SIXTH EDITION

ESSENTIALS OF
GEOLOGY

SIXTH EDITION

ESSENTIALS OF
GEOLOGY

STEPHEN MARSHAK

UNIVERSITY OF ILLINOIS

W. W. NORTON & COMPANY
NEW YORK | LONDON

W. W. Norton & Company has been independent since its founding in 1923, when William Warder Norton and Mary D. Herter Norton first published lectures delivered at the People's Institute, the adult education division of New York City's Cooper Union. The firm soon expanded their program beyond the Institute, publishing books by celebrated academics from America and abroad. By mid-century, the two major pillars of Norton's publishing program—trade books and college texts—were firmly established. In the 1950s, the Norton family transferred control of the company to its employees, and today—with a staff of four hundred and a comparable number of trade, college, and professional titles published each year—W. W. Norton & Company stands as the largest and oldest publishing house owned wholly by its employees.

Editor: Jake Schindel
Project Editor: Katie Callahan
Production Manager: Sean Mintus
Associate Editor: Rachel Goodman
Copy Editor: Chris Curioli
Managing Editor, College: Marian Johnson
Managing Editor, College Digital Media: Kim Yi
Digital Media Editor: Robert Bellinger
Associate Media Editors: Arielle Holstein and Gina Forsythe
Media Project Editor: Marcus Van Harpen
Editorial Assistant, Digital Media: Kelly Smith
Marketing Manager, Geology: Katie Sweeney
Design Director: Rubina Yeh
Designer: Jillian Burr
Director of College Permissions: Megan Schindel
Photography Editor: Trish Marx
Composition and page layout: MPS North America LLC
MPS Project manager: Jackie Strohl
Illustrations for the Second, Third, Fourth, and Fifth Editions: Precision Graphics / Lachina
Illustrations for the Sixth Edition: Stan Maddock and Joanne Brummett
Manufacturing: Transcontinental Interglobe—Beauceville, Quebec

Permission to use copyrighted material is included in the backmatter of this book.

Library of Congress Cataloging-in-Publication Data

Names: Marshak, Stephen, 1955- author.
Title: Essentials of geology / Stephen Marshak (University of Illinois).
Description: Sixth edition. | New York : W.W. Norton & Company, [2019] |
    Includes index.
Identifiers: LCCN 2018047804 | ISBN 9780393644456 (pbk.)
Subjects: LCSH: Geology—Textbooks.
Classification: LCC QE28 .M3415 2019 | DDC 551—dc23 LC record available at https://lccn.loc.gov/2018047804

W. W. Norton & Company, Inc., 500 Fifth Avenue, New York, NY 10110
wwnorton.com

W. W. Norton & Company Ltd., 15 Carlisle Street, London W1D 3BS

1 2 3 4 5 6 7 8 9 0

To Kathy, David, Emma, and Michelle

To Kelly, David, Emma, and Michelle

BRIEF CONTENTS

SPECIAL FEATURES

BRIEF CONTENT

CONTENTS

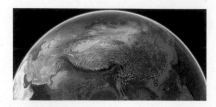

CHAPTER 15

Restless Realm: Oceans and Coasts · 498

CHAPTER 16

A Hidden Reserve: Groundwater · 530

CHAPTER 17

Dry Regions: The Geology of Deserts · 556

PREFACE

NARRATIVE THEMES

Why do earthquakes, volcanoes, floods, and landslides happen? What causes mountains to rise? How do beautiful landscapes develop? How have climate and life changed through time? When did the Earth form, and by what process? Where do we dig to find valuable metals, and where do we drill to find oil? Does sea level change? Do continents move? The study of geology addresses these important questions and many more. But from the birth of the discipline, in the late 18th century, until the mid-20th century, geologists considered each question largely in isolation, without pondering its relation to the others. This approach changed, beginning in the 1960s, in response to the formulation of two paradigm-shifting ideas that have since unified thinking about the Earth and its features. The first idea, called the *theory of plate tectonics*, states that the Earth's outer shell, rather than being static, consists of discrete plates that slowly move, relative to each other, so that the map of our planet continuously changes. Plate interactions cause earthquakes and volcanoes, build mountains, provide gases that make up the atmosphere, and affect the distribution of life on our planet. The second idea, the *Earth System concept*, emphasizes that the Earth's water, land, atmosphere, and living inhabitants are dynamically interconnected, so that materials constantly cycle among various living and nonliving reservoirs on, above, and within the planet. In the context of this idea, we have come to realize that the history of life is intimately linked to the history of the physical Earth, and vice versa.

Essentials of Geology, Sixth Edition, is an introduction to the study of our planet that uses the theory of plate tectonics, as well as the Earth System perspective throughout, to weave together a number of narrative themes, including:

1. The solid Earth, the oceans, the atmosphere, and life interact in complex ways.
2. Many important geologic processes involve the interactions of plates—pieces of the Earth's outer, relatively rigid shell.
3. The Earth is a planet formed, like other planets, from dust and gas. But, in contrast to other planets, the Earth is a dynamic place where new geologic features continue to form and old ones continue to be destroyed.
4. The Earth is very old—indeed, about 4.56 billion years have passed since its birth. During this time, the map of the planet and its surface features have changed, and life has evolved.
5. Internal processes (driven by the Earth's interior heat) and external processes (driven by heat from the Sun) interact at the Earth's surface to produce complex landscapes.

6. Geologic knowledge can help us to understand, and perhaps reduce, the danger of natural hazards, such as earthquakes, volcanoes, landslides, and floods.
7. Energy and mineral resources come from the Earth and are formed by geologic phenomena. Geologic study can help locate these resources and mitigate the consequences of their use.
8. Geology is a science, and the ideas of science come from observation, calculation, and experimentation by researchers—it is a human endeavor. Furthermore, geology utilizes ideas from physics, chemistry, and biology, so the study of geology provides an excellent opportunity for students to improve their overall science literacy.

These narrative themes serve as the book's take-home message, a message that hopefully, students will remember long after they finish their introductory geology course. In effect, the themes provide a mental framework on which students can organize and connect ideas, and can develop a modern, coherent image of our planet.

PEDAGOGICAL APPROACH

Educational research demonstrates that students learn best when they actively engage with a combination of narrative text and narrative art. Some students respond more to the words of a textbook, which help to organize information, provide answers to questions, fill in the essential steps that link ideas together, and develop a context for understanding ideas. Some students respond more to figures and photos, as images help students comprehend, visualize, and remember the narrative. And some respond best to active learning, an approach where students can practice their knowledge by putting ideas to work. *Essentials of Geology*, Sixth Edition, provides all three of these learning tools. The text has been crafted to be engaging, the art has been configured to tell a story, the chapters have been laid out to help students internalize key principles, and the online activities have been designed to both engage students and to provide active feedback. This book's narrative doesn't merely provide a dry statement of facts. Rather, it provides the story behind the story—the reasoning and observation that led to our current understanding, as well as an explanation of the processes that cause particular geologic phenomena.

Each chapter starts with a list of Learning Objectives that frame key pedagogical goals for each chapter. These objectives are revisited in the end-of-chapter Review Questions and in the Smartwork5 Online Activities. Take-Home Message panels, which include both a brief summary and a key question, appear at the end of each section to help students solidify key themes before proceeding to the next section. Throughout the chapter, brief Did You Ever Wonder? questions prompt students with real-life queries they may already have thought about—answers to which occur in the nearby text. See for Yourself panels guide students to visit spectacular examples of geologic features, using the power of Google Earth. They allow students to apply their newly acquired knowledge to the interpretation of real-world examples. In the ebook version of the text, these See for Yourself features are live links that "fly" students to the precise locations discussed. Each chapter concludes with a concise, two-page review that reinforces understanding and provides a concise study tool at the same time. Review Questions at the end of each chapter include two parts: the first addresses basic concepts; and the second, labeled as On Further Thought, stimulates critical thinking opportunities that invite students to think beyond the basics. Some of the questions use visuals from the chapter.

To enhance active learning opportunities, the Smartwork5 Online Activity System has been developed specifically for *Essentials of Geology*, Sixth Edition. Smartwork5 offers a wide range of visual exercises, including ranking, labeling, and sorting questions. Smartwork5 questions make the textbook art interactive, and they integrate the Narrative Art Videos, Animations, and Simulations that accompany the text. Questions are designed to give students answer-specific feedback when they are incorrect, coaching them towards developing a thorough understanding of the core concepts discussed in the book.

ORGANIZATION

The topics covered in this book have been arranged so that students can build their knowledge of geology on a foundation of overarching principles. To set the stage, the book starts by describing processes that led to the formation of the Earth, in the context of scientific cosmology. It then introduces the architecture of our planet, from surface to center. With this basic background, students are prepared to delve into plate tectonics theory. Plate tectonics appears early in the book so that students can relate the content of subsequent chapters to the theory. Knowledge of plate tectonics, for example, helps students understand the suite of chapters on minerals, rocks, and the rock cycle. Knowledge of plate tectonics and rocks together, in turn, provides a basis for studying volcanoes, earthquakes, and mountains. And with this background, students are prepared to see how the map of the Earth has changed throughout the vast expanse of geologic time, and how energy and mineral resources have developed. The book's final chapters address processes occurring at or near the Earth's surface, such as the flow of rivers, the evolution of coasts, and the carving of landscapes by glaciers. We also consider some problems that the Earth's surface processes can cause, such as landslides and floods. This part concludes with a topic of growing concern in society—global change, particularly climate change.

In addition to numbered chapters, the book contains several "interludes." These are, in effect, "mini-chapters" that focus on topics that are self-contained but are not broad enough to require an entire chapter. By placing selected topics in interludes, we can keep the numbered chapters reasonable in length, and can provide additional flexibility in sequencing topics within a course.

Although the sequence of chapters and interludes was chosen for a reason, this book is designed to be flexible, so that instructors can choose their own strategies for teaching geology. Therefore, each self-contained chapter reiterates relevant material where necessary. For example, if instructors prefer to introduce minerals and rocks before plate tectonics, they simply need to reorder the reading assignments. A low-cost, loose-leaf version of the book allows instructors to have students bring to class only the chapters that they need.

We have used a tiered approach in highlighting terminology in *Essentials of Geology*, Sixth Edition. Terminology, the basic vocabulary of a subject, serves an important purpose in simplifying the discussion of topics. For example, once students understand the formal definition of a mineral, the term can be used again in subsequent discussion without further explanation or redundancy. Too much new vocabulary, however, can be overwhelming. So we have tried to keep the book's Guide Terms (set in boldface and referenced at the end of each chapter for studying purposes) to a minimum. Other terms, less significant but still useful, appear in italics when presented, to provide additional visual focus for students as they read the chapters. We take care not to use vocabulary until it has been completely introduced and defined.

SPECIAL FEATURES OF THIS EDITION

Essentials of Geology, Sixth Edition, contains a number of new or revised features that distinguish it from all competing texts.

Narrative Art, What a Geologist Sees, and See for Yourself

It's difficult to understand many features of the Earth System without being able to see them. To help students visualize these and other features, this book is lavishly illustrated with figures that try to give a realistic context for the particular feature, without overwhelming students with too much extraneous detail. The talented artists who worked on the book have used the latest computer graphics software, resulting in the most sophisticated pedagogical art ever provided by a geoscience text. Many figures have been updated with an eye toward improving students' 3-D visualization skills. The figures have also been reconfigured to be more reader-friendly and intuitive. All of the plate tectonics figures have been revised in this Sixth Edition in order to provide students the clearest, most vibrant, and most accurate visual understanding of the Earth's interior dynamics.

In addition to the drawn art, the book also boasts over 1,000 stunning photographs from all around the world. Many of the photographs were taken by the author himself, in order to illustrate the exact concept under discussion. Where appropriate, photographs are accompanied by annotated sketches named

SEE FOR YOURSELF

USING GOOGLE EARTH

Visiting the SFY Field Sites Identified in the Text

There's no better way to appreciate geology then to see it first-hand in the field. Unfortunately, the great variety of geologic features that we discuss in this book can't be visited from any one locality. So, even if your class takes geology field trips during the semester, at most you'll see examples of just a few geologic settings. Fortunately, Google Earth makes it possible to address the challenge of seeing geology by allowing you to visit spectacular geologic field sites all over the world. In effect, you can take a virtual field trip anywhere, electronically, in a matter of seconds. In each chapter of this book, See for Yourself panels identify geologic sites that you can explore on your own personal computer (Mac or PC) using Google Earth software, or on your Apple or Android smartphone or tablet with the appropriate Google Earth app.

To get started, follow these three simple steps:

1. Check to see if Google Earth is installed on your personal computer, smartphone, or tablet. If not, download the free software from **https://www.google.com/earth**, or access the desktop version at **earth.google.com.** You can also download the app from the Apple or Android app store.
2. Each See for Yourself panel in the margin of the chapter provides a thumbnail photo of a geologically interesting site, as well as a very brief description of the site. The panel also provides the latitude and longitude of the site.
3. Open Google Earth and enter the coordinates of the site in the search window. As an example, let's find Mt. Fuji, a beautiful volcano in Japan. We note that the coordinates in the See for Yourself panel are as follows:

    Latitude   35°21'41.78"N
    Longitude   138°43'50.74"E

Type these coordinates into the search window of Google Earth as:

    35 21 41.78N, 138 43 50.74E

with the degree, minute, and second symbols left blank. When you click enter or return, your device will bring you to the viewpoint right above Mt. Fuji, as illustrated by the thumbnail above (on the left).

Google Earth contains many built-in and easy-to-use tools that allow you to vary the elevation, tilt, orientation, and position of your viewpoint, so that you can tour around the feature,

Vertical view, looking down.

Inclined view, looking north.

see it from many different perspectives, and thus develop a three-dimensional sense of the feature. In the case of Mt. Fuji, you'll be able to see its cone-like shape and the crater at its top. By zooming out to higher elevation, you can instantly perceive the context of the given geologic feature—for example, if you fly up into space above Mt. Fuji, you will see its position relative to the tectonic plate boundaries of the western Pacific. The thumbnail above (on the right) shows the view you'll see of the same location if you tilt your viewing direction and look north.

Need More Help?

If you're having trouble, please visit **digital.wwnorton.com/essgeo6**. There you will find a video showing how to download and install Google Earth, additional instructions on how to find the See for Yourself sites, links to Google Earth videos describing basic functions, and links to any hardware and software requirements. Also, notes addressing Google Earth updates will be available at this site.

We also offer a separate book—the *Geotours Workbook*, Second Edition (ISBN 978-1-324-00096-9), by Scott Wilkerson, Beth Wilkerson, and Stephen Marshak—that identifies additional interesting geologic sites to visit, provides active-learning exercises linked to the sites, and explains how you can create your own virtual field trips.

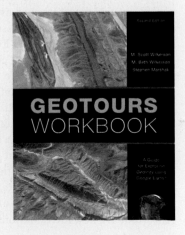

What a Geologist Sees. These figures allow students to see how geologists perceive the world around them and encourage students to start thinking like geologists.

Throughout the book, drawings and photographs have been integrated into narrative art, which has been laid out, labeled, and annotated to tell a story—the figures are drawn to teach! Subcaptions are positioned adjacent to relevant parts of each figure, labels point out key features, and balloons provide important annotation. Figure subparts are arranged to convey time progression, where relevant. The color schemes of drawings have been tied to those of relevant photos, so that students can easily relate features in the drawings to those in the photos. The author has also written and narrated over a dozen Narrative Art Videos, which bring the textbook art to life.

Google Earth provides an amazing opportunity for students to visit and tour important geologic sites wherever they occur. Throughout the book, we provide See for Yourself panels, which provide coordinates and descriptions of geologic features that students can visit at the touch of a finger or the click of a mouse. The adjacent box provides a quick guide for using these panels.

Featured Paintings—Geology at a Glance

Artist Gary Hincks has created or adapted spectacular two-page annotated paintings for most chapters. These paintings, called Geology at a Glance, integrate key concepts introduced in the chapters, visually emphasize the relationships between components of the Earth System, and allow students a way to review a subject . . . at a glance. Some of these paintings initially appeared in *Earth Story* (BBC Worldwide, 1998) and were conceived by the authors of that book, Simon Lamb and Felicity Maxwell. This Sixth Edition includes two new two-page spreads, drawn by artist Stan Maddock, illustrating the formation of minerals (Chapter 3) and the consequences of sea-level change (Chapter 19).

Enhanced Coverage of Current Topics

To ensure that *Essentials of Geology* continues to reflect the latest research discoveries and to help students understand geologic events that have been featured in current news, we have updated many topics throughout the book. For example, the Sixth Edition discusses the causes and lessons learned from recent natural disasters such as Hurricanes Harvey, Irma, and Maria (Chapter 15), and assesses the impact of recent earthquakes in Nepal, Japan, and Ecuador (Chapter 8). The Sixth Edition also includes updated coverage of the economics of oil and other energy resources, to clarify the difference between conventional and unconventional reserves (Chapter 12). These topics, along with expanded discussion of climate change and its impacts (Chapter 19), highlight the relevance of physical geology concepts and phenomena to students' lives today.

Other notable new content in the Sixth Edition includes a revision of the paleomagnetism discussion that makes this topic more accessible (Chapter 2); new coverage of mantle structure, using the results of the EarthScope experiment (Interlude D); new introductions to phylogenetics, ecosystems, and paleoecology (Interlude E); and an intensive revision of the explanation of the Coriolis force and other atmospheric concepts (Chapter 15), using text and figures developed in collaboration with atmospheric scientist Robert Rauber (University of Illinois) for the First Edition of a separate book, *Earth Science* (by Marshak and Rauber, W. W. Norton, 2016). These revisions ensure that students have access to the most contemporary and accurate explanations of these important, but complex, topics.

NEW GUIDED LEARNING EXPLORATIONS

Guided Learning Explorations are topical online activities that coach students through three carefully arranged stages: foundational concept review; application questions featuring geologic data, video and animation clips, and interactive simulations; and exploratory questions that have students interpret real-world sites using videos created with Google Earth.

Highly visual, interactive questions provide formative feedback every time a student clicks a possible answer choice, guiding them towards mastery of the concept at hand.

Guided Learning Explorations are built on 15 major topics taught in an introductory geology course, such as plate tectonics, volcanic hazards, and groundwater. These activities are available with every new book or ebook purchase, and can be set up to work with your campus LMS.

SMARTWORK5 ONLINE ACTIVITIES

Smartwork5

Smartwork5 is Norton's tablet-friendly, online activity platform. Both the system and its physical geology content were designed with the feedback of hundreds of instructors, resulting in unparalleled ease of use for students and instructors alike.

Smartwork5 features easy-to-deploy, highly visual assignments that provide students with answer-specific feedback. Students get the coaching they need to work through the assignments, while instructors get real-time assessment of student progress via automatic grading and item analysis. The question bank features a wide range of higher-order questions such as ranking, labeling, and sorting. All of the Narrative Art Videos, Animations, and Interactive Simulations are integrated directly into Smartwork5 questions—making them assignable. Smartwork5 also contains What a Geologist Sees questions that take students to sites not mentioned in the book, so they can apply their knowledge just as a geologist would. In addition, Smartwork5 offers reading quizzes for each chapter and Geotours—Guided Inquiry Activities using Google Earth.

Based on instructor feedback, Smartwork5 offers three formats to help students use what they have learned:

› Chapter **Reading Quizzes**, designed to help students prepare for lecture
› Chapter **Activities**, consisting of highly visual exercises covering all chapter Learning Objectives
› **Geotours Worksheets**—guided inquiry activities that use Google Earth

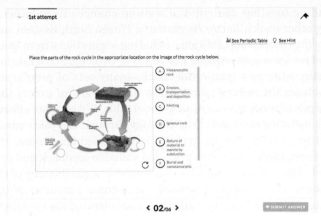

Smartwork5 features a variety of question types to get students working hands-on with geologic concepts.

Smartwork5 is fully customizable, meaning that instructors can add or remove questions, create assignments, write their own questions, or modify the questions in Smartwork5. Easy and intuitive tools allow instructors to filter questions by chapter, section, question type, and learning objective.

All Smartwork5 content is written by geology instructors. For the Sixth Edition, Smartwork5 authors include Heather Lehto of Angelo State University, Tobin Hindle of Florida Atlantic University, Christine Clark of Eastern Michigan University, and Jacqueline Richard of Delgado Community College.

MEDIA AND ANCILLARIES

Animations, Simulations, and Videos

Essentials of Geology, Sixth Edition, provides a rich collection of new Animations, developed by Alex Glass of Duke University, working with Heather Cook of California State University—San Marcos. These Animations illustrate geologic processes in a consistent style and with a 3-D perspective. Interactive Simulations allow students to control variables and see the resulting output. The newest Simulations are designed to help students understand basic terminology.

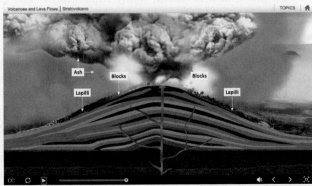

Animations illustrate geologic processes.

Narrative Art Videos, written and narrated by Stephen Marshak, bring both textbook art and supplementary field photos to life. And a robust suite of over 100 Real-World Video Clips illustrate key processes, concepts, and natural phenomena.

All the videos, animations, and interactive simulations are free, require no special software, and are available in a variety of settings to offer ultimate flexibility for instructors and students: on the Norton Digital Landing Page (digital.wwnorton.com/essgeo6); in LMS-compatible coursepacks; integrated into Smartwork5 questions; linked to the ebook; and as linked resources in the new Interactive Instructor's Guide that accompanies the text (iig.wwnorton.com/essgeo6/full).

Mobile-Ready Ebook

Essentials of Geology, Sixth Edition, is available in a new format perfect for tablets and phones. Within the ebook, art expands for a closer look; links send the reader to geologic locations in Google Maps; Animations and Videos link out from each chapter; and pop-up key terms allow for quick review. It's also easy to highlight, take notes, and search the text.

The *Geotours Workbook*

Created by Scott Wilkerson, Beth Wilkerson, and Stephen Marshak, the *Geotours Workbook,* Second Edition, provides active-learning opportunities that take students on virtual field trips to see outstanding examples of geology at locations around the world using Google Earth. Arranged by topic, the questions in the *Geotours Workbook* have been designed for auto-grading, and they are available as worksheets, both in print format (these come free with the book and include complete user instructions and advanced instruction) or electronically with automatic grading through Smartwork5 or your campus LMS. The *Geotours Workbook* also provides instructions that will allow instructors or students to make their own geotours. Request a sample copy to preview each worksheet.

Lecture PowerPoints and Image Files

Norton provides a variety of electronic presentations of art and photographs in the book to enhance the classroom experience. These include:

› Lecture PowerPoints—Designed for instant classroom use, these slides utilize photographs and line art from the book in a form that has been optimized for use in the PowerPoint environment. The art has been relabeled and resized for projection. Lecture PowerPoints also include supplemental photographs. For the Sixth Edition, the Lecture PowerPoints were revised by Karen McNeal of Auburn University.

› Labeled and Unlabeled Image Files—These include all art from the book, formatted as JPEGs, pre-pasted into PowerPoints. We offer one set in which all labeling has been stripped for use in quizzes and clicker questions, and one set in which the labeling has been retained. Individual JPEGs are also available for download.

> Quarterly Update PowerPoints—Norton offers a quarterly update service that provides new PowerPoint slides, with instructor support, covering recent geologic events. Monthly updates are authored by Paul and Nathalie Brandes of Lone Star College—Montgomery.

Instructor's Manual and Test Bank

The *Instructor's Manual*, prepared by Nick Soltis of Auburn University, is designed to help instructors prepare lectures and exams. It contains detailed Learning Objectives, Chapter Summaries, and complete answers to the end-of-chapter Review Questions and On Further Thought Questions for every chapter and interlude.

The *Test Bank*, written by Steven Petsch of the University of Massachusetts, has been revised not only to correlate with this new Edition, but to provide greater, more rounded assessment than ever before. Expert accuracy checkers have ensured that every question in the *Test Bank* is scientifically reliable and truly tests students' understanding of the most important topics in each chapter, so that the questions can be assigned with confidence.

Interactive Instructor's Guide—https://iig.wwnorton.com/essgeo6/full

New for the Sixth Edition, the Interactive Instructor's Guide is a dynamic, searchable, online resource that provides all instructor resources in one place. With content tagged by book chapter and section, learning objective, and keyword, instructors can find what they need, when they need it—whether a Real-World Video, an in-class activity idea, or the Lecture PowerPoints for the chapter.

LMS Coursepacks

Available at no cost to professors or students, Norton Coursepacks bring high-quality Norton digital media into a new or existing online course. Coursepacks contain ready-made content for your campus LMS. For *Essentials of Geology*, Sixth Edition, content includes the full suite of animations, simulations, and videos keyed to core figures in each chapter; the *Test Bank*; Reading Quizzes authored by Tim Cope of DePauw University, with accuracy checked by Scott Werts of Winthrop University; new European Case Studies; Geotour Questions; Vocabulary Flashcards; and links to the ebook.

ACKNOWLEDGMENTS

Many people contributed to the long and complex process of bringing this book from the concept stage to the shelf in the first place, and now to the continuous effort of improving the book and keeping it current. Textbooks are, by definition, always a work in progress.

Developing and revising this book is done in partnership with my wife, Kathy. She carries out the immense task of pulling together content and writing changes introduced in my other books, *Earth: Portrait of a Planet*, Sixth Edition, and *Earth Science*, First Edition, including suggestions from users and reviewers, to produce *Essentials of Geology*, Sixth Edition. Kathy edits new text, cross-checks many sets of proofs, and manages the never-ending inflow and outflow of proofs that perpetually occupy our dining-room table. Without her efforts, the updating of *Essentials of Geology* through the years would not be possible. We are grateful to our daughter, Emma, and our son, David, who have provided valuable feedback and several of the photos—and who also served as scale in many of the photos. They allowed "the book" to become a member of our household, for more than 20 years, and tolerated the overabundance of geo-stops on family trips.

Kathy and I are very grateful to all of the staff of W. W. Norton & Company for their incredible, continuing efforts during the development of this book and its companions, since the first contract was signed in 1992. It has been a privilege to work with an employee-owned company that is willing to collaborate so closely with its authors. In particular, I would like to thank Jake Schindel, the geology editor at Norton, who has injected new enthusiasm and ideas into the project, working steadfastly to bring order to the chaos of juggling multiple titles at once. His skill in editing, ability to oversee many moving parts, and his friendly reminders of deadlines, led this book to completion on an accelerated production schedule. Katie Callahan, the book's project editor, did an amazing job of guiding the book through production. She joined the Norton geology team for this latest component of a very lengthy and complicated project, and has remained both calm and supportive throughout the process. We are also grateful to Thom Foley, who served as project editor for all previous editions and who kindly provided advice throughout, and brought the book over the finish line when Katie moved to other projects. Rob Bellinger continues to keep the technology component of the book at the cutting edge, by introducing new web tools and overseeing the development of the new Guided Learning Explorations and Smartwork5 for the book. Katie Sweeney has done wonders as the marketing manager for the book, by helping to determine how to meet the needs of adopters worldwide. I also wish to thank the previous editors of this book. Eric Svendsen ably oversaw the Fourth and Fifth Editions and introduced many innovations. The late Jack Repcheck, served as the editor for the first three editions of the book. Jack suggested many ideas that strengthened the book, and his instincts about what works in textbook publishing brought the book to the attention of a wider geological community than we ever thought possible. He will always be remembered as an understanding friend and a fountain of sage advice.

Moving a new edition of *Essentials of Geology* from concept to completion involves a large team of professionals. Joanne Brummett and Stan Maddock, artists in Champaign, Illinois, have created beauty and enhanced pedagogy with the new illustrations that they have rendered for this Edition, along with the past work anchored at Precision Graphics, set the bar for the quality of art in geology textbooks. Stan established the initial style of the book's art and developed innovative ways of visualizing geologic phenomena. Trish Marx has done a fantastic job with the Herculean task of finding, organizing, and crediting photographs. Jillian Burr creatively developed a clean and friendly page design. I am also grateful to Rob Bellinger, Cailin Barrett Bressack, Liz Vogt, Kim Yi, Marcus Van Harpen, Leah Clark, Francesca Olivo, Arielle Holstein, Gina Forsythe, and Kelly Smith for their innovative approach to ancillary and e-media development. Thanks also go to Marcus Van Harpen, Lizz Thabeet, and Mateus Teixeira for their work on the tablet and mobile e-books, and to production manager Sean Mintus, who coordinated the back-and-forth between the publisher and various vendors and suppliers. Chris Curioli did an excellent job as copy editor of this Sixth Edition and Associate Editor Rachel Goodman provided consistent editorial support and trouble-shooting throughout the process of making this book.

The six editions of this book and its cousins, *Earth: Portrait of a Planet* and *Earth Science*, have benefited greatly from input by expert reviewers for specific chapters, by general reviewers of the entire book, and by comments from faculty and students who have used the book and were kind enough to contact the author or the publisher with suggestions and corrections. We gratefully acknowledge the contributions of the individuals listed below, who have provided invaluable input into this and past editions either through comments or reviews. I apologize if I've inadvertently left anyone off the list.

Jack C. Allen, *Bucknell University*
David W. Anderson, *San Jose State University*
Martin Appold, *University of Missouri, Columbia*
Mary Armour, *York University*
Philip Astwood, *University of South Carolina*
Eric Baer, *Highline University*
Victor Baker, *University of Arizona*
Julie Baldwin, *University of Montana*
Miriam Barquero-Molina, *University of Missouri*
Sandra Barr, *Acadia University*
Keith Bell, *Carleton University*
Mary Lou Bevier, *University of British Columbia*
Jim Black, *Tarrant County College*
Daniel Blake, *University of Illinois*
Andy Bobyarchick, *University of North Carolina, Charlotte*
Ted Bornhorst, *Michigan Technological University*
Michael Bradley, *Eastern Michigan University*
Mike Branney, *University of Leicester, UK*
Sam Browning, *Massachusetts Institute of Technology*
Bill Buhay, *University of Winnipeg*
Caroline Burberry, *University of Nebraska, Lincoln*
Rachel Burks, *Towson University*
Peter Burns, *University of Notre Dame*
Katherine Cashman, *University of Oregon*

Cinzia Cervato, *Iowa State University*
George S. Clark, *University of Manitoba*
Kevin Cole, *Grand Valley State University*
Patrick M. Colgan, *Northeastern University*
Amanda Colosimo, *Monroe Community College*
Peter Copeland, *University of Houston*
John W. Creasy, *Bates College*
Norbert Cygan, *Chevron Oil, retired*
Michael Dalman, *Blinn College*
Peter DeCelles, *University of Arizona*
Carlos Dengo, *ExxonMobil Exploration Company*
Meredith Denton-Hedrick, *Austin Community College, Cypress Creek*
John Dewey, *University of California, Davis*
Charles Dimmick, *Central Connecticut State University*
Robert T. Dodd, *Stony Brook University*
Holly Dolliver, *University of Wisconsin, River Falls*
Glen Dolphin, *University of Calgary*
Missy Eppes, *University of North Carolina, Charlotte*
Eric Essene, *University of Michigan*
David Evans, *Yale University*
James E. Evans, *Bowling Green State University*
Susan Everett, *University of Michigan, Dearborn*
Dori Farthing, *State University of New York, Geneseo*
Mark Feigenson, *Rutgers University*
Grant Ferguson, *St. Francis Xavier University*
Eric Ferré, *Southern Illinois University*
Leon Follmer, *Illinois Geological Survey*
Nels Forman, *University of North Dakota*
Bruce Fouke, *University of Illinois*
David Furbish, *Vanderbilt University*
Steve Gao, *University of Missouri*
Grant Garvin, *John Hopkins University*
Christopher Geiss, *Trinity College, Connecticut*
Bryan Gibbs, *Richland Community College*
Gayle Gleason, *State University of New York, Cortland*
Patrick Gonsoulin-Getty, *University of Connecticut*
Cyrena Goodrich, *Kingsborough Community College*
William D. Gosnold, *University of North Dakota*
Lisa Greer, *William & Mary College*
Steve Guggenheim, *University of Illinois, Chicago*
Henry Halls, *University of Toronto, Mississuaga*
Bryce M. Hand, *Syracuse University*
Anders Hellstrom, *Stockholm University*
Tom Henyey, *University of South Carolina*
Bruce Herbert, *Texas A & M University*
James Hinthorne, *University of Texas, Pan American*
Paul Hoffman, *Harvard University*
Curtis Hollabaugh, *University of West Georgia*
Bernie Housen, *Western Washington University*
Mary Hubbard, *Kansas State University*
Paul Hudak, *University of North Texas*
Melissa Hudley, *University of North Carolina, Chapel Hill*
Warren Huff, *University of Cincinnati*

Neal Iverson, *Iowa State University*
Charles Jones, *University of Pittsburgh*
Donna M. Jurdy, *Northwestern University*
Thomas Juster, *University of Southern Florida*
H. Karlsson, *Texas Tech*
Daniel Karner, *Sonoma State University*
Dennis Kent, *Lamont Doherty / Rutgers*
Charles Kerton, *Iowa State University*
Susan Kieffer, *University of Illinois*
Jeffrey Knott, *California State University, Fullerton*
Ulrich Kruse, *University of Illinois*
Robert S. Kuhlman, *Montgomery County Community College*
Lee Kump, *Pennsylvania State University*
David R. Lageson, *Montana State University*
Robert Lawrence, *Oregon State University*
Heather Lehto, *Angelo State University*
Scott Lockert, *Bluefield Holdings*
Leland Timothy Long, *Georgia Tech*
Craig Lundstrom, *University of Illinois*
John A. Madsen, *University of Delaware*
Jerry Magloughlin, *Colorado State University*
Scott Marshall, *Appalachian State University*
Kyle Mayborn, *Western Illinois University*
Jennifer McGuire, *Texas A&M University*
Judy McIlrath, *University of South Florida*
Paul Meijer, *Utrecht University, Netherlands*
Aric Mine, *California State University, Fresno*
Jamie Dustin Mitchem, *California University of Pennsylvania*
Alan Mix, *Oregon State University*
Marguerite Moloney, *Nicholls State University*
Otto Muller, *Alfred University*
Kristen Myshrall, *University of Connecticut*
Kathy Nagy, *University of Illinois, Chicago*
Pamela Nelson, *Glendale Community College*
Wendy Nelson, *Towson University*
Robert Nowack, *Purdue University*
Charlie Onasch, *Bowling Green State University*
David Osleger, *University of California, Davis*
Bill Patterson, *University of Saskatchewan*
Eric Peterson, *Illinois State University*
Ginny Peterson, *Grand Valley State University*
Stephen Piercey, *Laurentian University*
Adrian Pittari, *University of Waikato, New Zealand*
Lisa M. Pratt, *Indiana University*
Eriks Puris, *Portland Community College*

Mark Ragan, *University of Iowa*
Robert Rauber, *University of Illinois*
Bob Reynolds, *Central Oregon Community College*
Joshua J. Roering, *University of Oregon*
Randye Rutberg, *Hunter College*
Eric Sandvol, *University of Missouri*
William E. Sanford, *Colorado State University*
Jeffrey Schaffer, *Napa Valley Community College*
Roy Schlische, *Rutgers University*
Sahlemedhin Sertsu, *Bowie State University*
Anne Sheehan, *University of Colorado*
Roger D. Shew, *University of North Carolina, Wilmington*
Doug Shakel, *Pima Community College*
Norma Small-Warren, *Howard University*
Donny Smoak, *University of South Florida*
David Sparks, *Texas A&M University*
Angela Speck, *University of Missouri*
Larry Standlee, *University of Texas, Arlington*
Tim Stark, *University of Illinois*
Seth Stein, *Northwestern University*
David Stetty, *Jacksonville State University*
Kevin G. Stewart, *University of North Carolina, Chapel Hill*
Michael Stewart, *University of Illinois*
Don Stierman, *University of Toledo*
Gina Marie Seegers Szablewski, *University of Wisconsin, Milwaukee*
Barbara Tewksbury, *Hamilton College*
Thomas M. Tharp, *Purdue University*
Kathryn Thornbjarnarson, *San Diego State University*
Robert Thorson, *University of Connecticut*
Basil Tikoff, *University of Wisconsin*
Spencer Titley, *University of Arizona*
Robert T. Todd, *Stony Brook University*
Torbjörn Törnqvist, *University of Illinois, Chicago*
Jon Tso, *Radford University*
James Tyburczy, *Arizona State University*
Stacey Verardo, *George Mason University*
Barry Weaver, *University of Oklahoma*
John Werner, *Seminole State College of Florida*
Alan Whittington, *University of Missouri*
John Wickham, *University of Texas, Arlington*
Lorraine Wolf, *Auburn University*
Christopher J. Woltemade, *Shippensburg University*
Jackie Wood, *Delgado Community College, City Park*
Kerry Workman-Ford, *California State University, Fresno*

THANKS!

I am very grateful to the faculty who have selected *Essentials of Geology* for their classes, and to the students and other readers who engage so energetically with the book. I particularly appreciate the questions and corrections from readers that help to improve the book and keep it as accurate as possible. I always welcome comments and can be reached at smarshak @illinois.edu.

Stephen Marshak

Geology, perhaps more than any other department of natural philosophy, is a science of contemplation. It demands only an enquiring mind and senses alive to the facts almost everywhere presented in nature.

—SIR HUMPHRY DAVY (BRITISH SCIENTIST, 1778–1829)

ABOUT THE AUTHOR

Stephen Marshak is a Professor Emeritus of Geology at the University of Illinois, Urbana-Champaign, where he also served as the Director of the School of Earth, Society, and Environment. He holds an A.B. from Cornell University, an M.S. from the University of Arizona, and a Ph.D. from Columbia University. Steve's research interests lie in structural geology and tectonics, and he has participated in field projects on a number of continents. Steve loves teaching and has won his college's and university's highest teaching awards. He also received the 2012 Neil Miner Award from the National Association of Geoscience Teachers (NAGT), for "exceptional contributions to the stimulation of interest in the Earth sciences." In addition to research papers and *Essentials of Geology,* Steve has authored *Earth: Portrait of a Planet,* and has co-authored *Earth Science; Laboratory Manual for Introductory Geology; Earth Structure: An Introduction to Structural Geology and Tectonics;* and *Basic Methods of Structural Geology.*

SIXTH EDITION

ESSENTIALS OF
GEOLOGY

AND JUST WHAT IS GEOLOGY?

By the end of this prelude, you should be able to . . .

A. describe the scope and applications of geology.
B. explain the foundational themes of modern geologic study.
C. demonstrate how geologists employ the scientific method.

D. provide a basic definition of the theory of plate tectonics.
E. explain what geologists mean by the Earth System concept.
F. name the main layers of the Earth's interior.

P.1 In Search of Ideas

We arrived in the late-night darkness, at a campsite in western Arizona. Here in the desert, so little rain falls over the course of a year that hardly any plants can survive, and rocks crop out as jagged ledges on many hills. Under the dry sky, there's no need for tents, so we could sleep under the stars. At dawn, the red rays of the first sunlight made the hill near our campsite start to glow, and we could see our target, a prominent ledge of rusty-brown rock that formed a shelf at the top of the hill. To reach it, though, we'd have to climb a steep slope littered with jagged boulders.

After a quick breakfast, we loaded our day packs with water bottles and granola bars, slathered on a layer of sunscreen, and set off toward the slope. It was the breezeless morning of what was going to be a truly hot day, and we wanted to gain elevation before the Sun rose too high in the sky. After a tiring hour finding our way through the boulder obstacle course, we reached the base of the ledge and decided to take a break before ascending to the peak. But just as we leaned to rest our backs against a rock, we heard an unnerving vibration. Somewhere nearby, too close for comfort, a rattlesnake shook an urgent warning with its tail. Rest would have to wait, and we scrambled up the ledge. It was the right choice, for the view from the top of the surrounding landscape was amazing **(Fig. P.1a)**. But the rocks beneath our feet were even more amazing. Close up, we could see curving ribbons of light and dark layers. The ledge preserved the story of a distant age in our planet's past when the rock we now stood on was kilometers below ground level and was able to flow like soft plastic, ever so slowly **(Fig. P.1b)**. We now set to the task of figuring out what it all meant.

Geologists—scientists who study the Earth—explore many areas, including remote deserts, high mountains, damp rainforests, frigid glaciers, and deep canyons **(Fig. P.2)**. Such efforts can strike people in other professions as a strange way to make a living. This sentiment underlies a description by the Scottish poet Walter Scott (1771–1832) of geologists at work: "Some rin uphill and down dale, knapping the chucky stanes to pieces wi' hammers, like sae mony road-makers run daft—they say it is to see how the world was made!" Indeed—to see

FIGURE P.1 Geologic exploration provides beautiful views and mysteries to solve.

(a) A view of the western Arizona desert is not just beautiful—it holds clues to the Earth's past and to the changes taking place today.

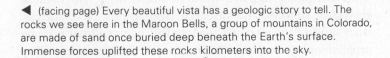

◀ (facing page) Every beautiful vista has a geologic story to tell. The rocks we see here in the Maroon Bells, a group of mountains in Colorado, are made of sand once buried deep beneath the Earth's surface. Immense forces uplifted these rocks kilometers into the sky.

(b) The contortions of the rock layers speak of a time when the rock flowed like soft plastic.

FIGURE P.2 Geologists explore many environments.

(a) Cliff exposures in the desert of Utah.

(b) The shore in Massachusetts.

(c) A rainforest in Peru.

(d) Mountains in Alaska.

how the world was made, to see how it continues to evolve, to find its resources, to protect against its natural hazards, and to predict what its future may bring. These are the questions that have driven geologists to explore the Earth, on all continents and in all oceans, from the equator to the poles, and everywhere in between.

Geologic discovery continues today. But while some geologists continue to work in the field with hammers and hand lenses, others have moved into laboratories where they employ sophisticated electronic instruments to analyze microscopic quantities of Earth materials or detect the configuration of layers underground, use satellites to detect the motions of continents or the stability of volcanoes, and use high-performance computers to locate earthquakes or analyze the flow of underground water. For over two centuries, geologists have pored over the Earth in search of ideas to explain the processes that form and change our planet. In this Prelude, we look at the questions geologists ask and have tried to answer. You'll see that the results of this work are not just of academic interest but have implications for society.

P.2 **Why Study Geology?**

Geology, or *geoscience*, is the study of the Earth. Not only do geologists address fundamental questions such as the formation and composition of our planet, the causes of earthquakes and ice ages, and the evolution of life, but they also address practical problems such as how to keep pollution out of groundwater, how to find oil and minerals, and how to avoid landslides.

> **Did you ever wonder...**
> will an earthquake happen near where you live?

The fascination of geology attracts many to careers in this science. Hundreds of thousands of geologists work for energy, mining, water, engineering, and environmental companies, while a smaller number work in universities, government geological surveys, and research laboratories. Nevertheless, since most of the students reading this book will not become professional geologists, it's fair to ask the question, "Why should people, in general, study geology?"

First, geology may be one of the most practical subjects you can learn. When a news report begins, "Scientists say," and then continues, "an earthquake occurred today off Japan," or "landslides will threaten the city," or "chemicals from a toxic waste dump will ruin the town's water supply," or "there's only a limited supply of oil left," or "the floods of the last few days are the worst on record," the scientists that the report refers to are geologists. In fact, ask yourself the following questions, and you'll realize that geologic phenomena and materials play major roles in daily life:

> Do you live in a region threatened by landslides, volcanoes, earthquakes, or floods **(Fig. P.3)**?
> Are you worried about the price of energy or whether there will be a war in an oil-supplying country?
> Do you ever wonder where the copper in your home's wires comes from? Or the lithium in the battery of your cell phone?
> Have you seen fields of green crops surrounded by desert and wondered where the irrigation water comes from?
> Have you worried about the consequences of deforestation?
> Would you like to buy a dream house on a beach?
> Are you following news stories about how radioactive waste can migrate underground into a river?

Clearly, all citizens of the 21st century, not just professional geologists, need to make decisions and understand news reports addressing Earth-related issues. A basic understanding of geology will help you do so.

Second, the study of geology gives you an awareness of the planet. As you will see, the Earth is a complicated place, where living organisms, oceans, atmosphere, and solid rock interact with one another in a great variety of ways. Geologic research reveals the Earth's antiquity and demonstrates how the planet has changed profoundly during its existence. What our ancestors considered to be the center of the Universe has become, with the development of geologic perspective, our "island in space" today. And what people believed to be an unchanging orb that originated at the same time as humanity is now considered to be a dynamic planet that existed long before people did and that continues to evolve.

Third, the study of geology puts the accomplishments and consequences of human civilization in a broader context. View the aftermath of a large earthquake, flood, or hurricane, and it's clear that the might of natural geologic phenomena greatly exceeds the strength of human-made structures. But watch a bulldozer clear a swath of forest, a dynamite explosion remove the top of a hill, or a prairie field disappear beneath a housing development, and it's clear that people can change the face of the Earth at rates often exceeding those of natural geologic processes.

Finally, when you finish reading this book, your view of the world may be forever colored by geologic curiosity. If you walk in the mountains, you may remember the many forces that shape and reshape the Earth's surface. If you hear about a natural disaster, you may think about the processes that brought it about. And if you go on a road trip, the rock exposures along the highway should no longer be gray, faceless cliffs, but will present complex puzzles of texture and color telling a story of the Earth's long history.

P.3 **Themes of This Book**

A number of narrative themes appear (and reappear) throughout this text. These themes, listed below, can be viewed as this book's overall take-home message.

> *Geology is a synthesis of many sciences:* The study of geology can help you understand physical science in general, for geology applies many of the basic concepts of physics and chemistry to the interpretation of visible phenomena. As you learn about the Earth, you'll also be learning about the behavior of matter and energy and about the nature of chemical reactions. Also, readers who pursue teaching careers will find that geological examples can help students to develop STEM (science-technology-engineering-math) learning skills.
> *The Earth has an internal structure:* The Earth isn't homogeneous inside, but rather it consists of concentric layers. From center to surface, our planet has a *core, mantle,* and *crust.* We live on the surface of the crust, where it meets the atmosphere and the oceans.

FIGURE P.3 Human-made cities cannot withstand the vibrations of a large earthquake. These apartment buildings collapsed during an earthquake in Turkey.

> *The outer layer of the Earth consists of moving plates:* In the 1960s, geologists recognized that the crust, together with the uppermost part of the underlying mantle, forms a 100- to 150-km-thick semi-rigid shell, called the **lithosphere**, that surrounds a softer part of the mantle called the **asthenosphere**. Distinct boundaries separate this shell into discrete pieces, called **plates**, which move very slowly relative to one another. Geologists recognize three kinds of plate boundaries, based on the relative motion of one plate with respect to the other across the plate boundary **(Fig. P.4)**. The theory that describes this movement and its consequences is called the **theory of plate tectonics**, and it serves as the foundation for understanding most geologic phenomena. Plate movements and interactions yield earthquakes, volcanoes, and mountain ranges and cause the map of the Earth's surface to change very slowly over time.

> *We can picture the Earth as a complex system:* The Earth is not static. Rather our planet's interior, solid surface, oceans, atmosphere, and life all interact with one another dynamically in many ways. Geologists refer to this interconnected web of interacting realms of materials and processes as the **Earth System** (**Geology at a Glance**, pp. 8–9).

> *The Earth is a planet:* Despite the uniqueness of the Earth System, the Earth is a planet, formed like the other planets of the Solar System. But because of the way the Earth System operates, our planet differs from others by having plate tectonics, an oxygen-rich atmosphere, a liquid-water ocean, and abundant life.

> *Internal and external processes drive geologic phenomena:* **Internal processes** are those driven by heat from inside the Earth. Plate movement serves as an example. Because plate movements cause mountain building, earthquakes, and volcanoes, we consider all of these phenomena to be internal processes. **External processes** are those driven by energy coming to the Earth from the Sun. The heat produced by this energy drives the movement of air and water, which grinds and sculpts the Earth's surface and transports the debris to new locations, where it accumulates. As we'll see, **gravity**—the pull that one mass exerts on another—plays an important role in both internal and external processes. The interaction between internal and external processes forms and shapes the mountains, canyons, beaches, and plains of our planet.

> *The Earth is very old:* Geologic data indicate that the Earth formed about 4.56 billion years ago—plenty of time for geologic processes to generate and destroy landscapes, for life forms to evolve and go extinct, and for the map of

Did you ever wonder... if a map of the Earth's surface today looks like a map of the surface 200 million years ago?

FIGURE P.4 A simplified map of the Earth's plates. The arrows indicate the direction each plate is moving, and the length of the arrow indicates plate velocity relative to the Earth's interior. The longer the arrow, the faster the motion.

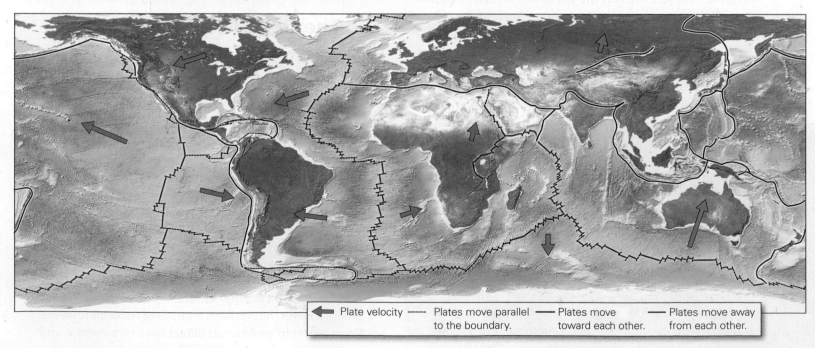

⟵ Plate velocity   ------ Plates move parallel to the boundary.   —— Plates move toward each other.   —— Plates move away from each other.

the planet to change. There is time to build mountains and grind them down, many times over. The Earth has a history, and it extends far into the past. **Geologic time** represents the duration of this history.

> *The geologic time scale divides the Earth's history into intervals:* To refer to specific portions of geologic time, geologists developed the **geologic time scale (Fig. P.5)**. The last 541 million years comprise the *Phanerozoic Eon*, and all time before that makes up the *Precambrian*. The Precambrian can be further divided into three main intervals named, from oldest to youngest: the *Hadean*, the *Archean*, and the *Proterozoic Eons*. The Phanerozoic Eon, in turn, can be divided into three main intervals named, from oldest to youngest: the *Paleozoic*, the *Mesozoic*, and the *Cenozoic Eras*.

> *Geologic phenomena affect society:* Volcanoes, earthquakes, landslides, floods, groundwater, energy sources, and mineral reserves are of vital interest to every inhabitant of this planet. Therefore, throughout this book we emphasize the linkages among geology, the environment, and society.

> *Physical aspects of the Earth System interact with life processes:* All life on this planet depends on such physical features as the composition of soil; the temperature, humidity, and composition of the atmosphere; and the flow of surface and subsurface water. And life in turn affects and alters physical features. For example, the oxygen in the Earth's atmosphere comes from photosynthesis, a life activity in plants. Without the physical Earth, life could not exist, but without life, this planet's surface might have become a frozen wasteland like that of Mars or a cloud-enshrouded oven like that of Venus.

> *The Earth has changed dramatically in many ways over geologic time and continues to change:* The landscape that you see outside your window today is not what you would have seen a thousand, a million, or a billion years ago. Over Earth history, the planet's surface, composition of the atmosphere, and sea level have all changed. Also, continents move relative to one another. Change continues today, and aspects of the Earth System are changing faster than ever before because of human activity.

> *Most of the resources that we use come from geologic materials:* Modern society uses vast quantities of oil, gas, coal, metal, concrete, clay, fertilizer, and other materials that all come from the Earth **(Fig. P.6)**.

> *Science comes from observation, and people make scientific discoveries:* Science does not consist of subjective guesses or arbitrary dogmas, but rather of a consistent set of objective statements resulting from the application of the scientific method **(Box P.1)**.

FIGURE P.5 The geologic time scale.

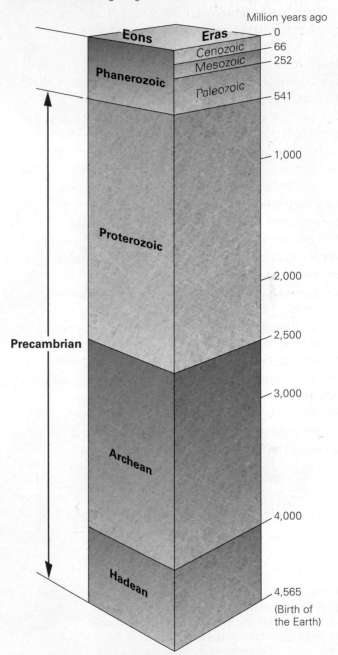

(a) The scale has been divided into eons and eras.

One thousand years ago = 1 **Ka**
(Ka stands for kilo-annum)

One million years ago = 1 **Ma**
(Ma stands for mega-annum)

One billion years ago = 1 **Ga**
(Ga stands for giga-annum)

(b) Abbreviations for time units.

The Earth System

Thunderhead

External energy

Mountain uplift

Lightning

Sun

Continental glacier

Rain and snow

Ocean

City

Desert

Rocky coastline

Valley

Arid mountains

Mining

Lakes

Field pattern

Deciduous forest

Forested mountains

Beach

Shark

Tropical rainforest

Coral reef

When you stand on the surface of the Earth, you can see the wondrous ways in which components of the Earth System interact. The *geosphere* consists of the solid part of our planet. You see it wherever you see exposed rock, sediment, or soil. Most of it lies underground, in the internal layers of our planet. The *hydrosphere* consists of all liquid water at or near the surface of the Earth. It fills oceans, lakes, underground pores, and occurs as gas in the atmosphere. The *cryosphere* consists of frozen water, mostly in glaciers. The *biosphere* consists of living organisms, from bacteria to whales. The *atmosphere* is the envelope of gas that encircles the planet. Flow in the air and sea transfer heat and water around the planet.

Internal energy rising from the interior and external energy coming to the Earth from the Sun keep the Earth System dynamic, so that materials cycle from component to component over time. Human society is having a growing impact on the Earth System, by extracting resources, building and farming on its surface, and emitting waste.

Internal energy

Cirrus clouds

Jet stream

Moon

Aurora

Wind system

Ice and snow

Coniferous forest

Evaporation

Volcanic islands

Industrial pollution

Cold surface current

Surface waters

Delta

Swamps

Twilight zone

Warm surface current

Abyssal zone

Whale

Seafloor

Bacteria and plankton

Giant squid

Deep-sea current

Black smokers

The Scientific Method

Sometime during the past 200 million years, a large block of rock or metal, which had been orbiting the Sun, slammed into our planet. It made contact at a site in what is now the central United States, a landscape of flat cornfields. The impact of this block, a *meteorite*, released more energy than a nuclear bomb. A cloud of shattered rock and dust blasted skyward, and once-horizontal layers of rock from deep below the ground sprang upward and tilted on end beneath the gaping crater left by the impact. When the event was over, the land surface looked radically different—a layer of debris surrounded and partially filled the crater at the impact site. Later in Earth history, running water and blowing wind wore down this jagged scar and carried away the debris. Then, about 15,000 years ago, sand, gravel, and mud carried by a vast sheet of ice—a glacier—buried what remained, hiding it entirely from view **(Fig. BxP.1)**. Wow! So much history beneath a cornfield. How do we know this? It takes scientific investigation.

The movies often portray science as a dangerous tool, capable of creating Frankenstein's monster, and scientists as nerdy characters with thick glasses and poor taste in clothes. In reality, **science** refers simply to the use of observation, experiment, and calculation to explain how nature operates, and scientists are people who study and try to understand natural phenomena. Scientists guide their work using the **scientific method**, a sequence of steps for systematically analyzing scientific problems in a way that leads to verifiable results. Let's see how geologists employed the scientific method to come up with the meteorite-impact story.

> ❯ *Recognizing the problem:* Any scientific project, like any detective story, begins by identifying a mystery. The cornfield mystery came to light when water drillers discovered that limestone, a rock typically made of shell fragments, lies just below the 15,000-year-old glacial sediment. In surrounding regions, the rock beneath the glacial sediment consists instead of sandstone, a rock made of cemented-together sand grains. Since limestone can be used to build roads, make cement, and produce the agricultural lime used in treating soil, workers stripped

FIGURE BxP.1 An ancient meteorite impact excavates a crater and permanently changes rock beneath the surface.

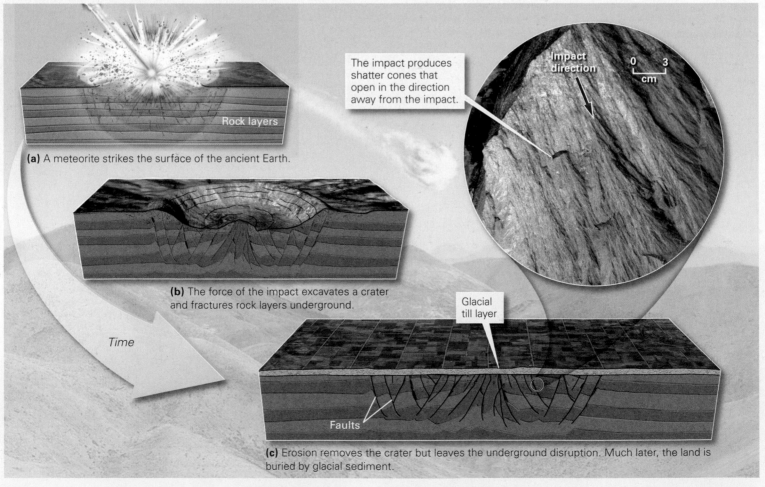

(a) A meteorite strikes the surface of the ancient Earth.

Rock layers

The impact produces shatter cones that open in the direction away from the impact.

Impact direction

0   3
cm

(b) The force of the impact excavates a crater and fractures rock layers underground.

Time

Glacial till layer

Faults

(c) Erosion removes the crater but leaves the underground disruption. Much later, the land is buried by glacial sediment.

off the glacial sediment and dug a quarry to excavate the limestone. They were amazed to find that rock layers exposed in the quarry were tilted steeply and had been shattered by large cracks. In the surrounding regions, all rock layers are horizontal like the layers in a birthday cake, the limestone layer lies underneath a sandstone layer, and the rocks contain relatively few cracks. When curious geologists came to investigate, they soon realized that the geologic features of the land just beneath the cornfield presented a problem to be solved. What phenomena had brought limestone up close to the Earth's surface, had tilted the layering in the rocks, and had shattered the rocks?

> *Collecting data:* The scientific method proceeds with the collection of observations or clues that point to an answer. Geologists studied the quarry and determined the age of its rocks, measured the orientation of the rock layers, and documented (made a written or photographic record of) the fractures that broke up the rocks.

> *Proposing hypotheses:* A scientific **hypothesis** is merely a possible explanation, involving only natural processes, that can explain a set of observations. Scientists propose hypotheses during or after their initial data collection. In this example, the geologists working in the quarry came up with two alternative hypotheses: either the features in this region resulted from a volcanic explosion, or they were caused by a meteorite impact.

> *Testing hypotheses:* Because a hypothesis is just an idea that can be either right or wrong, scientists try to put hypotheses through a series of tests to see if they work. The geologists at the quarry compared their field observations with published observations made at other sites of volcanic explosions and meteorite impacts, and they studied the results of experiments designed to simulate such events. If the geologic features visible in the quarry were the result of volcanism, the quarry should contain rocks formed by the freezing of molten rock erupted by a volcano. But no such rocks were found. If, however, the features were produced by an impact, the rocks should contain **shatter cones**, tiny cracks that fan out from a point. Shatter cones can be overlooked, so the geologists returned to the quarry specifically to search for them and found them in abundance. The impact hypothesis passed the test!

Our description of the scientific method is somewhat idealized, however, because sometimes serendipity works into the process, and scientists make discoveries by chance. Also, because we can't travel through time, we can't always completely test all geologic hypotheses.

Theories are scientific ideas supported by an abundance of evidence; they have passed many tests and have failed none. Scientists are much more confident in the correctness of a theory than of a hypothesis. Continued study in the quarry eventually yielded so much evidence for impact that the impact hypothesis came to be viewed as a theory. You may notice that in everyday conversation, people commonly use the word *theory* as a synonym for an untested or barely tested speculation. In scientific discussion, the word has a much more restricted meaning.

Scientists continue to test theories over a long time. Successful theories withstand these tests and are supported by so many observations that they become part of a discipline's foundation. However, some theories may eventually be disproved and replaced by better ones.

In a few cases, scientists have been able to devise concise statements that completely describe a specific relationship or phenomenon. Such statements, called **scientific laws**, apply without exception for a given range of conditions. Newton's law of gravitation serves as an example—it is a simple mathematical expression that always defines the invisible pull exerted by one mass on another. Note that scientific laws do not in themselves explain a phenomenon, and in this way they differ from theories. For example, the law of gravity does not explain why gravity exists, but the theory of evolution does provide an explanation of why evolution occurs.

FIGURE P.6 Workers excavate limestone in a quarry near Chicago. This rock commonly consists of shells and shell fragments, and can be used in the production of concrete.

Every scientific idea must be tested thoroughly and should be used only when supported by documented observations. Furthermore, scientific ideas do not appear out of nowhere; they are the result of human efforts **(Fig. P.7)**. Wherever possible, this book shows where geologic ideas came from and tries to answer the question, "How do we know that?"

As you read this book, please keep these themes in mind. Don't view geology as a list of words to memorize, but rather as an interconnected set of concepts to digest. Most of all, enjoy yourself as you learn about what may be the most fascinating planet in the Universe.

FIGURE P.7 Geologists work in many environments.

(a) A geologist explores a rock exposure.

(b) The crew of a research ship lowers instruments into the ocean.

(c) Computers aid in processing immense amounts of data.

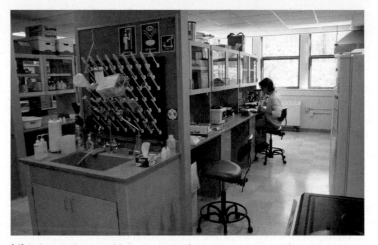

(d) Laboratories provide an opportunity to carry out experiments on Earth materials.

ANOTHER VIEW In this view of central New Zealand from an elevation of about 400 km, we see many components of the Earth System—air, water, ice, rock, life, and human activity.

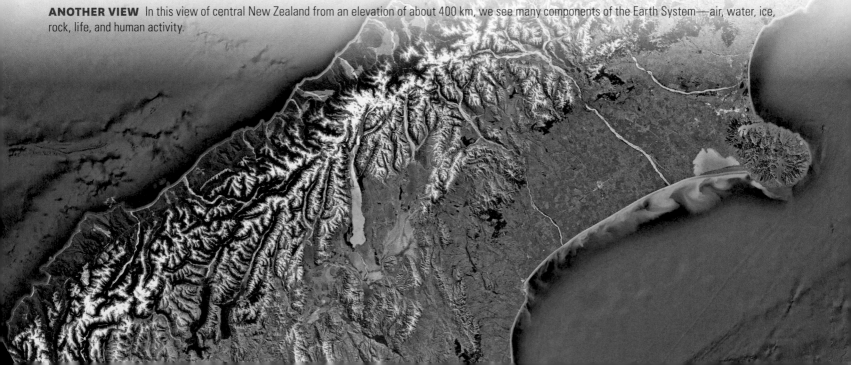

Prelude Review

PRELUDE SUMMARY

> Geologists are scientists who study the Earth. They search for the answers to the mysteries of our home planet, from why volcanoes explode to where we can find diamonds.

> Geologic study can involve field exploration, laboratory experiments, high-tech measurements, and calculations with computers.

> Geologic research not only provides answers to academic questions such as how the Earth formed, but also addresses practical problems like how to find resources and to avoid landslides. Many people pursue careers as geologists.

> A set of themes underlies geologic thinking. For example, the Earth's outer shell consists of moving plates whose interactions produce earthquakes, volcanoes, and mountains; the Earth is very old; and interacting realms of material on the planet comprise the Earth System.

> Research in geology can be guided by the scientific method.

GUIDE TERMS

asthenosphere (p. 6)
Earth System (p. 6)
external process (p. 6)
geologic time (p. 7)
geologic time scale (p. 7)

geologist (p. 3)
geology (p. 4)
gravity (p. 6)
hypothesis (p. 11)
internal process (p. 6)

lithosphere (p. 6)
plate (p. 6)
science (p. 10)
scientific law (p. 11)
scientific method (p. 10)

shatter cone (p. 11)
theory (p. 11)
theory of plate tectonics
   (p. 6)

REVIEW QUESTIONS

The letters following each Review Question refer to the corresponding Learning Objective from the Chapter Opener.

1. What are some of the practical applications of geology? (**A**)

2. Explain the difference between internal processes and external processes. (**B**)

3. How would the Earth's atmosphere differ if life didn't exist? (**B**)

4. Explain the difference between a hypothesis and a theory, in the context of science. (**C**)

5. What are the main layers of the Earth's interior? (**F**)

6. What is the basic premise of the theory of plate tectonics? What are the arrows shown on the map? (**D**)

7. What do geologists mean by the statement: the Earth is a complex system? (**E**)

8. What are the sources of data that geologists can use to understand the Earth? (**C**)

9. What are the major subdivisions of geologic time? Which time unit is longer, the Precambrian or the Paleozoic? (**B**)

10. This mine truck carries 100 tons of coal. Where does this resource, and others like it, come from? (**B**)

CHAPTER 1

THE EARTH IN CONTEXT

By the end of this chapter, you should be able to . . .

A. characterize how people's perceptions of the Earth's place in the Universe have changed over the centuries.

B. explain modern concepts concerning the basic architecture of our Universe and its components.

C. outline the premises of the expanding Universe and Big Bang theories for the formation of our Universe.

D. explain the nebula theory, a scientific model that describes how stars and planets form.

E. describe the nature of the magnetic field and atmosphere that surround our planet.

F. list the distinct interacting realms within the Earth System.

G. distinguish the internal layers (crust, mantle, and core) of the Earth.

H. explain the relationship between the lithosphere and the asthenosphere.

> I believe everyone should have a broad picture of how the Universe operates and our place in it.
> It is a basic human desire. And it also puts our worries in perspective.
>
> STEPHEN HAWKING (British cosmologist, 1942–2018)

1.1 Introduction

Sometime in the distant past, perhaps more than 50,000 generations ago, our ancestors developed the capacity for complex, conscious thought. This amazing ability, which distinguishes our species from all others, brought with it the gift of curiosity, an innate desire to understand and explain the workings of all our surroundings—of our *Universe*. Questions that we ask about the Universe differ little from questions a child asks of a new friend: Where do you come from? How old are you? Such musings first spawned legends in which heroes used supernatural powers to mold the planets and sculpt the landscape. Eventually, researchers applied scientific principles to **cosmology,** the study of the overall structure and history of the Universe.

In this chapter, we begin with a brief introduction to the principles of *scientific cosmology*. We explain the basic architecture of the Universe, introduce the Big Bang theory for the formation of the Universe, and discuss the nebular theory for the birth of the Solar System. Finally, we characterize our home planet by building an image of its surroundings, surface, and interior. This introductory high-speed tour of the Earth provides a context for the remainder of this book.

1.2 An Image of Our Universe

STARS, GALAXIES, AND BEYOND

Think about the mysterious spectacle of a clear night sky. What are the objects that sparkle up there? How far away are they? How do they move? How are they arranged? In addressing such questions, ancient philosophers learned to distinguish *stars* (points of light whose locations remain fixed, relative to each other) from *planets* (tiny spots of light that move relative to the backdrop of stars). Over the centuries, two schools of thought developed concerning how to explain the configuration of stars and planets and their relationships

to the Earth, Sun, and Moon. The first school advocated a **geocentric model (Fig. 1.1a)**, in which the Earth sits motionless at the center of the Universe while the Moon and the planets whirl around it, all within a revolving globe of stars. The second school advocated a **heliocentric model (Fig. 1.1b)**, in which the Sun lies at the center of the Universe while the Earth and other planets orbit around it.

FIGURE 1.1 Contrasting views of the Universe, as drawn by artists hundreds of years ago.

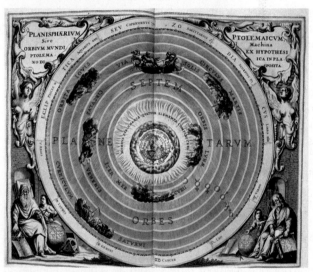

(a) Ptolemy's geocentric image of the Universe puts the Earth at the center.

(b) The heliocentric image of the Universe shows the Sun at the center, as envisioned by Copernicus.

◄ (facing page) The Hubble Space Telescope can see into space from its perch in Earth orbit above the atmosphere. In this photo, taken through the telescope, we see gas and dust in Nebula S106 (3,300 light-years from the Earth), a birthplace of new stars.

BOX 1.1 CONSIDER THIS...

The Basics of Matter, Force, and Energy

MATTER, ATOMS, AND MOLECULES

The material substance of the Universe consists of *matter*—it takes up space and you can feel it. We refer to the amount of matter in an object as its *mass*, so an object with greater mass contains more matter. *Density* refers to the amount of mass occupying a given volume of space—a cubic centimeter of a denser material contains more mass than does a cubic centimeter of a less-dense material.

An **element** is matter that cannot be subdivided into other components with different properties. If you were to keep subdividing an element into smaller and smaller pieces that have the same properties, you would end up with an **atom**, the smallest piece of an element that has the properties of the element. To picture how tiny atoms are, keep in mind that about 5 to 10 trillion atoms fit within the area of the period at the end of this sentence. We refer to elements by abbreviations, called *chemical symbols*, such as H (hydrogen), O (oxygen), Si (silicon), and Fe (iron).

Atoms themselves can be subdivided. Specifically, a single atom can contain three types of *subatomic particles*: *protons* that have a positive charge, *neutrons* that have a neutral charge, and *electrons* that have a negative charge. (Simplistically, the term "charge" refers to the electrical behavior of the particle.) Protons and neutrons, which are about the same size, stick together in a dense ball, the **nucleus**, at the center of an atom—the "glue" that holds particles together in a nucleus is called a *nuclear bond*. Electrons are much smaller than protons or neutrons, and swirl around the nucleus in an *electron cloud*—it's the outer edge of the electron cloud that defines the surface of an atom **(Fig. Bx1.1a)**.

An atom of one element differs from an atom of another in terms of the number of protons in its nucleus. We specify this quantity as the **atomic number** of the element. Hydrogen atoms have an atomic number of 1, helium has an atomic number of 2, and iron has an atomic number of 26. With the exception of the most common form of hydrogen, all nuclei contain neutrons. The sum of the number of protons plus the number of neutrons roughly equals the **atomic mass** of an atom. All atoms, except for hydrogen, have neutrons, so for all atoms except hydrogen, the atomic mass is more than the atomic number.

Atoms tend not to exist in isolation, but rather attach to other atoms, connected by an invisible "glue" called a *chemical bond*. Scientists use the word **molecule** for a particle composed of a combination of two or more atoms bonded together. Some molecules contain atoms of only the same element, whereas others contain atoms of two or more different elements. A substance in which molecules consist of two or more elements is a *compound*. We can specify the composition of a molecule by providing its *chemical formula*, a recipe which uses symbols to represent names and proportions of elements in a compound. For example, hydrogen gas consists of H_2 molecules, oxygen gas of O_2 molecules, water of H_2O molecules, and methane of CH_4 molecules. A molecule of methane contains one carbon atom and four hydrogen atoms. Note that the smallest piece of a compound that has the characteristics of the compound is a molecule, but that not all molecules are compounds. (Molecules that contain two or more atoms of the same element are not considered to be compounds.)

In everyday life, we see matter in four forms, called *states of matter* **(Fig. Bx1.1b)**. In a **gas**, atoms and/or molecules are free to move with respect to one another, so a gas can expand to fill a volume that it occupies. In a **liquid**, atoms and/or molecules are loosely connected, so a liquid can flow and assume the shape of its container, but it will not change its volume as it flows. In a **solid**, atoms and/or molecules are strongly connected, so a solid can retain its shape for a long time. Under very high temperatures, plasma, a fourth state of matter, forms—*plasma* is a gas in which the electrons have been stripped from atoms.

For millennia, the geocentric model garnered the most followers, due to the influence of an Egyptian mathematician, Ptolemy (100–170 C.E.), who developed equations that appeared to predict the wanderings of the planets in the context of the model. During the Middle Ages (ca. 476–1400 C.E.), church leaders in Europe adopted the geocentric model as dogma, because it justified the comforting thought that humanity's home occupies the most important place in the Universe, the center.

Then came the Renaissance. In 15th-century Europe, bold thinkers spawned a new age of exploration and scientific discovery. Thanks to the efforts of Nicolaus Copernicus (1473–1543) and Galileo Galilei (1564–1642), people gradually came to realize that the Earth and planets did indeed orbit the Sun, so the Earth could not possibly sit at the center of the Universe, and that orbits are elliptical, not cylindrical. As the Renaissance progressed, Isaac Newton (1642–1727) established the foundations of physics and optics, and others began to understand fundamental phenomena of chemistry. In the light of this work, the language of modern science came into focus. Using this language, we now define the **Universe** as all of space and the matter and energy

FORCE AND ENERGY

Matter in the Universe does not sit still. Objects spin, they move from one place to another, and they pull on or push against each other. Isaac Newton showed that one mass can exert a *force* on another mass, and by doing so, change the velocity (speed and/or direction) of the mass. A change in velocity is called an *acceleration*, so Newton wrote this relationship as a scientific law: $F = ma$. In this equation, F is force, m is mass, and a is acceleration.

Forces can be applied in two ways. A *contact force* happens when one mass touches another—for example, you apply a contact

FIGURE Bx1.1 The nature of atoms and matter.

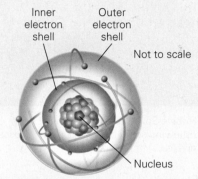

(a) An image of an atom with a nucleus orbited by electrons.

force when you strike a ball with a bat and send the ball careening away. A *field force* happens when an object applies a force to another object without touching it. Examples of field forces include *gravity*, which is the attraction that one mass exerts on another, and *magnetism*, which is the attraction or repulsion that a magnet or electric current exerts on magnetic or electrically charged materials.

Simplistically, *work*, in the jargon of physics, is the consequence of a force acting on an object over a distance. For example, when you lift a weight by 1 meter, against the pull of gravity, you have done work. If you lift the same weight by 2 meters, you have done twice as much work. You also do work when you heat a pot of water and the water turns to steam that rises into the air. In other words, work happens when a mass has changed its position or state.

Physicists refer to the ability or capacity of a system (a specified region of space, of any size, and all it contains) to do work as **energy**. A moving object, light, magnetism, and gravity all can cause matter to change in some way, and thus can do work, so they provide energy. We can distinguish between *kinetic energy*, the energy of motion, and *potential energy*, the energy stored in a mass. A flying ball has kinetic energy, which can do work when it strikes a window and shatters the glass. A boulder sitting on a hillslope has potential energy, because gravitational force is pulling on it, so it can move (bounce downhill) if the dirt beneath it gives way. Similarly, a liter of gasoline contains potential energy, in that if it burns, it can cause a car to move.

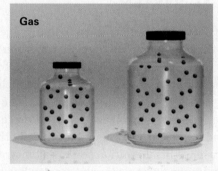

(b) The three common states of matter. Solids keep their shape, liquids conform to the shape of their container without changing density, and gases expand to fill their container.

within it. **Box 1.1** provides a brief refresher on the language for discussing energy.

As telescopes and other instruments improved so that astronomers could see and measure features progressively farther into space, the interpretation of stars evolved. Although it looks like a point of light, a **star** is actually an immense ball of incandescent gas that emits intense heat and light. Gravity holds stars together in immense groups called **galaxies**. Our Sun, together with over 300 billion other stars, makes up the *Milky Way Galaxy*. From our vantage point on Earth, the Milky Way looks like a hazy band **(Fig. 1.2a)**, but if we could

view the Milky Way from a great distance, it would resemble a flattened spiral with great curving arms slowly swirling around a glowing, disk-like center **(Fig. 1.2b)**. Presently, our Sun lies near the outer edge of one of these arms, orbiting the center of the Milky Way once every 250 million years. Astronomers estimate that as many as a trillion galaxies constitute the visible Universe **(Fig. 1.2c)**. Clearly, human understanding of the Earth's place in the Universe has evolved radically over the past few centuries. Today, we realize that neither the Earth, nor the Sun, nor even the Milky Way occupies the center of the Universe—and everything is in motion.

FIGURE 1.2 A galaxy may contain about 300 billion stars.

(a) The Milky Way on a clear night. The "haze" actually consists of millions of faraway stars. A comet appears in the lower right.

(b) A spiral galaxy that looks like the Milky Way, as viewed from the top.

(c) A Hubble Space Telescope view of deep space showing some of the billions of galaxies in the Universe.

WELCOME TO THE NEIGHBORHOOD: OUR SOLAR SYSTEM

Our Sun's gravitational pull holds on to many objects that, together with the Sun, comprise the **Solar System (Fig. 1.3a)**. Most of the mass of the Solar System—99.8%, to be exact—resides in the Sun itself. The remaining 0.2% includes a great variety of objects, the largest of which are the planets. Astronomers define a **planet** as an object that orbits a star, is roughly spherical, and has "cleared its neighborhood of other objects." The last phrase in this definition sounds a bit strange at first, but it merely implies that a planet's gravity has pulled in all particles of matter that had once been in its orbit. According to this definition, formalized in 2005, our Solar System includes eight planets—Mercury, Venus, Earth, Mars, Jupiter, Saturn, Uranus, and Neptune. Until 2005, astronomers considered one more object, Pluto, to be a planet. But Pluto does not fit the modern definition of a planet because it has not cleared its orbit of other objects, so it has been dropped from the roster. The eight planets orbit the Sun in the same direction and more or less in the same plane, called the **ecliptic (Fig. 1.3b)**. Our Solar System is not alone in hosting planets. Astronomers have located thousands of *exoplanets*, planets that orbit stars other than our Sun.

Planets in our Solar System differ radically from one another in both size and composition. The *inner planets* (Mercury, Venus, Earth, and Mars), the ones closer to the Sun, are relatively small. We call them the **terrestrial planets** because,

like the Earth, they consist of a mantle of rock surrounding a core of metal. The *outer planets* (Jupiter, Saturn, Uranus, and Neptune) are known as the **giant planets**, or Jovian planets. The largest, Jupiter, contains 318 times as much mass as the Earth and accounts for about 71% of the non-solar mass in the Solar System. The overall composition of giant planets differs markedly from that of terrestrial planets. Although rock and metal lie at their centers, most of the mass of Jupiter and Saturn consists of hydrogen and helium gas. These are known as *gas-giant* planets. Uranus and Neptune also contain hydrogen and helium gas, but in addition, they host large volumes of water, ammonia, and methane ice and are known as *ice-giant* planets. Note that researchers use the word *ice* to mean any solid compound that could evaporate relatively easily under conditions found at the Earth's surface.

In addition to the planets, the Solar System contains a great many smaller objects. A **moon** is a sizable body locked in orbit around a planet. All but two planets (Mercury and Venus) have moons in varying numbers—the Earth has 1, Mars has 2, and Jupiter has at least 67. Some moons, such as the Earth's Moon, are large and spherical, but many are small and have irregular shapes. **Asteroids** are rocky or metallic objects, with diameters ranging from 1 cm to about 930 km. Millions of asteroids occupy a belt between the orbits of Mars and Jupiter. About a trillion bodies of ice lie outside the orbit of Neptune. Most of these icy bodies are tiny, but a few (including Pluto) are spheres that have diameters as large as 2,300 km and are known as *dwarf planets*. The gravitational pull of planets has

FIGURE 1.3 The relative sizes and positions of planets in the Solar System.

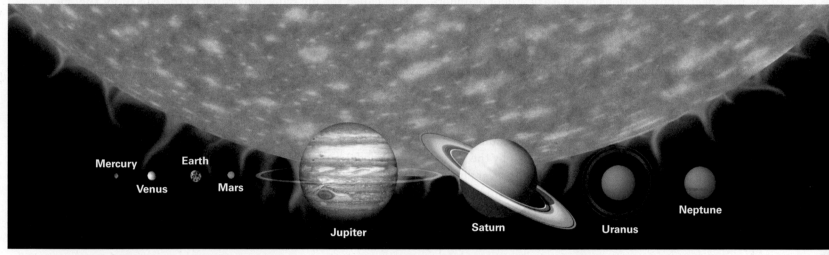

(a) Relative sizes of the planets. All are much smaller than the Sun, but the gas-giant planets are much larger than the terrestrial planets. Jupiter has a diameter about 11.2 times greater than that of the Earth.

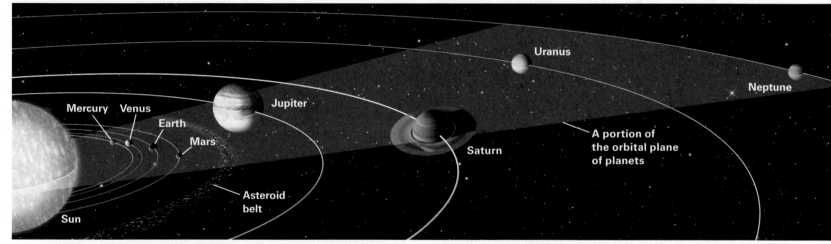

(b) Relative positions of the planets. This figure is not to scale. If the Sun in this figure was the size of a large orange, the Earth would be the size of a sesame seed 15 meters away. Note that all planetary orbits lie roughly in the same plane.

sent some of the icy objects on paths that take them into the inner part of the Solar System, where they heat up and release long tails of gas and dust—these objects are **comets**.

TAKE-HOME MESSAGE

People once thought the Earth lay at the center of the Universe. Now it's clear that it is one of eight planets orbiting our Sun, one of 300 billion stars of the spiral-shaped Milky Way Galaxy. Hundreds of billions of galaxies speckle the visible Universe.

QUICK QUESTION How do the inner planets differ from the outer planets?

1.3 Forming the Universe

A KEY CLUE: DISCOVERING THAT THE UNIVERSE EXPANDS

We stand on a planet, in orbit around a star, speeding through space on the arm of a galaxy. Beyond our galaxy lie hundreds of billions of other galaxies.

Where did all this matter come from, and when did it first form? For most of human history, a scientific

Did you ever wonder . . .
if galaxies move?

solution to these questions seemed intractable. But in the 1920s, unexpected observations opened a new door of understanding. Astronomers such as Edwin Hubble (1889–1953), after whom the Hubble Space Telescope was named, braved many a frosty night beneath the open dome of a mountain-top observatory in order to aim telescopes into deep space. These researchers were searching for distant galaxies. At first, they documented only the location and shape of newly discovered galaxies, but eventually they also began to document the motion of the galaxies, which they could deduce by studying characteristics of the light emitted by the galaxies. The results yielded a surprise that would forever change humanity's perception of the Universe. They found that all distant galaxies are moving away from the Earth, no matter which direction they looked. Furthermore, the farther away the galaxy lies, the faster it's moving away.

Around 1929, Hubble realized what these observations meant—the volume of the whole Universe must be increasing! This idea came to be known as the **expanding Universe theory**. To picture the expanding Universe, imagine a ball of bread dough with raisins scattered throughout. As the dough bakes and expands into a loaf, each raisin moves away from its neighbors, in every direction **(Fig. 1.4)**. Note how, in the finished loaf, raisins that were farther apart to start with moved more, relative to each other, than did raisins that were closer together to start with. Therefore, the velocity at which raisins move relative to each other depends on the distance between them, just as the velocity at which galaxies move away from each other depends on the distance between them.

THE BIG BANG

Hubble's ideas marked a revolution in cosmological thinking. Now we picture the Universe as an expanding bubble, in which galaxies race away from each other at incredible speeds. This image immediately triggers a key question of cosmology: Did the expansion begin at some specific time in the past? If it did, then that instant would mark the beginning of the physical Universe. Most astronomers have concluded that expansion

FIGURE 1.4 A raisin-bread analogy for the expanding Universe. As the dough expands, each raisin moves farther away from the others.

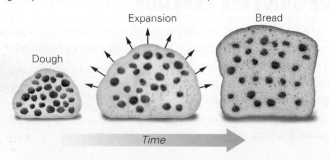

Dough

Expansion

Bread

Time

did indeed begin at a specific time, with a cataclysmic explosion called the Big Bang. According to the **Big Bang theory**, all matter and energy—everything that now constitutes the Universe—was initially packed into an infinitesimally small point. The point exploded and the Universe began, according to current estimates, about 13.8 billion years ago (Ga).

Of course, no one actually saw the Big Bang happen. But by combining clever calculations with careful observations, researchers have developed a scientific model of how the Universe evolved following the Big Bang **(Fig. 1.5a)**. According to this model, the Universe was so dense and so hot during the first instant of its existence that it consisted entirely of energy. At this time, not even the subatomic particles that make up atoms could exist. Within a few seconds, however, the Universe had expanded enough and had cooled enough that protons (hydrogen nuclei) and electrons could begin to form. **(Box 1.2** provides a brief refresher on heat and temperature.) By the time the Universe reached an age of 3 minutes, when its diameter had grown to about 53 million kilometers and its temperature had fallen below 1 billion degrees, nuclear bonds could attach hydrogen nuclei together to form helium nuclei. Note that the process by which the nuclei of atoms merge to form new, larger nuclei is called **nuclear fusion (Fig. 1.5b)**. Researchers refer to the formation of new atomic nuclei in the first few minutes of existence as **Big Bang nucleosynthesis**. This process happened very rapidly and produced only very small atoms, meaning those with atomic numbers of 4 or less. In fact, all of the matter of the Universe—almost entirely in the form of hydrogen and helium nuclei (at a ratio of 3:1) and electrons—had come into existence by the end of the first 5 minutes.

Eventually, the Universe cooled enough for electrons to form clouds around nuclei. Then, chemical bonds could bind atoms together in molecules. Most notably, hydrogen atoms bonded to form molecular hydrogen (H_2). As the Universe expanded and cooled further, atoms and molecules slowed down and accumulated into patchy clouds called **nebulae**. The earliest nebulae of the Universe consisted almost entirely of hydrogen and helium gas. The initial expansion of the Universe took place very rapidly. This *inflationary epoch* lasted less than a second. After that, the Universe expanded at a less rapid rate. Recent research suggests that its expansion has begun to accelerate.

BIRTH OF THE FIRST STARS

When the Universe reached its 200 millionth birthday, it contained immense, slowly swirling nebulae of hydrogen and helium gas separated by vast voids of empty space. Empty space is a **vacuum**, a region in which the density of matter is extremely low. In the regions between nebulae, the vacuum of space may contain less than 1 atom or molecule per cubic

FIGURE 1.5 The concept of the expanding Universe, and of fusion.

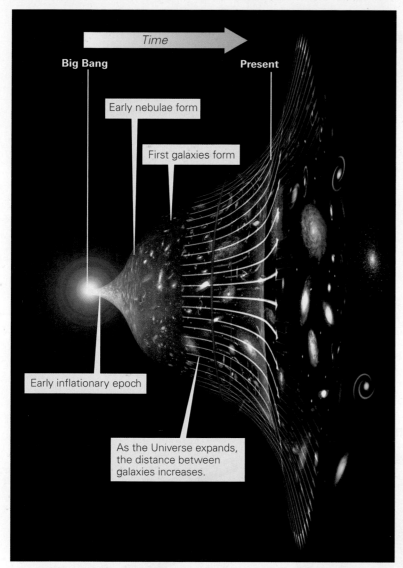

(a) After the Big Bang, the Universe started to expand. It has done so ever since.

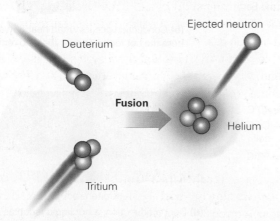

(b) Fusion happens when colliding particles fuse to form a new, larger nucleus. Deuterium and tritium are versions of hydrogen. These nuclei can collide to form a helium nucleus.

meter—by comparison, the air you breathe at sea level contains on the order of 300,000,000,000,000,000,000,000 molecules per cubic meter. The Universe could not remain this way forever, though, because of the invisible but persistent pull of gravity.

All matter exerts gravitational attraction—a type of force—on its surroundings. As Isaac Newton first pointed out, the amount of attraction depends on the amount of mass (the larger the mass, the greater its attraction) and on the distance between the masses (gravitational force increases as objects get closer to each other). Somewhere in the young Universe, gravity produced by an initially more massive region of a nebula began to suck in surrounding matter and, in a grand example of the rich getting richer, grew in mass and therefore in density. As progressively more matter accumulated, it formed into a relatively dense clump. Within this clump, gravity began to pull mass inwards, and hydrogen atoms got close enough for chemical bonding to take place, resulting in formation of hydrogen molecules (see Box 1.1). For reasons discussed in more advanced books, these molecules could more efficiently radiate heat outward. This process cooled the clump somewhat, allowing inward collapse to proceed, so that the density of the clump increased. As it became denser, its gravitational pull became stronger, so it pulled in more matter, and it collapsed more. As such *runaway collapse* took place, the swirling motion of the nebula transformed into spin around an axis, so the once cloud-like nebula evolved into a bulbous plate-like shape called an **accretion disk**. Most material gathered into the central ball of the disk, but faster- moving material fell onto the flattened cloud and remained in orbit around the central ball. As runaway collapse continued, the central ball became so hot that it glowed, and at this time it became a **protostar**.

Notably, the process of runaway collapse, when a protostar was produced out of a nebula consisting only of hydrogen and helium, could happen only where clumps in a nebula contained 50 to 1,000 times as much mass as our Sun. Why? When matter squeezes into a progressively smaller space, its temperature increases dramatically (see Box 1.2). Heat causes gas to expand. This means that a ball of gas can collapse only until the outward push caused by the hot gas balances the inward pull of gravity. Only in a situation where the nebula contains an immense amount of mass can the inward gravitational pull of the mass overcome the outward push, permitting the gas to become dense enough and hot enough to start glowing.

A protostar continues to grow, by pulling in successively more matter, until its core becomes so dense, and its temperature so high (about 10,000,000°C), that hydrogen nuclei within it move incredibly fast and slam together so forcefully that fusion takes place and new helium nuclei form. Fusion reactions produce huge amounts of energy, so when they begin, the protostar "ignites" to becomes a nuclear

BOX 1.2 SCIENCE TOOLBOX...

Heat, Temperature, and Heat Transfer

In everyday English, the words *heat* and *temperature* may seem interchangeable, but in scientific discussion, they have different, distinct definitions. Scientists refer to the total kinetic energy (energy of motion) contained in matter due to the motion of its particles as **thermal energy**, or **heat**. A jar of gas that has been warmed in an oven contains more heat than the same-sized jar that has been sitting in a refrigerator.

In contrast, the **temperature** of a material is a number that represents the average velocity of particles moving within the material. These movements can include the vibration of atoms or molecules in place and the physical movement of atoms or molecules from one location to another. The faster the particles vibrate or move, the higher the temperature. Adding heat to a material causes its temperature to rise. For example, when you place a jar of gas on a hot stove, heat moves into the jar so that the temperature of the gas rises because the gas molecules start to move faster. If, however, you place the jar in a cold refrigerator, heat moves out of the jar, and the gas molecules slow down.

When we say that one object is hotter or colder than another, we are describing its temperature. *Temperature* is a measure of warmth relative to some standard and represents the average kinetic energy of atoms in the material. Scientists use the *centigrade scale* (also known as the Celsius scale) for calibrating changes in temperatures in the metric system of measurement. In the United States, weather reports typically give temperatures using the *Fahrenheit scale* of the English system of measurement. Water freezes at 0°C, or 32°F, and it boils at 100°C, or 212°F. Note that a single degree in the centigrade scale represents a larger temperature change than does a single degree in the Fahrenheit scale.

Heat can be transferred from one object, place, or material to another. When heat is added to a substance, the substance warms, which means that its molecules start to vibrate or move more rapidly. When a substance cools, the motion of its molecules slows. There are four ways in which *heat transfer* takes place in the Earth System.

Radiation takes place when electromagnetic waves transmit energy into a body or out of a body **(Fig. Bx1.2a)**. For example, when sunlight strikes the ground during the day, radiative heating takes place. Similarly, when heat rises from the ground at night, the air warms and the ground cools.

Conduction takes place when you stick the end of an iron bar in a fire **(Fig. Bx1.2b)**. The iron atoms at the fire-licked end of the bar start to vibrate more energetically; they gradually incite atoms farther up the bar to start jiggling, and

these atoms in turn set atoms even farther along in motion. In this way, heat slowly flows along the bar until you feel it with your hand.

Convection takes place when you set a pot of water on a stove **(Fig. Bx1.2c)**. The heat from the stove warms the water at the base of the pot by making the molecules of that water vibrate faster and move around more. As a consequence, the density of the water at the base of the pot decreases. As you heat a liquid, the atoms move away from one another and the liquid expands. For a time, cold water remains at the top of the pot. But eventually, the warmer, less-dense water at depth becomes buoyant relative to the colder, denser water higher up. In a gravitational field, a buoyant material rises if the material above it is weak enough to flow out of the way. Since liquid water can flow easily, the hot water rises. When this happens, cold water sinks to take its place. Thus, during convection, the actual flow of the material itself carries heat. The trajectory of flow defines *convective cells*.

Advection happens when a fluid that flows through cracks and pores within a solid material carries heat with it **(Fig. Bx1.2d)**. The heat brought by the fluid conductively heats up the adjacent solid that the fluid passes through. Advection takes place, for example, where molten rock rises through the crust beneath a volcano and heats up the crust in the process.

FIGURE Bx1.2 The four processes of heat transfer.

(a) Radiation from sunlight warms the Earth.

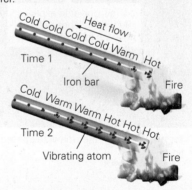

(b) Conduction occurs when heat flows from the hot region toward the cold region.

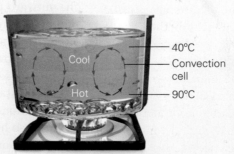

(c) Convection takes place when moving fluid carries heat with it. Hot fluid rises while cool fluid sinks, setting up a convection cell.

(d) During advection, a hot liquid (such as molten rock) rises into cooler material, and heat then conducts from the hot liquid into the cooler material.

furnace, and the ball becomes a true star. The first true star may have formed about 800 million years after the Big Bang, and when this happened, the first starlight pierced the new-born Universe **(Fig 1.6)**. Star formation happened again and again in the young Universe, and soon many first-generation stars had come into existence. The stellar wind produced by the new star blew away the gas rings surrounding it.

First-generation stars tended to be very massive, probably more than 100 times the mass of the Sun. Astronomers have shown that the larger the star, the hotter it burns and the faster it runs out of fuel and dies. A huge star may survive only a few million to a few tens of millions of years before it runs out of fuel, collapses, explodes, and becomes a **supernova**. The reason for this name is that, to modern observers on the Earth, supernovae appear as "new" stars in the night sky, remaining bright for only a short time (days to weeks) before disappearing from sight; *nova* is Latin for new. Not long after the first generation of stars formed, the Universe began to be peppered with the first generation of supernovae.

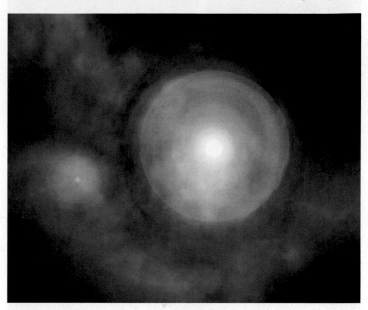

FIGURE 1.6 The first generation of stars formed from massive nebulae containing only hydrogen and helium. Here we see a new star igniting.

TAKE-HOME MESSAGE

All distant galaxies are moving away from us, an observation that requires the Universe to be expanding. According to the Big Bang theory, this expansion, and thus the beginning of our Universe, began with a cataclysmic explosion at about 13.8 Ga. According to the nebular theory, atoms formed during the Big Bang collected into nebulae, which, due to gravity, collapsed into dense balls that began to produce energy by nuclear fusion reactions. These objects were the first stars.

QUICK QUESTION What elements made up most of the nebulae in the very young Universe?

1.4 Forming the Solar System

WE ARE ALL MADE OF STARDUST

The nebulae from which the first-generation stars formed consisted almost entirely of hydrogen and helium, the lightest atoms (atoms with a very small nucleus), because only these atoms were generated by Big Bang nucleosynthesis. In contrast, the Universe of today contains 92 naturally occurring elements. Where did the other elements come from? In other words, how did heavier elements with larger atomic numbers (such as carbon, sulfur, silicon, iron, gold, and uranium) appear? Physicists have shown that intermediate-weight elements form by fusion reactions during the life cycle of stars, a process called **stellar nucleosynthesis**. Because of stellar nucleosynthesis, we can consider stars to be "element

factories," constantly fashioning larger atoms out of smaller atoms. Most heavy atoms, those with atomic numbers greater than that of iron, require even hotter environments to form than generally occurs within a star. In fact, most very heavy atoms form during the inconceivably hot conditions of a supernova explosion, a process called *supernova nucleosynthesis*.

How do the atoms formed in stars get into space? Some escape during the star's lifetime, simply by moving fast enough to overcome the star's gravitational pull. The stream of atoms emitted from a star during its lifetime is a **stellar wind (Fig. 1.7a)**. Some escape only when a star dies. Large stars, as we've noted, blast an immense amount of matter into space during supernova explosions **(Fig. 1.7b)**. A small or medium-sized star (like our Sun) does not explode, but rather balloons to become a *red giant*, when it starts to run out of fuel. Eventually, when the star runs out of fuel entirely, its interior collapses, leaving its outer shells of gas to drift out into space. Once ejected into space, atoms from stars, collapsed red giants, and supernova explosions form new nebulae or mix back into existing nebulae.

Did you ever wonder . . .
where the atoms in your body first formed?

When the first generation of stars died, they left a legacy of new, heavier elements that mixed with residual hydrogen and helium from the Big Bang. A second generation of stars and associated planets formed out of these compositionally more diverse nebulae. When these stars died, they contributed heavier elements to third-generation stars. Succeeding generations contain a progressively greater proportion of

FIGURE 1.7 Element factories in space.

(a) A photo of solar (stellar) wind streaming into space.

(b) This expanding cloud of gas, ejected into space from a supernova explosion whose light reached the Earth in 1054 C.E., is called the Crab Nebula.

heavier elements. Because not all stars live for the same duration of time, at any given moment the Universe contains many different generations of stars. Our Sun may be a third-, fourth-, or fifth-generation star. The mix of elements we find on the Earth includes relics of primordial gas from the Big Bang as well as the disgorged guts of dead stars. Think of it—the atoms that make up your body once resided inside a star!

FORMING THE SOLAR SYSTEM: THE NEBULAR THEORY

Earlier in this chapter, we described how the first stars formed from nebulae of atoms formed during the Big Bang. But we delayed our discussion of how the planets and other objects in our Solar System originated until we had discussed the production of heavier atoms such as carbon, silicon, iron, and uranium, because terrestrial planets consist predominantly of these elements. Now that we've discussed stars as element factories, we can consider the **nebular theory**, a scientific explanation of the origins of our Solar System, including the Sun, and the planets, moons, asteroids, and comets that orbit the Sun.

The process began in a nebula that formed billions of years after the Big Bang. This nebula contained hydrogen and helium, as well as larger atoms formed in stars or supernovae. Some of these atoms remained independent, but others bonded to each other and formed molecules. Over time, clusters of molecules including atoms of heavier elements stuck together, resulting in the growth of tiny particles of ice and dust. These particles played a very important role in Solar System formation, because they could disperse heat outward much more efficiently than could gases of independent atoms or molecules. Without this heat dispersal, the inward collapse of the solar nebula would not have proceeded to the stage of forming a star. An outward-directed *thermal pressure* (a push due to heating a gas) resulting from compression of the nebula would have stopped its inward collapse before matter had squeezed together into a body that was dense enough for nuclear-fusion reactions to begin. Heat dispersal by particles decreased the thermal pressure, allowing collapse of the solar nebula to proceed all the way, until the density of a star had been achieved and our Sun became a nuclear furnace. Note that first-generation stars of the newborn Universe were so massive that they could undergo inward collapse to achieve stellar densities even without heat dispersal by ice or dust particles.

As we've seen, during collapse, motion of the material in a nebula transformed from random swirling into rotation around an axis, to form an accretion disk. According to the nebular theory, the central ball of our accretion disk became the Sun, while the flattened outer part became a **protoplanetary disk**, the region in which planets would eventually form (Fig. 1.8a).

The protoplanetary disk of our Solar System contained all 92 elements, some as isolated atoms and some bonded to others in molecules. Geologists divide the material formed from these atoms and molecules into two classes. **Volatile materials**—such as hydrogen, helium, methane, ammonia, water, and carbon dioxide (CO_2)—can exist as gas at the Earth's surface. In the pressure and temperature conditions of space, all volatile materials remain in a gaseous state closer to the Sun. But beyond a

FIGURE 1.8 Nebular theory for planet formation.

Nebula

Time

Frost line

Protoplanetary disk

Rings of planetesimals

The eight planets

(a) Planets form in the protoplanetary disk that evolves from a nebula containing 92 elements.

(b) The grainy texture of this meteorite fragment may resemble the texture of a small planetesimal.

of refractory dust, whereas the outer portions accumulated large quantities of volatile materials and ice.

How did the dusty, icy, and gassy rings of the protoplanetary disk transform into planets? Even before the proto-Sun ignited, the material of the surrounding rings began to clump and bind together, due to gravity and electrical attraction, forming a series of concentric rings around the proto-Sun. In these rings, matter was denser, and particles got close enough to each other to attract. First, soot-sized particles merged to form sand-sized grains. These then stuck together to form grainy basketball-sized blocks, which in turn collided. If the collision was slow, blocks stuck together or simply bounced apart. If the collision was fast, one or both of the blocks shattered, producing smaller fragments that recombined later. Eventually, enough blocks coalesced to form **planetesimals (Fig. 1.8b)**, bodies whose diameter exceeded about 1 km. Because of their mass, the planetesimals exerted enough gravitational attraction to pull in other objects that were nearby. Planetesimals behaved like vacuum cleaners, sucking in small pieces of dust and ice as well as smaller planetesimals that lay in their orbit, and in the process they grew progressively larger. Eventually, victors in the competition to attract mass grew into **proto-planets**, bodies approaching the size of today's planets. Once a protoplanet succeeded in incorporating virtually all the debris within its orbit, it became a full-fledged planet. Probably occurred...

Early stages in the planet-forming process suggest that it curred very quickly—some computer m... from the dust may have taken less than a million y... Planets may and gas stage to the large plane... million years. have grown from planetesim... disk consisted In the inner orbits, where... formed, whereas in mostly of refractory d... significant amounts the Earth) compose... metallic protoplanets the outer part of... of volatiles r...

distance called the *frost line*, some volatiles can freeze into ice. (Recall that we do not limit use of the word *ice* to solid water alone.) **Refractory materials** are those that melt only at high temperatures, and they condense to form solid soot-sized particles of dust in the coldness of space. As the proto-Sun began to form, the inner part of the disk became hotter, causing volatile elements to evaporate and drift to the outer portions of the disk. Thus, the inner part of the disk consisted predominantly became surrounded by thick shells of gas or ice and evolved into the gas-giant and ice-giant planets. Fragments of materials that were not incorporated in planets remain today as asteroids and comets.

When did the Solar System form? Using techniques introduced in Chapter 10, geologists have found that special types of meteorites thought to be leftover planetesimals formed at 4.57 Ga, and they consider that date to be the birth date of the Solar System. This date means that the Solar System formed about 9 billion years after the Big Bang and thus is only about a third as old as the Universe.

WHY ARE PLANETS ROUND?

Small planetesimals were jagged or irregular in shape, and asteroids today have irregular shapes. Planets and larger moons, on the other hand, are more or less spherical. Why? The rock composing a small planetesimal is cool enough and strong enough that the force of gravity cannot cause the rock inside the object to flow, so irregularities can survive. But once a planetesimal grows beyond a certain critical size (about 300 to 500 km in diameter), its interior becomes warm enough and soft enough to flow in response to the pull of gravity. As a consequence, protrusions are pulled inward toward the center, and the planetesimal takes on a special shape—a sphere—in which mass is evenly distributed around the center, so the force of gravity can be nearly the same at all points on its surface (**Geology at a Glance**, pp. 28–29).

DIFFERENTIATION OF THE EARTH

When planetesimals started to form, they had a fairly homogeneous distribution of material throughout, because the smaller pieces from which they formed all had much the same composition and collected together in no particular order. But large planetesimals (about 500 km in diameter) did not stay homogeneous for long, because they began to heat up inside. The heat came primarily from three sources: the heat produced during collisions (similar to the heat produced when you bang on a nail with a hammer and they both get warm); the heat produced when matter squeezed into a smaller volume (see Box 1.2); and the heat produced from the decay of radioactive elements. In bodies whose temperature rose sufficiently to cause internal melting, denser metals (mostly iron) separated out and sank to the center of the body, whereas relatively lighter rocky materials remained in a shell surrounding the center. By this process, called **differentiation**, protoplanets and large planetesimals developed internal layering early in their history **(Fig. 1.9)**. As we will see later in this chapter, the Earth's central ball of metal constitutes the body's *core* and the rocky shell surrounding it is its *mantle*. Recent studies suggest that the terrestrial planets had differentiated by about 4.56 Ga.

TAKE-HOME MESSAGE

Heavier elements formed in stars and supernovae added to gases in nebulae from which new generations of stars formed. Planets formed from rings of dust and ice orbiting the stars, so we are all formed of stardust. When they became large enough, planets differentiated, with denser materials sinking to the center.

QUICK QUESTION Why do the giant planets orbit farther from the Sun than do the terrestrial planets?

FIGURE 1.9 Differentiation of the Earth's interior.

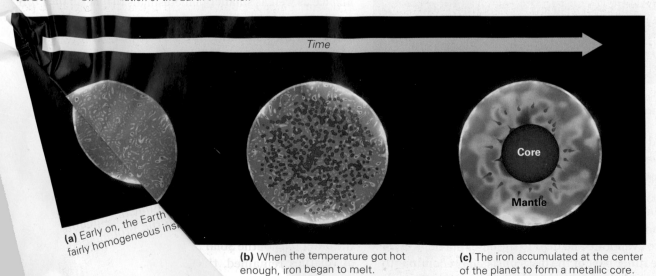

(a) Early on, the Earth fairly homogeneous ins

(b) When the temperature got hot enough, iron began to melt.

(c) The iron accumulated at the center of the planet to form a metallic core.

1.5 The Earth's Moon and Magnetic Field

The Earth does not orbit the Sun all on its own. Rather, the near space around the Earth encompasses two important entities—the Moon and the magnetic field.

THE MOON AND ITS FORMATION

As seen from Earth, the Moon glows an eerie white. Its radius of 1,737 km is about a quarter of the Earth's. Even with the naked eye, you can see that its surface differs from that of the Earth. The whiter areas, the lunar highlands, have been pockmarked by countless craters, whereas the darker areas, the maria (singular: mare), are smoother and less cratered **(Fig. 1.10)**. Gravitational interaction between the Earth and the Moon causes the same side of the Moon to face the Earth, so humans did not know what the far side of the Moon looked like until satellites and astronauts sent back pictures.

> **Did you ever wonder . . .**
> if the Moon is as old as the Earth?

How did the Moon form? Most geologists favor a model in which the Moon resulted when a protoplanet, perhaps comparable in size to Mars, collided with the newborn Earth. In the process, the colliding body disintegrated and evaporated, along with a large part of the Earth's mantle (see Geology at a Glance). A ring of debris formed around the remaining, now molten Earth. Gravity quickly caused this

FIGURE 1.10 The Earth's moon has distinct maria (dark areas) and lighter-colored, intensively cratered highlands.

material to coalesce into the Moon. Not all moons in the Solar System necessarily formed in this manner. Some may have been independent protoplanets or comets that were captured by a larger planet's gravity.

THE EARTH'S MAGNETIC FIELD

A **magnetic field**, in a general sense, is the region affected by the force emanating from a magnet. Magnetic force, which grows progressively stronger closer to the magnet, can attract or repel another magnet and can cause charged particles to move. The Earth has a magnetic field, which is largely a **dipole**, meaning that it has two different ends—a north pole and a south pole—just like the magnetic field around a familiar bar magnet **(Fig. 1.11a, b)**. When you bring two bar magnets close to one another, their opposite poles attract (pull) and their like poles repel (push). By convention, we represent the orientation of a magnetic dipole by an arrow that points from the south pole to the north pole, and we represent the magnetic field of a magnet by a set of invisible magnetic field lines that curve through the space around the magnet and enter the magnet at its poles. Arrowheads along these lines point in a direction to complete a loop. Magnetized needles, such as iron filings or compass needles, when placed in a field, align with the field lines.

If we represent the Earth's magnetic field as emanating from an imaginary bar magnet in the planet's interior, the north pole of this bar lies near the south geographic pole of the Earth, whereas the south pole of the bar lies near the north geographic pole. (The *geographic poles* are the places where the Earth's axis of rotation intersects the planet's surface.) By convention, geologists and geographers refer to the magnetic pole closer to the north geographic pole as the *north magnetic pole*, and the magnetic pole closer to the south geographic pole is called the *south magnetic pole*. This way, the north-seeking end of a compass points toward the north geographic pole, since opposite ends of a magnet attract.

The Sun's stellar wind, known as the *solar wind*, interacts with the Earth's magnetic field, distorting it into a huge teardrop pointing away from the Sun. The solar wind consists of dangerous, high-velocity charged particles. Fortunately, the magnetic field deflects most (but not all) of the particles, so that they do not reach the Earth's surface. In this way, the magnetic field acts like a shield against the solar wind; the region inside this magnetic shield is called the **magnetosphere (Fig. 1.11c)**. At distances of 3,000 to 10,500 km out from the Earth, the *Van Allen radiation belts*, named for an American physicist who first recognized them in 1959, have developed. These belts, which lie well within the magnetosphere, trap solar wind particles as well as cosmic rays (nuclei of atoms emitted from supernova explosions) that were moving so fast that they were able to

Forming the Planets and the Earth-Moon System

2. Gravity pulls gas and dust inward to form a protoplanetary disk. Eventually, a glowing ball—the proto-Sun—forms at the center of the disk.

1. Forming the Solar System, according to the nebular theory: A nebula forms from hydrogen and helium left over from the Big Bang, as well as from heavier elements that were produced by fusion reactions in stars or during explosions of stars.

6. Gravity reshapes the proto-Earth into a sphere. The interior of the Earth differentiates into a core and mantle.

5. Forming the planets from planetesimals: Planetesimals grow by continuous collisions. Gradually, an irregularly shaped proto-Earth develops. The interior heats up and becomes soft.

7. Soon after the Earth forms, a protoplanet collides with it, blasting debris that forms a ring around the Earth.

8. The Moon forms from the ring of debris.

3. "Dust" (particles of refractory materials) concentrates in the inner rings, while "ice" (particles of volatile materials) concentrates in the outer rings. Eventually, the dense ball of gas at the center of the disk becomes hot enough for fusion reactions to begin. When it ignites, it becomes the Sun.

4. Dust and ice particles collide and stick together, forming planetesimals.

9. Eventually, the atmosphere develops from volcanic gases. When the Earth becomes cool enough, moisture condenses and rains fill the oceans. Some gases may be added by passing comets.

FIGURE 1.11 A magnetic field permeates the space around the Earth. It can be represented by a bar magnet.

(a) A bar magnet produces a magnetic field. Magnetic field lines point into the "south pole" and out from the "north pole."

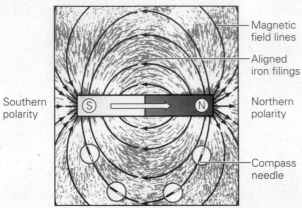

Magnetic field lines

Aligned iron filings

Southern polarity

Northern polarity

Compass needle

(b) We can represent the Earth's magnetic field by an imaginary bar magnet inside.

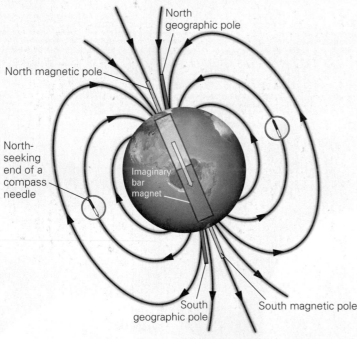

North geographic pole

North magnetic pole

North-seeking end of a compass needle

Imaginary bar magnet

South geographic pole

South magnetic pole

(c) The Earth behaves like a magnetic dipole, but the field lines are distorted by the solar wind. The Van Allen radiation belts trap charged particles.

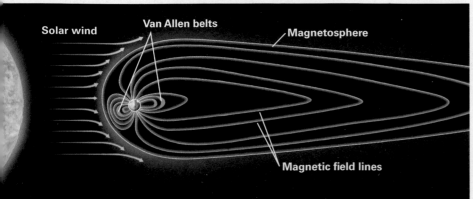

Solar wind

Van Allen belts

Magnetosphere

Magnetic field lines

FIGURE 1.12 The Earth, as seen from space, has an overall bluish glow. You can distinguish land, sea, clouds, ice, and vegetation.

penetrate the weaker outer part of the magnetic field. Some charged particles make it past the Van Allen belts and flow along magnetic field lines to the polar regions of the Earth.

TAKE-HOME MESSAGE

A collision of a protoplanet with the Earth soon after formation of the Earth may have yielded the material from which the Moon formed. The Earth produces a magnetic field that deflects solar wind.

QUICK QUESTION What is a magnetic pole?

1.6 **The Earth System**

So far in this chapter, we've discussed how the Universe, and then the Solar System, formed. Now let's focus on our home planet and develop an image of the Earth's overall architecture. To develop this image, pretend that we are explorers from space rocketing toward the Earth for a first visit. As our spacecraft approaches the Earth, we admire its beautiful bluish glow **(Fig. 1.12)**. At an elevation of about 10,000 km (6,200 miles) above the surface, we slow the spacecraft and go into orbit. The Earth almost fills our field of view, and we can see that its surface and surroundings consist of several distinct realms, which we begin to explore. These realms, as well as the interactions among the realms over time, comprise the **Earth System** (see Geology at a Glance, pp. 8–9).

THE ATMOSPHERE

At an elevation of 10,000 km, our instruments detect a slight increase in the density of gas outside. This change marks the edge of the Earth's **atmosphere**, the mixture of gas known as *air* that surrounds the planet **(Fig. 1.13a)**. The weight of

overlying atmosphere pushes down on the atmosphere below, so both the density of air and the *air pressure* (the amount of push that the air exerts on material around it) increases progressively closer to the surface **(Fig. 1.13b)**.

We can specify pressure in units of force per unit area. One such unit, an *atmosphere* (abbreviated atm), has a value of 1.03 kilograms per square centimeter (14.7 pounds per square inch),

Did you ever wonder . . .
how thick our atmosphere is?

FIGURE 1.13 Characteristics of the atmosphere that envelops the Earth.

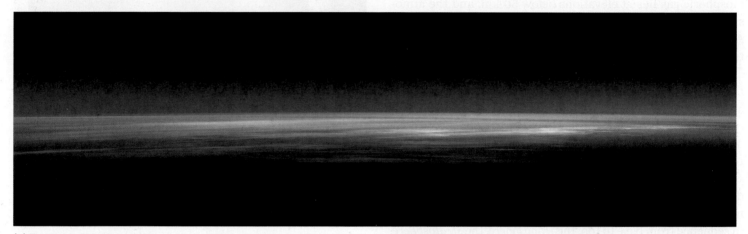

(a) The haze of the atmosphere fades up into the blackness of space.

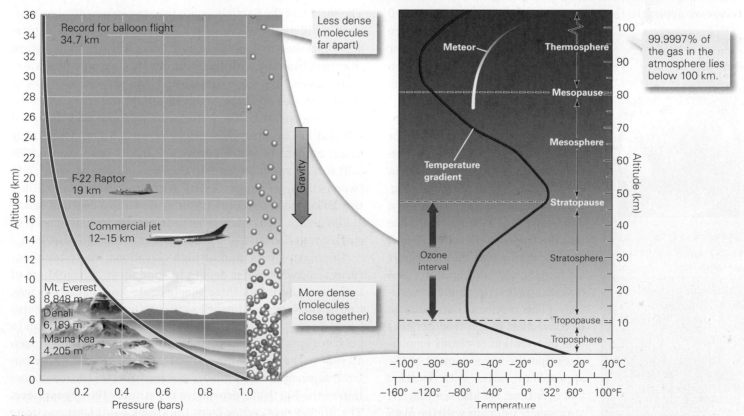

(b) Molecules pack together more tightly at the base of the atmosphere, so atmospheric pressure changes with elevation.

(c) The atmosphere can be divided into several distinct layers. We live in the troposphere.

which represents the average air pressure at sea level. In scientific discussions, another unit, the *bar*, is commonly used for defining pressure—an atmosphere and a bar are almost the same: 1 atm = 1.01 bar.

The highest permanent town in the world sits at an elevation of about 5.1 km. That's because above this elevation, air doesn't contain enough oxygen molecules to maintain long-term human survival. In fact, air pressure (and therefore, density) decreases by half for every 5.6 km that you rise above sea level, so at the peak of Mt. Everest, the highest point of the planet (8.85 km above sea level), air pressure is only 0.3 bars. Because of the decrease in density with elevation, 99% of atmospheric gas lies at elevations below 50 km, and the atmosphere becomes barely detectable at elevations above 120 km. At elevations of between 700 and 10,000 km, the atmosphere transitions into the vacuum of interplanetary space. By analyzing samples of dry air from the Earth's atmosphere, we find that overall, it currently consists of 78% nitrogen (N_2) and 21% oxygen (O_2). The remaining 1% consists of a variety of trace gases, including argon, carbon dioxide, neon, methane, ozone, carbon monoxide, and sulfur dioxide. Humidity, or water content, of air varies with time and location.

The character of the atmosphere changes with increasing distance from the Earth's surface. Because of these changes, atmospheric scientists divide the atmosphere into layers named, in sequence from base to top: the *troposphere*, the *stratosphere*, the *mesosphere*, and the *thermosphere* (**Fig. 1.13c**). The term *exosphere* applies to the transition from air to space above 700 km. Boundaries between layers are named for the underlying layer. For example, the boundary between the troposphere and the overlying stratosphere is the *tropopause*.

As we circle the Earth, we can see the swirls and streaks of *clouds*, mists of tiny water droplets or hazes of tiny ice crystals that float in the air of the troposphere (see Fig. 1.12). The complex distribution and shapes of clouds emphasize that the troposphere is in constant motion, with *winds* that transport air horizontally and updrafts and downdrafts that transport air vertically. (This motion occurs because of convection in the troposphere.) The distribution of clouds emphasizes that the water content of the atmosphere varies with location and over time. We also see that the atmosphere interacts with the magnetosphere. The charged particles

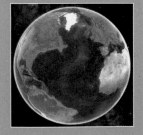
FIGURE 1.14 The aurorae are an atmospheric phenomenon.

(a) Charged particles flow toward the Earth's magnetic poles.

(b) Particles cause gases in the atmosphere to glow, forming colorful aurorae in polar skies.

that stream along the Earth's magnetic field lines toward the poles interact with gases in the atmosphere and causes them to glow, creating spectacular *aurorae*, pulsating sheets of color, in the atmosphere (**Fig. 1.14**).

THE HYDROSPHERE AND CRYOSPHERE

The surface of the Earth beneath the atmosphere displays a variety of different textures and colors. Our first task as explorers is to make a map of the planet. What features should go on this map? Starting with the most obvious, we note that water covers about 70% of the surface and land covers about 30% of the surface (see Fig. 1.12). About 98% of the water resides in the *oceans*, regions where the water contains about 3.5% dissolved salt and has an average depth of about 4.5 km. A relatively minor amount of water occurs as "fresh" (less than 0.05% salt) surface water on land, in the form of lakes, streams, and swamps. A larger amount exists as **groundwater**, water filling tiny holes and cracks underground, down to a depth of several kilometers. Geologists refer to the realm including oceans, surface water on land, and groundwater as the **hydrosphere**.

In polar regions, and at high elevations where temperatures are colder, water occurs in solid form, as ice. Much of this ice has built into thick sheets or streams, called *glaciers*, that last all year and slowly flow in response to gravity. Ice also forms a relatively thin sheet over the surface of the ocean in polar regions, and near-surface groundwater remains frozen all year in the form of permanently frozen ground, or *permafrost*, adjacent to many glacier-covered areas. The frozen portion of the Earth's hydrosphere constitutes the **cryosphere**. The prefix *cryo–* comes from the Greek word *kryo*, meaning very cold or ice-like.

THE GEOSPHERE

The solid Earth, from the surface to the center, constitutes the **geosphere**. As we've seen, *land* refers to regions where the surface of the geosphere is exposed to the atmosphere. The *seafloor* is the surface portion submerged beneath the oceans. Today's Earth includes several large areas of land (continents) and thousands of smaller ones (islands). As we'll see later in the book, the continents that we observe today have not always existed. At times in the past, they were merged into vast supercontinents or were divided into smaller pieces. Also, because sea level rises and falls over Earth history, some regions that are now dry land were once submerged by shallow seas, and some regions that are now submerged were once dry. Similarly, because glaciers grow and shrink over Earth history, some areas of dry land today were once covered by glaciers.

The surface of the geosphere is not perfectly smooth **(Fig. 1.15)**. From our spacecraft, we can see that the land surface elevation ranges from a few hundred meters below sea level (the shore of the Dead Sea) to 8.85 km above sea level (the peak of Mt. Everest). We refer to this variation in land surface elevation, defining the shape of the land surface, as **topography**.

Our instruments also allow us to study **bathymetry**, variations in the depth of the seafloor. Along the edges of many continents, we observe *continental shelves*, regions where water depths are generally less than 200 m. The seafloor descends down to broad flat regions, called *abyssal plains*, at depths of between 4 and 5 km. Elongate submarine mountain belts, known as *mid-ocean ridges*, where water decreases to less than 2.5 km deep, serve to separate abyssal plains from one another. Here and there, individual submarine mountains, *seamounts*, protrude from the seafloor, and some of these rise above the surface as *sea islands*. The deepest parts of the oceans occur in elongate *trenches*. Water in trenches can reach depths of 11 km. Note that the total vertical distance between the deepest point in the sea and the highest point on land is 19.8 km, only about 0.3% of the Earth's radius (6,371 km).

What does the surface of the geosphere consist of? Our exploratory probes make a preliminary analysis and find the following **Earth materials**:

> *Organic chemicals:* A carbon-containing compound that either occurs in living organisms or has characteristics that resemble those of compounds in living organisms is called an **organic chemical**.
> *Minerals:* A solid, natural substance in which atoms are arranged in an orderly pattern is a **mineral**. A coherent sample of a mineral that grew to its present shape is a **crystal**. Most minerals are *inorganic* chemicals, meaning that they are not organic.
> *Glass:* A solid in which atoms are not arranged in an orderly pattern is called **glass**.

FIGURE 1.15 This map of the Earth shows variations in elevation on both the land surface and the seafloor. Darker blues are deeper water in the oceans. Greens are lower elevation on land.

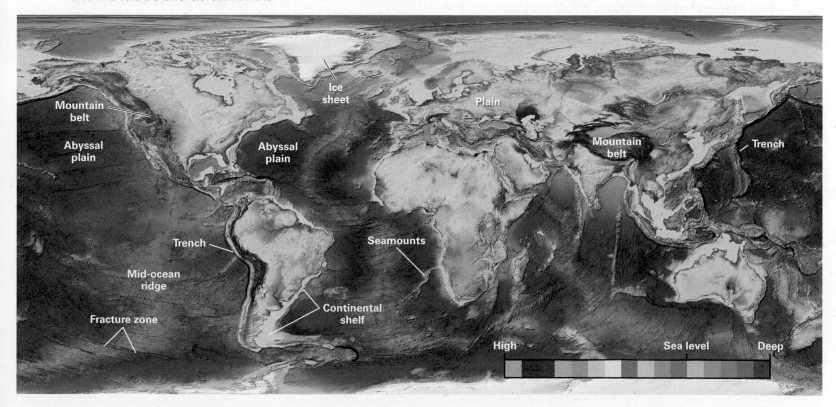

> *Melts:* A **melt** forms when solid materials become hot and transform into liquid. *Molten rock* is a type of melt. Geologists distinguish between *magma,* molten rock beneath the Earth's surface, and *lava,* molten rock that has flowed out onto the Earth's surface.

> *Rocks:* A coherent aggregate of mineral crystals or grains, or a mass of natural glass, is called a **rock**. Geologists recognize three main groups of rocks:
> *Igneous rock:* A rock that forms when molten rock cools and freezes solid, either underground or at the surface of the Earth.
> *Sedimentary rock:* A rock that forms either from fragments that broke off pre-existing rock and were then cemented together, or from minerals that precipitate out of a water solution at or near the Earth's surface.
> *Metamorphic rock:* A rock that forms when pre-existing rock undergoes changes in response to changes in temperature and pressure.

> *Grains:* We use the term **grain** either for an individual crystal within an igneous or metamorphic rock or for an individual fragment derived from a once-larger mineral sample or rock body.

> *Sediment:* An accumulation of loose (unconsolidated) grains, meaning grains that have not been cemented together, makes up **sediment**. Gravel and sand are types of sediment.

> *Metals:* A solid composed entirely of metal atoms (such as iron, aluminum, copper, and tin) is a **metal**. Metal can be stretched into wires or flattened into sheets; it tends to be shiny and can conduct electricity. An *alloy* is a mixture containing more than one type of metal.

> *Volatiles:* As noted earlier, a material that can transform into gas at the conditions found at the Earth's surface can be called a **volatile**.

Of the 92 naturally occurring elements that make up the Earth, 93% of the Earth's mass consists of only four—iron,

FIGURE 1.16 The proportions of major elements making up the mass of the whole Earth. Iron and oxygen dominate.

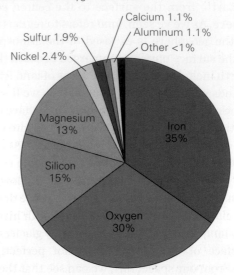

oxygen, silicon, and magnesium **(Fig. 1.16)**. Most of the rock in the geosphere consists of *silicate minerals*, which contain silica (SiO_2), either alone or bonded to other elements. Not surprisingly, rocks composed of silicate minerals are known as **silicate rocks**, which are the most common rocks on Earth. We can distinguish among igneous silicate rocks based on their chemical composition, as defined by the percentage of silica (SiO_2) that they contain relative to the percentage of iron oxide and magnesium oxide that they contain. This important ratio determines the density of rocks; rocks with more silica are less dense than rocks with more iron oxide or magnesium oxide. In sequence, from greatest to least proportion of silica, these types are known as *felsic, intermediate, mafic,* and *ultramafic*. To simplify our discussion of the Earth's layers, we name four common igneous rock types here: (1) *granite,* a felsic rock with large grains; (2) *basalt,* a mafic rock with small grains; (3) *gabbro,* a mafic rock with large grains; and (4) *peridotite,* an ultramafic rock with large grains **(Fig. 1.17)**.

FIGURE 1.17 These are examples of materials that compose the Earth. The left four are types of silicate rocks—granite has the most silica, peridotite the least. The image on the right is a metal alloy.

Granite	Basalt	Gabbro	Peridotite	Iron-nickel alloy
density = 2.7 g/cm³	density = 2.9 g/cm³	density = 3.0 g/cm³	density = 3.3 g/cm³	density = 7.5 g/cm³

Increasing density

THE BIOSPHERE

We call the aggregate of all living organisms, as well as the portion of the Earth in which living organisms exist, the **biosphere**. Although most familiar life inhabits the Earth's surface, the biosphere extends from several kilometers below the Earth's surface, where temperatures approach the boiling point of water, to several kilometers above the surface, where temperatures become low enough to freeze water. Note that life, as we know it, survives only under conditions in which liquid water can exist. The biosphere overlaps with portions of the geosphere, hydrosphere, and atmosphere.

Even viewed from space, we can see that the biosphere varies in character both spatially and temporally. Looking down, we can distinguish forests, fields, grassland, swamps, and tundra. Based on color changes of land surfaces over the year, we can determine that organisms grow, and then go dormant, during different seasons. And, of course, we realize from our survey that at least one species can modify the landscape substantially, by building networks of highways and clusters of buildings, by plowing, digging pits, building mounds, cutting canyons, and damming rivers. Based on the patterns of light at night, we can get a sense of variations in population density and degrees of technological development **(Fig. 1.18)**.

ENERGY AND EXCHANGES IN THE EARTH SYSTEM

As we'll see throughout this book, the Earth System stays in constant motion. Continents drift, mountains rise, volcanoes spew out lava, rivers and glaciers flow, and the atmosphere and oceans circulate. During these events, materials may transfer within or among different realms of the Earth System, and in some cases, they may end up back where they began, defining a *natural cycle*. The energy that drives the movement of rock, air, and water in the Earth System comes from three sources—internal energy, external energy, and gravity.

The term **internal energy** refers to heat (thermal energy) stored or produced inside the Earth. Much of this heat is primordial, originating in processes that formed the Earth in the first place. Some processes continue to work. Collisions of planetesimals converted kinetic energy to thermal energy; compression of matter into a smaller space elevated its temperature; the sinking of iron blobs to form the core generated heat when the blobs rubbed against their surroundings as they sank. Each time a radioactive atom inside the Earth decays, it generates heat. Over time, the Earth's internal energy can cause rock to melt, which can trigger volcanism. Internal energy keeps portions of the Earth's interior soft enough to flow, allowing pieces of its relatively rigid shell to move relative to one another, and these movements, in turn, lead to mountain building.

External energy reaches the Earth's surface from outside. Most of this energy comes in the form of radiation from the Sun, warming the atmosphere, the oceans, and the land surface. External energy, acting together with gravity, causes convection of the atmosphere and hydrosphere, resulting in winds, waves, and currents. Therefore, it also causes transfer of water from the oceans to the land surface, feeding rivers and glaciers that flow over and grind away or **erode** the land surface.

FIGURE 1.18 The lights of the Earth as seen at night from space. This image gives a sense of the distribution of human society, and of variations in industrialization.

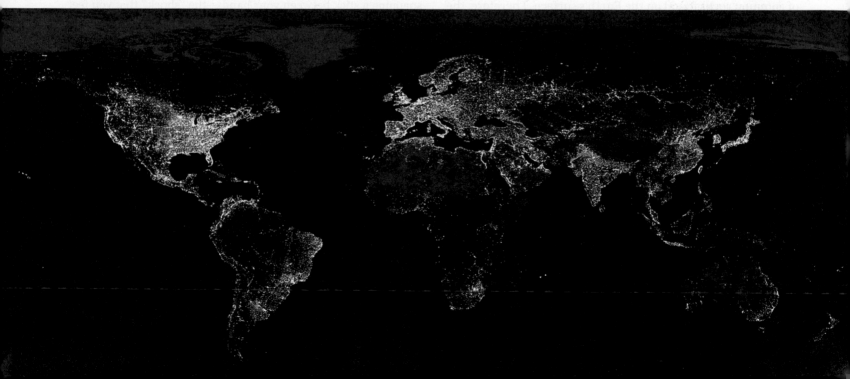

1.7 Looking Inward—Discovering the Earth's Interior

Having mapped the Earth's surface from our spacecraft, we now have a clear understanding that the planet is a variable and dynamic place. Our final task is to determine what's inside. We begin by describing the Earth's interior based on information we can detect from outside the planet. Then, we refine the image by placing instruments on the Earth's surface.

CLUES FROM EXAMINING THE EARTH'S SHAPE AND DENSITY

By measuring the gravitational pull of the Earth and the planet's volume, we can calculate its *average density*. A quick comparison shows that the average density exceeds the densities of common rocks found on the Earth's surface. Therefore, the interior of the Earth must contain denser material than its outermost layer.

In fact, we find that the overall mass of the Earth is so great that the planet must contain a large amount of dense metal. And, since the Earth has almost the shape of a sphere, the metal must be concentrated near the center—otherwise, centrifugal force due to the spin of the Earth on its axis would pull the equator outward and the planet would become somewhat disk-shaped. (To picture why, consider that when you swing a hammer, your hand feels more force if you hold the end of the light wooden shaft, rather than the heavy metal head.) Further, the interior must be mostly solid, because if it weren't, tidal forces would cause the land surface to rise and fall much more than it does in response to the Moon's gravitational pull. With these key observations on hand, we conclude that the Earth resembles a hard-boiled egg, in that it has three principal layers: a not-so-dense *crust* (like an eggshell), a denser

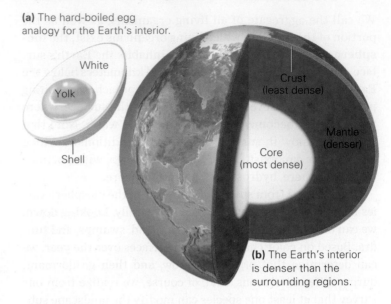

FIGURE 1.19 An early image of the Earth's internal layers.

(a) The hard-boiled egg analogy for the Earth's interior.

White

Yolk

Shell

Crust (least dense)

Mantle (denser)

Core (most dense)

(b) The Earth's interior is denser than the surrounding regions.

solid *mantle* in the middle (like an egg white), and a very dense *core* (the egg yolk) **(Fig. 1.19)**.

CLUES FROM THE STUDY OF EARTHQUAKES: REFINING THE IMAGE OF THE INTERIOR

To understand how studying earthquakes can provide information about the Earth's interior, we first need to understand what an earthquake is. Most earthquakes happen when rock within the outer portion of the Earth suddenly breaks along a *fault* (a fracture on which slip occurs), and generates energy that travels through the surrounding rock outward from the break in the form of vibrations, known as *seismic waves* or earthquake waves **(Fig. 1.20a)**. You can simulate this process, on a small scale, by snapping a stick between your hands—the "shock" that you feel is energy that propagated along the stick from the break to your hands **(Fig. 1.20b)**. When seismic energy reaches the Earth's surface, it causes the surface to vibrate (move up and down or back and forth). Geologists use the term **earthquake** for both the sudden movement that generates vibrations, and the ground shaking that results when these vibrations reach the Earth's surface.

Study of seismic waves reveals the character of the Earth's insides, much as ultrasound measurements help doctors study the insides of a patient. Seismic waves travel at different velocities through different materials, and they refract (bend) or reflect when they reach boundaries between different materials. To study these waves, we need to place *seismometers*, instruments that can detect seismic waves, at various sites on the Earth's surface. Using methods described in Chapter 8 and Interlude D, records made by our seismometers allow us to pinpoint the depth of boundaries between layers, to identify sublayers within the major layers, and to determine whether layers are solid or liquid.

FIGURE 1.20 Faulting and earthquakes.

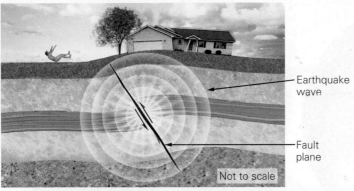

Earthquake wave

Fault plane

Not to scale

(a) When the rock inside the Earth suddenly breaks and slips, forming a fracture called a fault, it generates shock waves that pass through the Earth and shake the surface.

(b) Similarly, snapping a stick generates vibrations that pass through the stick.

PRESSURE AND TEMPERATURE INSIDE THE EARTH

The mass of the overlying rock layer increases with depth, so the pressure within the Earth also increases progressively with depth. In solid rock, the pressure at a depth of 1 km is about 300 atm, so a mine tunnel requires enough support to prevent collapse. At the Earth's center, pressure probably reaches about 3,600,000 atm.

Did you ever wonder . . .
how hot it gets at the center of this planet?

Temperature also increases with depth in the Earth. Even on a cool winter's day, if we could descend down into the crust, we would find that at almost 4 km below the surface, the depth of the deepest gold mine on Earth, the surrounding rock is 60°C. Humans exposed to these temperatures can survive no more than 10 minutes, without air-conditioning. We refer to the rate of change in temperature with depth as the **geothermal gradient**. In the upper part of the crust, the geothermal gradient averages between 20°C and 30°C per kilometer. At greater depths, the gradient decreases, so that 35 km below the surface of a continent, the temperature reaches 400°C to 700°C, and at the mantle-core boundary it is about 3,500°C. No one

will ever be able to measure the temperature at the Earth's center, but calculations suggest it may exceed 4,700°C, close to the Sun's surface temperature of 5,500°C.

TAKE-HOME MESSAGE

Measurements of the Earth's mass, shape, and tidal response led to the conclusion that the Earth has three principal internal layers—the crust, the mantle, and the core. Study of earthquake waves passing through the interior has refined this image.

QUICK QUESTION What is the geothermal gradient?

1.8 Basic Characteristics of the Earth's Layers

At this point, we leave our fantasy spacecraft and describe the basic properties of the Earth's interior, based on studies that geologists have undertaken during the past 150 years of research. We can determine the composition of the crust by directly analyzing rock samples collected from the surface. But to determine the composition of deeper layers, we must rely on other sources of data. Even the deepest drillhole, a 12-km-deep penetration in northern Russia, represents only a pinprick in comparison to the Earth's diameter of 6,371 km **(Fig. 1.21)**. In some localities, the magma that rises beneath volcanoes carries chunks of the mantle up to the surface, and researchers can collect these chunks for study. Also, meteorites bring samples to the Earth of material that was once in the interiors of differentiated protoplanets **(Box 1.3)**. The internal composition of these protoplanets likely resembles that of the Earth, so meteorites effectively provide pieces of materials resembling those inside our planet. We will be using information about the Earth's interior in the next chapter's discussion of plate tectonics, geology's grand unifying theory. Later, after we

SEE FOR YOURSELF...

Meteor Crater

LATITUDE
35°1'37.18" N

LONGITUDE
111°1'20.17" W

Look down from 5.7 km.

Some craters can be found on the Earth. Meteor Crater, Arizona, is 1.1 km across.

FIGURE 1.21 A modern view of the Earth's interior layers.

(a) By studying earthquake waves, geologists produced a refined image of the Earth's interior, in which the mantle and core are subdivided.

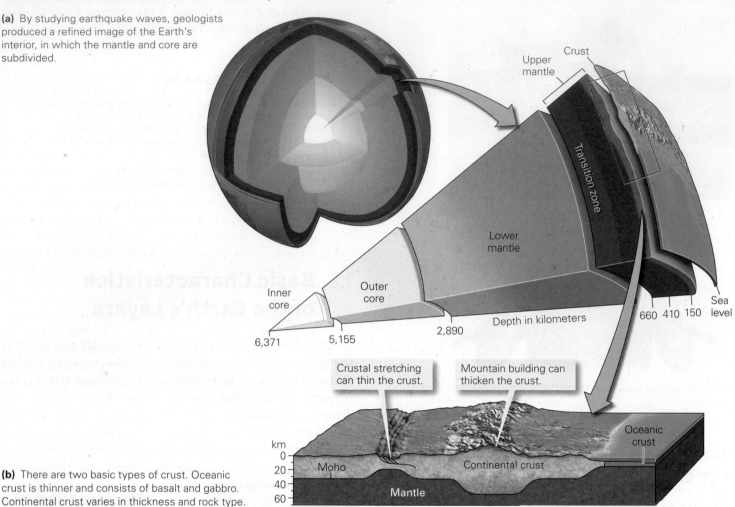

(b) There are two basic types of crust. Oceanic crust is thinner and consists of basalt and gabbro. Continental crust varies in thickness and rock type.

THE CRUST

have thoroughly discussed earthquakes and seismic waves, we will add further details to our image of the interior in Interlude D.

When you look at the surface of the Earth, you are standing on top of its outermost layer, the **crust**. This layer overlaps with the biosphere and is the source of all of humanity's resources. How thick is this all-important crust? Or, in other words, what is the depth to the crust-mantle boundary? An answer came from the studies of Andrija Mohorovičić (1857–1936), a researcher working in Zagreb, Croatia. In 1909, he discovered that the velocity of earthquake waves suddenly increased at a depth of tens of kilometers beneath the Earth's surface, and he suggested that this increase was caused by an abrupt change in the properties of rock (see Interlude D for details). Later studies showed that this change can be found all around our planet, though it occurs at different depths in different locations.

In particular, the area of change is deeper beneath continents than beneath oceans. Geologists now identify the change as the base of the crust, and they refer to it as the **Moho** in Mohorovičić's honor. The relatively shallow depth of the Moho, as compared to the radius of the Earth, emphasizes that the crust is very thin indeed. In fact, the crust is only 0.1% to 1.0% of the Earth's radius, so if the Earth were the size of a balloon, the crust would be about the thickness of the balloon's skin.

The crust does not consist simply of cooled mantle, like the skin on chocolate pudding, but rather, it's made up of rocks that differ in composition (chemical makeup) from mantle rock. *Oceanic crust*, which underlies the seafloor, is only 7 to 10 km thick—at highway speeds (100 km per hour), you could drive a distance equal to the thickness of the oceanic crust in about 5 minutes. The top of the oceanic crust is a blanket, generally less than 1 km thick, of sediment containing clay and tiny shells that settled out of the sea like snow. Beneath this blanket, the oceanic crust consists of a layer of basalt and, below that, a layer of gabbro.

BOX 1.3 CONSIDER THIS...

Meteors and Meteorites

Collisions between objects from space and the Earth happened frequently during the planet's early history. But even today collisions with such objects continue, and nearly 1 million kilograms of material (rock, metal, dust, and ice) fall to the Earth every year. The vast majority of this colliding material consists of fragments derived from comets and asteroids that were sent careening into the path of the Earth after billiard ball–like collisions with each other out in space or because the gravitational pull of a passing planet deflected their orbit. Some of the material, however, consists of chips of the Moon or Mars, ejected into space when large objects collided with those bodies.

Astronomers refer to any object that enters the Earth's atmosphere from space as a *meteoroid*. Meteoroids are moving at speeds of 20 to 75 km per second when they reach an altitude of 150 km. Below this altitude, compression of the air in front of a meteoroid makes the air hot enough to melt and vaporize the meteoroid's surface. This process produces a streak of bright, glowing gas in the meteoroid's wake. The glowing streak, an atmospheric phenomenon, is a **meteor**, also known, commonly though incorrectly, as a falling star **(Fig. Bx1.3a)**. Most visible meteors completely vaporize by an altitude of about 30 km. But dust-sized ones may slow down sufficiently to float to Earth, and some that are fist-sized or bigger can survive the heat of entry to reach the surface of the planet. In some cases, meteoroids explode in brilliant fireballs.

Objects that strike the Earth are called **meteorites**. Although almost all meteorites are small and do not cause notable damage on the Earth, a very few have smashed through houses, dented cars, and bruised people. During the longer term of Earth history, some catastrophic collisions have left huge craters **(Fig. Bx1.3b)**.

Researchers recognize three basic classes of meteorites: *iron* (made of iron-nickel alloy), *stony* (made of rock), and *stony iron* (rock embedded in a matrix of metal). Of all known meteorites, about 93% are stony and 6% are iron **(Fig. Bx1.3c)**. Researchers have concluded that most stony meteorites and all iron meteorites are fragments of planetesimals that had differentiated into a metallic core and a rocky mantle before being shattered by collisions with other planetesimals. Most of these appear to be 4.56 to 4.54 Ga. A few stony meteorites were derived from planetesimals that never differentiated into core and mantle sectors, and these are as old as 4.57 Ga—the oldest Solar System materials ever measured.

FIGURE Bx1.3 Meteors and meteorites.

(a) A shower of meteors over Hong Kong in 2001.

The meteorite that formed the crater was 50 m across.

(b) The Barringer meteor crater in Arizona, formed about 50,000 years ago, is 1.1 km in diameter.

(c) Examples of stony meteorites (left) and iron meteorites (right).

TABLE 1.1 Average Relative Abundances of Elements in the Earth's Crust

Element	Symbol	% by Weight	% by Volume	% by Atoms
Oxygen	O	46.3	93.8	60.5
Silicon	Si	28.0	0.9	20.5
Aluminum	Al	8.1	0.8	6.2
Iron	Fe	5.5	0.5	1.9
Calcium	Ca	3.4	1.0	1.9
Magnesium	Mg	2.8	0.3	1.4
Sodium	Na	2.4	1.2	2.5
Potassium	K	2.3	1.5	1.8
All others	—	1.2	>0.1	3.3

Continental crust generally has a thickness of 35 to 40 km (four to five times the thickness of oceanic crust), but its thickness varies significantly. Where continental crust has been stretched horizontally, it can be as thin as 25 km, and where it has been squeezed horizontally, it can be up to 70 km thick. Continental crust contains a great variety of rock types, ranging from mafic to felsic in composition. On average, upper continental crust has a felsic (granite-like) to intermediate composition—so the density of continental crust overall is less than that of oceanic crust. Notably, oxygen is the most abundant element in the crust, with silicon coming in second, because most rocks contain abundant silica, a compound with the formula SiO_2 **(Table 1.1)**.

THE MANTLE

The **mantle** of the Earth, a layer that ranges in thickness from 2,820 to 2,890 km, lies between the crust and the core. In terms of volume, it is the largest part of the Earth. In contrast to the crust, the mantle consists entirely of an ultramafic (dark and dense) rock called peridotite. This means that peridotite, though rare at the Earth's surface, is actually the most abundant rock in our planet! Seismic-wave velocities change at a depth of 410 km and again at a depth of 660 km in the mantle. Based on this observation, geologists divide the mantle into two sublayers: the *upper mantle*, down to a depth of 660 km, and the *lower mantle*, from 660 km down to about 2,890 km. The lower part of the upper mantle, the interval between 410 and 660 km deep, is called the *transition zone*. In this zone, seismic-wave velocities increase in a series of steps due to abrupt changes in the character of minerals making up mantle rock.

Almost all of the mantle is solid rock. Melt, which occurs in films or bubbles between grains, accounts for just a few percent of its volume, and generally occurs only at a depth of 100 to 150 km. But even though it's mostly solid, mantle rock is so hot that it's soft enough to flow. Rock, like wax, softens as it warms up. This flow, however, takes place extremely slowly, at a rate of less than 15 cm a year. So "soft" here does not mean liquid—it simply means that over long periods of time mantle rock is *plastic* and can change shape without breaking. Because of its softness, the mantle can undergo convection very slowly. During this convection, warmer mantle rock rises, while cooler mantle rock sinks.

THE CORE

Early calculations suggested (incorrectly) that the **core**, the central part of the Earth, had the same density as gold, so for many years people held the fanciful hope that vast riches lay at the heart of our planet. Alas, geologists eventually concluded that the core consists of a far less glamorous material—*iron alloy* (iron mixed with nickel and lesser amount of oxygen, silicon, and sulfur). Studies of seismic waves led geologists to divide the core into two parts (see Fig. 1.21): the *outer core* (between 2,890 and 5,155 km deep) and the *inner core* (from a depth of 5,155 km down to the Earth's center at 6,371 km). The outer core's iron alloy exists as a liquid because the temperature in the outer core is so high that even the great pressures squeezing the region cannot keep atoms locked into a solid framework. The liquid iron alloy's rapid convective flow generates the Earth's magnetic field.

The inner core, with a radius of about 1,220 km, consists of a solid iron alloy (see Fig. 1.17) that may reach a temperature of over 4,700°C. Even though it is hotter than the outer core, the inner core's alloy remains solid because it endures even greater pressure, enough to keep atoms locked together tightly in solid crystals.

THE LITHOSPHERE AND THE ASTHENOSPHERE

So far, we have identified three major layers inside the Earth (crust, mantle, and core) whose compositions differ from each other. Seismic waves travel at different velocities through these layers. An alternative way of thinking about Earth layers comes from studying the degree to which the material of each layer can flow. In this context, we distinguish between *rigid materials*, which can bend or break but cannot flow easily, and *plastic materials*, which are relatively soft and can flow without breaking.

The outer 100 to 150 km of the Earth behaves as a relatively rigid material. In other words, it's effectively a shell that cannot flow. This outer layer, called the **lithosphere**, consists of the crust plus the relatively cool uppermost part of the upper

FIGURE 1.22 A block diagram of the lithosphere emphasizing the difference between continental and oceanic lithosphere.

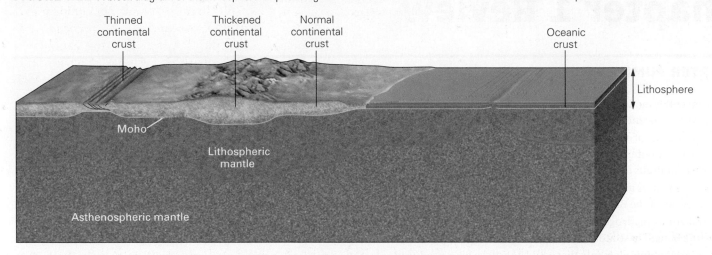

mantle **(Fig. 1.22)**. We refer to this portion of the upper mantle as the *lithospheric mantle*. Note that the terms lithosphere and crust are not the same—the crust represents only the top layer of the lithosphere. The lithosphere lies on top of the **asthenosphere**, the portion of the mantle in which rock behaves like soft plastic and can flow. The boundary between the lithosphere and asthenosphere occurs where the temperature reaches about 1,280°C, for when temperatures are higher than this value, mantle rock becomes soft enough to flow. Note that the asthenosphere generally lies below a depth of 100 to 150 km. We can't assign a specific depth to the base of the asthenosphere because all of the mantle below 150 km can flow, but for convenience, some geologists equate the base of the asthenosphere with the top of the transition zone. As we begin our study of plate tectonics in the next chapter, we'll see that plates consist of lithosphere, and that the movement of plates can take place only because of the ability of the asthenosphere to flow.

TAKE-HOME MESSAGE

The Earth's outermost layer, the crust, is very thin; oceanic and continental crust differ in composition. Most of the Earth's mass lies in the mantle. A metallic core lies at the planet's center. The crust and the outermost mantle together comprise the rigid lithosphere, while the asthenosphere is the inner mantle whose material can flow.

QUICK QUESTION Is the entire core solid?

ANOTHER VIEW Astronomers are discovering many planetary systems around other nearby stars in our galaxy. This is an artist's conception of what the nearest planetary system, Epsilon Eridani, might look like. This example has an asteroid belt and comets.

Chapter 1 Review

CHAPTER SUMMARY

> The geocentric model placed the Earth at the center of the Universe. The heliocentric model placed the Sun at the center.

> The Earth is one of eight planets orbiting the Sun. The Solar System lies on the outer edge of the Milky Way Galaxy. The Universe contains hundreds of billions of galaxies.

> Most astronomers agree that the Universe has been expanding since it began at the Big Bang, about 13.8 billion years ago.

> The first atoms (hydrogen and helium) developed within minutes of the Big Bang. These atoms formed vast gas nebulae.

> The Earth contains elements that could only have been produced during the life cycle of stars.

> Gravity causes nebulae to coalesce into disks with bulbous centers. The central ball of this accretion disk became a protostar. Eventually, the ball became so hot and dense that fusion reactions began and it became a true star.

> Planets developed from the rings of gas and dust surrounding newborn stars. These condensed into planetesimals that then merged to form protoplanets, and finally true planets.

> The Moon formed from debris ejected when a protoplanet collided with the Earth in the very young Solar System.

> Large protoplanets differentiate into a core and mantle and assume a near-spherical shape.

> The Earth has a magnetic field that shields it from solar wind.

> The atmosphere consists of 78% N_2, 21% O_2, and 1% other gases. Air pressure decreases with elevation.

> The surface of the Earth hosts land (30%) and ocean (70%).

> Earth materials include organic chemicals, minerals, glasses, rocks, metals, melts, and volatiles. Geologists recognize three classes of rocks.

> The Earth's interior can be divided into three distinct layers: the very thin crust, the rocky mantle, and the metallic core.

> Pressure and temperature increase with depth in the Earth. The rate of temperature increase is the geothermal gradient.

> The crust varies in thickness from 7 to 10 km (beneath the oceans) to 25 to 70 km (beneath the continents). Oceanic crust is mafic in composition, whereas average upper continental crust is felsic to intermediate. The mantle is composed of ultramafic rock. The core consists of iron alloy.

> The mantle can be subdivided into an upper mantle and a lower mantle. The core can be subdivided into the liquid outer core and a solid inner core.

> The crust plus the upper part of the mantle constitute the lithosphere, a rigid shell. The lithosphere lies over the asthenosphere, in which mantle can flow.

GUIDE TERMS

accretion disk (p. 21)
asteroid (p. 18)
asthenosphere (p. 41)
atmosphere (p. 31)
atom (p. 16)
atomic mass (p. 16)
atomic number (p. 16)
bathymetry (p. 33)
Big Bang nucleosynthesis (p. 20)
Big Bang theory (p. 20)
biosphere (p. 35)
comet (p. 19)
core (p. 40)
cosmology (p. 15)

crust (p. 38)
cryosphere (p. 32)
crystal (p. 33)
differentiation (p. 26)
dipole (p. 27)
Earth materials (p. 33)
Earth System (p. 31)
earthquake (p. 36)
ecliptic (p. 18)
element (p. 16)
energy (p. 17)
erosion (p. 35)
expanding Universe theory (p. 20)
external energy (p. 35)

galaxy (p. 17)
gas (p. 16)
geocentric model (p. 15)
geosphere (p. 33)
geothermal gradient (p. 37)
giant planet (p. 18)
glass (p. 33)
grain (p. 34)
groundwater (p. 32)
heat (p. 22)
heliocentric model (p. 15)
hydrosphere (p. 32)
internal energy (p. 35)
liquid (p. 16)
lithosphere (p. 40)

magnetic field (p. 27)
magnetosphere (p. 27)
mantle (p. 40)
melt (p. 34)
metal (p. 34)
meteor (p. 39)
meteorite (p. 39)
mineral (p. 33)
Moho (p. 38)
molecule (p. 16)
moon (p. 18)
nebula (p. 20)
nebular theory (p. 24)
nuclear fusion (p. 20)
nucleus (p. 16)

GEOTOURS *THIS CHAPTER'S GEOTOURS WORKSHEET (A) FEATURES QUESTIONS AND GOOGLE EARTH SITES ON:*

> Nebular supernovae   > Spiral galaxies   > Meteorite impacts

organic chemical (p. 33)
planet (p. 18)
planetesimal (p. 25)
protoplanet (p. 25)
protoplanetary disk (p. 24)
protostar (p. 21)

refractory material (p. 25)
rock (p. 34)
sediment (p. 34)
silicate rock (p. 34)
Solar System (p. 18)
solid (p. 16)

star (p. 17)
stellar nucleosynthesis (p. 23)
stellar wind (p. 23)
supernova (p. 23)
temperature (p. 22)
terrestrial planet (p. 18)

thermal energy (p. 22)
topography (p. 33)
Universe (p. 16)
vacuum (p. 20)
volatile (p. 34)
volatile materials (p. 24)

REVIEW QUESTIONS

The letters following each Review Question refer to the corresponding Learning Objective from the Chapter Opener.

1. Contrast the geocentric and heliocentric Universe concepts. **(A)**

2. What is the ecliptic, and why are the orbits of the planets within the ecliptic? Why is Pluto no longer considered to be a planet? **(B)**

3. Explain the expanding Universe theory. **(C)**

4. What is the Big Bang, and when did it occur? **(C)**

5. Describe the steps in the formation of the Solar System, according to the nebular theory. **(D)**

6. Why is the Earth round? **(D)**

7. What is the Earth's magnetic field? Draw a representation of the field on a piece of paper. What causes aurorae? **(E)**

8. What is the Earth's atmosphere composed of? Why would you die of suffocation if you were to jump from a plane at an elevation of 12 km without an oxygen tank? **(E)**

9. What is the proportion of land area to sea area on the Earth? **(F)**

10. Describe the major categories of materials constituting the Earth. On what basis do geologists distinguish among different kinds of silicate rocks? **(F)**

11. Identify the principal layers of the Earth in the figure. **(G)**

12. How do temperature and pressure change with increasing depth in the Earth? **(H)**

13. What is the Moho? Describe the differences between continental crust and oceanic crust. **(G)**

14. What is the mantle composed of? Is there any melt in it? **(G)**

15. What is the core composed of? How do the inner and outer cores differ? Which produces the magnetic field? **(G)**

16. What is the difference between lithosphere and asthenosphere? Identify each in the figure here. **(H)**

17. At what depth does the lithosphere-asthenosphere boundary occur? Is this above or below the Moho? Is the asthenosphere entirely liquid? **(H)**

ON FURTHER THOUGHT

18. Are all stars that we see today considered to be first-generation stars? What is the evidence for your answer? **(D)**

19. Popular media sometimes imply that the crust floats on a "sea of magma." Is this a correct image? **(H)**

ONLINE RESOURCES

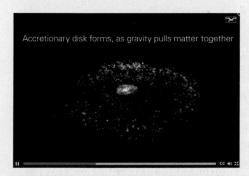

Accretionary disk forms, as gravity pulls matter together

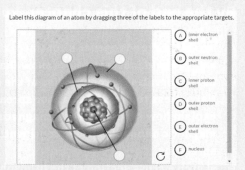

Label this diagram of an atom by dragging three of the labels to the appropriate targets.

- (A) inner electron shell
- (B) outer neutron shell
- (C) inner proton shell
- (D) outer proton shell
- (E) outer electron shell
- (F) nucleus

Videos This chapter features videos showing how our Solar System and the Earth formed, including coverage of nebular theory, proto-Earth, the formation of the Moon, and the beginning of the atmosphere.

Smartwork5 This chapter features visual labeling exercises on topics such as atomic structures and our Solar System.

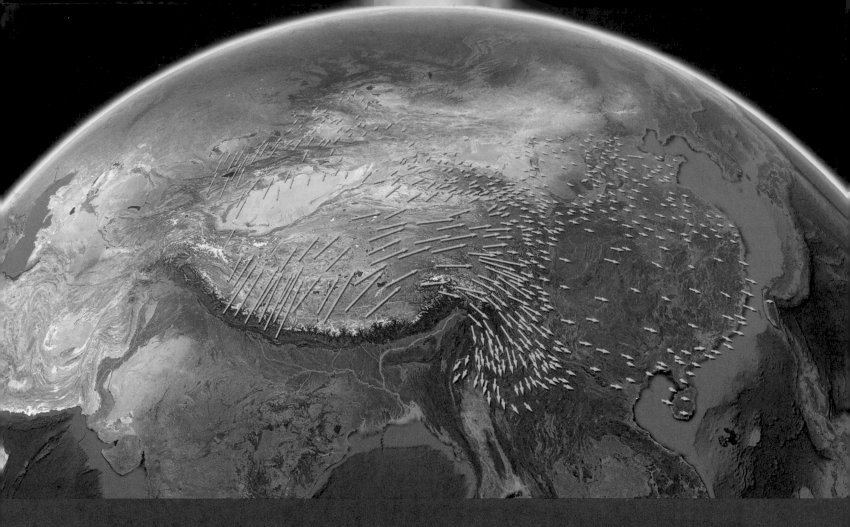

CHAPTER 2

THE WAY THE EARTH WORKS: PLATE TECTONICS

By the end of this chapter, you should be able to . . .

A. explain Alfred Wegener's continental-drift hypothesis and the evidence he used to show that it takes place.

B. list observations from studies of the ocean floor that led Harry Hess to propose seafloor spreading.

C. explain how studies of paleomagnetism and marine magnetic anomalies prove continental drift and seafloor spreading.

D. sketch a cross section of the lithosphere, and contrast oceanic and continental lithosphere.

E. use a map of earthquakes to locate plate boundaries and triple junctions.

F. distinguish among the three types of plate boundaries, and characterize geologic features associated with each.

G. discuss rifting, continental collision, and hot-spot formation, and show where these processes happen today.

H. characterize the processes driving plate motion, the rates at which this motion takes place, and how rates can be measured.

> It is only by combing the information furnished by all the earth sciences that we can hope to determine "truth" here.
>
> ALFRED WEGENER (German geophysicist and meteorologist, 1880–1930)

2.1 Introduction

In September 1930, 15 explorers led by a German meteorologist, Alfred Wegener, set out across the endless snowfields of Greenland to resupply two weather observers stranded at a remote camp. The observers had been planning to spend the long polar night recording wind speeds and temperatures on Greenland's polar plateau. Wegener was well known, not only to researchers studying climate but also to geologists. Some 15 years earlier, he had published a small book, *The Origin of the Continents and Oceans*, in which he had dared to challenge geologists' long-held assumption that the continents had remained fixed in position through all of Earth history.

Wegener thought, instead, that the continents had once fit together like pieces of a giant jigsaw puzzle, making one vast *supercontinent*. This supercontinent, which he named *Pangaea* (pronounced pan-JEE-ah; Greek for all land), later fragmented into separate continents that drifted apart, moving slowly to their present positions **(Fig. 2.1)**. This phenomenon came to be known as *continental drift*.

Wegener presented many observations that he believed proved that continental drift had occurred, but he met with strong resistance. At a widely publicized 1926 geology conference in New York City, a crowd of celebrated American professors scoffed, "What force could possibly be great enough to move the immense mass of a continent?" Wegener's writings didn't provide a good answer, so despite all the supporting

FIGURE 2.1 Alfred Wegener and his model of continental drift.

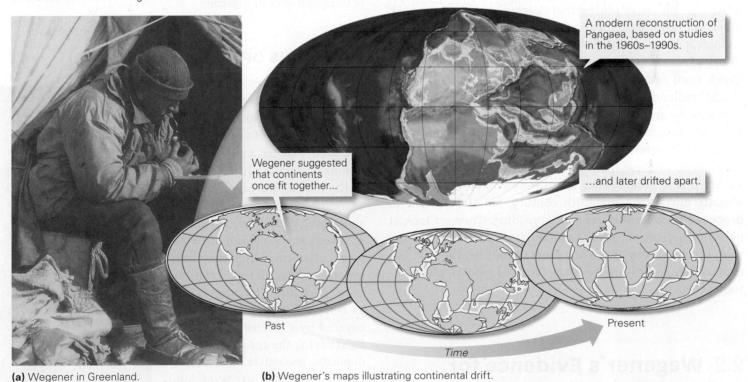

A modern reconstruction of Pangaea, based on studies in the 1960s–1990s.

Wegener suggested that continents once fit together...

...and later drifted apart.

Past

Present

Time

(a) Wegener in Greenland.

(b) Wegener's maps illustrating continental drift.

◄ (facing page) Measurements with GPS indicate that India is moving north relative to Asia. Such measurements allow us to "see" plate movements. Each yellow arrow indicates the velocity of the point at the end of the arrow. Note that much of the eastern Tibetan Plateau is moving to the east.

observations he had provided, most of the group rejected continental drift. Four years later, Wegener faced his greatest challenge—survival. Sadly, he lost. On October 30, 1930, Wegener and one companion reached the observers, dropped off enough supplies to last the winter, and set out on the return trip the next day, but they never made it home.

Had Wegener survived to old age, he would have seen his hypothesis become the foundation of a scientific revolution. Today, geologists accept Wegener's basic conclusion and take for granted that the map of the Earth constantly changes as continents waltz around this planet's surface, variously combining and breaking apart over geologic time.

The revolution began in 1960, when an American geologist, Harry Hess (1906–1969), proposed that as continents drift apart, new ocean floor forms between them by a process that his contemporary, Robert Dietz (1914–1995), named *seafloor spreading*. Hess and others suggested that continents move toward each other when the old ocean floor between bends down and sinks back down into the Earth's interior, a process now called *subduction*. By 1968, geologists had developed a fairly complete model encompassing continental drift, seafloor spreading, and subduction. In this model, the Earth's lithosphere, its outer, relatively rigid shell, consists of about 20 distinct pieces, or **plates**, that slowly move relative to each other. Because we can confirm this model using many observations, it has gained the status of a theory, now called the *theory of plate tectonics*, or simply **plate tectonics**, from the Greek word *tekton*, which means builder; plate movements "build" regional geologic features. Geologists view plate tectonics as the grand unifying theory of geology, because it can successfully explain a great many geologic phenomena.

In this chapter, we introduce the observations that led Wegener to propose his continental drift hypothesis. Next, we learn how observations about the seafloor, made by geologists during the mid-20th century, led Harry Hess to propose the concept of seafloor spreading. Then we look at paleomagnetism, the record of Earth's magnetic field in the past, because it provides a key proof of continental drift. We conclude by describing the many facets of modern plate tectonics theory.

2.2 Wegener's Evidence for Continental Drift

Wegener suggested that a vast supercontinent, Pangaea, existed until near the end of the Mesozoic Era (the interval of geologic time that lasted from 252 to 66 million years ago). He suggested that Pangaea then broke apart, and the landmasses

moved away from each other to form the continents we see today. Let's look at some of Wegener's arguments and see what led him to formulate this hypothesis of **continental drift**.

THE FIT OF THE CONTINENTS

Almost as soon as maps of the Atlantic coastlines became available in the 1500s, scholars noticed the fit of the continents (see Fig. 2.1b). The northwestern coast of Africa looks like it could tuck in snugly against the eastern coast of North America, and the bulge of eastern South America could nestle tightly into the indentation of southwestern Africa. Australia, Antarctica, and India could all connect to the southeast of Africa, while Greenland, Europe, and Asia could pack against the northeastern margin of North America. In fact, all the continents could be joined, with remarkably few overlaps or gaps, to produce Pangaea. Wegener concluded that the fit was too good to be coincidence and, therefore, that the continents of today did once fit together.

LOCATIONS OF PAST GLACIATIONS

Glaciers are rivers or sheets of ice that flow across the land surface. As a glacier flows, it carries sediment grains of all sizes (clay, silt, sand, pebbles, and boulders). Grains protruding from the base of the moving ice carve scratches, called *striations*, into the substrate. When the ice melts, sediment stays behind in a deposit called *till*, which may bury striations. The till and striations at a particular location serve as evidence that the region was covered by a glacier in the past. By studying the age of glacial till deposits, geologists have determined that large areas of land were covered by glaciers during time intervals called *ice ages*. One of these ice ages occurred from about 280 to 260 Ma (million years ago), near the end of the Paleozoic Era.

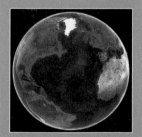

Wegener was an Arctic climate scientist by training, so it's no surprise that he had a strong interest in glaciers. He knew that glaciers form at high (polar) latitudes today, so he was bothered by the observation that sediments and striations indicative of late Paleozoic glaciation occurred in southern South America, southern Africa, southern India, Antarctica, and southern Australia (**Fig. 2.2a**). These places are now widely separated and, with the exception of Antarctica, do not currently lie in cold polar regions (**Fig. 2.2b**). To Wegener's amazement, all late Paleozoic glaciated areas lie adjacent to each other on his map of Pangaea (**Fig. 2.2c**). Furthermore, when he plotted the orientation of glacial striations, they all pointed roughly outward from a location in southeastern Africa. In other words, Wegener determined that the distribution of glaciations at the end of the Paleozoic could easily be explained if the continents had been united in Pangaea, with the southern part of Pangaea lying beneath the center of a huge ice cap, but not if the continents had always been in their present positions.

THE DISTRIBUTION OF CLIMATIC BELTS

If the southern part of Pangaea lay near the South Pole at the end of the Paleozoic, then during this same time interval, southern North America, southern Europe, and northwestern Africa would have straddled the equator and would have had tropical or subtropical climates. Wegener searched for evidence that this was so by studying descriptions of sedimentary rocks that were formed at this time, because the material making up these rocks can reveal clues to the past climate. For example, in the swamps and jungles of tropical regions, thick deposits of plant material accumulate, and when deeply buried, this material transforms into coal (see Chapter 12). And, in the clear, shallow seas of tropical regions, large reefs develop. Finally, subtropical regions, on either side of the tropical belt, contain deserts, an environment in which sand dunes form and salt from evaporating seawater or salt lakes accumulates. Wegener speculated that the distribution of late Paleozoic coal, reef, sand-dune, and salt deposits could define climate belts on Pangaea.

FIGURE 2.2 Evidence for Pangaea based on the distribution of glacial features.

(a) Glacial striations of late Paleozoic age exposed on the southern coast of Australia.

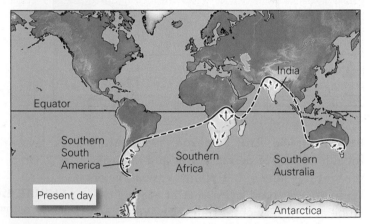

(b) The distribution of late Paleozoic glacial deposits and striations on present-day Earth are hard to explain.

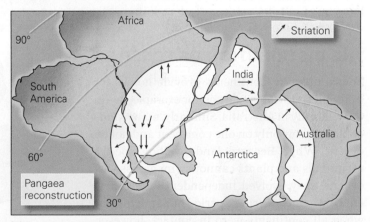

(c) On a map of Pangaea, areas with glacial deposits fit together within a southern polar cap.

Sure enough, in the belt of Pangaea that Wegener expected to be equatorial, late Paleozoic sedimentary rock layers include abundant coal and the relicts of reefs. In the portions of Pangaea that Wegener predicted would

FIGURE 2.3 Each symbol shows the location of an environment in which a distinctive type of sediment accumulates. The locations define climate belts, at appropriate latitudes, on a map of Pangaea.

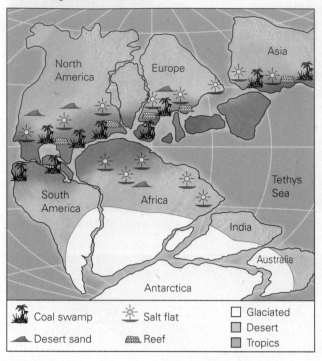

- 🌴 Coal swamp
- 🔆 Salt flat
- ⛰ Desert sand
- 🏛 Reef
- ☐ Glaciated
- ◻ Desert
- ◼ Tropics

be subtropical, late Paleozoic sedimentary rock layers include relics of desert dunes and deposits of salt. On a present-day map of our planet, exposures of these ancient rock layers scatter around the globe at a variety of latitudes. On Wegener's Pangaea, the exposures align in continuous bands that occupy appropriate latitudes **(Fig. 2.3)**.

THE DISTRIBUTION OF FOSSILS

Today, different continents provide homes for different species. Kangaroos, for example, live in the wild only in Australia. Similarly, many kinds of plants grow only on one continent and not on others. Why? Because land-dwelling species of animals and plants cannot swim across vast oceans; they evolved independently on different continents. During a period of Earth history when all continents were in contact, however, land animals and plants could have migrated among many continents.

With this concept in mind, Wegener plotted occurrences of fossils representing land-dwelling species that lived during the late Paleozoic and early Mesozoic Eras (between

300 and 210 Ma). He found that such fossils exist on continents now separated by oceans **(Fig. 2.4)**. Wegener argued, therefore, that the distribution of fossils requires that the continents were adjacent to one another in the late Paleozoic and early Mesozoic.

MATCHING GEOLOGIC UNITS

Art historians can recognize a Picasso painting, and architects know what makes a building look "Victorian." Similarly, geoscientists can identify distinctive assemblages of rocks. Wegener found that the same types of Precambrian rock assemblages occurred on the eastern coast of South America and the western coast of Africa, regions now separated by an ocean **(Fig. 2.5a)**. If the continents had been joined to produce Pangaea in the past, then these matching rock groups would have been adjacent to each other and therefore could have composed continuous blocks or belts. Wegener also noted that features of the Appalachian Mountains of the United States and Canada closely resemble mountain belts in southern Greenland, Great Britain, Scandinavia, and northwestern Africa, regions that would have lain adjacent to each other in Pangaea **(Fig. 2.5b, c)**. Wegener thus demonstrated that not only did the coastlines of continents match, so too did the rocks adjacent to the coastlines.

FIGURE 2.4 The distribution of fossil localities shows that Mesozoic land-dwelling or coastal organisms occurred on continents that were adjacent in Pangaea.

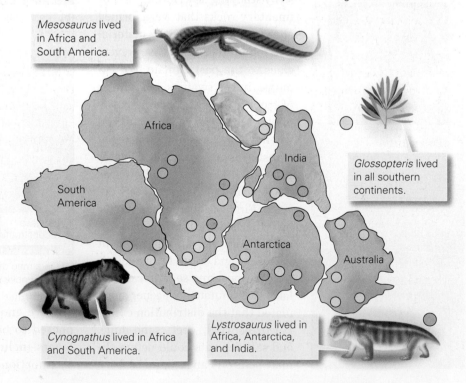

Mesosaurus lived in Africa and South America.

Glossopteris lived in all southern continents.

Cynognathus lived in Africa and South America.

Lystrosaurus lived in Africa, Antarctica, and India.

FIGURE 2.5 Further evidence of continental drift: rocks on different sides of the ocean match.

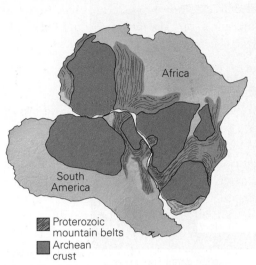

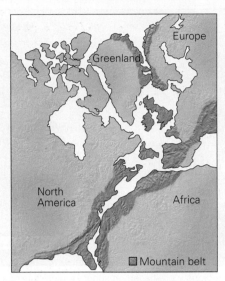

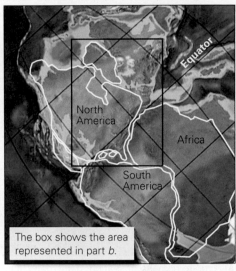

(a) Distinctive belts of rock in South America would align with similar ones in Africa if the Atlantic Ocean didn't exist.

Proterozoic mountain belts
Archean crust

(b) If the Atlantic didn't exist, Paleozoic mountain belts on both coasts would be adjacent.

Mountain belt

(c) A modern reconstruction showing the positions of mountain belts in Pangaea. Modern continents are outlined in white.

The box shows the area represented in part *b*.

CRITICISM OF WEGENER'S IDEAS

Wegener's model of a supercontinent that later broke apart explained the distribution of ancient glaciers, coal, sand dunes, rock assemblages, and fossils. Clearly, he had compiled a strong circumstantial case for continental drift. But as we noted earlier, he could not adequately explain how or why continents drifted. He left for Greenland having failed to convince his peers, and he died with no clue that his ideas, after lying dormant for decades, would be reborn as the basis of the broader theory of plate tectonics.

In effect, Wegener was ahead of his time. It would take an additional 30 years of research before geologists obtained sufficient data to test his hypotheses properly. Collecting such data required instruments and techniques that did not exist in Wegener's day. Between 1930 and 1960, geologists learned how to determine the age of rocks and how to "see" the ocean floor. In the next section, we describe some of the key results of this work.

TAKE-HOME MESSAGE

In the early 20th century, Alfred Wegener argued that the continents had once been connected in a supercontinent, Pangaea, that later broke up to produce smaller continents that drifted apart. He showed that the matching shapes of coastlines, as well as the distribution of ancient glaciers, climate belts, fossils, and rock units, make better sense if Pangaea existed. But Wegener couldn't convince his peers about continental drift.

QUICK QUESTION Why were Wegener's peers skeptical of continental drift?

2.3 The Discovery of Seafloor Spreading

NEW IMAGES OF SEAFLOOR BATHYMETRY

Military needs during World War II gave a boost to seafloor exploration, for as submarine fleets grew, navies required detailed information about *bathymetry*, or depth variations. The invention of *sonar* (echo sounding) permitted such information to be gathered quickly. Echo sounding works on the same principle that a bat uses to navigate and find insects. A ship emits a sound pulse that travels down through the water, bounces off the seafloor, and returns up as an echo through the water to a receiver on the ship. Because sound waves travel at a known velocity, the time between the sound's emission and the echo's detection indicates the distance between the ship and the seafloor. (Recall that Velocity = Distance/Time, so

SEE FOR YOURSELF...

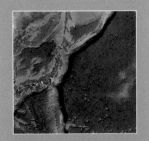

Trenches

LATITUDE
35°37'59.86" N

LONGITUDE
145°36'12.78" E

Look down from 5,500 km. Zoom in and use the elevation tool to measure trench depth.

Trenches of the western Pacific Ocean, near Japan.

FIGURE 2.6 Bathymetric features of the ocean floor.

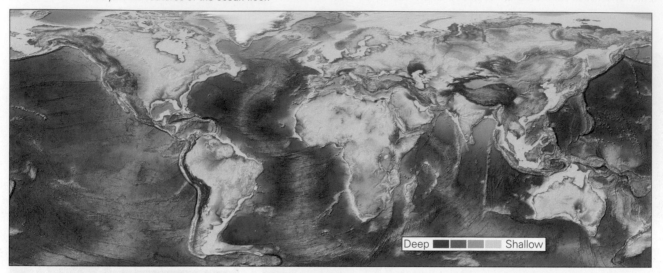

(a) A modern image of seafloor bathymetry. (Older versions were based on sonar studies; this one uses satellite data.) The colors indicate water depth.

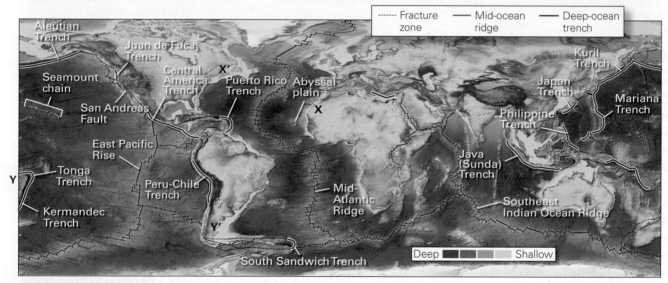

(b) On this map, major bathymetric features have been annotated with symbols.

Distance = Velocity × Time.) As the ship travels, observers can obtain a continuous record of the depth of the seafloor. The resulting cross section showing depth plotted as a function of location is called a *bathymetric profile*. By cruising back and forth across the ocean many times, investigators obtain a series of bathymetric profiles, and from these they construct *bathymetric maps* of the seafloor. (Geologists can now produce such maps rapidly using satellite data.) Bathymetric maps define the shape of the seafloor and reveal several important features **(Fig. 2.6)**.

> *Mid-ocean ridges:* The floor beneath all major oceans includes *abyssal plains*, which are broad, relatively flat regions of the ocean that lie at a depth of about 4 to 5 km below sea level, and **mid-ocean ridges**, submarine mountain ranges whose peaks lie only about 2 to 2.5 km below

sea level **(Fig. 2.7a)**. Geologists call the crest of a mid-ocean ridge the *ridge axis*. Mid-ocean ridges are roughly symmetrical—bathymetry on one side of the axis is nearly a mirror image of bathymetry on the other side.

> *Deep-sea trenches:* Along much of the perimeter of the Pacific Ocean, and in a few additional localities, the seafloor reaches depths greater than 5 km. These deep areas define elongate troughs that are referred to as **trenches** **(Fig. 2.7b, c)**. Trenches border **volcanic arcs**, curving chains of active volcanoes.

> *Fracture zones:* Surveys reveal that narrow bands of vertical cracks and broken-up rock cut the ocean floor of mid-ocean ridges. These **fracture zones** lie roughly at right angles to mid-ocean ridges and interrupt the continuity of the ridge axis **(Fig. 2.7d)**.

FIGURE 2.7 Examples of distinctive bathymetric features.

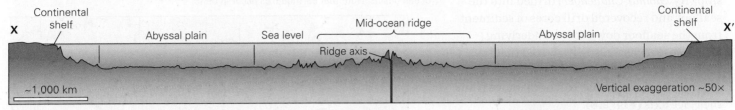

(a) A bathymetric profile across the North Atlantic Ocean (X to X' on Figure 2.6b), showing the Mid-Atlantic Ridge.

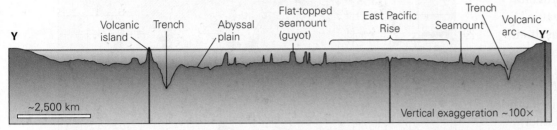

(b) A bathymetric profile across the South Pacific Ocean (Y to Y' on Figure 2.6b), showing trenches and arcs. Note that the bathymetric feature called the East Pacific Rise is a mid-ocean ridge.

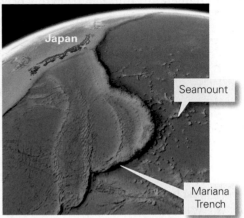

(c) The Mariana Trench in the western Pacific defines a distinct curve.

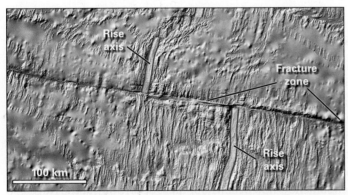

(d) A bathymetric map of a fracture zone along the Mid-Atlantic Ridge. The ridge axis is highlighted.

(e) A bathymetric map of seamounts on the Pacific Ocean floor, northwest of Hawaii.

> *Seamount chains:* Numerous volcanic islands poke up from the ocean floor; for example, the Hawaiian Islands lie in the middle of the Pacific. In addition to islands that rise above sea level, sonar has detected many **seamounts**, isolated submarine mountains, which were once volcanoes but no longer erupt **(Fig. 2.7e)**. Volcanic islands and seamounts typically occur in chains, but in contrast to the volcanic arcs that border deep-sea trenches, only the island at one end of seamount and island chains remains capable of erupting today. In island arcs, all of the volcanoes are capable of erupting.

Did you ever wonder...
why Hawaii rises above the middle of the ocean?

NEW OBSERVATIONS ON THE NATURE OF OCEANIC CRUST

By the mid-20th century, geologists had discovered many important characteristics of the seafloor, which led them to realize that oceanic crust differs markedly from continental crust, and that bathymetric features of the ocean floor provide clues to the origin of the crust. Specifically:

> A layer of sediment composed of clay and the tiny shells of dead plankton covers much of the ocean floor. This layer becomes progressively thicker away from the mid-ocean ridge axis. But even at its thickest, the sediment layer is too thin to have been accumulating for the entirety of

Earth history. Of note, when, in the 1960s, a ship (RV *Glomar Challenger*) drilled into the seafloor and recovered drill cores of sediment from the seafloor down to the underlying basalt, researchers also discovered that the sediment layer just above the basalt gets progressively older with increasing distance from the ridge axis (**Fig. 2.8**).

> The composition of the oceanic crust is fundamentally different from that of continental crust. Beneath its sediment cover, oceanic crust bedrock consists primarily of basalt—it does not contain the great variety of rock types found on continents.

> *Heat flow*, the rate at which heat rises from the Earth's interior up through the crust, is not the same everywhere in the ocean floor. Rather, more heat rises beneath mid-ocean ridges than elsewhere. This observation led researchers to speculate that molten rock rises into the crust just below a mid-ocean ridge axis.

> When maps showing the distribution of earthquakes in oceanic regions became available in the years after World War II, it became clear that earthquakes do not occur randomly, but rather in distinct **seismic belts** (**Fig. 2.9**). Some belts follow trenches, some follow mid-ocean ridge axes, and others lie along portions of fracture zones. Since earthquakes define locations where rocks break and move, geologists realized that these bathymetric features are places where motion of one part of the Earth, relative to the adjacent part, is taking place.

> The ridge axis of some mid-ocean ridges is marked by a narrow (a few kilometers wide), elongate trough hundreds of meters deeper than its borders. In this regard, the bathymetry of a mid-ocean ridge resembles the topography of the East African rift valley, a place where the crust of Africa is stretching and breaking apart.

FIGURE 2.8 Drilling into the sediment layer of the ocean floor confirms that the basal sediment in contact with basalt gets older the farther away it is from the ridge. The sediment at location A is older than the sediment at location D.

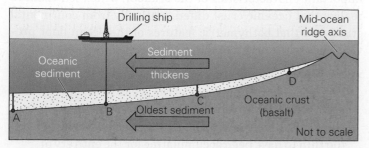

FIGURE 2.9 A 1953 map showing the distribution of earthquake locations in the ocean basins. Note that earthquakes occur in belts.

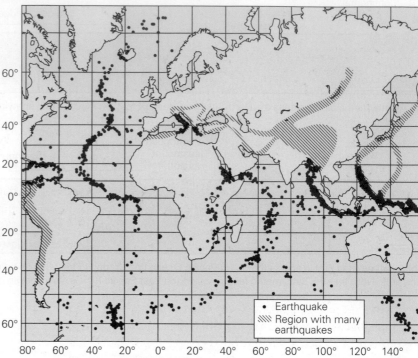

HARRY HESS AND HIS "ESSAY IN GEOPOETRY"

In the late 1950s, Harry Hess studied the observations described above and made an important deduction. He realized that, because the sediment layer on the ocean floor was thin overall, the ocean floor might be much younger than the continents. Also, because sediment thickens progressively away from a mid-ocean ridge axis, the ridges themselves are likely younger than the deeper parts of the ocean floor. If this is so, then somehow new ocean floor must be forming at the ridges, so an ocean basin could be getting wider with time. But how? The association of earthquakes with mid-ocean ridges hinted that the seafloor is cracking and splitting apart at the ridge, and the similarity in form to the East African rift valley suggested that the region is stretching. The discovery of high heat flow along mid-ocean ridge axes provided the final piece of the puzzle, for it suggested the presence of molten rock beneath the ridges.

In 1960, Hess brought these ideas together and suggested that magma rises upward at mid-ocean ridges, and that this material solidifies to form the basalt of oceanic crust (**Fig. 2.10**). The new seafloor then moves away from the ridge axis, leading to the widening of the ocean basin, a process we

Did you ever wonder...
if the distance between New York and Paris changes?

FIGURE 2.10 Harry Hess's early concept of seafloor spreading (1962). Hess implied, incorrectly, that only the crust moved. We will see that this sketch is an oversimplification and contains errors.

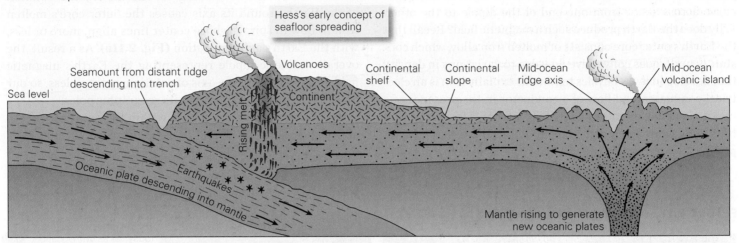

now call **seafloor spreading**. Hess realized that old ocean floor must be consumed somewhere, otherwise the Earth would be expanding, so he suggested that deep-ocean trenches might be places where the seafloor sinks back into the mantle, a process now known as *subduction*. Hess suggested that earthquakes at trenches are evidence of this movement, but he didn't understand how the movement took place. Other geologists were coming to similar conclusions at about the same time.

Hess and his contemporaries realized that the seafloor-spreading hypothesis instantly provided the long-sought explanation of how continental drift occurs. Continents passively move apart as the seafloor between them spreads at mid-ocean ridges, and they passively move together as the seafloor between them sinks back into the mantle, or subducts, at trenches. (As we will see later, geologists now realize that it is the whole lithosphere that moves, not just the crust.) Understanding seafloor spreading and subduction proved to be an important step on the route to plate tectonics—the ideas seemed so good that Hess referred to his paper describing them as an "essay in geopoetry." But first, the idea needed to be tested, and other key discoveries would have to take place before the whole theory of plate tectonics could come together.

TAKE-HOME MESSAGE

New observations about seafloor bathymetry, sediment cover, heat flow, and seismicity led to Hess's proposal of seafloor spreading—new seafloor forms at mid-ocean ridges and then moves away from the ridge axis, so ocean basins can get wider with time. As this happens, old ocean floor sinks back into the mantle by subduction.

QUICK QUESTION How do seafloor spreading and subduction provide an explanation for how continents move?

2.4 Paleomagnetism—Proving Continental Drift and Seafloor Spreading

The publication of Hess's hypothesis that ocean basins grow by seafloor spreading suddenly made Wegener's hypothesis that continents drift seem like a real possibility. But, in order for both hypotheses to be accepted as theory, there had to be proof of movement. A method to directly measure continental movement on a short time scale did not exist in 1962. Fortunately, a discovery made by Chinese sailors more than 1,500 years ago provided a basis for measuring such movement on a long time scale. The sailors had discovered that a piece of lodestone, when suspended from a thread, points in a northerly direction and can help guide a voyage. Lodestone exhibits this behavior because it consists of *magnetite*, an iron-rich mineral that, like a compass needle, aligns with the Earth's *magnetic field lines*. While not as magnetic as lodestone, several rock types contain trace amounts of magnetite or other magnetic minerals, so they behave overall like weak magnets. The study of such magnetic behavior led to the realization that rocks can preserve **paleomagnetism**, a record of the Earth's magnetic field in the past. An understanding of paleomagnetism provided proof of both continental drift and seafloor spreading. As a foundation for introducing paleomagnetism, we first provide further detail on the basic nature of the Earth's magnetic field.

THE EARTH'S MAGNETIC FIELD: SOME FURTHER DETAILS

As we learned in Chapter 1, we can represent the Earth's magnetic field by a *magnetic dipole*, an imaginary arrow that points from the north magnetic pole to the south magnetic pole and passes through the center of the Earth. Magnetic field lines curve across space from one end of the dipole to the other. Why does the Earth produce such a magnetic field? Recall that the Earth's outer core consists of molten iron alloy, which constantly undergoes convective flow due to variations in density that develop in it. As physics textbooks explain, iron is an electrical conductor, and flow of a conductor in the presence of a magnetic field generates an electric current. This current, in turn, maintains the magnetic field. Put in the terminology of physics, the outer core behaves like a *self-sustaining dynamo*.

Convective flow in the outer core does not take place in random directions. According to one model, the rotation of the Earth around its axis causes the outer core's molten iron to flow in spirals whose center lines align, more or less, with the Earth's axis of rotation **(Fig. 2.11a)**. As a result, the overall magnetic dipole representing the Earth's magnetic field also aligns with the axis of rotation, more or less, so our planet's magnetic poles lie near its geographic poles.

Over time, convection spirals within the outer core probably change their shape and orientation, causing the Earth's magnetic dipole to wobble and its magnetic poles to move a

FIGURE 2.11 Features of the Earth's magnetic field.

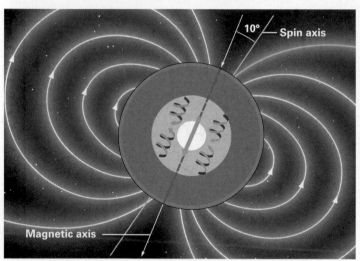

(a) The magnetic axis is not parallel to the axis of rotation. The field is generated by flow in the outer core.

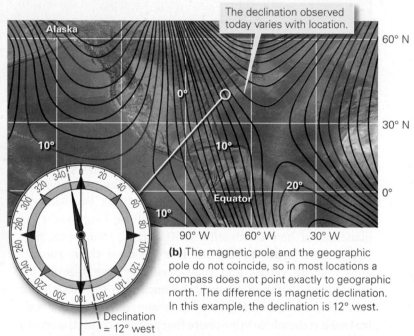

(b) The magnetic pole and the geographic pole do not coincide, so in most locations a compass does not point exactly to geographic north. The difference is magnetic declination. In this example, the declination is 12° west.

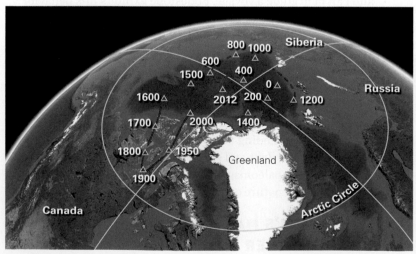

(c) A simplified map showing the changing position of the north magnetic pole over the past 2,000 years. Before about 1600, the position was not as well constrained, so the path is dashed.

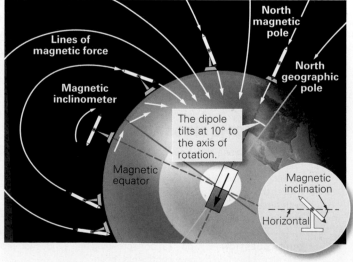

(d) The Earth's magnetic field lines curve, so the tilt of a magnetic needle changes with latitude. This tilt is the magnetic inclination.

little. As a result, at any given time, the magnetic dipole and the Earth's axis of rotation are not exactly parallel. This situation means that at any given time, a compass needle doesn't point to geographic north and doesn't trend exactly parallel to a line of longitude. The angle between the direction in which a compass needle points and a line of longitude at a given location is called the **magnetic declination (Fig. 2.11b)**. Today, the magnetic axis and the axis of rotation differ by several degrees, and the north magnetic pole lies a few hundred kilometers away from "true north" (*geographic north*). This distance changes every year because the magnetic pole trundles across the Arctic Ocean toward Siberia at a rate of 50 to 60 km per year. Significantly, a map depicting the North Pole's changing position over the past 2,000 years reveals that the pole follows a fairly random path that stays north of the Arctic Circle **(Fig. 2.11c)**. Because of this randomness, geologists assume that, averaged over thousands of years, the Earth's magnetic poles and geographic poles roughly coincide.

As we've seen, invisible magnetic field lines curve through space around the Earth **(Fig. 2.11d)**. Looking more closely, we see that these lines are horizontal (parallel to the Earth's surface) at the magnetic equator and vertical (perpendicular to the Earth's surface) at the magnetic poles. At mid-latitudes, they tilt at an angle to the Earth's surface. Geologists refer to the angle between the magnetic field and the surface of the Earth at a given location as the **magnetic inclination**. The magnetic field has an inclination of 0° at the magnetic equator and 90° at the magnetic pole. If you make a *magnetic inclinometer* by placing a magnetic needle on a horizontal axis so it can pivot up and down, you can observe magnetic inclination directly. (Regular compasses don't show inclination because their needles have been balanced to compensate for a downward pull.)

WHAT IS PALEOMAGNETISM?

In the early 20th century, when researchers developed instruments that could measure the very weak magnetic field produced by some kinds of rock, they made a surprising discovery. The orientation of the dipole representing the magnetic field of an ancient rock does not necessarily point to the present-day magnetic pole, as a compass needle does **(Fig. 2.12a)**. Geologists interpreted this observation to mean that these rocks preserve *paleomagnetism*.

Subsequent work revealed that paleomagnetism develops in rocks for different reasons. For example, in an igneous rock such as basalt, paleomagnetism forms when the rock solidifies and cools from melt **(Fig. 2.12b)**. Why? In hot melt, thermal energy causes iron-bearing molecules to vibrate, spin, and move about randomly, so at any given instant, each iron-bearing molecule's dipole points in a different direction. The fields produced by these differently oriented dipoles cancel one another out. As the melt cools and solid rock starts to

FIGURE 2.12 Paleomagnetism and how it can form.

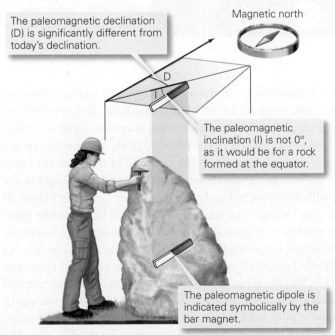

The paleomagnetic declination (D) is significantly different from today's declination.

Magnetic north

The paleomagnetic inclination (I) is not 0°, as it would be for a rock formed at the equator.

The paleomagnetic dipole is indicated symbolically by the bar magnet.

(a) A geologist finds an ancient rock sample at a location on the equator, where declination today is 0°. The orientation of the rock's paleomagnetism is different from that of today's field.

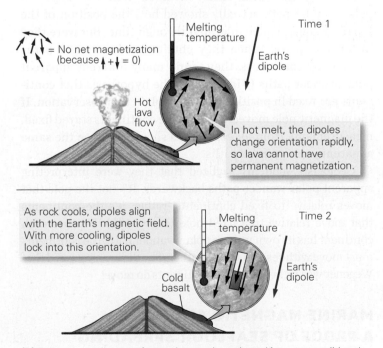

= No net magnetization (because ↑ + ↓ = 0)

Melting temperature

Time 1

Earth's dipole

Hot lava flow

In hot melt, the dipoles change orientation rapidly, so lava cannot have permanent magnetization.

As rock cools, dipoles align with the Earth's magnetic field. With more cooling, dipoles lock into this orientation.

Melting temperature

Time 2

Cold basalt

Earth's dipole

(b) Paleomagnetism can form when melt cools and becomes solid rock.

form, molecular motion slows down, so the dipoles, like tiny compass needles, align with the Earth's magnetic field. Finally, when crystals of solid magnetite grow and cool, the dipoles lock into alignment, yielding a measureable magnetic field that won't change as long as the rock remains sufficiently cool. In a sedimentary rock, paleomagnetism can form when magnetic minerals grow in the spaces between sediment grains; as these minerals grow, their dipoles align with the Earth's magnetic field.

APPARENT POLAR WANDER—A PROOF THAT CONTINENTS MOVE

Why doesn't the paleomagnetic dipole in ancient rocks point to the present-day magnetic field? When geologists first tried to answer this question, they assumed that continents stayed fixed in position, so they concluded that the positions of the Earth's magnetic poles in the past were different than they are today. They introduced the term **paleopole** to refer to the position the Earth's magnetic north pole had in the past. With this concept in mind, they set out to track the change in position of the paleopole over time. To do this, they measured the paleomagnetism in a succession of rocks of different ages from the same general location, and they plotted the associated paleopole positions on a map **(Fig. 2.13)**. The successive positions of dated paleopoles trace out a curving line that came to be known as an **apparent polar-wander path**.

At first, the researchers assumed that the apparent polar-wander path actually showed how the position of the Earth's magnetic pole migrated through time. But were they in for a surprise! When they obtained polar-wander paths from many continents, they found many different apparent polar-wander paths **(Fig. 2.14a)**. The hypothesis that continents are fixed in position cannot explain this observation. If the magnetic pole moved while all the continents stayed fixed, measurements from all continents should produce the same apparent polar-wander paths.

Geologists suddenly realized that they were interpreting apparent polar-wander paths backwards. It's not the pole that moves relative to fixed continents; rather, it's the continents that move relative to a fixed pole **(Fig. 2.14b)**. Because each continent has its own unique polar-wander path, the continents must move with respect to each other. The discovery proved that Wegener was right all along—continents do move!

MARINE MAGNETIC ANOMALIES: A PROOF OF SEAFLOOR SPREADING

Discovering Magnetic Reversals Through human history, the north-seeking end of a compass needle has always pointed

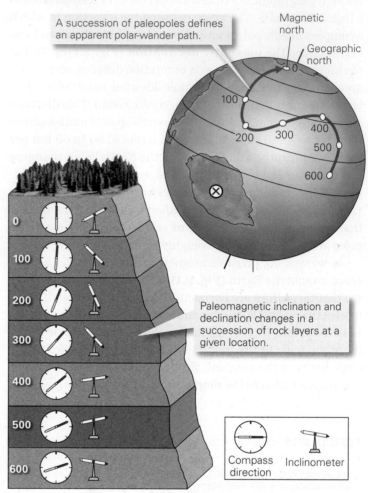

FIGURE 2.13 Production of an apparent polar-wander path. Measurements of paleomagnetism in a succession of rock layers on Continent X define an apparent polar-wander path relative to the continent. (Numbers are Ma.)

A succession of paleopoles defines an apparent polar-wander path.

Magnetic north

Geographic north

Paleomagnetic inclination and declination changes in a succession of rock layers at a given location.

Compass direction   Inclinometer

toward the north magnetic pole, and geologists once assumed that it had done so for all of geologic time. But research on paleomagnetism led to another big surprise. Geologists discovered locations where the polarity of the magnetism in one layer of rock is opposite to the polarity of the adjacent layer. In other words, the dipole measured in one layer points in the opposite direction to the dipole measured in the next layer. At first, they thought that such *reversals* could be caused by lightning strikes or chemical reactions. But eventually, examples of reversals convinced them that the polarity of the Earth's magnetic field itself really does flip every now and then, and quickly. During some intervals of time, the Earth has **normal polarity**, the same polarity that it has today, with the north magnetic pole near the north geographic pole. During other intervals of time, the Earth has **reversed polarity**, opposite to that of today, with the north magnetic pole near the south geographic pole. Note that when a polarity reversal takes

FIGURE 2.14 Interpretation of apparent polar-wander paths.

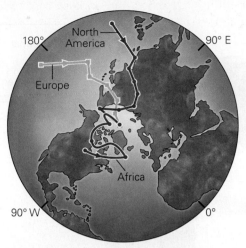

(a) The apparent polar-wander path of North America is not the same as that of Europe or Africa.

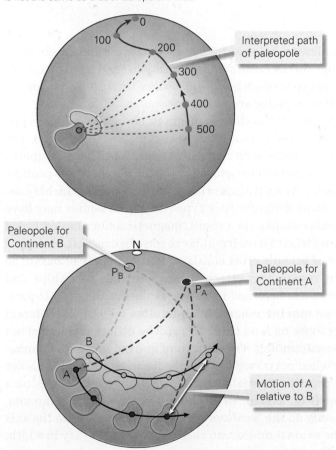

(b) If continents are fixed, then the pole moves relative to the continent. If the pole is fixed, then the continent must drift relative to the pole.

place, only the direction of the dipole changes—the Earth itself does not turn upside down **(Fig. 2.15)**.

At first, researchers had no way of knowing when these reversals took place. This situation changed when geologists developed techniques for measuring the *numerical ages* of rocks in years (see Chapter 10). By recording the

paleomagnetic polarity preserved in a succession of basalt layers, each of which had been dated numerically, researchers established a **magnetic-reversal chronology**, a chart showing when reversals happened and, therefore, how much time occurred between reversals **(Fig. 2.16)**. Their initial chart extended back 4.5 million years. The researchers referred to a major interval of a given polarity as a **chron**, with short-duration intervals of opposite polarity known as **subchrons**. They named chrons after famous researchers and subchrons after localities.

Interpreting "Seafloor Stripes" During the years when some researchers were figuring out the magnetic-reversal chronology, others were measuring subtle variations in the strength of the magnetic field with location. Using a *magnetometer*, an instrument that measures magnetic field strength, they found that in some localities, the field is stronger than they would have expected it would be if produced only by circulation in the outer core, whereas in others, the field is slightly weaker than expected. Researchers coined the term **magnetic anomaly** for

FIGURE 2.15 Magnetic polarity reversals.

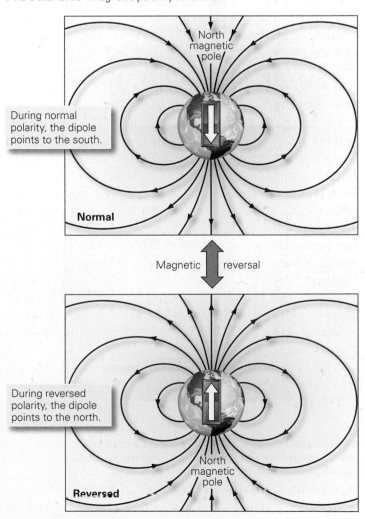

During normal polarity, the dipole points to the south.

Normal

Magnetic reversal

During reversed polarity, the dipole points to the north.

Reversed

FIGURE 2.16 The chronology of magnetic reversals.

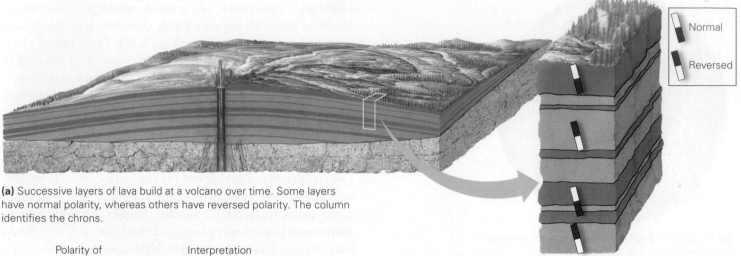

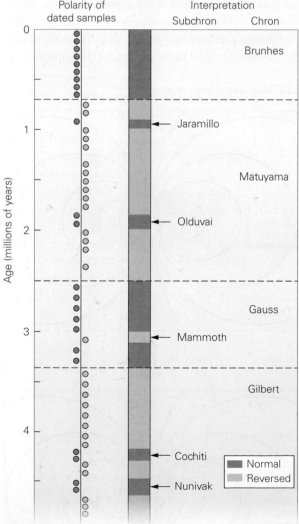

(a) Successive layers of lava build at a volcano over time. Some layers have normal polarity, whereas others have reversed polarity. The column identifies the chrons.

(b) Observations of such layers led to the production of a reversal chronology with named polarity intervals, or chrons. Subchrons occur with chrons.

the difference between the expected strength and the observed strength of the Earth's field at a location. Field strengths greater than expected are *positive anomalies*, whereas field strengths less than expected are *negative anomalies*.

On land, the pattern of magnetic anomalies depends primarily on the composition of rock below the ground. For example, rocks with more iron tend to be more magnetic, so a magnetometer placed over them records a positive anomaly. As we'll discuss later in this book, continents contain many different rock types, and rock bodies may have irregular shapes. As a result, magnetic anomalies recorded on land tend to have irregular shapes. In contrast, the upper layer of oceanic crust consists almost entirely of basalt, so when researchers attached magnetometers to ships and towed them back and forth across the oceans, they expected that **marine magnetic anomalies** might look different from those on land **(Fig. 2.17a)**. The difference turned out to be astounding. The pattern of marine magnetic anomalies, when portrayed by coloring positive anomalies darker and negative anomalies lighter, resembles the stripes on a zebra **(Fig. 2.17b)**. In addition, the pattern isn't random. Not only do the "seafloor stripes" trend parallel to the axis of the nearest mid-ocean ridge, but they also vary in width, and the pattern of stripes on one side of a mid-ocean ridge looks like the mirror image of the pattern on the other side **(Fig. 2.17c)**.

At first, researchers were baffled by marine magnetic anomalies. Eventually, however, they realized that the pattern of stripes close to a mid-ocean ridge resembles the pattern of chrons and subchrons in the magnetic-reversal chronology

FIGURE 2.17 The discovery of marine magnetic anomalies.

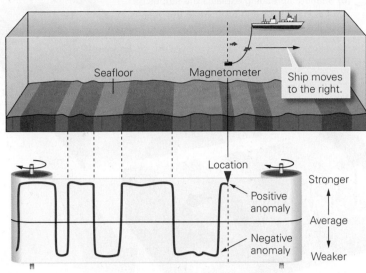

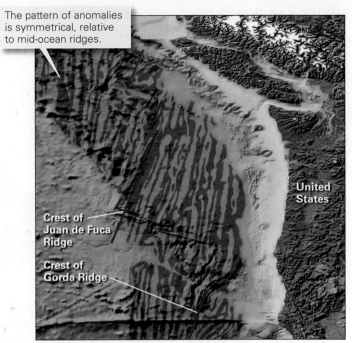

The pattern of anomalies is symmetrical, relative to mid-ocean ridges.

(a) A ship towing a magnetometer detects changes in the strength of the magnetic field on the seafloor. On a paper record, intervals of stronger magnetism (positive anomalies) alternate with intervals of weaker magnetism (negative anomalies).

(b) A map showing areas of positive anomalies (dark) and negative anomalies (light) off the west coast of North America. The pattern of anomalies resembles zebra stripes.

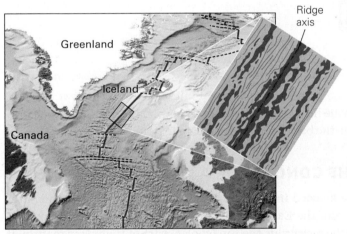

(c) Magnetic anomalies are symmetrical relative to the mid-ocean ridge.

produced by the Earth's dipole to produce a stronger signal. A negative anomaly develops where the basalt formed during a time of reversed polarity. The magnetic force produced by the reversed-polarity paleomagnetism in the basalt subtracts from the force produced by the Earth's dipole, so the magnetometer detects a weaker signal than expected **(Fig. 2.18a)**. As oceanic crust moves away from the ridge axis, it carries its paleomagnetic character with it. Seafloor spreading along a segment of ridge takes place at a fairly constant rate, so the relative widths of the stripes correspond to the relative durations of polarity intervals **(Fig. 2.18b, c)**.

chart: the relative widths of the stripes are exactly the same as the relative durations of the chrons and subchrons. What does the stripe-like pattern of marine magnetic anomalies mean?

In 1963, researchers realized that the striped pattern develops because magnetic reversals are occurring while seafloor spreading takes place. Specifically, the polarity of basalt in the oceanic crust depends on the polarity of the Earth's magnetic field at the time the crust forms. A positive anomaly develops where the seafloor basalt formed when the Earth had normal polarity. The magnetic force produced by the normal-polarity paleomagnetism in this basalt adds to the force

TAKE-HOME MESSAGE

Certain rocks preserve paleomagnetism, a record of the past position of the Earth's magnetic poles relative to a rock. By measuring paleomagnetism in rocks of different ages from various continents, geologists could show that continents must move relative to one another. The Earth's magnetic field reverses polarity every now and then, causing positive and negative anomalies over the seafloor. The discovery of such marine magnetic anomalies proves that seafloor spreading takes place.

QUICK QUESTION Why do geologists insert the word *apparent* in front of *polar-wander path*?

FIGURE 2.18 The progressive development of magnetic anomalies.

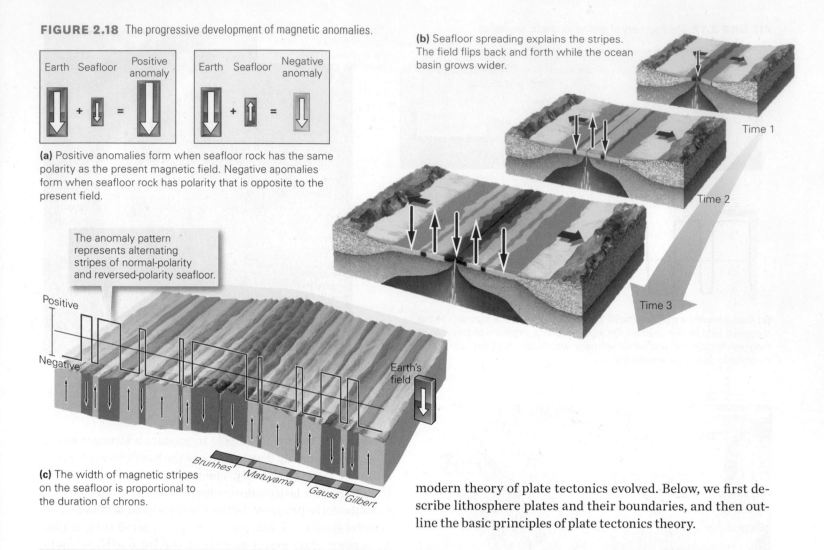

(a) Positive anomalies form when seafloor rock has the same polarity as the present magnetic field. Negative anomalies form when seafloor rock has polarity that is opposite to the present field.

The anomaly pattern represents alternating stripes of normal-polarity and reversed-polarity seafloor.

Positive

Negative

Earth's field

Brunhes Matuyama Gauss Gilbert

(c) The width of magnetic stripes on the seafloor is proportional to the duration of chrons.

(b) Seafloor spreading explains the stripes. The field flips back and forth while the ocean basin grows wider.

Time 1

Time 2

Time 3

modern theory of plate tectonics evolved. Below, we first describe lithosphere plates and their boundaries, and then outline the basic principles of plate tectonics theory.

2.5 **What Do We Mean by Plate Tectonics?**

The paleomagnetic proof of continental drift and the discovery and proof of seafloor spreading set off a scientific revolution in geology in the 1960s and 1970s. Geologists realized that many of their existing interpretations of global geology, based on the premise that the positions of continents and oceans remain fixed in position through time, were simply wrong! Researchers dropped what they were doing to study the broader implications of continental drift and seafloor spreading. It became clear that these phenomena required that the Earth's outer shell be divided into distinct, relatively rigid *plates* that moved relative to one another. New studies clarified the meaning of a plate, defined the types of plate boundaries, constrained plate motions, related plate motions to earthquakes and volcanoes, showed how plate interactions can explain mountain belts and seamount chains, and outlined the history of past plate motions. From these, the

THE CONCEPT OF A LITHOSPHERE PLATE

We learned in Chapter 1 that geoscientists divide the outer part of the Earth into two layers. The **lithosphere** consists of the crust plus the top (cooler) part of the upper mantle. It behaves relatively rigidly, meaning that when a force pushes or pulls on it, it does not flow but rather bends or breaks **(Fig. 2.19a)**. The lithosphere sits on a relatively soft, or "plastic," layer called the **asthenosphere**, composed of warmer (greater than 1,280°C) mantle that can flow very slowly when acted on by a force. As a result, the asthenosphere convects, like water in a pot, though much more slowly.

Continental lithosphere and oceanic lithosphere differ markedly in their thicknesses. On average, continental lithosphere has a thickness of 150 km, whereas old oceanic lithosphere has a thickness of about 100 km **(Fig. 2.19b)**. (For reasons discussed later in this chapter, new oceanic lithosphere at a mid-ocean ridge is much thinner.) Recall that the crustal part of continental lithosphere ranges from 25 to 70 km thick and consists largely of relatively low-density felsic and intermediate rock (granite). In contrast, the crustal part

FIGURE 2.19 The nature and behavior of the lithosphere.

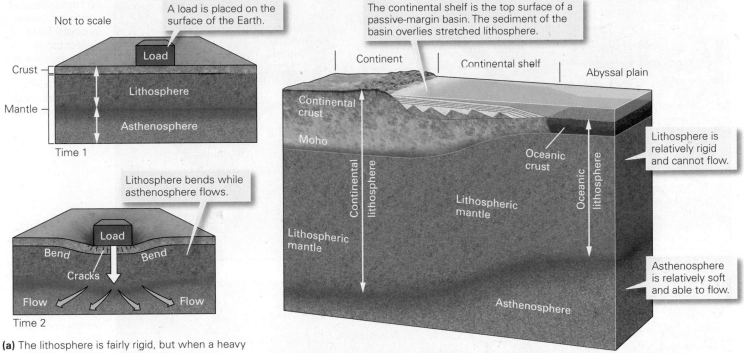

(a) The lithosphere is fairly rigid, but when a heavy load builds on its surface, the surface bends down. This can happen because the "plastic" asthenosphere can flow out of the way.

(b) The lithosphere consists of the crust plus the uppermost mantle. It is thicker beneath continents than beneath oceans.

of oceanic lithosphere is only 7 to 10 km thick and consists of relatively dense mafic rock (basalt and gabbro). The mantle part of both continental and oceanic lithosphere consists of very dense ultramafic rock (peridotite). Because of these differences, the surface of continental lithosphere lies at a higher elevation than does the surface of the oceanic lithosphere—that's why distinct ocean basins exist.

The lithosphere forms the Earth's relatively rigid shell. But unlike the shell of a chicken egg, the lithospheric shell contains a number of major breaks, which separate it into distinct pieces. As noted earlier, we call the pieces *lithosphere plates*, or simply *plates*. The breaks between plates are known as **plate boundaries**. Geoscientists distinguish 12 major plates and several microplates.

THE BASIC PRINCIPLES OF PLATE TECTONICS THEORY

With the background provided above, we can summarize plate tectonics theory concisely as follows:

> The Earth's lithosphere is divided into plates that move relative to each other. As a plate moves, its internal area remains mostly, but not perfectly, rigid and intact.

> Rock along plate boundaries undergoes intense deformation (cracking, sliding, bending, stretching, and

squashing) as the plate grinds or scrapes against its neighbors or pulls away from its neighbors.

> As plates move, so do continents that form part of the plates.

> Because of plate tectonics, the map of the Earth's surface constantly changes.

IDENTIFYING PLATE BOUNDARIES

How do we recognize the location of a plate boundary? The answer becomes clear from looking at a map showing the locations of earthquakes **(Fig. 2.20a, b)**. Recall from Chapter 1 that earthquakes are vibrations caused by shock waves that are generated where rock breaks and suddenly slips along a fault. The *epicenter* marks the point on the Earth's surface directly above the *focus*, the location at which rock broke or slipped to cause the earthquake. Earthquake epicenters do not speckle the globe randomly, like buckshot on a target. Rather, the majority occur in relatively narrow, distinct belts. These earthquake belts, or *seismic belts*, define the position of plate boundaries because the fracturing and slipping that occurs along plate boundaries generates

Did you ever wonder...
why earthquakes don't occur everywhere?

FIGURE 2.20 The locations of plate boundaries and the distribution of earthquakes.

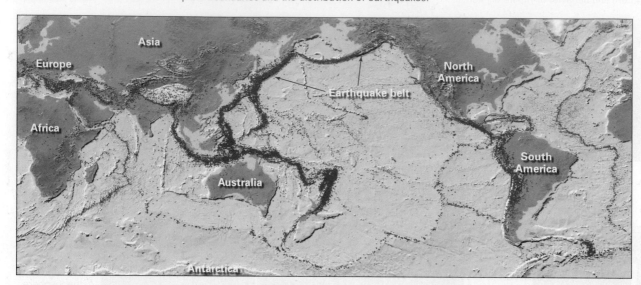

(a) The locations of most earthquakes (red dots) fall in distinct bands that correspond to plate boundaries. Relatively few earthquakes occur in the stabler plate interiors.

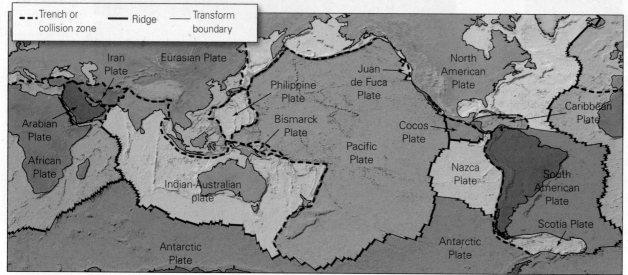

(b) A map of major plates shows that some consist entirely of oceanic lithosphere, whereas others consist of both continental and oceanic lithosphere. Active continental margins lie along plate boundaries; passive margins do not.

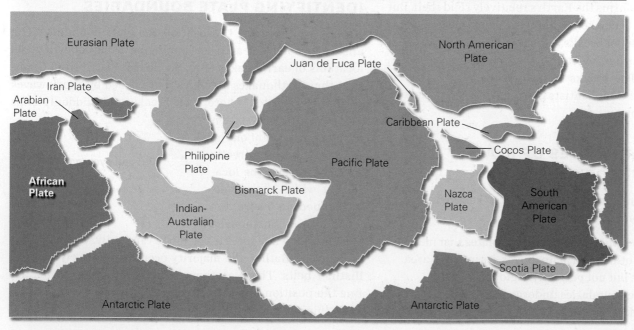

(c) An exploded view of the plates emphasizes the variation in their shape and size.

earthquakes. *Plate interiors*, regions away from the plate boundaries, remain relatively earthquake-free because they do not accommodate as much movement. While earthquakes serve as the most definitive indicators of a plate boundary, other prominent geologic features also develop along plate boundaries, as you will learn by the end of this chapter.

Note that some plates consist entirely of oceanic lithosphere, whereas some plates consist of both oceanic and continental lithosphere. Also, not all plates are the same size and some plate boundaries follow continental margins, the boundary between a continent and an ocean, but others do not (**Fig. 2.20c**). For this reason, we distinguish between **active margins**, which are plate boundaries, and **passive margins**, which are not plate boundaries. Earthquakes are common at active margins, but not at passive margins. Along passive margins, continental crust is thinner than in continental interiors. Thick (10 to 15 km) accumulations of sediment, the fill of a *passive-margin basin,* cover this thinned crust. The surface of this sediment layer is a broad, shallow (less than 500 m deep) region called the *continental shelf,* home to the major fisheries of the world.

Geologists define three types of plate boundaries, based on the relative motions of the plates on either side of the boundary (**Fig. 2.21**). A boundary at which two plates move apart from each other is a *divergent boundary*. A boundary at which two plates move toward each other so that one plate sinks beneath the other is a *convergent boundary*. And a boundary at which two plates slide sideways past each other is a *transform boundary*. Each type of boundary looks and behaves differently from the others, as we will see in the next three sections.

2.6 Divergent Boundaries and Seafloor Spreading

At a **divergent boundary**, or *spreading boundary*, two oceanic plates move apart (diverge) by the process of seafloor spreading. Note that an open space does not develop between diverging plates. Rather, as the plates move apart, new oceanic lithosphere forms continually along the divergent boundary (**Fig. 2.22a**). This process takes place at submarine mountain ranges called mid-ocean ridges, so geologists commonly refer to a divergent boundary as a *mid-ocean ridge*, or simply as a *ridge*.

To characterize a divergent boundary more completely, let's look at one mid-ocean ridge in detail (**Fig. 2.22b**). The Mid-Atlantic Ridge extends from the waters between

FIGURE 2.21 The three types of plate boundaries are distinguished by the nature of relative plate movement.

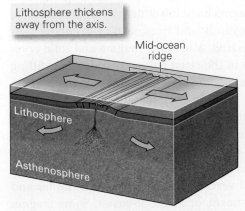

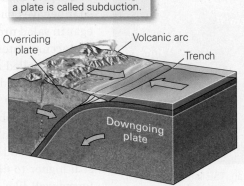

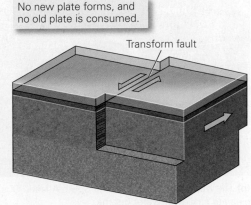

(a) At a *divergent boundary,* two plates move away from the axis of a mid-ocean ridge. New oceanic lithosphere forms.

(b) At a *convergent boundary,* two plates move toward each other; the downgoing plate sinks beneath the overriding plate.

(c) At a *transform boundary,* two plates slide past each other on a vertical fault surface. There is no up or down motion at such boundaries.

FIGURE 2.22 Divergent boundaries are delineated by mid-ocean ridges, where seafloor spreading occurs.

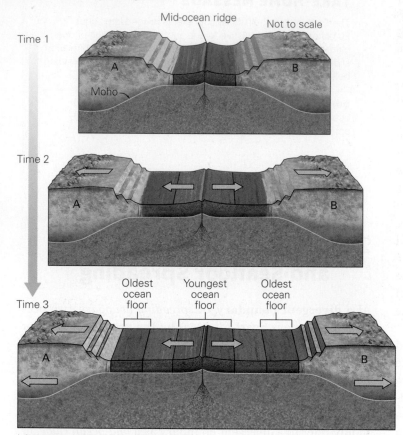

Mid-ocean ridge

Not to scale

Time 1

A B

Moho

Time 2

A B

Oldest ocean floor | Youngest ocean floor | Oldest ocean floor

Time 3

A B

(a) During seafloor spreading, the ocean floor gets wider, and continents on either side move apart. New oceanic crust forms at the ridge axis. Rising asthenosphere melts beneath the axis.

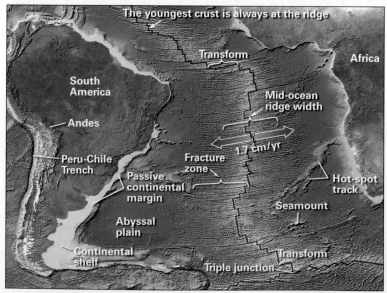

The youngest crust is always at the ridge

Transform

Africa

South America

Mid-ocean ridge width

Andes

1.7 cm/yr

Peru-Chile Trench

Fracture zone

Passive continental margin

Hot-spot track

Seamount

Abyssal plain

Continental shelf

Transform

Triple junction

(b) The Mid-Atlantic Ridge in the South Atlantic Ocean. The lighter shades of blue are shallower water depths.

northern Greenland and northern Scandinavia southward across the equator to the latitude of the southern tip of South America. It rises about 2 km above the depth of the Atlantic abyssal plains, so its crest lies at water depths of 2 to 2.5 km. The centerline of the mid-ocean ridge is called the *ridge axis*. Typically, a 1-km-deep elongate trough, the *median valley*, follows the trace of the ridge axis. Geologists have found that the formation of new oceanic crust takes place on or beneath the floor of the median valley. A series of steep scarps form the walls of the trough. Outside of the trough, seafloor slopes overall from the ridge toward the abyssal-plain, reaching abyssal-plain depth at a distance of about 500 to 800 km from the ridge axis. The Mid-Atlantic Ridge has a symmetrical shape, more or less, in that its eastern half looks like a mirror image of its western half.

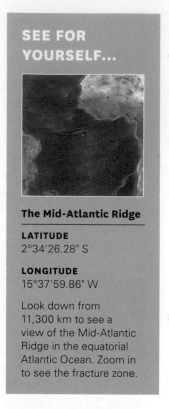
HOW DOES OCEANIC CRUST FORM AT A MID-OCEAN RIDGE?

As seafloor spreading takes place, hot asthenosphere rises beneath the ridge and begins to melt, and molten rock, or *magma*, forms. (We will explain why this magma forms in Chapter 4.) Magma has a lower density than solid rock, so it behaves buoyantly and rises, as oil rises above vinegar in salad dressing. A mush of magma and solid crystals accumulates in the crust below the ridge axis, filling a region called a *magma chamber*. Some of the magma mush cools and solidifies completely along the side of the chamber to make the coarse-grained, mafic igneous rock called gabbro **(Fig. 2.23a)**. (Igneous rocks, in general, form when a melt cools, so molecules lock together to produce solid crystals.) Some of the liquid magma rises still higher to fill vertical cracks, where it solidifies and forms wall-like sheets, or *dikes*, of basalt. Some magma makes it all the way to the surface of the seafloor at the

FIGURE 2.23 Geologic activity at a mid-ocean ridge.

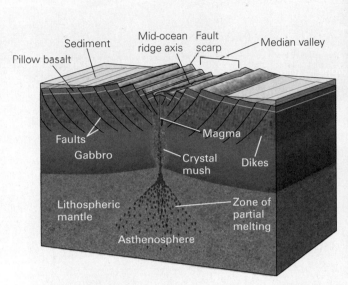

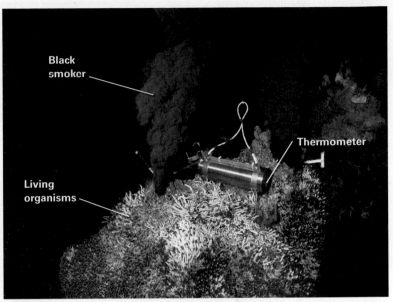

(a) Beneath a mid-ocean ridge lies a magma chamber. Gabbro forms on the sides of the magma chamber. Basalt dikes protrude upward.

(b) A column of superheated water gushing from a vent (known as a black smoker) along the ridge. Bacteria, shrimp, and worms live around the vent.

ridge axis and spills out of small submarine volcanoes as lava, which cools in blob- or pillow-like shapes to form a layer of *pillow basalt*. The gabbro, basalt dikes, and pillow basalt together become new crust.

Of note, the magma beneath a mid-ocean ridge heats seawater that has seeped into the crust, producing hot, mineralized water that rises and spews out from small chimneys, called **black smokers** because the rising water looks like a cloud of dark smoke. The dark color comes from a suspension of tiny mineral grains that precipitate in the water the instant that the hot water cools **(Fig. 2.23b)**. As they accumulate, these minerals build the chimney of a black smoker.

As soon as it forms, new oceanic crust moves away from the ridge axis, and when this happens, more magma rises from below, so still more crust forms. In other words, like a vast, moving conveyor belt, magma from the mantle rises to the Earth's surface at the ridge, solidifies to form oceanic crust, and then moves laterally away from the ridge. Because all seafloor forms at mid-ocean ridges, the youngest seafloor occurs at the ridge axis, and seafloor becomes progressively older away from the ridge. In the Atlantic Ocean, the oldest seafloor lies adjacent to the passive continental margins **(Fig. 2.24)**. The oldest seafloor on our planet underlies the western Pacific Ocean; this crust formed at about 200 Ma.

The tension (stretching force) applied to newly formed solid crust as spreading takes place breaks the crust, resulting in the formation of faults. Movement (slip) on the faults causes divergent-boundary earthquakes and produces numerous cliffs, or scarps, that trend parallel to the ridge axis.

HOW DOES THE LITHOSPHERIC MANTLE FORM AT A MID-OCEAN RIDGE?

So far, we've seen how oceanic crust forms at mid-ocean ridges. How does the *lithospheric mantle* (mantle part of the oceanic lithosphere) form? This part consists of the cooler uppermost layer of the mantle, in which temperatures are less than about 1,280°C. At the ridge axis, such temperatures occur almost at the base of the crust, because of the presence of rising hot asthenosphere and hot magma, so lithospheric mantle beneath the ridge axis effectively doesn't exist. But as the newly formed oceanic crust moves away from the ridge axis, the crust and the uppermost mantle directly beneath it gradually cool by losing heat to the ocean above. As soon as mantle rock (peridotite) cools below 1,280°C, it becomes, by definition, part of the lithosphere.

As oceanic lithosphere continues to move away from the ridge axis, it continues to cool, so the lithospheric mantle, and therefore the oceanic lithosphere as a whole, grows progressively thicker **(Fig. 2.25a)**. Note that this process doesn't change the thickness of the oceanic crust, for the crust formed

FIGURE 2.24 This map of the world shows the age of the seafloor. Note that the seafloor grows older with increasing distance from the ridge axis.

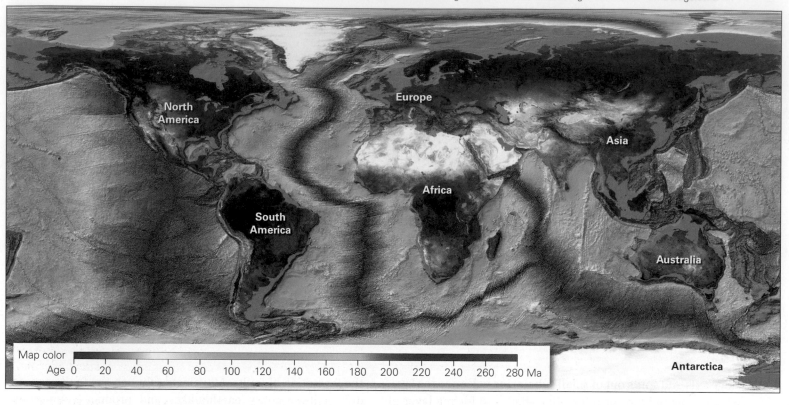

FIGURE 2.25 Changes accompanying the aging of lithosphere.

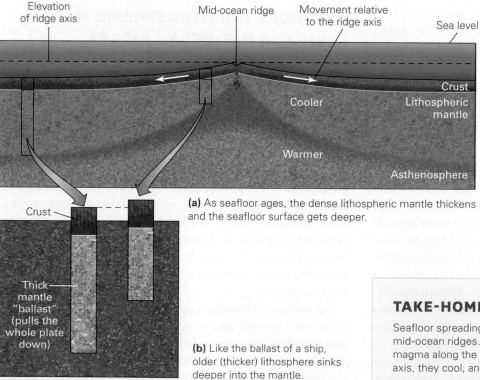

(a) As seafloor ages, the dense lithospheric mantle thickens and the seafloor surface gets deeper.

(b) Like the ballast of a ship, older (thicker) lithosphere sinks deeper into the mantle.

entirely at the ridge axis. The rate at which cooling and lithospheric thickening occur decreases progressively with increasing distance from the ridge axis. In fact, by the time the lithosphere is about 80 million years old, it has just attained its maximum thickness. As lithosphere thickens and gets cooler and denser, it sinks down into the asthenosphere, like a ship taking on ballast **(Fig. 2.25b)**. So the ocean becomes deeper over older ocean floor than over younger ocean floor. That's why abyssal plains are deeper than mid-ocean ridges.

TAKE-HOME MESSAGE

Seafloor spreading occurs at divergent boundaries, defined by mid-ocean ridges. New oceanic crust solidifies from basaltic magma along the ridge axis. As plates move away from the axis, they cool, and the lithospheric mantle forms and thickens.

QUICK QUESTION What are black smokers, and why do they form?

FIGURE 2.26 During the process of subduction, oceanic lithosphere sinks back into the deeper mantle.

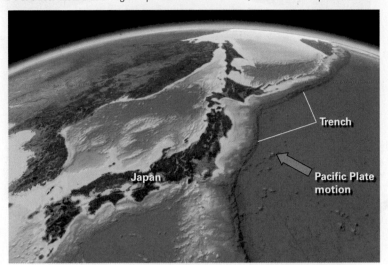

(a) The Pacific Plate subducts underneath Japan.

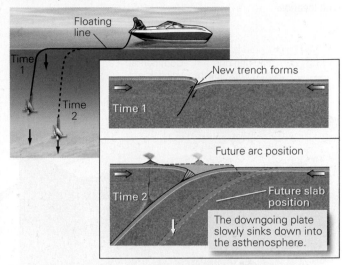

(b) Sinking of the downgoing plate resembles sinking of an anchor attached to a rope.

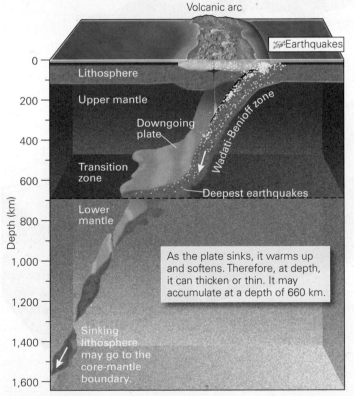

(c) A belt of earthquakes (dots) defines the position of the downgoing plate in the region above a depth of about 660 km. Sometimes plates "pile up" at a depth of 660 km.

2.7 **Convergent Boundaries and Subduction**

At a **convergent boundary**, two plates, at least one of which is oceanic, move toward one another. But rather than butting each other like angry rams, one oceanic plate bends and

sinks down into the asthenosphere beneath the other plate. Geologists refer to the sinking process as **subduction**, so convergent boundaries are also known as *subduction zones*. Because subduction at a convergent boundary "consumes" old ocean lithosphere, geologists also refer to convergent boundaries as *consuming boundaries*, and because they are delineated by deep-ocean trenches, they are sometimes simply called *trenches* **(Fig. 2.26a)**. The amount of oceanic plate consumption worldwide, averaged over time, equals the amount of seafloor spreading worldwide, so the surface area of the Earth remains constant through time.

Subduction occurs for a simple reason: oceanic lithosphere, once it has aged at least 10 million years, is denser than the underlying asthenosphere. Therefore, it can sink through the asthenosphere if given an opportunity. Where it lies flat on the surface of the asthenosphere, oceanic lithosphere can't sink. However, once the end of the convergent plate bends down and slips into the mantle, it continues downward like an anchor falling to the bottom of a lake **(Fig. 2.26b)**. As the lithosphere sinks, asthenosphere flows out of its way, just as water flows out of the way of a sinking anchor. But unlike water, the asthenosphere can flow only very slowly, so oceanic lithosphere can sink only very slowly, at a rate of less than about 15 cm per year. (To visualize the difference, imagine how much faster a coin can sink through water than it can through honey.) Because of the asthenosphere's high viscosity, the broad horizontal interior of an oceanic plate can't just sink straight down because there's too much resistance to flow in the underlying asthenosphere.

While the *downgoing plate*, the plate that has been subducted, must be composed of oceanic lithosphere, the *overriding plate*, which does not sink, can consist of either oceanic or

FIGURE 2.27 The nature of a convergent-plate margin varies with location.

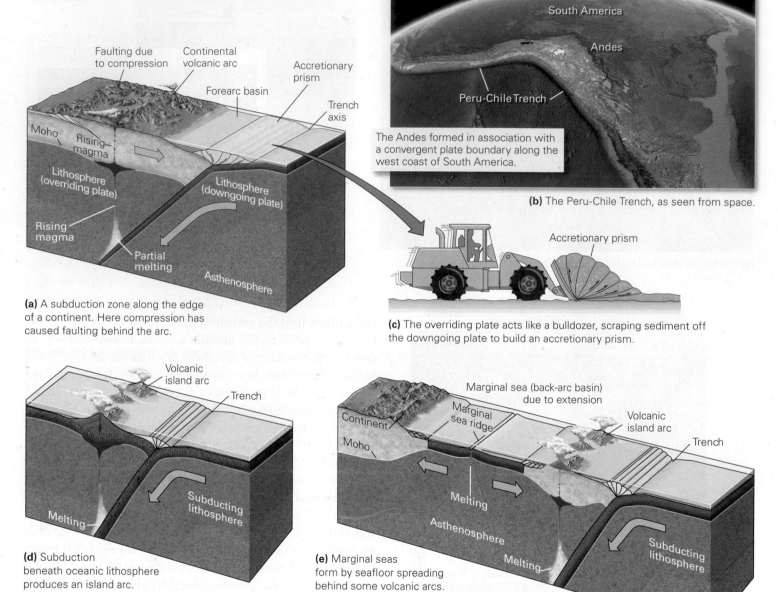

(a) A subduction zone along the edge of a continent. Here compression has caused faulting behind the arc.

The Andes formed in association with a convergent plate boundary along the west coast of South America.

(b) The Peru-Chile Trench, as seen from space.

(c) The overriding plate acts like a bulldozer, scraping sediment off the downgoing plate to build an accretionary prism.

(d) Subduction beneath oceanic lithosphere produces an island arc.

(e) Marginal seas form by seafloor spreading behind some volcanic arcs.

continental lithosphere. Continental lithosphere cannot be subducted, because it is too buoyant; its low-density crust acts like a life preserver keeping the continent afloat. Because continental crust cannot be completely subducted, some continental crust has survived for more than 3.8 billion years, much longer than the oldest oceanic lithosphere.

EARTHQUAKES AND THE FATE OF SUBDUCTED PLATES

At convergent boundaries, the downgoing plate grinds along the base of the overriding plate, a process that generates large earthquakes. These earthquakes occur fairly close to the Earth's surface, so some of them cause massive destruction in coastal cities. But earthquakes also happen in downgoing plates at greater depths. In fact, geologists have detected earthquakes within downgoing plates down to a depth of 660 km. The band of earthquakes in a downgoing plate is called a **Wadati-Benioff zone**, after its two discoverers **(Fig. 2.26c)**.

At depths greater than 660 km, conditions change and earthquakes no longer occur. Downgoing plates, however, do continue to sink below a depth of 660 km—they just do so without generating earthquakes. In fact, the lower mantle may be a graveyard for old subducted plates.

GEOLOGIC FEATURES OF A CONVERGENT BOUNDARY

To become familiar with the various geologic features that occur along a convergent boundary, let's look at an example, the boundary between the western coast of the South American Plate and the eastern edge of the Nazca Plate (a portion of the Pacific Ocean floor). A deep-ocean trench, the Peru-Chile Trench, delineates this boundary (Fig. 2.27a, b). Such trenches form where the plate bends as it starts to sink into the asthenosphere.

In the Peru-Chile Trench, as the downgoing plate slides under the overriding plate, sediment (clay and plankton) that had settled on the surface of the downgoing plate, as well as sand that fell into the trench from the shores of South America, gets scraped up and incorporated in a wedge-shaped mass known as an **accretionary prism** (Fig. 2.27c). An accretionary prism forms in basically the same way as a pile of snow or sand gathers in front of a plow, and like snow, the sediment tends to be squashed and contorted.

A chain of volcanoes known as a *volcanic arc* develops on the edge of the overriding plate behind the accretionary prism. As we will see in Chapter 4, the magma that feeds these volcanoes forms just above the surface of the downgoing plate where the plate reaches a depth of about 150 km below the Earth's surface. If the volcanic arc forms where an oceanic plate subducts beneath continental lithosphere, the resulting chain of volcanoes grows on the continent and forms a *continental volcanic arc*. (In some cases, the plates squeeze together across a continental arc, causing a belt of faults to form behind the arc.) If, however, the volcanic arc grows where one oceanic plate subducts beneath another oceanic plate, the resulting volcanoes form a chain of islands known as a *volcanic island arc* (Fig. 2.27d). A *back-arc basin* exists either where subduction happens to begin offshore, trapping ocean lithosphere behind the arc, or where stretching of the lithosphere behind the arc leads to the formation of a small spreading ridge behind the arc (Fig. 2.27e).

TAKE-HOME MESSAGE

At a convergent boundary, an oceanic plate sinks into the mantle beneath the edge of another plate. A volcanic arc and a trench delineate such plate boundaries, and earthquakes happen along the contact between the two plates as well as in the downgoing slab. Volcanic arcs can form on the edge of a continent or as a chain of islands in the sea.

QUICK QUESTION Can continents be completely subducted?

2.8 Transform Boundaries

When researchers began to explore the bathymetry of mid-ocean ridges in detail, they discovered that mid-ocean ridges are not long, uninterrupted lines, but rather consist of short segments that appear to be offset laterally from each other by narrow belts of broken and irregular seafloor (Fig. 2.28a). These belts, or *fracture zones*, lie roughly at right angles to the ridge segments, intersect the ends of the segments, and extend beyond the ends of the segments (see Fig. 2.7c). Originally, researchers assumed that the entire length of each fracture zone was a fault, and that slip on a fracture zone had displaced segments of the mid-ocean ridge sideways, relative to each other. In other words, they imagined that a mid-ocean ridge initiated as a continuous, fence-like line that only later was broken up by faulting. But when data on the exact locations of earthquakes along mid-ocean ridges became available, it was clear that this model could not be correct. Earthquakes don't occur along the portions of fracture zones that extend beyond the ends of ridge segments, out into the abyssal plains, so these portions of fracture zones are not actively slipping faults.

The nature of movement along fracture zones remained a mystery until a Canadian researcher, J. Tuzo Wilson, began to think about fracture zones in the context of the seafloor-spreading concept. Wilson proposed that fracture zones formed at the same time as the ridge axis itself, and that the ridge consisted of separate segments to start with. These segments were linked (not offset) by fracture zones. With this idea in mind, he drew a sketch map showing two ridge-axis segments linked by a fracture zone, and he drew arrows to indicate the direction that the ocean floor was moving, relative to the ridge axis, as a result of seafloor spreading (Fig. 2.28b). Look at Wilson's arrows. Clearly, the movement direction on the active portion of the fracture zone must be opposite to the movement direction that researchers originally thought occurred across the structure. Furthermore, in Wilson's model, slip occurs only along the segment of the fracture zone between the two ridge segments (Fig. 2.28c). Plates on opposite sides of the inactive part of a fracture zone move together, as one plate.

Wilson introduced the term **transform boundary,** or *transform fault*, for the actively slipping segment of a fracture zone between two ridge segments. At a transform boundary, one plate slides sideways, relative to its neighbor, on a vertical fault. As a result, the slip direction on the fault is horizontal (parallel to the Earth's surface), so no new plate forms, and no old plate is consumed, at a transform boundary.

FIGURE 2.28 The concept of transform faulting.

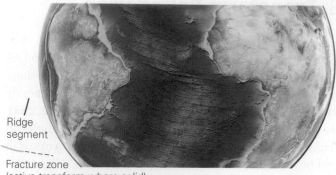

Ridge
segment

Fracture zone
(active transform where solid)

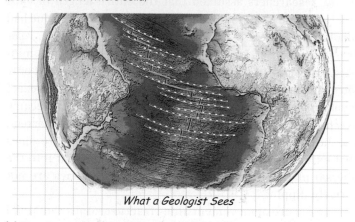

What a Geologist Sees

(a) Numerous transform faults segment the Mid-Atlantic Ridge.

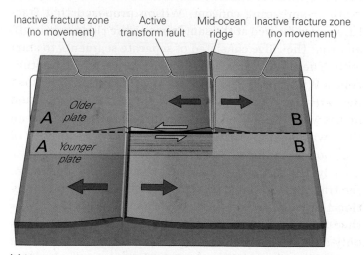

Inactive fracture zone (no movement)   Active transform fault   Mid-ocean ridge   Inactive fracture zone (no movement)

Older plate

A   B

A   B

Younger plate

(c) Note that only the segment of the fracture zone between the two ridge segments is active.

So far we've discussed only transforms along mid-ocean ridges. Not all transforms link ridge segments. Some, such as the Alpine fault of New Zealand, link trenches, while others link a trench to a ridge segment. Also, not all transform faults occur in oceanic lithosphere; a few cut across continental lithosphere. The San Andreas fault, for example, which cuts across California, defines part of the plate

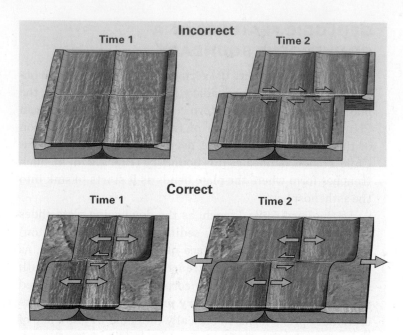

Incorrect

Time 1   Time 2

Correct

Time 1   Time 2

(b) A comparison of the old, incorrect model of transform faults with Wilson's new, correct model required by the seafloor-spreading hypothesis.

boundary between the North American Plate and the Pacific Plate—the portion of California that lies to the west of the fault (including Los Angeles) moves with the Pacific Plate, while the portion that lies to the east of the fault is part of the North American Plate **(Fig. 2.29)**.

TAKE-HOME MESSAGE

At transform boundaries, one plate slips sideways past another along a vertical fault. Thus, there is no production or destruction of lithosphere at a transform boundary. Most transform boundaries link segments of mid-ocean ridges, but some, such as the San Andreas fault, cut across continental crust.

QUICK QUESTION Why do earthquakes only occur on the portion of a fracture zone that links mid-ocean ridge segments?

2.9 Special Locations in the Plate Mosaic

TRIPLE JUNCTIONS

Geologists refer to a point where three plate boundaries intersect as a **triple junction**. We can name triple junctions after the types of boundaries that intersect them. For example, the triple junction formed where the Southwest Indian Ocean

FIGURE 2.29 The San Andreas fault—a continental transform boundary.

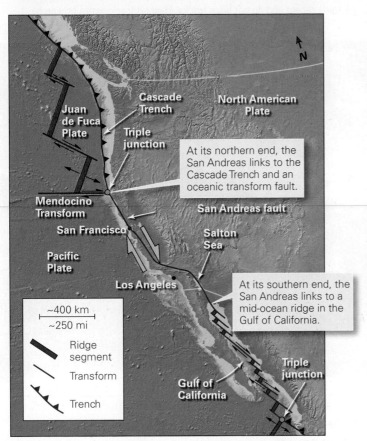

At its northern end, the San Andreas links to the Cascade Trench and an oceanic transform fault.

At its southern end, the San Andreas links to a mid-ocean ridge in the Gulf of California.

(a) The San Andreas fault is a transform plate boundary between the North American and Pacific Plates. The Pacific Plate is moving northwest, relative to the North American Plate.

(b) In southern California, the San Andreas fault cuts a dry landscape. The fault trace is in the narrow valley. The land has been pushed up slightly along the fault.

Ridge intersects two arms of the Mid–Indian Ocean Ridge is a ridge-ridge-ridge triple junction **(Fig. 2.30)**, and the triple junction north of San Francisco, where the Cascadia trench, the San Andreas fault, and the Mendocino fracture zone intersect, is a trench-transform-transform triple junction (see Fig. 2.29a).

FIGURE 2.30 A ridge-ridge-ridge triple junction occurs in the Indian Ocean.

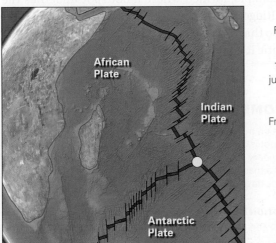

HOT SPOTS

The volcanoes of both volcanic arcs and mid-ocean ridges are *plate-boundary volcanoes*, in that they formed as a consequence of movement along the boundary. Not all volcanoes on Earth, however, are plate-boundary volcanoes. Worldwide, geoscientists have identified about 100 volcanoes that exist as isolated points. These are called *hot-spot volcanoes*, or simply **hot spots (Fig. 2.31)**.

What causes hot spots? In the early 1960s, J. Tuzo Wilson noted that active hot-spot volcanoes, examples that are erupting or may erupt in the future, occur at the end of a chain of extinct hot-spot volcanoes, meaning volcanoes that will never erupt again. This configuration is different from that of volcanic arcs along convergent boundaries—at volcanic arcs, all of the volcanoes are active. With this image in mind, Wilson suggested that the active volcano represents the present-day

FIGURE 2.31 The dots represent the locations of selected hot-spot volcanoes. The red lines represent hot-spot tracks. The most recent volcano (dot) is at one end of this track. Some of these volcanoes are extinct. Some hot spots are fairly recent and do not have tracks. Dashed tracks were broken by seafloor spreading.

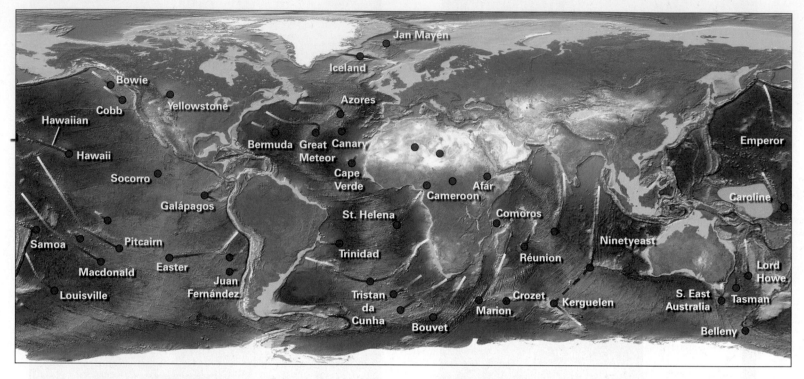

location of the heat source, whereas the chain of extinct volcanic islands represents locations on the plate that were once over the heat source but progressively moved off **(Fig. 2.32a)**. In other words, the position of the heat source causing a hot-spot volcano is fixed, relative to the moving plate above.

A few years after Wilson's proposal, researchers suggested that the heat source for hot spots is a **mantle plume**, a column of very hot rock rising up through the mantle to the base of the lithosphere **(Fig. 2.32b)**. In this model, still an area of active research, plumes originate deep in the mantle. Rock in the plume, though solid, is soft enough to flow, and it rises buoyantly because it is less dense than surrounding cooler rock. When the hot rock of the plume reaches the base of the lithosphere, it begins to melt (for reasons discussed in Chapter 4) and produces magma that seeps up through the lithosphere to the Earth's surface. The chain of extinct volcanoes delineates a *hot-spot track*, which forms when the overlying plate moves over a fixed plume. This movement slowly carries the volcano off the top of the plume, so that it becomes extinct. A new, younger volcano then grows over the plume.

We can apply Wilson's model of hot-spot track formation to the Hawaiian island chain and to its continuation, the Hawaiian-Emperor seamount chain. Volcanic eruptions occur today on the island of Hawaii, at the southeastern end of the chain. All of the other islands and seamounts to the northwest are remnants of inactive volcanoes that have sunk below

sea level and they get progressively older to the northwest **(Fig. 2.32c, d)**.

As we've noted, some hot spots lie within continents. For example, several have been active in the interior of Africa, and one now underlies Yellowstone National Park. The famous *geysers* (natural steam and hot-water fountains) of Yellowstone exist because hot magma, formed at the Yellowstone hot spot, lies not far below the surface of the park. A few hot spots lie on mid-ocean ridges. Where this happens, a volcanic island may protrude above sea level because the hot spot produces more magma than does a normal mid-ocean ridge. Iceland, for example, formed where a hot spot underlies the Mid-Atlantic Ridge. The extra volcanism of the hot spot built up the island so that it rises almost 3 km above other places on the Mid-Atlantic Ridge.

TAKE-HOME MESSAGE

A triple junction marks the point where three plate boundaries join. A hot spot is a point where volcanism occurs independently of plate-boundary movement. Hot spots may be due to melting at the top of a mantle plume. As a plate moves over a plume, a hot-spot track develops.

QUICK QUESTION Why is the volcano at the end of a hot-spot track the only one to be active?

FIGURE 2.32 The mantle plume hypothesis for the formation of hot-spot tracks.

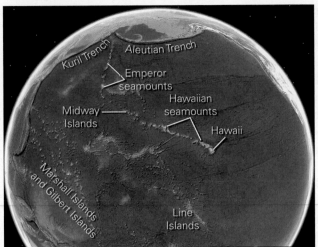

(a) A bathymetric map showing the hot-spot tracks of the Pacific Ocean.

A volcano forms on a moving plate above a mantle plume.

Active hot-spot volcano #1

Plate motion

Mantle plume

The first volcano moves off the plume and dies.

Extinct volcano #1   Active hot-spot volcano #2

Asthenosphere

Time

Seamount (remnant of volcano #1)   Extinct volcano #2   Active hot-spot volcano #3

Crust

Lithospheric mantle

Plate movement carries each successive volcano off the hot spot.

Lithosphere

Asthenosphere

(b) Progressive stages in the development of a hot-spot track, according to the plume model.

(c) According to the plume model, the Hawaiian island chain is a hot-spot track that formed as the Pacific Plate moved northwest relative to a plume.

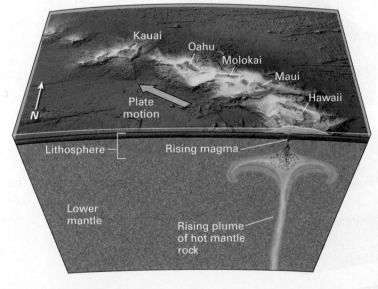

Kauai
Oahu
Molokai
Maui
Hawaii

N

Plate motion

Lithosphere   Rising magma

Lower mantle

Rising plume of hot mantle rock

(d) As a volcano moves off the hot spot, it gradually sinks below sea level due to sinking of the plate, erosion, and submarine landslides.

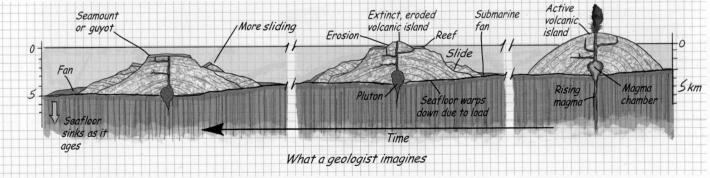

Seamount or guyot   More sliding   Extinct, eroded volcanic island   Submarine fan   Active volcanic island

Erosion   Reef

Slide

Fan

0   0

Pluton   Rising magma   Magma chamber

5   5 km

Seafloor sinks as it ages

Seafloor warps down due to load

Time

What a geologist imagines

2.10 How Do Plate Boundaries Form, and How Do They Die?

The configuration of plates and plate boundaries visible on our planet today has not existed for all of geologic history, and it will not exist indefinitely into the future. Because of plate motion, oceanic plates form and are later consumed, while continents merge and later split apart. How does a new divergent boundary come into existence, and how does an existing convergent boundary eventually cease to exist? Most new divergent boundaries form when a continent splits and separates into two continents. We call this process *rifting*. A convergent boundary ceases to exist when a piece of relatively buoyant crust, such as a continent or an island arc, moves into the subduction zone and, in effect, jams up the system. We call this process *collision*. Let's look at how these processes operate.

CONTINENTAL RIFTING

A **continental rift** is a linear belt in which continental lithosphere pulls apart **(Fig. 2.33)**. During the process, the lithosphere stretches horizontally and thins vertically, much like a piece of chewing gum that you pull between your fingers. Nearer the surface of the continent, where the crust is cold and brittle, stretching causes rock to break and faults to develop. Blocks of rock slip down the fault surfaces, leading to the formation of a low area, a *rift valley*, that gradually becomes buried by sediment eroded from its surroundings. Deeper in the crust, and in the underlying lithospheric mantle, rock is warmer and softer, so stretching takes place plastically, without breaking the rock. Geologists refer to the whole region that stretches as the rift, and the process of stretching as **rifting**.

When rifting takes place, hot asthenosphere rises beneath the rift and starts to melt, for reasons we describe in Chapter 4. Eruption of the molten rock produces volcanoes along the rift. If rifting continues for a long enough time, the continent breaks in two, a new mid-ocean ridge forms, and seafloor spreading begins. The relict of the rift underlies the continental shelf along a passive margin (see Fig. 2.19b). In some cases, however, rifting stops before the continent splits in two, and the rift, which fills with sediment, remains as a permanent scar in the crust.

A major rift, the East African Rift, extends in a north-south direction for over 3,500 km across the continent **(Fig. 2.34a)**.

FIGURE 2.33 During rifting, continental lithosphere stretches and thins. If rifting succeeds, a new mid-ocean ridge forms.

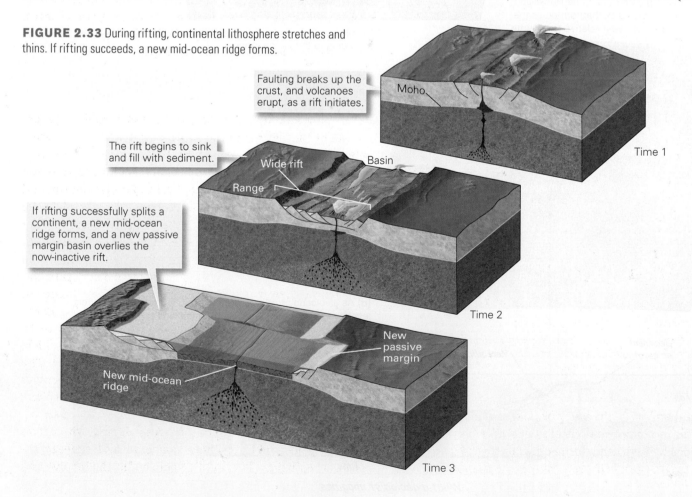

Faulting breaks up the crust, and volcanoes erupt, as a rift initiates.

Moho

Time 1

The rift begins to sink and fill with sediment.

Wide rift

Range

Basin

Time 2

If rifting successfully splits a continent, a new mid-ocean ridge forms, and a new passive margin basin overlies the now-inactive rift.

New passive margin

New mid-ocean ridge

Time 3

FIGURE 2.34 Examples of present-day crustal rifting.

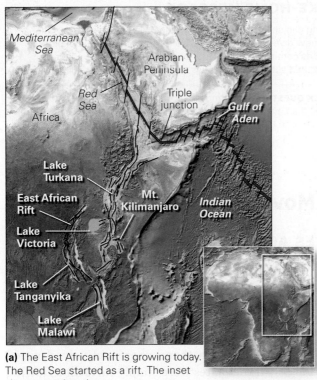

(a) The East African Rift is growing today. The Red Sea started as a rift. The inset shows map locations.

(b) An aerial photo of the northern end of the East African Rift, showing faults and volcanoes.

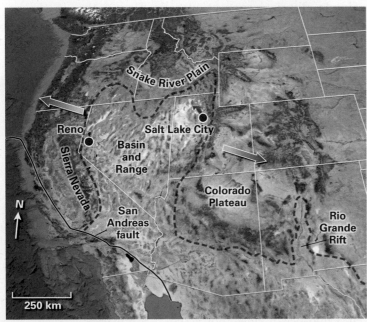

(c) The Basin and Range Province is a rift. Faulting bounds the narrow north-south-trending mountains, which are separated by basins. The arrows indicate the direction of stretching.

To astronauts in orbit, the rift looks like a giant gash in the crust. On the ground, it consists of a deep rift valley bordered on both sides by high cliffs formed by faulting (**Fig. 2.34b**). Along the length of the rift, several major volcanoes, such as snow-crested Mt. Kilimanjaro, smoke and fume above the savannah. At its north end, the rift joins the Red Sea Ridge and the Gulf of Aden Ridge at a triple junction. A much wider rift, the *Basin and Range Rift*, breaks up the landscape of the western United States (**Fig. 2.34c**). Here, movement on numerous faults tilted blocks of crust to form narrow mountain ranges, while sediment that eroded from the blocks filled the adjacent basins (the low areas between the ranges).

COLLISION

After the breakup of Pangaea, India was a small, separate continent that lay far to the south of Asia. As time passed, subduction consumed the ocean between India and Asia, and India moved northward, finally slamming into the southern margin of Asia at about 40 to 50 Ma. Continental crust, unlike oceanic crust, is too buoyant to subduct. So when India collided with Asia, the attached oceanic plate broke off and sank down into the deep mantle while India pushed hard into and partly under Asia, squeezing the rocks and sediment that once lay between the two continents into the 8-km-high welt that we now know of as the Himalayas. During this process, not only did the surface of the Earth rise, but the crust became thicker. In fact, the crust beneath a collisional mountain range can be up to 60 to 70 km thick, about twice the thickness of normal continental crust. The boundary between what was once two separate continents is called a **suture**; slivers of ocean crust may be trapped along a suture.

Geoscientists refer to the process during which two buoyant pieces of lithosphere converge and squeeze together as collision (**Fig. 2.35**). Some collisions involve two

FIGURE 2.35 Continental collision.

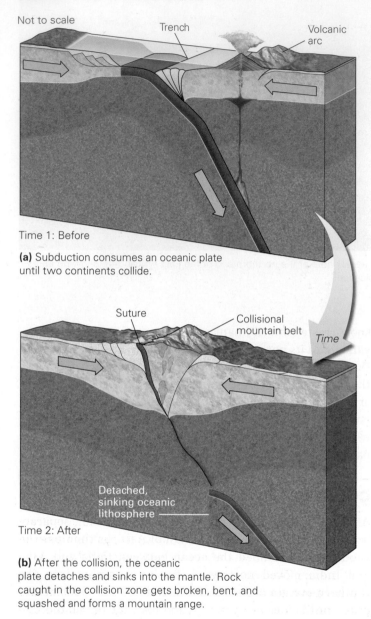

Not to scale

Trench

Volcanic arc

Time 1: Before

(a) Subduction consumes an oceanic plate until two continents collide.

Suture

Collisional mountain belt

Time

Detached, sinking oceanic lithosphere

Time 2: After

(b) After the collision, the oceanic plate detaches and sinks into the mantle. Rock caught in the collision zone gets broken, bent, and squashed and forms a mountain range.

continents, whereas some involve continents and an island arc. When a collision concludes, the convergent boundary that once existed between the two colliding pieces ceases to exist. Collisions yield some of the most spectacular mountains on the planet, such as the Himalayas and the Alps, and have yielded major mountain ranges in the past. These subsequently eroded away so that today we see only their relics. For example, the Appalachian Mountains in the eastern United States formed as a consequence of three collisions. After the last one, a collision between Africa and North America around 280 Ma, North America became part of the Pangaea supercontinent.

TAKE-HOME MESSAGE

Rifting can split a continent in two and can lead to the formation of a new divergent boundary. When two buoyant crustal blocks, such as continents and island arcs, collide, a mountain belt forms and subduction ceases.

QUICK QUESTION Do rifting and collision affect crustal thickness? If so, how?

2.11 **Moving Plates**

FORCES ACTING ON PLATES

We've now discussed the many facets of plate tectonics theory (**Geology at a Glance**, pp. 78–79). But to complete the story, we need to address a major question: What drives plate motion? When geoscientists first proposed plate tectonics, they thought the process occurred simply because convective flow in the asthenosphere actively dragged plates along, as if the plates were simply rafts on a flowing river. Thus, early images depicting plate motion showed simple elliptical convection cells—flow paths—in the asthenosphere (**Fig. 2.36a**). At first glance, this hypothesis looked pretty good. But, on closer examination, it became clear that a model of simple convection cells carrying plates on their backs can't explain the complex geometry of plate boundaries and the great variety of plate motions that we observe on the Earth, so it can't be correct. Researchers now prefer a model in which *convection*, along with *ridge-push force* and *slab-pull force*, all contribute to driving plates. Let's look at each of these in turn.

Convection affects plate motions in two ways. Recall that, at a mid-ocean ridge, hot asthenosphere rises and then cools to form oceanic lithosphere that slowly moves away from the ridge until, eventually, it sinks back into the mantle at a trench. Since the material forming the plate starts out hot, cools, and then sinks, we can view the plate itself

SEE FOR YOURSELF...

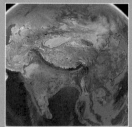

Himalayan Collision

LATITUDE
27°59'17.95" N

LONGITUDE
86°55'30.71" E

Look down from 7,000 km.

You're seeing the Himalayas. Zoom closer, and you'll be at the peak of Mt. Everest.

FIGURE 2.36 Plate-driving forces.

Old image

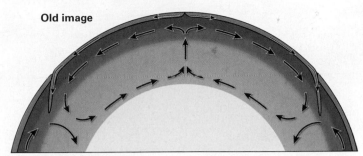

(a) The old, incorrect, image of simple convection cells in the asthenosphere.

Modern image

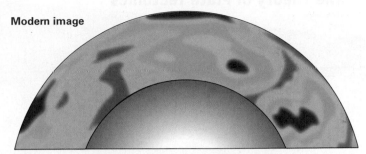

(b) The colors represent the difference between the observed velocity of seismic waves and the expected velocity. Geologists interpret red areas to be warmer, upwelling regions, and blue areas to be cooler, downwelling regions.

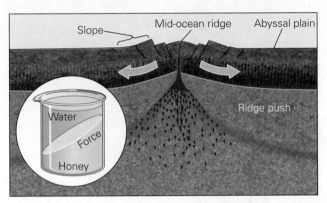

(c) Ridge push develops because the region of a rift is elevated. Like a wedge of honey with a sloping surface, the mass of the ridge pushes sideways.

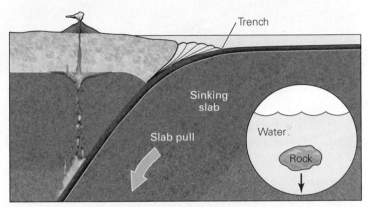

(d) Slab-pull force develops because lithosphere is denser than the underlying asthenosphere and sinks like a stone in water (though much more slowly).

as the top of a convection cell and plate motion as a form of convection. But in this view, convection is effectively a consequence of plate motion, not the cause.

Can convection actively cause plates to move? The answer may come from studies that demonstrate the interior of the mantle, beneath the plates, is indeed convecting on a very broad scale (see Interlude D). Specifically, geologists have found that in some places deeper, hotter asthenosphere is rising or *upwelling*, and in other places shallower, colder asthenosphere is sinking or *downwelling* **(Fig. 2.36b)**. Such asthenospheric flow probably does exert a force on the base of plates. But on a global scale, the pattern of upwelling and downwelling does not match the pattern of plate boundaries exactly. So asthenospheric flow may either speed up or slow down plates, depending on the orientation of the flow direction relative to the movement direction of the overlying plate.

Ridge-push force develops simply because the lithosphere of mid-ocean ridges lies at a higher elevation than that of the adjacent abyssal plains **(Fig. 2.36c)**. To understand ridge-push force, imagine that you have a glass containing a layer of water over a layer of honey. By tilting the glass momentarily and then returning it to its upright position,

you can create a temporary slope in the boundary between these substances. While the boundary has this slope, gravity causes the weight of elevated honey to push against the glass adjacent to the side where the honey surface lies at lower elevation. The geometry of a mid-ocean ridge resembles this situation, for the seafloor of a mid-ocean ridge is higher than the seafloor of abyssal plains. Gravity causes the elevated lithosphere at the ridge axis to push on the lithosphere that lies farther from the axis, making it move away. As lithosphere moves away from the ridge axis, during seafloor spreading, new hot asthenosphere rises to fill the gap. Note that the upward movement of asthenosphere beneath a mid-ocean ridge represents a consequence of seafloor spreading, not the cause.

Slab-pull force, the force that subducting, downgoing plates apply to oceanic lithosphere at a convergent margin, arises simply because lithosphere that was formed more than 10 million years ago is denser than asthenosphere, so it can sink into the asthenosphere **(Fig. 2.36d)**. Thus, once an oceanic plate starts to sink, it gradually pulls the rest of the plate along behind it, like an anchor pulling down the anchor line.

The Theory of Plate Tectonics

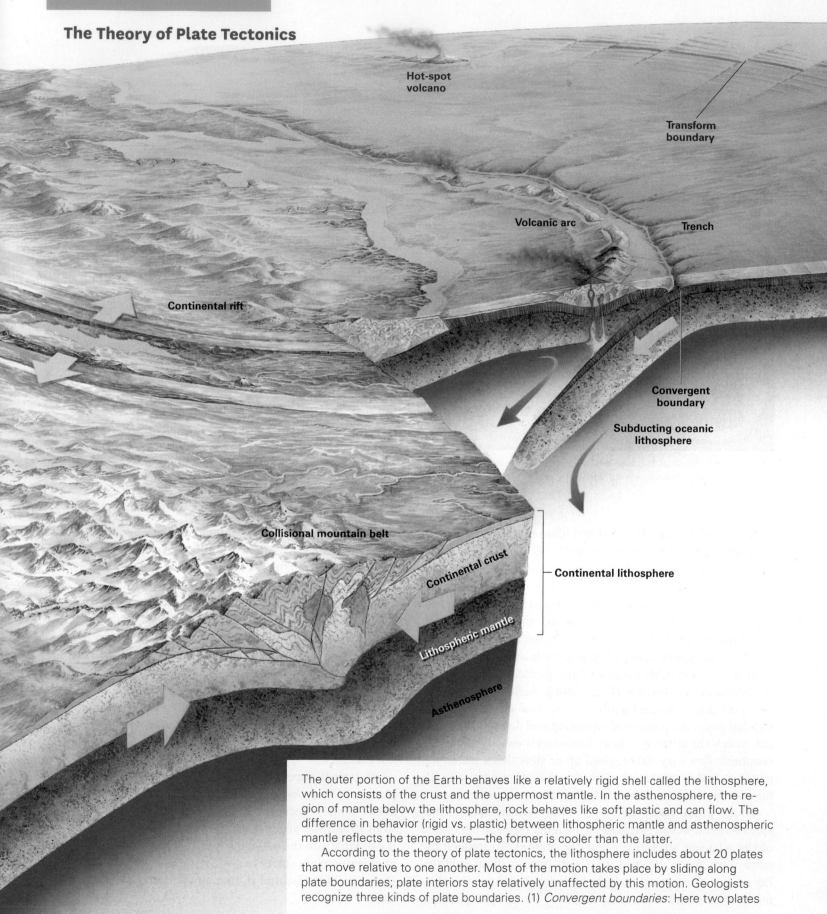

Hot-spot volcano

Transform boundary

Volcanic arc

Trench

Continental rift

Convergent boundary

Subducting oceanic lithosphere

Collisional mountain belt

Continental crust

Continental lithosphere

Lithospheric mantle

Asthenosphere

The outer portion of the Earth behaves like a relatively rigid shell called the lithosphere, which consists of the crust and the uppermost mantle. In the asthenosphere, the region of mantle below the lithosphere, rock behaves like soft plastic and can flow. The difference in behavior (rigid vs. plastic) between lithospheric mantle and asthenospheric mantle reflects the temperature—the former is cooler than the latter.

According to the theory of plate tectonics, the lithosphere includes about 20 plates that move relative to one another. Most of the motion takes place by sliding along plate boundaries; plate interiors stay relatively unaffected by this motion. Geologists recognize three kinds of plate boundaries. (1) *Convergent boundaries*: Here two plates

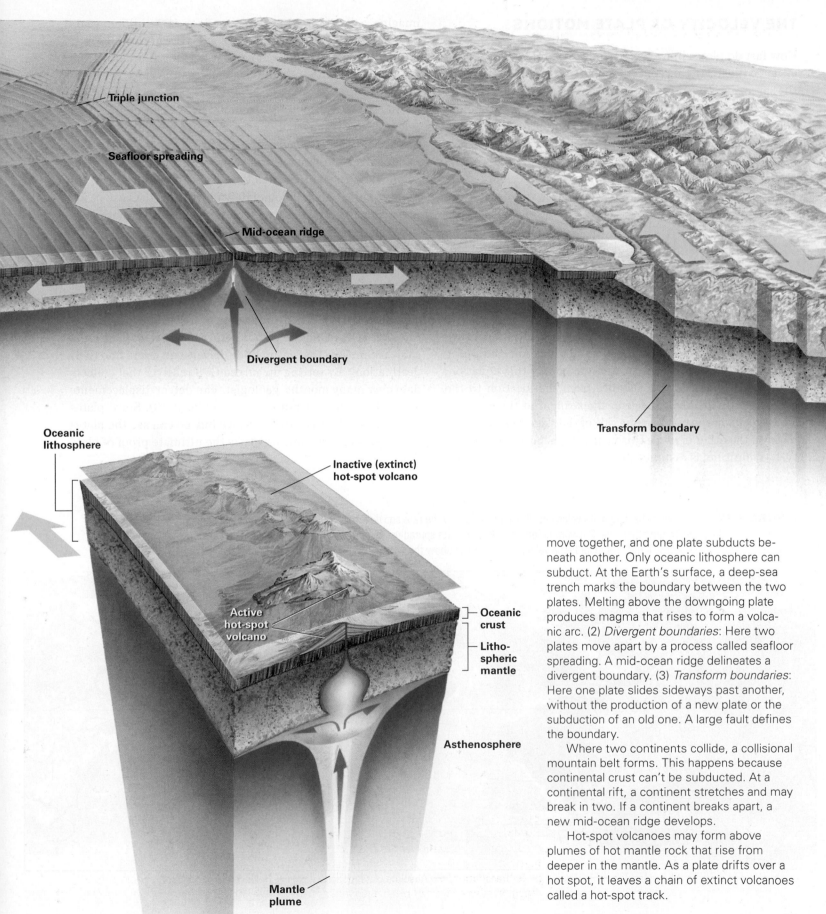

Triple junction

Seafloor spreading

Mid-ocean ridge

Divergent boundary

Transform boundary

Oceanic lithosphere

Inactive (extinct) hot-spot volcano

Active hot-spot volcano

Oceanic crust

Lithospheric mantle

Asthenosphere

Mantle plume

move together, and one plate subducts beneath another. Only oceanic lithosphere can subduct. At the Earth's surface, a deep-sea trench marks the boundary between the two plates. Melting above the downgoing plate produces magma that rises to form a volcanic arc. (2) *Divergent boundaries*: Here two plates move apart by a process called seafloor spreading. A mid-ocean ridge delineates a divergent boundary. (3) *Transform boundaries*: Here one plate slides sideways past another, without the production of a new plate or the subduction of an old one. A large fault defines the boundary.

Where two continents collide, a collisional mountain belt forms. This happens because continental crust can't be subducted. At a continental rift, a continent stretches and may break in two. If a continent breaks apart, a new mid-ocean ridge develops.

Hot-spot volcanoes may form above plumes of hot mantle rock that rise from deeper in the mantle. As a plate drifts over a hot spot, it leaves a chain of extinct volcanoes called a hot-spot track.

THE VELOCITY OF PLATE MOTIONS

How fast do plates move? It depends on your frame of reference. To illustrate this concept, imagine two cars speeding in the same direction down the highway. From the viewpoint of a tree along the side of the road, Car A zips by at 100 km an hour, while Car B moves at 80 km an hour. But relative to Car B, Car A moves at only 20 km an hour. Geologists use two different frames of reference for describing plate velocity. If we describe the movement of Plate A with respect to Plate B, then we are speaking about **relative plate velocity**. But if we describe the movement of both plates relative to a fixed location in the mantle below the plates, then we are speaking of **absolute plate velocity (Fig. 2.37)**.

> **Did you ever wonder...**
> whether we can really "see" continents drift?

To determine relative plate velocity, geoscientists measure the distance between the ridge axis and a location containing ocean floor of a specified age, and then apply this equation: plate velocity equals the distance from the location to the ridge axis, divided by the age of the seafloor at the location. The velocity of the plate on one side of the ridge relative to the plate on the other is twice this value. (Remember that velocity, by definition, is distance divided by time.) As an example, imagine a location where the seafloor is 1 million years old and lies 10 km from the ridge axis. Given that 10 km is 1,000,000 cm, the plate is moving at 1 cm per year away from the ridge axis. The spreading rate across the ridge is, therefore, 2 cm/year.

To estimate absolute plate motions, we can assume that the position of a hot spot does not change much over time. If this is so, then the track of hot-spot volcanoes on the plate moving over the plume provides a record of the plate's absolute velocity and indicates the direction of movement. (In reality, plumes are not completely fixed; geologists use other, more complex, methods to calculate absolute plate motions.)

Working from the calculations described above, geologists have determined that plate motions today occur at rates of 1 to 15 cm per year—about the rate that your fingernails grow. These rates, though small, can yield large displacements over geologic time. For example, at a rate of 10 cm per year, a plate can move 100 km in a million years! Can we detect such slow rates? Yes, by using the *global positioning system* (*GPS*), the same technology that automobile drivers can use to find their destinations. By setting up a fixed GPS receiver that collects data over many months, geologists can detect displacements as small as about 2 mm per year (**Fig. 2.38**). Since plates move at 5 to 75 times this rate, we indeed can see the plates move—this observation serves as the ultimate proof of plate tectonics.

FIGURE 2.37 Relative and absolute plate velocities. Black arrows show the rate and direction at which the plate on one side of a boundary is moving with respect to the plate on the other side. Outward-pointing arrows indicate spreading, inward-pointing arrows indicate subduction, and parallel arrows show transform motion. Arrow length represents the velocity. Red arrows show the velocity of the plates with respect to a fixed point in the mantle.

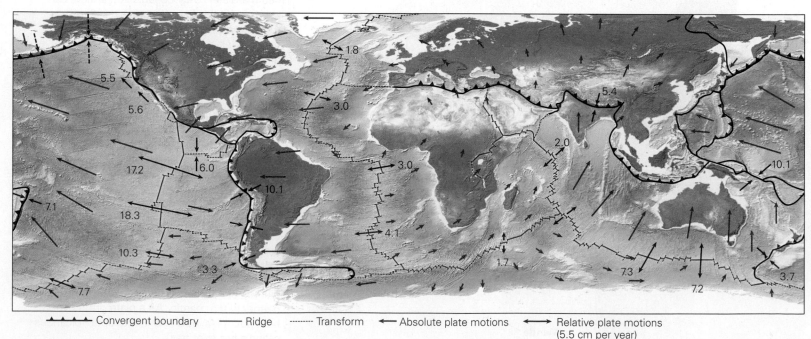

Taking into account many data sources that define the motion of plates, geologists have greatly refined the image of continental drift that Wegener tried so hard to prove nearly a century ago. We can now see how the map of our planet's surface has evolved radically during the past 400 million years (Fig. 2.39), and even before.

FIGURE 2.38 GPS measurements indicate that Turkey is moving west relative to Asia. Each yellow arrow indicates the velocity of the point at the end of the arrow. Red arrows give overall plate-movement direction.

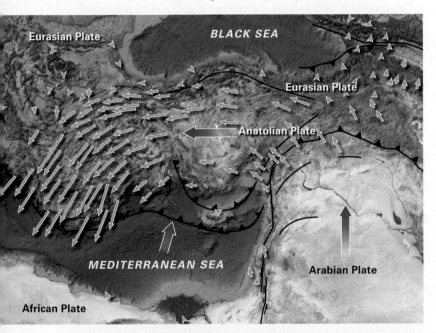

An example of a GPS station

TAKE-HOME MESSAGE

Plate motion takes place because plates are acted on by ridge push and slab pull, and are affected by asthenosphere convection. This motion takes place at rates of 1 to 15 cm per year. Relative motion specifies the rate that a plate moves relative to its neighbor, whereas absolute motion specifies the rate that a plate moves relative to a fixed point beneath the plate.

QUICK QUESTION How do GPS measurements serve as the ultimate proof of plate tectonics?

FIGURE 2.39 Due to plate tectonics, the map of the Earth's surface slowly changes. Here we see the assembly and the later breakup of Pangaea during the past 400 million years.

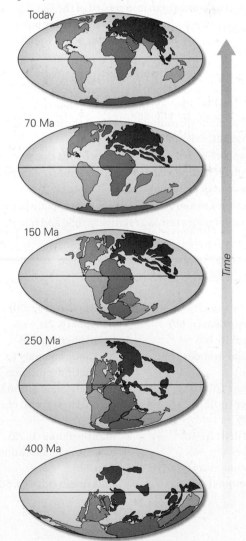

Chapter 2 Review

CHAPTER SUMMARY

> Wegener proposed that continents had once been joined together to form a single supercontinent (Pangaea) and then drifted apart. This idea is the continental-drift hypothesis.

> Wegener supported his hypothesis by studying the matching shapes of coastlines; the distribution of Paleozoic glaciers, climate belts, and fossil species; and the correlation of rock assemblages.

> Around 1960, Hess proposed the hypothesis of seafloor spreading, which states that new seafloor forms at mid-ocean ridges, then spreads symmetrically away from the ridge axis. Ocean floor sinks back into the mantle at deep-ocean trenches.

> Rocks retain paleomagnetism, a record of the Earth's magnetic field at the time the rocks formed. By measuring paleomagnetism, geologists could use studies of apparent polar-wander paths to prove continental drift.

> Geologists documented that the Earth's magnetic field reverses polarity every now and then. Magnetic reversals explain the existence of stripe-like marine magnetic anomalies.

> A proof of seafloor spreading came from the interpretation of marine magnetic anomalies.

> The lithosphere is broken into discrete plates that move relative to each other. Continental drift and seafloor spreading are manifestations of plate movement.

> Most earthquakes and volcanoes occur along plate boundaries; the interiors of plates remain relatively rigid and intact.

> The three types of plate boundaries—divergent, convergent, and transform—are distinguished from each other by the relative movement of plates.

> At divergent boundaries, which are marked by mid-ocean ridges, seafloor spreading produces new oceanic lithosphere.

> Convergent boundaries are marked by deep-ocean trenches and volcanic arcs. At convergent boundaries, oceanic lithosphere subducts beneath an overriding plate.

> Transform boundaries are marked by large faults along which one plate slides sideways past another.

> At triple junctions, three plate boundaries intersect.

> Hot spots are places where volcanism occurs at an isolated volcano. As a plate moves over the hot spot, the volcano moves off and dies, and a new volcano forms over the hot spot. Hot spots may be caused by mantle plumes.

> During rifting, continental lithosphere stretches and thins. If it finally breaks apart, a new mid-ocean ridge forms.

> Convergent boundaries cease to exist when a buoyant piece of crust (a continent or an island arc) moves into the subduction zone. When that happens, collision occurs.

> Ridge-push force and slab-pull force contribute to driving plate motions. Plates move at rates of about 1 to 15 cm per year. GPS satellite measurements can detect these motions.

GUIDE TERMS

absolute plate velocity (p. 80)
accretionary prism (p. 69)
active margin (p. 63)
apparent polar-wander
  path (p. 56)
asthenosphere (p. 60)
black smoker (p. 65)
chron (p. 57)
continental drift (p. 46)
continental rift (p. 74)
convergent boundary (p. 67)
divergent boundary (p. 63)

fracture zone (p. 50)
hot spot (p. 71)
lithosphere (p. 60)
magnetic anomaly (p. 57)
magnetic declination (p. 55)
magnetic inclination (p. 55)
magnetic-reversal chronology
  (p. 57)
mantle plume (p. 72)
marine magnetic anomaly
  (p. 58)
mid-ocean ridge (p. 50)

normal polarity (p. 56)
paleomagnetism (p. 53)
paleopole (p. 56)
passive margin (p. 63)
plate (p. 46)
plate boundary (p. 61)
plate tectonics (p. 46)
relative plate velocity (p. 80)
reversed polarity (p. 56)
ridge-push force (p. 77)
rifting (p. 74)
seafloor spreading (p. 53)

seamount (p. 51)
seismic belt (p. 52)
slab-pull force (p. 77)
subchron (p. 57)
subduction (p. 67)
suture (p. 75)
transform boundary
  (p. 69)
trench (p. 50)
triple junction (p. 70)
volcanic arc (p. 50)
Wadati-Benioff zone (p. 68)

GEOTOURS *THIS CHAPTER'S GEOTOURS WORKSHEET (B) FEATURES QUESTIONS AND GOOGLE EARTH SITES ON:*

> Divergent boundaries > Convergent boundaries > Transform boundaries > Hot spots > Seafloor spreading

REVIEW QUESTIONS

The letters following each Review Question refer to the corresponding Learning Objective from the Chapter Opener.

1. What was Wegener's continental drift hypothesis? What was his evidence? Why didn't other geologists accept Wegener's proposal of continental drift, at first? **(A)**

2. How do apparent polar-wander paths show that the continents have moved? **(C)**

3. Describe the hypothesis of seafloor spreading. **(B)**

4. Describe the pattern of marine magnetic anomalies across a mid-ocean ridge. How is this pattern explained? **(C)**

5. How did drilling into the seafloor contribute further proof of seafloor spreading? How did the seafloor-spreading hypothesis explain variations in ocean floor heat flow? **(C)**

6. What are the characteristics of a lithosphere plate? Can a single plate include both continental and oceanic lithosphere? **(D)**

7. How does oceanic lithosphere differ from continental lithosphere in thickness, composition, and density? **(D)**

8. How do we identify a plate boundary? **(E)**

9. Describe the three types of plate boundaries. Which type of plate boundary does the line labeled with yellow arrows show in the figure? **(F)**

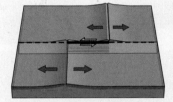

10. How does crust form along a mid-ocean ridge? **(D)**

11. Why is the oldest oceanic lithosphere less than 200 Ma? **(D)**

12. Identify the major geologic features of a convergent boundary in the figure shown. **(F)**

13. Why are transform plate boundaries required on an Earth with spreading and subducting plate boundaries? **(F)**

14. What is a triple junction? **(E)**

15. How is a hot-spot track produced, and how can hot-spot tracks be used to track the past motions of a plate? Which direction is the plate shown in the figure moving? **(G)**

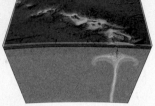

16. Describe the characteristics of a continental rift and give examples of where this process is occurring today. **(G)**

17. Describe the process of continental collision and give examples of where this process has occurred. **(G)**

18. Discuss the major forces that move lithosphere plates. **(H)**

19. Explain the difference between relative plate velocity and absolute plate velocity. **(H)**

ON FURTHER THOUGHT

20. Why are the marine magnetic anomalies bordering the East Pacific Rise in the Pacific Ocean wider than those bordering the Mid-Atlantic Ridge? **(C)**

21. The North Atlantic Ocean is 3,600 km wide. Seafloor spreading along the Mid-Atlantic Ridge occurs at 2 cm per year. When did rifting start to open the Atlantic? **(H)**

ONLINE RESOURCES

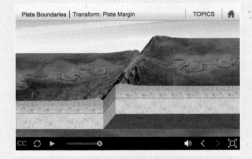

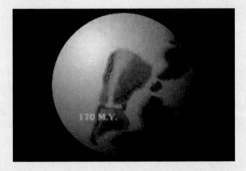

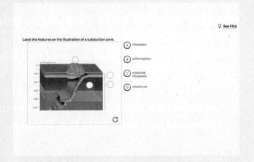

Animations This chapter features animations on the different types of plate boundaries.

Videos This chapter includes videos on plate motions over time, subduction trenches, and deep ocean volcanoes.

Smartwork5 This chapter features questions on topics including plate boundaries, plate formation, and plate movement.

CHAPTER 3

PATTERNS IN NATURE: MINERALS

By the end of this chapter, you should be able to . . .

A. explain why the term *mineral* has a very special meaning in a geologic context.

B. describe the processes by which minerals can form.

C. explain how geologists organize thousands of different minerals into relatively few classes.

D. specify which minerals are the most common ones on the Earth, and describe how they are classified.

E. identify common mineral specimens based on their properties.

F. distinguish gems from ordinary minerals, and describe how to produce the shiny facets of gems.

I died a mineral, and became a plant. I died as plant and rose to animal,
I died as animal and I was Man. Why should I fear?

JALAL-UDDIN RUMI (Persian mystic and poet, 1207–1273)

3.1 Introduction

In Greek legend, the god Dionysus, in a drunken rage, vowed to kill the next mortal that he saw. Just then, a beautiful young woman named Amethyst walked by, and Dionysus ordered two fearsome tigers to attack her. The goddess Artemis prevented a tragedy by changing Amethyst into a pure white statue made of quartz, a mineral that is much harder than the tigers' teeth. The statue was so beautiful that Dionysus sobered up and regretted his rashness. In remorse, he spilled his wine onto the statue as an offering, and the wine stained the quartz, turning it into purple amethyst (see chapter-opening photo). The word comes from the Greek *amethustos*, meaning not intoxicated. For centuries afterward, amethyst was thought of as an antidote for drunkenness. It isn't—amethyst has no effect on alcohol and can't prevent inebriation. But legend aside, amethyst is beautiful, one of many minerals that have been used in jewelry making for millennia.

Amethyst, the purple version of a common mineral, quartz, is one of about 4,000 minerals that *mineralogists*, people who specialize in the study of minerals, have identified so far. The vast majority of mineral types are rare. In fact, fewer than 50 are considered common rock-forming minerals of the Earth's crust. Most rocks that you pick up consist almost entirely of no more than six of these common minerals.

Why study minerals? Without exaggeration, we can say that minerals are the building blocks of our planet. To a geologist, almost any study of Earth materials depends on an understanding of minerals, for minerals make up most of the rocks and sediments comprising the Earth and its landscapes. Minerals are also important from a practical standpoint (see Chapter 12). Industrial minerals serve as the raw materials for manufacturing chemicals, concrete, and wallboard. Ore minerals are the source of valuable metals, such as copper and gold **(Fig. 3.1)**, and of powerful energy sources, such as uranium. Particularly beautiful forms of minerals—gems—delight the eye in jewelry. Unfortunately, though, some minerals pose environmental or health hazards. No wonder **mineralogy**, the study of minerals, fascinates professionals and amateurs alike.

◄ (facing page) This cluster of amethyst crystals, each about 1 cm across, precipitated from water passing through open spaces in a layer of basalt from southern Brazil.

This chapter, an introduction to mineralogy, begins with the geologic definition of a mineral. We then look at how minerals form and at the main characteristics that enable us to identify minerals. Finally, we describe the basic scheme that mineralogists use to classify minerals. Some of the discussions in this chapter utilize basic concepts from chemistry. If you are rusty on your understanding of these, please study **Box 3.1** before going further.

FIGURE 3.1 Copper ore is a useful mineral that serves as a source of copper metal.

The malachite in this partially polished chunk grew in a succession of layers.

5 cm

(a) Malachite, a type of copper ore ($Cu_2[CO_3][OH]_2$), contains copper plus other chemicals.

(b) The copper used for pots comes from copper ore.

BOX 3.1 SCIENCE TOOLBOX...

Some Basic Concepts from Chemistry

To describe minerals, we need to use several terms from chemistry. We introduce this vocabulary in an order that permits each successive term to use previous terms.

> *Element:* A pure substance that cannot be separated into other materials.

> *Atoms and their components:* The smallest piece of an element retaining the characteristics of the element is an atom. Atoms are so small that more than 5 trillion (5,000,000,000,000) could fit on the head of a pin. An atom consists of a nucleus surrounded by a cloud of orbiting electrons. A nucleus is a compact ball of protons and neutrons (except in hydrogen, whose nucleus contains only one proton and no neutrons). The mass of an electron is only about 1/1,836 that of a proton. Electrons have a negative charge, protons have a positive charge, and neutrons have a neutral charge.

> *Neutral atom:* An atom with the same number of electrons as protons is said to be neutral, in that it does not have an overall electrical charge.

> *Atomic number:* The number of protons in an atom of an element.

> *Atomic mass:* Approximately the number of protons plus neutrons in an atom of an element.

> *Ion:* An atom that is not neutral. An ion with an excess negative charge (it has more electrons than protons) is an *anion*, whereas an ion with an excess positive charge (it has more protons than electrons) is a *cation*. We indicate the charge with a superscript. For example, Cl^- has a single excess electron; Fe^{2+} is missing two electrons.

> *Chemical bond:* An attractive force that holds two or more atoms together **(Fig. Bx3.1)**. *Covalent bonds* form when atoms share electrons, and *ionic bonds* form when a cation and an anion (ions with opposite charges) get close together and attract each other. In materials with *metallic bonds*, some of the electrons can move freely.

> *Molecule:* Two or more atoms bonded together. The atoms may be of the same element or of different elements.

> *Compound:* A pure substance that can be subdivided into two or more elements. The smallest piece of a compound that retains the characteristics of the compound is a molecule.

> *State of matter:* The form of a substance, which reflects the degree to which the atoms or molecules comprising the matter are bonded together. In everyday experience, you see three states of matter—solid, liquid, and gas. There are more bonds in a solid than in a liquid, and more in a liquid than in a gas. Which state exists at a given location depends on pressure and temperature. A fourth state, plasma, exists only at very high temperatures.

> *Evaporation:* When atoms or molecules escape from the surface of a liquid and turn into gaseous form.

> *Freezing:* When a liquid turns into solid form.

> *Condensation:* When a gas turns into liquid form.

> *Chemical:* A general name used for a pure substance (either an element or a compound).

3.2 What Is a Mineral?

In everyday English, the word *mineral* has many uses. When you play the game 20 Questions, the word refers to anything that's not animal or vegetable, and if you read food ingredients, the word indicates certain nutrients that people need in order to be healthy. Geologists use the term in a very specific way. To a geologist, a **mineral** is a naturally occurring, solid, crystalline material, formed by geologic processes, that has a definable chemical composition. Almost all minerals are inorganic. Let's pull apart this mouthful of a definition and examine its meaning in detail.

> *Naturally occurring:* True minerals grow in nature, not in factories. In recent decades, chemists have learned how to manufacture materials that have characteristics virtually identical to those of real minerals. Such materials can be referred to as *synthetic minerals*.

> *Formed by geologic processes:* Traditionally, this phrase implied that minerals were the result only of solidification of molten rock or direct precipitation from a water solution, processes that did not involve living organisms. Increasingly, however, geologists recognize life's involvement in the Earth System. So geologists now consider solid, crystalline materials produced by organisms to be minerals, too. To avoid confusion, the term **biogenic mineral** may be used when discussing such materials.

> *Solid:* A mineral, like any matter in the solid state, can maintain its shape indefinitely, so it will not conform to the shape of its container. Minerals cannot be liquids (such as oil or water) or gases (such as air).

> *Crystalline material:* In a **crystalline material**, the atoms reside in an orderly, fixed pattern, locked in place by chemical bonds **(Fig. 3.2a)**. The three-dimensional geometric arrangement of atoms or ions that defines that pattern is called a **crystal structure**.

FIGURE Bx3.1 Types of chemical bonds.

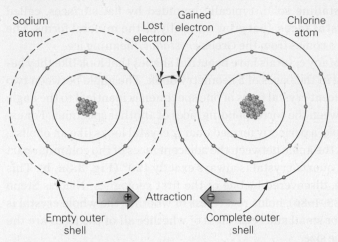

Sodium atom · Lost electron · Gained electron · Chlorine atom

Empty outer shell · ⊕ Attraction ⊖ · Complete outer shell

(a) An ionic bond forms between a positive ion of sodium (Na⁺) and chloride (Cl⁻), a negative ion of chlorine. When sodium gives up one electron to chlorine, so that both have filled shells, halite (NaCl) is produced.

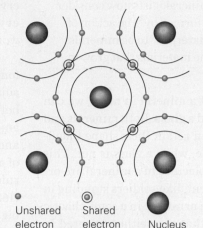

Unshared electron · Shared electron · Nucleus

(b) Covalent bonds form when carbon atoms share electrons so that all have filled electron shells.

(c) In metallically bonded material, nuclei and their inner shells of electrons float in a "sea" of free electrons. The electrons stream through the metal if there is an electrical current.

> *Chemical formula:* A shorthand recipe that itemizes the various elements in a chemical and specifies their relative proportions. For example, the formula for water, H_2O, indicates that water consists of molecules in which two hydrogens bond to one oxygen.

> *Chemical reaction:* A process that involves breaking or forming chemical bonds. Chemical reactions can break molecules apart or yield new molecules and/or isolated atoms. Energy is absorbed or released during a chemical reaction.

> *Mixture:* A combination of two or more elements or compounds that can be separated without a chemical reaction. For example, cereal composed of bran flakes and raisins is a mixture—you can separate the raisins from the flakes without destroying either.

> *Solution:* A type of material in which one chemical (the *solute*) dissolves in another (the *solvent*). In solutions, a solute may separate into ions during the process. For example, when salt (NaCl) dissolves in water, it separates into sodium (Na⁺) and chloride (Cl⁻) ions. In a solution, atoms or molecules of the solvent surround atoms, ions, or molecules of the solute.

> *Precipitate:* (noun) A compound that forms when ions in a liquid solution or a gas join together to make a solid that settles out of the solution; (verb) the process of forming solid grains by separation and settling from a solution. For example, when saltwater evaporates, solid salt crystals precipitate out of solution and form a precipitate of salt.

> *Definable chemical composition:* This phrase simply means that it's possible to write a chemical formula for a mineral. Some minerals contain only one element. For example, both diamond and graphite have the formula C because

FIGURE 3.2 The contrast between crystalline and noncrystalline solids.

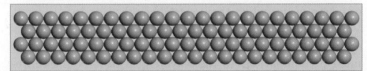

(a) A crystal contains an orderly arrangement of atoms.

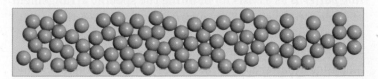

(b) Atoms in noncrystalline solids, such as glass, are not orderly.

they consist entirely of carbon. Most minerals, however, are compounds of two or more elements. For example, quartz has the formula SiO_2. It contains the elements silicon and oxygen in the proportion of one silicon atom for every two oxygen atoms. Some mineral formulas are complicated: for example, the formula for biotite is $K(Mg,Fe)_3(AlSi_3O_{10})(OH)_2$. According to this formula, the proportion of magnesium to iron can vary in biotite.

> *Inorganic:* To understand the meaning of *inorganic*, we must first remember what we mean by *organic*. Organic chemicals consist of molecules that (1) include carbon-carbon and/or carbon-hydrogen bonds and (2) either form in living organisms or have structures similar to those of chemicals that form in living organisms. Sugar ($C_{12}H_{22}O_{11}$), fat, plastic, propane, and protein, for example, are organic chemicals. Some organic chemicals contain only carbon and hydrogen, while others include other elements, such as oxygen, nitrogen, or phosphorus. Almost all minerals are inorganic,

in that they are not organic chemicals. But we need the qualifier "almost all" because mineralogists now consider a few dozen organic substances formed by "the action of geologic processes on organic materials" to be minerals. Examples include the crystalline materials that grow in ancient deposits of bat guano.

With the geologic definition of a mineral in mind, we can distinguish between a mineral and a **glass**. Both minerals and glasses are solids, in that they can retain their shape indefinitely, but a mineral is crystalline, while a glass is not. This means that the atoms, ions, or molecules in a mineral are ordered into a geometric arrangement, like soldiers standing in formation, but those in a glass are arranged in a semi-chaotic way, in small clusters or chains that are neither oriented in the same way nor spaced at regular intervals, like guests at a party **(Fig. 3.2b)**.

Let's examine some everyday materials and see if they are minerals, geologically. Is motor oil a mineral? No—it's an organic liquid. Is table salt a mineral? Yes—it's a natural solid crystalline compound with the formula NaCl. Is the hard material making up the shell of an oyster considered to be a mineral? Yes, it's a biogenic mineral—it has the same composition and structure as an inorganic mineral. Is rock candy a mineral? No. Even though it is solid and crystalline, it's not made by geologic processes and it consists of sugar (an organic chemical).

> **Did you ever wonder . . .**
> is rock candy a mineral?

TAKE-HOME MESSAGE

Minerals are solids with a crystalline structure (an orderly arrangement of atoms inside) and a definable chemical formula. They form by natural processes in the Earth System, and most are inorganic.

QUICK QUESTION Is Styrofoam a mineral? Why or why not?

3.3 Beauty in Patterns: Crystals and Their Structures

WHAT IS A CRYSTAL?

The word *crystal* brings to mind sparkling chandeliers, elegant wine goblets, and shiny jewels. But, as is the case with the word *mineral*, geologists have a more precise definition.

A **crystal** is a single, continuous (uninterrupted) piece of a crystalline solid, typically bounded by flat surfaces, called **crystal faces**, that grow naturally as the mineral forms. The word comes from the Greek *krystallos*, meaning ice.

Many crystals have beautiful shapes that look like they belong in the pages of a geometry book. The angle between two adjacent crystal faces of one specimen is identical to the angle between the corresponding faces of another specimen. For example, a perfectly formed quartz crystal looks like an obelisk, and the angle between the adjacent faces of the columnar part of a quartz crystal is always exactly 120° **(Fig. 3.3a, b)**. This rule, discovered by one of the first geologists, Nicolas Steno (1638–1686), holds regardless of whether the whole crystal is big or small and regardless of whether all of the faces are the same size.

Crystals come in a great variety of shapes, such as cubes, trapezoids, pyramids, octahedrons, hexagonal columns, blades, needles, columns, and obelisks **(Fig. 3.3c)**. Because a crystal has a regular geometric form, people have always considered them to be special, perhaps even a source of magical powers. For example, shamans of some cultures relied on talismans or amulets made of crystals, which supposedly brought power to their wearer or warded off evil spirits. Scientists have concluded, however, that crystals have no effect on health or mood. For millennia, crystals have inspired awe because of the way they sparkle, but such behavior is simply a consequence of how crystal structures interact with light.

LOOKING INSIDE A MINERAL

What do the insides of a mineral actually look like? This problem was the focus of study for centuries. An answer finally came from the work of a German physicist, Max von Laue, in 1912. He showed that an X-ray beam passing through a crystal breaks up into many tiny beams to make a pattern of dots on a screen **(Fig. 3.4a)**. Physicists refer to this phenomenon as *diffraction*. It occurs when waves interact with regularly spaced objects whose spacing resembles the wavelength of the waves—you can see diffraction of ocean waves when they pass through gaps in a seawall. Von Laue concluded that, for a crystal to cause diffraction, atoms within it must be regularly spaced, and the spacing must be comparable to the wavelength of X-rays. Eventually, Von Laue and others learned how to use *X-ray diffraction patterns* as a basis for defining the specific arrangement of atoms in crystals. This arrangement defines the crystal structure of a mineral.

If you've ever examined wallpaper, you've seen an example of a pattern **(Fig. 3.4b)**. Crystal structures contain one of nature's most spectacular examples of such a pattern.

FIGURE 3.3 Some characteristics of crystals.

A crystal face

A crystal lattice resembles intersecting points in scaffolding.

(a) A quartz crystal can resemble an obelisk. Inside, atoms are arranged in a specific geometric pattern, like the joint points in scaffolding.

Halite   Diamond   Staurolite   Quartz

Garnet   Stibnite   Calcite   Kyanite

(c) Crystals come in a variety of shapes, including cubes, prisms, blades, and pyramids. Some terminate at a point and some terminate with flat surfaces.

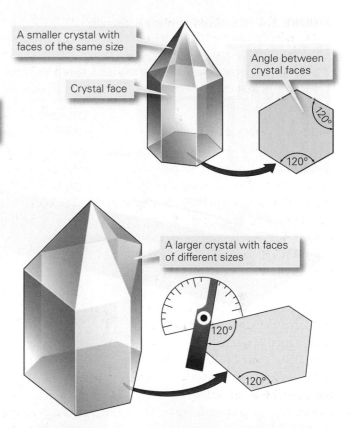

A smaller crystal with faces of the same size

Crystal face

Angle between crystal faces

120°

120°

A larger crystal with faces of different sizes

120°

120°

(b) Regardless of specimen size, the angle between two adjacent crystal faces is consistent in a particular mineral.

In crystals, the pattern is defined by the regular spacing of atoms and, if the crystal contains more than one element, by the regular alternation of atoms **(Fig. 3.4c)**. (Mineralogists refer to a three-dimensional geometry of points representing this pattern as a *crystal lattice*.) The arrangement of atoms in a crystal may control the shape of a crystal. For example, if atoms in a crystal pack into the shape of a cube, the crystal may have faces that intersect at 90° angles—galena (PbS) and halite (NaCl) have such a cubic shape. A crystal structure has *symmetry*, meaning that the shape of one part of the structure is the mirror image of the shape of a neighboring part. For example, if you were to cut a halite crystal or a water crystal

(snowflake) in half, and place the half against a mirror, it would look whole again **(Fig. 3.4d)**.

To illustrate crystal structures, we look at a few examples. Halite (rock salt) consists of oppositely charged ions that stick together because opposite charges attract. In halite, six chloride (Cl^-) ions surround each sodium (Na^+) ion in an overall arrangement of atoms that defines the shape of a cube **(Fig. 3.5a, b)**. Diamond, by contrast, is a mineral made entirely of carbon. In diamond, each atom bonds to four neighbors arranged in the form of a tetrahedron; some naturally formed diamond crystals have the shape of a double tetrahedron **(Fig. 3.5c)**. Graphite, another mineral composed entirely of carbon, behaves very differently from diamond. In contrast to diamond, graphite is so soft that we use it as the "lead" in a pencil. When a pencil moves across paper, tiny flakes of graphite peel off the pencil point and adhere to the paper. This behavior occurs because the carbon atoms in graphite are not arranged in tetrahedra, but rather occur in sheets **(Fig. 3.5d)**. The sheets are bonded to each other by weak bonds and thus can separate from each other easily. Two different minerals (such as diamond and graphite) that have the same composition but different crystal structures are called **polymorphs**.

FIGURE 3.4 Patterns and symmetry in minerals.

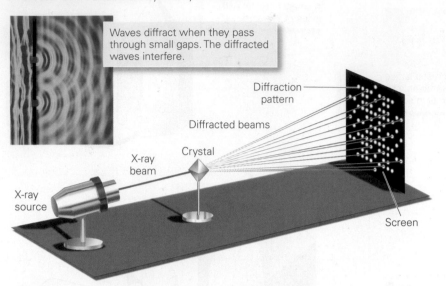

Waves diffract when they pass through small gaps. The diffracted waves interfere.

(a) Diffraction of an X-ray beam passing through a crystal produces a pattern of bright spots on a screen. The spots are due to interference of overlapping light waves.

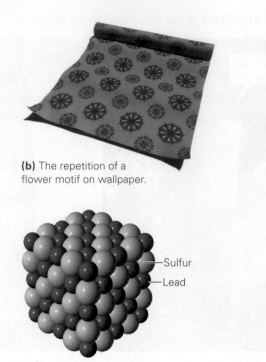

(b) The repetition of a flower motif on wallpaper.

(c) The repetition of alternating sulfur and lead atoms in the mineral galena (PbS).

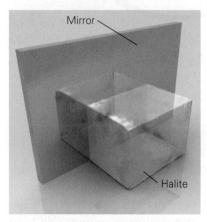

(d) Minerals display symmetry. One-half of a halite crystal or a water crystal is a mirror image of the other.

THE FORMATION AND DESTRUCTION OF MINERALS

New mineral crystals can form in several ways (**Geology at a Glance**, pp. 92–93): (1) *Solidification (freezing) of a melt* happens when a liquid cools and turns into a solid. Just as ice crystals form by solidification of water, a great variety of minerals form by solidification of molten rock. (2) *Precipitation from a water solution* takes place when dissolved ions bond together to form crystals that settle out of the water or grow out from the walls of a container. Salt crystals, for example, precipitate when saltwater evaporates (see Box 3.1). (3) *Solid-state diffusion* results from the movement of atoms or

ions through a solid to arrange into a new crystal structure. Garnets, for example, grow by diffusion in solid rock. During this process, they replace pre-existing minerals. (4) *Biomineralization* takes place when minerals grow at the interface between the physical and biological components of the Earth System. This process can happen because metabolic processes of some living organisms can cause minerals to precipitate either within their bodies, on their bodies, or immediately adjacent to their bodies. Shells produced by clams and oysters, for example, grow when these organisms extract ions from the water they live in. And (5) *precipitation directly from a gas* can occur around volcanic vents or around geysers, for at such locations volcanic gases or steam enter the atmosphere and cool, so some gas molecules of some elements are able to bind together. Some of the bright yellow sulfur deposits found in volcanic regions form in this way.

The first step in forming a crystal involves the chance formation of a seed, an extremely small crystal (**Fig. 3.6a**). Once the seed exists, other atoms in the surrounding material attach themselves to the face of the seed. As the crystal grows, crystal faces move outward but maintain the same orientation (**Fig. 3.6b**). The youngest part of the crystal always forms its outer edge. In the case of crystals formed by the solidification of a melt, atoms begin to attach to the seed when the melt becomes so cool that thermal vibrations can no longer break apart the attraction between the seed

FIGURE 3.5 The nature of crystalline structure in minerals.

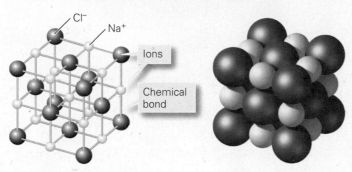

(a) A ball-and-stick model of halite portrays ions as balls, and chemical bonds as sticks.

(b) This packed-ball model gives a sense of how ions fit together in a crystal.

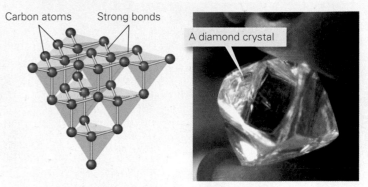

(c) In a diamond, carbon atoms are arranged in tetrahedra. All of the bonds are strong.

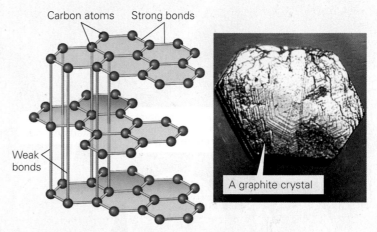

(d) Graphite consists of carbon atoms arranged in hexagonal sheets. The sheets are connected by weak bonds.

and the atoms in the melt. Crystals formed by precipitation from a solution develop when the solution becomes *saturated,* meaning the number of dissolved ions per unit volume of solution becomes so great that they can get close enough to each other to bond together.

As crystals grow, they develop their particular crystal shape, based on the geometry of their internal structure. The shape (for instance, needle-like or sheet-like) is defined by the relative dimensions of the crystal and by the angles between crystal faces. Typically, already-formed crystals act as obstacles preventing the unimpeded growth of new minerals. In such cases, the new minerals grow to fill the space that is available, so their shape does not reflect their crystal structures. Minerals without well-formed crystal faces are *anhedral* grains **(Fig. 3.6c)**. If a mineral's growth is unimpeded so that it displays well-formed crystal faces, then it is a *euhedral* crystal. The surface crystals of a *geode,* a mineral-lined cavity in rock, may be euhedral **(Fig. 3.6d)**.

A mineral can be destroyed by melting, dissolution, or some other chemical reaction. *Melting* involves heating a mineral to a temperature at which thermal vibration of the atoms or ions in the lattice break the chemical bonds holding them to the lattice. The atoms or ions then separate, either individually or in small groups, to move around again freely. *Dissolution* occurs when you immerse a mineral in a solvent, such as water. Atoms or ions then separate from the crystal face and are surrounded by solvent molecules. *Chemical reactions* can destroy a mineral when it comes in contact with reactive materials; for example, iron-bearing minerals react with air and water to form rust. *Microbial metabolism* in the environment can also destroy minerals. In effect, some microbes can "eat" certain minerals; the microbes use the energy stored in the chemical bonds that hold the atoms of the mineral together as their source of energy for metabolism.

TAKE-HOME MESSAGE

The crystal structure of minerals is defined by a regular geometric arrangement of atoms that has symmetry. Minerals can form by solidification of a melt, by precipitation from a water solution or a gas, or by rearrangement of atoms in a solid.

QUICK QUESTION How are minerals destroyed in nature?

3.4 How Can You Tell One Mineral from Another?

Amateur and professional mineralogists alike get a kick out of recognizing minerals. They might hover around a display case in a museum and name specimens without bothering to look at the labels. How do they do it? The trick lies in

GEOLOGY AT A GLANCE

Mineral Formation

Solidification from a melt happens when molten rock, such as lava erupted from a volcano, cools down. Different minerals grow in succession.

Eventually, all the melt solidifies.

At first, a few crystals form. They remain surrounded by melt.

A mafic melt starts to cool.

Temperature decreases over time

Diffusion can happen in a solid rock, though very slowly. During the process, atoms migrate through the crystal. New minerals, such as garnet, can grow in the rock.

Fumerole

Sulfur crystals

Precipitation of minerals from volcanic gas can also occur. Yellow sulfur crystals form this way.

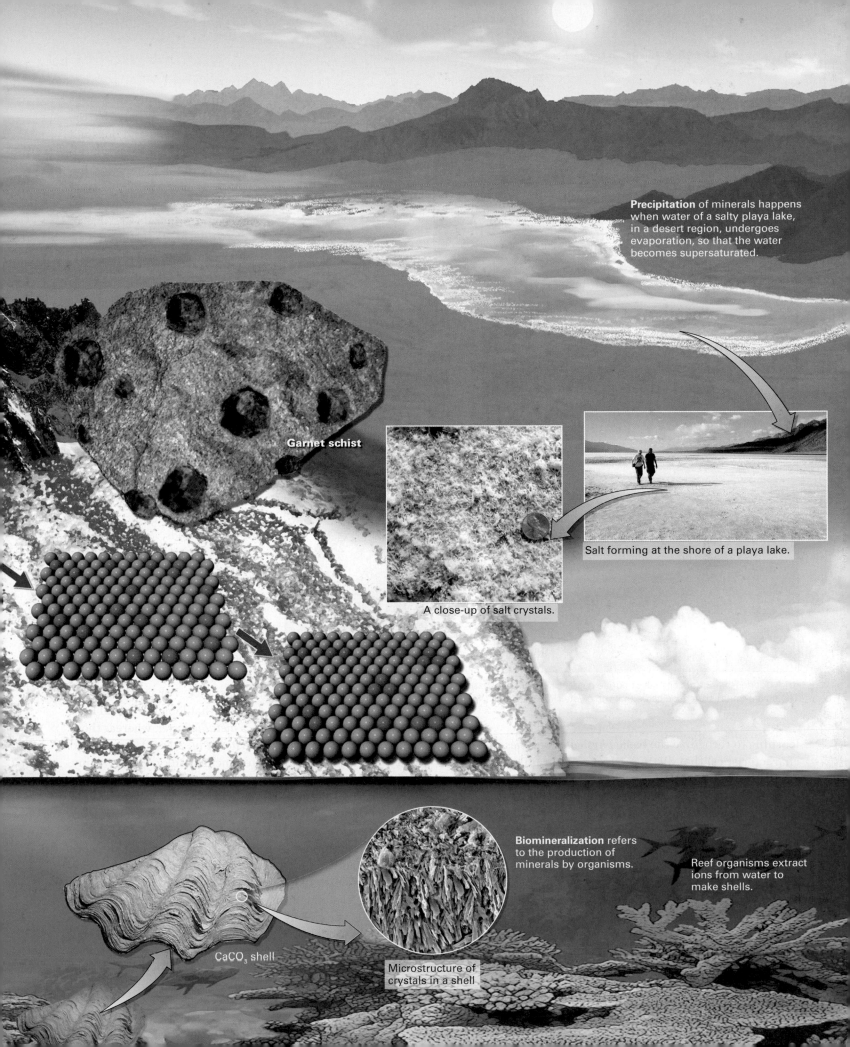

Garnet schist

Precipitation of minerals happens when water of a salty playa lake, in a desert region, undergoes evaporation, so that the water becomes supersaturated.

A close-up of salt crystals.

Salt forming at the shore of a playa lake.

Biomineralization refers to the production of minerals by organisms.

Reef organisms extract ions from water to make shells.

CaCO$_3$ shell

Microstructure of crystals in a shell

FIGURE 3.6 The growth of crystals.

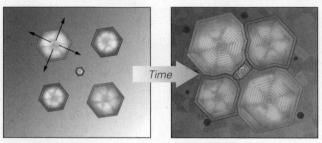

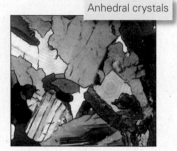

Ions attach to the crystal face.

(a) New crystals nucleate and begin to precipitate out of a water solution. As time progresses, they grow into the open space.

Anhedral crystals

(b) New crystals grow outward from the central seed. As time passes, they maintain their shape until they interfere with each other.

(c) A crystal growing in a confined space will be anhedral.

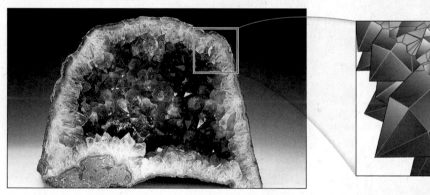

Euhedral crystals

(d) A geode from Brazil consists of purple quartz crystals (amethyst) that grew from the wall into the center. The enlargement sketch indicates that the crystals are euhedral.

learning to recognize the basic *physical properties* (material characteristics) that distinguish one mineral from another. Some physical properties, such as shape and color, can be seen from a distance. Others, such as hardness and magnetization, can be determined only by handling the specimen or by performing an *identification test* on it. Such tests include scratching the mineral by another object, placing it near a magnet, weighing it, tasting it, or placing a drop of acid on it. Let's examine some of the physical properties most commonly used in basic mineral identification.

Color: **Color** results from the way a mineral interacts with light. Sunlight contains the whole spectrum of colors, each with a different wavelength. A mineral absorbs certain wavelengths and reflects others—the color you see when looking at a specimen represents the wavelengths the mineral reflects and does not absorb. Certain minerals always have the same color, but many display a range of colors (**Fig. 3.7a**). Color variations in some minerals are due to the presence of impurities. For example, trace amounts of iron may give quartz a reddish color.

Streak: The **streak** of a mineral refers to the color of a powder produced by pulverizing the mineral. You can obtain a streak by scraping the mineral against an unglazed ceramic plate (**Fig. 3.7b**). The color of a mineral powder tends to be more consistent than the color of a whole crystal, so it provides a fairly reliable clue to a mineral's identity. Calcite, for example, always yields a white streak even though pieces of calcite may be white, pink, or clear.

Luster: **Luster** refers to the way a mineral surface scatters light. Geoscientists describe luster by comparing the appearance of the mineral with the appearance of a familiar substance. For example, minerals that look like metal have a *metallic luster,* whereas those that do not have a *nonmetallic luster* (**Fig. 3.7c, d**). Terms used for types of nonmetallic luster include silky, glassy, satiny, resinous, pearly, or earthy.

Hardness: **Hardness** is a measure of the relative ability of a mineral to resist scratching, and it therefore represents the resistance of bonds in the crystal structure to being broken. The atoms or ions in crystals of a hard mineral are more strongly bonded than those in a soft mineral. Hard minerals can scratch soft minerals, but soft minerals cannot scratch hard ones. Diamond, the hardest mineral known, can scratch anything, which is why it is used to cut glass. In the early 1800s, a mineralogist named Friedrich Mohs listed some minerals in sequence of relative hardness; a mineral with a hardness of 5 can scratch all minerals with a hardness of

FIGURE 3.7 Physical characteristics of minerals.

(a) Color is diagnostic of some minerals, but not all. For example, quartz can come in many colors.

(b) To obtain the streak of a mineral, rub it against a porcelain plate. The streak consists of mineral powder.

(c) Pyrite has a metallic luster because it gleams like metal.

(d) Feldspar has a nonmetallic luster.

(e) Crystal habit refers to the shape of the crystal. Wulfenite crystals are very thin, tabular plates.

(f) Kyanite crystals are bladed and columnar.

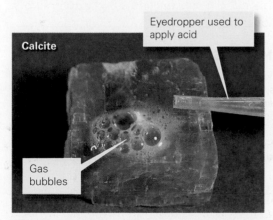

(g) Calcite reacts with hydrochloric acid to produce carbon dioxide gas.

(h) Magnetite is magnetic, so it can attract magnetized objects.

TABLE 3.1 Mohs Hardness Scale

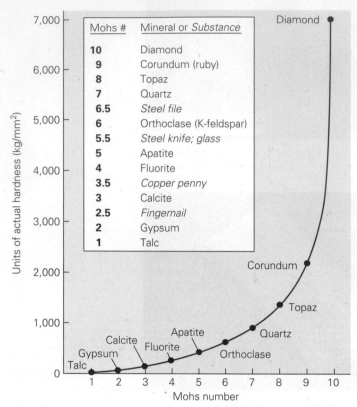

Mohs #	Mineral or *Substance*
10	Diamond
9	Corundum (ruby)
8	Topaz
7	Quartz
6.5	*Steel file*
6	Orthoclase (K-feldspar)
5.5	*Steel knife; glass*
5	Apatite
4	Fluorite
3.5	*Copper penny*
3	Calcite
2.5	*Fingernail*
2	Gypsum
1	Talc

Mohs numbers are relative—in reality, diamond is 3.5 times harder than corundum, as the graph shows.

5 or less. This list, the **Mohs hardness scale**, helps in mineral identification. To make the scale easy to use, it includes common items such as a fingernail, a penny, or a glass plate **(Table 3.1)**.

Specific gravity: **Specific gravity** represents the density of a mineral, as defined by the ratio between the weight of a volume of the mineral and the weight of an equal volume of water at 4°C. For example, one cubic centimeter of quartz has a weight of 2.65 grams, whereas one cubic centimeter of water has a weight of 1.00 gram. Thus, the specific gravity of quartz is 2.65. In practice, you can develop a feel for specific gravity by hefting minerals in your hands. A piece of galena (lead ore) feels heavier than a similar-sized piece of quartz.

Crystal habit: The **crystal habit** of a mineral refers to the shape of a single crystal with well-formed crystal faces, or to the character of an aggregate of many well-formed crystals that grew together as a group **(Fig. 3.7e)**. The habit depends on the internal arrangement of atoms in the crystal. When describing habit, mineralogists commonly compare the mineral

to a common geometric shape using adjectives such as cubic, prismatic, bladed, platy, or fibrous **(Fig. 3.7f; Box 3.2)**. The relative dimensions depend on relative rates of crystal growth in different directions. For example, crystals that grow rapidly in one direction but slowly in the other two directions become needle-like.

Special properties: Some minerals have distinctive properties that readily distinguish them from other minerals. For example, calcite ($CaCO_3$) reacts with dilute hydrochloric acid (HCl) to produce carbon dioxide (CO_2) gas **(Fig. 3.7g)**. Dolomite ($CaMg[CO_3]_2$) also reacts with acid, but not as strongly. Graphite makes a gray mark on paper, magnetite attracts a magnet **(Fig. 3.7h)**, halite tastes salty, and plagioclase has striations (thin parallel corrugations or stripes) on its surface.

Fracture and cleavage: Different minerals fracture (break) in different ways. These depend on the internal arrangement of atoms in the minerals. If a mineral breaks to form distinct planar surfaces that have a specific orientation in relation to the crystal structure, then we say that the mineral has cleavage, and we refer to each surface as a cleavage plane.

Cleavage forms in directions where the bonds holding atoms together in the crystal are weaker **(Fig. 3.8)**. Some minerals have one direction of cleavage. For example, mica has very weak bonds in one direction but strong bonds in the other two directions. Thus, it easily splits into parallel sheets; the surface of each sheet is a cleavage plane. Other minerals have two or three directions of cleavage that intersect at a specific angle. For example, halite has three sets of cleavage planes that intersect at right angles, so halite crystals break into little cubes. Materials with no cleavage at all (because bonding is equally strong in all directions) break either by forming irregular fractures or by forming conchoidal fractures **(Fig. 3.9a)**. *Conchoidal fractures* are smoothly curving, clamshell-shaped surfaces that typically form in glass. Cleavage planes are sometimes hard to distinguish from crystal faces **(Fig. 3.9b)**.

TAKE-HOME MESSAGE

The properties of minerals (such as color, streak, luster, crystal habit, hardness, specific gravity, cleavage, magnetism, and reaction with acid) are a manifestation of the crystal structure and chemical composition of minerals and can be used for mineral identification.

QUICK QUESTION Which minerals react with acid to produce CO_2 bubbles?

FIGURE 3.8 The nature of mineral cleavage and fracture.

(a) Mica has one strong plane of cleavage and splits into sheets.

(b) Pyroxene has two planes of cleavage that intersect at 90°.

(c) Amphibole has two planes of cleavage that intersect at 60°.

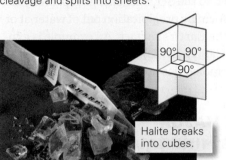

(d) Halite has three mutually perpendicular planes of cleavage.

(e) Calcite has three planes of cleavage, one of which is inclined.

(f) Diamond has four planes of cleavage, each inclined to the others.

3.5 Organizing Your Knowledge: Mineral Classification

Just about every object you come across in daily life has been classified in some way, because classification schemes help organize information and streamline discussion. Biologists, for example, classify animals into groups based on how they feed their young and on the architecture of their skeletons, and botanists classify plants according to the way they reproduce and by the shape of their leaves. In the case of minerals, a good means of classification eluded researchers until it became possible to determine the chemical makeup of minerals. A Swedish chemist, Baron Jöns Jacob Berzelius (1779–1848), analyzed minerals and noted chemical similarities among many of them. Berzelius, along with his students, established that most minerals can be classified by specifying the principal anion (negative atom) or anionic group (negative molecule) within the mineral (see Box 3.1). Using this approach, it's possible to divide the 4,000 known minerals into a relatively small number of groups, or **mineral classes**. We now take a look at principal mineral classes, focusing especially on silicates, the class that constitutes most of the Earth's rock.

MINERAL CLASSES

Mineralogists distinguish several principal classes of minerals. Here are some of the major ones.

> *Silicates:* The fundamental component of most silicates in the Earth's crust is the SiO_4^{4-} anionic group. A well-known example, quartz (see Fig. 3.7a), has the formula SiO_2. We will learn more about silicates in the next section.

> *Sulfides:* Sulfides consist of a metal cation bonded to a sulfide anion (S^{2-}). Examples include galena (PbS) and pyrite (FeS_2; see Fig. 3.7c).

> *Oxides:* Oxides consist of metal cations bonded to oxygen anions. Typical oxide minerals include hematite (Fe_2O_3; see Fig. 3.7b) and magnetite (Fe_3O_4; see Fig. 3.7h).

> *Halides:* The anion in a halide is a halogen ion (such as chloride [Cl^-] or fluoride [F^-]), an element from the second column from the right in the periodic table (a periodic table is provided at the back of the book). Halite, or rock salt (NaCl; see Fig. 3.8d), and fluorite (CaF_2) are common examples.

FIGURE 3.9 The different ways minerals can break.

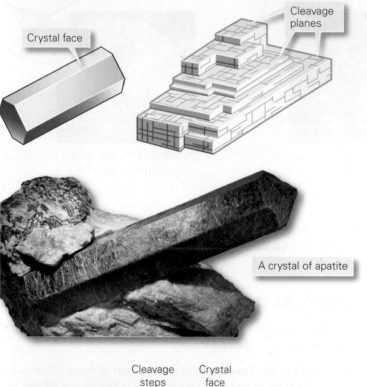

(a) Minerals without cleavage can develop irregular or conchoidal fractures.

(b) How do you distinguish between crystal faces and cleavage planes? A crystal face is a single surface, whereas cleavage planes can be repeated.

> *Carbonates:* In carbonate minerals, CO_3^{2-} serves as the anionic group. Elements such as calcium or magnesium bond to this group. The two most common carbonates are calcite ($CaCO_3$; see Fig. 3.8e) and dolomite ($CaMg[CO_3]_2$).

> *Native metals:* Native metals consist of pure masses of a single metal. The metal atoms are bonded by metallic bonds (see Box 3.1). Copper and gold, for example, may occur as native metals.

> *Sulfates:* Sulfates consist of a metal cation bonded to the SO_4^{2-} anionic group. Many sulfates form by precipitation out of water at or near the Earth's surface. An example is gypsum ($CaSO_4 \cdot 2H_2O$). Note that in gypsum, water molecules bond to calcium sulfate.

SILICATES: THE MAJOR ROCK-FORMING MINERALS

Silicate minerals, or **silicates**, make up over 95% of the continental crust and almost 100% of the oceanic crust. Further, nearly all of the Earth's mantle consists of silicates. Thus, silicates are the most common minerals of the Earth. As we've noted, silicates in the Earth's crust and upper mantle contain the SiO_4^{4-} anionic group. In this group, four oxygen atoms surround a single silicon atom, thereby defining the corners of a tetrahedron, a pyramid-like shape with four triangular faces **(Fig. 3.10a)**. We refer to this anionic group as the **silicon-oxygen tetrahedron** (or, informally, as the silica tetrahedron), and it acts, in effect, as the building block of silicate minerals.

Mineralogists distinguish among several groups of silicate minerals based on the way in which silica tetrahedra are arranged and bonded together **(Fig. 3.10b)**. The extent of bonding, in turn, determines the degree to which tetrahedra share oxygen atoms. Note that the number of shared oxygens determines the ratio of silicon (Si) to oxygen (O) in the mineral. Here are the groups, in order from fewer shared oxygens to more shared oxygens:

> *Isolated tetrahedra:* In this group, the tetrahedra do not share any oxygen atoms. The attraction between tetrahedra and positive ions holds these minerals together. Examples include olivine, a glassy green mineral, and garnet.

> *Single chains:* In a single-chain silicate, the tetrahedra link to form a chain by sharing two oxygen atoms. The most common of the many different types of single-chain silicates are pyroxenes.

BOX 3.2 CONSIDER THIS...

Asbestos and Health—When Crystal Habit Matters

There are many types of asbestos minerals, but they all share a key characteristic—all have a fibrous crystal habit **(Fig. Bx3.2a)**. Thus, samples of asbestos consist of clusters of needle-like crystals **(Fig. Bx 3.2b)**. For years, asbetos has been incorporated into floor tiles, roof shingles, brake pads, fireproof clothes, and insulation. Its popularity derived from the fact that because it's a mineral, it doesn't burn, and because it's fibrous, it can be woven into other materials. Its presence in other materials adds strength because fibers have strong bonds along their length.

When intact, asbestos-bearing materials are not especially hazardous. But in the form of dust, during mining, or during construction or demolition, the fibers can be dangerous to human health. If inhaled, they lodge in the lungs where they cause irritation and may trigger cancer. Over the years, many lawsuits involving asbestos have taken place, and its use has been banned since the mid-1980s.

Today, building owners must pay for *asbestos abatement* during renovation projects that involve tearing out old asbestos-bearing materials. During abatement, workers handling asbestos must wear protective gear **(Fig. Bx 3.2c)**, and the space in which they're working must be sealed off from its surroundings. Debate continues as to whether all abatement is necessary. In some cases, asbestos-bearing materials that have been covered over by paint or other materials, or that remain undisturbed, may not be problematic.

FIGURE Bx3.2 Asbestos has characteristics that can make it both useful and hazardous.

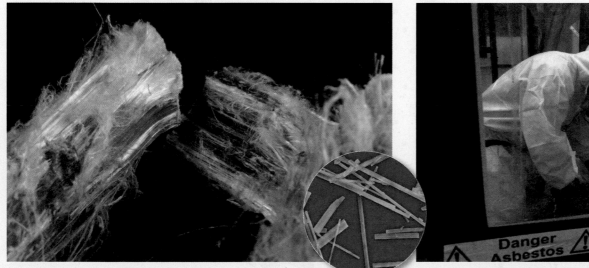

(a) Asbestos is a fibrous mineral.

(b) At high magnification, asbestos fibers look like tiny needles.

(c) To remove asbestos requires protective gear and for the area to be sealed off.

> *Double chains:* In a double-chain silicate, the tetrahedra link by sharing two or three oxygen atoms. Amphiboles are the most common type.

> *Sheet silicates:* All the tetrahedra in this group share three oxygen atoms and therefore link to form two-dimensional sheets. Other ions, and in some cases water molecules, fit between the sheets in some sheet silicates. Because of their structure, sheet silicates have cleavage in one direction and occur in books of very thin sheets. This group includes micas and clays (which occur only in extremely tiny flakes).

> *Framework silicates:* In a framework silicate, each tetrahedron shares all four oxygen atoms with its neighbors, so the tetrahedra are configured in a three-dimensional structure. Examples include feldspar and quartz. The two most common feldspars are plagioclase, which tends to be white, gray, or blue; and orthoclase (also called potassium feldspar, or K-feldspar), which tends to be pink.

FIGURE 3.10 The structure of silicate minerals.

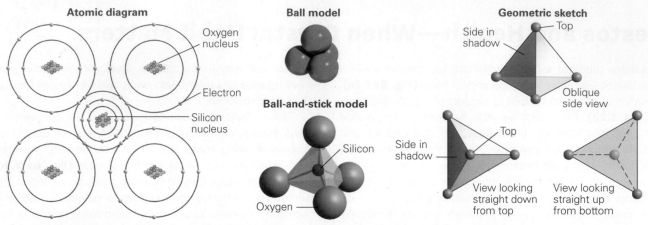

(a) The fundamental building block of a silicate mineral is the silicon-oxygen tetrahedron. Oxygen atoms occupy the corners of the tetrahedron, and silicon lies at the center. Geologists portray the tetrahedron in a number of different ways.

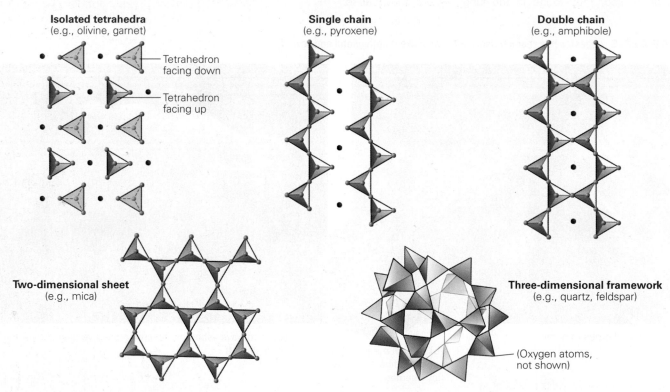

(b) The classes of silicate minerals differ from one another by the way in which the silicon-oxygen tetrahedra are linked. Where the tetrahedra link, they share an oxygen atom. Oxygen atoms are shown in blue. Positive ions (not shown) occupy spaces between tetrahedra.

TAKE-HOME MESSAGE

The 4,000 known minerals can be organized into a relatively small number of classes based on chemical makeup. Most minerals are silicates, which contain silicon-oxygen tetrahedra arranged in various ways.

QUICK QUESTION What is the principal anionic group in carbonate minerals?

3.6 Something Precious— Gems!

Mystery and romance follow famous gems. Consider the stone now known as the Hope Diamond, recognized by name the world over **(Fig. 3.11)**. No one knows who first dug it out of the ground **(Box 3.3)**. Was it mined in the 1600s or was

BOX 3.3 CONSIDER THIS...

Where Do Diamonds Come From?

Diamonds consist of carbon, which typically accumulates only at or near the Earth's surface. Experiments demonstrate that the pressures needed to form diamond are so extreme that, in nature, they generally occur only at depths of around 150 km below the Earth's surface. Nowadays, engineers can duplicate these conditions in the laboratory, so corporations manufacture several tons of synthetic diamonds a year.

How does carbon get down to depths of 150 km? Geologists speculate that subduction or collision carries carbon-containing rocks and sediments down to the depth at which it transforms into diamond beneath continents. But if diamonds form at great depth, how do they return to the surface? Some diamonds rise when rifting cracks the continental crust and causes a small part of the underlying mantle to melt. Magma generated during this process rises to the surface, bringing the diamonds with it. Near the surface, the magma solidifies to form an igneous rock called *kimberlite*, named for Kimberley, South Africa. Diamonds brought up with the magma are embedded as crystals in solid kimberlite (**Fig. Bx3.3a**). Much of the world's diamond supply comes from mines in this rock. But some sources occur in deposits of sediment formed from the breakdown and erosion of kimberlite that had been exposed at the surface. Rivers and glaciers may transport diamond-bearing sediments far from their original bedrock source. Currently, diamonds are being mined in both open-pit mines and underground mines (**Fig. Bx3.3b, c**).

Not all natural diamonds are valuable; value depends on color and clarity. Diamonds that contain imperfections (cracks, or specks of other material) or are dark gray in color are not used for jewelry. These stones, called *industrial diamonds*, can be used as abrasives. Gem-quality diamonds come in a range of sizes. Jewelers measure the size of these gems using carats, where 1 carat equals 200 mg (0.2 g). In English units of measurement, 1 ounce equals 142 carats. The largest diamond ever found, a stone called the Cullinan Diamond, was discovered in South Africa in 1905 and weighed 3,106 carats (621 g) before being cut. A 910-carat diamond, expected to sell for over $40 million, came out of a South African mine in 2018. By comparison, the diamond on a typical engagement ring weighs less than 1 carat. Gem-quality diamonds are actually more common than you might expect—suppliers stockpile the stones in order to avoid flooding the market and lowering the price.

FIGURE Bx3.3 Diamond occurrences.

(a) A kimberlite sample containing a diamond.

A diamond embedded in solid kimberlite.

(b) A diamond mine in northern Canada.

(c) Modern equipment excavates kimberlite in an underground mine near Kimberley, South Africa.

FIGURE 3.11 The Hope Diamond, now on display at the Smithsonian Institution in Washington, DC.

SEE FOR YOURSELF...

Kimberley Diamond Mine

LATITUDE
28°44'17.06" S

LONGITUDE
24°46'30.77" E

Look straight down from 13 km.

The field of view shows the town of Kimberley, South Africa, and its inactive diamond mine. The mine looks like a circular pit. You can also see the tailings pile of excavated rock debris.

Did you ever wonder...
where diamonds come from and how they form?

it stolen off an ancient religious monument? What we do know is that in the 1600s, a French trader named Jean-Baptiste Tavernier obtained a large (112.5 carats, where 1 carat = 200 mg), rare blue diamond in India, perhaps from a Hindu statue, and carried it back to France. King Louis XIV bought the diamond and had it fashioned into a jewel of 68 carats. This jewel vanished during a burglary in 1762. Perhaps it was lost forever—perhaps not. In 1830, a 44.5-carat blue diamond mysteriously appeared on the jewel market for sale. Henry Hope, a British banker, purchased the stone, which then became known as the Hope Diamond. It changed hands several times until 1958, when a famous New York jeweler named Harry Winston donated it to the Smithsonian Institution in Washington, DC, where it now sits behind bulletproof glass in a heavily guarded display.

What makes stones such as the Hope Diamond so special that people risk life and fortune to obtain them? What is the difference

between gemstones, gems, and other minerals? A **gemstone** is a mineral that has special value because it is rare and people consider it beautiful. A **gem**, or jewel, is a finished stone ready to be set in jewelry. Jewelers distinguish between *precious stones* (such as diamond, ruby, sapphire, and emerald), which are particularly rare and expensive, and *semiprecious stones* (such as topaz, tourmaline, aquamarine, and garnet), which are less rare and less expensive. All precious stones are transparent crystals, though most have some color. The category of semiprecious stones also includes opaque or translucent minerals such as lapis, malachite (see Fig. 3.1a), and opal.

In everyday language, pearls and amber may also be considered gemstones. Unlike diamonds and garnets, which form inorganically in rocks, pearls form in living oysters when the oyster extracts calcium and carbonate ions from water and precipitates them around an impurity, such as a sand grain, embedded in its body. Thus, pearls are a result of biomineralization. Most pearls used in jewelry today are "cultured" pearls, made by artificially introducing round sand grains into oysters in order to stimulate production of round pearls. Amber also forms as a consequence of organic processes—it consists of fossilized tree sap. But because amber consists of organic compounds that are not arranged in a crystal structure, it does not meet the definition of a mineral.

In some cases, gemstones are merely pretty and rare versions of more common minerals. For example, ruby is a special version of the common mineral corundum, and emerald is a special version of the common mineral beryl **(Fig. 3.12a)**. As for the beauty of a gemstone, this quality lies basically in its color and, in the case of transparent gems, its "fire"—the way the mineral bends and internally reflects the light passing through it, and disperses the light into a spectrum. Fire makes a diamond sparkle more than a similarly cut piece of glass can.

Gemstones form in many ways. Some solidify from a melt, some form by diffusion, some precipitate out of a water solution in

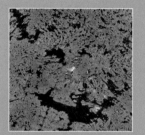

SEE FOR YOURSELF...

Ekati Diamond Mine, Canada

LATITUDE
64°43'15.44" N

LONGITUDE
110°36'56.27" W

Look straight down from 100 km.

The Ekati Diamond Mine is in a remote, largely uninhabited region of the Northwest Territories of Canada. Prospectors found diamond pipes here in the early 1990s after a 20-year search. The mine opened in 1998 and within 10 years had produced more than 40 million carats (8,000 kg) of diamonds. Zoom down to 10 km to see details of the mining operation.

FIGURE 3.12 Cutting gemstones.

Non-gem-quality beryl

Rough emerald

Cut emerald

(a) Emerald is a green, transparent variety of the mineral beryl.

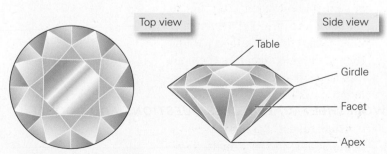

Lap

Doping arm

Goniometer (to adjust angle)

Gemstone

Cooling water supply

Grinding surface on a spinning lap

(b) A faceting machine. The shiny faces of a gem are made by grinding the stone on a lap.

Top view

Side view

Table

Girdle

Facet

Apex

(c) There are many different "cuts" for a gem. Here we see the top and side views of a brilliant-cut diamond.

cracks, and some are a consequence of the chemical interaction of rock with water near the Earth's surface. Several types of gems come from *pegmatites,* particularly coarse-grained rocks formed by the solidification of steamy melt.

Most gems used in jewelry are "cut" stones, meaning that they are not raw crystals right from the ground, but rather have been faceted. The smooth **facets** on a gem are ground and polished surfaces made with a faceting machine **(Fig. 3.12b)**. Facets are not the natural crystal faces of the mineral, nor are they cleavage planes, though gem cutters sometimes make the facets parallel to cleavage directions and will try to break a large gemstone into smaller pieces by splitting it on a cleavage plane. A faceting machine consists of a doping arm, a device that holds a stone in a specific orientation, and a lap, a rotating disk covered with a wet paste of grinding powder and water. The gem cutter fixes a gemstone to the end of the doping arm and positions the arm so that it holds the stone against the moving lap. The movement of the lap grinds a facet. After completing a facet, the gem cutter rotates the arm by a specific angle, lowers the stone, and grinds another facet. The geometry of the facets defines the cut of the stone. Different cuts have names, such as brilliant, French, star, and pear. Grinding facets is a lot of work—a typical engagement-ring diamond with a brilliant cut has 57 facets **(Fig. 3.12c)**!

> **Did you ever wonder . . .**
> how jewelers make the facets on a jewel?

TAKE-HOME MESSAGE

Gemstones are particularly rare and beautiful minerals. The gems or jewels found in jewelry have been faceted using a lap—the facets are not natural crystal faces or cleavage surfaces. The fire of a jewel comes from the way it reflects light internally.

QUICK QUESTION What's the difference between a facet on a gem and a crystal face?

Chapter 3 Review

CHAPTER SUMMARY

> Minerals are naturally occurring, solid substances, formed by geologic processes, with a definable chemical composition and an internal structure characterized by an orderly arrangement of atoms, ions, or molecules in a crystalline lattice. Most minerals are inorganic.

> Biogenic minerals, such as those in clam shells, are produced by organisms through a process called biomineralization.

> In the crystalline lattice of minerals, atoms occur in a specific pattern—one of nature's finest examples of ordering.

> Minerals can form by solidification of a melt, precipitation from a water solution, diffusion through a solid, metabolism of organisms, or precipitation from a gas.

> Mineralogists have identified about 4,000 different types of minerals. Each has a name and distinctive physical properties (such as color, streak, luster, hardness, specific gravity, crystal habit, cleavage, magnetism, and reactivity with acid).

> The unique physical properties of a mineral reflect its chemical composition and crystal structure. By observing physical properties, you can identify minerals.

> The most convenient way to classify minerals is to group them according to their chemical composition. Mineral classes include silicates, oxides, sulfides, sulfates, halides, carbonates, and native metals.

> Silicate minerals are the most common minerals on the Earth. The silicon-oxygen tetrahedron, a silicon atom surrounded by four oxygen atoms, serves as the fundamental building block of silicate minerals.

> Groups of silicate minerals are distinguished from each other by the ways in which the silicon-oxygen tetrahedra that constitute them are linked.

> Gemstones are minerals known for their beauty and rarity. The facets on cut gems used in jewelry are made by grinding and polishing the stones with a faceting machine.

GUIDE TERMS

biogenic mineral (p. 86)
color (p. 94)
crystal (p. 88)
crystal face (p. 88)
crystal habit (p. 96)
crystal structure (p. 86)

crystalline material (p. 86)
facet (p. 103)
gem (p. 102)
gemstone (p. 102)
glass (p. 88)
hardness (p. 94)

luster (p. 94)
mineral (p. 86)
mineral classes (p. 97)
mineralogy (p. 85)
Mohs hardness scale (p. 96)
polymorph (p. 89)

silicate (p. 98)
silicon-oxygen tetrahedron (p. 98)
specific gravity (p. 96)
streak (p. 94)

 GEOTOURS *THIS CHAPTER'S GEOTOURS WORKSHEET (C) FEATURES QUESTIONS AND GOOGLE EARTH SITES ON:*

> Rare Earth elements   > Physical properties of minerals   > Diamond mines and conflict diamonds   > Mineral reactions after coal mining

REVIEW QUESTIONS

The letters following each Review Question refer to the corresponding Learning Objective from the Chapter Opener.

1. What is a mineral, as geologists understand the term? How is this definition different from the everyday usage of the word? **(A)**

2. Why is glass not a mineral? **(A)**

3. Salt is a mineral, but the plastic making up an inexpensive pen is not. Why not? **(A)**

4. Describe several ways that mineral crystals can form. **(B)**

5. Why do some minerals occur as euhedral crystals, whereas others occur as anhedral grains? Are the crystals shown in this image anhedral or euhedral? **(B)**

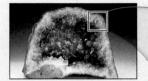

6. List and define the principal physical properties used to identify a mineral. Which minerals react with acid to produce CO_2? **(E)**

7. How can you determine the hardness of a mineral? What is the Mohs hardness scale? **(E)**

8. How do you distinguish cleavage surfaces from crystal faces on a mineral? How does each type of surface form? **(E)**

9. What is the prime characteristic that geologists use to separate minerals into classes? **(C)**

10. What is a silicon-oxygen tetrahedron? What is the anionic group that occurs in carbonate minerals? **(C)**

11. On what basis do mineralogists organize silicate minerals into distinct groups? **(D)**

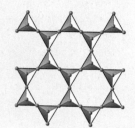

12. Which type of silicate mineral structure does the diagram show? **(D)**

13. What is the relationship between the way in which silicon-oxygen tetrahedra bond in micas and the characteristic cleavage of micas? **(E)**

14. Why are some minerals considered gemstones? How do you make the facets on a gem? **(F)**

ON FURTHER THOUGHT

15. Compare the chemical formula of magnetite with that of biotite. Considering that iron is a relatively heavy element, which mineral has the greater specific gravity? **(E)**

16. Imagine that you find two milky white crystals, each about 2 cm across. One consists of plagioclase and the other of quartz. How can you determine which is which? **(E)**

17. Could you use crushed calcite to grind facets on a diamond? Why or why not? **(E)**

18. Could a diamond precipitate out of a salt lake? **(F)**

ONLINE RESOURCES

 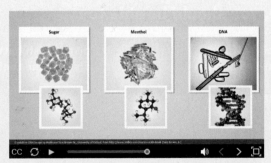 

Animations This chapter features animations on the formation, classification, and composition of various types of minerals.

Smartwork5 This chapter includes visual matching and labeling exercises designed to help students better understand crystal structure, how crystals grow, and the physical characteristics of minerals.

INTERLUDE A

INTRODUCING ROCKS

By the end of this interlude, you should be able to . . .

A. provide a geologic definition of rock.

B. explain the basis that geologists use to classify rocks into three groups.

C. recognize and describe key characteristics used to describe rocks.

D. distinguish between clastic and crystalline textures.

E. discuss the tools that can be used to study rocks.

A.1 **Introduction**

During the 1849 gold rush in the Sierra Nevada of California, only a few lucky individuals actually became rich. The rest of the "forty-niners" either slunk home in debt or took up less-glamorous jobs in new boom towns such as San Francisco. These towns grew rapidly, and soon people from the American west coast were demanding large quantities of manufactured goods from east-coast factories. Making the goods was not a problem, but getting them to California meant either a stormy voyage around the southern tip of South America or a trek with stubborn mule teams through the deserts of Nevada and Utah. The time was ripe to build a railroad linking the east and west coasts of North America, so with much fanfare, a consortium of companies set to work in 1863. The Union Pacific surveyed a route that crossed the Sierra Nevada, and as the Civil War raged, the company transported thousands of Chinese laborers across the Pacific and set them to work chipping ledges around, and blasting tunnels through, the range's towering peaks. Sadly, untold numbers of laborers died of frostbite and exhaustion or from mistimed blasts, landslides, and avalanches.

Through their efforts, the railroad laborers certainly gained an intimate knowledge of how rock feels and behaves—it's solid, heavy, and hard! They also found that some rocks break easily into layers but others do not, and that some rocks are dark-colored while others are light-colored. Like anyone who looks closely at rock exposures, they realized that rocks are not just gray, featureless masses, but rather come in a great variety of colors, textures, and configurations.

Why are there so many distinct types of rocks? The answer is simple: rocks can form in many different ways and from many different materials. Because of the relationship between rock types and the process of formation, rocks provide a historical record of geologic events, and they give insight into interactions among components of the Earth System. We devote the next few chapters to a discussion of rocks and a description of how rocks form. To provide a general introduction to these chapters, this interlude provides the geologic definition of the term *rock*, describes the basic components of rock, and characterizes the three principal classes or groups of rocks. We also describe a few of the methods that geologists use to study rocks.

◀ (facing page) A cliff of weathered limestone rises from the shore of New Caledonia. Limestone is one of many rock types making up the Earth's crust.

A.2 **What Is Rock?**

To geologists, a **rock** is a coherent, naturally occurring solid that consists of an aggregate of minerals or, less commonly, a body of glass. Let's take this definition apart to understand its components.

> *Coherent:* A rock holds together, so it must be broken to be separated into smaller pieces. As a result of its coherence, rock can form cliffs or can be carved into sculptures. A pile of unattached mineral grains does not constitute a rock.
> *Naturally occurring:* Geologists consider only naturally occurring materials to be rocks. Manufactured materials, such as concrete and brick, are not rocks.
> *An aggregate of minerals or a body of glass:* The vast majority of rocks consist of an aggregate (a collection) of many mineral grains or crystals, attached to one another. Some rocks contain only one kind of mineral, whereas others contain several different kinds. A few rock types consist of glass.

What holds rock together? Grains in rock stick together to form a coherent mass either because they are bonded by natural **cement**, mineral material that precipitates from water and fills the space between grains **(Fig. A.1a)**, or because they interlock with one another like pieces of a jigsaw puzzle **(Fig. A.1b)**. Rocks whose grains are stuck together by cement are called **clastic rocks**, whereas rocks whose crystals interlock with one another are called **crystalline rocks**. A glassy rock can be coherent either because it originated as a continuous mass (that is, it does not contain separate grains) or because it formed when separate glass grains welded together while still hot.

At the surface of the Earth, rock occurs either as broken chunks (pebbles, cobbles, or boulders; see Chapter 6) that have moved by falling down a slope or by being transported in ice, water, or wind, or as **bedrock**, meaning rock that remains attached to the Earth's crust. Geologists refer to an exposure of bedrock as an **outcrop**. An outcrop may be a rounded knob out in a field, a ledge forming a cliff or ridge, a stream cut (where running water has cut down into bedrock), a road cut, a rail cut, or the wall of an excavation **(Fig. A.2)**. To inhabitants of cities, forests, or farmland, outcrops of bedrock may be unfamiliar, since bedrock may be completely covered by vegetation, loose sand and gravel, water, asphalt, concrete, or buildings. Outcrops are particularly rare in regions such as the midwestern United States, where, during the past 2 million years, melting ice-age glaciers buried bedrock under thick deposits of debris (see Chapter 18).

FIGURE A.1 Rocks, aggregates of mineral grains and/or crystals, can be clastic or crystalline.

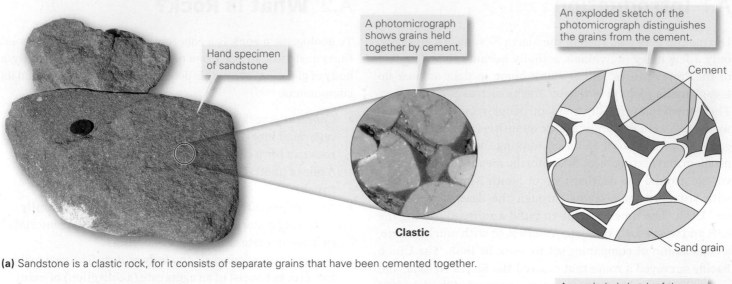

Hand specimen of sandstone

A photomicrograph shows grains held together by cement.

An exploded sketch of the photomicrograph distinguishes the grains from the cement.

Cement

Clastic

Sand grain

(a) Sandstone is a clastic rock, for it consists of separate grains that have been cemented together.

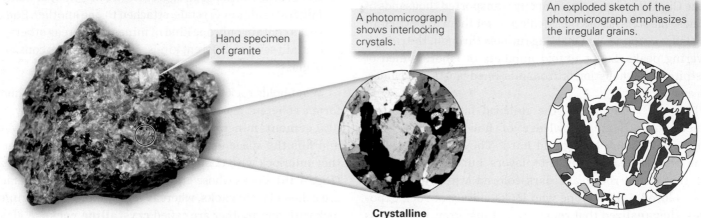

Hand specimen of granite

A photomicrograph shows interlocking crystals.

An exploded sketch of the photomicrograph emphasizes the irregular grains.

Crystalline

(b) Granite is a crystalline rock, for it consists of interlocking crystals that grew together.

A.3 The Basis of Rock Classification

Beginning in the 18th century, geologists struggled to develop a sensible way to classify rocks, for like miners, they realized that not all rocks are the same. Classification schemes help us to organize information and to remember significant details about materials. Also, they help us to recognize similarities and differences among materials. By the end of the 18th century, most geologists had accepted a *genetic scheme* for classifying rocks, based on an interpretation of the origin (genesis) of rocks. Using this approach, which we still use today, geologists recognize three basic groups: (1) **igneous rocks**, which form by the freezing (solidification) of molten rock **(Fig. A.3a)**; (2) **sedimentary rocks**, which form either by the cementing together of fragments (grains) that had

broken off pre-existing rocks, or by the precipitation of mineral crystals out of water solutions at or near the Earth's surface **(Fig. A.3b)**; and (3) **metamorphic rocks**, which form when pre-existing rocks change character in response to a change in temperature, pressure, or chemical environment **(Fig. A.3c)**. Metamorphic change occurs in the solid state, which means that it does not require melting. In the context of modern plate tectonics theory, different rock types form in different geologic settings, as we will discuss in succeeding chapters **(Fig. A.4)**.

Each of the three groups contains many individual rock types, distinguished from one another by physical characteristics, such as the following:

> *Grain size and shape:* The dimensions of individual *grains* (here used to mean either fragments or crystals) vary greatly. Some grains are so small that they can't be seen without a microscope, whereas others are as big as a car or larger. Grain shape also helps identify rock type; in some

FIGURE A.2 Examples of rock exposures.

(a) Large cliffs form the face of this outcrop in arid New Mexico.

(b) A small outcrop, mostly hidden by trees, in Illinois.

(c) A stream cut in New York where water stripped away soil and vegetation.

(d) To produce a level grade for a highway in Maryland, engineers excavated this road cut.

rocks grains are **equant** (meaning they have the same dimensions in all directions), whereas in others the grains are **inequant** (meaning the dimensions are not the same in all directions) **(Fig. A.5)**.

> *Composition:* In a broad sense, rock is a mass of chemicals. The term **rock composition** refers to the proportions of chemicals that make up the rock. The proportions of chemicals, in turn, affect the proportions of different minerals that make up the rock.

> *Texture:* This term refers to the arrangement of grains in a rock, that is, the way grains connect to one another and whether or not inequant grains are aligned parallel to each other. The concept of rock texture will become easier to grasp as we look at different examples of rocks in the following chapters.

> *Layering:* Some rock bodies appear to contain distinct layering, defined either by bands of different compositions or textures or by the alignment of inequant grains so that they trend parallel to each other. Different types of layering occur in different kinds of rocks. For example, the layering in sedimentary rocks is called **bedding (Fig. A.6a)**, whereas the layering in metamorphic rocks is called **metamorphic foliation (Fig. A.6b)**.

Each distinct rock type has a name. Some names reflect the dominant mineral making up the rock, some are derived from the name of the region where the rock was first discovered or occurs in abundance, some come from ancient legends, some from a root word of Latin or Greek origin, and some from a traditional name used by people in an area where the rock can be found. Many rock names date back to antiquity, but some were assigned only in recent decades. In this book we will introduce only about 30 out of hundreds of rock names.

FIGURE A.3 Examples of the three major classes of rocks.

(a) Lava (molten rock that has reached the Earth's surface) freezes quickly to form igneous rock. Here, the molten tip of a brand-new flow still glows red. Older flows are already solid.

(b) Sand, formed from grains eroded from these rock cliffs, collects on the beach. If buried and turned to rock, it becomes layers of sandstone, like that making up the cliffs.

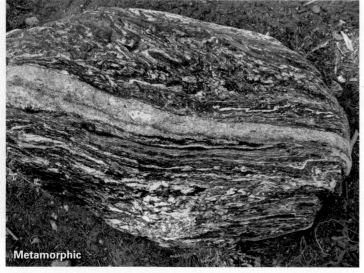

(c) Metamorphic rock forms when pre-existing rocks endure changes in temperature and pressure and/or are subjected to shearing, stretching, or squashing.

A.4 **Studying Rock**

OUTCROP OBSERVATIONS

The study of a rock begins by observing the rock in an outcrop. If the outcrop is big enough, such an examination will reveal relationships between the rock you're interested in and the rocks around it and will allow you to detect layering. Geologists carefully record observations about an outcrop, and then, using a *rock hammer*, may break off a **hand specimen**, a fist-sized piece, that they can examine more closely with a magnifying glass or *hand lens* **(Fig. A.7)**. Observation with a hand lens enables geologists to identify sand-sized or larger grains, and may enable them to describe the rock's texture.

FIGURE A.4 A cross section illustrating various geologic settings in which rocks form.

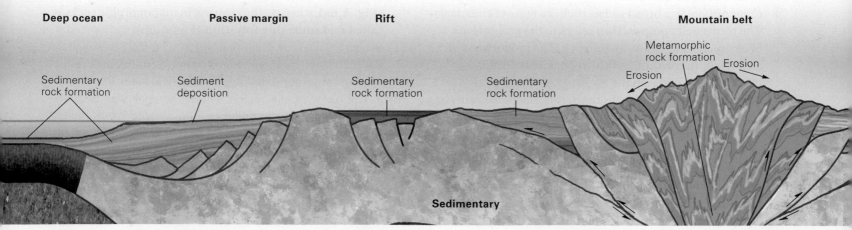

FIGURE A.5 Describing grains in rock.

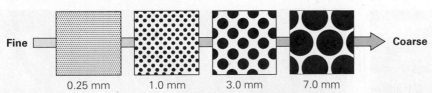

(a) Geologists define grain size by using this comparison chart.

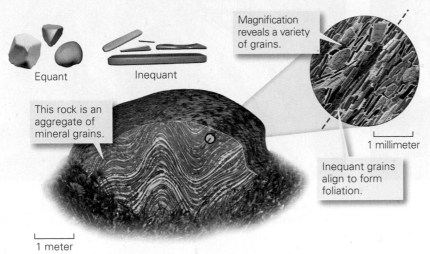

Equant

Inequant

Magnification reveals a variety of grains.

This rock is an aggregate of mineral grains.

1 millimeter

Inequant grains align to form foliation.

1 meter

(b) Grains in rock come in a variety of shapes. Some are equant, whereas some are inequant. In this example of metamorphic rock, inequant grains align to define a foliation.

THIN-SECTION STUDY

Geologists often must examine rock composition and texture in minute detail in order to identify a rock and develop a hypothesis for how it formed. To do this, they take a specimen back to the lab and cut it with a rock saw to make a small rectangular block, which is glued to a glass slide. The block is then ground away until all that remains is a very thin slice (about 0.03 mm thick, the thickness of a human hair) mounted on the glass slide. They then study this **thin section** with a petrographic microscope (*petro* comes from the Greek word for rock). A *petrographic microscope* differs from an ordinary microscope in that it illuminates the thin section with transmitted polarized light. This means that the illuminating light beam first passes through a special polarizing filter that makes all the light waves in the beam vibrate in the same plane. Then the light passes up through the thin section and finally through another polarizing filter. An observer, therefore, looks through the thin section as if it were a window. When illuminated with transmitted polarized light, and viewed through two polarizing filters, each type of mineral grain displays a unique suite of colors **(Fig. A.8)**. The specific color the observer sees depends on both the identity of the grain and its orientation with respect to the waves of polarized light.

The brilliant colors and strange shapes in a thin section, viewed in polarized light, rival the beauty of an abstract painting or stained glass.

By examining a thin section with a petrographic microscope, geologists can identify most of the minerals that make up the rock and can describe the ways in which grains connect to each other. To convey the formation visible in the thin section, they can attach a camera to the microscope eyepiece to take a **photomicrograph**.

HIGH-TECH ANALYTICAL EQUIPMENT

Beginning in the 1950s, high-tech electronic instruments became available that enabled geologists to examine rocks on an even finer scale than is possible with a petrographic mi-

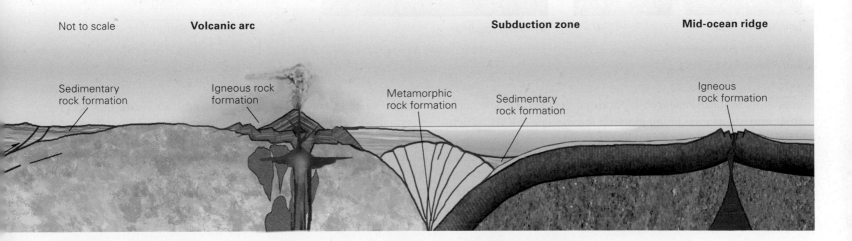

Not to scale

Volcanic arc

Subduction zone

Mid-ocean ridge

Sedimentary rock formation

Igneous rock formation

Metamorphic rock formation

Sedimentary rock formation

Igneous rock formation

FIGURE A.6 Layering in rock.

(a) Bedding in sedimentary rock, here defined by alternating layers of coarser and finer grains, as exposed on a cliff in Utah.

(b) Foliation in this outcrop of metamorphic rock near Mecca, California, is defined by alternating light and dark layers. The color of a layer depends on the minerals that make up that layer.

FIGURE A.7 Studying rocks in the field.

(a) A rock hammer.

(b) A hand specimen.

FIGURE A.8 Photomicrograph of a thin section in polarized light.

500 μm

croscope. Modern research laboratories typically boast such instruments as a *scanning electron microscope* (SEM), which can image the surface of a rock chip at extremely high magnification and can map the distribution of elements in the chip; an *electron microprobe*, which can focus a beam of electrons on a small part of a grain to produce a signal that defines the chemical composition of the mineral; a *mass spectrometer*, which analyzes the proportions of atoms with different atomic weights contained in a rock; and an *X-ray diffractometer*, which identifies minerals by measuring how X-ray beams interact with crystals. Such instruments, in conjunction with optical examination, can provide geologists with highly detailed characterizations of rocks. This information can in turn help them understand how the rocks formed and where they came from, so the study of rocks serves as a basis for deciphering Earth history.

Interlude A Review

INTERLUDE SUMMARY

> Rock is a coherent, naturally occurring solid, consisting of an aggregate of minerals or of a body of glass. Nonglassy rocks can be classified as crystalline or clastic.

> Bedrock consists of rock that remains connected to the underlying crust. Outcrops, exposures of bedrock at the Earth's surface, can be natural or human-made.

> Geologists classify a rock as igneous, sedimentary, or metamorphic based on how the rock formed.

> A variety of characteristics prove helpful in describing rocks. Examples include: grain size, shape, composition, texture, and the nature of layering.

> Geologists make thin sections, very thin slices of rock, to study details of rock texture.

> Hand lenses, microscopes (after making thin sections), and sophisticated electronic equipment help geologists interpret the origin of rocks.

GUIDE TERMS

bedding (p. 109)	equant (p. 109)	metamorphic rock (p. 108)	sedimentary rock (p. 108)
bedrock (p. 107)	hand specimen (p. 110)	outcrop (p. 107)	thin section (p. 111)
cement (p. 107)	igneous rock (p. 108)	photomicrograph (p. 111)	
clastic rock (p. 107)	inequant (p. 109)	rock (p. 107)	
crystalline rock (p. 107)	metamorphic foliation (p. 109)	rock composition (p. 109)	

REVIEW QUESTIONS

The letters following each Review Question refer to the corresponding Learning Objective from the Chapter Opener.

1. How do geologists define the term *rock*? Can a brick be considered a rock? Explain your answer. **(A)**

2. Explain the difference between a clastic and a crystalline rock. **(D)**

3. Give examples of different kinds of rock outcrops. Can you find outcrops everywhere? Explain your answer. **(A)**

4. On what basis do geologists define rocks into three classes? What are these classes? **(B)**

5. Distinguish between an equant and an inequant grain. **(C)**

6. Give two examples of types of layering that can occur in rock. Which type appears in the photo? **(C)**

7. What are thin sections, how are they examined, and what do they allow you to see? **(E)**

8. Name examples of high-tech equipment that can be used to study rocks. What extra information can geologists learn by using such equipment? **(E)**

ONLINE RESOURCES

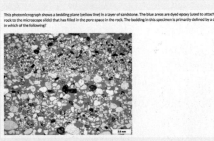

Animations This interlude features animations on distinguishing rock groups and conducting rock analysis.

Smartwork5 This interlude features visual identification questions on rock classifications and the three main rock groups.

CHAPTER 4

UP FROM THE INFERNO: MAGMA AND IGNEOUS ROCKS

By the end of this chapter, you should be able to . . .

A. distinguish between magma and lava, and explain the chemical distinctions among different types of molten rock.

B. describe the special places in the Earth where magma forms.

C. explain why melt moves to locations where it solidifies, and how solidification takes place.

D. describe and classify different kinds of igneous rocks.

E. discuss where and why igneous activity happens, in the context of plate tectonics theory.

> Granite—it seems inevitable to begin with granite, even though so many people have ended with it, lying under those glossy pinkish slabs labeled in gold or black. . . .
>
> JACQUETTA HAWKES (British archaeologist and writer, 1910–1996)

4.1 Introduction

Every now and then, incandescent molten rock (or *melt*) fountains or spills out of the ground on the Big Island of Hawaii, which hosts volcanoes **(Fig. 4.1a)**. Formally defined, a **volcano** is a *vent* or opening from which melt that originates inside the Earth emerges onto the planet's surface, or rises into the air during an episode called a *volcanic eruption*. The word volcano also applies to the hill or mountain built from the products of an eruption. In May 2018 in Hawaii, a particularly intense eruption began. The melt that fountained from the ground produced molten rivers—hundreds of meters across and up to 15 m thick—that eventually destroyed hundreds of homes.

Geologists refer to melt underground as **magma**, and melt that has emerged at the surface as **lava**. At a volcano, some lava pools around the vent, while some moves downslope as a syrupy red-yellow stream called a *lava flow* **(Fig. 4.1b)**. Near the vent, lava from a Hawaiian volcano has a temperature of about 1,150°C, and it can move swiftly, cascading over escarpments at speeds of 10 to 60 km per hour. At the base of the volcano, the lava flow slows but continues to advance, engulfing roads, houses, or vegetation in its path **(Fig. 4.1c)**. As lava cools, its surface darkens and forms a hardened rind that occasionally cracks open to reveal the hot, sticky mass oozing within. Finally, the lava flow stops moving entirely and, within days or weeks, cools through and through. As a result, the once red-hot melt becomes a hard, dark-gray rock **(Fig. 4.1d)**. Not all magma makes it to the surface—some solidifies underground. We refer to any rock formed by the solidifying (freezing) of a melt as an **igneous rock**. Considering the fiery heat of the melt from which igneous rocks solidify, their name—from the Latin *ignis*, meaning fire—makes sense. Igneous rocks make up the entire oceanic crust and much of the continental crust, so you can find them exposed at many places on the Earth's surface.

It may seem strange to speak of "freezing" in regard to forming rock. Most people think of freezing as a change from liquid water to solid ice at a temperature below 0°C (32°F). Nevertheless, the transformation of liquid melt to solid igneous rock represents the same phenomenon, solidification of a liquid. Freezing of molten rock takes place at considerably higher temperatures than does freezing of water. Igneous rocks freeze at temperatures between 650°C and 1,100°C, depending on their chemical makeup. To put such temperatures in perspective, keep in mind that a home oven attains a maximum temperature of only 260°C (500°F).

Geologists distinguish between two main categories of igneous rock, based on where it solidified **(Fig. 4.2a)**. Rock that forms from lava that freezes above ground, in contact with air or water, after it erupts or *extrudes*, is **extrusive igneous rock (Fig. 4.2b)**. Examples include rock that solidified within a lava flow, as well as rock made from cemented-together fragments of **pyroclastic debris** (from the Greek word *pyro*, meaning fire) that blasted out of a volcano and into the air. Pyroclastic debris occurs in many sizes, from very fine particles, called **volcanic ash**, to coarser chunks. The spectacle of a volcano erupting may give the impression that igneous rock forms exclusively at the Earth's surface. In fact, a vastly greater volume of igneous rock forms by solidification of magma underground, after it has pushed its way, or *intruded*, into pre-existing *wall rock*. Geologists refer to the rock formed by solidification of magma as **intrusive igneous rock (Fig. 4.2c)**.

A great variety of igneous rocks exist on the Earth. To understand why, we first discuss the process of forming magma. Next, we examine phenomena that cause magma to rise from depth, factors that determine how easily it flows, and conditions under which magma or lava freezes. Finally, we consider the scheme that geologists use to classify igneous rocks. We will see that, by examining features of an igneous rock, we can characterize the context in which the rock formed.

4.2 Why Does Magma Form, and What Is It Made Of?

The popular image that the crust, the Earth's outer shell, floats on a sea of molten rock is simply not correct. In fact, geologists knew by the end of the 19th century that the Earth's interior was mostly solid, for if it weren't, spinning on its axis would flatten the Earth more than it does (see Chapter 1). Yet the occurrence of both extrusive and intrusive igneous rocks of many ages, and of erupting

> **Did you ever wonder . . .**
> whether the Earth's crust floats on a magma sea?

◀ (facing page) The "Elephant Rocks" of Missouri are boulders of reddish granite that formed from magma that oozed into the crust almost 1.5 billion years ago.

FIGURE 4.1 Formation and evolution of lava flows.

Smoke comes from burning vegetation.

STOP

(a) A view of Hawaii, looking north. Recent lava flows are darker than other areas.

(c) At a distance from the vent, the lava has completely crusted over with new rock, but the interior of the flow remains molten.

Lava fountain

Active lava flow

Solidified lava flow

(b) In this helicopter view from June 2018, 100-m-high lava fountains feed a river of new lava that flows on top of an already solid lava flow from a few days earlier. The lava covered forest, grassland, and buildings.

(d) Eventually, the flow cools completely and becomes a layer of new rock. This flow engulfed a road on Hawaii.

volcanoes today, means that melting to produce magma does take place in some locations. Let's consider the processes that cause melting and the nature of magma produced by melting.

CAUSES OF MELTING

What is the source of the heat that can cause magma to form? As we discussed in Chapter 1, much of this heat is a relict of our planet's formation. As the proto-Earth grew, sources of heat included the compression of mass into a smaller volume,

the sinking of iron to form the core, the impact of meteorites, and the decay of radioactive elements. The Earth remains hot inside, even after 4.56 billion years, because heat produced by decaying radioactive elements in the crust, as well as heat released by the core as it slowly solidifies, replaces much of the heat lost to space by radiation at the Earth's surface. But even though the Earth's interior continues to be very hot, most of the crust and the mantle remain in solid form. The immense pressure produced by the weight of overlying rock prevents molecules from moving freely relative to one another, which

FIGURE 4.2 The intrusive and extrusive realms.

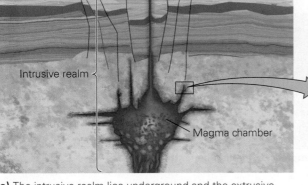

(a) The intrusive realm lies underground and the extrusive realm lies above ground. Lava flows, as well as various types of explosive eruptions, all produce extrusive rocks.

(b) Extrusive rocks include lava flows and pyroclastic layers.

(c) An intrusion of basalt (dark rock) cuts across an earlier intrusion of granite (light rock).

would allow rock to turn into liquid. Magma, therefore, forms only in special places, where conditions trigger melting of pre-existing solid rock. Below, we describe these conditions, and briefly note the geologic settings, in the context of plate tectonics, where melting takes place. We'll wait until the end of this chapter to associate specific rock types with each setting.

Melting Due to Decompression Because pressure prevents melting, even in a very hot rock, a decrease in pressure can trigger melting as long as the rock remains hot **(Fig. 4.3a)**. This process, called *decompression melting*, takes place where hot mantle rock rises slowly. As the rock moves up, its pressure becomes less (due to the decrease in overburden) while its temperature remains nearly unchanged (rock acts as an excellent insulator). As we'll see, upward movement causes decompression melting in mantle plumes, beneath rifts, and beneath mid-ocean ridges **(Fig. 4.3b)**.

Melting Due to Addition of Volatiles Magma also forms at locations where volatiles mix with hot mantle rock. Recall that *volatiles* are substances, such as water (H_2O) and carbon dioxide (CO_2), that evaporate relatively easily. When volatiles mix with hot, dry rock, they react with minerals and break chemical bonds, so that the rock begins to melt,

a process known as *flux melting* **(Fig. 4.4a)**. In other words, adding volatiles decreases a rock's melting temperature. As we'll see, volatiles seep into hot asthenosphere in the region just above subducting oceanic lithosphere, causing volcanism at convergent plate boundaries.

Melting Due to Heat Transfer When very hot magma from the mantle rises into the crust, the heat it brings raises the temperature of the surrounding crustal rock. In some cases, the rise in temperature may be sufficient to cause the crustal rock to begin melting, a process called *heat-transfer melting*. To picture the process, imagine injecting hot fudge into ice cream—the fudge transfers heat to the ice cream, raises its temperature, and causes it to melt **(Fig. 4.4b)**.

THE MAJOR TYPES OF MELT

All molten rocks (magma or lava) contain *silica*, a compound of silicon and oxygen (SiO_2). They also contain varying proportions of other elements, including aluminum (Al), calcium (Ca), sodium (Na), potassium (K), iron (Fe), and magnesium (Mg). Because molten rock is a liquid, its molecules do not lie in an orderly crystalline lattice, but rather they occur in clusters or chains that can move with respect to one another.

FIGURE 4.3 Decompression melting.

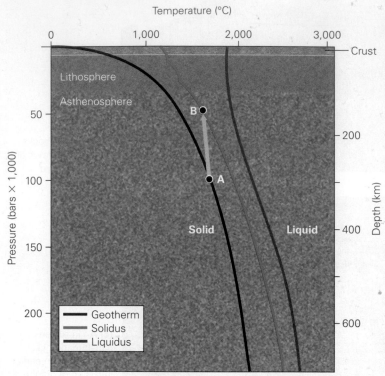

(a) Decompression takes place when the pressure acting on hot rock decreases. As this graph of pressure and temperature conditions in the Earth shows, when rock rises from point A to point B, the pressure decreases a lot, but the rock cools only a little, so the rock begins to melt.

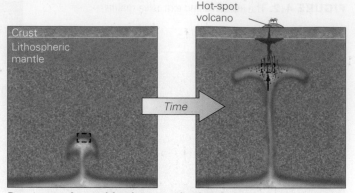

Decompression melting in a mantle plume

Decompression melting beneath a rift

Decompression melting beneath a mid-ocean ridge

(b) The conditions leading to decompression melting occur in several different geologic environments. In each case, a volume of hot asthenosphere (outlined by dashed lines) rises to a shallower depth, and magma (red dots) forms.

Geologists distinguish between "dry" melts, which do not contain volatiles, and "wet" melts, which do. Wet melts include up to 15% dissolved volatiles, including water, carbon dioxide, nitrogen (N_2), hydrogen (H_2), and sulfur dioxide (SO_2). These volatiles come out of the Earth at volcanoes in the form of gas. Usually, water makes up about half of the gas erupting at a volcano. Thus, molten rock contains not only the molecules that constitute solid minerals in rocks, but also the molecules that become water and air.

Molten rocks differ from one another in terms of the proportions of chemicals that they contain **(Table 4.1)**. Geologists distinguish among four major compositional types depending, overall, on the proportion of silica (SiO_2) relative to the sum of magnesium oxide (MgO) and iron oxide (FeO or Fe_2O_3) in the melt. *Mafic melts* contain a relatively high proportion of magnesium oxide and iron oxide compared to silica—the *ma–* in the word stands for magnesium, and the *–fic* comes from the Latin word for iron. Ultramafic melts have an even higher proportion of magnesium oxide and iron oxide, relative to silica. *Felsic (or silicic) melts* have a fairly high proportion of silica, compared to magnesium oxide and iron oxide. *Intermediate melts* get their name because their composition is partway between that of mafic and felsic melts.

TABLE 4.1 The Four Categories of Magma

Magma Type	Weight Percent of Silica*
Felsic (or silicic)	66–76%
Intermediate	52–66%
Mafic	45–52%
Ultramafic	38–45%

*Weight percent means the proportion of the magma's weight that consists of silica (SiO_2).

Why do melts of so many compositions form in the Earth? Several factors play a role:

> *Source-rock composition:* The composition of a melt reflects the composition of the solid from which it was derived. Not all melts form from the same source rock, so not all melts have the same composition.

FIGURE 4.4 Flux melting and heat-transfer melting.

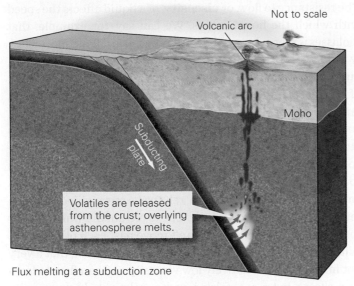

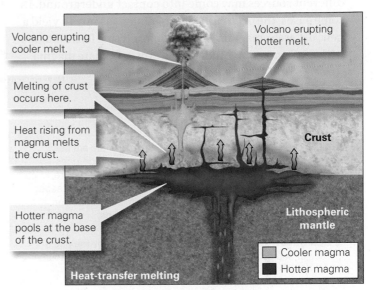

(a) When volatiles enter hot mantle rock above a subducting plate, flux melting takes place.

(b) When rising magma brings heat up with it, heat-transfer melting can take place in the overlying or surrounding rock.

> *Partial melting:* Under the temperature and pressure conditions that occur in the Earth, only 2% to 30% of a source rock can melt to produce magma at a given location. The temperatures at sites of magma production simply never get high enough to melt the entire source rock, and magma tends to migrate away from the site of melting before all of the original rock has melted. Geologists refer to the process by which only part of an original rock melts to produce magma as **partial melting (Fig. 4.5a)**. Magmas formed by partial melting are more felsic than the source rock from which they were derived because more silica enters the liquid, as melting begins, than remains behind in the still-solid source. For example, partial melting of an ultramafic rock produces a mafic magma.

> *Assimilation*: As magma sits underground before solidifying completely, it may incorporate chemicals dissolved from the wall rock or from blocks that detached from the wall and sank into the magma **(Fig. 4.5b)**. This process is called contamination or **assimilation**.

FIGURE 4.5 Phenomena that can affect the composition of magma.

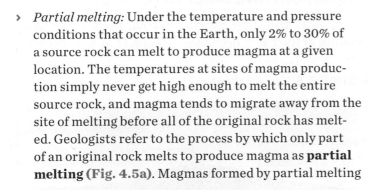

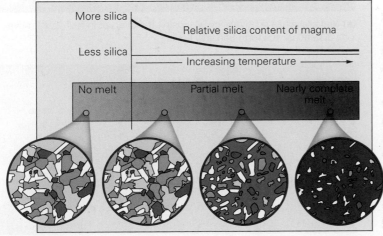

(a) Partial melting: The first-formed melt will be richer in silica than the original rock. As melting continues, magma becomes increasingly mafic.

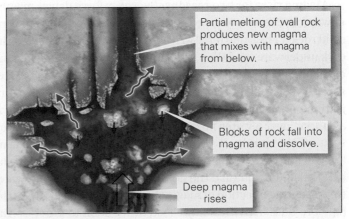

(b) Assimilation: Heat provided by magma partially melts wall rock; the new magma may then mix with the original magma. Also, blocks of wall rock can dissolve in the original magma, and that wall rock may chemically react with the magma.

› *Magma mixing*: Magmas formed in different locations from different sources may come into contact underground. In some cases, the originally distinct magmas mix to yield a new, different magma. For example, mixing felsic magma with mafic magma could produce intermediate magma.

TAKE-HOME MESSAGE

Though the Earth is hot inside, the crust and mantle are solid, except in special places where pre-existing rock undergoes melting. Melting can be triggered by a decrease in pressure, addition of volatiles, and/or injection of hot magma from deeper below. Geologists classify magma based on its composition, specifically, the proportion of silica that it contains.

QUICK QUESTION Why doesn't a magma formed by melting a particular source rock have the same composition as the source rock?

4.3 Movement and Solidification of Molten Rock

WHY DOES MAGMA RISE?

If magma stayed put once it formed, new igneous rocks would not develop in or on the crust. But it doesn't stay put—magma tends to move upward, away from the site of melting. In some cases it reaches the Earth's surface and erupts at a volcano. Movement of magma serves an important role in the Earth System, because materials from deeper parts of the Earth flow upward to provide the raw material from which new rocks, the atmosphere, and oceans form.

Magma rises for two reasons. *Buoyancy* force drives magma upward because molten rock is less dense than surrounding solid rock. (The same force drives a cork upward, if you hold the cork underwater and then let it go.) Pressure within the magma also contributes to driving magma upward. This *magma pressure* develops due to the weight of overlying rock and to the injection of new magma. To picture why, imagine that you place a balloon over a water nozzle, fill the balloon to capacity, then turn off the nozzle. If you squeeze the balloon (representing the weight of overlying rock), you increase the pressure in the balloon, and if you turn the nozzle on again to add more water (representing injection of magma), you increase pressure in the balloon. Both activities may cause the balloon to burst, so the water gets pushed out into the surrounding area. Similarly, if magma pressure increases, magma gets pushed into overlying crust, or even to the Earth's surface.

WHAT CONTROLS THE SPEED OF FLOW?

The resistance to flow, or **viscosity**, of a liquid affects the speed with which the liquid moves. We can say, for example, that molasses is more viscous than water because it flows more slowly than water. All molten rock is more viscous than molasses, but not all molten rock has the same viscosity. The viscosity of a given melt depends on its temperature, volatile content, and silica content. Specifically, hotter melt is less viscous than cooler melt, because thermal energy breaks bonds and allows atoms or molecules to move more easily. A wet melt is less viscous than a dry melt, because reactions with volatiles also tend to break apart silicate molecules. Finally, mafic melt is less viscous than felsic melt because relatively more silicon-oxygen tetrahedra occur in felsic melt. These tetrahedra link together to make long molecules that can't move past each other easily. With such relationships in mind, it's not surprising that a very hot mafic lava has relatively low viscosity and can flow to form thin sheets, but a cool felsic lava has relatively high viscosity and clumps up into a bulbous mound (**Fig. 4.6**).

TRANSFORMING MELT INTO ROCK

When melts rise, they eventually cool, because the temperature of the Earth decreases upward. If magma becomes

FIGURE 4.6 Viscosity affects lava behavior.

(a) Mafic lava has relatively low viscosity. It can erupt in fountains, move long distances, and form thin lava flows.

(b) Felsic to intermediate lava is very viscous. When it erupts, it may form a mound-like lava dome around the volcano's vent.

trapped underground as an intrusion, it slowly loses heat to the surrounding wall rock, drops below its freezing temperature, and solidifies underground. If magma reaches the Earth's surface and extrudes as lava, it cools because it comes in contact with much cooler air or water.

The time it takes for a magma to cool depends on how fast it can transfer heat into its surroundings. To see why, think about the process of cooling coffee. If you spill coffee on a table, it cools quickly because it loses heat directly to the cold air. For the same reason, lava in an extrusive environment cools relatively quickly. In contrast, if you pour hot coffee into an insulated thermos bottle and seal it, the coffee stays hot for hours, because insulation slows the transfer of heat to the air outside. Like a thermos bottle, wall rock acts as insulation, so magma in an intrusive environment cools slowly.

Not all magma trapped in the intrusive realm cools at the same rate. Three factors control the cooling time of such magma:

> *The depth of intrusion:* Magma intruded deep in the crust, where hot wall rock surrounds it, cools more slowly than does magma intruded into cold wall rock near the ground surface.

> *The shape and size of a magma body:* Heat escapes from magma at an intrusion's surface, so the greater the surface area for a given volume of intrusion, the faster it cools. As a result, a body of magma shaped like a pancake cools faster than one shaped like a melon **(Fig. 4.7a, b)**. And because the ratio of surface area to volume increases as size decreases, a body of magma the size of a car cools faster than one the size of a ship.

> *The presence of circulating groundwater:* Water passing through magma absorbs and carries away heat, much like the coolant that flows around an automobile engine. The presence of groundwater in wall rock, therefore, can accelerate cooling.

CHANGES IN MOLTEN ROCK DURING COOLING: FRACTIONAL CRYSTALLIZATION

If you cool a tray of water to a temperature of 0°C, crystals of ice start to grow, and if you keep the temperature cold enough for long enough, the water in the tray becomes water ice, a crystalline solid composed entirely of H_2O. Molten rock can freeze, too, but many different minerals form, because unlike pure water, a melt contains many different chemicals.

FIGURE 4.7 Factors that affect the freezing of molten rock.

Faster cooling	Slower cooling
Effect of size — For a given shape, a smaller volume cools faster.	
Effect of shape — For a given volume, a pancake shape—with its greater surface area—cools faster.	

(a) The shape and size of an intrusion affect the ratio of surface area to volume.

Small drops cool very quickly.

A thin flow cools quickly.

Cooler wall rock

Increasing temperature

Warmer wall rock

A large blob cools slowly.

A shallow sheet cools faster than a deep sheet.

(b) The cooling rate of molten rock depends on the size and shape of the magma or lava body, and on its depth.

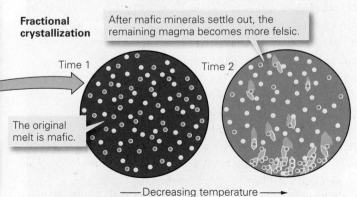

Fractional crystallization

After mafic minerals settle out, the remaining magma becomes more felsic.

Time 1   Time 2

The original melt is mafic.

← Decreasing temperature →

(c) The process of fractional crystallization results in a progressive change in magma composition during freezing. Blue dots represent mafic components, and yellow dots represent felsic components.

Early-formed minerals tend to be relatively mafic, so their growth preferentially removes iron and magnesium from the magma. As a consequence, the remaining magma becomes more felsic **(Fig. 4.7c)**. This process of sequential crystal formation, as cooling takes place, is known as **fractional crystallization**. If an originally mafic magma freezes before much fractional crystallization has occurred, a mafic igneous rock forms. Freezing of the magma that remains after much fractional crystallization has occurred, however, yields a felsic igneous rock. **Box 4.1** provides further details about the process.

WHERE DO MELTS COOL?

Magma rising into the crust may accumulate to form a relatively large volume, known as a **magma chamber**,

underground. Typically, magma chambers do not contain melt only. Rather, they hold a "mush" consisting of solid, but hot, crystals mixed with melt. Very high magma pressures can develop in a magma chamber, and these pressures can push magma into underground cracks, sometimes all the way to the surface. Sometimes magma solidifies into rock within the magma chamber, sometimes it solidifies in underground cracks that tap into the magma chamber, and sometimes it cools at the surface. Let's consider these igneous rock-forming settings more closely.

Extrusive Igneous Settings Different volcanoes extrude molten rock in different ways. Some volcanoes erupt streams of low-viscosity lava that flood down the flanks of the volcano and then cover broad swaths of the countryside. When this

FIGURE 4.8 Examples of eruptions and extrusive volcanic materials.

(a) This volcano is producing lava flows and fountains.

(b) A stack of over 50 thin lava flows, capped by debris, is visible inside Mt. Vesuvius, Italy.

(c) This volcanic explosion produced two styles of ash eruption.

(d) A close-up of a layer of ash containing lapilli.

FIGURE 4.9 Igneous sills and dikes, examples of tabular intrusions.

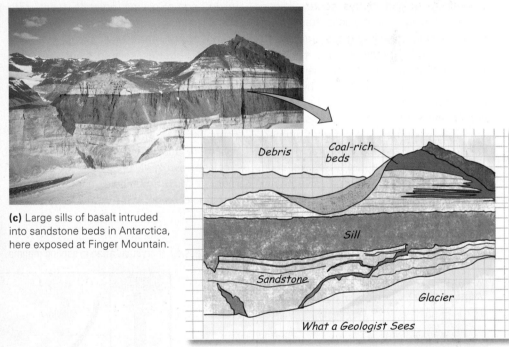

Dike cuts across layers.

Sill pushes between layers.

Layers of sandstone

Intrusive contact

If all the sandstone were removed, the intrusions would look like this (before erosion).

(a) Dikes cut across pre-existing layering. Sills are parallel to pre-existing layering.

(b) A wall-like intrusion cutting into pre-existing igneous rock is also called a dike.

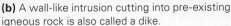

(c) Large sills of basalt intruded into sandstone beds in Antarctica, here exposed at Finger Mountain.

Debris   Coal-rich beds

Sill

Sandstone

Glacier

What a Geologist Sees

lava freezes, it forms a relatively thin lava flow. Such flows may cool in days to months. In contrast, some volcanoes erupt viscous masses of lava that pile into rubbly domes. And still others erupt explosively, sending clouds of pyroclastic debris soaring skyward, and/or avalanches of ash tumbling down the sides of the volcano. Which type of eruption occurs depends largely on a magma's composition and volatile content. Mafic lavas tend to have low viscosity and spread in broad, thin flows **(Fig. 4.8a, b)**. Volatile-rich felsic lavas tend to erupt explosively and form thick ash and debris deposits **(Fig. 4.8c, d)**. Chapter 5 describes the products of extrusive eruptions, and the causes for their differences, in more detail.

Intrusive Igneous Settings Geologists distinguish among different types of intrusions on the basis of their shape. *Tabular intrusions*, or *sheet intrusions*, are roughly planar and have

a fairly uniform thickness. They form when magma injects into cracks underground. Most are from centimeters to tens of meters thick, and tens of meters to tens of kilometers long. A **dike** is a tabular intrusion that cuts across pre-existing layering (bedding or foliation), whereas a **sill** is a tabular intrusion that injects between layers, and, therefore, parallels layering **(Fig. 4.9)**. In places where tabular intrusions cut across rock that does not have layering, a nearly vertical, wall-shaped tabular intrusion is called a dike, and a nearly horizontal, tabletop-shaped tabular intrusion is called a sill. When an intrusion starts to inject between layers but then domes upward, it yields a blister-shaped intrusion known as a **laccolith**. **Plutons** are blob-shaped intrusions that range in size from tens of meters across to tens of kilometers across **(Fig. 4.10a, b)**. They typically form when magma in a magma chamber solidifies. Intrusion of numerous plutons in a region

BOX 4.1 CONSIDER THIS . . .

Bowen's Reaction Series

In the 1920s, Norman L. Bowen began a series of laboratory experiments designed to determine the sequence in which silicate minerals crystallize from a melt. First, Bowen melted powdered mafic igneous rock by raising its temperature to about 1,280°C. Next, he cooled the melt just enough to cause part of it to solidify. Then he quenched the remaining melt by submerging it quickly in cold mercury. *Quenching*, which in this context means sudden cooling to form a solid, transformed the remaining liquid into glass, trapping the earlier-formed crystals within it. Bowen made a thin section of the sample and identified the mineral crystals by using a microscope, and he analyzed the chemical composition of the resulting glass.

After experiments at different temperatures, Bowen was able to characterize the sequence of mineral-producing reactions that took place during progressive cooling of a mafic magma (**Fig. Bx4.1a**). Geologists now refer to this sequence as **Bowen's reaction series,** in his honor.

Let's examine the sequence more closely. In a cooling mafic melt, olivine and calcium-rich plagioclase form first. The plagioclase reacts with the remaining melt, and new plagioclase, containing more sodium (Na), grows. This new plagioclase may replace or grow around the earlier-formed plagioclase. Meanwhile, some olivine crystals react with the remaining melt to produce pyroxene. Some of the early olivine and pyroxene crystals become isolated from the melt, effectively extracting iron and magnesium, so the remaining melt becomes progressively enriched with silica.

As the melt continues to cool, plagioclase continues to form. Later-formed plagioclase has more sodium than earlier-formed plagioclase, pyroxene crystals react with the melt to form amphibole, and then some amphibole reacts with the remaining melt to form biotite. All the while, some newly formed crystals become isolated and don't exchange atoms with the melt—some crystals may, in fact, sink and accumulate at the bottom of the melt. As more crystals form, the remaining melt becomes progressively more felsic. Finally, at temperatures between 650°C and 850°C, only about 10% of the melt remains, and this melt has a high silica content, so when the last melt finally freezes, only quartz, potassium feldspar (K-feldspar), and muscovite can form from it.

Note that the reaction series has two tracks. The *discontinuous reaction series* refers to the sequence of olivine, pyroxene, amphibole, biotite, K-feldspar/muscovite/quartz, in that each step yields a different kind of silicate mineral. The *continuous reaction series* refers to the sequence from calcium-rich to sodium-rich plagioclase, in that each step yields a different version of the same mineral (**Fig. Bx4.1b**).

FIGURE Bx4.1 Bowen's reaction series indicates the succession of crystallization in cooling magma.

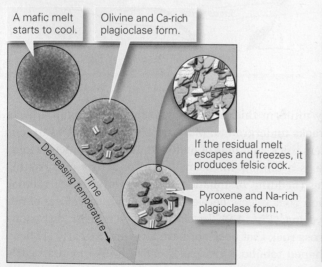

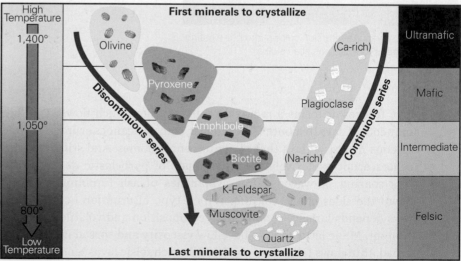

(a) With decreasing temperature, fractional crystallization begins and the composition of the remaining magma becomes more felsic.

(b) This chart displays the discontinuous and continuous reaction series. Rocks formed from minerals at the top of the series are mafic, whereas rocks formed from the bottom of the series are felsic.

FIGURE 4.10 Igneous plutons are blob-shaped intrusions.

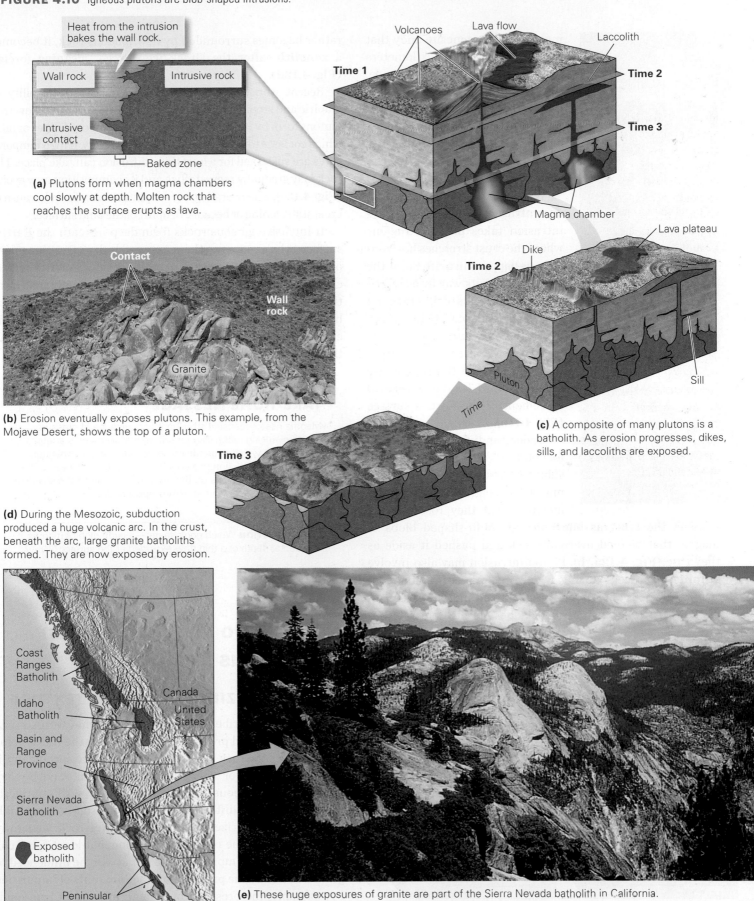

Heat from the intrusion bakes the wall rock.

Wall rock

Intrusive rock

Intrusive contact

Baked zone

(a) Plutons form when magma chambers cool slowly at depth. Molten rock that reaches the surface erupts as lava.

Volcanoes

Lava flow

Laccolith

Time 1

Time 2

Time 3

Magma chamber

Contact

Wall rock

Granite

(b) Erosion eventually exposes plutons. This example, from the Mojave Desert, shows the top of a pluton.

Lava plateau

Dike

Time 2

Pluton

Sill

(c) A composite of many plutons is a batholith. As erosion progresses, dikes, sills, and laccoliths are exposed.

Time

Time 3

(d) During the Mesozoic, subduction produced a huge volcanic arc. In the crust, beneath the arc, large granite batholiths formed. They are now exposed by erosion.

Coast Ranges Batholith

Canada

Idaho Batholith

United States

Basin and Range Province

Sierra Nevada Batholith

Exposed batholith

Peninsular Batholith

Present day

(e) These huge exposures of granite are part of the Sierra Nevada batholith in California.

makes a vast composite body that may be hundreds of kilometers long and tens of kilometers wide. The resulting immense mass of igneous rock is called a **batholith** (**Fig. 4.10c–e**). The boundary between wall rock and any igneous intrusion defines an *intrusive contact*.

Where does the space for igneous intrusions come from? Dike intrusion takes place in regions where the crust stretches horizontally, as happens in a rift. So, as the magma forces its way into a vertical crack, the walls of the crack can move apart sideways (**Fig. 4.11a**). Sill intrusion occurs near the surface of the Earth, so pressure from magma can push the rock above the sill upward, leading to vertical movement of the Earth's surface (**Fig. 4.11b**).

Understanding how the space for plutons develops remains a subject of research. Some plutons may have originated as *diapirs*, meaning that they rose upward through the crust as buoyant, light-bulb-shaped blobs of magma that pierced overlying rock and pushed it aside as they rose (**Fig. 4.12a, b**). Pluton intrusion may also involve **stoping**, a process during which magma assimilates wall rock, and blocks of wall rock break off and sink into the magma (**Fig. 4.12c**). If a stoped block does not melt entirely, but

rather becomes surrounded by new igneous rock, it becomes a **xenolith**, after the Greek word *xeno*, meaning foreign (**Fig. 4.12d**).

Recent work has questioned the general applicability of diapiric and stoping models, leading to the alternative view that plutons form by intrusion of several superimposed dikes or sills, which coalesce to become a single, massive body. If high temperatures are sustained for a long time, diffusion can take place. The rock may gradually recrystallize and its composition may evolve (**Fig. 4.12e**). The overall space for plutons may form because of crustal stretching or because of uplift of the land surface.

If intrusive igneous rocks form deep beneath the Earth's surface, why can we see them exposed today? Over long periods of geologic time, mountain building slowly uplifts belts of crust. Erosion by water, wind, and ice can gradually strip away the thick, overlying rock and expose the intrusive rock that has formed below. Some intrusive rocks exposed in mountain cliffs today solidified kilometers to tens of kilometers below the surface.

TAKE-HOME MESSAGE

Magma rises because it's buoyant and because of pressure due to overlying rocks. The rate of melt movement is affected by viscosity, which depends on composition and temperature. When molten rock enters a cooler environment, it freezes. The rate of cooling depends on the environment and on the shape of the magma body. Different names apply to different shapes of intrusions.

QUICK QUESTION Which cools faster—a large blob of magma intruded at depth or a thin flow extruded at the surface? Why?

4.4 How Do You Describe an Igneous Rock?

CHARACTERIZING COLOR AND TEXTURE

If you had to describe a rock to a friend, what words might you use? You would probably start by noting the rock's color. Overall, does the rock look dark or light? More specifically, is it gray, pink, white, or black? Describing color may not be easy, because some igneous rocks contain many visible mineral grains, each with a different color; but even so, you'll probably be able to characterize the overall hue of the rock. Generally, the color reflects the rock's composition, but identifying color isn't always so simple, because it may also be influenced by grain size and by the presence of trace amounts of impurities. For example, the presence of a small amount of iron oxide gives rock a reddish tint.

FIGURE 4.11 Making room for igneous dikes and sills.

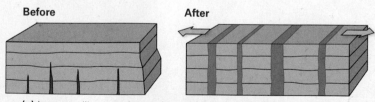

(a) Igneous dikes can form where the crust is stretched sideways.

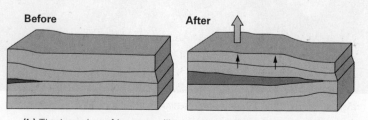

(b) The intrusion of igneous sills pushes the Earth's surface upward.

FIGURE 4.12 Making room for igneous plutons.

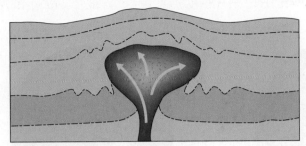

(a) During diapiric rise, a blob-shaped mass of magma forces its way up through wall rock, which plastically deforms to move out of the way.

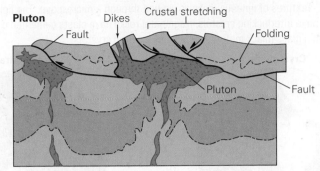

(b) Faulting associated with crustal stretching may accommodate the emplacement of a diapir.

Xenolith

Time

(c) Magma pushes up into cracks and breaks off blocks of wall rock. The blocks may be incorporated in the melt by stoping.

The white rock (granite) is intruding into the dark wall rock.

Xenolith

(d) A xenolith of darker rock surrounded by light-colored granite.

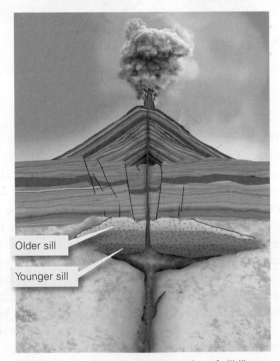

Older sill

Younger sill

(e) Plutons may intrude as a succession of sill-like sheets, whose composition evolves.

Next, you would probably characterize the rock's texture. A description of *igneous texture* specifies whether the rock consists of interlocking crystals, stuck-together fragments, or solid glass. If the rock consists of crystals or fragments, a description of texture also specifies the grain size. Here are the common terms for defining texture:

> *Crystalline texture:* When a melt solidifies, minerals in some rocks grow and interlock like pieces of a jigsaw puzzle. Geologists refer to such rocks as **crystalline igneous rocks (Fig. 4.13a).** The crystals interlock because the rock does not solidify instantly. Rather, different crystals grow at different rates and at different

FIGURE 4.13 Textures of igneous rocks as viewed through a microscope. The field of view is about 3 mm. Crystalline rocks have interlocking crystals, fragmental rocks have clasts cemented or welded together, and glassy rocks contain glass (black material) as well as isolated crystals.

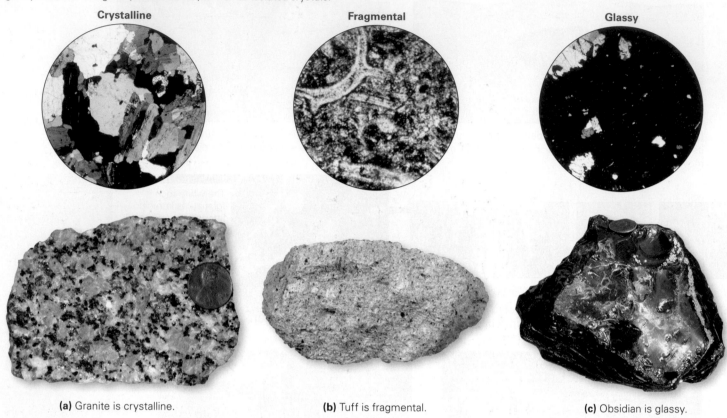

(a) Granite is crystalline.　　　**(b)** Tuff is fragmental.　　　**(c)** Obsidian is glassy.

times and, as the crystals grow, they interfere with each other. For example, a faster-growing crystal may surround a slower-growing crystal partly or even entirely, and later-forming crystals fill in gaps between earlier-forming crystals. Subcategories of crystalline igneous rocks are based on the size of the crystals. Coarse-grained (*phaneritic*) rocks have crystals large enough to identify with the naked eye. The crystals of fine-grained (*aphanitic*) rocks are too small to identify with the naked eye. *Porphyritic rocks* have larger crystals (called *phenocrysts*) surrounded by a mass of fine crystals (called *groundmass*).

› *Fragmental texture:* **Fragmental igneous rocks** form from pyroclastic debris and consist of chunks or shards that are packed together, welded together, or cemented together after they have solidified **(Fig. 4.13b)**.

› *Glassy texture:* Rocks made of a solid mass of glass or of tiny crystals surrounded by glass are **glassy igneous rocks (Fig. 4.13c)**. Glassy rocks typically fracture conchoidally (fractures are curved, like a clam shell).

The texture of a nonfragmental igneous rock largely reflects its cooling rate. Why? Recall from Chapter 3 that mineral crystals grow when atoms diffuse (move) through melts and attach to crystal seeds. At high temperatures, seeds constantly form and dissolve because heat causes diffusion to occur rapidly. Only some seeds are "successful" enough to grow into small crystals, and even most of these end up dissolving in the melt again. Relatively few crystals grow large enough to survive until the melt cools to a level at which crystals stop dissolving. If a melt cools very rapidly, it can become solid before seeds have grown into crystals, so the resulting rock has a glassy texture. If the melt cools rapidly, but not rapidly enough to form glass, it will develop a fine-grained (aphanitic) texture. Many crystals in the resulting rocks will have grown from seeds, but none of those crystals will have had enough time to grow large. If a melt cools slowly, successful crystals have time to grow large before the rock solidifies completely. As the large crystals are growing, new seeds and small crystals do form, but because the melt remains hot, they tend to dissolve into the melt again before they can grow large.

FIGURE 4.14 Crystalline igneous rocks are classified based on composition and texture.

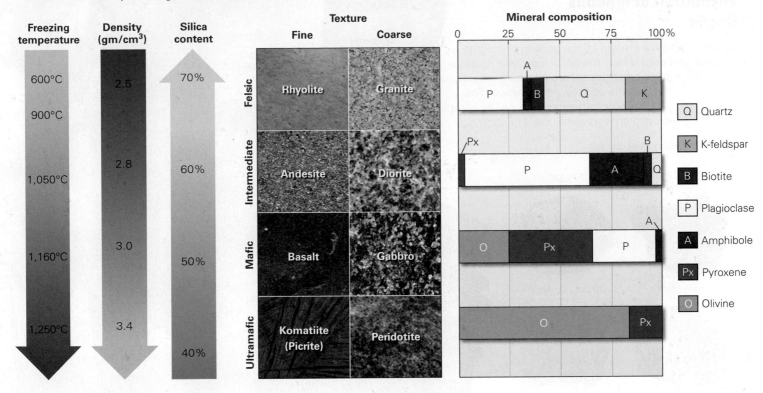

Because of the relationship between cooling rate and texture, lava flows, dikes, and sills tend to be composed of fine-grained rock. In contrast, plutons tend to be composed of coarse-grained rock. Plutons that intrude into hot wall rock at great depth cool particularly slowly and tend to have larger crystals than plutons that intrude into cool wall rock at shallow depth. Porphyritic rocks form when a melt cools in two stages. First, the melt cools at depth slowly enough for phenocrysts to form. Then the melt erupts, and the remainder cools quickly, so fine-grained groundmass forms around the phenocrysts.

There is, however, an exception to the standard cooling rate-grain size relationship. A very coarse-grained igneous rock called *pegmatite* doesn't form by slow cooling. Pegmatite, which can contain crystals up to tens of centimeters across, typically occurs in dikes. Because dikes generally cool relatively quickly, the coarseness of this rock may seem surprising. Researchers have shown that pegmatite becomes coarse because it forms from water-rich melts in which atoms can diffuse so rapidly that large crystals can grow very quickly.

CLASSIFYING IGNEOUS ROCKS

Melts have a variety of compositions and can freeze to form igneous rocks in many different environments above and below the surface of the Earth. We can classify igneous rocks according to their texture and composition. Studying a rock's texture tells us about the rate at which it cooled, as we've seen, and therefore about the environment in which it formed (**Geology at a Glance**, p. 130). The rock's composition tells us about the original source of the magma and the way in which the melt evolved before finally solidifying. Below, we introduce some of the more important igneous rock types.

Did you ever wonder . . .
how the black glass once used for arrowheads formed?

Types of Crystalline Igneous Rocks The scheme for classifying the principal types of crystalline igneous rocks is fairly simple. We distinguish among the different compositional classes—*ultramafic, mafic, intermediate,* or *felsic*—on the basis of silica content, and we distinguish among the different textural classes according to whether the grains are coarse or fine (**Fig. 4.14**). As indicated by the center section of Figure 4.14, contrasts in chemical composition mean that different rock types contain different groups of minerals. Note that *basalt, andesite,* and *rhyolite* could have come from the same magmas as *gabbro, diorite,* and *granite,* respectively. But the three fine-grained rock types cooled quickly

Formation of Igneous Rocks

Igneous rocks are formed by the cooling of magma underground or of lava at the Earth's surface. Igneous rocks that solidify underground are intrusive, whereas those that solidify at the surface are extrusive. The type of igneous rock that forms depends on the composition of the melt and on the environment of cooling.

Stratified volcanic tuff

Increasing silica content

MAFIC	FELSIC
Scoria (glassy)	Obsidian (glassy)
Basalt (fine grained)	Rhyolite (fine grained)
Gabbro (coarse grained)	Granite (coarse grained)

Fast cooling

Slow cooling

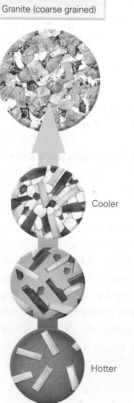

In an extrusive environment, melt may cool quickly, so extrusive rocks tend to be fine-grained or may even have a glassy texture. Melt that explodes into the air forms ash and other pyroclastic debris. Rocks formed from this debris have a fragmental texture.

In an intrusive environment, magma can cool slowly, so larger crystals can grow.

Minerals in an igneous rock crystallize in succession as the melt cools. They interlock to produce a crystalline texture.

Cooler

Hotter

EXTRUSIVE ENVIRONMENT

Lava flow

Pyroclastic flow

Dike swarm

Sills

Lava dome

Ring dikes

Volcanic neck

Laccolith

Irregular stock

INTRUSIVE ENVIRONMENT

Pluton

Magma chamber

FIGURE 4.15 Examples of igneous rocks, arranged by grain size and composition.

Fine grained

Rhyolite

Andesite

Basalt

Coarse grained

Granite

Diorite

Gabbro

Felsic

Mafic

Increasing silica content

Types of Glassy Igneous Rocks Glassy texture develops more commonly in felsic igneous rocks because the high concentration of silica inhibits the easy growth of crystals. But basaltic and intermediate lavas can form glass if they cool rapidly enough. In some cases, a rapidly cooling lava freezes while it still contains a high concentration of gas bubbles—these bubbles remain as open holes known as **vesicles**. Geologists distinguish among several kinds of glassy rocks.

> *Obsidian* is a mass of solid, felsic glass. It tends to be black or brown (see Fig. 4.13c). Because it breaks conchoidally, sharp-edged pieces split off its surface when you hit a sample with a hammer. Pre-industrial people worldwide used such pieces for arrowheads, scrapers, and knife blades.
> *Tachylite* is a relatively rare, vesicle-free mafic glass.
> *Pumice* is a felsic volcanic rock that contains abundant (75% to 90% of the rock's volume) tiny vesicles, each surrounded by a thin screen of glass. With far more open space than solid glass, it can look like a sponge, and some specimens can actually float on water, like Styrofoam **(Fig. 4.16a)**. Pumice forms from quickly cooling, volatile-rich frothy lava.
> *Scoria* is a mafic volcanic rock with numerous vesicles (more than about 30%). Generally, the bubbles in scoria are bigger than those in pumice, and the rock looks darker, overall **(Fig. 4.16b)**.

in lava flows or near-surface dikes and sills, while the three coarse-grained rock types cooled more slowly, in plutons. As a rough guide, the color of an igneous rock indicates its composition—mafic rocks tend to be black or dark gray, intermediate rocks tend to be lighter gray or greenish gray, and felsic rocks tend to be light tan to pink or maroon **(Fig. 4.15)**.

FIGURE 4.16 Some igneous rocks contain an abundance of vesicles, gas bubbles that were frozen into the rock as it cooled.

(a) Pumice is a felsic glassy rock with tiny vesicles.

(b) Scoria is a mafic glassy rock with many vesicles.

Types of Pyroclastic Igneous Rocks When volcanoes erupt explosively, they spew out clots or droplets of lava, as well as glass shards (the broken-up walls of vesicles in pumice), larger fragments of pumice, and other broken-up chunks of recently formed igneous rocks. As we've noted, geologists refer to accumulations of such fragments as pyroclastic debris. When pyroclastic debris becomes consolidated into a solid mass, due either to still-hot clasts welding together during accumulation, or to cementation by minerals precipitating from groundwater long after accumulation, it becomes a **pyroclastic rock**. Pyroclastic rocks have a fragmental texture, and we can distinguish among rock types based on grain size. *Tuff*, for example, consists mostly of volcanic ash; larger fragments of pumice may be mixed in with the ash (see Figs. 4.8 and 4.13b).

TAKE-HOME MESSAGE

Geologists divide igneous rocks into three general categories, based on texture—crystalline rocks have interlocking crystals, glassy rocks have no crystalline structure, and pyroclastic rocks have a fragmental texture. Crystalline rocks can be classified based on composition and on whether they are fine- or coarse-grained. The color of a rock may provide a clue to its composition.

QUICK QUESTION What kind of rock forms from felsic magma cooled in a large pluton at depth?

4.5 Plate-Tectonic Context of Igneous Activity

Earlier in this chapter, we pointed out that melting occurs only in special locations where conditions lead to decompression, addition of volatiles, or heat transfer. The conditions that lead to melting and, therefore, to igneous activity can develop in four geologic settings **(Fig. 4.17)**: (1) at hot spots; (2) along volcanic arcs bordering oceanic trenches; (3) along mid-ocean ridges; and (4) within continental rifts. Let's look more carefully at melting and igneous rock production of these settings, in the context of plate tectonics theory. In Chapter 5, we'll add to the story by discussing the types of volcanic eruptions associated with different settings.

PRODUCTS OF HOT SPOTS

As we learned in Chapter 2, most (but not all) researchers associate hot-spot igneous activity with mantle plumes—columns or streams of hot mantle rock rising from deeper in the mantle. According to the plume hypothesis, the plume itself does not consist of magma. Rather, it's composed of solid rock that is hot enough to be relatively soft so that it can flow plastically, at rates of a few centimeters a year. When the hot rock of a plume reaches the base of the lithosphere, decompression causes partial melting within the hot rock. Partial melting of ultramafic rock (peridotite) generates mafic (basaltic) magma. At oceanic hot spots, mafic magma erupts at the surface and solidifies to form basalt. At continental hot spots, part of the mafic magma erupts to form basalt, but some transfers heat to the continental crust, which itself then partially melts, producing felsic magmas that erupt to yield rhyolite.

PRODUCTS OF SUBDUCTION

A chain of volcanoes, called a **volcanic arc** (or just an *arc*), forms on the overriding plate adjacent to the deep-ocean trenches that mark convergent plate boundaries (see Chapter 2). *Continental volcanic arcs* grow on continental crust in locations where oceanic lithosphere subducts beneath continental lithosphere. *Volcanic island arcs* grow on oceanic crust where one oceanic plate subducts beneath another. Beneath volcanic arcs, a variety of intrusions—plutons, dikes, and sills—develop, to be exposed only later by erosion of the overlying volcanoes. In some localities, arc-related igneous activity produces huge batholiths.

How does subduction trigger melting? Some minerals in oceanic crust rocks contain volatile compounds. At shallow depths, these volatiles are bonded to other elements within mineral crystals. But when subduction carries crust down into the hot asthenosphere, the crust warms up, and at a depth of about 150 km, it becomes so hot that volatiles separate and diffuse up into the overlying hot ultramafic rock (peridotite) of the asthenosphere as molecules of H_2O or, to a lesser extent, CO_2. Addition of these volatiles causes flux melting of peridotite. As we have seen, only part of the original ultramafic rock actually melts, since silica preferentially goes into the melt. So the process yields mafic magma, some of which rises to form basaltic sills and dikes in the crust, and some travels all the way to the surface to extrude as basaltic lava.

In continental volcanic arcs, not all mantle-derived basaltic magma rises directly to the surface. Some gets trapped at the base of the continental crust, and some resides in magma

FIGURE 4.17 The tectonic setting of igneous rocks.

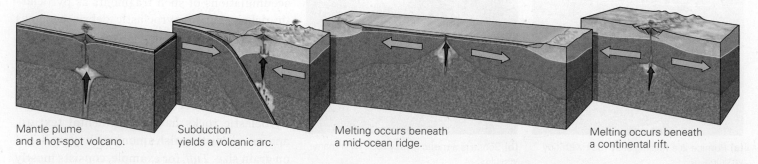

Mantle plume and a hot-spot volcano.

Subduction yields a volcanic arc.

Melting occurs beneath a mid-ocean ridge.

Melting occurs beneath a continental rift.

chambers deep in the crust. Fractional crystallization, as well as assimilation, may cause some of the magma to become progressively more felsic. But in addition, this very hot mantle-derived magma transfers heat into the adjacent continental crust, which causes partial melting of the continental crust. Some of this new melt becomes assimilated in the rising melt, and some rises on its own. And, as the magma cools, it undergoes fractional crystallization. Because much of the continental crust has a mafic to intermediate composition, the resulting magma is intermediate to felsic. As this magma rises, it either cools higher in the crust to form plutons in magma chambers, or it rises to the surface to erupt as lava. These processes (heat-transfer melting, assimilation, and fractional crystallization) contribute to the abundance of granite, rhyolite, diorite, and andesite at continental arcs.

FORMING IGNEOUS ROCKS AT MID-OCEAN RIDGES

The entire oceanic crust, a 7- to 10-km-thick layer of basalt and gabbro that covers 70% of the Earth's surface, forms at mid-ocean ridges (divergent plate boundaries). We can say, therefore, that most igneous rocks present at the Earth's surface are a product of seafloor spreading. Igneous activity happens at mid-ocean ridges because, as seafloor spreading takes place, oceanic lithosphere plates drift away from the ridge. Hot asthenosphere rises to keep the resulting space filled. As asthenosphere rises, it undergoes partial melting due to decompression. Partial melting of mantle peridotite yields mafic

magma which, as noted in Chapter 2, rises into the crust and collects in a magma chamber. Some cools slowly in the magma chamber to form massive gabbro, while some intrudes upward to fill vertical cracks that appear as newly formed crust splits apart (see Fig. 2.23a). Magma that cools in the cracks forms basalt dikes, and magma that rises as far as the seafloor, and extrudes as lava, forms pillow-basalt flows.

FORMING IGNEOUS ROCKS AT RIFTS

Rifts are places where continental lithosphere stretches horizontally. As a result of this stretching, the lithosphere also thins vertically (see Fig. 2.33). As this process takes place, the asthenosphere undergoes decompression. Partial melting due to decompression produces basaltic magma, which rises into the crust. Some magma makes it to the surface and erupts as basalt. However, some becomes trapped in crustal magma chambers and transfers heat to the crust. The resulting partial melting of the crust yields felsic magmas that erupt explosively as rhyolite. As a result, volcanic rocks in a rift generally include lava flows of basalt and layers of rhyolitic tuff.

LARGE IGNEOUS PROVINCES (LIPs)

In many places on the Earth, especially voluminous quantities of mafic magma have erupted or intruded (**Fig. 4.18**). Some of these regions occur along the margins of continents, some in the interior of oceanic plates, and some in the interiors of continents. The largest of these, the Ontong Java

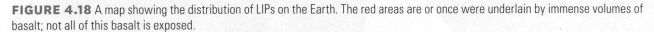

FIGURE 4.18 A map showing the distribution of LIPs on the Earth. The red areas are or once were underlain by immense volumes of basalt; not all of this basalt is exposed.

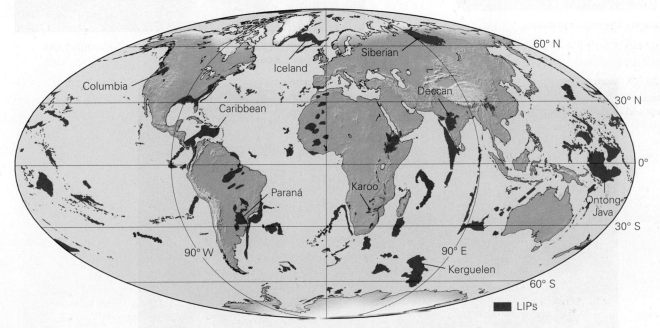

Oceanic Plateau of the western Pacific, covers an area of about 5,000,000 km² of the seafloor and has a volume of about 50,000,000 km³. It's no surprise that such a huge volume of igneous rock is called a **large igneous province (LIP)**. The term LIP has also been applied to huge eruptions of felsic ash.

Mafic LIPs may form when the bulbous head of a mantle plume first reaches the base of the lithosphere. More partial melting can occur in a plume head than in normal asthenosphere, because temperatures are higher in a plume head. Thus, a very large quantity of particularly hot mafic magma forms in the plume head, and when this magma reaches the surface, huge amounts of lava spew out of the ground. If the plume head lies beneath a rift, thinning of the lithosphere causes even more decompression, which in turn leads to even more melting **(Fig. 4.19a)**.

On land, the hot mafic lava that erupts in LIPs has such low viscosity that it can flow tens to hundreds of kilometers across the landscape, forming a vast sheet of basalt. The process may repeat hundreds of times, yielding thick stacks of thin, broad lava flows known as **flood basalts**. Flood basalts make up the bedrock of the Columbia River Plateau in Oregon and Washington **(Fig. 4.19b, c)**, the Paraná Plateau in southeastern Brazil, the Karoo region of southern Africa, and the Deccan region of southwestern India.

TAKE-HOME MESSAGE

The formation of igneous rocks can be understood in the context of plate tectonics. Flux melting happens in the mantle above subducting plates, and decompression melting takes place in the mantle within hot spots and under both mid-ocean ridges and rifts. Melting of the mantle produces basalt. Other types of igneous rocks most commonly form where rising mantle magma interacts with continental crustal rock, and heat-transfer melting takes place.

QUICK QUESTION Which tectonic setting has yielded the most igneous rocks in the past 200 million years?

FIGURE 4.19 Flood basalts form when vast quantities of low-viscosity mafic lava "flood" over the landscape and freeze into a thin sheet. Accumulation of successive flows builds a flat-topped plateau.

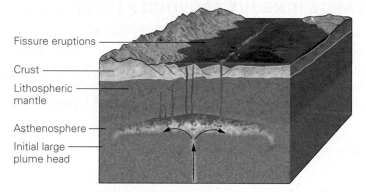

(a) The plume model for forming flood basalts.

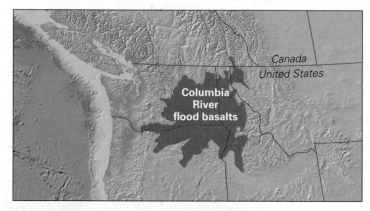

(b) Flood basalts underlie the Columbia Plateau in Washington and Oregon, the dark area of this map.

(c) The flood basalts of the Columbia Plateau form the layers exposed in Palouse Canyon, Washington.

ANOTHER VIEW A cliff of basalt (a thick sill) towers above a group of geology students in a park in Edinburgh, Scotland. After studying this exposure in the late 18th century, James Hutton became the first researcher to understand that these rocks formed from the cooling of an igneous intrusion.

Chapter 4 Review

CHAPTER SUMMARY

> Magma is liquid rock (melt) under the Earth's surface. Lava is melt that has erupted from a volcano at the Earth's surface.

> Magma forms when hot rock partially melts. This process occurs when pressure decreases (decompression), when volatiles (such as water) diffuse into hot rock, and when heat transfers from hot magma into adjacent rock.

> Magma occurs in a range of compositions: felsic (silicic), intermediate, mafic, and ultramafic. Composition depends partly on the original composition of the source rock and partly on the way the magma evolves.

> During partial melting, only part of the source rock melts, so magma tends to be more felsic.

> Magma rises because of its buoyancy and because of pressure within the magma.

> Magma viscosity (resistance to flow) depends on its composition. Felsic magma is more viscous than mafic magma.

> Geologists distinguish between extrusive igneous rocks, formed from lava that erupts from a volcano, and intrusive igneous rocks that develop from magma that freezes inside the Earth.

> Lava may solidify to form flows or domes, or it may explode into the air to form pyroclastic debris.

> Magma may accumulate in magma chambers underground.

> Intrusive igneous rocks form when magma intrudes into pre-existing rock below the Earth's surface. Blob-shaped intrusions are called plutons. Sheet-like intrusions that cut across layering in wall rock are dikes, and sheet-like intrusions that form parallel to layering in wall rock are sills. Huge intrusions composed of numerous plutons are known as batholiths.

> The rate at which intrusive magma cools depends on the depth of intrusion, on the size and shape of the magma body, and on groundwater circulation. Cooling time controls igneous rock texture.

> Lava extruded at the surface cools faster than lava intruded underground.

> Crystalline igneous rocks are classified according to grain size and composition. Glassy igneous rocks are classified according to composition and the presence of gas bubbles. Fragmental igneous rocks are classified by grain size.

> The origin of igneous rocks can be understood in the context of plate tectonics. Magma forms at continental or island volcanic arcs along convergent boundaries due to flux melting. Igneous rocks form at hot spots due to decompression melting or heat transfer. Igneous rocks form at rifts as a result of decompression melting or heat transfer. Igneous rocks form along mid-ocean ridges because of decompression melting.

GUIDE TERMS

assimilation (p. 119)
batholith (p. 126)
Bowen's reaction series (p. 124)
crystalline igneous rock (p. 127)
dike (p. 123)
extrusive igneous rock (p. 115)
flood basalt (p. 134)

fractional crystallization (p. 122)
fragmental igneous rock (p. 128)
glassy igneous rock (p. 128)
igneous rock (p. 115)
intrusive igneous rock (p. 115)
laccolith (p. 123)

large igneous province (LIP) (p. 134)
lava (p. 115)
magma (p. 115)
magma chamber (p. 122)
partial melting (p. 119)
pluton (p. 123)
pyroclastic debris (p. 115)
pyroclastic rock (p. 131)

sill (p. 123)
stoping (p. 126)
vesicle (p. 131)
viscosity (p. 120)
volcanic arc (p. 132)
volcanic ash (p. 115)
volcano (p. 115)
xenolith (p. 126)

 GEOTOURS *THIS CHAPTER'S GEOTOURS WORKSHEET (D) FEATURES QUESTIONS AND GOOGLE EARTH SITES ON:*

> Batholiths and laccoliths   > Dikes

REVIEW QUESTIONS

The letters following each Review Question refer to the corresponding Learning Objective from the Chapter Opener.

1. How does the process of freezing magma resemble that of freezing water? How is it different? **(A)**

2. What sources produce heat inside the Earth? How did the first igneous rocks on the planet form? **(B)**

3. Describe the three processes responsible for the formation of magmas. **(C)**

4. Why can we find so many different types of magmas? **(A)**

5. Why do magmas rise from depth to the surface of the Earth? **(B)**

6. What factors control the viscosity of a melt? **(C)**

7. What factors control the cooling rate of a magma within the crust? **(C)**

8. How does grain size reflect the cooling rate of a magma? Did the pictured rock come from rapidly or slowly cooling magma? **(D)**

9. What does the mixture of grain sizes in a porphyritic igneous rock indicate about its cooling history? **(D)**

10. Why does magma form at a convergent boundary? **(E)**

11. What process in the mantle may be responsible for causing hot-spot volcanoes to form? **(E)**

12. Describe how magmas are produced at continental rifts. **(E)**

13. Why does melting take place beneath the axis of a mid-ocean ridge? **(E)**

14. What is a large igneous province (LIP)? **(E)**

ON FURTHER THOUGHT

15. The Cascade volcanic arc of the northwestern United States is only about 800 km long. The Andean arc of western South America is several thousand kilometers long. Look at a map showing the Earth's plate boundaries, and explain why the Andean arc is so much longer than the Cascade arc. **(E)**

16. Polished rock slabs have long been a popular material for building facades, or outside surfaces, because of their durability. Architects and designers use the generic term "granite" for such rock if it contains silicate minerals, even if the rock does not have the proportions of minerals found in a true granite. Imagine that you're an architect, and your clients want the outside surface of their new building to be made of black, coarse-grained stone. What rock type (using a geologic classification) would you recommend that they use? **(D)**

17. This photograph shows a nearly horizontal layer of basalt that now crops out as a cliff along the Hudson River, across from New York City. It is parallel to layers of sandstone above and below, and is younger than all of those layers. The basalt intruded at 190 Ma, as Pangaea was breaking apart. What is this body of basalt, and in what geologic setting did it form? **(E)**

ONLINE RESOURCES

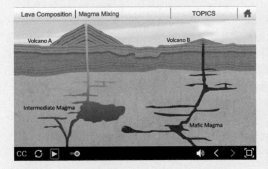

Animations This chapter features animations on partial melting, fractional crystallization, assimilation, and magma mixing.

Videos This chapter features a video on the topic of partial melting.

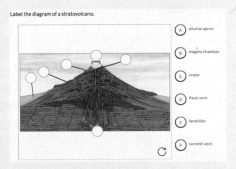

Smartwork5 Features include visual and labeling exercises on the processes involved in magma formation, the creation of various igneous rocks, and volcanic activity.

CHAPTER 5

THE WRATH OF VULCAN: VOLCANIC ERUPTIONS

By the end of this chapter, you should be able to . . .

A. distinguish among the different types of volcanoes.

B. explain how volcanic eruptions produce such a variety of materials, including lava, pyroclastic debris, and gases.

C. explain why some eruptions yield streams of lava while others produce catastrophic explosions.

D. describe how the type of eruption reflects the character of lava and the geologic setting.

E. assess the many hazards that eruptions pose to life and environment.

F. interpret clues to impending eruptions.

> Glowing waves rise and flow, burning all life on their way, and freeze into black, crusty rock which adds to the height of the mountain and builds the land, thereby adding another day to the geologic past. . . . I became a geologist forever, by seeing with my own eyes: the Earth is alive!
>
> HANS CLOOS (German geologist, 1886–1951), on seeing the eruption of Mt. Vesuvius in Italy

5.1 Introduction

Every few hundred years, one of the hills on Vulcano, an island in the Mediterranean Sea off the western coast of Italy, rumbles and spews out molten rock, glassy cinders, and dense "smoke" (actually a mixture of gas, ash, and tiny liquid droplets). Ancient Romans thought that such **eruptions**, episodes when volcanoes extrude lava or pyroclastic debris, happened when Vulcan, the god of fire, fueled his forges beneath the island to manufacture weapons for the other gods. Geologic study suggests, instead, that eruptions take place when hot magma, formed by melting inside the Earth, rises through the crust and emerges at the surface. No one believes the Roman myth anymore, but the island's name evolved into the English word **volcano**, which geologists use to designate either an erupting vent through which molten rock reaches the Earth's surface or the hill or mountain built from the products of eruptions.

On the main peninsula of Italy, another volcano, Mt. Vesuvius, towers over the Bay of Naples. Nearly 2,000 years ago, a prosperous Roman resort and trading town named Pompeii sprawled at the foot of Vesuvius. One morning in 79 C.E., earthquakes signaled the mountain's awakening. At 1:00 P.M. on August 24, a dark mottled cloud, streaked by lightning, billowed up above Mt. Vesuvius's summit. The cloud spread over Pompeii, filling the air with fine ash and choking fumes, and turned day into night **(Fig. 5.1a)**. Blocks and pellets of rock fell like hail, collecting on streets, plazas, and tile roofs—eventually, the heavy rubble began to crush buildings. People frantically rushed to escape, but for most it was too late. Suddenly, a turbulent current of scalding ash and pumice fragments surged down the flank of the volcano and swept over Pompeii. By the next day the town, along with its neighbor, Herculaneum, had vanished beneath a 6-m-thick gray-black blanket of debris. This covering protected the towns' ruins so well that when archaeologists excavated them 1,800 years later, they found an amazingly complete record of Roman daily life **(Fig. 5.1b, c)**. They also discovered strange open spaces in the debris. When they filled the spaces with

plaster and then removed the surrounding ash, the archaeologists realized that they had produced casts of Pompeii's unfortunate inhabitants, their bodies twisted in agony or huddled in despair, at the moment of death **(Fig. 5.1d)**.

Clearly, volcanoes are unpredictable and dangerous. Volcanic activity can build a towering mountain, or it can blast one apart. The materials a volcano produces can provide the fertile soil and mineral deposits that enable a civilization to thrive, or they can provide a rain of destruction that can snuff one out. To characterize volcanic activity, this chapter sets ambitious goals. We first build on Chapter 4 by looking more closely at the products of volcanic eruptions and at the basic components of volcanoes. Then we consider the different kinds of volcanic eruptions on the Earth, and why they occur where they do. Finally, we review the hazards posed by volcanoes, the efforts made by geoscientists to predict eruptions and help minimize the damage they cause, the possible influence of eruptions on climate and civilization, and the nature of volcanoes elsewhere in the Solar System.

5.2 The Products of Volcanic Eruptions

The drama of a volcanic eruption transfers materials from inside the Earth to our planet's surface. Products of an eruption come in three forms—lava flows, pyroclastic debris, and gas. Note that we use the term **lava flow** both for a molten, moving mass of lava and for the solid layer of rock that forms when the lava freezes.

LAVA FLOWS

Sometimes lava races down the side of a volcano like a fast-moving, incandescent stream, sometimes it oozes like a sticky but scalding paste, and sometimes it builds into a rubble-covered mound. Clearly, not all lava behaves in the same way, so not all lava flows look the same. Why?

The character of a lava reflects its *viscosity* (resistance to flow), and lava viscosity, in turn, depends primarily on chemical composition, for silica in lava tends to bond together into large molecules that can't move easily. Therefore, basaltic lava is less viscous than andesitic lava, which in turn is less

◀ (facing page) The drama of an eruption, in this case, at the peak of the Santiaguito Volcano, in Guatemala. Behind a cloud of ash (tiny glass shards mixed with fragmented rock), molten rock, which has risen from the mantle, spurts out explosively from the summit.

FIGURE 5.1 The eruption of Vesuvius buried Pompeii and nearby Herculaneum in 79 C.E.

(a) In this 1817 painting, the British artist J. M. W. Turner depicted the cataclysmic explosion.

(b) Streets and buildings of Pompeii are well preserved.

You can see ruts carved into streets by chariots.

Pre-79 C.E. profile

(c) Excavations exposed the ruins of Pompeii, shown here with Vesuvius in the distance. The dashed line shows the volcano's profile prior to its eruption.

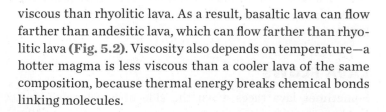

(d) A plaster cast of Pompeii residents who had been buried in ash.

viscous than rhyolitic lava. As a result, basaltic lava can flow farther than andesitic lava, which can flow farther than rhyolitic lava **(Fig. 5.2)**. Viscosity also depends on temperature—a hotter magma is less viscous than a cooler lava of the same composition, because thermal energy breaks chemical bonds linking molecules.

Basaltic Lava Flows Basaltic (mafic) lava has very low viscosity when it first emerges from a volcano, because it contains relatively little silica and a high temperature. Therefore, on the steep slopes near the summit of a volcano, mafic lava can move at speeds up to 30 km per hour **(Fig. 5.3a)**, but it slows down to walking pace after it starts to cool. Most basaltic flows measure less than a few kilometers long, but some extend tens to hundreds of kilometers from the source **(Fig. 5.3b)**.

FIGURE 5.2 The characteristics of a lava flow depend on its viscosity.

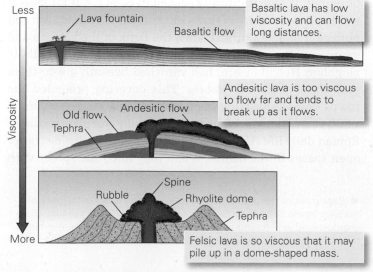

Less

Lava fountain

Basaltic flow

Basaltic lava has low viscosity and can flow long distances.

Viscosity

Old flow
Tephra

Andesitic flow

Andesitic lava is too viscous to flow far and tends to break up as it flows.

Rubble

Spine

Rhyolite dome

Tephra

More

Felsic lava is so viscous that it may pile up in a dome-shaped mass.

FIGURE 5.3 Features of basaltic lava flows. Basaltic lava has low viscosity and thus can flow long distances.

(a) A fast-moving flow coming from Mt. Etna, Sicily.

(b) A basaltic lava flow covers a highway on the Big Island of Hawaii.

(c) In a lava tube, still-molten lava flows beneath a crust of solid basalt.

(d) A drained lava tube exposed in a road cut on Hawaii.

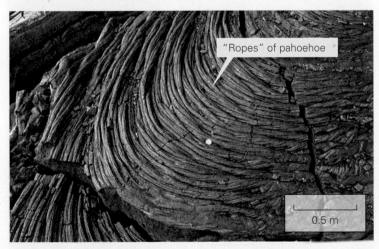

(e) Pahoehoe from a recent lava flow on Hawaii. Note the coin for scale.

(f) The rubbly surface of an a'a' flow, Sunset Crater, Arizona.

FIGURE 5.3 (continued)

(g) Pillow basalt develops when lava erupts underwater. Later uplift may expose pillows above sea level, as in this outcrop in Cyprus.

(h) Columnar jointing develops when the interior of a flow cools and cracks. This example is Devils Postpile in California.

Although all the lava of a flow moves when it first emerges, rapid cooling causes the surface to solidify after the flow has moved away from the source. The surface shell serves as insulation, so the hot interior of the flow can remain liquid and continue to move. As time progresses, more of the interior solidifies. Eventually, molten lava moves only through a tunnel-like passageway, or **lava tube**, within the flow—the largest of these may be tens of meters in diameter **(Fig. 5.3c)**. In some cases, lava tubes drain and eventually become empty tunnels **(Fig. 5.3d)**.

The surface texture of a basaltic lava flow depends, in part, on the timing of the lava's freezing relative to its movement. Basalt flows whose surfaces are warm and pasty wrinkle into smooth, glassy, rope-like ridges—geologists use the Hawaiian word **pahoehoe** (pronounced pa-HOY-hoy) for such flows **(Fig. 5.3e)**. If a moving lava flow becomes too viscous to contort into pahoehoe ropes, the solid surface of the lava breaks up into a jumble of sharp, angular fragments, creating a rubbly flow also called by its Hawaiian name, **a'a'** (pronounced AH-ah) **(Fig. 5.3f)**. Footpaths made by people living in basaltic volcanic regions follow the smooth surface of pahoehoe rather than the foot-slashing surface of a'a'.

Basaltic flows that erupt underwater look very different from those that erupt on land. Lava cools much more quickly in water, so submarine basaltic lava can travel only a short distance before its surface freezes, producing a glass-encrusted blob, or *pillow* **(Fig. 5.3g)**. The glass rind of a pillow momentarily stops the flow's advance, but within

minutes the pressure of the lava squeezing into the pillow breaks the rind, and a new blob of lava squirts out, freezes, and produces another pillow. Geologists refer to a flow made from many of these blobs as **pillow lava**.

Even after a flow stops moving and has become solid, it remains warm. As the last of its heat radiates outward, the rock in a lava flow contracts, because like many solids, rock shrinks as it loses heat. During this final cooling, the shrinkage causes the solid flow to crack, forming natural fractures called *joints*. In some cases, the pattern of jointing outlines polygonal columns with roughly hexagonal cross sections. This type of fracturing is called **columnar jointing** **(Fig. 5.3h)**. Of note, such joints can also form in dikes and sills.

Andesitic and Rhyolitic Lava Flows
Because of its greater viscosity, intermediate-composition (andesitic) lava cannot flow as easily as

FIGURE 5.4 This rhyolite dome formed about 650 years ago in Panum Crater, California. Tephra (cinders) accumulated around the vent.

basaltic lava. When erupted, andesitic lava first forms a mound above the vent. This mound advances relatively slowly down the volcano's flank at about 1 to 5 m a day, in a lumpy flow with a bulbous snout. Typically, andesitic flows are less than a few kilometers long. Because the lava is so viscous, the surface can solidify as it moves, forming angular chunks that may tumble down in advance of the flow. The whole flow, called *blocky lava*, looks like a jumble of rubble. If the flow extrudes onto a steep slope, the blocks may tumble down the side of a volcano while still glowing red.

Rhyolitic lava is the most viscous type of lava because it is the most felsic and the coolest. Therefore, it tends to accumulate either above the vent in a *lava dome* **(Fig. 5.4)**, or in short and bulbous flows rarely more than 1 to 2 km long. Sometimes rhyolitic lava freezes while still in the vent and then pushes upward as a column-like *spire* up to 100 m above the vent. Rhyolitic flows have broken and blocky surfaces.

VOLCANIC GASES AND AEROSOLS

Most magma contains dissolved gases such as water, carbon dioxide, and sulfur dioxide (H_2O, CO_2, and SO_2). In fact, up to 9% of a magma may consist of gaseous components—generally, lavas with more silica contain a greater proportion of gas. Volcanic gases come out of solution when the magma approaches the Earth's surface and pressure decreases, just as bubbles come out of solution in a soda when you pop the bottle top off. Commonly, water emitted by a volcano condenses into a liquid that dissolves the sulfur-bearing gases to form **aerosols** (tiny droplets) of sulfuric acid.

In low-viscosity magma, gas bubbles can rise more quickly than the magma moves, and thus most reach the surface of the magma and enter the atmosphere before the lava does. As a result, for a while, some volcanoes may continue to erupt large quantities of gas, without much lava **(Fig. 5.5a)**. Lava may still contain gas bubbles as it starts to flow. When the lava freezes, the bubbles are trapped in the rock and become holes called **vesicles (Fig. 5.5b)**. In high-viscosity magmas, gas has trouble escaping because bubbles can't push through the sticky lava.

VOLCANICLASTIC DEPOSITS

On a mild day in February 1943, as Dionisio Pulido prepared to sow the fertile soil of his field, 330 km west of Mexico City, an earthquake jolted the ground. Earthquakes had occurred with increasing frequency in recent days, but this time, to Dionisio's amazement, the surface of his field visibly bulged upward by a few meters and then cracked. Ash and sulfurous fumes

> **Did you ever wonder . . .**
> whether anyone has ever seen a brand-new volcano form?

FIGURE 5.5 The gas component of volcanic eruptions.

(a) A volcano in Alaska erupting large quantities of steam.

(b) Gas bubbles frozen in lava produce vesicles, as in this block from Sunset Crater, Arizona.

rose into the air, and the farmer fled. When he returned the following morning, his field lay buried beneath a 40-m-high mound of gray cinders—Dionisio had witnessed the birth of a new volcano, Paricutín. During the next several months, Paricutín erupted continuously, at times blasting clots of lava into the sky like fireworks. By the following year, it had become a steep-sided cone more than 300 m high. Nine years later, when the volcano ceased erupting, its products covered 25 km².

The description of Paricutín's eruption, like that of Vesuvius at the beginning of this chapter, emphasizes that volcanoes can erupt fragmental igneous material. Geologists use either the term **tephra** or the term **pyroclastic debris** (from the Greek *pyro*, meaning fire) for unconsolidated fragments resulting from an eruption. Tephra can include many different materials: specks of glass and pieces of rock formed from drops or clots of molten lava that solidified in midair or on the ground soon after falling, chunks of recently solidified lava or pumice blasted out of the volcano's vent, and chunks formed by the forceful fragmentation of pre-existing volcanic rock surrounding the volcano's vent. A broader term, *volcaniclastic debris*, includes not only tephra, but also debris that tumbled down the flank of a volcano in landslides long after an eruption, debris that mixed with water to form a flowing muddy slurry, and debris originally formed at a volcano but later carried and sorted by streams. Let's look at these components in more detail—you'll see that different debris types form in association with different kinds of eruptions.

Pyroclastic Debris from Basaltic Eruptions Basaltic magma rising in a volcano may contain dissolved gas. As such magma approaches the surface, the gas comes out of solution and forms bubbles. When the bubbles reach the vent of the volcano, they expand and burst, producing enough pressure to drive columns or sprays of basaltic lava upward in dramatic **lava fountains**, from which clots of lava rain down onto the volcano from the fountains **(Fig. 5.6a)**.

The solidified pea-sized to golf-ball-sized fragments of glassy lava and scoria, informally called *cinders*, are a type of **lapilli**, from the Latin word for little stones. Rarely, flying droplets stretch into long strands that freeze as glassy threads called *Pele's hair*, after the Hawaiian goddess of volcanoes; the droplets freeze into tiny streamlined glassy beads known as *Pele's tears*. Larger blobs of soft lava that squirt out of a vent and then solidify become streamlined **bombs (Fig. 5.6b)**. Eruptions may break apple- to refrigerator-sized fragments of already-solid lava from the walls of the vent and eject them as chunky, angular **blocks (Fig. 5.6c)**. As a result, the pyroclastic debris surrounding a basaltic volcanic vent commonly includes blocks, bombs, and cinder-like lapilli.

Pyroclastic Debris from Andesitic and Rhyolitic Eruptions Andesitic and rhyolitic lavas are more viscous than basalt and are generally richer in gas. As we'll see, eruptions of these lavas tend to be explosive and to eject large quantities of pyroclastic debris **(Fig. 5.7a)**. Debris ejected

FIGURE 5.6 Pyroclastic debris from basaltic eruptions.

(b) A bomb has a smooth, streaked surface.

(a) A fountain of basaltic lapilli spouts from a vent on Hawaii.

(c) Blocks and lapilli on the flank of a Hawaiian volcano.

FIGURE 5.7 Pyroclastic debris from andesitic and rhyolitic eruptions.

(a) Pyroclastic debris billowing from the 2008 eruption of Chaitén in Chile.

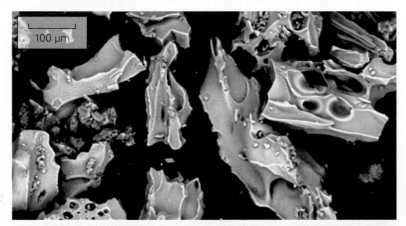

(b) Electron photomicrograph of ash.

(c) Pumice lapilli.

(d) Accretionary lapilli.

(e) A pyroclastic flow rushes down the flank of Mt. Merapi, Indonesia, in 2006.

(f) Water-soaked volcanic debris slid down the side of this volcano in Nicaragua.

(g) A lahar fills a riverbed in New Zealand after an eruption in 2007.

during explosive eruptions includes the following: **ash**, with particles less than 2 mm in diameter, made from glass shards formed either when frothy lava explosively breaks up during an eruption and the glassy walls of bubbles separate into small pieces **(Fig. 5.7b)**, or when pre-existing volcanic rock undergoes pulverization; *pumice lapilli*, angular pumice fragments formed when frothy lava breaks into marble-sized chunks **(Fig. 5.7c)**; and *accretionary lapilli*, small, snowball-like spheres of ash formed when ash mixes with water in the air and then sticks together, collecting more ash as it falls **(Fig. 5.7d)**.

Some pyroclastic debris that erupts from an exploding volcano shoots upward into the air. Some, however, rushes down the flank of the volcano in an avalanche-like current known as a **pyroclastic flow (Fig. 5.7e)**. The debris may have a temperature between 200°C and 450°C, so this type of flow was once called a *nuée ardente* (French for glowing cloud). Tephra from andesitic or rhyolitic eruptions (ash, or ash mixed with lapilli) becomes **tuff** when buried and transformed into coherent rock. Tuff from material that rains out of an ash cloud is *air-fall tuff*, whereas a sheet of tuff deposited by a pyroclastic flow is an *ignimbrite*. Tuff generally forms when debris accumulates while it is still hot enough to weld together, or when its minerals grow in the tephra due to reaction with groundwater after deposition.

Volcanic Debris Flows and Lahars In cases where volcanoes are covered with snow and ice, or are drenched with rain, water can mix with debris to form a *volcanic debris flow* that moves downslope like a sheet of wet concrete **(Fig. 5.7f)**. If ash-rich debris becomes very wet, it becomes a muddy, rapidly flowing slurry called a **lahar**, which can travel for tens of kilometers **(Fig. 5.7g)**. When debris flows and lahars stop moving, they form a layer consisting of volcanic debris suspended in ashy mud.

TAKE-HOME MESSAGE

Volcanoes erupt lava, pyroclastic debris, and gases. The character of a lava flow—whether it has low viscosity and spreads over a large area or has high viscosity and builds a mound over the vent—depends largely on its composition. Pyroclastic debris includes pumice, lapilli, blocks, and bombs. Some may fall over the countryside like snow, but some surges down the flank of a volcano as a pyroclastic flow.

QUICK QUESTION How can lava travel tens of kilometers or more from a volcanic vent without freezing?

5.3 Structure and Eruptive Style

VOLCANIC ARCHITECTURE

As magma rises from depth, it doesn't always travel all the way to the surface but may accumulate underground in a **magma chamber**, a large, irregularly shaped zone. Inside a magma chamber, melt, together with crystals, forms a crystal mush. It may solidify to form a pluton, inject into cracks or between layers to form dikes or sills, or rise along a pathway, called a *conduit*, to the Earth's surface and erupt from a volcano. A conduit may resemble a vertical pipe or chimney, or it may take the form of a crack called a **fissure (Fig. 5.8)**. The products of an eruption, or a succession of eruptions, build into a hill or mountain known formally as a *volcanic edifice*, but informally, simply as a volcano. Typically, a circular, bowl-like depression, called a **crater**, occurs at the top of a volcano. Craters, which may be as large as 1 km across and 200 m deep, form either during eruption, as material accumulates around the summit vent, or just after eruption, as the summit collapses into the drained conduit.

During some major eruptions, a large magma chamber beneath a volcano may drain suddenly and produce a **caldera**, a circular or elliptical depression that's much larger than a crater. Typically, a caldera has steep walls and a fairly flat floor and may be partially filled with ash or solidified lava, or submerged by water **(Fig. 5.9)**. Examples range from a few kilometers to tens of kilometers across, and they may be several hundred meters deep. Later eruptions can build new volcanic edifices within a caldera. For example, Wizard Island grew on the floor of the Crater Lake caldera and now protrudes from the lake (see Fig. 5.9d). The caldera collapsed between 6,000 and 8,000 years ago, and Wizard Island last erupted about 4,600 years ago.

Geologists distinguish among several different shapes of subaerial (above sea level) volcanic edifices. **Shield volcanoes** are broad, gentle domes whose shape resembles a soldier's shield lying on the ground **(Fig. 5.10a)**. They form when the products of eruption have low viscosity and can't build into a mound at the vent, but rather flow easily and spread out as thin sheets over large areas. **Cinder cones**, also known as *scoria cones*, consist of cone-shaped piles of basaltic lapilli and blocks **(Fig. 5.10b)**. The steep slopes of a cinder cone represent the *angle of repose* for lapilli, meaning the maximum slope that the loose fragments can sustain before sliding down. (You can build a similar-shaped pile by pouring dry sand from a bucket onto the ground.) **Stratovolcanoes**, also called *composite volcanoes*, are large volcanoes (up to 3 km high and 25 km across) that consist of interleaved layers

FIGURE 5.8 Crater eruptions and fissure eruptions come from conduits of different shapes.

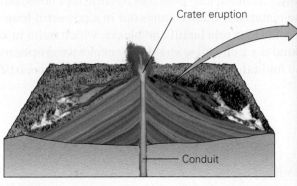

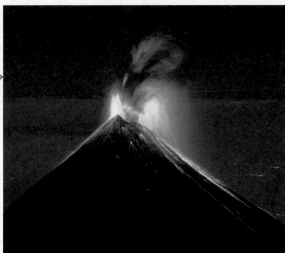

(a) At a crater eruption, lava spouts from a chimney-shaped conduit. Such eruptions typically happen at the summit of a volcano.

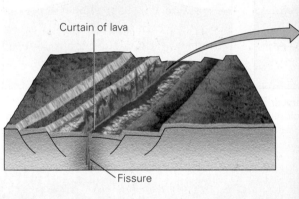

(b) At a fissure eruption, lava comes out in a curtain, along the length of a crack.

of lava, tephra, and volcaniclastic debris **(Fig. 5.10c)**. The prefix *strato–* was coined to emphasize that such volcanoes have internal layering, or stratification. Their shape, exemplified by Japan's Mt. Fuji, serves as the classic image that most people have of a volcano. If you look closely, you'll note that stratovolcanoes generally have steeper slopes near the summit and gentler slopes near the base. This is because the upper part builds up out of successive layers of tephra, deposited at its angle of repose and coated by lava flows, whereas the lower part consists of debris flows, lahars, and landslide deposits.

WILL IT FLOW OR WILL IT BLOW?

Concept of Eruptive Style Kilauea, a volcano on Hawaii, produces rivers of lava that cascade down the volcano's flanks. Mt. St. Helens, a volcano near the Washington-Oregon border, exploded catastrophically in 1980 and blanketed the surrounding countryside with tephra. Clearly, volcanoes erupt in different ways. Geologists refer to the character of an eruption as *eruptive style*.

Geologists distinguish between two end-member categories of eruptions, based on eruptive style. During an **effusive eruption** (from the Latin word meaning pour out), low-viscosity lava spills or fountains steadily from a vent or fissure. The lava may collect in a lava lake around the vent or it may flow downslope **(Fig. 5.11a)**. The 2018 eruption of Kilauea on the Big Island of Hawaii is an example of an effusive eruption. During this event, fountains of molten rock poured from vents along fissures. These fountains fed rivers of lava that flowed on land for almost 20 km before spilling into the sea. In places, individual flows were more than 15 m thick. During an **explosive eruption**, in contrast, pyroclastic debris blasts forcefully into the air **(Fig. 5.11b, c)**. We can identify distinct types of explosive eruptions based on the violence of the explosion and on the nature of the pyroclastic debris that the explosion produces, as we now see. Traditionally, each type was named after a specific example (**Geology at a Glance**, pp. 154–155).

Styles of Explosive Eruptions Ancient Romans referred to the island of Stromboli as "the lighthouse of the Mediterranean" because it has erupted about every 10 to 20 minutes throughout recorded history, and the red-hot clots that it ejects trace out glowing arcs of light in the night sky. Occasionally, Stromboli also produces basaltic lava flows, but most of the material it erupts comes out in a powerful fountain that produces scoria lapilli and blocks, which build into a cone around the vent. Somewhat larger explosive eruptions emit both a fountain of lava and a dense plume of pyroclastic debris.

FIGURE 5.9 The formation of volcanic calderas.

Time

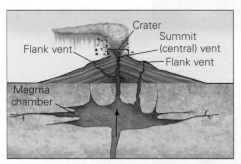
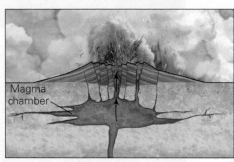
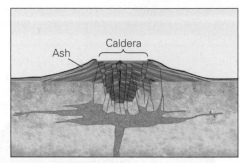

(a) As an eruption begins, the magma chamber inflates with magma. There can be a central vent and one or more flank vents.

(b) During an eruption, the magma chamber drains, and the central portion of the volcano collapses downward.

(c) The collapsed area becomes a caldera. Later, a new volcano may begin to grow within the caldera.

(d) This caldera in Oregon formed about 7,700 years ago. Afterward, it filled with water to become Crater Lake. Wizard Island, protruding from the lake, is a small volcano that grew on top of the caldera floor.

FIGURE 5.10 Volcanoes of different shapes.

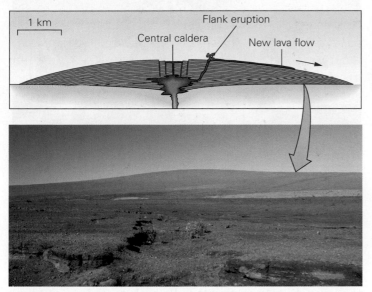

(b) A cinder cone in Arizona, with a lava flow covering the land surface to its right.

(a) A shield volcano in Hawaii, made from successive flows of low-viscosity basalt, has very gentle slopes.

Mt. Fuji, a composite volcano in Japan, last erupted in 1707.

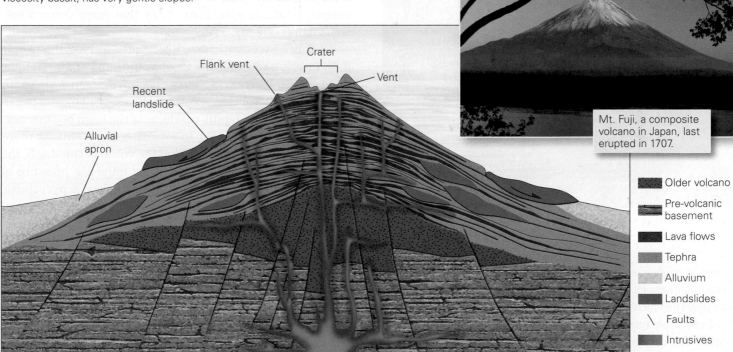

	Older volcano
	Pre-volcanic basement
	Lava flows
	Tephra
	Alluvium
	Landslides
\\	Faults
	Intrusives

(c) A composite volcano consists of layers of tephra and lava. Volcanic debris flows and ash avalanches modify slopes and contribute to the development of a classic cone-like shape.

Even larger explosions happen in felsic or andesitic volcanoes, where the rising magma tends to be very viscous and rich in gas. We've noted that gas forms bubbles as lava rises. During the final stages of its rise, as magma slowly moves up into the volcano itself, the magma may partly solidify into glass and trap the bubbles. When overlying material erupts, the pressure pushing down on a magma decreases. As external pressure decreases, because bubbles can't expand,

their internal gas pressure increases. If cracks occur in the glass screens between bubbles, the gas escapes and suddenly has room to expand. The sudden expansion blasts out any remaining melt and pulverizes the glass that had surrounded the bubbles, as well as pre-existing rock and debris that had formed the volcanic edifice. The resulting pyroclastic debris shoots out of the volcano like a shotgun blast, or like champagne spouting from a newly opened shaken bottle.

FIGURE 5.11 Effusive eruptions compared to explosive eruptions.

(a) A 1986 effusive eruption on Hawaii.

Sea surface

(b) An explosive eruption of a subsea volcano near Tonga.

Convective cloud

(c) The explosive eruption of Mt. Pinatubo in the Philippines.

Such an explosion, awesome in its power and catastrophic in its consequences, can eject cubic kilometers of debris and may destroy the volcano **(Box 5.1)**.

Let's look more closely at the structure of the eruptive cloud. The debris from an explosive eruption (ash, lapilli, and blocks) shoots skyward in a vertical column **(Fig. 5.12a, b)**. But eruptive force can push the material only so high. The huge plume of ash that rises to stratospheric heights above a large explosion does so by becoming a billowing *convective cloud* **(Fig. 5.12c)**. This means that the warm mixture of volcanic ash, gas, and air, which is less dense than the surrounding, cooler air, rises skyward buoyantly. The resulting plume resembles a mushroom cloud above a nuclear explosion. Coarse-grained ash from the cloud settles close to the volcano, whereas fine ash gets carried farther away. In fact, high-elevation winds may transport fine ash around the globe. Not all of the debris ejected by an explosive eruption ends up in

the convective cloud; gravity pulls a substantial amount back down around the vent. This phenomenon, known as *column collapse*, produces pyroclastic flows that surge down a volcano's flanks (see Fig. 5.7e).

What is a pyroclastic flow like? Sadly, in 1902 the people of St. Pierre, a town on the Caribbean island of Martinique, found out. St. Pierre was a busy port town, about 7 km south of the peak of Mt. Pelée, a volcano. When the volcano began emitting steam and lapilli, residents of the town became nervous and debated whether to evacuate. Meanwhile, a rhyolite spire rose from the conduit of the volcano. On May 8, the spire suddenly cracked, releasing the immense pressure of gases in the partially solidified magma below. A pyroclastic flow swept down Pelée's flank. Partly riding on a cushion of air, this flow reached speeds of 300 km per hour and slammed into St. Pierre. Within moments, all the town's buildings had been flattened and all but two of its 28,000 inhabitants were dead of incineration or asphyxiation **(Fig. 5.12d)**. Similar eruptions have happened more recently on the nearby island of Montserrat, but with a much smaller death toll because of timely evacuation.

Relation of Volcanic Type to Eruptive Style Note that the type of volcano (shield, cinder cone, or stratovolcano) depends on its eruptive style. Volcanoes with only low-viscosity lava eruptions become shield volcanoes, those that generate small pyroclastic eruptions due to fountaining basaltic lava yield cinder cones, and those that alternate between lava and pyroclastic eruptions become stratovolcanoes. Large explosions yield calderas and blanket the surrounding countryside with ash and/or ignimbrites. Why do such contrasts in eruptive style exist? Eruptive style depends on the viscosity and gas content of the magma in the volcano. These characteristics, in turn, depend on the composition and temperature of

FIGURE 5.12 The components of a large explosive volcanic eruption.

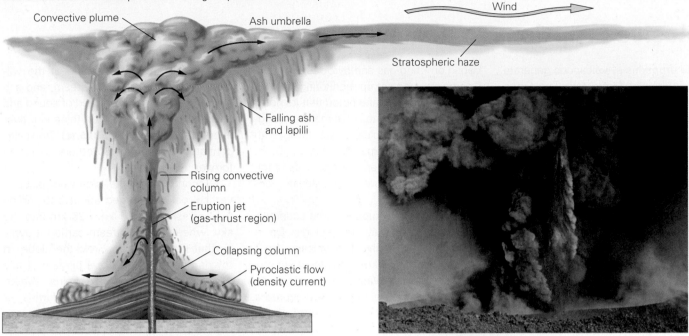

Convective plume

Ash umbrella

Wind

Stratospheric haze

Falling ash
and lapilli

Rising convective
column

Eruption jet
(gas-thrust region)

Collapsing column

Pyroclastic flow
(density current)

(a) A Plinean eruptive column contains several components.

(b) The eruption jet at an eruption of Mt. Etna, Italy.

(c) The mushroom cloud of the 1989 eruption of Redoubt Volcano, Alaska.

(d) The aftermath of the 1902 eruption of Mt. Pelée.

the magma and on the environment (subaerial or submarine) where the eruption occurs, as we discuss below.

Supervolcanoes The largest explosive volcanic eruption ever observed during human history took place in 1815. This eruption, which blasted the top off Mt. Tambora, a volcano on an island in Indonesia, ejected about 145 km³ of debris. The geologic record shows that much larger explosions have happened in prehistory (see Fig. Bx5.1a). In fact, during the past few million years, a number of eruptions

have ejected over 1,000 km³ of debris. Geologists now refer to the source of an incomprehensibly huge eruption as a **supervolcano**. After such an eruption, the area of the volcano typically collapses to form a large caldera. The largest known supervolcano, which erupted about 73,000 B.C.E. in Indonesia, produced 2,800 km³ of debris and left behind a caldera now filled with Lake Toba. Supervolcanic eruptions have happened at least twice in the area where Yellowstone Park now lies—in the aftermath of one of these explosions, a 72-km-diameter caldera remained.

BOX 5.1 CONSIDER THIS. . .

Explosive Eruptions to Remember

Explosive eruptions of volcanoes generate enduring images of destruction. The historical record shows a vast range in the volume of debris erupted. Notably, the largest eruption in historic times (Tambora in 1815) was small compared to a superexplosion that took place about 600,000 years ago in what is now Yellowstone National Park, Wyoming **(Fig. Bx5.1a)**. Let's look at two examples of explosions.

Mt. St. Helens, a snow-crested stratovolcano in the Cascades of the northwestern United States, had not erupted since 1857. However, geologic evidence suggested that the mountain had a violent past, punctuated by many explosive eruptions. On March 20, 1980, an earthquake announced that the volcano was awakening once again. A week later, a crater 80 m in diameter burst open at the summit and began emitting gas and pyroclastic debris. Geologists who set up monitoring stations to observe the volcano noted that its north side was beginning to bulge markedly, suggesting that the volcano was filling with magma and was expanding like a balloon. Their concern that an eruption was imminent led local authorities to evacuate people in the area.

The climactic eruption came suddenly. At 8:32 A.M. on May 18, geologist David Johnston, monitoring the volcano from a distance of 10 km, shouted over his two-way radio, "Vancouver, Vancouver, this is it!" An earthquake had triggered a huge landslide that caused 3 km³ of the volcano's weakened north side to slide away. The sudden landslide released pressure on the magma in the volcano, causing a sudden and violent expansion of gases that blasted through the side of the volcano **(Fig. Bx5.1b)**. Rock, steam, and ash screamed north at the speed of sound and flattened a forest and everything in it over an area of 600 km² **(Fig. Bx5.1c)**. Tragically, Johnston, along with 60 others, vanished forever.

Seconds after the sideways blast, a vertical column carried about 540 million tons of ash (about 1 km³) 25 km into the sky, where the jet stream carried it away so that it was able to circle the globe. In towns near the volcano, a blizzard of ash choked roads and buried fields. Water-saturated ash formed viscous slurries, or lahars, that flooded river valleys, carrying away everything in their path. When the eruption was finally over, the once cone-shaped peak of Mt. St. Helens had disappeared—the summit now lay 440 m

FIGURE Bx5.1 Examples of explosive eruptions.

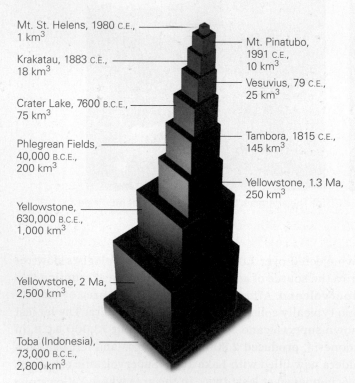

Mt. St. Helens, 1980 C.E., 1 km³

Krakatau, 1883 C.E., 18 km³

Crater Lake, 7600 B.C.E., 75 km³

Phlegrean Fields, 40,000 B.C.E., 200 km³

Yellowstone, 630,000 B.C.E., 1,000 km³

Yellowstone, 2 Ma, 2,500 km³

Toba (Indonesia), 73,000 B.C.E., 2,800 km³

Mt. Pinatubo, 1991 C.E., 10 km³

Vesuvius, 79 C.E., 25 km³

Tambora, 1815 C.E., 145 km³

Yellowstone, 1.3 Ma, 250 km³

(a) The relative amounts of pyroclastic debris (in km³) ejected during major explosive eruptions.

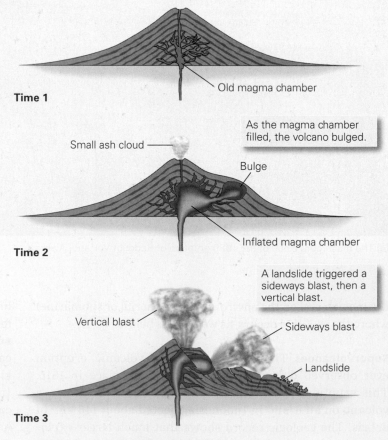

Old magma chamber

Time 1

Small ash cloud

As the magma chamber filled, the volcano bulged.

Bulge

Inflated magma chamber

Time 2

A landslide triggered a sideways blast, then a vertical blast.

Vertical blast

Sideways blast

Landslide

Time 3

(b) Stages during the eruption of Mt. St. Helens, 1980.

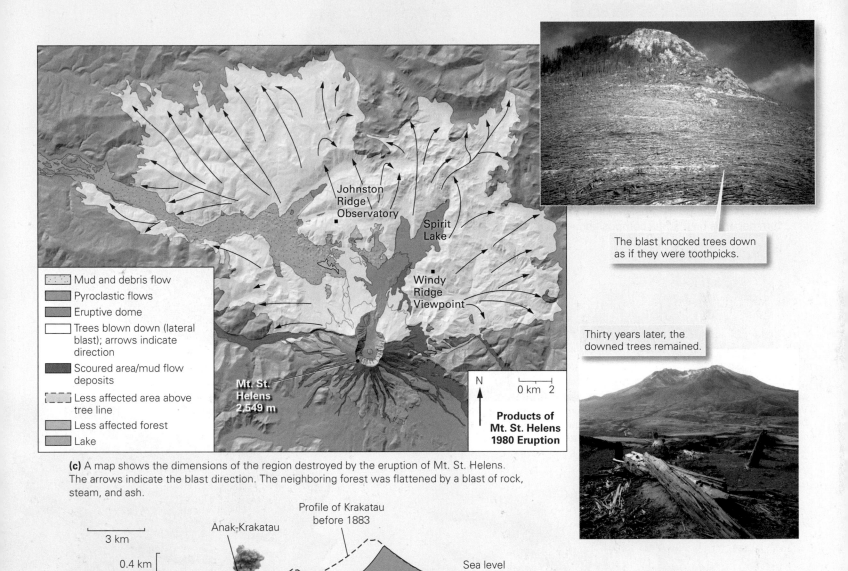

The blast knocked trees down as if they were toothpicks.

Thirty years later, the downed trees remained.

(c) A map shows the dimensions of the region destroyed by the eruption of Mt. St. Helens. The arrows indicate the blast direction. The neighboring forest was flattened by a blast of rock, steam, and ash.

(d) Profile of Krakatau before and after the eruption. Note that a new volcano (Anak-Krakatau) has formed.

lower, and the once snow-covered mountain was a gray mound with a large gouge in one side. The volcano came alive again in 2004, but did not explode.

In 1883, an even greater explosive eruption happened in Krakatau (Krakatoa), a volcano in Indonesia between Java and Sumatra that had built a 9-km-long island rising 800 m above the sea. On May 20, the island began to erupt with a series of large explosions, yielding ash that settled as far as 500 km away. Smaller explosions continued through June and July,

and steam and ash rose from the island, forming a huge black cloud that rained ash into the surrounding straits. Ships sailing by couldn't see where they were going, and their crews had to shovel ash off the decks.

Krakatau's demise came at 10 A.M. on August 27, perhaps when the volcano cracked and the magma chamber suddenly flooded with seawater. The resulting blast, 5,000 times greater than the Hiroshima atomic bomb explosion, could be heard as far as 4,800 km away, and subaudible

sound waves traveled around the globe seven times. Giant sea waves pushed out by the explosion slammed into coastal towns, killing over 36,000 people. Near the volcano, a layer of ash up to 40 m thick accumulated. When the air finally cleared, Krakatau was gone, replaced by a submarine caldera some 300 m deep **(Fig. Bx5.1d)**. All told, the eruption shot 20 km³ of rock into the sky. Some ash reached elevations of 27 km. Because of this ash, people around the world could view spectacular sunsets during the next several years.

Volcanoes

Beneath a volcano, magma rises to fill an open or cracked region of crust and forms a magma chamber. Some of the magma erupts at a surface vent. Once molten rock has erupted at the surface, it is called lava. Some lava spills down the side of the volcano in lava flows. Some may fountain out of a vent to form scoria fragments, which pile up in a cone around the vent. Eruptions may eject larger chunks as blocks or bombs.

The nature of eruptions depends on the viscosity of the magma, which in turn depends on magma composition.

Cinder cone

Caldera

Vulcanian eruptions occur when a buildup of gas and magma explodes.

Strombolian crater explosions frequently burst through thinly crusted lava.

Hawaiian fountain explosions are caused by escaping gas.

Flank vent

Eroded cone

Lava cone

Lava flow

Sills

Dikes

Cinder cones

Lava pavement (cracked/broken)

Plinean explosions shoot a huge column of pyroclastic debris up to 50 km into the atmosphere. The ash fall rains down and the column collapses back around the vent, traveling overland as a pyroclastic flow.

Lava flow
50 km

Mud flow
150 km

Pyroclastic flow
200 km

Ash fall
2500 km

The distance volcanic hazards can travel from an eruption.

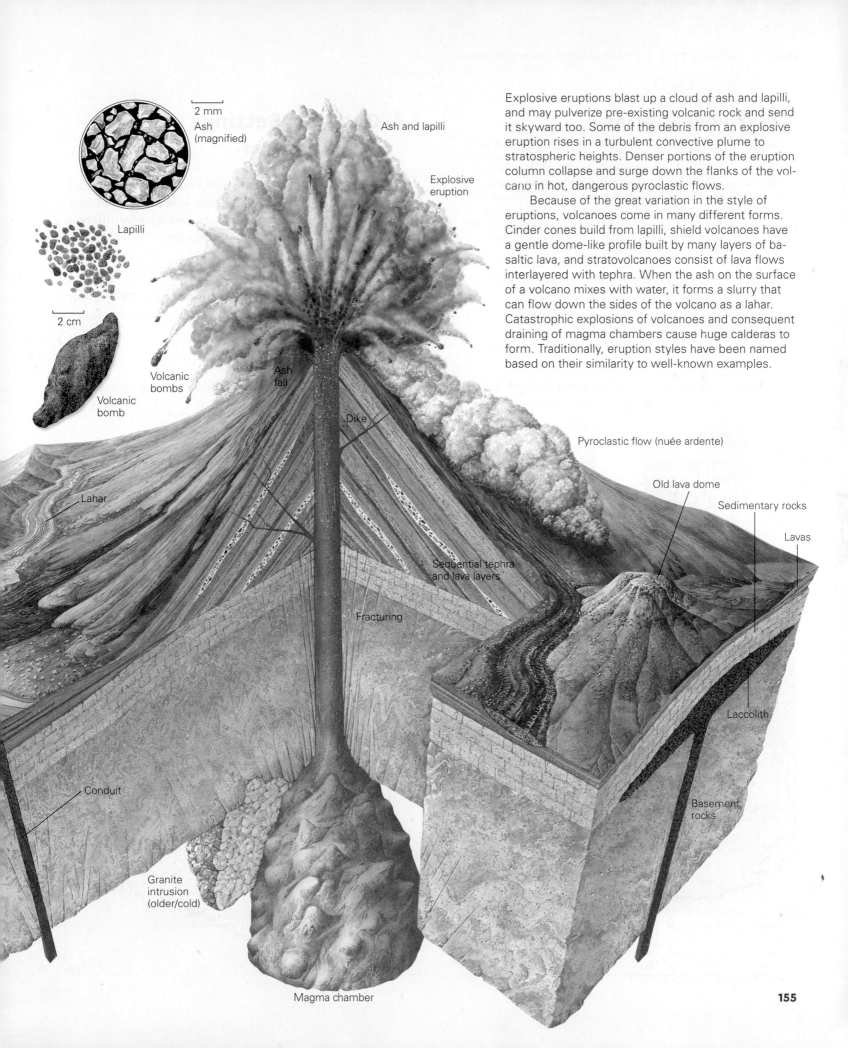

2 mm
Ash (magnified)

Lapilli

2 cm

Volcanic bomb

Volcanic bombs

Ash and lapilli

Explosive eruption

Ash fall

Dike

Lahar

Sequential tephra and lava layers

Fracturing

Pyroclastic flow (nuée ardente)

Old lava dome

Sedimentary rocks

Lavas

Laccolith

Conduit

Granite intrusion (older/cold)

Basement rocks

Magma chamber

Explosive eruptions blast up a cloud of ash and lapilli, and may pulverize pre-existing volcanic rock and send it skyward too. Some of the debris from an explosive eruption rises in a turbulent convective plume to stratospheric heights. Denser portions of the eruption column collapse and surge down the flanks of the volcano in hot, dangerous pyroclastic flows.

Because of the great variation in the style of eruptions, volcanoes come in many different forms. Cinder cones build from lapilli, shield volcanoes have a gentle dome-like profile built by many layers of basaltic lava, and stratovolcanoes consist of lava flows interlayered with tephra. When the ash on the surface of a volcano mixes with water, it forms a slurry that can flow down the sides of the volcano as a lahar. Catastrophic explosions of volcanoes and consequent draining of magma chambers cause huge calderas to form. Traditionally, eruption styles have been named based on their similarity to well-known examples.

TAKE-HOME MESSAGE

At a volcano, lava rises from a magma chamber and erupts from chimney-like conduits or from crack-like fissures. Geologists distinguish between effusive (lava-dominated) and explosive eruptions. Low-viscosity basalt lava flows build shield volcanoes. Fountaining basalt spatters tephra to build cones. Successive eruptions of pyroclastic debris and lava build stratovolcanoes, and explosions produce calderas. A large explosive eruption can produce a convecting cloud of ash that rises to stratospheric heights, as well as devastating pyroclastic flows.

QUICK QUESTION Why do cinder cones have steeper slopes than shield volcanoes?

5.4 Geologic Settings of Volcanism

Volcanic eruptions occur along plate boundaries, and at hot spots and in rifts **(Fig. 5.13)**. We'll now look at these settings in the context of plate tectonics theory, and see why different styles of volcanoes form in different settings.

MID-OCEAN RIDGES

Products of mid-ocean ridge volcanism cover 70% of our planet's surface. We don't generally see this volcanic activity,

FIGURE 5.13 Volcanoes of the world. The map shows the distribution of volcanoes around the world, and the diagrams illustrate basic geologic settings in which volcanoes form, in the context of plate tectonics theory.

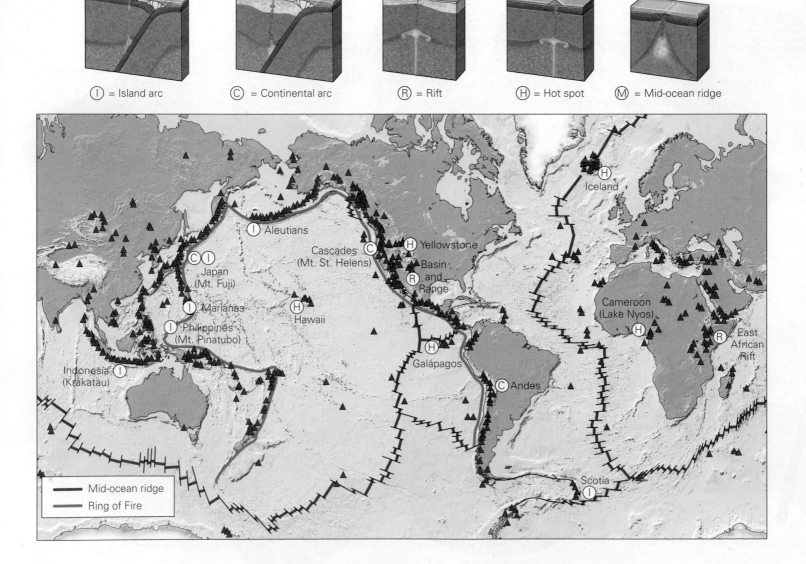

FIGURE 5.14 Igneous activity at plate boundaries.

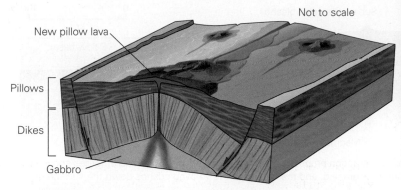

(a) Mounds of pillow basalt erupt along fissures.

(b) Pillow basalt on the seafloor along the Juan de Fuca Ridge.

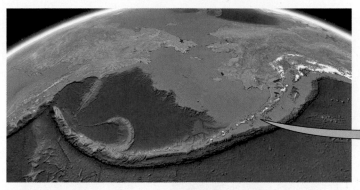

(c) The Aleutian arc forms the northern edge of the Pacific Plate. It displays a distinct curvature.

(d) An oblique view of the Aleutian arc, as seen looking northwest.

however, because the ocean hides most of it beneath a blanket of water. Mid-ocean ridge volcanoes develop along fissures parallel to the ridge axis. Each one turns on and off in a time scale measured in tens to hundreds of years. When active, they erupt basalt, which, because it cools so quickly underwater, forms pillow-lava mounds **(Fig. 5.14a, b)**. Water that heats up as it circulates through the crust near the magma chamber bursts out of hydrothermal (hot-water) vents, known as *black smokers*, along these mounds.

CONVERGENT BOUNDARIES

Most subaerial volcanoes on the Earth lie on the edge of an overriding plate along a convergent plate boundary. Geologists refer to the chain of volcanoes along a given convergent boundary as a volcanic arc, because many of these chains have an arc-shaped trace in map view. Convergent boundaries border over 60% of the Pacific Ocean, so a 20,000-km-long chain of volcanic arcs, known as the *Ring of Fire*, has developed along the rim of the ocean. Typically, individual volcanoes in volcanic arcs lie about 50 to 100 km apart **(Fig. 5.14c, d)**. Some of these volcanoes grow on oceanic crust and become *island arcs*, such as the Marianas of the western Pacific. Others grow on continental

crust, building *continental arcs* such as the Cascade Range of Washington and Oregon or the Andes of South America. Because of the variety of magma types produced beneath continental arcs, such arcs produce large stratovolcanoes.

CONTINENTAL RIFTS

Due to the diversity of magmas that form beneath rifts, rifts can host both basaltic fissure eruptions, in which curtains of lava fountain up or linear chains of cinder cones develop, and explosive rhyolitic volcanoes. In some places, eruptions build stratovolcanoes, such as Mt. Kilimanjaro in Africa.

OCEANIC HOT-SPOT VOLCANISM

When a hot-spot volcano forms on oceanic lithosphere, basaltic magma erupts at the surface of the seafloor. First, such submarine eruptions yield an irregular mound of pillow lava, eventually growing above the sea surface. After the volcano has emerged from the sea, effusive eruptions of basalt yield a broad shield shape with gentle slopes **(Fig. 5.15a)**. As the volcano grows, portions of it succumb to the pull of gravity and slip seaward, creating large submarine slumps.

FIGURE 5.15 Oceanic hot-spot volcanoes.

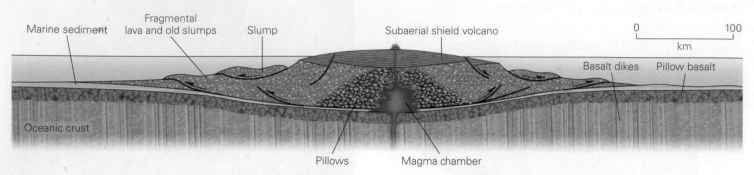

(a) The interior structure of an oceanic hot-spot volcano is complicated. Initially, eruption produces pillow basalts. When the volcano emerges above sea level, it becomes a shield volcano. The margins of the island frequently undergo slumping, and the weight of the volcano pushes down the surface of the lithosphere. The Hawaiian Islands exemplify this architecture.

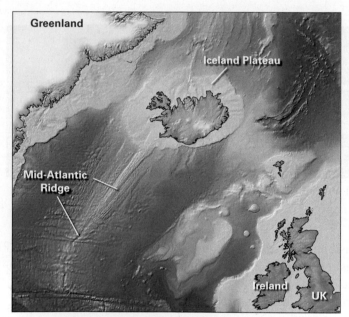

(b) A bathymetric map shows that Iceland sits atop a huge plateau straddling the Mid-Atlantic Ridge. Light blue is shallower water; dark blue is deeper.

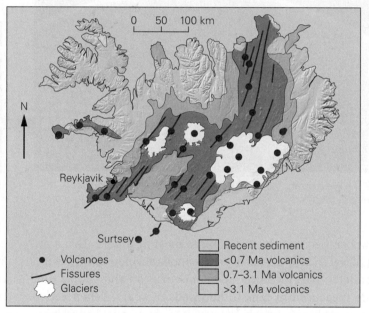

(c) A geologic map of Iceland shows that the youngest volcanoes occur in the central rift, effectively the on-land portion of the Mid-Atlantic Ridge.

Hawaii and other volcanic islands within the Pacific Ocean basin serve as examples of oceanic hot-spot volcanoes. Because the Pacific Plate moves relative to their sources, these active hot-spot volcanoes lie at the ends of chains of now-dead volcanoes and seamounts, defining the hot-spot track (see Figure 2.32). Iceland is one of a few places on the Earth where a hot spot lies at the crest of a mid-ocean ridge. Because of the hot spot, vastly more magma erupts at Iceland than at other places along the Mid-Atlantic Ridge, and igneous activity has built a broad submarine plateau **(Fig. 5.15b)**. Iceland straddles a divergent plate boundary, which means that it is being stretched apart, and it has been cut up by faulting. Indeed, the central part of the island is a narrow rift, marking the trace of the Mid-Atlantic Ridge axis,

where the youngest volcanic rocks of the island were extruded **(Fig. 5.15c)**. Fissure eruptions along faults yield curtains of lava. Eruptions along faults can also build linear chains of small cinder cones. Not all volcanic activity on Iceland occurs subaerially—some eruptions take place under glaciers and melt large amounts of ice. When the meltwater bursts through the edge of the glacier it becomes a devastating flood, called a *jokulhlaup* in Icelandic.

CONTINENTAL HOT-SPOT VOLCANISM

Yellowstone National Park lies at the northeast end of a string of calderas, the oldest of which erupted 16 million years ago **(Fig. 5.16a, b)**. Recent and ongoing activity beneath the

Earth's surface has yielded fascinating landforms, volcanic rock deposits, and geysers. Eruptions at the Yellowstone hot spot differ from those in Hawaii in an important way: some eruptions at Yellowstone yield basaltic lava, but unlike those of Hawaii, others yield rhyolitic pyroclastic debris. Some of the rhyolitic eruptions are the products of supervolcanoes. The most recent of these, 630,000 years ago, produced an immense caldera that filled with pyroclastic debris and sent up gigantic convective clouds of ash. Some ash fell as far east as the Mississippi River **(Fig. 5.16c)**. The name of

FIGURE 5.16 Hot-spot volcanic activity in Yellowstone National Park.

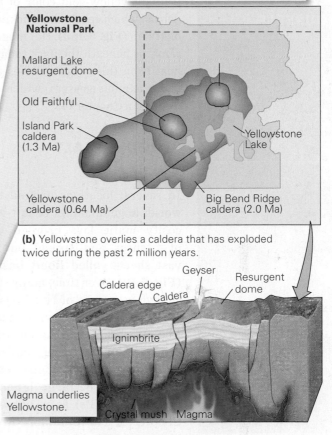

A resurgent dome is a bulge formed when a magma chamber inflates.

(b) Yellowstone overlies a caldera that has exploded twice during the past 2 million years.

Magma underlies Yellowstone.

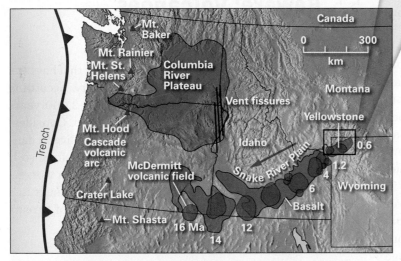

(a) Yellowstone lies at the end of a continental hot-spot track. Progressively older calderas follow the Snake River Plain. The blue arrow indicates plate motion.

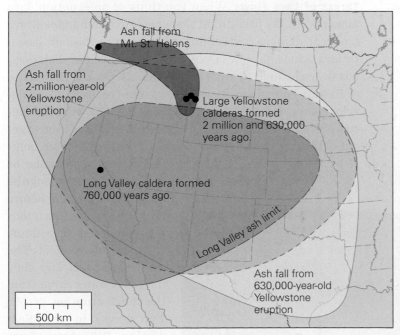

(c) The ash produced by explosions of the Yellowstone calderas covered vast areas—much more than did the Mt. St. Helens eruption.

(d) Felsic tuffs form the colorful walls of Yellowstone Canyon.

Yellowstone Park comes from the brilliant yellow-gold color of the debris, which reached a thickness of 400 m **(Fig. 5.16d)**. Hot magma remains beneath the park today—heat from this magma drives the park's famous geysers, which we discuss in Chapter 16.

FLOOD-BASALT VOLCANISM

In several locations around the world, huge amounts of low-viscosity mafic lava erupted out of fissures and spread out in vast sheets called **flood basalt** **(Fig. 5.17)**. Over time, many successive eruptions of flood basalt can build up a broad plateau. The aggregate volume of rock produced by these eruptions can be so great that geologists consider them to be **large igneous provinces (LIPs)**, as we noted in Chapter 4. The Columbia River Plateau, of Washington and Oregon (see Fig. 4.19), serves as an example of a LIP. About 15 million years ago, the region of the plateau hosted huge volcanic fissures yielding enough lava to build a 3.5-km-thick layer, covering an area of 220,000 km². Even larger flood-basalt provinces occur in eastern Siberia (an occurrence known as the Siberian Traps), the Deccan Plateau of India, the Paraná Plateau of Brazil, and the Karoo Plateau of southern Africa.

FIGURE 5.17 Flood-basalt layers exposed on the wall of a canyon in Idaho.

5.5 Beware: Volcanoes Are Hazards!

Like earthquakes, volcanoes are natural hazards that have the potential to cause great destruction to humanity. According to one estimate, volcanic eruptions in the last 2,000 years have caused about a quarter of a million deaths. Considering the rapid expansion of cities, far more people live in dangerous proximity to volcanoes today than ever before, so if anything, the hazard posed by volcanoes has gotten worse—imagine if a large explosive eruption were to occur next to a major city today. Let's look at the different kinds of threats posed by volcanic eruptions.

HAZARDS DUE TO ERUPTIVE MATERIALS

Threat of Lava Flows When you think of an eruption, lava flows may be the first threat that comes to mind. Indeed, lava flows can damage real estate, and on many occasions, basaltic lava, which has low viscosity and can flow long distances, has overwhelmed towns **(Fig. 5.18a–d)**. On the Big Island of Hawaii, recent lava flows have covered roads, housing developments, and vehicles. Lava from the 2018 eruption destroyed hundreds of homes. Although people have time to get out of the way of such flows, they might have to watch helplessly from a distance as an advancing flow engulfs their homes. Even before the lava touches it, a building will burst into flame from the intense heat. The most disastrous lava flow in recent times came from the 2002 eruption of Mt. Nyiragongo in the East African Rift. Lava flows traveled almost 50 km and

FIGURE 5.18 Hazards due to lava and ash from volcanic eruptions.

Lava Flows

(a) A lava flow reaches a house in Hawaii and sets it on fire.

(b) Lava from Mt. Etna threatens a town and olive grove in Sicily.

(c) Residents rescue household goods after a lava flow filled the streets of Goma, along the East African Rift.

(d) This empty school bus was engulfed by lava in Hawaii.

Pyroclastic Debris

(e) A pyroclastic flow rushes down the slopes of the Soufrière Hills volcano of Montserrat.

(f) A blizzard of ash fell from the cloud erupted by Mt. Pinatubo in the Philippines.

(g) Lapilli falls from an eruption in Iceland.

Lahar

(h) A lahar submerges farmland in Colombia.

flooded the streets of the Congolese city of Goma, encasing them in a 2-m-thick layer of basalt. The flows destroyed almost half the city.

Threat of Pyroclastic Flows Pyroclastic flows can move extremely fast (100 to 300 km per hour) and are so hot (500°C to 1,000°C) that they represent a profound hazard to humans and the environment **(Fig. 5.18e)**. Even relatively small examples, such as the flow that struck St. Pierre on Martinique, can flatten towns and devastate fields despite leaving only a few centimeters of ash and lapilli behind. People caught in the direct path of such flows may be incinerated, and even those protected from the ash itself may die from inhaling toxic, superhot gases.

Threat of Falling Ash and Lapilli During a large pyroclastic eruption, ash and lapilli erupt into the air, then later fall back to the ground **(Fig. 5.18f, g)**. Close to the volcano, lapilli tumble out of the sky and can accumulate to form a blanket up to several meters thick. The mass, especially when saturated with rainwater, causes roofs and power lines to collapse. Winds can carry fine ash over a broader region. In the Philippines, for example, typhoon winds during the 1991 eruption of Mt. Pinatubo covered a 4,000-km^2 area with ash. An ash fall buries crops, coats the leaves of trees, and may spread toxic chemicals that poison the soil. Ash also insidiously infiltrates machinery, causing moving parts to wear out.

Threat of High Elevation to Aircraft Fine ash from an eruption that rises to stratospheric heights can be dangerous to airplanes. Like a sandblaster, sharp, angular shards of ash abrade turbine blades, greatly reducing engine efficiency. The ash, along with sulfuric acid aerosols, scores windows and damages the fuselage. Also, when heated inside a jet engine, the ash melts, yielding a liquid that coats interior parts of the engine and freezes into glass, coating temperature sensors, which falsely indicate that the engines are overheating, so that they automatically shut down.

Encounters between airliners and high-altitude ash have led to terrifying incidents. In 1982, a British Airways 747 flew through the ash cloud above a volcano in Java. The windshield turned opaque and all four engines shut down. For 13 minutes, the plane silently glided earthward, dropping from its initial 11.5-km altitude. The pilot frantically tried to restart the engines to no avail and prepared to ditch at sea. Finally, at 3.7 km, the engines cooled sufficiently and suddenly roared back to life, and the plane headed to Jakarta for an emergency landing. There, without functioning instruments, the pilot squinted out an open side window to see the runway and land safely. A similar event happened to a KLM 747 that

encountered the eruptive cloud of Redoubt Volcano in Alaska in 1989 (see Fig. 5.12c).

Because of the lessons learned from such incidents, the 2010 eruption of the volcano Eyjafjallajökull in Iceland had a profound impact on air traffic. All told, the eruption sent only about 0.25 km^3 of pyroclastic material up into the air. But, at the time of the eruption, the jet stream, a high-altitude current of rapidly moving air, was passing over Iceland. This dispersed the volcanic ash throughout European air space. Because of concern about possible damage to planes, officials shut down almost all air traffic across Europe for 6 days. The closure directly cost airlines $200 million per day, disrupted travel plans for countless passengers, and halted air shipment of everything from television sets to flowers, thus affecting economies worldwide.

OTHER HAZARDS RELATED TO ERUPTIONS

Threat of the Blast The explosion of a volcano, like the blast of a bomb, flattens everything in its path. After the explosion of Mt. St. Helens, for example, the once-towering trees of the forest around the volcano were stripped of bark and needles and lay scattered over the hillslopes like matchsticks (see Box 5.1). Were such a blast to strike a city, the consequences would be catastrophic.

Threat of Landslides and Lahars Eruptions commonly trigger large landslides along a volcano's flanks. The debris, composed of ash and solidified lava that erupted earlier, can move both fast (as much as 250 km per hour) and far. During the eruption of Mt. St. Helens, 8 billion tons of debris took off down the mountainside, careered over a 360-m-high ridge, and tumbled down a river valley, until the last of it finally came to rest more than 20 km from the volcano.

Mixing volcanic ash and other debris with water produces a lahar that resembles wet concrete. Because lahars are denser and more viscous than clear water, they pack more force than clear water and can carry away everything in their path. The lahars of Mt. St. Helens, for example, traveled along existing drainages more than 40 km from the volcano, at speeds of 5 to 50 km per hour. When they had passed, they left a gray and barren wake of mud, boulders, broken bridges, and crumpled houses, as if a giant knife had scraped across the landscape. Perhaps the most destructive lahar of recent times accompanied the eruption of the snow-crested Nevado del Ruiz in Colombia on the night of November 13, 1985. The lahar surged down a valley like a 40-m-high wave, hitting the sleeping town of Armero, 60 km from the volcano. Within the town, 90% of the buildings

FIGURE 5.19 The CO_2 disaster at Lake Nyos.

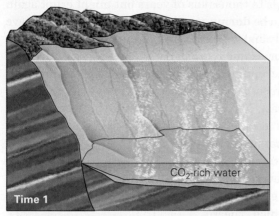 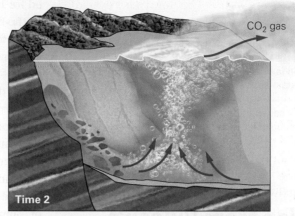

(a) Carbon dioxide dissolved in the colder bottom water. When a landslide or wind disturbed the lake, the saturated water rose. The CO_2 came out of solution and, in gas form, flowed out of the crater.

vanished, replaced by a 5-m-thick layer of mud, which now entombs the remains of 25,000 people **(Fig. 5.18h)**.

Threat of Earthquakes and Tsunamis Earthquakes accompany almost all major volcanic eruptions, for the movement of magma breaks rocks underground. Such earthquakes may trigger landslides on the volcano's flanks and can cause buildings to collapse and dams to rupture, even before the eruption itself begins.

Where explosive eruptions occur in the sea or along a coast, the blast and the underwater collapse of a caldera, or the sudden slip of giant landslides on the flank of the volcano, triggered by the eruption, can generate huge sea waves, or tsunamis, tens of meters high. Most of the 36,000 deaths attributed to the 1883 eruption of Krakatau were due not to ash or lava, but rather to tsunamis that slammed into nearby coastal towns.

Threat of Gas We have already seen that volcanoes erupt not only solid material but also large quantities of gases. The hydrogen sulfide in volcanic gas has a rotten-egg smell, and along with acidic aerosols and other gases, produces a yellowish, choking, caustic cloud that is dangerous to breathe. Occasionally, tourists visiting Japan's Mt. Aso, a volcano known for the clouds of volcanic gas rising from its summit, have inhaled too much of the gas and have fainted; over the years, a few have died.

Carbon dioxide, unlike the sulfurous gases and aerosols that we've just described, can be a deadly threat, as illustrated by the catastrophe that happened near Lake Nyos in 1986. Lake Nyos is a small, deep lake that fills the crater of a volcano in Cameroon, western Africa. Though only 1 km across, the lake reaches a depth of over 200 m. Carbon dioxide gas slowly bubbles out of cracks in the floor of the crater and dissolves

Volcanic Gas

(b) The CO_2 suffocated cattle on the slopes below the volcano.

in the stable layer of cool water at the bottom of the lake. Apparently, by August 21, 1986, the denser, cool bottom-water layer had become supersaturated in carbon dioxide. On that day, perhaps because of a landslide, the layer was disturbed, and the cold water was forced upward **(Fig. 5.19a)**. As the water rose, it underwent decompression. As a result, the CO_2 gas came out of solution, violently bubbled upward, and rose from the lake. Since it's denser than air, the invisible CO_2 flowed down the flank of the volcano and spread out over the countryside for a distance of about 23 km before dispersing. Although it is not toxic, CO_2 does not provide the oxygen that is essential for metabolism and oxidation. Sadly, when the gas cloud engulfed the village of Nyos, it quietly put out cooking fires and suffocated 1,742 sleeping people, as well as thousands of cattle **(Fig. 5.19b)**.

TAKE-HOME MESSAGE

Volcanoes can be dangerous! The lava flows, ash clouds, pyroclastic flows, explosions, mudflows (lahars), landslides, earthquakes, and tsunamis that can be produced during eruptions can destroy towns and farmland. Even the gas emitted by a volcano can lead to catastrophe.

QUICK QUESTION Why can a lahar do so much more damage than a similarly sized flood of clear water?

5.6 Protection from Vulcan's Wrath

Volcanic eruptions are a natural hazard of extreme danger. Can anything be done to protect lives and property from this danger? The answer is yes. Below, we first examine the evidence that geologists can use to determine whether a volcano has the potential to erupt. We then consider the suite of observations that may allow geologists to predict the timing of an impending eruption.

ACTIVE, DORMANT, AND EXTINCT VOLCANOES

Geologists refer to volcanoes that are erupting, have erupted recently, or are likely to erupt within a time frame of decades as **active volcanoes**, and consider vents that have not erupted for hundreds to thousands of years but might erupt again in the future to be **dormant volcanoes**. Volcanoes that have shut off entirely and will never erupt again are called **extinct volcanoes**. For example, Hawaii's Kilauea is an active volcano, for it is erupting currently and has erupted numerous times during recorded history. In contrast, Mt. Fuji in Japan last erupted in 1708 but has not had activity for over 300 years. Subduction, however, continues along the eastern edge of Japan, which means that Mt. Fuji could erupt in the future, so geologists consider it to be dormant. Devils Tower, in eastern Wyoming, formed when magma intruded beneath the region about 40 million years ago **(Fig. 5.20)**. No volcanism can happen near Devils Tower now, for the geologic conditions that caused volcanism there have not existed for almost 40 million years. We can say, therefore, that any volcanoes that once erupted in the Devils Tower area are now extinct. In fact, they've been extinct so long that they've completely eroded away.

The condition of a volcano's surface, whether it has undergone erosion or has been covered with vegetation, serves as one clue to whether a volcano should be classified as active, dormant, or extinct. Specifically, the surface of an active volcano tends to be relatively free of mature vegetation, because not enough time has passed since lava or ash covered the surface for new vegetation to grow. Also, an active volcano's surface tends to host relatively few gullies, because erosion hasn't yet had time to carve into the surface. Notably, the rate of erosion depends greatly on the composition of a volcano's surface—cinders, for example, erode rapidly, whereas lava flows can survive for a longer time. In contrast, a dormant volcano may be covered with forest (below the treeline), and may have undergone so much erosion that its surface hosts many deep gullies **(Fig. 5.21)**. Typically, extinct volcanoes have been eroded so deeply that none of the edifice remains, and exposures of intrusive rocks that once lay deep beneath the edifice are now exposed at the Earth's surface.

> **Did you ever wonder . . .**
> whether a volcano could erupt beneath London, England?

PREDICTING ERUPTIONS

Predicting the time of an eruption at a given volcano decades or even years in

FIGURE 5.20 Devils Tower, Wyoming, formed as an intrusion into sedimentary rocks beneath a volcano. It solidified hundreds of meters below the surface of the Earth and has been exposed by erosion. Cooling produced spectacular columnar joints.

The vertical lines are columnar joints.

FIGURE 5.21 The shape of a volcano changes as it is eroded.

An active volcano is a smooth cone.

Erosion carves gullies into the volcano.

Eventually, only hills of volcanic rock remain.

Time

advance is impossible. The only basis for constraining a long-term prediction comes from determining the *recurrence interval* (the average time between eruptions) for the volcano. Geologists calculate recurrence intervals by measuring the ages of erupted layers comprising the volcano. For example, Mt. Fuji has erupted about 65 times in the last 10,000 years, so it has an eruptive recurrence interval of about 150 years. Note that a recurrence interval does not indicate periodicity—some of Fuji's eruptions were decades apart, while others were centuries apart. The recurrence interval just gives us a sense of the probability of an eruption during a time period.

Some volcanoes send out distinct warning signals announcing that an eruption may take place within a time frame of days or months, for as magma squeezes into the magma chamber, it causes a number of changes that geologists can measure:

> *Earthquake activity:* Movement of magma generates vibrations in the Earth. When magma flows into a volcano, rocks surrounding the magma chamber crack, and blocks slip with respect to one another. Such cracking and shifting cause earthquakes at depths between 1 and 7 km beneath a volcano.

> *Changes in heat flow:* The presence of hot magma increases the local *heat flow,* the amount of heat passing upward through rock. In some cases, the increase in heat flow melts snow or ice on the volcano, triggering floods and lahars even before an eruption occurs.

> *Changes in shape:* As magma fills the magma chamber inside a volcano, it pushes outward and can cause the surface of the volcano to bulge; the same effect happens when you blow air into a balloon. Geologists now use laser sighting, accurate tiltmeters, surveys using global positioning systems, and a technique known as *InSAR* (interferometric synthetic aperture radar), which uses radar beams emitted by satellites to measure elevation changes in order to detect any alteration in a volcano's shape due to the rise of magma **(Fig. 5.22)**.

> *Increases in gas and steam emission:* Even though magma remains below the surface, gases bubbling out of the magma, together with steam formed as the magma heats groundwater, percolate upward through cracks in the Earth and rise from the volcanic vent. So an increase in the volume and composition of gas emissions, or the birth of new hot springs may indicate that magma has entered the ground below.

FIGURE 5.22 To produce this InSAR map of Mt. Longonot, Kenya, a satellite measures the elevation at two different times (2004 and 2006). If the land has warped between measurement times, interference of the radar-beam waves, when superimposed, appears as color-spectrum bands. Each rainbow-like set indicates 3 cm of uplift, so the total uplift is 9 cm.

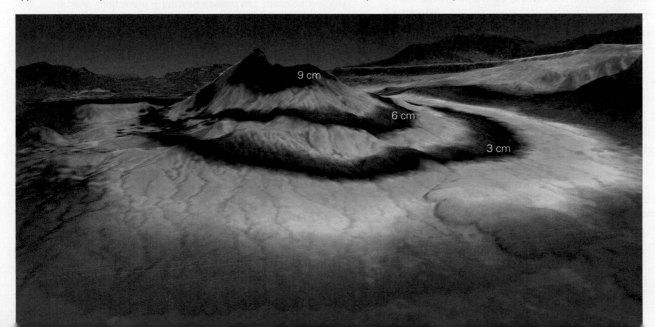

9 cm

6 cm

3 cm

FIGURE 5.23 A volcanic-hazard assessment map for Mt. Rainier, Washington, showing the regions that might be affected by flows and lahars. Note that lahars may travel long distances down river valleys.

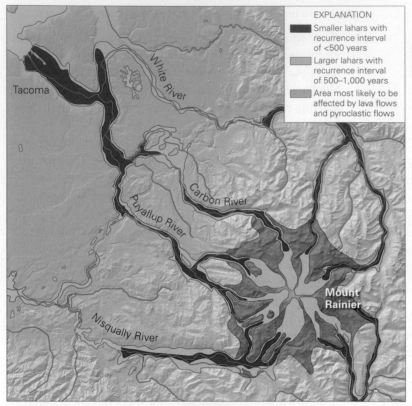

EXPLANATION

■ Smaller lahars with recurrence interval of <500 years

■ Larger lahars with recurrence interval of 500–1,000 years

■ Area most likely to be affected by lava flows and pyroclastic flows

MITIGATING VOLCANIC HAZARDS

Danger Assessment Maps and Evacuation Let's say that a particular active volcano has the potential to erupt in the near future. What can geologists do to prevent the loss of life and property? First, they compile a *volcanic-hazard assessment map* (Fig. 5.23). Such maps show areas that lie in the path of potential lava flows, lahars, or pyroclastic flows and should therefore be evacuated if a volcano starts to erupt. In the case of Mt. St. Helens, hundreds of lives were saved in 1980 by timely evacuation, but in the case of Mt. Pelée in 1902, thousands of lives were lost because warning signs were ignored.

Diverting Flows Sometimes people have used direct force to change the direction of a lava flow, or even to stop it. For example, during a 1669 eruption of Mt. Etna, an active volcano on the Italian island of Sicily, basaltic lava formed a glowing orange river that began to spill down the side of the mountain. When the flow approached the town of Catania, 16 km from the summit, 50 townspeople protected by wet cowhides boldly hacked through the

Did you ever wonder . . . whether people can redirect a lava flow?

chilled side of the flow to create an opening through which the lava could exit. By doing so, they hoped to cut off the supply of lava feeding the end of the flow near their homes. Their strategy worked, and the flow began to ooze through the new hole in its side. But unfortunately, the diverted flow began to ooze toward the neighboring town of Paterno. Seeing the threat, 500 men from Paterno ran up to the flow and chased away the Catanians who had been keeping the diversion hole open. Eventually, the hole closed, and the flow resumed its path toward Catania, sadly destroying part of the town. More recently, in 1983 and 1992, people have tried more aggressive techniques to divert Etna's flows—they've used high explosives to blast breaches in the flanks of flows and used bulldozers to build dams and channels of rubble to divert flows (**Fig. 5.24a**).

Inhabitants of Heimaey, a small volcanic island off the south shore of Iceland, used a particularly creative approach to stop a flow before it completely overran a harbor town. Between February and July 1973, workers both on land and on ships in the harbor used powerful pumps to spray millions of cubic meters of cold seawater onto the flow, hoping to freeze it in its tracks (**Fig. 5.24b**). The lava boiled away the water, and clouds of steam engulfed the town during the operation, but the pumpers persisted, and probably succeeded in preventing the flows from advancing as far as they might have otherwise. When the pumps were turned off, the lava was covered by over 200,000 tons of salt, precipitated from the evaporating seawater.

TAKE-HOME MESSAGE

Volcanoes don't erupt continuously and don't last forever, so we can distinguish among active, dormant, and extinct volcanoes. Once a volcano ceases to erupt, erosion destroys its eruptive shape. Geologists can provide near-term predictions of eruptions, so people can take precautions. In some cases, people have actually been able to divert flows.

QUICK QUESTION What does a volcanic-hazard assessment map show?

5.7 Effect of Volcanoes on Climate and Civilization

In 1783, Benjamin Franklin was living in Europe, serving as the American ambassador to France. The summer of that year seemed to be unusually cool and hazy. Franklin,

FIGURE 5.24 Efforts to divert lava flows away from inhabited locations.

(a) Workers spraying a lava flow to solidify it and using a bulldozer to build an embankment to divert it on the flanks of Mt. Etna.

(b) Firefighters pumping 6 million cubic meters of water on a lava flow on the island of Heimaey, Iceland, in an effort to freeze it and stop it.

who was an accomplished scientist as well as a statesman, couldn't resist seeking an explanation for this phenomenon, and learned that in June of that year, a huge volcanic eruption had taken place in Iceland. He wondered if the "smoke" from the eruption had prevented sunlight from reaching the Earth, thus causing the cooler temperatures. Franklin reported this idea at a meeting, and by doing so, may well have been the first scientist ever to suggest a link between eruptions and climate.

Franklin's idea seemed to be confirmed in 1815, when Mt. Tambora in Indonesia exploded, ejecting more than 145 km³ of ash and pumice into the air. The sky became hazy, and stars dimmed by a full magnitude. Temperatures dipped so low in the northern hemisphere that 1816 became known as "the year without a summer." The unusual weather of that year influenced artists and writers. For example, memories of fabulous sunsets and the hazy glow of the sky inspired the luminous atmospheric quality that made the landscape paintings of the English artist J. M. W. Turner so famous (see Fig. 5.1a). Mary Shelley, trapped in her house by bad weather, wrote *Frankenstein*, with its numerous scenes of gloom and doom. Similar eruption-triggered global temperature drops also followed the 1883 eruption of Krakatau and the 1991 eruption of Pinatubo.

How can a volcanic eruption produce these cooling effects? When a large explosive eruption takes place, ash and aerosols enter the stratosphere. In only about 2 weeks, winds distribute the ash and aerosol cloud around the planet. Because they circulate above the weather and do not get washed away by rainfall, the ash and aerosol may remain suspended for many months to years. The resulting haze keeps the

Sun's heat from reaching the Earth—in the case of Pinatubo's eruption, researchers were able to document that the event temporarily diminished the amount of sunlight reaching the Earth by about 10%, thereby affecting global temperatures **(Fig. 5.25a)**. Unlike greenhouse gases, such as CO_2, volcanic haze does not prevent heat from escaping.

The largest explosive eruption to happen in the last million years of the Earth's history blasted apart the Toba Volcano in Indonesia about 75,000 years ago. The explosion sent out huge quantities of ash, covering much of southern Asia like a snowfall. In addition, the eruption injected a huge dose of aerosols into the atmosphere. Not only did the aerosols and suspended ash decrease the solar radiation reaching the Earth, but the white ash on the ground reflected back into space some of the radiation that did reach the planet. As a result, the atmosphere may have remained cool for as long as 1,000 years, long enough to drive many species of organisms to extinction. Some anthropologists have speculated that after the cooling event, fewer than 10,000 humans remained on the Earth.

In more recent millennia, volcanic eruptions may have led to the demise of civilizations. For example, archaeological research suggests that the Minoan culture, which thrived in the eastern Mediterranean during the Bronze Age, disappeared during the century following explosive eruptions of the Santorini Volcano in 1645 B.C.E. **(Fig. 5.25b)**. All that remains of Santorini today is a huge caldera whose rim projects above sea level as the Greek island Thera. Ash clouds, tsunamis, and earthquakes generated by Santorini may have disrupted their daily lives so severely that the Minoan people decided to move elsewhere.

FIGURE 5.25 Volcanic eruptions may affect climate.

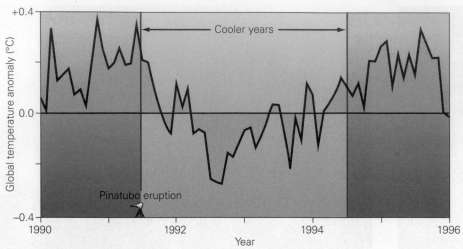

(a) The large injection of dust and aerosols into the atmosphere caused the average global temperature to drop for a few years after the eruption of Mt. Pinatubo in 1991.

(b) The eruption of Santorini in 1645 B.C.E. may have contributed to the demise of the Minoans, who left behind ruins of elaborate palaces.

TAKE-HOME MESSAGE

The ash, gases, and aerosols produced by explosive eruptions can be blown around the globe in the stratosphere. This material can cause significant global cooling.

QUICK QUESTION How might large explosive eruptions have affected the course of human history?

5.8 Volcanoes on Other Planets

We conclude this chapter by looking beyond the Earth, for our planet is not the only object in the Solar System to host volcanic eruptions. We can see the effects of volcanic activity on our nearest neighbor, the Moon, just by looking up on a clear night. The broad darker areas of the Moon, the **maria** (singular *mare*, after the Latin word for sea), consist of flood basalts that erupted more than 3 billion years ago **(Fig. 5.26a)**. Geologists propose that the flood basalts formed after huge meteors collided with the Moon, blasting out giant craters. Crater formation decreased the pressure in the Moon's mantle so that it underwent partial melting, producing basaltic magma that eventually rose to the surface through fissures and filled the craters.

Did you ever wonder . . . if the Earth is the only planet with volcanoes?

The surfaces of both Mars and Venus display distinct volcanic edifices, some of which have calderas at their peaks **(Fig. 5.26b)**. In fact, the largest known mountain in the Solar System, Olympus Mons **(Fig. 5.26c)**, is an extinct shield volcano on Mars. The base of Olympus Mons spans a distance of 600 km, and its peak rises 27 km above the surrounding plains.

Active volcanism currently occurs on Io, one of the many moons of Jupiter. Cameras in the *Galileo* spacecraft have recorded these volcanoes in the act of spraying plumes of sulfur gas into space **(Fig. 5.26d)** and have tracked immense, moving lava flows. Different colors of erupted material make the surface of this moon resemble a pizza. Researchers have proposed that the heat driving this volcanic activity is due to tides, in that the gravitational pull exerted by Jupiter and by other moons alternately stretches and then squeezes Io, generating sufficient friction to keep Io's mantle hot. Geologists have also detected eruptions from moons of Saturn **(Fig. 5.26e, f)**.

TAKE-HOME MESSAGE

Space exploration reveals that volcanism occurs not only on the Earth, but has also left its mark on other terrestrial planets and on the moons of giant planets.

QUICK QUESTION Since most volcanic activity on the Earth is a result of plate tectonics, what does the lack of volcanic activity on the Moon tell us about whether plate tectonics happens on the Moon?

FIGURE 5.26 Volcanism on other planets and moons in the Solar System.

(a) Maria of the Moon were seas of basaltic lava.

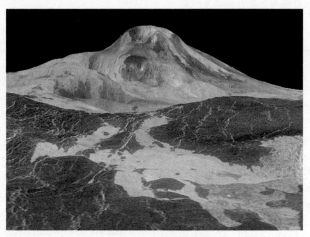

(b) A volcano rises above the plains of Venus. Lava flows cover part of the land in the foreground.

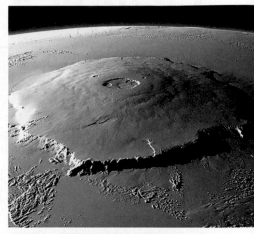

(c) Olympus Mons rises 27 km above the surface of Mars. It's the largest volcano in our Solar System.

(d) An active volcano erupts sulfur on Io, a moon of Jupiter.

(e) Large cracks at the southern end of Enceladus, a moon of Saturn, erupt water vapor.

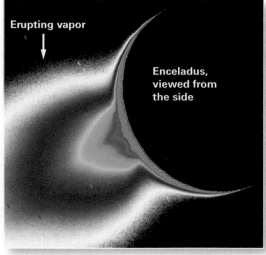

(f) The *Cassini* spacecraft detected gas erupting from Enceladus.

ANOTHER VIEW Molten rock spills into the sea at the end of a lava flow produced by the 2018 eruption on Hawaii. The heat of the lava instantly turns the water that it touches into steam.

Chapter 5 Review

CHAPTER SUMMARY

> Volcanoes are vents at which molten rock (lava), pyroclastic debris, gas, and aerosols erupt at the Earth's surface.

> The characteristics of a lava flow depend on its viscosity, which in turn depends on its temperature and composition.

> Basaltic lava can flow great distances. Pahoehoe flows have smooth, ropy surfaces, whereas a'a' flows have rough, rubbly surfaces. Andesitic and rhyolitic lava flows tend to pile into mounds at a volcano's vent.

> Pyroclastic debris includes powder-sized ash, marble-sized lapilli, and apple- to refrigerator-sized blocks and bombs. Some debris falls from the air, and some settles from pyroclastic flows.

> The summit of an erupting volcano usually includes a crater. Collapse of a volcano, when a large magma chamber drains, can yield a much larger bowl-shaped depression called a caldera.

> A volcano's shape depends on the style of eruption. Shield volcanoes are broad, gentle domes, and cinder cones are steep-sided, symmetric hills. Stratovolcanoes (composite volcanoes) can become quite large and consist of alternating layers of pyroclastic debris and lava.

> The type of eruption depends on the lava's viscosity and gas content. Effusive eruptions produce flows, and explosive eruptions produce pyroclastic debris. Typically, the former are basaltic, and the latter are andesitic or rhyolitic.

> Different kinds of volcanoes form in different geologic settings, as explained by plate tectonics theory.

> Volcanic eruptions pose many hazards: lava flows overrun roads and towns, ash falls blanket the landscape, pyroclastic flows incinerate towns and fields, landslides and lahars bury the land surface, earthquakes topple structures and rupture dams, tsunamis wash away coastal towns, and gases suffocate people and animals.

> Eruptions can be predicted by earthquake activity, changes in heat flow, changes in shape of the volcano, and changes in the volume of gas and steam.

> We can minimize the consequences of an eruption by using volcanic-hazard assessment maps and by making evacuation plans.

> Immense flood basalts cover portions of the Moon. The largest known volcano, Olympus Mons, towers over the surface of Mars. Satellites have documented evidence for eruptions on moons of Jupiter and Saturn.

GUIDE TERMS

a'a' (p. 142)
active volcano (p. 164)
aerosol (p. 143)
ash (p. 146)
block (p. 144)
bomb (p. 144)
caldera (p. 146)
cinder cone (p. 146)
columnar jointing (p. 142)
crater (p. 146)

dormant volcano (p. 164)
effusive eruption (p. 147)
eruption (p. 139)
explosive eruption (p. 147)
extinct volcano (p. 164)
fissure (p. 146)
flood basalt (p. 160)
lahar (p. 146)
lapilli (p. 144)

large igneous province (LIP) (p. 160)
lava flow (p. 139)
lava fountain (p. 144)
lava tube (p. 142)
magma chamber (p. 146)
mare (p. 168)
pahoehoe (p. 142)
pillow lava (p. 142)

pyroclastic debris (p. 144)
pyroclastic flow (p. 146)
shield volcano (p. 146)
stratovolcano (p. 146)
supervolcano (p. 151)
tephra (p. 144)
tuff (p. 146)
vesicle (p. 143)
volcano (p. 139)

GEOTOURS THIS CHAPTER'S GEOTOURS WORKSHEET (E) FEATURES QUESTIONS AND GOOGLE EARTH SITES ON:

> Shield volcanoes   > Stratovolcanoes   > Cinder cone volcanoes

REVIEW QUESTIONS

The letters following each Review Question refer to the corresponding Learning Objective from the Chapter Opener.

1. Describe three different kinds of material that can erupt from a volcano. Identify them on the figure. **(B)**

2. How does pyroclastic flow differ from a lahar? **(B)**

3. Why do some volcanic eruptions consist mostly of lava flows, whereas others are explosive and do not produce lava flows? **(C)**

4. Distinguish among different types of explosive eruptions. **(D)**

5. Describe the activity in the mantle that leads to hot-spot eruptions. **(D)**

6. How do continental-rift eruptions form flood basalts? **(D)**

7. What is a large igneous province (LIP)? **(D)**

8. Identify some of the major volcanic hazards, and explain how they develop. **(E)**

9. Contrast shield volcanoes, stratovolcanoes, and cinder cones. Which is pictured in the diagram? How are these differences explained by the composition of their lavas and other factors? **(A)**

10. How do geologists predict impending volcanic eruptions? **(F)**

11. Distinguish among active, dormant, and extinct volcanoes. **(F)**

12. Explain how steps can be taken to protect people from the effects of eruptions. **(E)**

ON FURTHER THOUGHT

13. The Long Valley caldera, near the Sierra Nevada range in California, resulted from an explosion about 700,000 years ago that produced a huge ignimbrite called the Bishop Tuff. About 30 km to the northwest lies Mono Lake, with an island in the middle. Hot springs can be found along the lake. You can see the lake on Google Earth at Latitude 37°59″56.58″ N, Longitude 119°2″18.20″ W. Explain the origin of Mono Lake and its island. Is it a volcanic hazard? **(B, E)**

14. Why don't large eruptions on Hawaii cause global cooling? **(E)**

15. Mt. Fuji is a 3.6-km-high stratovolcano in Japan formed as a consequence of subduction. Find it using Google Earth at Latitude 35°21″46.72″ N, Longitude 138°43″49.38″ E. Why do andesites erupt here? Very little andesite occurs on the Mariana Islands. Why? **(D)**

16. Using information you can find on the Web, calculate the recurrence interval of eruptions at Mt. Vesuvius. **(F)**

ONLINE RESOURCES

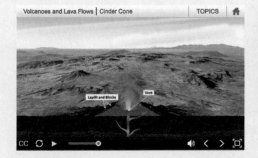

Animations This chapter features animations covering the various types of volcanoes and lava flows, and the formation and activity of hot spots.

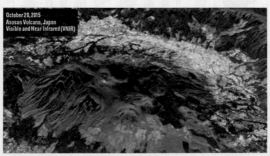

Videos This chapter features videos on monitoring volcanoes and on lava activity, flows, and hazards.

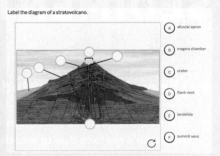

Smartwork5 Questions include visual and labeling exercises on the processes involved in the structure and eruptive styles of various types of volcanoes, and assessments of the hazards volcanoes present.

A SURFACE VENEER: SEDIMENTS AND SOILS

By the end of this interlude, you should be able to . . .

A. describe the changes that rocks undergo at and near the Earth's surface due to weathering.

B. explain how weathering produces sediment.

C. recognize the basic components and layers of a soil.

D. appraise factors that affect the character and thickness of soil.

B.1 Introduction

In the 1950s, the government of Egypt decided to build the Aswan High Dam to trap water of the Nile River in a huge reservoir before the water could reach the Mediterranean Sea. In the process of identifying a good site for the dam's foundation, geologists discovered that the present-day Nile River flows on the surface of a 1,500-m-thick layer of gravel, sand, and mud that fills what was once a canyon as large as the Grand Canyon (Fig. B.1). How could the river once have carved a canyon this deep, and why did the canyon later fill with debris?

The origin of the pre-Nile canyon remained a mystery until the summer of 1970, when geologists began to study the material that underlies the floor of the Mediterranean Sea. They expected to find layers consisting of the shells of *plankton* (tiny floating organisms) that had settled out of the water or out of clay that rivers had carried to the sea. To their surprise, however, they also found a 2-km-thick layer of halite and gypsum. Such minerals, types of salts, precipitate from seawater when the seawater dries up. In order to yield a salt layer that is 2 km thick, the entire Mediterranean must have dried up almost completely several times, with the sea refilling after each drying event. This discovery solved the mystery of the pre-Nile canyon. When the Mediterranean Sea dried up, the Nile River flowed down into a deep lowland, and in the process, it carved a canyon. Later, when the sea refilled with water, the river could no longer cut down, and the flooded canyon filled with sand and gravel brought in from upstream.

Why did the Mediterranean Sea dry up? Only 10% of its water comes from rivers, so for the Mediterranean Sea to remain full, water must flow in from the Atlantic Ocean through the Strait of Gibraltar. About 6 million years ago, the northward-drifting African Plate collided with the European Plate, forming a natural dam separating the Mediterranean from the Atlantic. At times when global sea level dropped, water stopped flowing over this dam from the Atlantic, and the Mediterranean evaporated. The salt that had been dissolved in its water precipitated to form a solid deposit of halite and gypsum on the floor of the resulting basin, and the pre-Nile canyon formed. At times when sea level rose, water flooded in from the Atlantic into the Mediterranean, filling the basin again. About 5.5 million years ago, the Mediterranean rose to its present level, and gravel, sand, and mud carried by the Nile River accumulated in the pre-Nile canyon.

◀ (facing page) On the coast of Heimaey Island, off the south coast of Iceland, aprons of sediment collect at the base of lava cliffs. At the shore, waves break up this sediment to make black sand. In the island's wet climate, soils form quickly on the sediment and grass takes root in the soil. Sediment and soil form a veneer that hides bedrock.

FIGURE B.1 The Nile River flows on top of a sediment-filled canyon. This canyon formed when water evaporated from the Mediterranean, and the river cut down to the level of the basin floor.

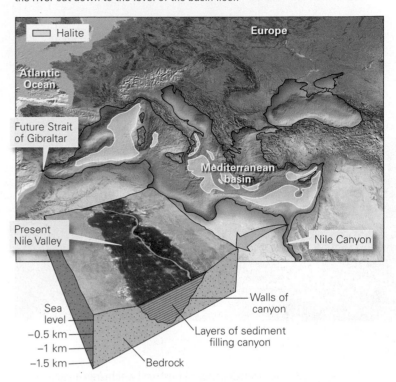

Geologists refer to the kinds of deposits just described—sand, mud, gravel, halite and gypsum layers, and shell fragments—as sediment. **Sediment**, broadly defined, consists of loose fragments of rocks or minerals broken off bedrock, mineral crystals that precipitate directly out of water, and shells or shell fragments. The materials in sediments ultimately came from the *weathering* (physical and chemical breakdown) of pre-existing rock. Sediments form a veneer, or *cover*, over bedrock (Fig. B.2). This cover's thickness ranges from 0 km, in places where bedrock crops out at the Earth's surface, to 20 km, in rapidly sinking basins. Some sediments transform into *soil*, which may serve as a substrate for plants. In this interlude, we look at how weathering produces sediment and how soils form and evolve. Note that geologists refer to any loose debris (sediment or soil) as *regolith*.

B.2 Weathering: Forming Sediment

Rock exposed at the Earth's surface sooner or later disintegrates and crumbles away, due to weathering. In a geologic context, **weathering** refers to the combination of phenomena

FIGURE B.2 A layer of unconsolidated sediment (sand, clay, and cobbles), topped by dark soil, overlies bedrock in this outcrop along the coast of western Ireland.

that corrode and break up solid rock, eventually transforming it into loose debris. The process also produces ions that dissolve in surface water and groundwater. Weathered rock may look discolored, or rough, compared to unweathered or *fresh rock* **(Fig. B.3)**. Just as a plumber can unclog a drain by using physical force (with a plumber's snake) or by causing a chemical reaction (with a dose of liquid drain opener), nature can attack rocks with two types of weathering: physical and chemical.

PHYSICAL WEATHERING

Physical weathering, also called *mechanical weathering*, breaks intact rock into unconnected grains or chunks known as **clasts**, which come in a range of sizes **(Table B.1)**. Geologists may refer to an accumulation of clasts as *detritus*. Many phenomena contribute to physical weathering, as we now describe.

Jointing A joint is a natural crack in rock—the formation of a joint separates one piece of rock into two separate pieces. Almost all rock outcrops contain joints. For example, large granite plutons typically split into onion-like sheets along

FIGURE B.3 The contrast between fresh and weathered granite in an Arizona outcrop. Weathered granite can break apart. Grains fall off and collect as regolith at the base of the outcrop. The inset photos show how weathering visibly changes the rock.

TABLE B.1 Clasts Are Classified by Grain Diameter

Boulders	More than 256 mm
Cobbles	Between 64 mm and 256 mm
Pebbles	Between 2 mm and 64 mm
Sand	Between 1/16 mm and 2 mm
Silt	Between 1/256 mm and 1/16 mm
Clay	Less than 1/256 mm

exfoliation joints that lie parallel to the mountain face, and sedimentary rock layers tend to break into rectangular blocks bounded by joints on the sides and layer boundaries above and below **(Fig. B.4a, b)**. Jointing effectively breaks bedrock into many separate blocks which, when exposed on a slope, eventually tumble downslope, fragmenting into smaller pieces as they fall. The resulting chunks may collect in an apron of *talus*, the rock rubble at the base of a slope, or may be carried away by rivers or glaciers at the base of

FIGURE B.4 Joints (natural cracks) break bedrock into blocks and sheets, which can tumble down a slope.

(a) Vertical joints break beds of sedimentary rock into blocks that fall to the base of this 100-m-high cliff in Ireland.

(b) Exfoliation joints on this granite mountain slope in the Sierra Nevada break the rock into onion-like sheets.

(c) Recently formed talus, collecting at the base of a cliff in the Canadian Rockies.

Sandstone

Granite

Bedding

Vertical joints

Exfoliation joints

Different rock types display different types of joints.

a cliff (**Fig. B.4c**). Joints form for a variety of reasons (see Chapter 9). For example, when rock that had been buried deeply in the crust rises toward the Earth's surface, as **erosion** (the breaking off and removal of rock or sediment) strips away overburden, the rock becomes cooler and the pressure squeezing it decreases. This change causes the rock to change shape only slightly, but enough to cause hard rock to crack.

Frost Wedging Freezing water can burst pipes and shatter bottles because water expands when it freezes and pushes the walls of the container apart. The same phenomenon happens in rock. When water trapped in a joint freezes, it forces the joint open and may cause the joint to grow. Such *frost wedging* helps break blocks free from intact bedrock (**Fig. B.5a**).

Salt Wedging In arid climates, dissolved salt in groundwater precipitates in open pore spaces in rocks, forming crystals that push apart the surrounding grains. This process, called *salt wedging*, weakens rock so that when exposed to wind and rain, the rock disintegrates into separate grains. The same phenomenon happens along the seacoast, where salt spray percolates into rock and then dries (**Fig. B.5b**).

Root Wedging Have you ever noticed how the roots of a tree can break up a sidewalk? As roots grow, they apply pressure to their surroundings and can push joints open in a process known as *root wedging* (**Fig. B.5c**).

FIGURE B.5 Wedging is one type of physical (mechanical) weathering.

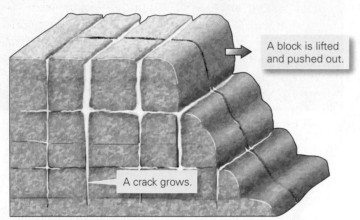

A block is lifted and pushed out.

A crack grows.

(a) Frost wedging occurs when the water that fills cracks freezes; it expands and wedges the cracks open.

(b) Salt wedging in rock on a beach cliff in Scotland yields "honeycomb weathering".

Tree growing in a joint.

Eventually, the blocks tumble to the base of the cliff.

(c) Root wedging pushes open a joint, slowly separating a block from the cliff.

Thermal Expansion When the heat of an intense forest fire bakes a rock, the outer layer of the rock expands. On cooling, the layer contracts, which generates force in the rock sufficient to break off the outer part of the rock. Recent research suggests that the heat of the Sun's rays sweeping across dark rocks in a desert may, over time, cause cobbles on the ground surface to fracture into thin slices.

Animal Attack Animal life also contributes to physical weathering, for burrowing creatures, from earthworms to gophers, can move rock fragments. In the past century, humans have become perhaps the most energetic agent of physical weathering on the planet, for when we excavate quarries, foundations, mines, or roadbeds by digging and blasting, we shatter and displace large volumes of rock that might otherwise have remained intact for millions of years.

CHEMICAL WEATHERING

Up to this point, we've taken the plumber's-snake approach to breaking up rock. Now, let's look at the liquid-drain-opener approach. **Chemical weathering** refers to the many chemical reactions that alter or destroy minerals when rock comes in contact with water solutions or air. Common reactions involved in chemical weathering include the following:

> *Dissolution:* Water serves as a solvent, so when it flows over or through rock, it slowly dissolves minerals. This process of *dissolution* affects primarily salts and carbonate minerals **(Fig. B.6)**, but even silicate minerals can dissolve slightly.
> *Hydrolysis:* During *hydrolysis*, water reacts chemically with minerals and breaks them down to form other minerals (*lysis* means loosen in Greek). For example, hydrolysis reactions transform feldspar and many other silicate minerals into clay.
> *Oxidation: Oxidation* reactions in rocks transform iron-bearing minerals, such as biotite and pyrite, into a rusty-brown mixture of various iron-oxide and iron-hydroxide minerals. In effect, oxidation causes iron-bearing rocks to "rust."
> *Hydration: Hydration*, the absorption of water into the crystal structure of minerals, causes some minerals, such as certain types of clay, to swell. Such expansion weakens rock.

Not all minerals undergo chemical weathering at the same rates. Some weather in a matter of months or years, whereas others remain unweathered for millions of years. For example, when a granite undergoes chemical weathering, most of its minerals transform into clay due to hydrolysis. But the quartz it contains does not change, so we can say that quartz is *resistant* to chemical weathering.

Until fairly recently, geoscientists tended to think of chemical weathering as a strictly inorganic chemical reaction, occurring entirely independently of life forms.

FIGURE B.6 Examples of chemical weathering.

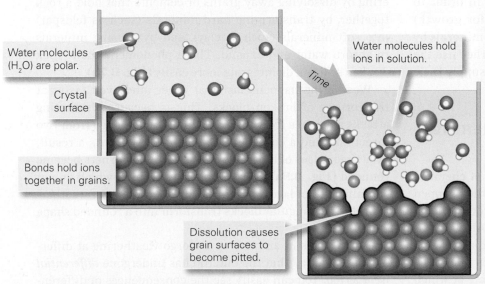

Water molecules (H$_2$O) are polar.

Crystal surface

Bonds hold ions together in grains.

Time

Water molecules hold ions in solution.

Dissolution causes grain surfaces to become pitted.

(a) Dissolution occurs when water molecules pluck ions off grain surfaces.

Water seeping into joints in limestone produced troughs.

(b) Dissolution of limestone on an outcrop surface in Ireland.
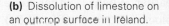

FIGURE B.7 Physical and chemical weathering processes work together to break down rock.

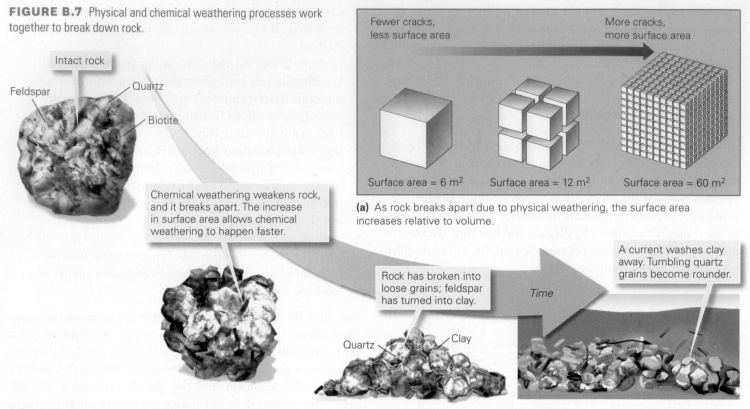

(a) As rock breaks apart due to physical weathering, the surface area increases relative to volume.

(b) Chemical weathering weakens rock, so it breaks apart. As this happens, the surface area increases, so chemical weathering happens still faster. Eventually, the rock completely disaggregates to form sediment. Here, the weathering of granite produces both quartz sand and clay.

But researchers now realize that organisms can play a major role in the chemical-weathering process. For example, the roots of plants, fungi, and lichens secrete organic acids that help dissolve minerals in rocks in order to extract *nutrients* (essential elements needed for growth) from the minerals. Some microbes literally eat minerals by plucking molecules off of a mineral's surface—they use the energy from the molecules' chemical bonds to supply their own life force.

PHYSICAL AND CHEMICAL WEATHERING WORKING TOGETHER

So far we've looked separately at the processes of chemical and physical weathering, but in the real world they happen together, aiding one another in disintegrating rock to form sediment (**Geology at a Glance**, pp. 180–181). Physical weathering speeds up chemical weathering. To understand why, keep in mind that chemical-weathering reactions take place at a material's surface. As a result, the overall rate at which chemical weathering occurs depends on the ratio of surface area to volume—the greater the surface area, the faster the volume of the whole material can chemically weather. When jointing (physical weathering) breaks a large block of rock

into smaller pieces, the surface area increases, so chemical weathering happens faster (**Fig. B.7a**).

Similarly, chemical weathering speeds up physical weathering by dissolving away grains or cements that hold a rock together, by transforming hard minerals (such as feldspar) into soft minerals (such as clay), and by causing minerals to absorb water and expand. These phenomena make rock weaker, so it can disintegrate more easily (**Fig. B.7b**).

Weathering happens faster at edges, and even faster at the corners of broken blocks. This is because weathering attacks a flat face from only one direction, an edge from two directions, and a corner from three directions. As a result, over time, edges of blocks become blunt and corners become rounded (**Fig. B.8a**). In rocks such as granite, which do not contain distinct layers or fabrics that can influence weathering rates, rectangular blocks transform into a rounded shape (**Fig. B.8b**).

When rocks in an outcrop undergo weathering at different rates, we say that the outcrop has undergone *differential weathering*. You can easily see the consequences of differential weathering if you walk through a graveyard. The inscriptions on some headstones are sharp and clear, but those on other stones have become blunted or have even disappeared (**Fig. B.8c**). That's because the minerals in different stones

FIGURE B.8 Spheroidal weathering and differential weathering.

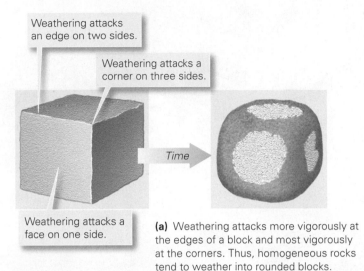

Weathering attacks an edge on two sides.

Weathering attacks a corner on three sides.

Time

Weathering attacks a face on one side.

(a) Weathering attacks more vigorously at the edges of a block and most vigorously at the corners. Thus, homogeneous rocks tend to weather into rounded blocks.

(b) At this location in Nevada, we can spot the granite outcrops by looking for light-colored, spheroidally weathered rocks.

(c) Inscriptions on a granite headstone (left) last for centuries, but those on a marble headstone (right) may weather away in decades. These gravestones are in the same cemetery and are about the same age.

Weak shale

Strong sandstone

(d) Sawtooth weathering profiles develop in sequences of alternating strong and weak layers on this exposure in New Mexico. Weak layers are indented, whereas strong layers protrude.

have different resistances to weathering. Granite, an igneous rock with a high content of resistant quartz, retains inscriptions the longest. But marble, a metamorphic rock composed of soluble calcite, dissolves away rapidly in acidic rain. As a result of differential weathering, cliffs composed of layers of different rock types develop a stair-step or sawtooth shape (Fig. B.8d).

B.3 Soil

If you've ever had the chance to dig in a garden, you've seen firsthand that the material in which flowers grow looks and feels different from beach sand or potter's clay. We call the

material in a garden *dirt* or, more technically, soil. **Soil** consists of rock or sediment that has been modified by physical and chemical interaction with organic material, rainwater, and organisms at or just below the Earth's surface over time. Soil represents one of our planet's most valuable resources, for without it there could be no forests or grasslands, and no horticulture, agriculture, forestry, or ranching.

HOW DOES SOIL FORM?

Three processes play a role in soil formation. First, chemical and physical weathering produce loose debris and ions in solution. The resulting detritus consists of mineral grains

Weathering, Sediment, and Soil Production

Glacial erosion

River erosion

Weathered granite

Cliff retreat

Glacial deposition

Limestone dissolution

New boulders tumble from a cliff in New Mexico.

Silt collects along a stream in Indiana.

Wind

Tectonic processes raise the land surface above sea level. Once exposed, rock interacts with air and water and undergoes chemical and physical weathering, ultimately
breaking down to produce sediment. Convection in the atmosphere generates wind,
rain, and snow. Flowing water, ice, and air erode and transport sediment to sites of
deposition. Leaching by downward-percolating rainwater, along with the addition of
organic material, produces soil.

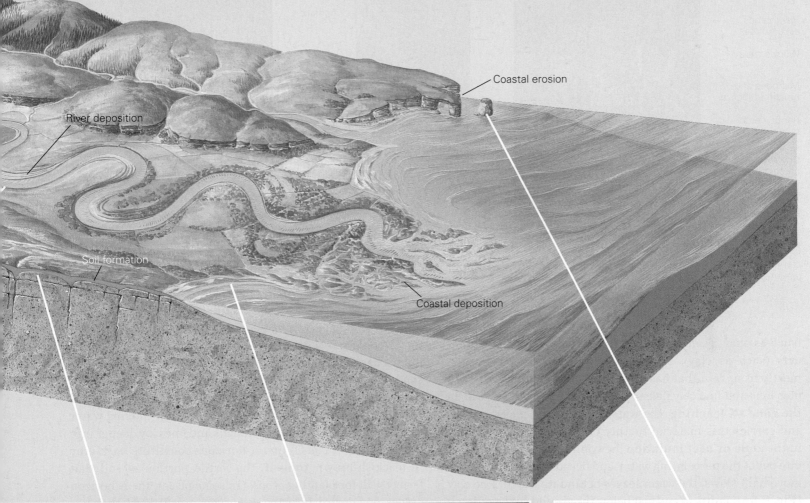

River deposition

Coastal erosion

Soil formation

Coastal deposition

Soil forms on bedrock of
chalk in southern England.

Waves move sand on a
beach in Brazil.

Erosion carves the
coastal cliffs of Ireland.

FIGURE B.9 Formation of soil horizons.

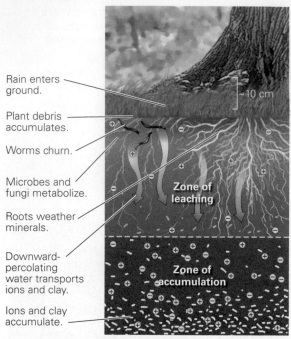

Rain enters ground.

Plant debris accumulates.

Worms churn.

Microbes and fungi metabolize.

Roots weather minerals.

Downward-percolating water transports ions and clay.

Ions and clay accumulate.

Zone of leaching

Zone of accumulation

~10 cm

(a) Soil horizons develop as water percolates downward, carrying ions and clay with it, and as organisms interact with the soil.

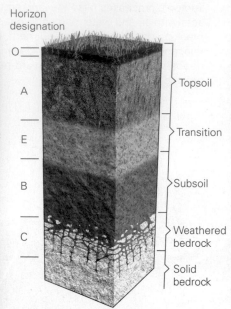

Horizon designation

O

A

E

B

C

Topsoil

Transition

Subsoil

Weathered bedrock

Solid bedrock

(b) Distinct soil horizons develop, each with a characteristic composition and texture.

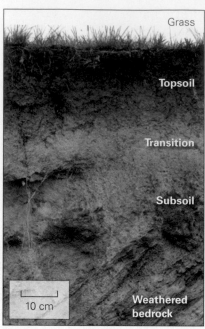

Grass

Topsoil

Transition

Subsoil

Weathered bedrock

10 cm

(c) Soil horizons exposed on the wall of a gully in eastern Brazil.

(such as quartz sand or silt), as well as new weathering products (such as clay). Second, downward-percolating water, mostly from rain that has seeped into the ground, redistributes ions and fine clay flakes. Closer to the ground surface, in the **zone of leaching**, the water extractions, picks up clay, and carries this material further downward. Deeper down, in the **zone of accumulation**, new mineral crystals precipitate out of the percolating water, and because the rate of water movement slows, the water leaves behind its load of fine clay **(Fig. B.9a)**. Third, at and just below the ground surface, detritus interacts with organisms. Organic material such as dead leaves, roots, and animals, along with waste materials from animals, mixes into the inorganic minerals of the detritus. Microbes, fungi, and insects decompose this organic material, breaking large and complex organic chemicals down into simpler ones. Some of these organisms also absorb nutrients from inorganic minerals or release acids that dissolve minerals. Finally, plant roots or burrowing animals (insects, worms, and gophers) churn and break up the debris that may have clumped together.

Because different soil-forming processes operate at different depths, soils typically develop distinct zones, known as **soil horizons**, arranged in a vertical sequence called a **soil profile (Fig. B.9b, c)**. Let's look at an idealized soil profile, from top to bottom, using a soil formed in a temperate forest

as our example. The highest horizon is the *O-horizon* (the prefix stands for organic), so called because it consists almost entirely of *humus* (plant debris) with barely any mineral matter. Below the O-horizon, we find the *A-horizon*, in which humus has decayed further and has mixed with clay, silt, and sand. Water percolating down through the A-horizon causes further chemical weathering reactions to occur, yielding ions in solution and new clay minerals. Downward-moving water eventually carries soluble chemicals and fine clay deeper into the subsurface. The O- and A-horizons constitute dark-gray to blackish-brown **topsoil**, the fertile portion of soil that farmers till for planting crops. (In some places, the A-horizon grades downward into the *E-horizon*, a soil level that has undergone substantial leaching but has not yet mixed with organic material.) Ions and clay accumulate in the next level, the *B-horizon*, or **subsoil**. Note from our description that the O-, A-, and E-horizons make up the zone of leaching, whereas the B-horizon makes up the zone of accumulation. Finally, at the base of a soil profile we find the *C-horizon*, which consists of material derived from the substrate that's been chemically weathered and broken apart, but has not yet undergone leaching or accumulation. The C-horizon grades downward into unweathered bedrock or into unweathered sediment.

Farmers, foresters, and ranchers are aware that the soil in one locality can differ greatly from the soil in another in terms of composition, thickness, and texture. Indeed, crops that grow well in one type of soil may wither and die in another. Such diversity exists because the character of a soil—its texture, nutrient content, and thickness—depends on several soil-forming factors **(Fig. B.10)**.

> *Climate:* Climate is the most important factor in determining the nature of soil development. The aspects of climate that have significant impact on weathering are the amount and distribution of rainfall and temperature. Large amounts of rainfall and warm temperatures accelerate chemical weathering and cause most soluble elements to be leached. Small amounts of rainfall and cool temperatures result in slower rates of weathering and leaching, so soils take a long time to develop and can retain unweathered minerals and soluble components.

> *Substrate composition:* Some soils form on basalt, some on granite, some on volcanic ash, and some on recently deposited quartz silt. These different substrates consist of different materials, so the soils formed on each end up with different chemical compositions. Also, substrates have different resistances to erosion—thicker soils develop on less-resistant material.

> *Slope steepness:* A thick soil can accumulate under land that lies flat. But on a steep slope, weathered rock may wash away or tumble downslope before it can evolve into soil. All other factors being equal, soil thickness increases as slope angle decreases.

> *Wetness:* Depending on the details of local topography and on the depth below the surface at which groundwater occurs, some soil becomes wetter than other soil in the same region. Wet soils tend to contain more organic material than do dry soils.

> *Time:* Because soil formation takes time, a younger soil tends to be thinner and less developed (with less distinct horizons) than an older soil. The rate of soil formation varies greatly with environment.

> *Local ecosystem:* The type and quantity of organisms living in a region influence the character of the organic content of soil. For example, where plants grow, plant debris can mix into soil, and roots may bind soil and prevent it from washing away. Where microbes and maggots thrive, decomposition happens more rapidly to provide organic compounds in soil. If decomposition does not take place, dead organic matter simply dries out, turns to dust, and blows away.

SOIL CLASSIFICATION

Soil scientists worldwide have struggled to develop a rational scheme for soil classification. Not all schemes utilize the same criteria, and even today there's not worldwide agreement on which scheme works best. In the United States, a country with many mid-latitude climates, soil scientists frequently use the *U.S. Comprehensive Soil Classification System,* which

FIGURE B.10 Factors that control the character of soil.

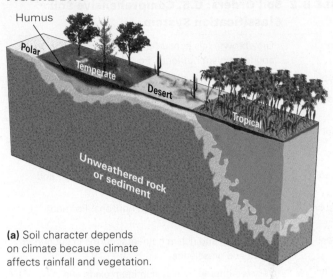

(a) Soil character depends on climate because climate affects rainfall and vegetation.

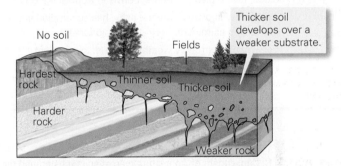

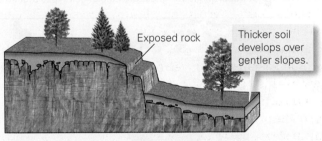

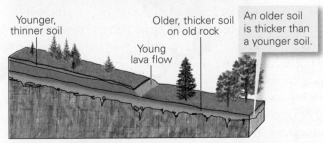

(b) Soil character also depends on the strength of the substrate, the steepness of the slope, and the length of time that soil has been forming.

distinguishes among 12 **soil orders** based on the physical characteristics and environment of soil formation **(Table B.2; Fig. B.11)**. Canadian soil scientists use a different scheme that focuses on soils that develop in cooler, high-latitude climates.

TABLE B.2 Soil Orders: U.S. Comprehensive Soil Classification System

Alfisol	Gray/brown, has subsurface clay accumulation and abundant plant nutrients. Forms in humid forests.
Andisol	Forms in volcanic ash.
Aridisol	Low in organic matter, has carbonate horizons. Forms in arid environments.
Entisol	Has no horizons. Formed very recently.
Gelisol	Underlain with permanently frozen ground.
Histosol	Very rich in organic debris. Forms in swamps and marshes.
Inceptisol	Moist, has poorly developed horizons. Formed recently.
Mollisol	Soft, black, and rich in nutrients. Forms in subhumid to subarid grasslands.
Oxisol	Very weathered, rich in aluminum oxide and iron oxide, low in plant nutrients. Forms in tropical regions.
Spodosol	Acidic, low in plant nutrients, ashy, has accumulations of iron and aluminum. Forms in humid forests.
Ultisol	Very mature, strongly weathered soils, low in plant nutrients.
Vertisol	Clay-rich soils capable of swelling when wet and shrinking and cracking when dry.

Let's look at the differences among a few examples of soil types. In deserts, where there is very little rainfall and sparse vegetation, an *aridisol* forms **(Fig. B.12a)**. (In older classifications, these were known as pedocal soils.) Aridisols have no O-horizon, because there is so little organic material, and the A-horizon is thin. Soluble minerals, specifically calcite, that would be washed away entirely if there were more rainfall, accumulate instead in the B-horizon. In fact, capillary action may bring calcite up from deeper down as water evaporates at the ground surface. The calcite locally cements clasts together in the B-horizon to form a rock-like mass called *caliche* or *calcrete*.

In temperate environments, an *alfisol* forms—this soil has an O-horizon, and because of moderate amounts of rainfall, materials leached from the A-horizon accumulate in the B-horizon **(Fig. B.12b)**. (In older classifications, these were known as pedalfer soils.)

In a tropical climate, *oxisols* develop. (Older soil classifications used the term *laterite* for an oxisol.) Because of the heat and humidity at the ground surface, organic matter rots away quickly and relatively little humus accumulates, so the A-horizon of an oxisol will be thin or may even be absent. Abundant water from rainfall percolates into the ground, causing all reactive minerals in the soil to undergo chemical

FIGURE B.11 U.S. Department of Agriculture map of soil types around the world.

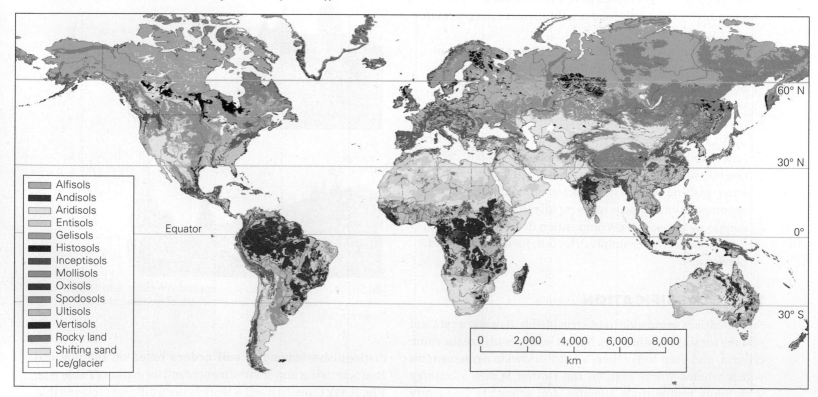

Legend:
- Alfisols
- Andisols
- Aridisols
- Entisols
- Gelisols
- Histosols
- Inceptisols
- Mollisols
- Oxisols
- Spodosols
- Ultisols
- Vertisols
- Rocky land
- Shifting sand
- Ice/glacier

Equator

60° N
30° N
0°
30° S

0 2,000 4,000 6,000 8,000
km

FIGURE B.12 Examples of soil classification.

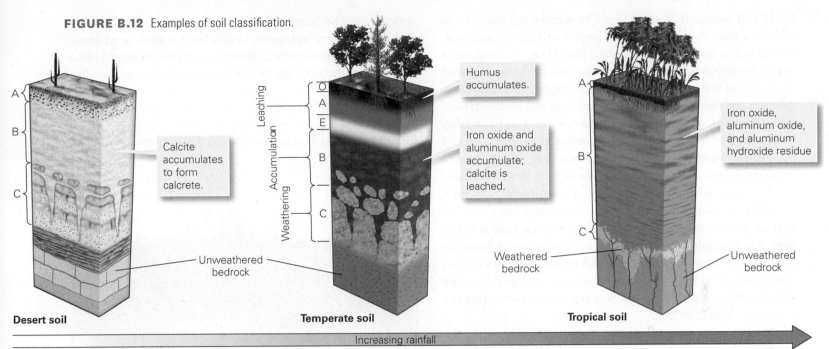

A

B — Calcite accumulates to form calcrete.

C

Unweathered bedrock

Desert soil

Leaching / Accumulation / Weathering

O
A
E
B
C

Humus accumulates.

Iron oxide and aluminum oxide accumulate; calcite is leached.

Unweathered bedrock

Temperate soil

A

B — Iron oxide, aluminum oxide, and aluminum hydroxide residue

C

Weathered bedrock

Unweathered bedrock

Tropical soil

Increasing rainfall

(a) Aridisol forms in deserts. Rainfall is so low that no O-horizon forms, and soluble minerals accumulate in the B-horizon.

(b) Alfisol forms in temperate climates. An O-horizon forms, and less-soluble materials accumulate in the B-horizon.

(c) Oxisol forms in tropical climates where percolating rainwater leaches all soluble minerals, leaving only iron- and aluminum-rich residues.

weathering, producing ions and clay that flush downward. As a consequence, the A- and B-horizons contain substantial residues of less-soluble minerals such as iron oxide, aluminum oxide, aluminum hydroxide, and also abundant clay. The iron oxide means that oxisols tend to be rusty red in color **(Fig. B.12c)**. Because the A-horizon, if it exists, may have a color similar to the B-horizon, horizons are not always obvious in oxisols. In some cases, oxisols become so hard, due to cementation by oxide minerals, that they can be chopped into blocks and used as bricks.

SOIL EROSION

As we have seen, soils take time to form, and soils capable of supporting crops or forests are a natural resource worthy of protection **(Fig. B.13)**. However, human activities such as agriculture, overgrazing, and clear-cutting have

FIGURE B.13 Soils provide a substrate for the growth of vegetation.

(a) This soil in a temperate realm hosts a forest. It displays a good O-horizon.

(b) Recently tilled soil of an Illinois farm field.

led to soil destruction. Crops rapidly remove nutrients from soil, so if they are not replaced, the soil will not contain sufficient nutrients to maintain plant life. When natural plant cover disappears, the surface of the soil becomes exposed to wind and water. Forces such as the impact of falling raindrops or the rasping of a plow can break up soil at the surface, with the result that it can wash away in water or blow away as dust. When this happens, **soil erosion**, the removal of soil by running water or by wind, takes place **(Fig. B.14)**. Human activities can increase rates of soil erosion by 10 to 100 times, so that the erosion rate far exceeds soil-formation rate. Droughts worsen the situation, so in some localities, erosion carries away as much as 6 tons of soil from an acre of land per year. For example, during the 1930s a succession of droughts killed off so much vegetation in the American plains of Oklahoma, Texas, and adjacent states that wind stripped the land of soil and caused devastating dust storms, transforming the region into the *Dust Bowl*.

The consequences of *rainforest destruction* have particularly profound effects on soil. In an established rainforest, lush growth provides sufficient organic debris so that trees can grow. But if the forest has been logged or cleared for agriculture, the humus disappears rapidly, leaving laterite that contains few nutrients. Crop plants consume whatever nutrients remain so rapidly that the soil soon becomes infertile, useless for agriculture and unsuitable for regrowth of rainforest trees.

FIGURE B.14 As a tractor plows the soil, dust rises and blows away.

ANOTHER VIEW A laterite exposed along a dirt road in New Caledonia displays the typical bright red color of tropical soils.

Interlude B Review

INTERLUDE SUMMARY

> Sediment consists of loose fragments derived from pre-existing rock, precipitated from water, or formed by the breakup of shells.

> Rock at or near the Earth's surface weathers over time.

> Physical weathering breaks larger rock bodies into smaller pieces.

> Chemical weathering involves a variety of chemical reactions that dissolve or alter minerals.

> Soil develops over time when water percolates down through detritus or weathering bedrock. Soil can be modified by interactions with organisms. The character of soil depends on the composition of its source material, as well as on climate and time.

> Soil serves as an essential resource to society, for it provides the basis for agriculture. Various phenomena, some caused by humans, can lead to the loss of soil.

GUIDE TERMS

chemical weathering (p. 177)
clast (p. 174)
erosion (p. 176)
physical weathering (p. 174)

sediment (p. 173)
soil (p. 179)
soil erosion (p. 186)
soil horizon (p. 182)

soil order (p. 183)
soil profile (p. 182)
subsoil (p. 182)
topsoil (p. 182)

weathering (p. 173)
zone of accumulation (p. 182)
zone of leaching (p. 182)

REVIEW QUESTIONS

The letters following each Review Question refer to the corresponding Learning Objective from the Chapter Opener.

1. Explain the difference between physical and chemical weathering. (**A**)

2. What processes can cause originally solid rock to break into pieces? Which type does the figure show? (**B**)

3. Why do sawtooth weathering profiles develop? (**B**)

4. What are the various reactions that can contribute to chemical weathering? (**A**)

5. Why doesn't weathering take place on the Moon? (**A**)

6. Explain the process of soil formation. (**C**)

7. Why do soils develop distinct horizons? Label the horizons on the figure. (**C**)

8. What factors determine the character (e.g., thickness, texture, types of horizons, etc.) of a soil? (**D**)

9. How does a soil that forms in a tropical climate differ from one that forms in an arid climate? (**D**)

10. Explain why soil erosion has been exacerbated by human activity. (**D**)

ONLINE RESOURCES

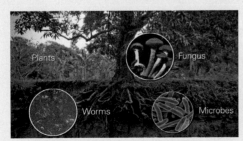

Videos This interlude features a video on how soil is formed and the role of organisms in this process.

The honeycomb-like network of ridges and depressions in these sandstone blocks is characteristic of physical weathering caused by exposure to seawater. Which of the following best explains how seawater causes this type of physical degradation?

Smartwork5 This interlude features questions on topics including physical weathering, chemical weathering, and soil-forming processes.

CHAPTER 6

PAGES OF THE EARTH'S PAST: SEDIMENTARY ROCKS

By the end of this chapter, you should be able to . . .

A. distinguish among various classes of sedimentary rocks.

B. explain how clastic sedimentary rocks form, and how to recognize and name major types.

C. describe the role of life in the production of rocks such as limestone and coal.

D. produce a model to illustrate how the layering (bedding) in sedimentary rock develops.

E. recognize the shapes and textures preserved in sedimentary rocks that reflect depositional environments.

F. discuss why thick accumulations of sedimentary rock can be found only in certain locations.

In every grain of sand there is a story of Earth.

RACHEL CARSON (American marine biologist and conservationist, 1907–1964)

6.1 Introduction

In this day when Google Earth can take you to every nook and cranny of our planet's surface at the touch of a phone screen or computer mouse, it's hard to imagine a world in which vast regions remained blank on a map. But only a century ago, that was the state of affairs that members of the 1910–1913 British Antarctic Expedition faced. Led by Robert Falcon Scott, a team of explorers from the expedition set out hoping to be the first to reach the South Pole. The early part of the journey took them over the Ross Ice Shelf, a broad plain of ice not far above sea level. But to reach the pole, they had to haul their heavy sledges up the Beardmore Glacier, a river of ice that had cut its way down through the rugged Transantarctic Mountains. The cliffs overlooking the glacier expose layer upon layer of a light-colored grainy rock. One of the expedition's members, Edward Wilson, served as the team's geologist, and during the journey he collected specimens of these rocks.

Scott, Wilson, and the others succeeded in reaching the South Pole on January 17, 1912. But when they arrived, they found to their profound disappointment that a Norwegian explorer, Roald Amundsen, had beaten them there by 34 days.

On their return journey, all of the British explorers perished in the blizzards and cold of the southernmost continent. When rescuers eventually came upon Scott's last campsite, they found Wilson's specimens, some of which contained fossils of *Glossopteris* (see Fig. 2.4), the fossil whose distribution Alfred Wegener would use as evidence of continental drift.

What are the grainy, layered rocks that Wilson collected? They are a type of sedimentary rock called sandstone. Formally defined, **sedimentary rock** is rock that forms at or near the surface of the Earth in one of several ways: by the cementing together of loose *clasts* (fragments or grains) that had been produced by physical or chemical weathering of pre-existing rock; by the growth of mounds of shells; by the cementing together of shells and shell fragments; by the accumulation and subsequent alteration of organic matter derived from living organisms; or by the precipitation of minerals directly from surface-water solutions. Layers, or *beds*, of sedimentary rock are like the pages of a book, recording tales of ancient events and environments on the ever-changing face of the Earth. They occur only in the upper part of the crust, and form a *cover* that buries the underlying *basement* of igneous and/or metamorphic rock **(Fig. 6.1)**.

FIGURE 6.1 In this close-up of the inner gorge of the Grand Canyon, we see a sedimentary-rock blanket that forms a cover over an older basement of metamorphic and igneous rock.

(a) The contact between the sedimentary cover and the underlying basement lies at the top of the inner gorge.

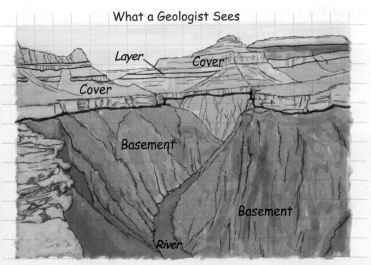

(b) A geologist's sketch emphasizes the contact. Here the basement consists of metamorphic and igneous rock.

◀ (facing page) Sedimentary strata exposed in Utah. The cliff at the top consists of sandstone beds, and the slope below consists of shale beds.

In Interlude B, we introduced the concept of weathering and showed how it attacks solid rock to break it down into ions and loose sediment grains. In this chapter, we see how these materials can be buried and transformed into sedimentary rock. We also describe various specific types of sedimentary rocks and show how geologists use the study of sedimentary rocks to characterize the Earth System's history. Finally, we discuss special settings, called *sedimentary basins*, in which particularly thick successions of sedimentary rock accumulate.

6.2 Classes of Sedimentary Rocks

Geologists divide sedimentary rocks into four major classes, based on their mode of origin. (1) **Clastic sedimentary rock** forms from cemented-together **clasts**, solid fragments and grains broken off of pre-existing rocks (the word comes from the Greek *klastos*, meaning broken); (2) **biochemical sedimentary rock** consists of shells; (3) **organic sedimentary rock** consists of carbon-rich relicts of plants or other organisms; and (4) **chemical sedimentary rock** is made up of minerals that precipitated directly from water solutions. Let's now look at each of these major classes in more detail.

CLASTIC SEDIMENTARY ROCKS

Formation Nine hundred years ago, a thriving community of Native Americans inhabited the high plateau of Mesa Verde,

Colorado. In hollows beneath huge overhanging ledges, they built multistory stone-block buildings that have survived to this day **(Fig. 6.2)**. Clearly, the blocks are solid and durable—they are, after all, rock. But if you were to rub your thumb along one, it would feel gritty, and small grains of quartz would break free and roll under your thumb, for the block consists of quartz sand grains cemented together. Geologists call such rock **sandstone**.

Sandstone is an excellent example of clastic sedimentary rock. It consists of loose clasts that have been stuck together to form a solid mass. The clasts, or *grains*, can consist of individual minerals (such as grains of quartz or flakes of clay) or of chunks of rock (such as pebbles of granite). Production of a clastic sedimentary rock involves five steps **(Fig. 6.3)**.

> *Weathering:* Clasts form by disintegration of bedrock into separate grains due to physical and chemical weathering.
> *Erosion:* **Erosion** refers to the combination of processes that separate rock or *regolith* (surface debris) from its substrate. Erosion involves abrasion, falling, plucking, scouring, and dissolution, and can be caused by moving air, water, or ice.
> *Transportation:* Gravity, wind, water, or ice can carry sediment. The ability of a medium to carry sediment depends on the medium's viscosity and velocity. Solid ice can transport sediment of any size, regardless of how slowly the ice moves. Very fast-moving, turbulent water can transport coarse fragments (cobbles and boulders), as well as finer ones; moderately fast-moving water can carry only sand and gravel; and slow-moving water carries only

FIGURE 6.2 The cliff dwellings nestled beneath a ledge at Mesa Verde, Colorado, are made of sandstone blocks. The inset shows the grainy character of the rock.

Sandstone
layer

2 cm

FIGURE 6.3 The five steps in clastic sedimentary rock formation.

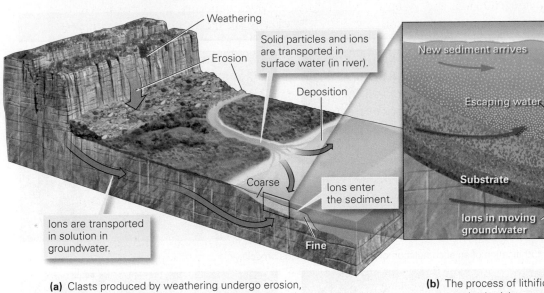

(a) Clasts produced by weathering undergo erosion, transportation, and deposition. Dissolved ions may eventually become cement.

(b) The process of lithification takes place during progressive burial.

finer-grained sediments, such as silt and clay. Strong winds can move sand and dust, but gentle breezes carry only dust.

> *Deposition:* **Deposition** is the process by which sediment settles out of the transporting medium. Sediment settles out of wind or moving water when these fluids slow down, because as the velocity decreases, the fluids no longer have the ability to carry sediment. Sediment carried by ice accumulates when the ice melts.

> *Lithification:* Geologists refer to the transformation of loose clasts into solid rock as **lithification**, a process that takes place in two stages. It begins with **compaction**, when the weight of overburden squeezes air or water out from between grains, so the grains can fit together more tightly. It ends with **cementation**, when minerals (commonly quartz or calcite) precipitate from groundwater and fill the remaining spaces between clasts, to form a *cement* that binds grains together.

Classification Say that you pick up a clastic sedimentary rock and want to describe it in enough detail that, from your words alone, another person can picture the rock. What characteristics should you mention? Geologists find the following characteristics most useful.

> *Clast size:* Size refers to the diameter of fragments or grains making up a rock. The names used for clast size, listed in order from coarsest to finest, are: boulder, cobble, pebble, sand, silt, and mud (see Table B.1 in Interlude B).

> *Clast composition:* Composition refers to the makeup of clasts in sedimentary rock. Clasts may be composed of rock fragments or individual mineral grains. Sand typically consists mostly of quartz grains.

> *Angularity and sphericity:* Angularity indicates the degree to which clasts have smooth surfaces, or have sharp corners and edges **(Fig. 6.4a)**. Sphericity, in contrast, refers to how closely the shape of a clast resembles a sphere.

FIGURE 6.4 Grain characteristics and their evolution with increasing transport.

Angularity

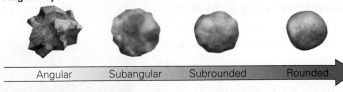

Angular   Subangular   Subrounded   Rounded

(a) Individual clasts tend to become more rounded (spherical) and smoother.

Sorting

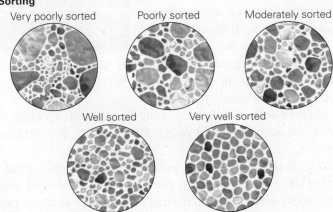

Very poorly sorted   Poorly sorted   Moderately sorted

Well sorted   Very well sorted

(b) If transport sifts grains, carrying smaller ones farther and leaving coarser ones behind, grains in a sediment tend to be the same size.

> *Sorting:* **Sorting** of clasts indicates the proportion of clasts in a rock that are the same size. *Well-sorted* sediment consists entirely of clasts that are the same size, whereas *poorly sorted* sediment contains a mixture of clast sizes (**Fig. 6.4b**).

> *Character of cement:* Not all clastic sedimentary rocks contain the same kind of cement. In some, the cement consists predominantly of quartz, whereas in others, it consists predominantly of calcite.

With these characteristics in mind, we can distinguish among several common types of clastic sedimentary rocks, listed in **Table 6.1**. This table provides common rock names; specialists sometimes use more precise names based on more complex classification schemes.

The size, angularity, sphericity, and sorting of clasts depends on the *transporting medium* (water, ice, or wind) that carries the clasts and, in the case of water or wind, on both the velocity of the medium and the distance of transport. The composition of the clasts depends on the composition of rock from which the clasts were derived and on the degree of chemical weathering that the clasts have undergone. Therefore, the type of clasts that accumulate in a sedimentary deposit varies with location. To see how, let's follow the fate of rock fragments as they gradually move from a cliff face in the mountains via a river to the seashore. Different kinds of sediment develop along the route. Each kind, if buried and lithified, would yield a different type of sedimentary rock.

To start the discussion of sedimentary rock types, imagine that large blocks of granite tumble off a cliff and slam into other blocks already at the bottom. The impact shatters the blocks, producing a pile of angular clasts. If these fragments were to be cemented together before being transported very

FIGURE 6.5 Different kinds of clasts lithify into different kinds of sedimentary rocks.

(a) Lithification of an accumulation of angular clasts yields breccia.

(b) Layers of river gravel lithify into conglomerate.

Alluvial fan

(c) Sediment deposited in an alluvial fan, close to its source, can be feldspar rich. Lithification of this sediment yields arkose.

far, the resulting rock would be **breccia** (**Fig. 6.5a**). Later, a storm causes the clasts to be carried away by a turbulent river. In the moving water, clasts bang into each other and into the riverbed, a process that shatters them into still smaller pieces and breaks off their sharp

Did you ever wonder...
where beach sand comes from?

FIGURE 6.5 (*continued*)

Sediment ——— Lithification ———▶ Sedimentary rock

(d) Layers of beach or dune sand lithify into sandstone. In the image on the right, note the pin for scale.

Sandstone

Shale

(e) Layers of mud, exposed beneath marsh grass, lithify to form shale. Here the thin-bedded shale is interbedded with sandstone.

If the gravel stays put for a long time, it undergoes chemical weathering. As a consequence, cobbles and pebbles break apart into individual mineral grains, eventually producing a mixture of quartz, feldspar, and clay. Clay is so fine that flowing water easily picks it up and carries it downstream, leaving sand containing a mixture of quartz and some feldspar grains—this sediment, if buried and lithified, becomes **arkose (Fig. 6.5c)**. Over time, feldspar grains in sand continue to weather into clay so that gradually, during successive events that wash the sediment downstream, the sand loses feldspar and ends up being composed almost entirely of durable quartz grains. Some of the sand may make it to the sea, where waves carry it to beaches, and some may end up in desert dunes. This sediment, when buried and lithified, becomes quartz sandstone **(Fig. 6.5d)**. Meanwhile, some of the silt and clay accumulates in the flat areas bordering streams, regions called *floodplains* (see Chapter 14) that become submerged only during floods. The rest settles in a wedge, called a *delta*, at the mouth of the river, in lagoons, or in *mudflats* along the shore. When lithified, the silt becomes **siltstone**, and the mud becomes **shale**, if finely laminated, or **mudstone**, if fairly massive **(Fig. 6.5e)**.

edges. As the clasts are carried downstream, they gradually become rounded pebbles and cobbles. When the river water slows, these clasts stop moving and form a mound or bar of gravel. Burial and lithification of these rounded clasts produces **conglomerate (Fig. 6.5b)**.

TABLE 6.1 Classification of Clastic Sedimentary Rocks

Clast Size	Clast Character	Rock Name (Alternate Name)
Coarse to very coarse (> 2 mm)	Rounded pebbles and cobbles Angular clasts Large clasts in muddy matrix	**Conglomerate** **Breccia** **Diamictite**
Medium to coarse (0.07–2 mm)	Sand-sized grains › Quartz grains only › Quartz and feldspar sand › Sand-sized rock fragments › Sand and rock fragments in a clay-rich matrix	**Sandstone** › Quartz sandstone (quartz arenite) › Arkose › Lithic sandstone › Wacke (informally called graywacke)
Fine (0.004–0.06 mm)	Silt-sized clasts	**Siltstone**
Very fine (< 0.004 mm)	Clay and/or very fine silt	**Shale,** if it breaks into platy sheets **Mudstone,** if it doesn't break into platy sheets

FIGURE 6.6 The formation of carbonate rocks (limestone).

(a) In this modern coral reef, corals produce shells. If buried and preserved, these shells become limestone.

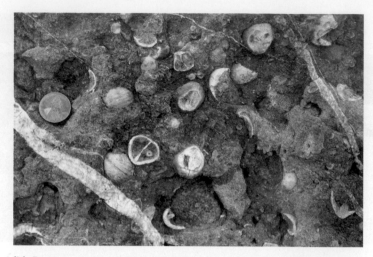

(b) Fossil shells and shell fragments of brachiopods protrude from an outcrop of ~415-Ma limestone in New York.

(c) This road cut near Kingston, New York, exposes beds of limestone. The vertical stripes are drillholes.

BIOCHEMICAL SEDIMENTARY ROCKS

The Earth System involves many interactions between living organisms and the physical planet. Numerous organisms have evolved the ability to extract dissolved ions from seawater to make solid shells. When the organisms die, the solid material in their shells survives. This material, when lithified, comprises *biochemical sedimentary rock*. Geologists recognize several distinct types of biochemical sedimentary rocks.

Biochemical Limestone A snorkeler gliding above a reef sees an incredibly diverse community of coral and algae, around which creatures such as clams, oysters, snails (gastropods), and lampshells (brachiopods) live, and above which plankton float **(Fig. 6.6a)**. Although they look so different from each other, many of these organisms share an important characteristic: they make solid shells of calcium carbonate ($CaCO_3$). The $CaCO_3$ crystallizes either as calcite or aragonite. (These minerals have the same composition, but different crystal structures.) When the organisms die, the shells remain and may accumulate. Rock formed dominantly from this material is *biochemical limestone*. Since the principal compound making up limestone is $CaCO_3$, geologists refer to limestone as a type of **carbonate rock**.

Limestone comes in a variety of textures, because the material that forms it accumulates in a variety of ways. For example, limestone can originate from reef builders (such as corals) that grew in place, from shell debris that was broken up and transported, from carbonate mud, or from plankton shells that settled like snow out of water. Because of this variety, geologists distinguish among *fossiliferous limestone*, consisting of visible fossil shells or shell fragments **(Fig. 6.6b)**; *micrite*, consisting of very fine carbonate mud; and *chalk*, consisting of plankton shells. Experts recognize many other types as well.

Typically, limestone is a massive light-gray to dark-bluish-gray rock that breaks into chunky blocks. It doesn't look much like a pile of shell fragments **(Fig. 6.6c)**. That's because several processes change the texture of the rock over time, after it has been buried deeply. For example, water passing through the rock not only precipitates cement but also dissolves some carbonate grains and causes new ones to grow.

FIGURE 6.7 Examples of biochemical and organic sedimentary rocks.

(a) Chert is a biochemical sedimentary rock. This bedded chert developed on the deep seafloor by the deposition of plankton that secrete silica shells.

Sandstone and shale

Coal layer

(b) Coal is an organic sedimentary rock. It is deposited in layers (beds), just like other kinds of sedimentary rocks.

Biochemical Chert If you walk beneath the northern end of the Golden Gate Bridge in California, you will find outcrops of reddish, almost porcelain-like rock occurring in 3- to 15-cm-thick layers **(Fig. 6.7a)**. Hit it with a hammer, and the rock cracks to form smooth, spoon-shaped (conchoidal) fractures. This biochemical chert is made from cryptocrystalline quartz (*crypto* is Greek for hidden), consisting of quartz grains that are too small to be seen without the extreme magnification of an electron microscope. This chert formed from the shells of silica-secreting plankton that accumulated on the seafloor. Gradually, after burial, the shells dissolved, forming a silica-rich gel. Chert then formed when the gel solidified.

ORGANIC SEDIMENTARY ROCKS

We've seen how the mineral shells of organisms ($CaCO_3$ or SiO_2) can accumulate and lithify to become biochemical sedimentary rocks. What happens to the "guts" of the organisms—the cellulose, fat, carbohydrate, protein, and other organic compounds that make up living matter? Commonly, organic debris gets eaten by other organisms or decays at the Earth's surface. But in some environments, the organic debris settles along with other sediment and is eventually buried. When lithified, organic-rich sediment becomes *organic sedimentary rock*. Since the dawn of the industrial revolution in the early 19th century, organic sedimentary rock has provided the fuel of modern industry and transportation, for organic chemicals can burn to produce energy. Coal and oil shale are the two most common types of organic sedimentary rocks.

Coal is a black, combustible rock containing between 40% and 90% carbon. The remainder consists of clay and quartz. Typically, the carbon in coal occurs in large, complex organic molecules. As discussed further in Chapter 12, coal forms when plant remains have been buried deeply enough and long enough for the material to become compacted and to lose significant amounts of volatiles (hydrogen, water, CO_2, and ammonia). As the volatiles seep away, the concentration of carbon increases **(Fig. 6.7b)**.

Oil shale contains not only clay but also between 15% and 75% organic material in a form called *kerogen*. The kerogen in oil shale comes from the fats and proteins that made up the living part of plankton or algae. If the tiny organisms settle in an environment where they do not immediately rot away or get eaten, they mix with the clay minerals in mud. When the mud gets buried and lithified, to form shale, the organic material transforms into kerogen. The presence of organic material colors oil shale black.

CHEMICAL SEDIMENTARY ROCKS

The colorful terraces, or mounds, that grow around the vents of hot-water springs; the immense layers of salt that underlie the floor of the Mediterranean Sea; the smooth, sharp point of an ancient arrowhead—these materials all have something in common. They all consist of rock formed primarily by the precipitation of minerals from water solutions. We call such rocks *chemical sedimentary rocks*. Typically they have a crystalline texture, partly formed during their original precipitation and partly when, at a later time, new crystals grow at

FIGURE 6.8 The formation of evaporite deposits.

(a) In lakes with no outlet, tiny amounts of salt brought in by streams stay behind as the water evaporates. When the water evaporates entirely, a white crust of salt remains.

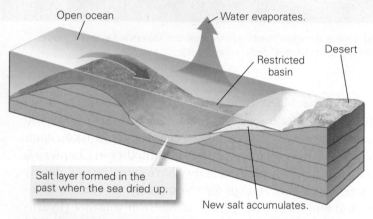

(b) Salt precipitation can also occur along the margins of a restricted marine basin if saltwater evaporates faster than it can be resupplied.

the expense of old ones through a process called *recrystallization*. In some examples, crystals are coarse, whereas in others, they are too small to see. Geologists distinguish among many types of chemical sedimentary rocks, primarily on the basis of composition.

Evaporites—The Products of Saltwater Evaporation In 1965, two daredevil drivers in jet-powered cars battled to be the first to set the land speed record of 600 mph. On November 7, the *Green Monster* peaked at 576.127 mph; but eight days later, the *Spirit of America* reached 600.601 mph. Traveling at such speeds, a driver must maintain an absolutely straight line; any turn will catapult the vehicle out of control. Thus, high-speed trials take place on extremely long and flat racecourses. Not many places can provide such conditions—but the Bonneville Salt Flats of Utah do. The salt flats formed when an ancient salt lake evaporated. Under the heat of the Sun, the water turned to vapor and drifted up into the atmosphere, while the salt that had been dissolved in the water stayed behind.

Salt precipitation occurs where saltwater becomes supersaturated, meaning that it can't keep all the dissolved ions that it contains in solution. This situation happens because evaporation removes water from a water body faster than the rate at which new water enters. This process takes place in desert lakes and along the margins of restricted seas (**Fig. 6.8**). For thick deposits of salt to form, large volumes of water must evaporate. Because salt deposits form as a consequence of evaporation, geologists refer to them as **evaporites**. The specific type of salt minerals comprising an evaporite depends on the amount of evaporation. When 80% of the water evaporates, gypsum forms; and when 90% of the water evaporates, halite precipitates.

Travertine (Chemical Limestone) **Travertine** is a rock composed of crystalline calcium carbonate ($CaCO_3$) that precipitates directly from groundwater that has seeped out at the ground surface either in hot- or cold-water springs, or on the walls of caves. What causes this precipitation? It happens, in part, when the groundwater degasses, meaning that some of the carbon dioxide that had been dissolved in the groundwater bubbles out of solution, for removal of carbon dioxide encourages the precipitation of carbonate. Precipitation also occurs when water evaporates, thereby increasing the concentration of carbonate. Various species of microbes live in the environments in which travertine accumulates, so biological activity may also contribute to the precipitation process. Travertine produced at springs forms terraces and mounds that are meters or even hundreds of meters thick, such as those at Mammoth Hot Springs (**Fig. 6.9a**). Travertine also grows on the walls of caves where groundwater seeps out (**Fig. 6.9b**). In cave settings, travertine builds up beautiful and complex growth forms called *speleothems* (see Chapter 16). Travertine

FIGURE 6.9 Examples of travertine (chemical limestone) deposits.

Terrace of new travertine

2 m

(a) Travertine accumulates in terraces at Mammoth Hot Springs in Yellowstone National Park, Wyoming.

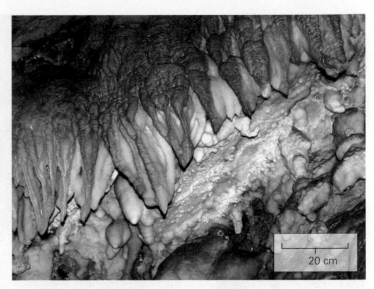

20 cm

(b) Travertine speleothems form as calcite-rich water drips from the ceiling of Timpanogos Cave in Utah.

(c) This slice of travertine forms a decorative panel for a building's interior wall.

tends to have distinctive layering, making it a popular stone for construction **(Fig. 6.9c)**.

Dolostone Another carbonate rock, *dolostone*, differs from limestone in that it contains the mineral dolomite ($CaMg[CO_3]_2$), which contains equal amounts of calcium and magnesium. Where does the magnesium come from? Most dolostone forms by a chemical reaction between solid calcite and magnesium-bearing groundwater. This change may take place beneath lagoons along a shore soon after the limestone formed, or a long time later, after the limestone has been buried deeply.

Chemically Precipitated Chert A community of Native Americans, the Onondaga, once lived in eastern New York State. Here, outcrops of limestone contain layers or *nodules* (lenses or lumps) of black chert, or flint **(Fig. 6.10a)**. Because of the way this chert breaks, Onondaga artisans could fashion sharp-edged tools (arrowheads and scrapers) from it, so they collected it for their own toolmaking industry and for use in trade. Unlike the deep-sea (biochemical) chert described earlier, the chert collected by the Onondaga formed when microscopic quartz crystals gradually replaced calcite crystals within a body of limestone long after the limestone was deposited. Geologists call such material *replacement chert*.

> **Did you ever wonder . . .**
> how flint used for arrowheads first formed?

Chert comes in many colors (black, white, red, brown, green, gray), depending on the impurities it contains. *Petrified wood* is chert that forms when silica-rich sediment, such as ash from a volcanic eruption, buries trees. The silica dissolves in groundwater and then later precipitates as microcrystalline quartz within wood, gradually replacing the wood's cellulose **(Fig. 6.10b)**. The chert deposit retains the shape of the wood cells and the growth rings within it. Some chert, known as *agate*, precipitates in concentric rings inside hollows in a rock and ends up with a striped appearance, caused by variations in the content of impurities incorporated in the chert **(Fig. 6.10c)**.

FIGURE 6.10 Examples of chert that precipitated in place.

(a) Replacement chert forms as layers of nodules between tilted limestone beds in New York.

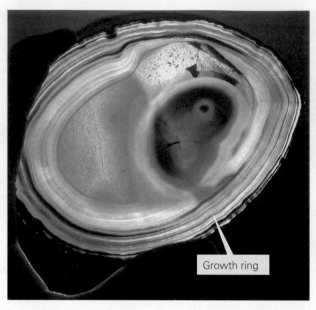

(c) A thin slice of Brazilian agate, lit from the back, shows growth rings.

(b) This 14-cm-diameter log of petrified wood from Wyoming formed about 50 Ma.

TAKE-HOME MESSAGE

Geologists distinguish among many types of sedimentary rocks based on the mode of formation. Clastic sedimentary rocks consist of grains weathered and eroded from pre-existing rock that are then transported, deposited, and lithified; clastic rocks can be classified by grain size. Biochemical sedimentary rocks, such as limestone, consist of the shells of organisms, and organic sedimentary rocks, such as coal, form from the organic remains of organisms. Chemical sedimentary rocks precipitate from water solutions.

QUICK QUESTION Do all sedimentary rocks have the same composition? Why or why not?

6.3 Sedimentary Structures

Geologists use the term **sedimentary structure** to describe the layering of sedimentary rocks, surface features on layers formed during deposition, and the arrangement of grains within layers.

BEDDING AND STRATIFICATION

Let's start by introducing the language geologists use for discussing sedimentary layers. A single layer of sediment or sedimentary rock with a recognizable top and bottom is called a **bed**; the boundary between two beds is a *bedding plane*. Several beds together constitute **strata** (singular stratum, from the Latin *stratum*, meaning pavement), and the overall arrangement of sediment into a sequence of beds is *bedding*, or *stratification*. In some outcrops, stratification can be quite

FIGURE 6.11 Bedding in sedimentary rocks.

FIGURE 6.11 Bedding in sedimentary rocks.

A layer of silt is deposited during normal river flow.

Silt

Basement

A layer of gravel is deposited during flood.

Gravel

Later, another layer of silt accumulates.

Silt

Gravel

Time

After burial, the sediment turns to beds of rock.

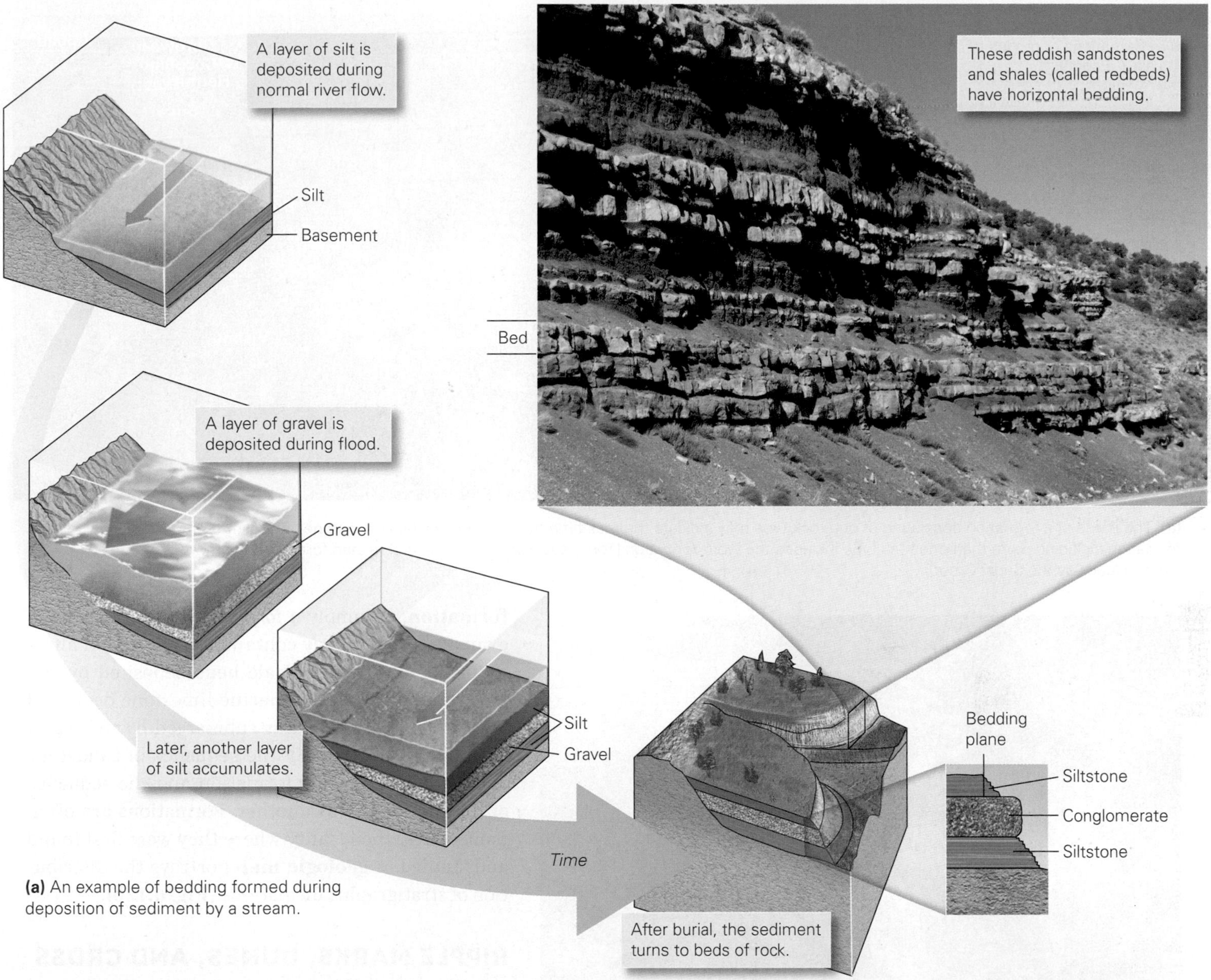

(a) An example of bedding formed during deposition of sediment by a stream.

(b) Beds of sedimentary rock exposed along a road in Utah.

These reddish sandstones and shales (called redbeds) have horizontal bedding.

Bed

Bedding plane

Siltstone

Conglomerate

Siltstone

subtle. But commonly, successive beds have different colors, textures, and resistance to erosion, so bedding gives outcrops a striped appearance.

Why does bedding form? To find the answer, we need to think about how sediment accumulates. Changes in the climate, water depth, current velocity, or the sediment source control the type of sediment deposited at a location at a given time. For example, on a normal day, a slow-moving river may carry only silt, which collects on the riverbed. During a flood, the river flows faster and carries sand and pebbles, so a layer of sandy gravel forms over the silt layer. Then, when the flooding stops, more silt buries the gravel. If this succession of

sediments becomes lithified and later exposed for us to see, it appears as alternating beds of siltstone and sandy conglomerate **(Fig. 6.11)**.

During geologic time, the overall character of a depositional environment can change dramatically. For example, if sea level rises, and water submerges an area in which desert sand had been accumulating, layers of carbonate shells may be deposited over layers of sand—when lithified, the sand layers become beds of sandstone, and the shell layers become beds of limestone.

A sequence of strata that is distinctive enough to be traced as a unit across a fairly large region is called a **stratigraphic**

FIGURE 6.12 The concept of a stratigraphic formation as exemplified by the Grand Canyon.

The surface between two formations or groups is called a contact.

(a) The name of a formation consisting of one rock type may indicate the rock type (for instance, Kaibab Limestone). The name of a formation including more than one rock type includes the word *formation* (Toroweap Formation). Several related formations comprise a group (such as the Supai Group).

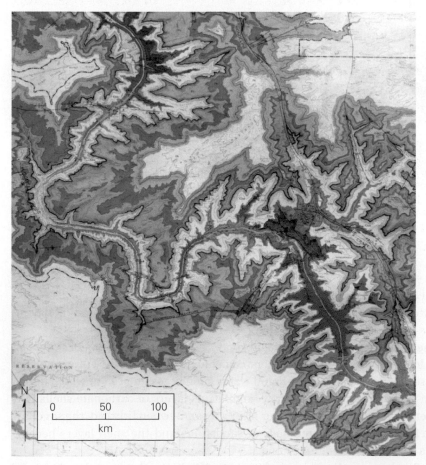

(b) A geologic map portrays the distribution of formations in a portion of the Grand Canyon. Each color band represents a specific formation.

formation, or simply a *formation* (**Fig. 6.12a**). For example, a region may contain a succession of alternating sandstone and shale beds deposited by rivers, overlaid by beds of marine limestone deposited later when the region was submerged by the sea. A geologist might identify the sequence of sandstone and shale beds as one formation and the sequence of limestone beds as another. Formations are often named after the locality where they were first found and studied. A **geologic map** portrays the distribution of stratigraphic formations (**Fig. 6.12b**).

RIPPLE MARKS, DUNES, AND CROSS BEDDING: CONSEQUENCES OF DEPOSITION IN A CURRENT

Many clastic sediments accumulate in moving fluids (wind, rivers, or waves). Fascinating sedimentary structures develop at the interface between the sediment and the fluid. These structures, called *bedforms*, develop at a given location and reflect such factors as the velocity of the flow and the size of the clasts. Though bedforms fall into many types, here we'll focus on only two—ripple marks and dunes. The growth of both produces cross bedding, a special type of lamination within beds.

Ripple marks are relatively small (generally no more than a few centimeters high), elongated

FIGURE 6.13 Ripple marks, a type of sedimentary structure, are visible on the surface of modern and ancient beds.

(a) Modern ripples exposed at low tide along a sandy beach on the shore of Cape Cod, Massachusetts.

(b) These 145-Ma ripples are preserved on a tilted bed of solid sandstone at Dinosaur Ridge, Colorado.

ridges that form on a bed surface at right angles to the direction of current flow. You can find ripples on modern beaches or streambeds, and others preserved on bedding planes of ancient rocks **(Fig. 6.13)**. A **dune** looks like a ripple, only it's much larger. For example, dunes on the bed of a stream may be tens of centimeters to several meters high, and wind-formed dunes in deserts may be tens of meters to over 100 meters high **(Fig. 6.14a)**.

If you examine a vertical slice cut into a ripple or dune, you will find distinct internal laminations that are inclined at an angle to the main sedimentary layer. Such laminations are called **cross beds**. To see how cross beds develop, imagine a current of air or water moving uniformly in one direction **(Fig. 6.14b, c)**. The current erodes and picks up clasts from the upstream part of the bedform and deposits them on the downstream or leeward face of the crest. Sediment builds up until gravity causes it to slip down the leeward face. With time, the dune or ripple builds in the downstream direction. The surface of the *slip face* establishes the shape of the cross beds. Eventually, a new cross-bedded layer builds out over a pre-existing one. The boundary between two successive layers is called the *main bedding*, and the internal inclined surfaces within the layer are called the *cross bedding* **(Fig. 6.14d)**.

TURBIDITY CURRENTS AND GRADED BEDS

Sediment deposited on a submarine slope tends to be unstable, so an earthquake or storm might disturb this sediment and cause it to slip downslope and mix with water to produce a murky, turbulent cloud. This cloud, which is denser than clear water, flows downslope like an underwater avalanche **(Fig. 6.15)**. Geologists refer to such a moving submarine sediment suspension as a *turbidity current*. Downslope, the turbidity current slows, and the sediment that it carried starts to settle out. Larger grains sink faster through a fluid than do finer grains, so the coarsest sediment settles out first. Progressively finer grains accumulate on top, with the finest sediment (clay) settling out last. This process forms a *graded bed*, meaning a layer of sediment in which grain size varies from coarse at the bottom to fine at the top. Geologists refer to a deposit from a turbidity current as a *turbidite*.

BED-SURFACE MARKINGS

A number of features develop on the surface of a bed as a consequence of events that happen during deposition or soon after, while the sediment layer remains soft. Such *bed-surface markings* include the following:

> *Mud cracks:* If a mud layer dries up after deposition, it cracks into roughly hexagonal plates that typically curl up at their edges. We refer to the openings between the plates as *mud cracks* **(Fig. 6.16)**.

> *Scour marks:* As currents flow over a sediment surface, they may erode small troughs, called *scour marks*, parallel to the current flow.

> *Fossils:* **Fossils** are relics of past life. Some fossils are shell imprints or footprints on a bedding surface (see Interlude E).

FIGURE 6.14 Cross bedding, a type of sedimentary structure within a bed.

(a) Large sand dunes formed in a strong wind.

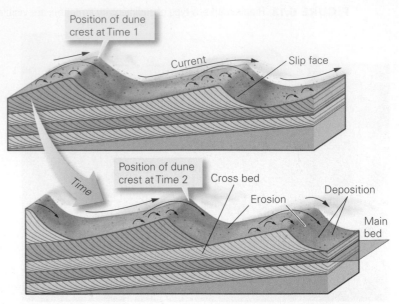

(b) Cross beds form as sand blows up the windward side of a dune or ripple and then accumulates on the slip face. With time, the dune crest moves.

(c) Slip face of a small sand dune, in Death Valley, California. The edges of the slip face are highlighted.

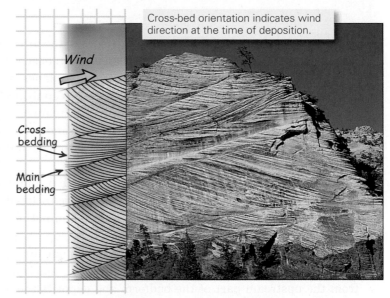

(d) A cliff face in Zion National Park, Utah, displays large cross beds formed between 200 and 180 Ma, when the region was a desert with large sand dunes.

Burial and lithification of bed-surface markings can preserve the markings.

WHY STUDY SEDIMENTARY STRUCTURES?

Sedimentary structures are not just a curiosity. Rather, they provide important clues that help geologists understand the *depositional environment* in which sediments accumulated. For example, the presence of ripple marks and cross bedding indicates that layers were deposited in a current, mud cracks indicate that the sediment layer was exposed to the air and dried out, and graded beds indicate deposition by turbidity currents. Also, fossil types can tell us whether sediment was deposited along a river or in the deep sea, because different species of organisms live in different environments. In the next section of this chapter, we examine depositional environments in greater detail.

FIGURE 6.15 The deposition of graded bedding by turbidity currents.

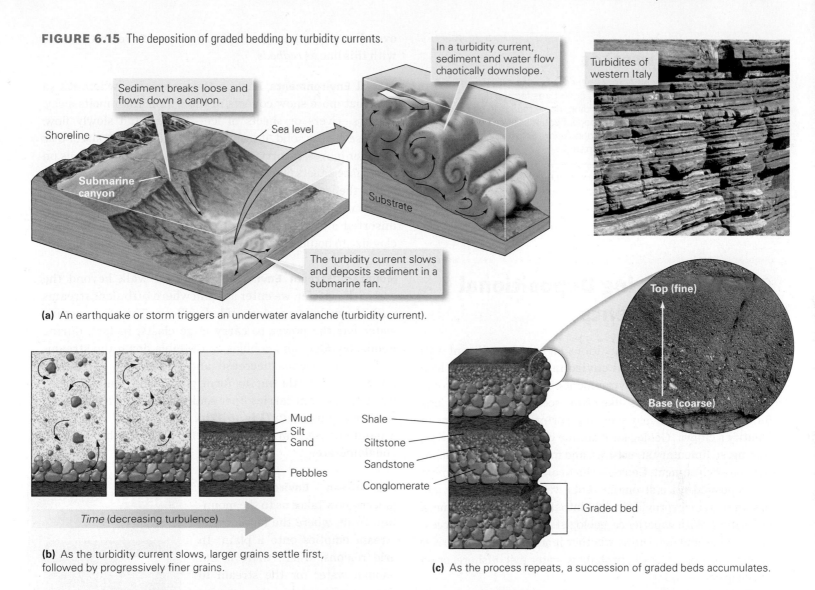

Sediment breaks loose and flows down a canyon.

Shoreline

Submarine canyon

Sea level

In a turbidity current, sediment and water flow chaotically downslope.

Substrate

Turbidites of western Italy

The turbidity current slows and deposits sediment in a submarine fan.

(a) An earthquake or storm triggers an underwater avalanche (turbidity current).

Mud
Silt
Sand

Pebbles

Time (decreasing turbulence)

(b) As the turbidity current slows, larger grains settle first, followed by progressively finer grains.

Top (fine)

Base (coarse)

Shale
Siltstone
Sandstone
Conglomerate

Graded bed

(c) As the process repeats, a succession of graded beds accumulates.

FIGURE 6.16 Mud cracks, a sedimentary structure formed by the drying out of mud.

(a) Mud cracks in red mud at Bryce Canyon, Utah. Note how the edges of the mud plates curl up.

(b) Mud cracks visible on the surface of a 410-Ma bed exposed on a cliff in New York.

6.4 Recognizing Depositional Environments

Geologists refer to the conditions in which sediment was deposited as a **depositional environment**. Examples include beach, glacial, and river environments. To identify depositional environments, geologists, like crime scene investigators, look for clues. Detectives may seek fingerprints and bloodstains to identify a culprit. Geologists examine grain size, composition, sorting, sedimentary structures, and fossils to identify a depositional environment. Geologic clues can tell us if the sediment was deposited by ice, strong currents, waves, or quiet water, and in some cases can provide insight into the climate at the time of deposition. With experience, geologists can examine a succession of beds and determine whether it accumulated on a river floodplain, along a beach, in shallow water just offshore, or on the deep ocean floor.

Let's now explore some examples of depositional environments and the sediments deposited in them, by imagining that we are taking a journey from the mountains to the sea, examining sediments as we go (**Geology at a Glance**, pp. 206–207). We will see that geologists distinguish among three basic categories of depositional environments: terrestrial, coastal, and marine.

TERRESTRIAL (NONMARINE DEPOSITIONAL) ENVIRONMENTS

We begin our exploration with *terrestrial depositional environments*, those that develop inland, far enough from the shoreline that they are not affected by ocean tides and waves. Terrestrial sediments, or *nonmarine sediments*, accumulate either on dry land or under and adjacent to freshwater. In some settings, oxygen in surface water or groundwater reacts with iron in the sediment to produce rust-like iron-oxide minerals, which give the sediment an overall reddish hue. Geologists informally refer to strata with this hue as *redbeds*.

Glacial Environments High in the mountains, where it's so cold that more snow collects in the winter than melts away, glaciers—rivers or sheets of ice—develop and slowly flow. Because ice is a solid, it can move sediment of any size. So as a glacier moves, it carries along all the sediment that falls on its surface or gets plucked from the ground at its base or sides. When the ice finally melts away, the sediment that had been in or on the ice accumulates as *glacial till* (**Fig. 6.17a**). Till is unsorted and unstratified—it contains clasts ranging from clay size to boulder size all mixed together.

Mountain Stream Environments As we walk beyond the end of the glacier, we enter a realm where turbulent streams rush downslope in steep-sided valleys. This fast-moving water has the power to carry large clasts; in fact, during floods, boulders and cobbles can tumble down the streambed. Where slopes decrease and water flow slows, the larger clasts settle out to form gravel and boulder beds, while the stream carries finer sediments like sand and mud farther downstream (**Fig. 6.17b**). Sedimentary deposits of a mountain stream would, therefore, include breccia and conglomerate.

Alluvial-Fan Environments Our journey now takes us to the mountain front, where the fast-moving stream empties onto a plain. In arid regions, where there is not enough water for the stream to flow continuously, the stream deposits its load of sediment near the mountain front, producing a wedge-shaped apron of gravel and sand called an **alluvial fan** (**Fig. 6.17c**). Deposition takes place here because when the stream pours from a canyon mouth and spreads out into multiple channels, friction causes the water to slow down, and slow-moving water does not have the power to move coarse sediment. Sediment in alluvial fans may accumulate close to the source, so it will not have undergone much chemical weathering. Thus, sand layers still contain feldspar grains, for these have

FIGURE 6.17 Examples of nonmarine depositional environments.

(a) Glacial till at the end of a glacier in France.

(b) Boulders and cobbles deposited by a mountain stream in Colorado.

(c) An alluvial fan in Death Valley, California.

(e) Deposits of an ancient river channel in Indiana. Note how the floor of the channel cuts across older strata. The geologist's sketch emphasizes the relationship.

Edge of photo
What a Geologist Sees
Younger floodplain deposits
Channel fill
Older floodplain deposits
(Talus)

(d) Sand dunes in Brazil.

(f) Laminated mud visible in a core of lake-bed sediment. The layers were bent by the coring process.

not yet weathered into clay. Alluvial-fan sediments become breccia, conglomerate, and arkose.

Desert Environments In very dry climates, few plants can grow and the ground surface lies exposed. Strong winds can move dust and sand. The dust gets carried away, and the resulting well sorted sand can accumulate in dunes. Thus,

thick layers of well-sorted sandstone, in which we can find large cross beds, are relics of desert sand-dune environments **(Fig. 6.17d)**.

River Environments In climates where streams flow, we find several distinctive depositional environments. Rivers transport gravel, sand, silt, and mud. The coarser sediment

GEOLOGY AT A GLANCE

The Formation of Sedimentary Rocks

Glacial environment

Estuary

Beach

Bar

Continental shelf

Coastal erosion

Turbidity current

Submarine fan

Deep-sea current

Redbeds

Bedding

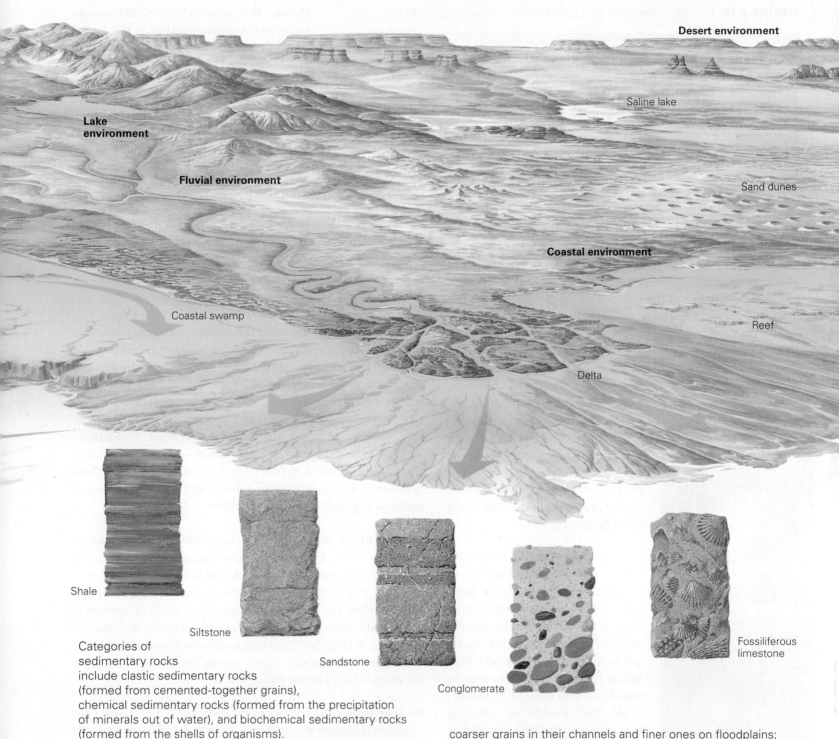

Lake environment

Desert environment

Saline lake

Fluvial environment

Sand dunes

Coastal environment

Coastal swamp

Reef

Delta

Shale

Siltstone

Sandstone

Conglomerate

Fossiliferous limestone

Categories of sedimentary rocks include clastic sedimentary rocks (formed from cemented-together grains), chemical sedimentary rocks (formed from the precipitation of minerals out of water), and biochemical sedimentary rocks (formed from the shells of organisms).

Clastic sedimentary rocks develop when grains (clasts) break off pre-existing rock by weathering and erosion and are transported to a new location by wind, water, or ice; the grains are deposited to create sediment layers, which are then lithified. We distinguish among types of clastic sedimentary rocks on the basis of grain size.

The character of a sedimentary rock depends on the composition of the sediment and on the environment in which it accumulated. For example, glaciers carry sediment of all sizes, so they leave deposits of poorly sorted till; streams deposit coarser grains in their channels and finer ones on floodplains; a river slows down at its mouth and deposits an immense pile of silt in a delta. Fossiliferous limestone develops on coral reefs. In desert environments, sand accumulates into dunes and evaporites precipitate in saline lakes. Offshore, submarine canyons channel avalanches of sediment, or turbidity currents, out to the deep seafloor.

Sedimentary rocks tell the history of the Earth. For example, the layering, or bedding, of sedimentary rocks is initially horizontal. So where we see layers bent or folded, we can conclude that the layers were deformed during mountain building.

FIGURE 6.18 A simple "Gilbert-type" delta formed where a stream enters a lake.

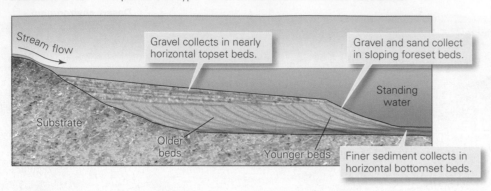

tumbles along the bed in the river's channel and collects in cross-bedded, rippled layers while the finer sediment drifts along, suspended in the water. This fine sediment settles out along the banks of the river, or on the *floodplain*, the flat land on either side of the river that water covers only during floods. On the floodplain, mud layers dry out between floods, so they develop mud cracks. River sediments lithify to form rippled sandstone, siltstone, and shale. Typically, coarser sediments occur in elongate bands, relics of river channels. Layers of fine-grained floodplain deposits surround the relict channels, so in cross section, the channel has a lens-like shape **(Fig. 6.17e)**. Geologists commonly refer to river deposits as *fluvial sediments*, from the Latin word *fluvius*, for river.

Lake Environments In temperate climates, where water remains at the surface throughout the year, lakes form. In the offshore portions of a lake, the deeper water is relatively quiet, and clay can settle out to form mud on the lake bed. When lithified, such laucustrine mud turns into shale **(Fig. 6.17f)**.

At the mouths of streams that empty into lakes, small deltas may form. A **delta** is a wedge of sediment that accumulates where moving water enters standing water. Deltas were so named because the map shape of some deltas resembles the Greek letter *delta* (Δ), as we discuss further in Chapter 14. In 1885, an American geologist named G. K. Gilbert showed that small lakeshore deltas contain three components **(Fig. 6.18)**: topset beds composed of gravel, foreset beds of gravel and sand, and silty bottomset beds.

COASTAL AND MARINE DEPOSITIONAL ENVIRONMENTS

Along the seashore, a variety of distinct coastal depositional environments occur; the character of each reflects the nature of the sediment supply and the climate. Marine environments start at the high-tide line and extend offshore, to include the deep seafloor. The type of sediment deposited at a location depends on the climate, water depth, and whether or not clastic grains are available.

Marine Delta Deposits After following the river downstream for a long distance, we reach its mouth, where it empties into the sea. Here, the river water stops flowing, so sediment settles out to build a delta of sediment out into the sea. Large marine deltas are much more complex than the lake examples that Gilbert studied. They include many different sedimentary environments, such as swamps, channels, floodplains, and submarine slopes. Sea-level changes may cause the positions of the different environments to move with time. Thus, deposits of an ocean-margin delta produce a great variety of sedimentary rock types **(Fig. 6.19a)**.

Coastal Beach Sands Now we leave the delta and wander along the coast. Oceanic currents transport sand along the coastline. The sand washes back and forth in the surf, so it becomes well sorted (waves winnow out silt and mud) and well rounded, and because of the back-and-forth movement of ocean water over the sand, the sand surface may become rippled **(Fig. 6.19b)**. So if you find well-sorted, medium-grained sandstone, perhaps with ripple marks, you may be looking at the remnants of a beach environment.

Shallow-Marine Clastic Deposits From the beach, we proceed offshore. In deeper water, where wave energy does not stir the seafloor, finer sediment can accumulate. Because the water here may be only a few meters to tens of meters deep, geologists refer to this depositional setting as a *shallow-marine environment*. Clastic sediments that accumulate in this environment tend to be fine-grained, well-sorted, well-rounded silt, and they are inhabited by a great variety of organisms such as mollusks and worms. So, if you see beds of siltstone and mudstone containing marine fossils, you may be looking at shallow-marine clastic deposits.

FIGURE 6.19 Examples of coastal depositional environments.

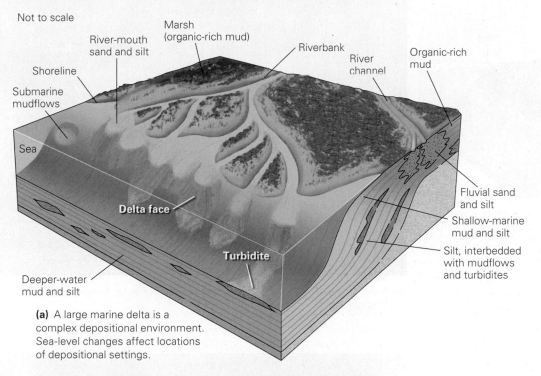

Not to scale

Shoreline · River-mouth sand and silt · Marsh (organic-rich mud) · Riverbank · River channel · Organic-rich mud · Submarine mudflows · Sea · Delta face · Turbidite · Deeper-water mud and silt · Fluvial sand and silt · Shallow-marine mud and silt · Silt, interbedded with mudflows and turbidites

(a) A large marine delta is a complex depositional environment. Sea-level changes affect locations of depositional settings.

(b) Waves along the coast sort beach sand.

Shallow-Water Carbonate Environments In shallow-marine settings, where relatively little sand and mud enter the water, warm, clear, nutrient-rich water can host an abundance of organisms with carbonate shells, which eventually become carbonate sediment **(Fig. 6.20)**. Beaches collect sand composed of shell fragments; lagoons are sites where carbonate mud accumulates; and reefs consist of coral and coral debris. Farther offshore from a reef, we can find a sloping apron of reef fragments. Shallow-water carbonate environments transform into various kinds of limestone.

Deep-Marine Deposits We conclude our journey by sailing far offshore. Along the transition between coastal regions and the deep ocean, turbidity currents deposit graded beds. In the deep-ocean realm, only fine clay and plankton provide a source for sediment **(Fig. 6.21)**. The clay eventually settles out onto the deep seafloor, forming deposits of finely laminated mudstones, and plankton shells settle to form chalk (from calcite shells) or chert (from silica shells). Consequently, deposits of mudstone, chalk, or bedded chert indicate a deep-marine origin.

TAKE-HOME MESSAGE

Different types of sedimentary rocks accumulate in different depositional environments. Thus, strata deposited along a river differ from strata deposited by ocean waves, by glaciers, or in the deep sea. By studying sedimentary rocks at a location, geologists can deduce environments that existed at the locality in the past.

QUICK QUESTION How can you distinguish sediment deposited in an alluvial fan from sediment deposited in a shallow-marine environment?

FIGURE 6.20 Reef environments for the deposition of carbonate sediments.

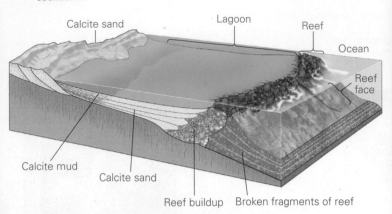

(a) Carbonate reefs form along shorelines in warm-water environments. In detail, reefs include many distinct depositional environments.

(b) A dramatic reef surrounds an island in the tropical Pacific. Deeper water is darker.

FIGURE 6.21 Examples of deep-marine sediment.

(a) These plankton shells, which make up some kinds of deep-marine sediment, are so small that they could pass through the eye of a needle.

(b) The chalk cliffs of southeastern England consist of plankton shells deposited on the seafloor tens of millions of years ago.

6.5 **Sedimentary Basins**

The sedimentary veneer on the Earth's surface varies greatly in thickness. If you stand in central Siberia or Canada, you will find yourself on igneous and metamorphic basement rocks that are over a billion years old—sedimentary rocks are nowhere in sight. Yet if you stand along the southern coast of Texas, you would have to drill through over 15 km of sedimentary beds before reaching igneous and metamorphic basement. Thick accumulations of sediment form only in special regions where the surface of the Earth's lithosphere sinks, providing space in which sediment can collect. Geologists use the term **subsidence** to refer to the process by which the surface of the lithosphere sinks, and the term **sedimentary basin** for the sediment-filled depression. In what geologic settings do sedimentary basins form? Plate tectonics theory can provide a key.

FIGURE 6.22 The geologic setting of sedimentary basins.

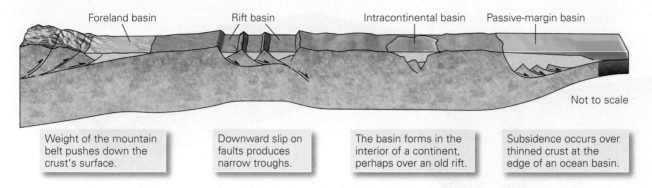

Foreland basin   Rift basin   Intracontinental basin   Passive-margin basin

Not to scale

Weight of the mountain belt pushes down the crust's surface.

Downward slip on faults produces narrow troughs.

The basin forms in the interior of a continent, perhaps over an old rift.

Subsidence occurs over thinned crust at the edge of an ocean basin.

CATEGORIES OF BASINS IN THE CONTEXT OF PLATE TECTONICS THEORY

Geologists distinguish among different kinds of sedimentary basins in the context of plate tectonics theory. Let's consider a few examples (**Fig. 6.22**).

> *Rift basins:* These form in continental rifts, regions where the lithosphere is stretching horizontally, and therefore thins vertically. As the rift grows, slip on faults drops blocks of crust down, producing low areas—narrow basins bordered by elongate mountain ridges. These basins fill with terrestrial sediment. In deserts, overlapping alluvial fans line the margins of the basins.

> *Passive-margin basins:* These form along the edges of continents that are not plate boundaries. They are underlain by stretched lithosphere, the remnants of a rift whose evolution successfully led to the formation of a mid-ocean ridge and subsequent growth of a new ocean basin. Passive-margin basins form because subsidence of stretched lithosphere continues long after rifting ceases. They fill both with sediment carried to the sea by rivers, and with carbonate rocks formed in coastal reefs. Passive-margin basins include some of the thickest accumulations of sediment on the Earth.

> *Intracontinental basins:* These develop in the interiors of continents, initially because of subsidence over a rift. They continue to subside in pulses, even hundreds of millions of years after they formed, for reasons that are not well understood.

> *Foreland basins:* These form on the continental side of a mountain belt because the forces produced during convergence or collision push large slices of rock up faults and onto the surface of the continent. The weight of these slices pushes down on the surface of the continent, producing a wedge-shaped depression adjacent to the mountain range

that fills with sediment eroded from the range. Fluvial and deltaic strata accumulate in foreland basins.

TRANSGRESSION AND REGRESSION

Sea-level changes, relative to the land surface, control the succession of sediments that we see in a sedimentary basin. At times during the Earth's history, sea level has risen by as much as a couple of hundred meters, yielding shallow seas that submerge the interiors of continents. At other times, relative sea level has fallen by hundreds of meters, exposing the continental shelves to air. Global sea-level changes may be due to a number of factors—including climate change—which control the amount of ice stored in polar ice caps and cause changes in the volume of ocean basins. Sea level at a specific location may also be due to local uplift or sinking of the land surface.

When relative sea level rises, the shoreline migrates inland—we call this process **transgression**. When relative sea level falls, the coast migrates seaward—we call this process **regression**. During transgression and regression, the positions of depositional environments migrate, so the depositional environment at a given location changes over time. These processes, acting over time, can lead to the formation of broad blankets of sediment. Note that the age of a given sediment layer deposited during a transgression or regression varies with location (**Fig. 6.23**).

DIAGENESIS

Earlier in this chapter we discussed lithification, by which sediment hardens into rock. Lithification is an aspect of a broader phenomenon called diagenesis. Geologists use the term **diagenesis** for all the physical, chemical, and biological processes that transform sediment into sedimentary rock and that alter characteristics of sedimentary rock after the rock has formed. This includes changes to a sedimentary rock

FIGURE 6.23 The concept of transgression and regression during deposition of sedimentary sequence.

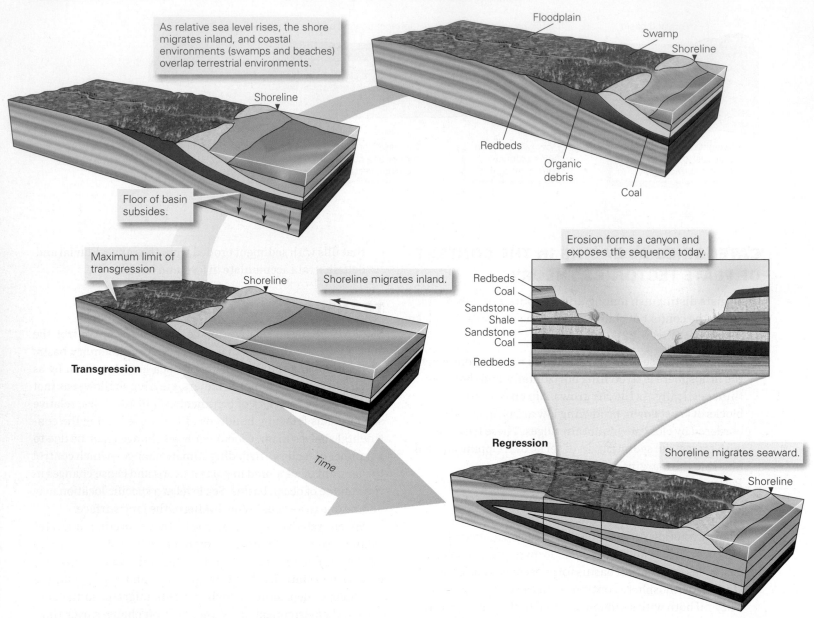

As relative sea level rises, the shore migrates inland, and coastal environments (swamps and beaches) overlap terrestrial environments.

Shoreline

Floodplain

Swamp

Shoreline

Redbeds

Organic debris

Coal

Floor of basin subsides.

Maximum limit of transgression

Shoreline

Shoreline migrates inland.

Transgression

Time

Erosion forms a canyon and exposes the sequence today.

Redbeds
Coal
Sandstone
Shale
Sandstone
Coal
Redbeds

Regression

Shoreline migrates seaward.

Shoreline

that happen when it reacts with groundwater underground, even long after lithification. Such reactions can cause existing cements to dissolve and new ones to precipitate, or may cause new minerals to grow in remaining pores.

As temperature and pressure increase still deeper in the subsurface, the changes that take place in rocks become more profound. At sufficiently high temperature and pressure, metamorphism begins, in that a new assemblage of minerals forms and in some cases mineral grains become aligned parallel to each other. The transition between diagenesis and metamorphism in sedimentary rocks is gradational and occurs between temperatures of 150°C and 250°C. In the next chapter, we enter the realm of true metamorphism.

TAKE-HOME MESSAGE

In certain geologic settings, the Earth's surface sinks (subsides) to form a depression that fills with sediment. The depression with its thick fill of sediment is a sedimentary basin. As sea level rises and falls, the coast, and therefore depositional environments, can migrate inland or offshore, respectively. Once sediment has been deposited, it undergoes various changes, known as diagenesis, in response to pressure and to interaction with groundwater.

QUICK QUESTION How thick can the fill of a sedimentary basin become?

ANOTHER VIEW Even from a distance, this cliff of horizontal redbeds along the shore of Lake Superior provides clues to the environment in which it was deposited over half a billion years ago. The red color means it was deposited in a terrestrial environment, probably in a fluvial setting.

Chapter 6 Review

CHAPTER SUMMARY

> Geologists recognize four major classes of sedimentary rocks. Clastic sedimentary rocks form from cemented-together grains that were first produced by weathering and were then transported, deposited, and lithified. Biochemical sedimentary rocks develop from the shells of organisms. Organic sedimentary rocks form from plant debris or the altered remains of plankton. Chemical sedimentary rocks precipitate directly from water.

> Formation of clastic rocks begins when grains erode from pre-existing rock. Moving water, air, or ice transport these grains to a site of deposition, where they accumulate. Lithification, involving compaction and cementation, converts loose sediment into rock.

> Clastic rocks are classified primarily by grain size. Other characteristics of grains (roundness, sorting, composition) help to distinguish sediment source and depositional environment.

> Sedimentary structures include bedding, cross bedding, graded bedding, ripple marks, dunes, and mud cracks. These features serve as clues to depositional settings.

> Biochemical and organic rocks form from materials produced by living organisms. Limestone consists predominantly of calcite; chert forms from silica, and coal forms from carbon.

> Evaporites consist of minerals precipitated from saline water.

> Different sediments accumulate in different depositional environments.

> Glaciers, streams, alluvial fans, deserts, rivers, lakes, deltas, beaches, shallow seas, and deep seas each accumulate a distinctive assemblage of sedimentary strata.

> Thick piles of sedimentary rocks collect in sedimentary basins, regions where the lithosphere subsided.

> Transgression occurs when sea level rises and the coastline migrates inland. Regression occurs when sea level falls and the coastline migrates seaward.

> Diagenesis involves processes leading to lithification and processes that alter sedimentary rock once it has formed.

GUIDE TERMS

alluvial fan (p. 204)
arkose (p. 193)
bed (p. 198)
biochemical sedimentary rock (p. 190)
breccia (p. 192)
carbonate rock (p. 194)
cementation (p. 191)
chemical sedimentary rock (p. 190)
clast (p. 190)
clastic sedimentary rock (p. 190)

coal (p. 195)
compaction (p. 191)
conglomerate (p. 193)
cross bed (p. 201)
delta (p. 208)
deposition (p. 191)
depositional environment (p. 204)
diagenesis (p. 211)
dune (p. 201)
erosion (p. 190)
evaporite (p. 196)
fossil (p. 201)

geologic map (p. 200)
lithification (p. 191)
mudstone (p. 193)
oil shale (p. 195)
organic sedimentary rock (p. 190)
regression (p. 211)
ripple mark (p. 200)
sandstone (p. 190)
sedimentary basin (p. 210)
sedimentary rock (p. 189)
sedimentary structure (p. 198)

shale (p. 193)
siltstone (p. 193)
sorting (p. 192)
strata (p. 198)
stratigraphic formation (p. 199)
subsidence (p. 210)
transgression (p. 211)
travertine (p. 196)

 GEOTOURS *THIS CHAPTER'S GEOTOURS WORKSHEET (F) FEATURES QUESTIONS AND GOOGLE EARTH SITES ON:*

> Sedimentary rocks exposed in the Grand Canyon, the Lewis Range, Death Valley, and Zion National Park

> Erosion and differential and physical weathering

> Regression and transgression

> Fluvial and arid depositional environments

> Modern and ancient depositional environments

REVIEW QUESTIONS

The letters following each Review Question refer to the corresponding Learning Objective from the Chapter Opener.

1. Describe how a clastic sedimentary rock forms from its unweathered parent rock. **(A, B)**

2. Explain how biochemical sedimentary rocks form. **(A)**

3. How do grain size, sorting, sphericity, and angularity change as sediments move downstream? **(B, E)**

4. Describe the two different kinds of chert. How are they similar? How are they different? **(A)**

5. What conditions lead to deposition of evaporites? **(E)**

6. How does dolostone differ from limestone, and how does dolostone form? **(C)**

7. What kinds of rock form in the environment shown in this diagram? **(C)**

8. What are cross beds, and how do they form? How can you read the current direction from cross beds? **(D)**

9. Describe how a turbidity current forms and moves. How does graded bedding form? **(D)**

10. Compare the deposits of an alluvial fan with those of a deep-marine environment. **(E)**

11. What kinds of sediments accumulate in river and delta systems? How can they be distinguished from glacial sediments? **(E)**

12. Why don't sediments accumulate everywhere? What types of tectonic conditions are required to create sedimentary basins? **(F)**

13. Compare the consequences of a transgression with those of a regression. Which process is depicted in the diagram? **(F)**

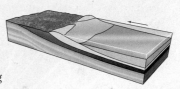

14. What changes take place during diagenesis? **(F)**

ON FURTHER THOUGHT

15. Recent exploration of Mars by robotic vehicles suggests that layers of sedimentary rock cover portions of the planet's surface. The layers contain cross bedding and relics of gypsum crystals. At face value, what do these features suggest about depositional environments on Mars in the past? **(D, E)**

16. The Gulf Coast of the United States is a passive-margin basin that contains a very thick accumulation of sediment. Drilling reveals that the base of the sedimentary succession in this basin consists of redbeds. These are overlain by a thick layer of evaporite, which in turn, is overlain by deposits composed predominantly of sandstone and shale. In some strata, the sandstone occurs in channels and contains ripple marks, and

the shale contains mud cracks. In other strata, the sandstone and shale contain fossils of marine organisms. The sequence contains hardly any conglomerate or arkose. Be a sedimentary detective and explain the succession of sediment in the basin. **(D)**

17. Examine the Bahamas (Latitude 23°58′40.98″ N, Longitude 77°30′20.37″ W) with Google Earth. Describe the depositional environments that you see. **(E)**

18. The bedrock of Florida consists mostly of shallow-marine limestone. What does this observation suggest about the nature of the Florida peninsula in the past? Keep in mind that Florida's land surface lies at less than 50 m above sea level. **(E)**

ONLINE RESOURCES

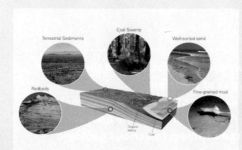

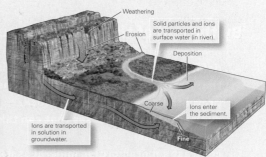

Videos This chapter features videos on the formation of sedimentary rocks, transgression and regression, and how this appears in the stratigraphic record.

Smartwork5 This chapter features questions on recognizing types of sedimentary rocks, structures, and environments.

CHAPTER 7

METAMORPHISM: A PROCESS OF CHANGE

By the end of this chapter, you should be able to . . .

A. define metamorphism, and characterize the changes that a rock undergoes during the transformation of a protolith into a metamorphic rock.

B. explain the key processes that can take place during metamorphism.

C. contrast a metamorphic mineral assemblage with the assemblage of a protolith, and contrast a metamorphic texture with the texture of a protolith.

D. recognize the occurrence of foliation, and distinguish among different types of foliation.

E. identify and name examples of metamorphic rock, and explain how these rocks differ from one another.

F. relate kinds of metamorphism to various geologic settings in the context of plate tectonics theory.

G. describe how the character of a metamorphic rock reflects the grade of metamorphism, and how metamorphic grade depends on temperature and pressure.

7.1 Introduction

Cool winds sweep across Scotland for much of the year. In this blustery climate, vegetation has a hard time taking hold, so the landscape provides countless outcrops of barren rock. During the mid-18th century, James Hutton, who would come to be known as the father of geology, examined these outcrops, hoping to learn how the rocks formed. Hutton found that many features in the outcrops resembled the products of present-day sediment deposition or of volcanic eruptions. So, he developed a sense for how sedimentary and igneous rock originated. But Hutton also found rock that contained minerals and textures quite different from those in sedimentary and igneous samples. He described this third, puzzling, kind of rock as "a mass of matter which had evidently formed originally in the ordinary manner . . . but which is now extremely distorted in its structure . . . and variously changed in its composition." This type is now known as metamorphic rock, from the Greek words *meta*, meaning change, and *morphe*, meaning form. In modern terms, a **metamorphic rock** is one that forms when a pre-existing rock, or **protolith**, undergoes a solid-state change in response to the modification of its environment. This process of change is called **metamorphism**.

Let's look more closely at the definition of metamorphic rock to understand what each component of the definition means. By *solid state*, we mean that a metamorphic rock does not form by solidification of magma—rocks that do, by definition, are igneous. By *change*, we mean that a metamorphic rock contains new minerals that did not occur in the protolith, and/or a new texture (arrangement of mineral grains) that differs from that of the protolith. And by *modification of environment*, we mean a change in temperature or pressure, an application of *stress* (a directed compression, stretching, or shear), or exposure to *hydrothermal fluids* (solutions of very hot water). Rocks undergo metamorphism when they are subjected to such *agents of metamorphism*, just as caterpillars undergo metamorphosis because of hormonal changes in their bodies.

Did you ever wonder . . .
if, once formed, rocks ever change?

◄ (facing page) A metamorphic rock exposed on a hillside in Scotland displays contorted foliation and contains minerals that formed at high pressure and temperature at depth in the crust.

From Hutton's day to the present, geologists have undertaken field studies, laboratory experiments, and theoretical calculations to better characterize metamorphism, and to understand the conditions that lead to metamorphism. In this chapter, we introduce the results of this work. We begin by explaining the causes of metamorphism and the basis for classifying metamorphic rocks. We conclude by discussing the geologic settings in which these rocks form. As you will see, the full story of metamorphism couldn't be told until the theory of plate tectonics became established.

7.2 Consequences and Causes of Metamorphism

WHAT IS A METAMORPHIC ROCK?

If someone were to put a rock on a table in front of you, how would you know that it is metamorphic? First, metamorphic rocks can possess **metamorphic minerals**, new minerals that grow in place within solid rock during metamorphism. In fact, metamorphism can produce a group of minerals that together make up a *metamorphic mineral assemblage*. Second, metamorphic rocks have **metamorphic texture**, defined by the arrangement of their mineral grains. That texture can be manifested by **metamorphic foliation**, which is characterized by the parallel alignment of platy minerals (such as mica) or by the presence of alternating light-colored and dark-colored layers.

Because of the changes that take place during metamorphism, a metamorphic rock can differ in appearance from its protolith as much as a butterfly can differ from a caterpillar. For example, metamorphism of red shale can yield a metamorphic rock consisting of aligned mica flakes, flattened quartz grains, and brilliant garnet crystals (**Fig. 7.1a**), and metamorphism of a fossiliferous limestone can yield a metamorphic rock consisting of large interlocking crystals of calcite (**Fig. 7.1b**).

PROCESSES THAT TAKE PLACE DURING METAMORPHISM

The process of forming metamorphic minerals and textures generally takes place very slowly, over thousands to millions

FIGURE 7.1 Metamorphism causes changes in mineral makeup and texture.

Protolith ⟶ Metamorphic rock

Foliation plane

(a) A specimen of red shale (left) contains clay, quartz, and iron oxide. When this rock undergoes intense metamorphism, it may change into a metamorphic rock (right) containing biotite, quartz, feldspar, and purple garnet.

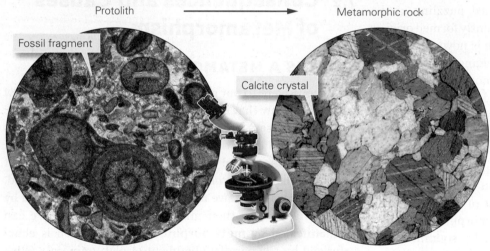

Protolith ⟶ Metamorphic rock

Fossil fragment

Calcite crystal

(b) A protolith of fossiliferous limestone changes into a metamorphic rock with a very different texture, as seen through a microscope.

of years, and involves several processes that sometimes occur alone and sometimes together. The most common processes include the following:

> *Recrystallization* changes the shape and size of grains without changing the identity of the mineral making up the grains **(Fig. 7.2a)**.
> A *phase change* transforms one mineral into another mineral (a polymorph) with the same composition but a different crystal structure. On an atomic scale, a phase change involves the rearrangement of atoms.
> *Metamorphic reaction*, or *neocrystallization* (from the Greek *neos*, for new) results in growth of new minerals that differ from those of the protolith **(Fig. 7.2b)**.

During neocrystallization, chemical reactions digest minerals of the protolith and cause new minerals to grow.

> *Pressure solution* happens where the surface of one mineral grain pushes against the surface of another, under conditions where a water film separates the grains. The grains preferentially dissolve at the surface of contact, and the resulting ions migrate away **(Fig. 7.2c)**. The ions may precipitate on the sides of grains that aren't being pushed together as tightly.
> *Plastic deformation* happens when a rock becomes warm enough to behave like soft plastic, so the minerals within it can change shape without breaking **(Fig. 7.2d)**.

In the following sections, we'll consider how various agents of metamorphism cause these processes to take place.

METAMORPHISM DUE TO HEATING

When you heat cake batter sufficiently, the batter transforms into a new material—cake. Similarly, when you heat a rock, its ingredients transform into a new material—metamorphic rock. Why? Think about what happens to atoms in a mineral grain when the grain warms. Heat causes the atoms to vibrate rapidly, stretching and bending chemical bonds that lock the atoms to their neighbors. If bonds stretch too far and break, atoms detach from their original neighbors, move slightly, and form new bonds with other atoms. Repetition of this process leads to *solid-state diffusion*. During diffusion, atoms either rearrange within grains or migrate into and out of grains. Diffusion allows recrystallization or neocrystallization to take place.

Metamorphism happens at temperatures between those at which diagenesis (sedimentary rock formation, see Chapter 6) occurs and those that cause melting. Roughly speaking, this means that most metamorphic rocks formed at temperatures between 250°C and 850°C.

FIGURE 7.2 Metamorphic processes as seen through a microscope.

Protolith ⟶ Metamorphic rock

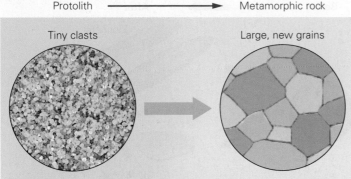

Tiny clasts   Large, new grains

(a) Mineral grains recrystallize to form new, interlocking grains of the same mineral. Typically, the new grains are larger.

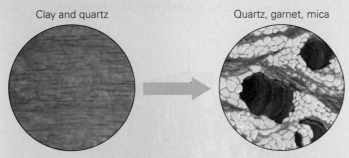

Clay and quartz   Quartz, garnet, mica

(b) Chemical reactions change the original assemblage of minerals into a new metamorphic assemblage of minerals.

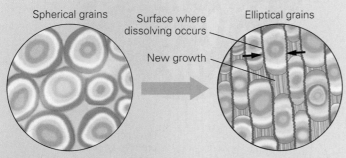

Spherical grains   Surface where dissolving occurs   Elliptical grains   New growth

(c) Pressure solution dissolves grains on the sides undergoing the highest pressure and precipitates new mineral material where the pressure is lower. Arrows indicate the squeezing direction.

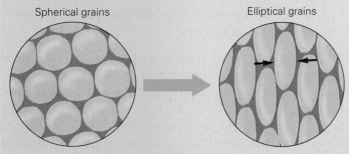

Spherical grains   Elliptical grains

(d) Plastic deformation changes the shape of grains—without breaking them—as a result of squeezing or shear at high temperatures.

METAMORPHISM DUE TO PRESSURE

As you swim underwater in a swimming pool, water squeezes or pushes against you equally from all sides—in other words, your body feels *pressure*. If you pull an air-filled balloon underwater in a lake, the balloon becomes smaller. Pressure can have a similar effect in minerals and can cause a material to collapse inward. Near the Earth's surface, minerals with relatively open crystal structures can be stable. However, if you subject these minerals to extreme pressure, the atoms in them pack together more closely, and denser minerals tend to form. Such transformations take place during phase changes or during neocrystallization.

CHANGING BOTH PRESSURE AND TEMPERATURE

So far, we've considered changes in temperature and pressure as separate phenomena. But in the Earth, temperature and pressure change together with increasing depth. Why? Rocks deeper in the crust lie beneath the weight of more overburden, so the pressure is greater. Also, rocks deeper in the crust endure higher temperature, due to the geothermal gradient. For example, at a depth of 15 km, temperature beneath a mountain belt reaches about 450°C, and pressure rises to about 4.5 kbar, whereas at a depth of 30 km, temperature reaches 720°C, and pressure rises to 9.0 kbar. Experiments and calculations show that a mineral's *stability*, meaning its ability to form and survive, depends on both temperature and pressure. Thus, as depth increases, the original mineral assemblage in a rock becomes unstable, and a new mineral assemblage becomes stable. As a consequence, the minerals that grow in a metamorphic rock at a depth of 15 km differ from those that grow in a metamorphic rock at 30 km.

COMPRESSION, SHEAR, AND DEVELOPMENT OF PREFERRED ORIENTATION

Put a ball of dough on a table top, lay a book on it, and press down on the book. You'll see the ball flatten into a pancake, oriented parallel to the table, because the downward push you apply exceeds the push provided by air in other directions **(Fig. 7.3a)**. If a material undergoes squeezing or stretching unequally from different sides, we say that it has been subjected to **differential stress**. In other words, under conditions of differential stress, the push or pull in one direction differs in magnitude from the push or pull in another direction. Note that differential stress differs from pressure—in the latter, the push is the same in all directions. (The more

FIGURE 7.3 Directed compression and shear change the shapes and sizes of mineral grains during metamorphism.

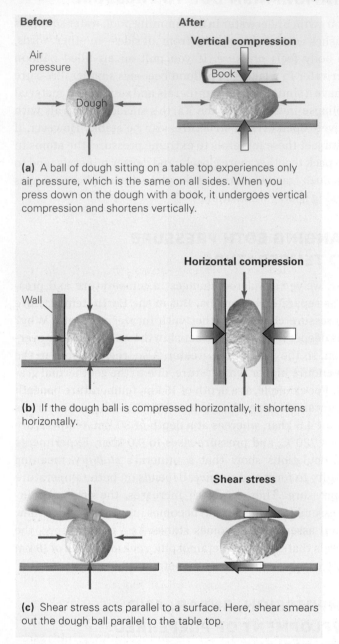

(a) A ball of dough sitting on a table top experiences only air pressure, which is the same on all sides. When you press down on the dough with a book, it undergoes vertical compression and shortens vertically.

(b) If the dough ball is compressed horizontally, it shortens horizontally.

(c) Shear stress acts parallel to a surface. Here, shear smears out the dough ball parallel to the table top.

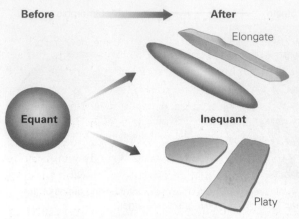

(d) Compression and shear can transform equant grains into inequant grains. Inequant grains can be elongate (cigar-shaped) or platy (pancake-shaped).

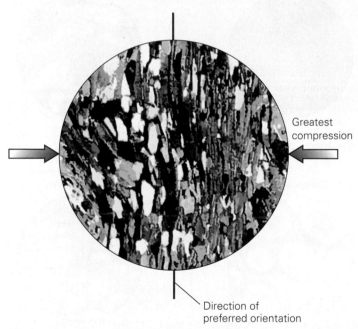

(e) In metamorphic rock, inequant grains may be aligned to form a preferred orientation. As seen through a microscope, the flat planes of grains are perpendicular to the greatest compression direction.

general term, stress, can refer to both differential stress and pressure.)

We distinguish between two kinds of differential stress:

> *Normal stress:* Normal stress pushes or pulls perpendicular to a surface. We call a push *compression* and a pull *tension.* Differential compression flattens a material in an orientation perpendicular to the greatest compression (**Fig. 7.3b**). Differential tension stretches a material in a direction parallel to the greatest pull.

> *Shear stress:* Shear stress, or *shear,* moves one part of a material sideways relative to another part. If you place a ball of dough on a table, set your hand on top of it, and

move your hand parallel to the table top, you're applying *shear* to the ball (**Fig. 7.3c**).

Compression and shear at metamorphic temperatures and pressures may cause a body of rock to change shape without cracking or breaking. During the process, a **preferred orientation** develops, in that pancake-shaped (platy) grains become aligned with one another, as do cigar-shaped (elongate) grains. Note that platy and elongate grains are examples of *inequant grains,* in that they have different dimensions in different directions (**Fig. 7.3d, e**); in contrast, *equant grains* have roughly the same dimensions in all directions.

THE ROLE OF HYDROTHERMAL FLUIDS

Metamorphic reactions commonly take place in the presence of hydrothermal fluids. We initially defined hydrothermal fluids simply as solutions of very hot water. In fact, they actually can include hot water, steam, and so-called supercritical fluid. (A *supercritical fluid* is a substance that forms under high temperatures and pressures and has characteristics of both liquid and gas.) Hydrothermal fluids react chemically with rock by dissolving, transporting, and providing ions. These fluids also provide water molecules that can become incorporated in minerals. In some cases, hydrothermal fluids passing through a rock during metamorphism pick up some ions of one element and drop off ions of another, thereby changing the overall chemical composition of the rock. When this happens, we say that the rock has undergone **metasomatism**.

TAKE-HOME MESSAGE

Metamorphism takes place in response to changes in temperature, pressure, application of compression and shear, or interaction with hydrothermal fluids. The process involves reactions that take place without melting and can produce new textures and new minerals.

QUICK QUESTION Can the overall composition of rock change during metamorphism?

7.3 Types of Metamorphic Rocks

Coming up with a way to classify and name the great variety of metamorphic rocks on the Earth hasn't been easy. After decades of debate, geologists have found it most convenient to divide metamorphic rocks into two fundamental classes: foliated rocks and nonfoliated rocks. Each class contains several rock types, as we now see.

FOLIATED METAMORPHIC ROCKS

To understand this group of rocks, we first need to describe the nature of **foliation** in more detail. The word comes from the Latin *folium*, for leaf. Geologists use *foliation* to refer to the parallel surfaces or layers that can occur in a metamorphic rock. Foliation can give a metamorphic rock a striped or streaked appearance in an outcrop, or can give it the ability to split into thin sheets. A rock has foliation either because its minerals have a preferred orientation or because different minerals concentrate in different layers so the rock displays alternating dark-colored and light-colored layers.

Foliated metamorphic rocks can be distinguished from one another according to their composition, their grain size, and the nature of their foliation. The most common types include:

> *Slate:* The finest-grained foliated metamorphic rock, **slate**, forms by metamorphism of shale or mudstone (rocks composed dominantly of clay) under relatively low pressures and temperatures. Slate contains a type of foliation called *slaty cleavage*. This allows slate to split into thin sheets, which make excellent roofing shingles **(Fig. 7.4a)**. Slaty cleavage develops when pressure solution removes portions of clay flakes that are not perpendicular to the compression direction, while clay flakes that are perpendicular to the compression direction grow. During the process, some flakes may passively rotate to be parallel with the plane of cleavage, pushed into the new orientation by compression. Slaty cleavage forms perpendicular to the direction of maximum compression. Therefore, horizontal compression of a sequence of shale beds produces vertical slaty cleavage, as well as folds **(Fig. 7.4b)**.

> *Phyllite:* **Phyllite** is a fine-grained metamorphic rock with a foliation caused by the preferred orientation of very fine-grained white mica. The word comes from the Greek *phyllon*, meaning leaf. The texture of phyllite gives it a silky sheen known as phyllitic luster **(Fig. 7.5a)**. Phyllite forms when slate undergoes metamorphism at a temperature high enough to cause neocrystallization and the growth of white mica.

> *Schist:* **Schist** is a medium- to coarse-grained metamorphic rock containing a type of foliation called *schistosity*. This fabric is defined by the preferred orientation of large mica flakes (typically, muscovite or biotite; **Fig. 7.5b**). Schist forms at a higher temperature than does phyllite, for under hotter conditions, mica crystals can grow larger.

> *Metaconglomerate:* Under the metamorphic conditions that produce slate, phyllite, or schist, a conglomerate metamorphoses into **metaconglomerate**. Typically, pressure solution and plastic deformation flatten pebbles and cobbles of such rock into pancake-like shapes. The alignment of these inequant clasts defines a foliation **(Fig. 7.5c)**.

> *Gneiss:* **Gneiss** is a compositionally layered metamorphic rock, composed of alternating dark-colored layers and light-colored layers. The former have a higher concentration of mafic minerals, whereas the latter have a higher concentration of felsic minerals. The layers range in thickness from millimeters to meters. Such compositional layering, or *gneissic banding*, gives gneiss a striped appearance **(Fig. 7.6a, b)**.

Did you ever wonder . . .
why slate makes such nice roofing shingles?

FIGURE 7.4 Slate is a foliated metamorphic rock that forms at relatively low temperatures and pressures.

(a) A block of slate splits easily along cleavage planes, which may be at a high angle to the bedding planes. Workers split slate to produce roof shingles that, when overlapped, make a watertight surface.

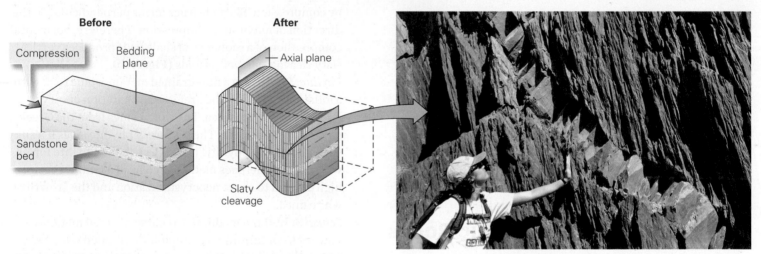

(b) Slaty cleavage forms in response to differential compression. In this example, beds also bend to form folds as slaty cleavage develops. The cleavage plane tends to be oriented parallel to the axial plane, an imaginary surface that, simplistically, divides the fold in half.

How does gneissic banding form? Some banding evolves directly from the original bedding in a rock. For example, metamorphism of a protolith consisting of alternating beds of sandstone and shale could produce a gneiss with alternating bands of quartzite and mica. Most gneissic banding forms when the protolith undergoes an extreme amount of shearing under conditions in which the rock can flow like soft plastic **(Fig. 7.6c)**. The shearing stretches and smears out any pre-existing compositional contrasts in the rock and transforms them into aligned sheets. Formation of banding may also develop by *metamorphic differentiation*, an incompletely understood process during which different metamorphic minerals grow preferentially in different layers **(Fig. 7.6d)**.

> *Migmatite:* Under high temperatures in the presence of water, gneiss may begin to melt, producing felsic magma

and leaving behind solid, relatively mafic metamorphic rock. If the melt freezes again before flowing out of the source area, a mixture called **migmatite**, consisting of igneous rock and relict metamorphic rock, forms. Plastic flow during the formation of migmatite can contort the felsic and mafic rocks into complex shapes.

NONFOLIATED METAMORPHIC ROCKS

Nonfoliated metamorphic rocks contain minerals that recrystallized or grew during metamorphism, but the rock overall has no foliation. The lack of foliation means either that metamorphism occurred in the absence of compression or shear, or that most of the new crystals are equant. Here, we describe some of the rock types that can occur without foliation.

FIGURE 7.5 Examples of foliated metamorphic rocks formed at high temperatures and pressures.

(a) During formation of phyllite, clay recrystallizes to form tiny mica flakes that reflect light, giving the rock a sheen.

(b) A schist contains coarse mica flakes, along with other metamorphic minerals.

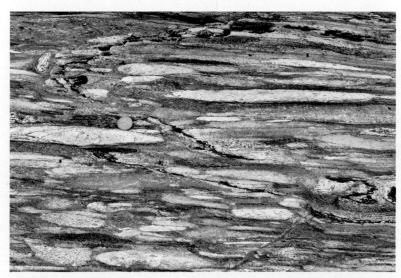

(c) In metaconglomerate, pebbles and cobbles flatten into a pancake shape without cracking.

> *Hornfels:* **Hornfels** is a fine-grained nonfoliated rock that contains a variety of metamorphic minerals. The specific mineral assemblage in a hornfels depends on the composition of the protolith and on the temperature and pressure of metamorphism. Inequant mineral grains of a hornfels, in contrast to those of a schist, are randomly oriented.

> *Quartzite:* Most **quartzite** forms by the metamorphism of pure quartz sandstone. During metamorphism, pre-existing quartz grains recrystallize, yielding new, larger grains. In the process, the distinction between cement and grains disappears, open pore space disappears, and the grains become interlocking. When quartzite fractures, cracks cut across grain boundaries—in contrast, cracks in a sandstone curve around grains. Quartzite looks glassier than sandstone and does not have the grainy, sandpaper-like surface characteristic of sandstone **(Fig. 7.7a)**. Quartzite can be white, gray, maroon, or green, depending on the impurities it contains.

> *Marble:* The metamorphism of limestone yields **marble**. During the formation of marble, calcite of the protolith recrystallizes, so fossil shells, pore space, and the distinction between grains and cement disappear. Therefore, marble typically consists of a fairly uniform mass of interlocking calcite crystals **(Fig. 7.7b)**. Sculptors love to work with marble because the rock is relatively soft and has a uniform texture that gives it the cohesiveness and homogeneity needed to fashion large, smooth, highly detailed sculptures. Like quartzite, marble comes in a variety of colors—white, pink, green, and black—depending on the impurities it contains. Michelangelo, one of the great Italian Renaissance artists, sought large, unbroken blocks of creamy white marble from quarries in northwestern Italy for his masterpieces **(Fig. 7.7c)**. Significantly, not all marble is nonfoliated. If the original protolith contained layers with different impurities, and shear caused the marble to flow plastically, the resulting marble displays complex color banding that makes it a prized decorative stone **(Fig. 7.7d)**.

DEFINING METAMORPHIC INTENSITY

The physical conditions under which metamorphism happens vary from place to place. For example, rocks carried to a great depth beneath a mountain range undergo more intense metamorphism than do rocks closer to the Earth's surface. Geologists use the term **metamorphic grade** in a somewhat informal way to indicate the intensity of metamorphism, meaning the amount or degree of metamorphic change. Metamorphic grade depends on the temperature and pressure at which metamorphism takes place.

FIGURE 7.6 The formation of gneiss, which takes place at high temperatures and pressures.

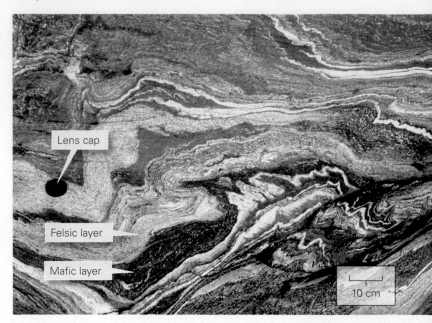

(a) A gneiss from northern Scotland. The light layers contain more felsic minerals, whereas the dark layers contain more mafic minerals.

(b) In this outcrop of gneiss from Brazil, the gneissic banding has been contorted by flow in the rock when it was at high temperature.

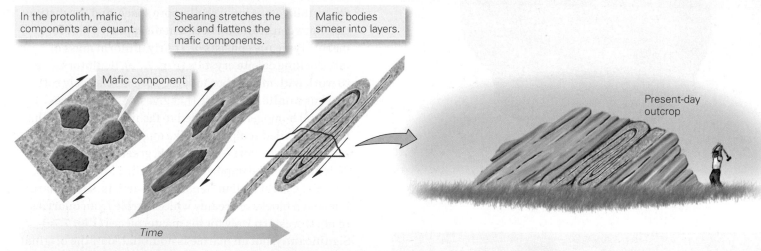

(c) Formation of gneiss, in some cases, involves extreme shear. Original contrasting rock types are smeared into parallel layers.

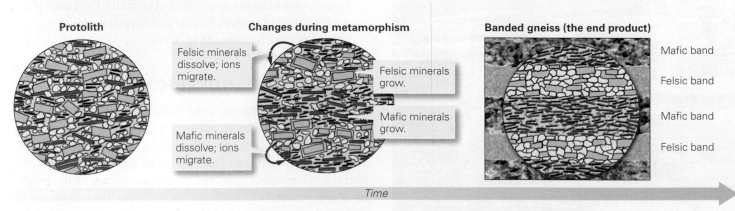

(d) Gneiss may also form by metamorphic differentiation, during which chemical reactions cause felsic and mafic minerals to grow in distinct, separate layers.

FIGURE 7.7 Examples of quartzite and marble, which are typically, but not always, nonfoliated metamorphic rocks.

Quartz sandstone—the protolith of quartzite.

(a) In this sandstone from Kuwait (left), the sand grains stand out. In contrast, this nonfoliated maroon quartzite from Wisconsin (right) looks glassy and breaks on smooth fractures that cut across grains.

In this unmetamorphosed limestone, fossils and bedding are visible.

Italian marble quarry sliced into a mountain

(b) In this unmetamorphosed limestone from New York (left), you can see fossils and shell fragments. Such grains are not visible in the white marble (right) exposed in an Italian quarry.

(c) Sculptors, such as Michelangelo, like to work with nonfoliated white marble due to its softness and uniform texture.

(d) This marble has color banding. The layers became contorted during metamorphism.

BOX 7.1 CONSIDER THIS...

Metamorphic Facies

In the early years of the 20th century, geologists working in Scandinavia, where erosion by glaciers left beautiful, nearly unweathered exposures of metamorphic rocks, came to realize that a metamorphic rock, in general, does not contain a hodgepodge of minerals formed at different times under different conditions, but rather contains a distinct set of minerals that grew in association with each other at a specific range of pressures and temperatures. It seemed that such a mineral assemblage more or less represents a condition of *chemical equilibrium*, meaning that the chemicals making up the rock had organized into a group of minerals that were—to anthropomorphize a bit—comfortable with one another and their surroundings, and thus did not feel the need to change further. The geologists realized that the metamorphic mineral assemblage in a rock depends both on the pressure and temperature conditions of metamorphism and on the composition of the protolith.

With the above concept in mind, the geologists defined a **metamorphic facies** as a set of metamorphic mineral assemblages indicative of a certain range of pressure and temperature. Each specific assemblage in a facies reflects a protolith composition. According to this definition, a given metamorphic facies includes several different kinds of rocks that vary in terms of chemical composition and, therefore, mineral content—but all the rocks of a given facies formed under roughly the same temperature and pressure conditions.

Geologists recognize eight metamorphic facies. We can represent the conditions under which different metamorphic facies formed on a pressure-temperature graph **(Fig. Bx7.1)**. Each area of the graph, labeled with a facies name, represents the approximate range of temperatures and pressures in which mineral assemblages characteristic of that facies grew. For example, a rock subjected to the pressure and temperature at Point A (4.5 kbar and 400°C) develops a mineral assemblage characteristic of the greenschist facies. As the

graph implies, the pressure and temperature conditions defining boundaries between facies cannot be precisely determined, so the transitions between facies are gradual.

We can also portray the geothermal gradients of different crustal regions on the graph. Beneath mountain ranges, for example, the geothermal gradient passes through the zeolite, greenschist, amphibolite, and granulite facies. In contrast, in the accretionary prism that forms at a subduction zone, temperature increases slowly with increasing depth, so blueschist assemblages can form. Note that rocks of the eclogite facies occur at great depth.

FIGURE Bx7.1 The common metamorphic facies. Boundaries between the facies are depicted as wide bands because they are gradational. Note that some amphibolite-facies rocks and all granulite-facies rocks form at pressure-temperature conditions to the right of the melting curve for wet granite. Such metamorphic rocks develop only if the protolith is dry.

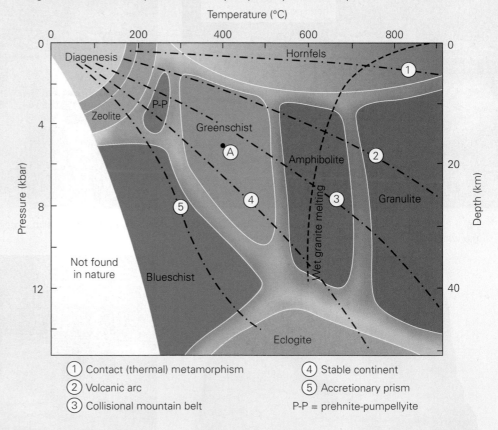

Temperature plays the dominant role in determining the extent of recrystallization and neocrystallization that take place during metamorphism. Metamorphic rocks that form at relatively low temperatures (between about 250°C and 400°C) are *low-grade rocks*, and metamorphic rocks

that form at relatively high temperatures (over about 600°C) are *high-grade rocks. Intermediate-grade rocks* form at temperatures between these two extremes **(Fig. 7.8a)**. As grade increases, recrystallization and neocrystallization tend to produce coarser grains and new mineral assemblages that

FIGURE 7.8 Intensity of metamorphism is indicated by metamorphic grade.

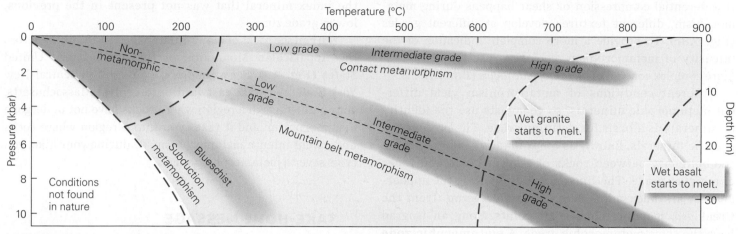

(a) This graph depicts the approximate temperatures and pressures of metamorphic grades. Different conditions occur in different geologic settings.

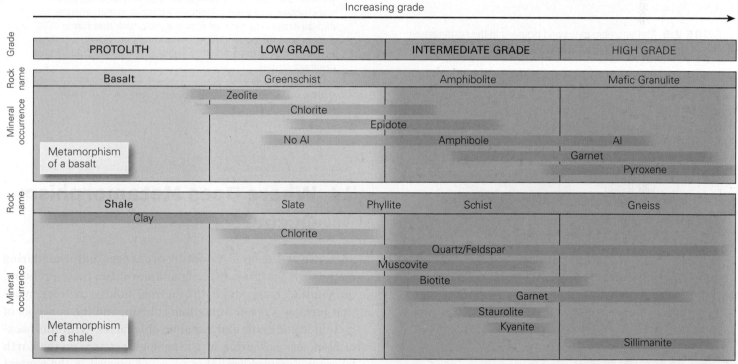

Increasing grade

(b) The metamorphic minerals that form in a given rock depend on metamorphic grade and protolith composition. This chart contrasts important metamorphic minerals that form from a basalt protolith with those formed from a shale protolith.

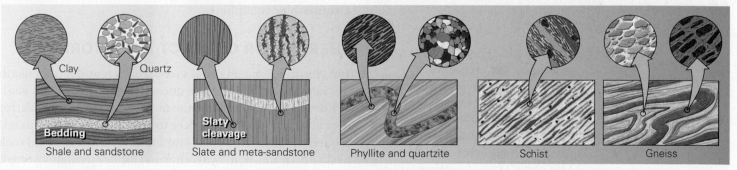

(c) Here we see the consequences of the progressive metamorphism of shale and sandstone during mountain building. Note how textures change with increasing grade.

are stable at higher temperatures and pressures (Fig. 7.8b). If differential compression or shear happens during metamorphism, different textures develop at different grades (Fig. 7.8c). To provide a more complete indication of the intensity of metamorphism, geologists use the somewhat more complex concept of *metamorphic facies* (Box 7.1).

Different conditions of metamorphism yield different metamorphic minerals, so geologists use the identity of minerals as a basis for defining grade. The presence of specific minerals, known as *index minerals*, can delineate the boundary between rocks of a lower grade and rocks of a higher grade. The line on a map marking the appearance of an index mineral is called an *isograd* (from the Greek *iso*, meaning equal); all points along an isograd have the same metamorphic grade. A **metamorphic zone**

is the region between two isograds—zones are named after the index mineral that was not present in the previous, lower-grade zone.

You can see rocks of many different grades by hiking across the Appalachian Mountains in the northeastern United States (Fig. 7.9). For example, if you start in central New York State and walk eastward, into central Massachusetts, your path starts in a region where rocks have not undergone metamorphism, and it takes you into a region where rocks underwent intense metamorphism, so during your hike, you cross several metamorphic zones.

FIGURE 7.9 Metamorphic zones and isograds in the northeastern United States. The minerals listed are index minerals.

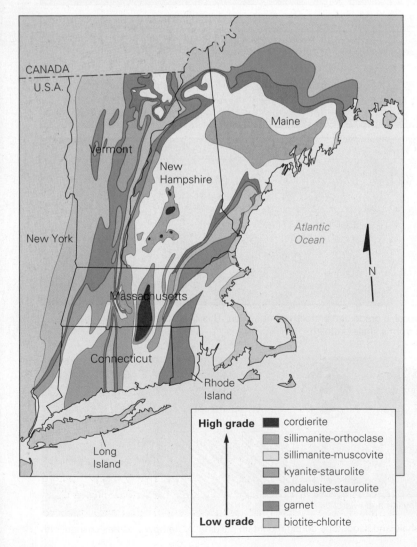

TAKE-HOME MESSAGE

Geologists divide metamorphic rocks into classes based on whether the rock contains foliation. Foliated rocks include slate, schist, and gneiss. Nonfoliated rocks include marble and quartzite. The type of metamorphic rock that forms depends on the conditions of metamorphism. Specific grades of metamorphic minerals form under specific temperature and pressure ranges.

QUICK QUESTION Why don't builders use gneiss to make roofing shingles?

7.4 Where Does Metamorphism Occur?

So far, we've discussed the nature of changes that occur during metamorphism, the causes of metamorphism (heat, pressure, differential stress, and hydrothermal fluids), the rock types that form as a result of metamorphism, and the concepts of metamorphic grade and metamorphic facies. With this background, let's now examine the geologic settings on the Earth where metamorphism takes place, as viewed in the context of plate tectonics theory (**Geology at a Glance**, pp. 232–233). We'll see that different geologic settings can produce different metamorphic conditions.

THERMAL OR CONTACT METAMORPHISM

Imagine a hot magma that rises from great depth beneath the Earth's surface and intrudes into cooler rock at a shallow depth. Heat flows from the magma into the wall rock, for heat always moves from hotter to colder materials. As a consequence, the magma cools and solidifies while the wall rock heats up. In addition, hydrothermal fluids circulate through both the intrusion and the wall rock. As a consequence

FIGURE 7.10 Geologic settings of metamorphism.

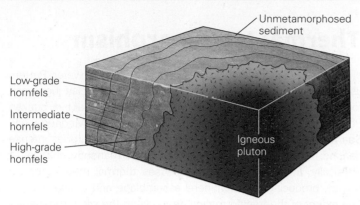

(a) Heat radiated from a large pluton can produce a metamorphic aureole, in which hornfels develops. Grade decreases progressively away from the pluton contact.

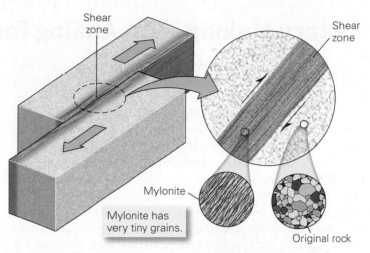

(b) Shearing of a rock under plastic conditions causes its original crystals to divide into tiny crystals without breaking, forming a mylonite.

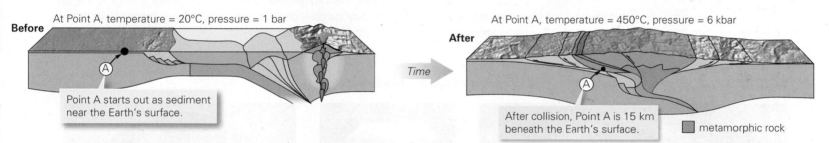

(c) Dynamothermal, or regional, metamorphism happens when one part of the crust shoves over another part, so that rocks once near the surface end up at great depth.

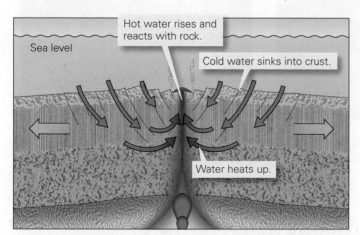

(d) The rising magma at a ridge axis heats water, which then rises. The hot water reacts with the crust and forms metamorphic minerals.

of the heat and hydrothermal fluids, the wall rock undergoes metamorphism, with the highest-grade rocks forming immediately adjacent to the pluton, where the temperatures are highest, and progressively lower-grade rocks forming farther away. The distinct belt of metamorphic rock that forms around an igneous intrusion is called a **metamorphic aureole** or *contact aureole* **(Fig. 7.10a)**. The width of an aureole depends on the size and temperature of the intrusion—larger and hotter intrusions produce wider aureoles.

Geologists refer to the local metamorphism caused by an igneous intrusion as **thermal metamorphism (Box 7.2)**, to emphasize that it develops in response to heat without a change in pressure and without differential compression or shear. This type of metamorphism is also called **contact metamorphism**, to emphasize that it develops adjacent to the contact between an intrusion with its wall rock. Because such metamorphism takes place without application of differential compression or shear, aureoles typically contain hornfels, a nonfoliated metamorphic rock. Contact metamorphism occurs anywhere that plutons intrude into the crust. In the context of plate tectonics theory, plutons intrude at convergent boundaries, in rifts, and during mountain building, so thermal metamorphism can occur in these settings.

BOX 7.2 CONSIDER THIS...

Pottery Making—An Analog for Thermal Metamorphism

A brick in the wall of an adobe house, an earthenware pot, a stoneware bowl, or a translucent porcelain plate may all be formed from the same lump of soft clay, scooped from the surface of the Earth and shaped by human hands **(Fig. Bx7.2)**. This pliable and slimy muck forms when water saturates a mixture of clay minerals and very fine quartz grains left by the chemical weathering of rock. Fine potter's clay for making white china contains a particular clay mineral called *kaolinite*, named after Kauling, a locality in China where it was originally discovered.

People in arid climates make adobe bricks by forming damp clay into blocks, which they then dry under the Sun. Such blocks can be used for construction only in arid climates, because if it rains frequently, the bricks rehydrate and turn back into a sticky muck. Drying clay under the Sun does not change the structure of the clay minerals.

To make a more durable material, workers place clay blocks in a kiln and bake ("fire") them at high temperatures. Heating transforms blocks of clay into hard bricks that are impervious to water. Potters use the same process to make jugs. In fact, fired clay jugs used for storing wine and olive oil have been found intact in sunken Phoenician ships that have rested on the floor of the Mediterranean Sea for thousands of years! Clearly, when clay undergoes intense heating, the process fundamentally and permanently changes it physically. In other words, firing causes thermal metamorphism of clay, producing a new mineral assemblage and a new texture. The extent of the transformation depends on the kiln temperature, just as the grade of metamorphic rock depends on the Earth's temperature at a particular time and place. Potters usually fire earthenware at about 1,100°C and stoneware, which is harder, at about 1,250°C. To produce porcelain—fine china—the clay must be heated to even higher temperatures, up to 1,400°C, for under these conditions the clay begins to melt. If the potter cools the slightly molten clay quickly, glass forms, which gives porcelain its translucent, vitreous appearance.

FIGURE Bx7.2 When metamorphosed, common mud becomes stronger and more durable.

(a) Mud can be shaped into blocks that, when dried, can be used to build houses like this one in Peru.

(b) Baking the mud turns it into much harder brick.

(c) At high temperatures, mud turns into porcelain, as in this plate from China.

BURIAL METAMORPHISM

As sediment becomes buried progressively deeper in a subsiding sedimentary basin, the pressure increases due to the weight of overburden, and the temperature increases due to the geothermal gradient. At depths greater than about 8 to 15 km, depending on the geothermal gradient, temperatures may be great enough for metamorphic reactions to begin, and low-grade metamorphic rocks form. Metamorphism due only to the consequences of very deep burial is called **burial metamorphism**.

DYNAMIC METAMORPHISM

Faults are surfaces on which one piece of crust slides, or shears, past another. Near the Earth's surface (in the upper 10 to 15 km) this movement can fracture rock, breaking it into angular fragments or even crushing it to a powder. But at greater depths, rock becomes warm enough to behave like soft plastic when shear takes place along the fault. During this process, the minerals in the rock recrystallize. We call this process **dynamic metamorphism**, because it occurs as a consequence only of shearing under metamorphic conditions,

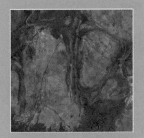
without requiring a change in temperature or pressure. The resulting rock, a *mylonite*, has a foliation that roughly parallels the fault **(Fig. 7.10b)**. Mylonites are very fine-grained, due to processes during dynamic metamorphism that replace larger crystals with a mass of very tiny ones. Dynamic metamorphism takes place wherever faulting occurs at depth in the crust. Consequently, mylonites can be found at all plate boundaries, in rifts, and in collision zones.

DYNAMOTHERMAL OR REGIONAL METAMORPHISM

During the development of mountain ranges, in response to either convergent-boundary tectonics or continental collision, regions of crust undergo compression, and large slices of continental crust slip up and over other portions of the crust along faults. As a consequence, rock that was once near the Earth's surface ends up at great depth beneath the mountain range **(Fig. 7.10c)**. In this environment, the protolith changes in three ways: (1) it heats up because of the geothermal gradient and because of igneous activity; (2) it endures greater pressure because of the weight of overburden; and (3) it undergoes compression and shearing. As a result of these changes, the protolith transforms into foliated metamorphic rock. The type of foliated rock that forms depends on the grade of metamorphism—slate forms at shallower depths, whereas schist and gneiss form at greater depths. Since the metamorphism we've just described involves not only heat but also compression and shearing, geologists call it **dynamothermal metamorphism**. Because such metamorphism affects a large region, we can also call it **regional metamorphism**.

HYDROTHERMAL METAMORPHISM AT MID-OCEAN RIDGES

Hot magma rises beneath the axis of a mid-ocean ridge, so when cold seawater sinks down into the oceanic crust through cracks along the ridge, it heats up and transforms into hydro-thermal fluid. This fluid then rises through the crust, near the ridge, causing **hydrothermal metamorphism** of ocean-floor basalt **(Fig. 7.10d)**. Eventually, the fluid escapes through vents back into the sea at black smokers (see Chapter 2).

METAMORPHISM IN SUBDUCTION ZONES

Blueschist is a relatively rare rock that contains an unusual blue-colored amphibole. Laboratory experiments indicate that formation of this mineral requires very high pressure but relatively low temperature. Such conditions cannot develop in continental crust, for at the high pressure needed to produce blue amphibole, the temperature in continental crust is also high (see Box 7.1). So, to understand where blueschist forms, we must determine where high pressures can develop at relatively low temperatures.

Plate tectonics theory provides the answer to this puzzle. Researchers found that blueschist develops only in the accretionary prisms that form at subduction zones (see Geology at a Glance, pp. 232–233). Because these prisms grow to be over 20 km thick, rock at the base of the prism endures very high pressure (due to the weight of overburden). But because subducted oceanic lithosphere beneath the prism is cool, temperatures at the base of the prism remain relatively low and blue amphibole can form. The blueschist develops a foliation because of the compression and shearing that takes place in an accretionary prism.

SHOCK METAMORPHISM

When large meteorites slam into the Earth, a pulse of extreme compression—a *shock wave*—propagates into the Earth away from the site of impact. The heat may be sufficient to melt or even vaporize rock at the impact site, and the extreme compression of the shock wave causes quartz in rocks below the impact site to undergo a phase change and become a more compact mineral called coesite. The process of changing rock in response to the passage of a shock wave is called **shock metamorphism**.

Environments of Metamorphism

Metamorphic rocks form when a pre-existing rock (a protolith) undergoes changes in texture and/or mineral content in the solid state in response to changes in temperature, changes in pressure, and/or differential stress. Metamorphism may also reflect interaction with

Regional Metamorphism in an Orogenic Belt

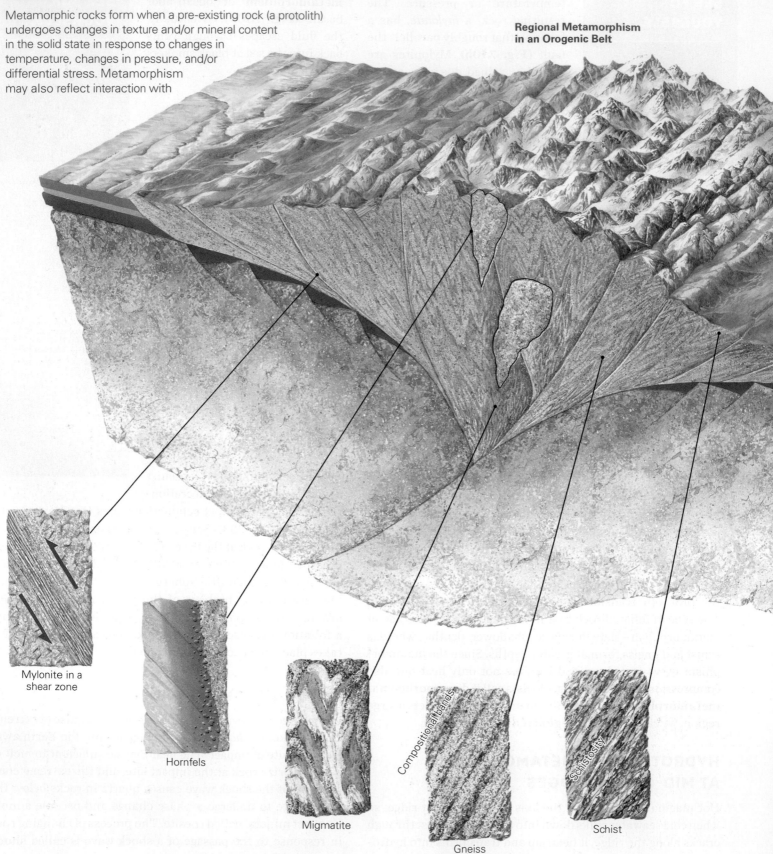

Mylonite in a shear zone

Hornfels

Migmatite

Gneiss

Compositional bands

Schist

Schistosity

Slate

Unmetamorphosed shale

Slaty cleavage

Relict bedding

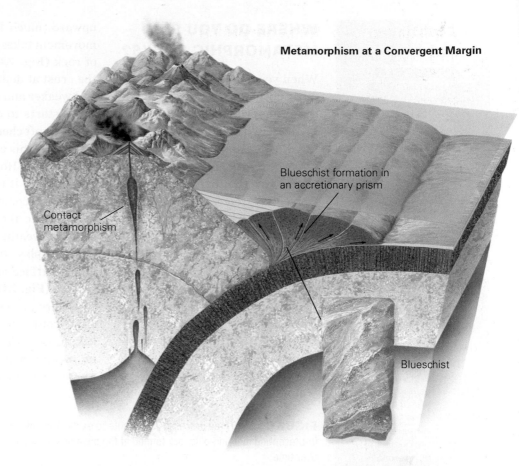

Contact metamorphism

Blueschist formation in an accretionary prism

Blueschist

hydrothermal fluids. Some metamorphic rocks are foliated (have metamorphic layering), whereas others are not. Foliation results when rock is compressed or sheared during metamorphism, causing mineral crystals to grow or rotate into alignment with one another. Dynamothermal (regional) metamorphism occurs during mountain building. Contact metamorphism, which takes place around an igneous intrusion, or pluton, is caused by the heat released by the pluton.

Geologists distinguish among metamorphic rocks according to the type of foliation and the mineral assemblage a rock contains. Hornfels is nonfoliated and forms as a result of contact metamorphism. Mylonite develops when shearing yields a foliation but not necessarily a change in types of minerals. Slate, which forms from shale, contains slaty cleavage; clay flakes in slate are typically aligned at an angle to the bedding plane. Schist contains coarse grains of mica (muscovite and/or biotite) aligned parallel to one another. Gneiss has compositional banding. (Migmatite forms when part of the rock melts, and thus it is a mixture of metamorphic and igneous rock.) Quartzite is composed predominantly of quartz (it is metamorphosed sandstone), whereas marble consists predominantly of calcite or dolomite (it is metamorphosed limestone or dolostone). Quartzite and marble are usually unfoliated.

The types of minerals and foliation in a metamorphic rock indicate the rock's grade. High-grade rocks, such as gneiss, form at higher temperatures and pressures, whereas low-grade rocks, such as shale, form at lower pressures and temperatures. Blueschist is an unusual metamorphic rock that develops under relatively high pressures but relatively low temperatures—the environment of an accretionary prism.

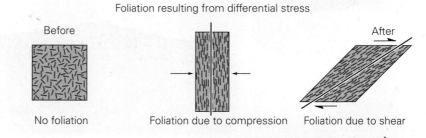

Foliation resulting from differential stress

Before

After

No foliation

Foliation due to compression

Foliation due to shear

Increasing temperature

Increasing pressure

Increasing metamorphic grade

Shale

Hornfels

Slate

Low grade

Schist

Gneiss

Migmatite

Blueschist

High grade

233

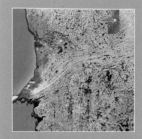
WHERE DO YOU FIND METAMORPHIC ROCKS?

When you stand on an outcrop of metamorphic rock, you are standing on material that once lay many kilometers beneath the surface of the Earth. How does metamorphic rock, formed at great depth, return to the Earth's surface? Geologists refer to the overall process by which deeply buried rocks end up back at the surface as **exhumation**.

Let's look at the specific processes that contribute to bringing metamorphic rocks from below a collisional mountain range back to the surface. First, as two continents push together, the rock caught between them squeezes upward, much like dough pressed in a vise—this upward movement takes place by slip on faults and by plastic-like flow of rock **(Fig. 7.11a)**. Second, as the mountain range grows, the crust at depth beneath it warms up and becomes softer and weaker and able to flow like soft plastic. Eventually, the range starts to collapse under its own weight, much like a block of soft cheese placed under the hot Sun **(Fig. 7.11b)**. As a result of this collapse, the upper part of the crust spreads out laterally. Horizontal stretching of the upper part of the crust causes it to become thinner in the vertical direction, and as it becomes thinner, the deeper part of the crust ends up closer to the surface. Third, erosion takes place at the surface—weathering, landslides, river flow, and glacial flow together play the role of a giant rasp, stripping away rock at the surface and exposing rock that was once below the surface **(Fig. 7.11c)**.

Where are metamorphic rocks exposed today? You can start your quest to find metamorphic rock outcrops by hiking into a mountain range. As we've seen, the process of mountain building produces and eventually exhumes metamorphic rocks, so the towering cliffs in the interior of

FIGURE 7.11 Three processes contribute to exhumation of metamorphic rock formed at depth. Here the red dot (representing metamorphic rock formed at the base of a mountain range) gets progressively closer to the surface over time.

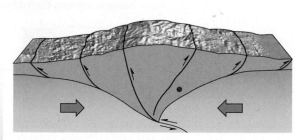
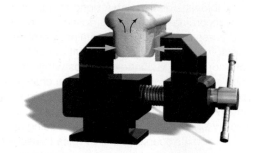

(a) As continents squeeze together, rock is pushed up like dough in a vise.

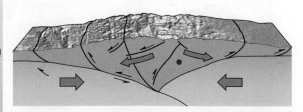
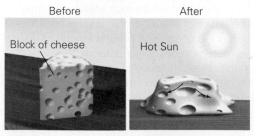

Before   After

Block of cheese   Hot Sun

(b) When rock at depth warms up and softens, the mountain belt collapses like cheese under the hot Sun.

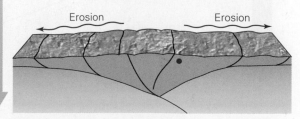
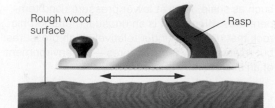

Erosion   Erosion

Rough wood surface   Rasp

(c) Erosion grinds away and removes rocks like a giant rasp.

a mountain range typically reveal schist, gneiss, quartzite, and marble **(Fig. 7.12a)**. Even after the peaks have eroded away, the record of mountain building at a location remains, in the form of a belt of metamorphic rock at the ground surface. Vast expanses of metamorphic rock crop out in continental shields. A **shield** is a broad region of long-lived, stable continental crust where sedimentary cover either was not deposited or has been eroded away so that Precambrian basement rocks are exposed **(Fig. 7.12b, c)**. The basement rocks were metamorphosed during a succession of Precambrian mountain-building events that led to the original growth of continents.

TAKE-HOME MESSAGE

Thermal (contact) metamorphism develops around an igneous intrusion, due to heat from the pluton. Dynamothermal (regional) metamorphism develops beneath mountain ranges where rock undergoes compression and shear at high temperatures and pressures. Erosion and uplift may eventually expose metamorphic rock in mountain ranges or continental shields.

QUICK QUESTION Could you find a layer of metamorphic rock between the layers of sedimentary rock in a sedimentary basin? Why or why not?

FIGURE 7.12 Examples of rock exposures consisting of Precambrian metamorphic rocks.

(a) This cliff, in the Wasatch Mountains of Utah, exposes gneiss, which has been intruded by granite.

(b) A photograph from an airplane window of the flat landscape of the eastern Canadian Shield.

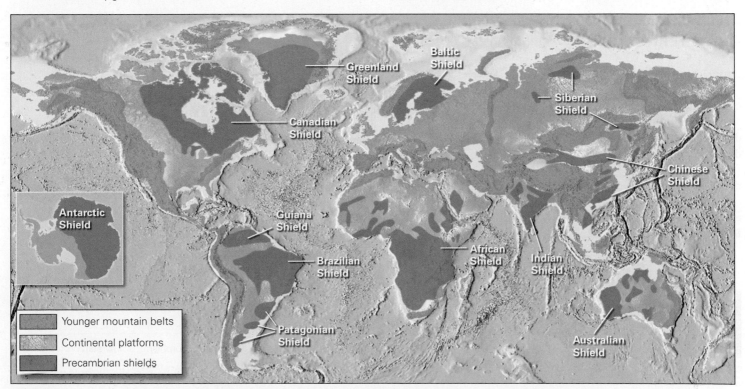

(c) A map showing the distribution of shields, areas where broad expanses of Precambrian crust, including Precambrian metamorphic rocks, crop out.

Chapter 7 Review

CHAPTER SUMMARY

> Metamorphism refers to changes in a rock that result in the formation of a metamorphic mineral assemblage and metamorphic texture in response to change in temperature or pressure, application of differential stress, or interaction with hydrothermal fluids (hot-water solutions).

> Metamorphism involves recrystallization, phase changes, metamorphic reactions (neocrystallization), pressure solution, or plastic deformation. If hydrothermal fluids bring in or remove elements, we say that metasomatism has occurred.

> Metamorphic foliation can be defined either by preferred orientation (alignment of inequant crystals) or by compositional banding. Preferred orientation develops where compression and shearing of a rock causes its inequant grains to align parallel with one another.

> Geologists separate metamorphic rocks into two classes, foliated rocks and nonfoliated rocks, depending on whether the rocks contain foliation.

> The class of foliated rocks includes slate, phyllite, metaconglomerate, schist, and gneiss. The class of nonfoliated rocks includes hornfels, quartzite, and marble. Migmatite, a mixture of igneous and metamorphic rock, forms under conditions where melting begins.

> Rocks formed at relatively low temperatures are known as low-grade rocks, whereas those formed at high temperatures are known as high-grade rocks. Intermediate-grade rocks develop between these two extremes. Different mineral assemblages form at different metamorphic grades.

> Geologists track the distribution of different grades of rock by looking for index minerals. Isograds indicate the locations at which index minerals first appear. A metamorphic zone is the region between two isograds.

> A metamorphic facies consists of a group of metamorphic mineral assemblages that develop under a specified range of temperature and pressure conditions.

> Thermal metamorphism (also called contact metamorphism) occurs in an aureole surrounding an igneous intrusion. Burial metamorphism develops at depth in a sedimentary basin. Dynamic metamorphism occurs along faults, where rocks undergo plastic shearing. Dynamothermal metamorphism (also called regional metamorphism) results when rocks undergo heating and shearing during mountain building. Hydrothermal metamorphism takes place due to the circulation of hot water in oceanic crust at mid-ocean ridges. Shock metamorphism happens during the impact of a meteorite.

> We find belts of metamorphic rocks in mountain ranges. Blueschist forms in accretionary prisms. Shields expose broad areas of Precambrian metamorphic rocks.

GUIDE TERMS

burial metamorphism (p. 230)
contact metamorphism (p. 229)
differential stress (p. 219)
dynamic metamorphism (p. 230)
dynamothermal metamorphism (p. 231)
exhumation (p. 234)
foliation (p. 221)

gneiss (p. 221)
hornfels (p. 223)
hydrothermal metamorphism (p. 231)
marble (p. 223)
metaconglomerate (p. 221)
metamorphic aureole (p. 229)
metamorphic facies (p. 226)
metamorphic foliation (p. 217)

metamorphic grade (p. 223)
metamorphic mineral (p. 217)
metamorphic rock (p. 217)
metamorphic texture (p. 217)
metamorphic zone (p. 228)
metamorphism (p. 217)
metasomatism (p. 221)
migmatite (p. 222)
phyllite (p. 221)
preferred orientation (p. 220)

protolith (p. 217)
quartzite (p. 223)
regional metamorphism (p. 231)
schist (p. 221)
shield (p. 235)
shock metamorphism (p. 231)
slate (p. 221)
thermal metamorphism (p. 229)

GEOTOURS *THIS CHAPTER'S GEOTOURS WORKSHEET (G) FEATURES QUESTIONS AND GOOGLE EARTH SITES ON:*

> Metamorphic zones   > Diverse metamorphic environments   > Protoliths   > Types of metamorphism   > Metamorphic foliation

REVIEW QUESTIONS

The letters following each Review Question refer to the corresponding Learning Objective from the Chapter Opener.

1. How do metamorphic rocks differ from igneous and sedimentary rocks? (**A**)

2. What two features can develop in a rock during metamorphism? (**C**)

3. What phenomena may be involved in transforming a protolith into metamorphic rock? (**B**)

4. What is metamorphic foliation, and how does it form? Identify the orientation of foliation on the thin section. (**D**)

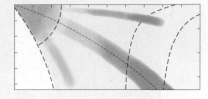

5. How does slate differ from phyllite? How does phyllite differ from schist? How does schist differ from gneiss? (**E**)

6. What conditions lead to the formation of a migmatite? (**E**)

7. Why doesn't hornfels contain foliation? Name two other nonfoliated metamorphic rocks. (**D**)

8. What does a metamorphic grade refer to, and how can it be determined? Identify the axes of the graph, and the areas representing conditions in which low-, intermediate-, and high-grade metamorphic rocks form. (**G**)

9. What does a metamorphic facies represent? (**G**)

10. Describe the geologic settings where thermal, dynamic, and dynamothermal metamorphism take place. What kind(s) of rock might you expect to find in each? (**F**)

11. Identify the aureole of highest-grade rock on the diagram. (**F**)

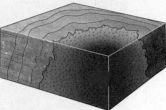

12. Why does metamorphism happen at the site of meteor impacts or along mid-ocean ridges? (**F**)

13. How does plate tectonics theory explain the combination of low-temperature but high-pressure minerals found in a blueschist? (**F**)

14. Where would you go if you wanted to find exposed metamorphic rocks? How did such rocks return to the surface of the Earth after undergoing metamorphism at depth in the crust? (**F**)

ON FURTHER THOUGHT

15. Do you think that you would be likely to find a broad region (hundreds of kilometers across and hundreds of kilometers long) in which the outcrops consist of high-grade hornfels? Why or why not? (Hint: Think about the causes of metamorphism and the conditions under which a hornfels forms.) (**B, E**)

16. Would we be likely to find gneiss and schist on the Moon? Why or why not? (**E, F**)

ONLINE RESOURCES

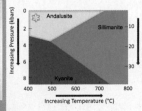

Animations This chapter features animations on the six major types of metamorphic changes, focusing on each process close up.

Smartwork5 This chapter features visual exercises on metamorphism and its effects.

Igneous rock forming, Hawaii

Sedimentary rock, Colorado

Metamorphic rock, California

INTERLUDE C

THE ROCK CYCLE

By the end of this interlude, you should be able to . . .

A. explain why rocks don't last forever in the Earth's crust, and why there are more younger rocks than older rocks at the surface of the Earth.

B. define the rock cycle, and illustrate the various paths through it.

C. relate the paths through the rock cycle to geologic settings, in the context of plate tectonics theory.

D. describe the source of the energy that drives Earth System processes, and how this energy plays a role in the rock cycle.

E. develop a model to explain the general concept of a cycle in the Earth System.

C.1 **Introduction**

"Stable as a rock." This familiar expression implies that rock, once formed, can last forever. In reality, however, components making up a rock may later be rearranged or moved elsewhere to form a new rock of the same class or even one of a different class. In some places, over the course of millions to billions of years, this process has happened many times. Geologists refer to this progressive transformation of Earth materials as the **rock cycle (Fig. C.1)**. In this interlude, we illustrate how to apply the concept of the rock cycle, and we conclude by showing how it represents one of many important cycles in the Earth System.

C.2 **Rock Cycle Paths**

PATHS REFLECT GEOLOGIC HISTORY

Different rock types form in different environments (**Geology at a Glance**, pp. 240–241). During the rock cycle, materials can transfer among these environments. By following the arrows in Figure C.1, you can see that there are many paths around or through the rock cycle. For example, igneous rock formed by solidification of a melt that rose from the mantle may undergo weathering and erosion to produce sediment. Later, burial and lithification transforms that sediment into new sedimentary rock. This sedimentary rock may, in turn,

FIGURE C.1 The stages of the rock cycle, showing various alternative paths.

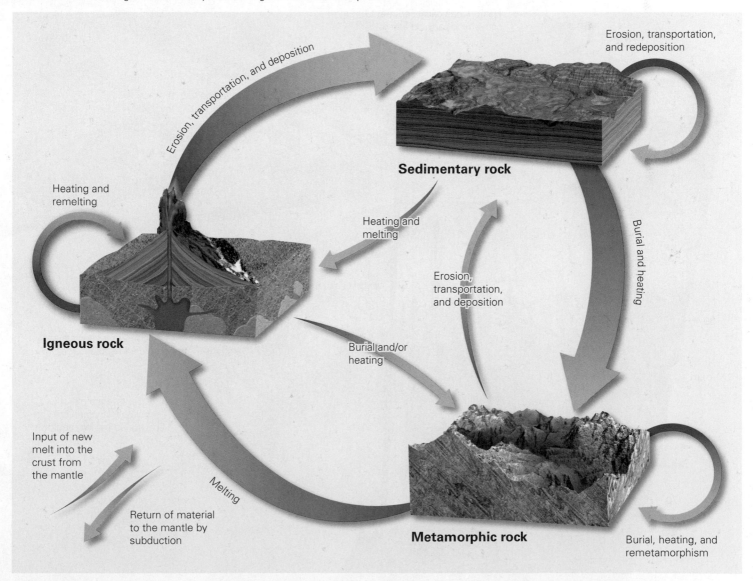

◄ (facing page) On our dynamic planet, the atoms in a rock of one class may, over time, be incorporated into a rock of another class, and then another. Such transformations comprise the rock cycle.

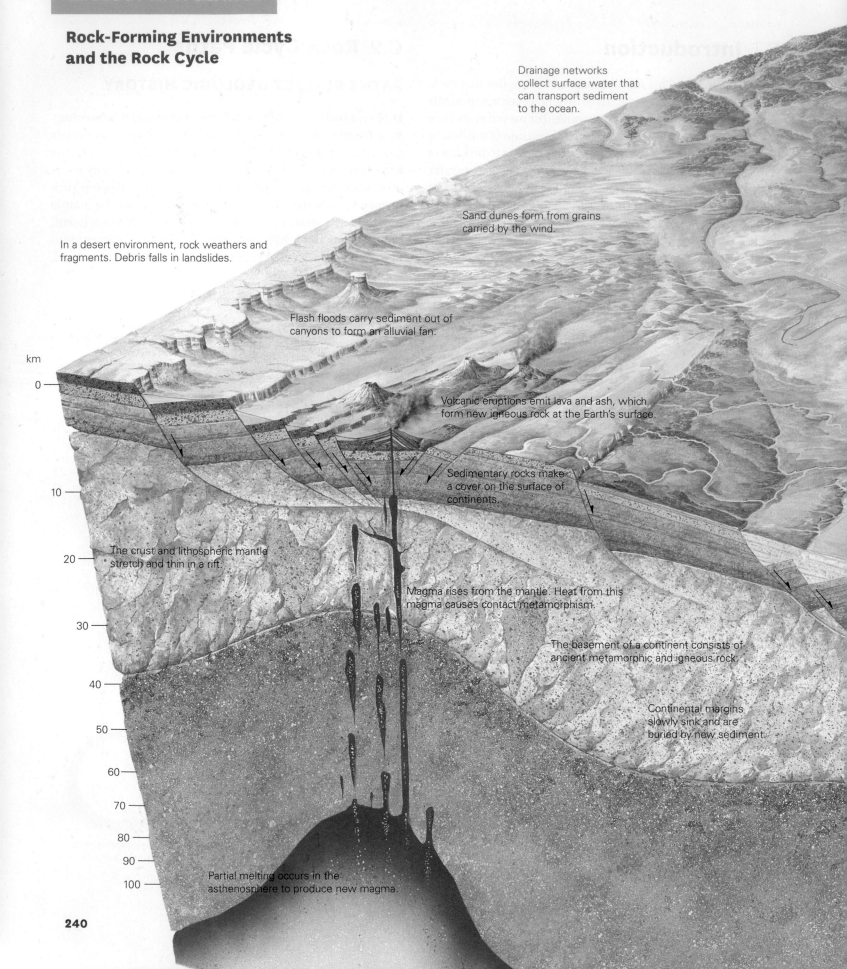

Rock-Forming Environments and the Rock Cycle

Drainage networks collect surface water that can transport sediment to the ocean.

In a desert environment, rock weathers and fragments. Debris falls in landslides.

Sand dunes form from grains carried by the wind.

Flash floods carry sediment out of canyons to form an alluvial fan.

Volcanic eruptions emit lava and ash, which form new igneous rock at the Earth's surface.

Sedimentary rocks make a cover on the surface of continents.

The crust and lithospheric mantle stretch and thin in a rift.

Magma rises from the mantle. Heat from this magma causes contact metamorphism.

The basement of a continent consists of ancient metamorphic and igneous rock.

Continental margins slowly sink and are buried by new sediment.

Partial melting occurs in the asthenosphere to produce new magma.

km
0
10
20
30
40
50
60
70
80
90
100

Glaciers erode rock and can transport sediment of all sizes.

In a region of continental collision, rocks that were near the surface are deeply buried and metamorphosed.

In humid climates, thick soils develop.

Magma that cools and solidifies underground forms igneous intrusions.

Along coastal plains, rivers meander. Sediment collects in the channel and floodplain.

Where a river enters the sea, sediment settles out to form a delta.

Reefs grow from calcite-secreting organisms. These will eventually turn into limestone.

Many different kinds of sediment accumulate along coastlines, building out a continental shelf.

Turbidity currents carry a cloud of sediment that settles to form a submarine fan.

The oceanic crust consists of igneous rocks formed at a mid-ocean ridge.

Fine clay and plankton shells settle on the oceanic crust.

Rocks form in many different environments. Igneous rocks develop where melt rises from depth and cools; sedimentary rocks form at or near the surface; and metamorphic rocks form deep underground. Because the Earth is dynamic—due to plate motion and the circulation of water, ice, and air—environments at a given location change over time, so atoms don't necessarily stay within one rock type for all of geologic time. This progressive shift of material over time, from being in one rock type to another, is the rock cycle. There are many paths through the rock cycle; which path (or succession of paths) an atom takes over time depends on the geologic setting.

241

become buried so deeply that it transforms into metamorphic rock. Extreme heating of the metamorphic rock might cause it to partially melt and produce new magma. This new magma might later solidify to form new igneous rock. We can express this path as follows:

Igneous → Sedimentary → Metamorphic → Igneous

Given local geologic conditions, a rock at one stage in the cycle could follow another path. For example, the metamorphic rock, once formed, could itself be uplifted and eroded to form new sediment and, later, new sedimentary rock, without melting. This path takes a shortcut through the cycle that we can illustrate as follows:

Igneous → Sedimentary → Metamorphic → Sedimentary

Likewise, the original igneous rock might be deeply buried before it could be eroded and could therefore undergo metamorphism directly, without first turning to sediment. The resulting metamorphic rock might then be uplifted and eroded to produce sediment that becomes sedimentary rock, defining another shortcut path:

Igneous → Metamorphic → Sedimentary

But not all steps have to yield a new rock type. For example, if a sedimentary rock, once formed, were uplifted and eroded to form new sediment that was then buried and lithified, its path would be

Sedimentary → Sedimentary

To get a clearer sense of how the rock cycle works, let's look at a case study of the rock cycle, in the context of plate tectonics theory.

C.3 A Case Study of the Rock Cycle

Material can enter the rock cycle when magma rises from the mantle. Suppose that magma reaches the Earth's surface at a hot-spot volcano and erupts as lava that solidifies to become basalt, an igneous rock (**Fig. C.2a**). Once at the surface, the basalt weathers as it interacts with wind, rain, and vegetation. The minerals that originally made up the basalt transform into new, tiny flakes of clay. Flowing water can pick up this clay and transport it downstream—if you've ever seen a brown-colored river, you've seen clay traveling to a site of deposition. Eventually, when the river reaches the sea, the water slows and the clay settles out.

If the clay accumulates as a deposit of mud on the slowly subsiding continental shelf (in this case, along the margin of Continent X), it will gradually become buried to a depth of several kilometers in a passive-margin basin. At this depth, the weight of the overburden packs the clay flakes tightly together, and minerals precipitating out of groundwater cement the flakes to each other, so eventually the mud turns into a sedimentary rock, shale. The shale can reside in the passive-margin basin for millions of years, until Continent X collides with Continent Y. When the two continents squeeze together, and mountains start to build, slip on faults transports the edge of Continent Y up and over the passive-margin basin of Continent X. The shale may end up at a depth of 20 km below the surface, where temperature, pressure, and stress metamorphose it into schist (**Fig. C.2b**).

The story's not over. During mountain building, erosion constantly grinds away the mountain range as uplift continues. The resulting *exhumation* may eventually expose the schist at the ground surface. On an outcrop, the schist itself undergoes erosion to form new sediment, which can be carried off and deposited elsewhere, where it will ultimately form a new sedimentary rock. Below the ground surface, some of the schist remains intact (**Fig. C.2c**). Imagine now that continental rifting takes place at the site of the former mountain range, and that the crust containing the schist begins to split apart. During rifting, injection of hot magma, rising from the mantle below, brings so much heat up into the crust that some of the schist partially melts, yielding a new felsic magma. This felsic magma rises to the surface of the crust and freezes into rhyolite, a new igneous rock (**Fig. C.2d**). In terms of the rock cycle, plate interactions have taken Earth materials through a full circle—the atoms that once resided in an igneous rock later resided in a sedimentary rock, then a metamorphic rock, and finally in a new igneous rock.

RATES OF TRANSFER

It's important to keep in mind that not all atoms pass through the rock cycle at the same rate. It's for this reason that we can find rocks of many different ages at the surface of the Earth. Some rocks survive for less than a million years, while others have lasted, virtually unchanged, for much of the Earth's history. For example, a rock in the Appalachian Mountains has passed through stages of the rock cycle many times during the past billion years, because during this time the eastern margin of North America was subjected to multiple events of rifting, basin formation, and mountain building. In contrast, geologists have found lava flows as old as 3 billion years in the shield areas of continents—these rocks have not gone through any further stages of the rock cycle since they were initially extruded. Research shows, however, that such long-lived rocks account for a very small proportion of the continental

FIGURE C.2 A case study of the rock cycle, in the context of plate tectonics theory.

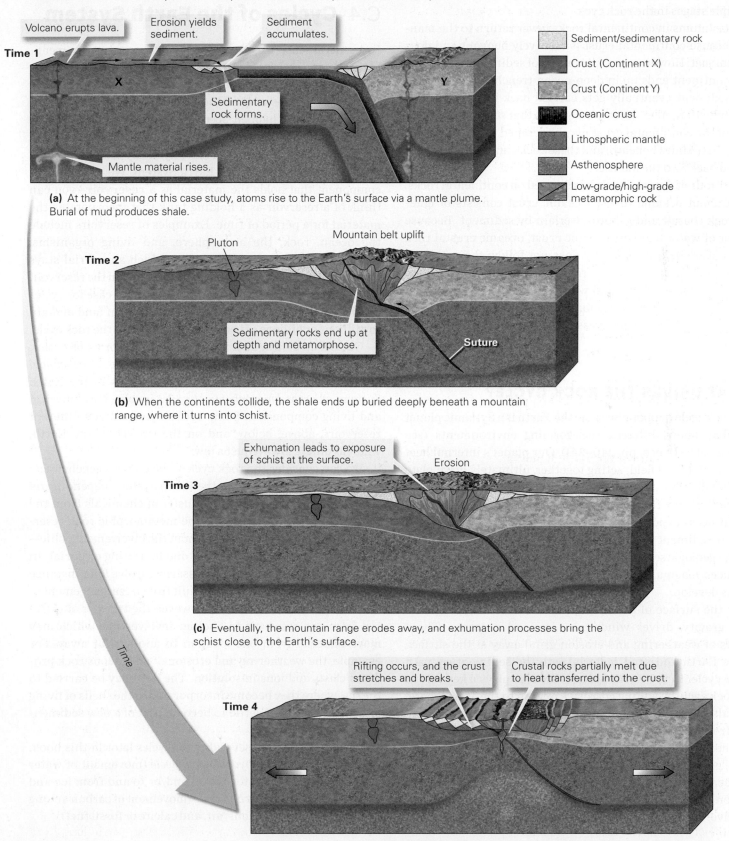

Time 1

Volcano erupts lava.

Erosion yields sediment.

Sediment accumulates.

Sedimentary rock forms.

Mantle material rises.

Sediment/sedimentary rock

Crust (Continent X)

Crust (Continent Y)

Oceanic crust

Lithospheric mantle

Asthenosphere

Low-grade/high-grade metamorphic rock

(a) At the beginning of this case study, atoms rise to the Earth's surface via a mantle plume. Burial of mud produces shale.

Pluton

Mountain belt uplift

Time 2

Sedimentary rocks end up at depth and metamorphose.

Suture

(b) When the continents collide, the shale ends up buried deeply beneath a mountain range, where it turns into schist.

Exhumation leads to exposure of schist at the surface.

Erosion

Time 3

(c) Eventually, the mountain range erodes away, and exhumation processes bring the schist close to the Earth's surface.

Rifting occurs, and the crust stretches and breaks.

Crustal rocks partially melt due to heat transferred into the crust.

Time 4

Time

(d) When rifting splits the continents apart, some of the schist melts, and its atoms become incorporated into magma again.

crust, and that most rocks of this crust have been through multiple stages in the rock cycle.

Most atoms in continental rocks never return to the mantle, because continental crust is relatively buoyant and does not subduct. However, a small amount of sediment that erodes off a continent ends up in deep-ocean trenches, and some of this sediment eventually gets carried back into the mantle by subduction. Also, research suggests that some rocks at the base of the continental crust may be scraped off as the downgoing plate slides beneath, and these rocks may also be transported back into the mantle.

Our tour of the rock cycle has focused on continental rocks. What about oceanic rocks? Oceanic crust consists of igneous rock (basalt and gabbro) overlain by sediment. Because a layer of water blankets oceanic crust, oceanic crustal rock does not erode and generally does not follow the path into the sedimentary loop of the rock cycle. But sooner or later, oceanic crust subducts, and when this happens, the rock of the crust undergoes metamorphism. As they sink, crustal rocks are subjected to progressively higher temperatures and pressures.

WHAT DRIVES THE ROCK CYCLE?

The rock cycle happens because the Earth is a dynamic planet that has many different rock-forming environments (see Geology at a Glance, pp. 240–241). Our planet's internal heat and gravitational field, acting together, ultimately drive plate movements and generate plume-associated hot spots. Plate interactions, in turn, cause the uplift of mountain ranges. This process exposes rock to weathering and erosion, which lead to sediment production. Plate interactions also generate the geologic settings in which pre-existing rock melts and produces magma, metamorphism occurs, and sedimentary basins develop.

At the surface of the Earth, heat from the Sun, together with gravity, drives wind, rain, ice, and currents. These agents of weathering and erosion grind away at the surface of the Earth and send material into the sedimentary path of the cycle. In the Earth System, life also plays a key role in the rock cycle by adding corrosive oxygen to the atmosphere. Life also directly contributes to weathering by extracting ions from water solutions and forming shells. In sum, external energy (solar heat), internal energy (the Earth's internal heat), gravity, and life all play roles in driving the rock cycle by keeping the mantle, crust, atmosphere, and oceans in motion and by making portions of the system chemically reactive.

C.4 Cycles of the Earth System

In a general sense, a **cycle**, from the Latin word *cyclus*, meaning circle, is a series of interrelated events or steps that occur in succession and can be repeated. During a *temporal cycle*—such as the phases of the Moon, the seasons of the year, or the tides—events happen according to a timetable, but the materials involved do not necessarily change. During a **mass-transfer cycle**, materials physically transfer among different components of the Earth System. We refer to each component that holds the material as a **reservoir**—you can think of a reservoir as a holding tank that incorporates the material for a period of time. Examples of reservoirs include the ocean, rock, the atmosphere, and living organisms. Geologists refer to the time during which a material stays within a given reservoir as its **residence time** in the reservoir. Residence times can vary from hours, as is the case for cycles involving transfer of volatile elements between land and air, to millions or billions of years, as is the case for the rock cycle.

Geologists distinguish between two categories of cycles, depending on whether or not they involve life. *Geochemical cycles* involve primarily physical components in the Earth System, whereas *biogeochemical cycles* involve both physical and living components. Some cycles involve many different reservoirs, above, below, and on the surface of the Earth, whereas others involve just a few.

We can consider the rock cycle to be either a geochemical or a biogeochemical cycle in the Earth System, depending on circumstances. For example, transfer of chemicals from the sedimentary rock reservoir to the metamorphic rock reservoir may be entirely physical, without the involvement of life—the process could happen simply due to heating or burial. In fact, such a transfer doesn't necessarily involve long-distance movement of atoms, but could result from a rearrangement of atoms in place. In some cases, however, the transfer of material from one reservoir to another involves life, which may move material from one location to another far away. For example, the weathering and erosion of an igneous rock produces clasts and ions in solution. The ions may be carried to the sea, where they become incorporated in the shells of living organisms that later settle to become part of a new sedimentary rock.

We'll be discussing several other cycles later in this book. Examples include the *hydrological cycle* (movement of water from sea to air to rain to stream and/or to and from ice and living tissue) and the *carbon cycle* (movement of carbon among organisms, water solutions, air, and calcite or fossil fuel).

Interlude C Review

INTERLUDE SUMMARY

> A given rock doesn't necessarily last forever. The atoms in a rock may, over time, be incorporated in different rock types. This transfer of atoms progressively from one rock type to another, over time, is the rock cycle.

> Not all atoms follow the same path through the rock cycle. For example, an igneous rock could later be eroded and turned into sediment, which could become a sedimentary rock, which might eventually be metamorphosed. Or the igneous rock could be metamorphosed directly.

> The rock cycle happens because the Earth is dynamic, and its internal and external sources of energy drive melting, uplift, faulting, weathering, erosion, and burial.

> The rock cycle is one of many cycles in the Earth System.

> Mass-transfer cycles involve transfer of material from one reservoir to another over time. Some of these cycles involve only physical processes, but some involve life.

GUIDE TERMS

cycle (p. 244)   reservoir (p. 244)   rock cycle (p. 239)
mass-transfer cycle (p. 244)   residence time (p. 244)

REVIEW QUESTIONS

The letters following each Review Question refer to the corresponding Learning Objective from the Chapter Opener.

1. Why don't rocks, once formed, last for all of the Earth's history? **(A)**

2. Define the rock cycle, and give three examples of paths through it. Label the processes associated with the arrows shown in the diagram. **(B)**

3. Have all rocks on the Earth passed through the rock cycle the same number of times? Explain your answer. **(C)**

4. Is there a rock cycle on the Moon? Why or why not? **(D)**

5. Explain the concept of a cycle in the context of the Earth System. **(E)**

6. What is a biogeochemical cycle? **(E)**

ONLINE RESOURCES

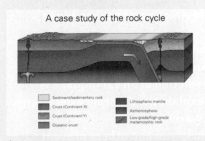

Videos This interlude features a video that explains the rock cycle, and includes a case study.

Smartwork5 This interlude features video questions on the processes and outcomes of the rock cycle

CHAPTER 8

A VIOLENT PULSE: EARTHQUAKES

By the end of this chapter, you should be able to . . .

A. describe an earthquake, and explain where the energy released during an earthquake comes from.

B. relate earthquakes to specific geologic settings, in the context of plate tectonics theory.

C. draw a sketch illustrating how a seismometer operates, and explain what the squiggles on a seismogram mean.

D. distinguish among the different kinds of seismic waves, and show how the arrival times of seismic waves can indicate where an earthquake occurred.

E. explain the difference between the intensity and magnitude of an earthquake, and how these indicators of earthquake size can be determined.

F. discuss the many ways in which earthquakes cause damage and injury.

G. distinguish between tsunamis and storm waves, and explain how large tsunamis can cause so much damage.

H. determine whether a prediction of an earthquake is worth listening to, and explain the difference between a prediction and an early warning.

I. identify steps that can help you and others prevent earthquake damage and avoid injury.

8.1 Introduction

It was mid-afternoon on March 11, 2011, and in many seaside towns along the eastern coast of Honshu, the northern island of Japan, fishing fleets unloaded their catch, factories churned out goods, shoppers browsed the stores, and office workers tapped at their computers, all unaware that their surroundings were about to change forever. This coast lies near a convergent boundary at which the Pacific Plate grinds beneath the edge of Japan and sinks back into the mantle. Averaged over time, the relative movement across this boundary takes place at a rate of about 8 cm per year. But the motion doesn't happen smoothly. Rather, for a while, rocks adjacent to the boundary quietly and subtly flex and warp. Then, suddenly, like a wooden stick that snaps after you've bent it too far, a measureable amount of movement takes place in a matter of seconds to minutes (**Fig. 8.1**). This movement occurs by slip on a **fault**, a fracture plane on which sliding takes place. On March 11, at 2:46 P.M., the "snap" began on a very large, gently sloping fault at a point located about 130 km east of Japan's coast, and about 24 km below the Earth's surface. Within a few minutes, slip had affected a vast area of the fault, as Japan lurched eastward by as much as several meters, relative to the Pacific Ocean floor. Because of the character of this particular movement, a region of seafloor above the fault abruptly rose vertically by as much as 30 cm. The stage had been set for a disaster.

The instant that the slip started on the fault east of Honshu, the bending that had been accumulating in rock adjacent to the fault, over decades to centuries, was released and the rock suddenly straightened out and began to vibrate, just as the two pieces of a bent stick straighten out and vibrate when the stick snaps. Simultaneously, some of the rock bordering the fault fractured. Growth of each new fracture yielded more vibrations. Effectively, the slip on the fault, with associated fracturing, released an immense amount of

◀ (facing page) In 2017, a large earthquake shook Mexico City. Sadly, the vibrations caused some buildings to collapse, with catastrophic consequences. Earthquakes like this one emphasize that our planet remains a dynamic place.

energy that instantly began to propagate through the crust in the form of waves. In solid rock, such waves race along at an average speed of 11,000 km per hour, 10 times the speed of sound in air. When the waves reached the surface of the Earth, they caused an episode of ground shaking—an **earthquake**.

So much energy was released by the resulting *Tōhoku earthquake* (named for the eastern province of Honshu) that the land surface lurched back and forth and bounced up and down for over a minute. People panicked, became disoriented, and even seasick—some lost their balance and crouched or fell, and some heard a dull rumbling or thumping. Bottles and plates flew off shelves and crashed to the floor, buildings twisted and swayed, ceilings and facades fell in a shower of debris, dust rose from the ground to create a fog-like mist, power lines stretched and sparked, and weaker buildings collapsed (**Fig. 8.2a**). In addition, several natural gas tanks and pipes broke, sending flammable vapors into the air—in

FIGURE 8.1 What happens during an earthquake?

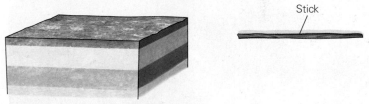

Stick

(a) Before deformation, rock layers are not bent.

The layer of rock bends elastically (exaggerated here).

(b) Before an earthquake, rock bends elastically, like a stick that you arch between your hands.

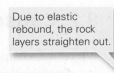
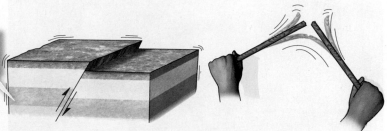

Due to elastic rebound, the rock layers straighten out.

(c) Eventually, the rock breaks, and sliding suddenly occurs on a fault. This break generates vibrations. You feel such vibrations when you break a stick.

247

FIGURE 8.2 The great Tōhoku earthquake and tsunami in Japan, 2011.

(a) Some buildings collapsed due to ground shaking.

(b) The tsunami filled harbors and spilled over seawalls.

(c) The tsunami washed over low coastal areas. In this photograph, the wave is starting to wash over a canal.

some places, the gas ignited in billows of flame that set fire to damaged buildings.

During the Tōhoku earthquake, damage due to the shaking itself, though significant, was not devastating because most buildings, due to Japan's stringent building codes, were sturdy enough to resist collapse. Unfortunately, even strong buildings could not resist what happened subsequent to the earthquake shock. The sudden displacement of the seafloor off Japan's coast had displaced the surface of the ocean and produced *tsunamis*. Such distinctive, very broad waves travel rapidly away from a location where the seafloor has been disturbed and water displaced. The first tsunami generated on March 11 reached land 20 to 80 minutes after the earthquake. As it approached, the tsunami locally grew to a height of over 10 m, so it was able to overtop seawalls and wash inland, picking up debris and sediment as it moved. In places, it submerged low-lying land several kilometers in from the shoreline **(Fig. 8.2b, c)**. In the end, the immensity of the devastation due to the Tōhoku earthquake and its aftermath was almost beyond comprehension.

Earthquakes are a fact of life on planet Earth—almost 1 million detectable earthquakes happen every year, most as a consequence of plate movement. They punctuate each step in the growth of mountains, the drift of continents, and the opening and closing of ocean basins. Fortunately, most cause no damage or casualties, either because they are too small or because they occur in unpopulated areas. But a few hundred earthquakes per year rattle the ground sufficiently to crack or topple buildings and injure their occupants, and every 5 to 20 years, on average, a great earthquake triggers a horrific calamity. In fact, during the past two millennia, building collapse, tsunamis, landslides, fires, and other phenomena caused by earthquakes have killed over 3.5 million people **(Table 8.1)**.

What geologic phenomena trigger earthquakes? Why do earthquakes take place where they do? How do they cause damage? Can we predict when earthquakes will happen or even prevent them from happening? **Seismologists** (from the Greek word *seismos*, for shock or earthquake), geoscientists who study earthquakes, have addressed many of these questions during the past century. In this chapter, we present some of the answers that they have obtained.

8.2 Causes of Earthquakes

Why do earthquakes happen? Ancient cultures offered a variety of explanations for **seismicity** (earthquake activity), most of which involved the restless movements of mythical animals (catfish, elephants, turtles) who supposedly lived underground. Today, we realize that instead, earthquakes happen when masses of rock suddenly vibrate. The energy released by

TABLE 8.1 Some Notable Earthquakes

Year	Location	Deaths
2017	Mexico City, Mexico	369
2011	Tōhoku, Japan (tsunami)	20,000
2011	Christchurch, New Zealand	180
2010	Haiti	230,000
2010	Concepción, Chile	1,000
2008	Sichuan, China	70,000
2005	Pakistan	80,000
2004	Sumatra (tsunami)	230,000
2003	Bam, Iran	41,000
2001	Bhuj, India	20,000
1999	Calaraca/Armenia, Colombia	2,000
1999	Izmit, Turkey	17,000
1995	Kobe, Japan	5,500
1994	Northridge, California	51
1990	Western Iran	50,000
1989	Loma Prieta, California	65
1988	Spitak, Armenia	24,000
1985	Mexico City	9,500
1983	Turkey	1,300
1978	Iran	15,000
1976	T'ang-shan, China	255,000
1976	Caldiran, Turkey	8,000
1976	Guatemala	23,000
1972	Nicaragua	12,000
1971	San Fernando, California	65
1970	Peru	66,000
1968	Iran	12,000
1964	Anchorage, Alaska (Good Friday)	131
1963	Skopje, Yugoslavia	1,000
1962	Iran	12,000
1960	Agadir, Morocco	12,000
1960	Southern Chile	6,000
1948	Turkmenistan, USSR	110,000
1939	Erzincan, Turkey	40,000
1939	Chillán, Chile	30,000
1935	Quetta, Pakistan	60,000
1932	Gansu, China	70,000
1927	Tsinghai, China	200,000
1923	Tokyo, Japan	143,000
1920	Gansu, China	180,000
1915	Avezzano, Italy	30,000
1908	Messina, Italy	160,000
1906	San Francisco	500
1896	Japan	22,000
1886	Charleston, South Carolina	60
1866	Peru and Ecuador	25,000
1811–12	New Madrid, Missouri (3 events)	Few
1783	Calabria, Italy	50,000
1755	Lisbon, Portugal	70,000
1556	Shen-shu, China	830,000

this process travels away from the source in the form of waves, known as **seismic waves** or *earthquake waves*, that can reach and shake the surface of the Earth. What causes the sudden vibration of rock? Earthquakes can occur due to the intrusion and flow of magma beneath a volcano, to underground nuclear tests, to landslides, or even to meteor impacts. Most earthquakes, though, are a consequence of faulting, as was the case for the Tōhoku event. Below, we first describe faults, and then show how movement on them generates seismic waves.

FAULTS IN THE CRUST

At first glance, a fault may look simply like a break or cut that slices across rock or sediment. But on closer examination, you may be able to see evidence of the sliding that occurred on a fault. For example, rock adjacent to the fault may be broken up into angular fragments or may be pulverized into tiny grains, due to the crushing and grinding that can accompany slip. The surface of a fault may be polished and grooved as if scratched by a rasp. In some localities, a fault cuts through a distinct *marker*: a sedimentary bed, an igneous dike, or a fence. Where this happens, the end of the marker on one side of the fault is offset relative to the end on the other side. Geologists refer to the distance between two ends of the offset marker, as measured along the fault surface in the direction of slip, as the fault's **displacement (Fig. 8.3)**. Most faults are completely underground, so we can't see the displacement across them when it happens. In fact, such faults may become visible only millions of years later, when exposed by the erosion of overlying rock. But some faults intersect and offset the ground surface, producing a step called a **fault scarp (Fig. 8.4a)**. The ground surface exposure of a fault, whether it appears due to erosion of a once-buried fault or due to displacement of the present-day ground surface by recent movement, is called a *fault line* or *fault trace*.

In the 19th century, miners who encountered faults in mine tunnels referred to the rock mass above a sloping fault plane as the *hanging wall*, because it hung over their heads, and the rock mass below the fault plane as the *footwall*, because it lay beneath their feet. The miners described the direction in which rock masses slipped on a sloping fault by specifying the *sense of slip*, the direction that the hanging wall moved in relation to the footwall. We still use the terms they developed (**Geology at a Glance**, pp. 252–253). Specifically, when the hanging wall slips down the slope or *dip* of the fault, it's a *normal fault* (see Fig. 8.4a), and when the hanging wall slips up the slope, it's a *reverse fault* or *thrust fault* (**Fig. 8.4b, c**). Note that reverse faults and thrust faults have the same sense of slip—they differ in their dip, in that the former have a steeper dip than the latter. A *strike-slip fault* is a near-vertical fracture on which slip occurs parallel to an imaginary horizontal line, called a strike line, on the fault plane—no up or down motion takes place on a strike-slip fault (**Fig. 8.4d**).

FIGURE 8.3 Examples of fault displacement on the San Andreas fault in California.

What a Geologist Sees

These ruptures formed where the fault broke the ground surface.

Dirt road

(a) A wooden fence built across the San Andreas fault was offset during the 1906 San Francisco earthquake. The displacement indicates strike-slip motion.

(b) An aerial photograph shows offset of a dirt road by slip on a fault in the Mojave Desert in 1999.

Faults are found in many locations—but don't panic! Not all are likely to be the source of earthquakes. We refer to a fault that has moved relatively recently, or might move in the future, as an *active fault*—if it generates earthquakes, news media sometimes refer to it as an "earthquake fault." A fault that last moved in the distant past and probably won't move again in the near future is an *inactive fault*.

FIGURE 8.4 The basic types of faults. Fault types are distinguished from one another by the direction of slip relative to the fault surface.

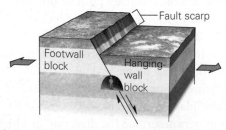

Fault scarp

Footwall block

Hanging-wall block

(a) Normal faults form during extension of the crust. The hanging wall moves down.

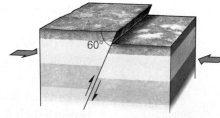

60°

(b) Reverse faults form during shortening of the crust. The hanging wall moves up and the fault is steep.

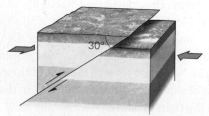

30°

(c) Thrust faults also form during shortening. The fault's slope is gentle (less than 30°).

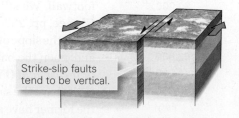

Strike-slip faults tend to be vertical.

(d) On a strike-slip fault, one block slides laterally past another, so no vertical displacement takes place.

GENERATING EARTHQUAKE ENERGY

How does faulting generate earthquakes? Seismologists have identified two causes for the vibrations of an earthquake. First, an earthquake can happen when a new fault forms by rupturing previously intact rock, and second, an earthquake can happen when a pre-existing fault suddenly slips again. Let's look more closely at these two situations.

Imagine that you grip each side of a brick-shaped block of rock with a clamp. Now, apply an upward push on one of the clamps and a downward push on the other. By doing so, you have applied a **stress** to the rock—simplistically, we can think of stress as a push, pull, or shear, as we discussed in Chapter 7. At first, the rock bends slightly but doesn't break **(Fig. 8.5a)**. In fact, if you were to stop applying stress at this stage, the rock would straighten out and return to its original shape. Geologists refer to such a phenomenon as **elastic behavior**—the same phenomenon happens when you stretch a spring and then let go. The change in shape due to elastic bending, stretching, or shortening is called *elastic strain*. Now repeat the experiment, but bend the rock even more.

FIGURE 8.5 A model representing the development of a new fault. The process can generate earthquake-like vibrations.

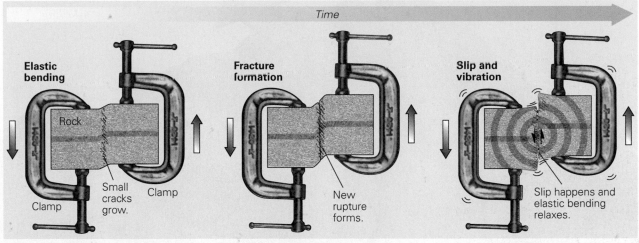

Time

Elastic bending

Rock

Small cracks grow.

Clamp

Clamp

Fracture formation

New rupture forms.

Slip and vibration

Slip happens and elastic bending relaxes.

(a) Imagine a block of rock gripped by two clamps. Move one clamp up, and the rock starts to bend. Small cracks develop along the bend.

(b) Eventually, the cracks link. When this happens, a throughgoing fracture forms.

(c) The instant that the fracture forms, the rock breaks into two pieces that slide past each other. The energy that is released generates vibrations.

If you bend the rock far enough, a number of small cracks start to form in the rock, typically in a diagonal zone. Eventually the cracks connect to one another to form a fracture (or rupture) that cuts across the entire block of rock **(Fig. 8.5b)**. The instant that such a throughgoing fracture forms, the block breaks in two, and the rock on one side of the fracture suddenly slides past the rock on the other side. When sliding occurs, the fracture has become a fault. Any elastic strain that had built up in the rock adjacent to the new fault gets released, so this rock straightens out, or rebounds **(Fig. 8.5c)**. Such fault formation in a previously intact rock releases energy and generates earthquake vibrations.

When a fault forms and starts to slip, it doesn't continue to slip forever, because **friction**, defined as the force that resists sliding on a surface, eventually slows and stops the movement. Why does friction exist? All real surfaces have bumps protruding from them. These bumps act like tiny anchors, snagging and digging into the opposing surface, eventually halting slip.

We've just seen how earthquakes can happen when intact rock ruptures. Earthquakes can also result from sudden slip on a pre-existing fault, for once a fault has formed, it remains weaker than surrounding, intact crust. Even though it's weaker, however, friction across a pre-existing fault resists sliding, so slip is not constant. As stress builds across a frictionally locked fault, rocks adjacent to the fault bend elastically, as they do prior to initiation of a new fault. When such stress becomes great enough that anchor-like bumps protruding from the walls of the fault either break off or scratch a furrow in the opposing wall, the fault slips again.

In sum, we see that earthquakes happen because stresses build up, causing rock to develop elastic strain until either intact rock breaks or a pre-existing fault reactivates. When slip takes place, the once-bent rocks adjacent to the fault rebound and vibrate back and forth until they regain their relaxed shape, thereby relieving the elastic strain. Geologists refer to this overall concept as the **elastic-rebound theory** of earthquake generation. Notably, the stress necessary to reactivate a fault tends to be less than the stress necessary to break intact rock, so most earthquakes probably represent slip on pre-existing faults. Geologists refer to such alternation between stress buildup and earthquake-generating slip events on an existing fault as **stick-slip behavior**.

Of note, the major earthquake, or **mainshock**, along a fault may be preceded by smaller ones, called **foreshocks**. These possibly result from the development or propagation of smaller cracks in the vicinity of what will be the major fault. In the days to months following a large earthquake, the region affected by the large earthquake will endure a series of **aftershocks**. The largest aftershock tends to be 10 times smaller than the main shock, and most are much smaller than that. Aftershocks happen because slip during the mainshock does not leave the fault in a perfectly stable configuration. For example, after the mainshock, bumps on one side of the fault surface may, in their new position, push so hard into the opposing side that they generate new stresses that are large enough to cause a small portion of the fault around the bump to slip again, or to cause slip on a nearby fault.

Faulting in the Crust

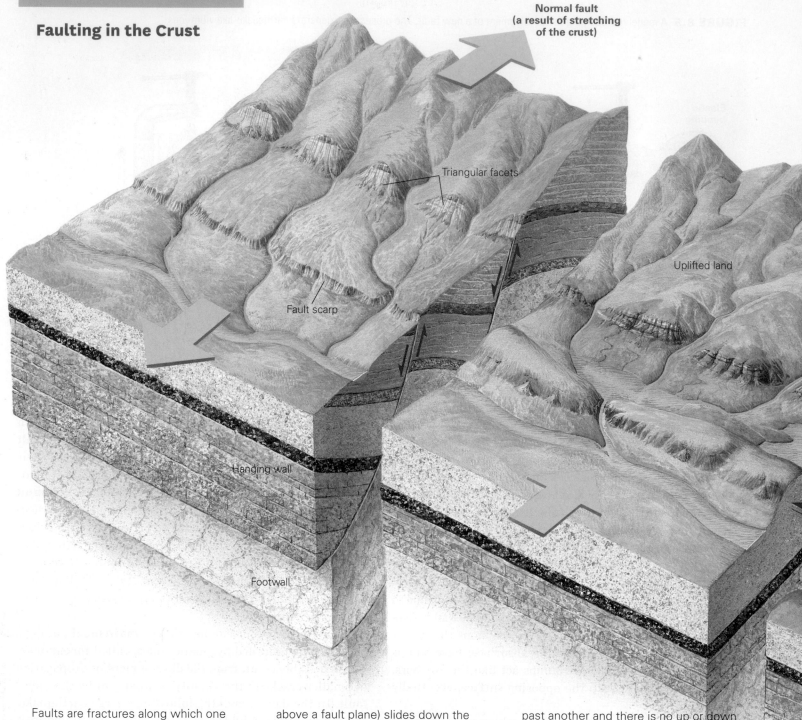

Normal fault
(a result of stretching
of the crust)

Triangular facets

Uplifted land

Fault scarp

Hanging wall

Footwall

Faults are fractures along which one block of crust slides past another block. Sometimes movement takes place slowly and smoothly, without earthquakes, but other times the movement happens suddenly, and rocks break as a consequence. The sudden breaking of rock sends seismic waves through the crust, creating vibrations at the Earth's surface—an earthquake.

Geologists recognize three types of faults. If the hanging-wall block (the rock

above a fault plane) slides down the fault's slope relative to the footwall block (the rock below the fault plane), the fault is a normal fault. (Normal faults form where the crust is being stretched apart, as in a continental rift.) If the hanging-wall block is being pushed up the slope of the fault relative to the footwall block, then the fault is a reverse fault. (Reverse faults develop where the crust is being compressed or squashed, as in a collisional mountain belt.) If one block of rock slides

past another and there is no up or down motion, the fault is a strike-slip fault. Strike-slip fault planes tend to be nearly vertical.

If a fault displaces the ground surface, it creates a ledge called a fault scarp. Where a fault scarp cuts a system of rivers and valleys, it truncates ridges and produces triangular facets. Strike-slip faults may offset ridges and streams sideways.

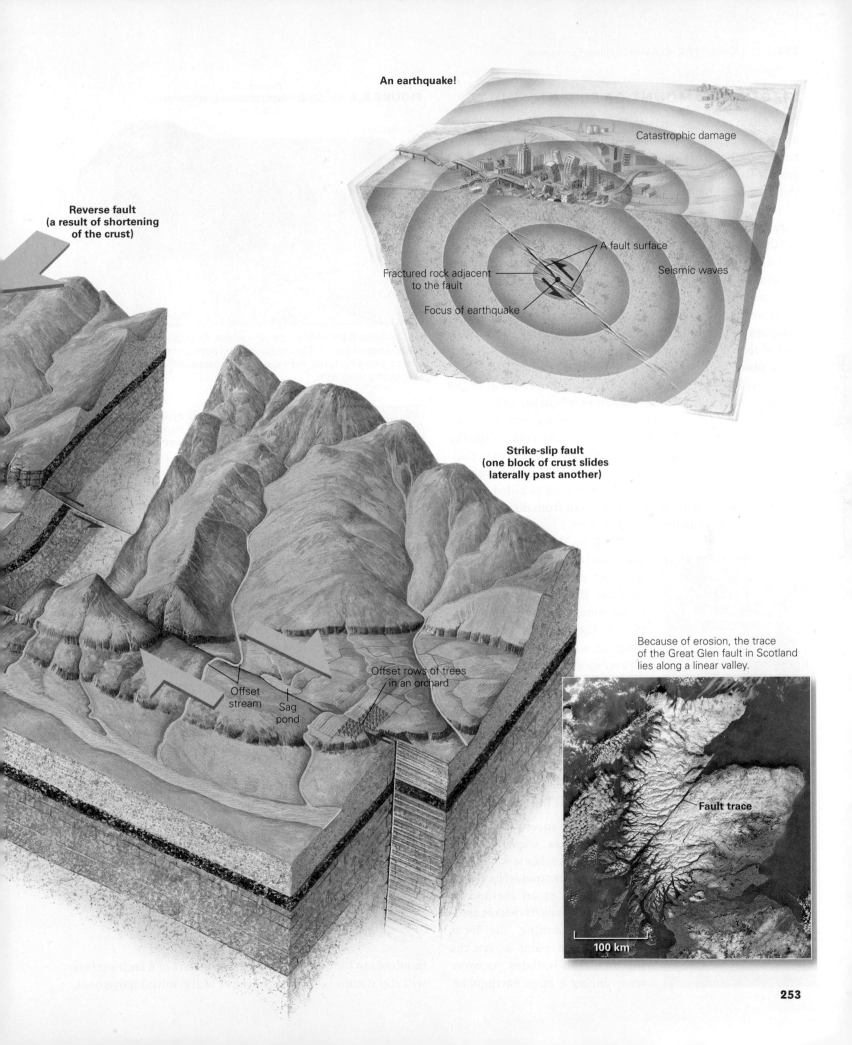

An earthquake!

Catastrophic damage

A fault surface

Fractured rock adjacent to the fault

Seismic waves

Focus of earthquake

Reverse fault (a result of shortening of the crust)

Strike-slip fault (one block of crust slides laterally past another)

Offset rows of trees in an orchard

Offset stream

Sag pond

Because of erosion, the trace of the Great Glen fault in Scotland lies along a linear valley.

Fault trace

100 km

SIZES AND AMOUNT OF SLIP ON FAULTS

How much of a fault surface slips during an earthquake? The answer depends on the size of the earthquake—generally, the larger the earthquake, the larger the slipped area and the greater the displacement. For example, the major earthquake that hit San Francisco, California, in 1906 ruptured a segment of the San Andreas fault that was 430 km long (measured parallel to the Earth's surface) by 15 km deep (measured perpendicular to the Earth's surface). Thus, the area that slipped was almost 6,500 km². During the 2011 Tōhoku earthquake, an area 300 km long by 100 km wide (30,000 km²) slipped.

The amount of slip varies along a fault—it tends to be greatest near the place where the slip begins and tends to die out progressively toward the ends of the slipped area. Beyond the end of the slipped area, the displacement is zero. The maximum observed displacement on very large earthquakes can be several meters. For example, a maximum of 30 m of slip happened during the 2011 Tōhoku earthquake, 12 m of slip during the 1964 Good Friday earthquake in southern Alaska, and 7 m during the 1906 San Francisco earthquake. Smaller earthquakes, such as the 1994 event that hit Northridge, California, resulted in only about 0.5 m of slip; even so, this earthquake toppled homes, ruptured pipelines, and killed 51 people. The smallest-felt earthquakes result from displacements measured in millimeters to centimeters.

Although the slip on an active fault during a human life span may not amount to much, over geologic time the accumulation can become significant. For example, if the earthquakes taking place on a strike-slip fault cause an average of 1 cm of displacement per year, the fault's movement will yield 10 km of displacement after 1 million years.

DEFINING THE FOCUS AND EPICENTER OF AN EARTHQUAKE

How can we specify where an earthquake has occurred? The location where seismic waves first begin to be generated is the **focus**, or *hypocenter*, of an earthquake (**Fig. 8.6a**). For earthquakes associated with faulting, the focus represents the point where the slip on a fault initiates. As we've seen, during a large earthquake,

FIGURE 8.6 Earthquake hypocenters and epicenters.

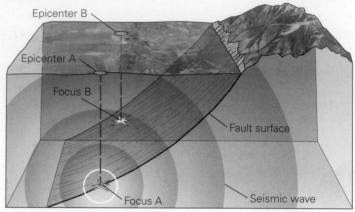

(a) The focus is the point on the fault where slip begins. Seismic energy starts radiating from it. The epicenter is the point on the Earth's surface directly above the focus. Earthquake A just happened; earthquake B happened a while ago.

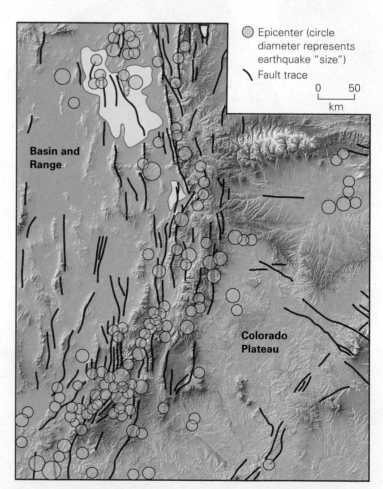

(b) A map of Utah showing the distribution of earthquake epicenters, recorded over several years. The map indicates that seismic activity happens mostly in a distinct belt following the boundary between the Basin and Range rift and the Colorado Plateau.

hundreds to thousands of square kilometers of a fault surface will slip within seconds or minutes of the initial movement.

Sometimes, the focus lies near the center of the slipped area—but not always. For example, in the case of the huge 2004 earthquake off Sumatra, the focus lay near one end of what would grow to become a slipped area more than 1,500 km long. Seismologists use a more general term, *seismic source*, in reference to the entire region where energy leading to an earthquake was released.

In continental crust, the focus of most earthquakes generally lies between 5 and 20 km depth, with a few lying as deep as 50 km. As we'll see, the foci of earthquakes in subducting oceanic lithosphere can lie as deep as 660 km below the surface. Overall, seismologists distinguish among three classes of earthquakes based on focal depth: *shallow-focus earthquakes* occur in the top 60 km of the Earth, *intermediate-focus earthquakes* take place between 60 and 300 km, and *deep-focus earthquakes* occur down to a depth of about 660 km. Earthquakes do not happen below 660 km.

We define the point on the surface of the Earth that lies directly above the focus as the earthquake's **epicenter**. We can portray the position of an epicenter as a dot on a map **(Fig. 8.6b)**. Generally, the relative size of the dot represents the "size" of the earthquake (specifically, the magnitude, as discussed later in this chapter). The color of the dot can indicate the depth of the focus.

TAKE-HOME MESSAGE

Most earthquakes happen when stress builds to cause either sudden formation of a new fault or slip on a pre-existing fault. When the slip takes place, elastically deformed rock on either side of the fault rebounds. This process generates seismic waves. Once formed, faults display stick-slip behavior in that stress builds until slip takes place and then builds again. The focus is the point within the Earth where slip begins, and the epicenter is the point on the Earth 's surface vertically above the focus.

QUICK QUESTION What are foreshocks and aftershocks?

8.3 Seismic Waves and Their Measurement

TYPES OF SEISMIC WAVES

As we've noted, the energy produced by slip on a fault moves through rock and sediment in the form of seismic waves. You can simulate such waves by simply holding one end of a wooden block with your hand while you strike the other end with a hammer—the energy transmitted by the head of the hammer to the block travels through the block as seismic waves, and causes the end of the block in contact with your hand to vibrate. Seismologists distinguish between two categories of seismic waves: **body waves** pass through the interior of the Earth, whereas **surface waves** travel along the Earth's surface. To picture the difference, imagine that the focus of an earthquake lies 20 km below the ground surface. The instant that faulting occurs, body waves propagate outward in all directions through solid rock, like a succession of concentric bubbles. Waves that propagate downward go deep into the interior of the Earth and will not reach the surface until they have traversed much of the planet. Those that go upward reach the top of the crust in a relatively short time, and their arrival generates surface waves that then propagate outward from the epicenter.

Body waves can cause a material to vibrate in two different ways **(Fig. 8.7a, b)**. **Compressional waves** cause particles of material to move back and forth parallel to the direction in which the wave itself moves. As a compressional wave passes, the material first compresses (squeezes together), then dilates (expands). To see this kind of motion in action, push on the end of a spring and watch as the little pulse of compression moves along the length of the spring. **Shear waves** cause particles of material to move back and forth perpendicular to the direction in which the wave itself moves. To see shear-wave motion, jerk the end of a rope up and down and watch how the up-and-down motion travels along the rope. Surface waves also vibrate in two different ways—some cause the ground surface to go up and down in rolling undulations, and some cause the surface to go back and forth sideways, like the movement of a snake **(Fig. 8.7c)**. Seismologists have assigned names to the different kinds of waves we've just described.

> *P-waves* (*P* stands for primary) are compressional body waves.
> *S-waves* (*S* stands for secondary) are shear body waves.
> *L-waves* (*L* stands for Love) are surface waves that cause the ground to shimmy back and forth.
> *R-waves* (*R* stands for Rayleigh) are surface waves that cause the ground to undulate up and down.

The different types of seismic waves travel at different velocities. P-waves travel the fastest. S-waves are the next fastest—an S-wave travels at about 60% of the speed of a P-wave. Surface waves are the slowest, so both Love waves and Rayleigh waves arrive substantially after the body waves have arrived.

SEISMOMETERS AND THE RECORD OF AN EARTHQUAKE

Researchers use a **seismometer** (traditionally called a seismograph) to measure the ground motion produced by an earthquake. Seismometers can be configured in two ways—a

FIGURE 8.7 Different types of earthquake waves.

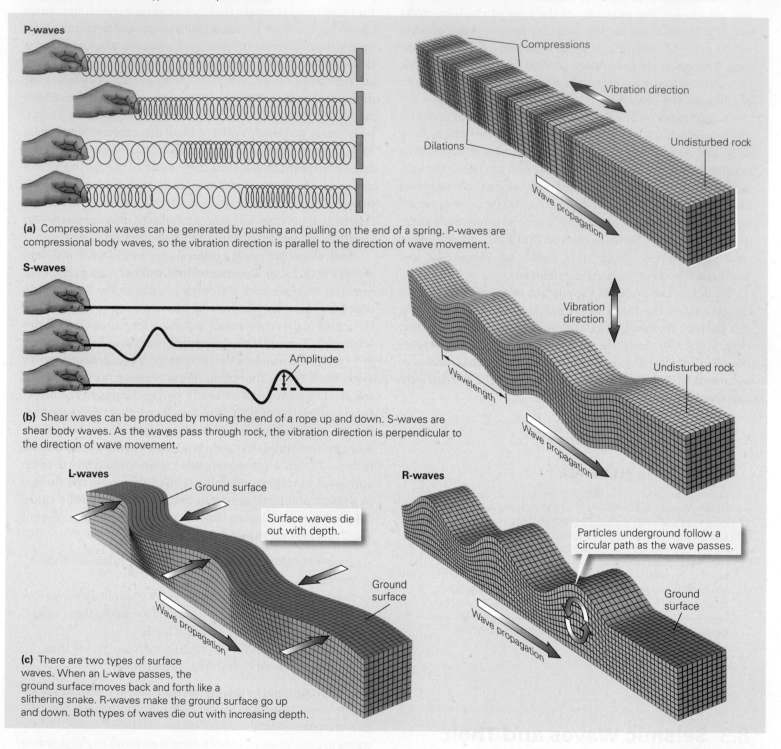

P-waves

(a) Compressional waves can be generated by pushing and pulling on the end of a spring. P-waves are compressional body waves, so the vibration direction is parallel to the direction of wave movement.

S-waves

(b) Shear waves can be produced by moving the end of a rope up and down. S-waves are shear body waves. As the waves pass through rock, the vibration direction is perpendicular to the direction of wave movement.

L-waves   **R-waves**

(c) There are two types of surface waves. When an L-wave passes, the ground surface moves back and forth like a slithering snake. R-waves make the ground surface go up and down. Both types of waves die out with increasing depth.

vertical-motion seismometer detects up-and-down ground motion, and a *horizontal-motion seismometer* detects back-and-forth ground motion. Let's examine how these instruments work.

The heart of a mechanical vertical-motion seismometer consists of a heavy weight suspended from a spring. The spring, in turn, hangs from a sturdy frame that has been bolted to the ground **(Fig. 8.8a)**. A horizontal bar connected to a pivot keeps the spring steady. A pen extends sideways from the weight and touches a revolving, paper-covered vertical cylinder that is also attached to the frame. When the ground is not moving, the pen traces out a straight reference

FIGURE 8.8 The basic operation of a seismometer.

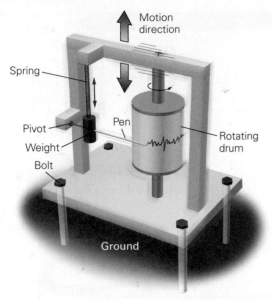

(a) A mechanical vertical-motion seismometer records up-and-down ground motion.

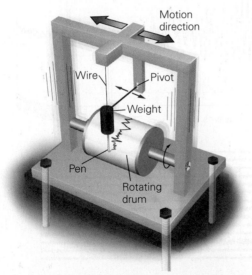

(b) A mechanical horizontal-motion seismometer records back-and-forth ground motion.

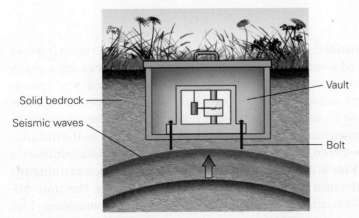

(c) Seismometers are bolted to bedrock in a protected shelter or vault.

line on the paper, as the cylinder turns. But when a seismic wave arrives and causes the ground surface to lurch up and down, the seismometer frame—along with the paper-covered cylinder—moves up and down. The weight, however, because of its *inertia* (the tendency of an object at rest to remain at rest), remains fixed in space. As the revolving paper-covered cylinder goes up and down with respect to the fixed pen, the line traced by the pen on the paper deflects away from the reference line. The deflection of the pen, therefore, represents the up-and-down movement of the ground. Because the paper cylinder revolves under the pen, the pen traces out curves that resemble waves. A mechanical horizontal-motion seismometer works on the same principle, except that the paper-covered cylinder is horizontal and the weight hangs from a vertical wire, and is connected to a pivot that prevents it from bouncing up and down (**Fig. 8.8b**). When the cylinder and the frame move sideways relative to the pen, the pen traces out waves.

In a modern electronic seismometer, the weight consists of a magnet that moves relative to a wire coil, thereby producing an electric signal that can be recorded digitally. Such seismometers are so sensitive that they can record ground movements of a millionth of a millimeter (only 10 times the diameter of an atom)—movements that people can't feel. Typically, seismologists place the instruments in an underground vault, preferably on bedrock, located away from traffic, urban noise, or swaying trees that could cause vibrations that are not due to earthquakes (**Fig. 8.8c**). The entire configuration comprises a *seismometer station*.

An earthquake record produced by a seismometer is called a **seismogram** (**Fig. 8.9**). At first glance, a typical seismogram looks like a messy squiggle of lines, but to a seismologist it contains a wealth of information. The horizontal axis on a seismogram represents time, and the vertical axis represents the *amplitude* (the size) of the seismic waves that arrived at the seismometer. The instant at which a seismic wave appears at a seismometer station is the *arrival time* of the wave. The first squiggles on the record represent P-waves, because P-waves travel the fastest. Next come the S-waves, and finally the surface waves (L-waves and R-waves). Typically, the surface waves have the largest amplitude and arrive over a relatively long interval of time.

FINDING THE EPICENTER

How do we find the location of an earthquake's epicenter? The answer to this question comes from comparing a seismogram recorded closer to the epicenter to one recorded farther from the epicenter (**Fig. 8.10a**). You will see that a given seismic wave takes less time to reach a nearby seismometer than it does to reach a distant one. Also, because different

FIGURE 8.9 The nature of seismograms.

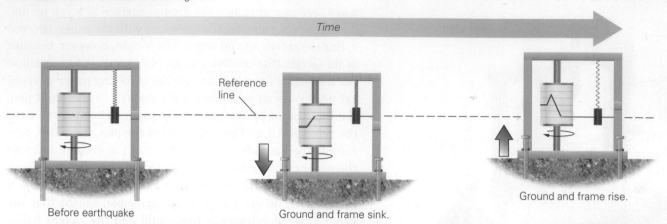

Time

Reference line

Before earthquake   Ground and frame sink.   Ground and frame rise.

(a) Before an earthquake, the pen traces a straight line. During an earthquake, the paper cylinder moves up and down while the pen stays in place.

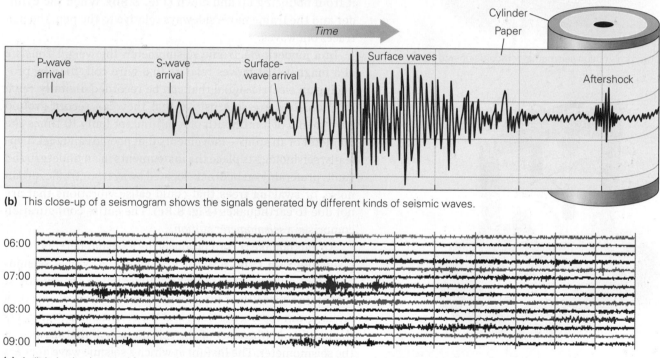

Time

Cylinder

Paper

P-wave arrival   S-wave arrival   Surface-wave arrival   Surface waves   Aftershock

(b) This close-up of a seismogram shows the signals generated by different kinds of seismic waves.

06:00

07:00

08:00

09:00

(c) A digital seismic record from a seismometer station in Arkansas. The space between vertical lines represents 1 minute. Colors have no meaning but make the figure more readable. Each color band represents the record of 15 minutes.

waves travel at different velocities, the difference between the arrival time of a faster wave and the arrival time of a slower wave increases as the distance between the epicenter and the seismometer increases. To picture why, imagine a car race— if one car travels faster than another, the distance between them increases as the race proceeds.

Researchers have found that measuring the difference between the P-wave arrival time and the S-wave arrival time provides an easy way to calculate the distance between the epicenter and a seismometer, for these waves are very

distinctive. We can represent the time delay between P-waves and S-waves by depicting a **travel-time curve** on a graph whose horizontal axis indicates distance from the epicenter and whose vertical axis indicates time. The curve shows the increase in time that it takes for an earthquake wave to move from its origin to a seismometer station, as the distance between the epicenter and the seismometer station increases **(Fig. 8.10b)**. To use travel-time curves for determining the distance to an epicenter, start by measuring the time difference between the P- and S-waves on a seismogram. This

FIGURE 8.10 The method for locating an earthquake epicenter.

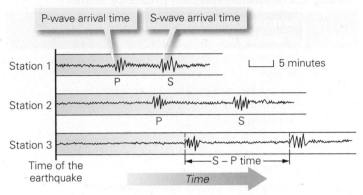

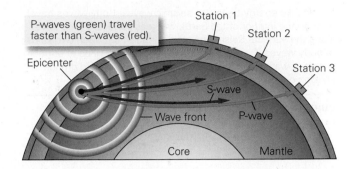

(a) Different types of seismic waves travel at different velocities. The greater the distance between the epicenter and the seismometer station, the longer it takes for earthquake waves to arrive, and the greater the delay between the P-wave and S-wave arrival times.

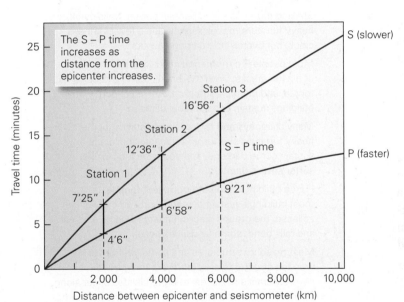

(b) We can represent the different arrival times of P-waves and S-waves on a graph of travel-time curves.

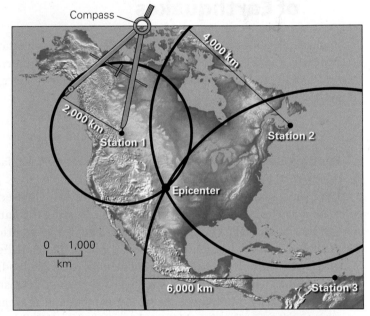

(c) If an earthquake epicenter lies 2,000 km from Station 1, we draw a circle with a radius of 2,000 km around the station at the scale of the map. We do the same for the other two stations. The intersection of the three circles defines the epicenter location.

quantity is called the *S − P* (pronounced "S minus P") *time*. Then draw a line segment on a piece of tracing paper to represent this amount of time, at the scale used for the vertical axis of the graph. Orient the line segment parallel to the time axis and move it back and forth until one end lies on the P-wave curve and the other end lies on the S-wave curve (this gives the S − P time). Extend the line down to the horizontal axis, and simply read off the distance to the epicenter from the seismic station.

The analysis of one seismogram indicates only the distance between the epicenter and the seismometer station—it does not indicate the direction from the station to the epicenter. To determine the map location of the epicenter,

it is necessary to *triangulate* by plotting the distance from the epicenter to three stations. For example, if we know that the epicenter lies 2,000 km from Station 1, 4,000 km from Station 2, and 6,000 km from Station 3, we can draw a circle around each station on a map, such that the radius of the circle represents the distance between the station and the epicenter at the scale of the map. The epicenter lies at the intersection of the three circles, for this is the only point at which the epicenter has the appropriate measured distance from all three stations **(Fig. 8.10c)**.

TAKE-HOME MESSAGE

Earthquake energy travels as seismic waves. Body waves travel through the interior of the Earth, whereas surface waves travel along the surface. Seismologists distinguish between two kinds of body waves (P-waves and S-waves) and two kinds of surface waves (L-waves and R-waves). Seismometers record ground shaking on a seismogram.

QUICK QUESTION How can you determine the location of an earthquake's epicenter?

8.4 Defining the Size of Earthquakes

Some earthquakes shake the ground violently, whereas others can barely be felt. Seismologists have developed two different scales—the intensity scale and the magnitude scale—to permit systematic composition of the sizes of earthquakes.

MERCALLI INTENSITY SCALE

Earthquake intensity at a locality on the Earth's surface refers to the degree of ground shaking at that locality. In 1902, an Italian scientist named Giuseppe Mercalli devised a scale for defining intensity based both on assessing the damage that the earthquake caused and on people's perception of the shaking. A version of this scale, called the **Modified Mercalli Intensity (MMI) scale**, continues to be used today (**Table 8.2**). On this scale, seismologists represent intensity at a location by a Roman numeral. An earthquake can have an intensity ranging from I (not destructive) to XII (highly destructive). Note that specifying earthquake intensity depends on subjective observations of damage, and how the shaking felt, not on a direct measurement with an instrument.

Significantly, the Mercalli intensity value varies with location for a given earthquake—intensity tends to be greater near the epicenter and to decrease progressively away from the epicenter. Why? First, the energy carried by a seismic wave decreases with increasing distance from the source. Rocks and sediment act as shock absorbers, so a vibration is less intense as it passes (much as a car's shock absorbers soften the effect of a bounce when a car goes over a bump). Second, some of the seismic energy passing through the Earth gets reflected back at layer boundaries. Therefore, it doesn't reach the seismometer. Seismologists draw contour lines on a map to illustrate zones where an earthquake has a certain intensity. Such maps show how intensity varies over a region for the earthquake (**Fig. 8.11**).

TABLE 8.2 Modified Mercalli Intensity Scale

MMI	Destructiveness (Perceptions of the Extent of Shaking and Damage)
I	Detected only by seismic instruments; causes no damage.
II	Felt by a few stationary people, especially in upper floors of buildings; suspended objects, such as lamps, may swing.
III	Felt indoors; standing automobiles sway on their suspensions; it seems as though a heavy truck is passing.
IV	Shaking awakens some sleepers; dishes and windows rattle.
V	Most people awaken; some dishes and windows break, unstable objects tip over; trees and poles sway.
VI	Shaking frightens some people; plaster walls crack, heavy furniture moves slightly, and a few chimneys crack, but overall little damage occurs.
VII	Most people are frightened and run outside; a lot of plaster cracks, windows break, some chimneys topple, and unstable furniture overturns; poorly built buildings sustain considerable damage.
VIII	Many chimneys and factory smokestacks topple; heavy furniture overturns; substantial buildings sustain some damage, and poorly built buildings suffer severe damage.
IX	Frame buildings separate from their foundations; most buildings sustain damage, and some buildings collapse; the ground cracks, underground pipes break, and rails bend; some landslides occur.
X	Most masonry structures and some well-built wooden structures are destroyed; the ground severely cracks in places; many landslides occur along steep slopes; some bridges collapse; some sediment liquifies; concrete dams may crack; facades on many buildings collapse; railways and roads suffer severe damage.
XI	Few masonry buildings remain standing; many bridges collapse; broad fissures form in the ground; most pipelines break; severe liquefaction of sediment occurs; some dams collapse; facades on most buildings collapse or are severely damaged.
XII	Earthquake waves cause visible undulations of the ground surface; objects are thrown up off the ground; there is complete destruction of buildings and bridges of all types.

EARTHQUAKE MAGNITUDE SCALES

When you hear a report of an earthquake disaster in the news, you will likely hear a phrase like, "An earthquake with a magnitude of 7.2 struck the city yesterday at 10:22 in the morning." What does this mean? **Earthquake magnitude** represents

FIGURE 8.11 This map shows Modied Mercalli Intensity contours for the 1886 Charleston, South Carolina, earthquake. Note that near the epicenter, ground shaking reached MMI of X, but in New York City, ground shaking reached MMI of II to III.

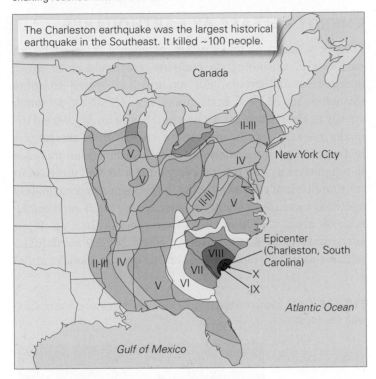

The Charleston earthquake was the largest historical earthquake in the Southeast. It killed ~100 people.

Because the Richter scale and all subsequent earthquake magnitude scales are logarithmic, an increase of one unit of magnitude represents a 10-fold increase in the maximum amplitude of ground motion. Thus, a magnitude 8 earthquake results in ground motion that is 10 times greater than that of a magnitude 7 earthquake, and 1,000 times greater than that of a magnitude 5 earthquake. Richter used 100 km as the reference distance—and since most seismic stations do not happen to lie exactly 100 km from the epicenter, he developed a simple chart to adjust for distance of the station from the epicenter **(Fig. 8.12)**.

FIGURE 8.12 Using the Richter magnitude scale.

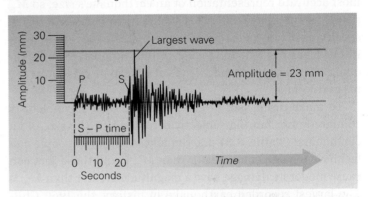

(a) To calculate the Richter magnitude from a seismogram, first measure the S – P time to determine the distance to the epicenter. Then measure the amplitude of the largest wave.

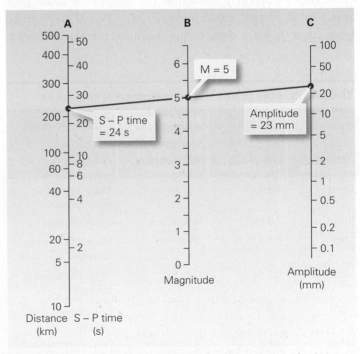

(b) Draw a line from the point on Column A representing the S – P time or the distance to the epicenter, to the point on Column C representing the wave amplitude. Read the Richter magnitude from Column B.

the amount of energy released from the seismic source, as indicated by the amplitude of ground shaking recorded by a seismometer. *Amplitude* means the amount of up-and-down or back-and-forth motion of the ground—the larger the amplitude, the greater the deflection of a seismometer pen or needle as it traces out a seismogram.

To calculate a magnitude, seismologists first measure the height of the largest deflections on a seismograph. Then, after they have determined the distance between the epicenter and the seismometer, they adjust the measurement to be representative of the deflections recorded by a seismometer that is positioned a certain distance, the *reference distance*, from the epicenter. This necessary adjustment takes into account the weakening of earthquake waves as they travel. Because of this adjustment, ideally seismologists obtain the same magnitude for an earthquake by using a measurement at any seismometer. In other words, a given earthquake should have only one magnitude number; in contrast to earthquake intensity, earthquake magnitude does not depend on distance from the epicenter.

In 1935, an American seismologist, Charles Richter, developed a logarithmic scale for defining earthquake magnitude. His scale, now known as the **Richter scale**, became so widely known that news reports often include wording such as, "The earthquake registered a 7.2 on the Richter scale."

The specific way that Richter calculated magnitude works well only for shallow earthquakes that are relatively close to the seismometer whose data go into the calculation. Because of the distance limitation, a number given by the Richter scale is now called a *local magnitude* (M_L).

These days, seismologists actually use several different scales to describe earthquake magnitude. Of these, the moment magnitude has become the most widely used. To calculate an earthquake's **moment magnitude** (M_W), seismologists measure the amplitudes of several different seismic waves, determine the dimensions of the slipped area on the fault, and estimate the displacement that occurred. The moment magnitude scale, like the Richter scale, is logarithmic. Seismologists consider the moment magnitude to be the most accurate representation of an earthquake's size, so M_W serves as the magnitude used in the historic record. When specifying magnitude, seismologists indicate the type of magnitude measurement and the magnitude number, so an earthquake with a moment magnitude of 6.5 can be called "an M_W 6.5 earthquake."

To make discussion of earthquakes easier, seismologists commonly use familiar adjectives to describe the size of an earthquake **(Table 8.3)**. Earthquakes that most people can feel have a magnitude greater than M_W 4, and those that can cause moderate damage have a magnitude greater than M_W 5. The largest recorded earthquake in history, the 1960 Chilean quake, registered as an M_W 9.5, and the catastrophic 2011 Tōhoku earthquake registered as an M_W 9.0. The news media sometimes incorrectly state that the magnitude scale goes from 1 to 10. In fact, earthquakes of M_W –1 or M_W –2 can be detected, if they are close to the seismometer. Furthermore, there is no defined upper limit to the magnitude scale. That said, seismologists estimate that an M_W 9.5 is about as big as an earthquake can get, given the known dimensions of faults.

ENERGY RELEASE BY EARTHQUAKES

To give a sense of the amount of energy released by an earthquake, seismologists compare earthquakes to other energy-releasing events. For example, according to some estimates, an M_W 5.3 earthquake releases about as much energy as the Hiroshima atomic bomb, and an M_W 9.0 earthquake releases significantly more energy than the largest hydrogen bomb ever detonated. Notably, though an increase in magnitude by one integer represents a 10-fold increase in the amplitude of ground shaking, it represents approximately a 32-fold increase in energy release. So an M_W 8 earthquake releases about 1 million times more energy than an M_W 4 earthquake **(Fig. 8.13)**. In fact, a single M_W 8.9 earthquake releases as much energy as the entire average global annual

FIGURE 8.13 Energy released by earthquakes increases dramatically with magnitude.

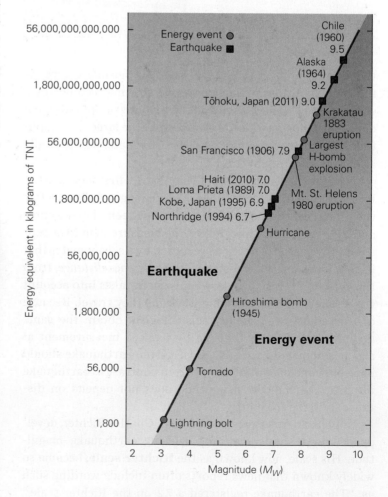

TABLE 8.3 Adjectives for Describing Earthquakes

Adjective	Magnitude	Approximate Maximum Intensity at Epicenter	Effects
Great	>8.0	X to XII	Major to total destruction
Major	7.0 to 7.9	IX to X	Great damage
Strong	6.0 to 6.9	VII to VIII	Moderate to serious damage
Moderate	5.0 to 5.9	VI to VII	Slight to moderate damage
Light	4.0 to 4.9	IV to V	Felt by most; slight damage
Minor	<3.9	III or smaller	Felt by some; hardly any damage

release of seismic energy coming from all other earthquakes combined! Fortunately, such large earthquakes occur much less frequently than small earthquakes. There are about 100,000 M_W 3 earthquakes every year, but an M_W 8 earthquake happens only about once or twice a year.

TAKE-HOME MESSAGE

We can specify earthquake size by intensity (based on a perception of damage caused and shaking felt) or by magnitude (a representation of energy released by the source, based on a measurement of ground motion on a seismogram). An increase of one magnitude in number represents a 10-fold increase in the amplitude of shaking, at a reference distance, and a 32-fold increase in energy release from the source. A great earthquake releases vastly more energy than a giant hydrogen bomb.

QUICK QUESTION Can you specify the "size" of an earthquake by giving just one intensity number? How about a single magnitude number?

8.5 Where and Why Do Earthquakes Occur?

Earthquakes do not take place everywhere on the globe. By plotting the distribution of earthquake epicenters on a map, seismologists find that most, but not all, earthquakes occur in elongate **seismic belts**, or localized **seismic zones** (Fig. 8.14). Most seismic belts correspond to plate boundaries, and earthquakes within these belts are called *plate-boundary earthquakes*. Earthquakes that occur away from plate boundaries are called *intraplate earthquakes*. Let's look at the characteristics of earthquakes in various geologic settings and learn why earthquakes take place where they do.

> **Did you ever wonder . . .**
> if an earthquake might happen near where you live?

FIGURE 8.14 A map of epicenters emphasizes that most earthquakes occur in distinct belts along plate boundaries.

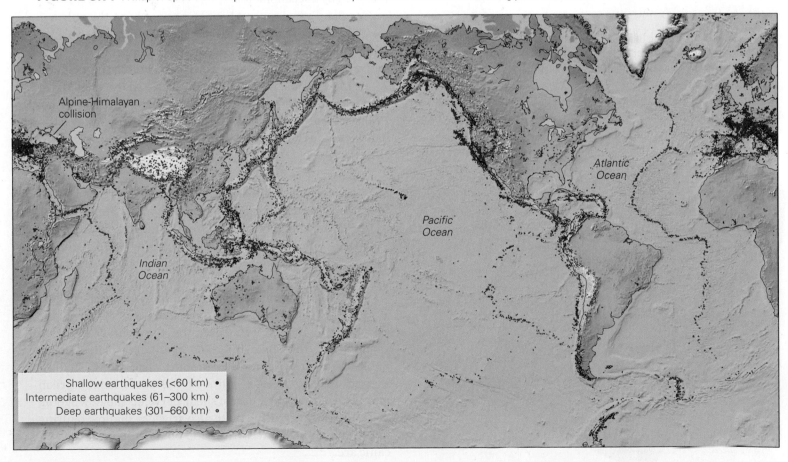

Shallow earthquakes (<60 km) •
Intermediate earthquakes (61–300 km) ◦
Deep earthquakes (301–660 km) ⊙

EARTHQUAKES AT PLATE BOUNDARIES

The majority of earthquakes happen on faults along plate boundaries, because the relative motion between plates causes slip on faults. We find different kinds of faulting at different types of plate boundaries.

Divergent-Boundary Seismicity   At divergent boundaries (mid-ocean ridges), two oceanic plates form and move apart. Divergent boundaries consist of spreading segments linked by transform faults. Therefore, two kinds of faults develop at divergent boundaries. Along spreading segments, stretching generates normal faults, whereas along transform faults, strike-slip displacement occurs **(Fig. 8.15)**. Earthquakes along mid-ocean ridges take place at depths of less than 10 km, making them shallow-focus earthquakes.

Transform-Boundary Seismicity   At transform boundaries, where one plate slides past another without producing or consuming oceanic lithosphere, most faulting results in strike-slip motion. The majority of transform faults in the world link segments of oceanic ridges, on the seafloor (see Fig. 8.15). But some, such as the San Andreas fault of California, the Alpine fault of New Zealand, and the Anatolian faults

FIGURE 8.15 The distribution of earthquakes at a mid-ocean ridge. Note that normal faults occur along the ridge axis, and strike-slip faults occur along active transform faults. Earthquakes don't take place along inactive fracture zones.

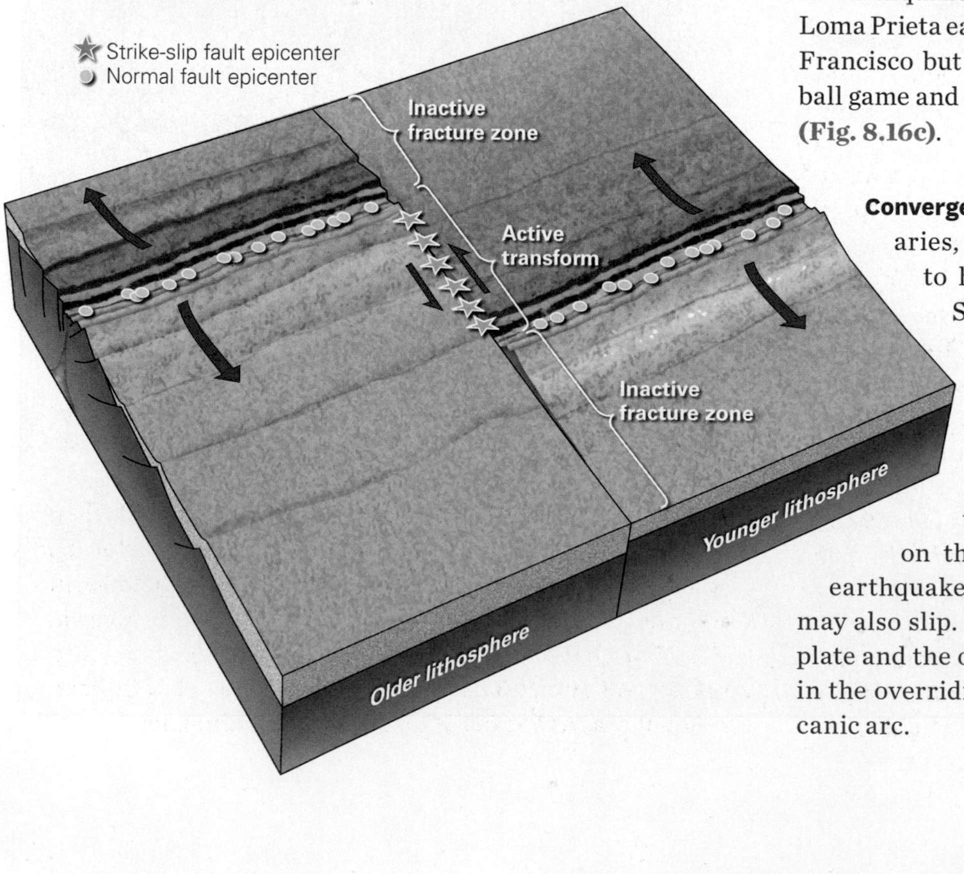

in Turkey, cut through continental lithosphere or volcanic arcs. All transform-fault earthquakes have a shallow focus, so the larger ones on land can cause disaster. This was the case when, in 2010, an M_W 7 earthquake struck Haiti, which sits astride the transform fault that delineates part of the plate boundary between the North American and Caribbean plates.

The San Francisco earthquake of 1906 serves as an example of a continental transform-fault earthquake that occurred on the San Andreas fault **(Fig. 8.16a)**. In the wake of the gold rush, San Francisco was a booming city with broad streets and numerous large buildings. But it was built on the transform boundary along which the Pacific Plate moves north at an average rate of 6 cm per year, relative to North America. The stick-slip behavior of the fault causes this movement to happen in sudden jerks, each of which causes an earthquake. At 5:12 A.M. on April 18, the fault near San Francisco slipped by as much as 7 m, and minutes later seismic waves struck the city. Witnesses watched in horror as the streets undulated, buildings swayed and banged together, laundry lines stretched and snapped, and weaker buildings collapsed. Fire followed soon after, consuming huge areas of the city **(Fig. 8.16b)**. Seismologists estimate that the earthquake would have registered as an M_W 7.9, had it been recorded by a seismometer.

The San Francisco earthquake has not been the only one to strike along the San Andreas and nearby related faults. Over a dozen major earthquakes have happened on these faults during the past two centuries, including the 1857 M_W 7.7 earthquake just north of Los Angeles, and the 1989 M_W 7.1 Loma Prieta earthquake, which occurred 100 km south of San Francisco but nevertheless shut down a World Series baseball game and caused the collapse of a double-decker freeway **(Fig. 8.16c)**.

Convergent-Boundary Seismicity   Convergent boundaries, where one plate subducts under another, tend to host several different kinds of earthquakes. Specifically, as the downgoing plate begins to subduct, it bends, causing normal faults to develop in the downgoing plate, seaward of the trench. Then, where it scrapes along the base of the overriding plate, large thrust faults form, defining the contact between the downgoing and overriding plates. Shear on these faults can produce disastrous, shallow earthquakes. Thrust faults in the accretionary prism may also slip. In some cases, shear between the downgoing plate and the overriding plate also triggers shallow faulting in the overriding plate within and on both sides of the volcanic arc.

FIGURE 8.16 The San Andreas fault system, a continental transform boundary.

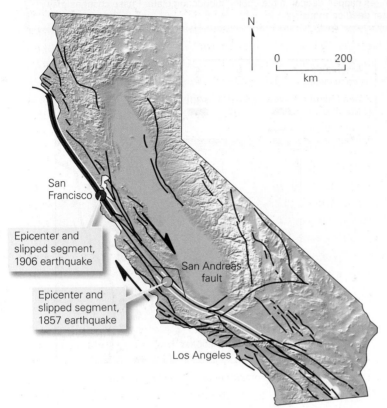

(a) The San Andreas fault system in California. Note that the system includes many faults in a 100-km-wide band.

(b) A street in San Francisco after the 1906 earthquake. The event caused fires to sweep through the city, adding to the damage and devastation.

(c) The 1989 Loma Prieta earthquake caused a two-level freeway in San Francisco, 100 km northwest of the epicenter, to collapse, as its support columns gave way.

In contrast to other types of plate boundaries, convergent boundaries can also host intermediate and deep earthquakes. These occur in the downgoing slab as it sinks into the mantle, defining the sloping belt of seismicity called a **Wadati-Benioff zone**, after the seismologists who first recognized it **(Fig. 8.17a)**. Earthquakes of this zone occur for several resons: in response to stresses caused by shear between the sinking lithosphere plate and surrounding asthenosphere; in response to the pull exerted by the deeper part of the plate on the shallower part as the plate sinks; due to stress generated by the resistance that the subducting plate encounters as it pushes into the mantle below; and due to volume changes that take place when olivine, a mineral in the plate, undergoes a phase change and collapses to form denser minerals under the extreme pressure that develops in deeply subducted lithosphere.

How are intermediate and deep earthquakes of a Wadati-Benioff zone even possible? Shouldn't the rock of a subducted plate at these depths be too warm and soft to break seismically? Seismologists have determined that rock is such a good insulator that the interior of a subducting plate actually remains cool enough to break seismically, down to the depths of intermediate- and deep-focus earthquakes.

Earthquakes in southern Mexico, southern Alaska, eastern Japan, the western coast of South America, the coast of Oregon and Washington, and along island arcs in the western Pacific serve as examples of convergent-boundary earthquakes. Some of these earthquakes are very large, and when they occur near populated areas, they can be devastating. Notable examples include the 1960 M_W 9.5 earthquake in Chile, the 1964 M_W 9.2 Good Friday earthquake in Alaska, the 1995 M_W 6.9 earthquake in Kobe, Japan **(Fig. 8.17b)**, the 2004 M_W 9.3 Sumatra earthquake, the 2010 M_W 8.8 Chilean earthquake, and the 2011 M_W 9.0 Tōhoku earthquake.

FIGURE 8.17 Convergent-boundary earthquakes.

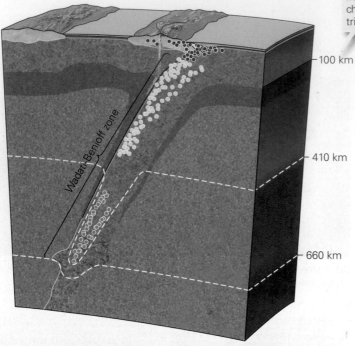

(a) At a convergent boundary, earthquakes occur along the contact between the two plates, as well as in the downgoing plate and overriding plate.

In the asthenosphere, phase changes take place at 410 km and 660 km. These changes happen deeper in the cooler subducting plate. Phase changes may trigger deep earthquakes.

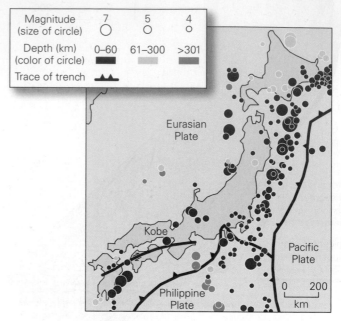

(b) Map of earthquakes in and near Japan.

EARTHQUAKES WITHIN CONTINENTS

We've seen that earthquakes occur along transform faults that cut across continents. They also take place where continental lithosphere undergoes stretching in rifts, where continents squeeze together in collisional zones, and where old faults within otherwise stable crust get reactivated, as we now see **(Fig. 8.18a)**.

Continental Rifts The stretching of crust at continental rifts generates normal faults. Active rifts today include the East African Rift, the Basin and Range Province (mostly in Nevada, Utah, and Arizona), and the Rio Grande Rift (in New Mexico). In all these places, shallow earthquakes occur, similar in nature to the earthquakes at mid-ocean ridges. But in contrast to mid-ocean ridges, seismic belts of continental rifts can be located under populated areas, so they can cause major damage.

Collision Zones Two continents collide when the oceanic lithosphere that once separated them has been completely subducted. Such collisions produced great mountain ranges, such as the Alps and the Himalayas. Most earthquakes in collision zones involve movement on thrust faults that accommodate crustal shortening.

An example of a collision-zone earthquake took place in April 2015. Compression resulting from the northward push of the Indian subcontinent into Asia caused an M_w 7.8 earthquake in Nepal, a country that encompasses a portion of the Himalayas **(Fig. 8.18b)**. The event occurred when a 7,000 km² portion of a large thrust fault, 15 km below the surface, suddenly slipped by up to 3 m. Ground shaking destroyed whole towns with poorly constructed homes. Tragically, thousands died and many more were injured. Landslides blocked rivers and ripped out roads, making hard-hit areas difficult to reach. Many monuments recognized as United Nations World Heritage sites crumbled completely. Aftershocks, the largest of which had M_w 6.6, rattled the area for weeks.

INTRAPLATE EARTHQUAKES

Some earthquakes occur in the interiors of plates and are not associated with plate boundaries, active rifts, or collision zones (see Fig. 8.18). These **intraplate earthquakes**, most with a shallow focus, account for only about 5% of the earthquake energy released in a year. What causes intraplate earthquakes? Most seismologists favor the idea that stress applied to the boundary of a plate can cause the interior of the plate to break suddenly at weak, pre-existing fault zones, which may initially have formed during the Precambrian.

FIGURE 8.18 The tectonic settings in which earthquakes occur in continental lithosphere.

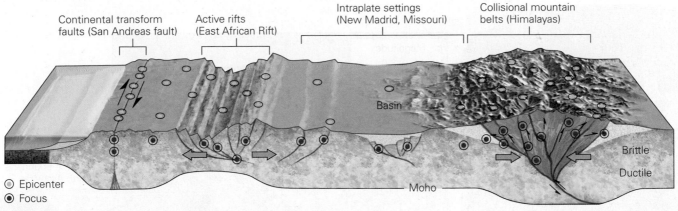

(a) Earthquakes occur along continental transforms, in rifts, within intraplate settings, and in collision zones.

(b) The 2015 Nepal earthquake is related to continental collision. It destroyed many historic buildings.

Intraplate seismic zones exist on all continents. In North America, such seismic zones occur near New Madrid, Missouri; Charleston, South Carolina; eastern Tennessee; and Montreal. An M_W 7.3 earthquake occurred near Charleston in 1886, ringing church bells up and down the coast and vibrating buildings as far away as Chicago. In Charleston itself, over 90% of the buildings were damaged, and 60 people died. In 2011, an M_W 5.9 earthquake struck central Virginia, abruptly reminding residents of the eastern United States that the region is not immune to seismicity. The tremor was felt from the Carolinas to New England.

The largest intraplate earthquakes to affect the continental United States happened in the early 19th century, in the New Madrid seismic zone, which lies near the Mississippi River in southernmost Missouri. During the winter of 1811–12, three M_W 7.0 to 7.4 earthquakes struck the region. Displacement of the ground surface temporarily reversed the flow of the Mississippi River and toppled cabins (**Fig. 8.19a**). The earthquakes resulted from slip on faults that underlie the Mississippi Valley (**Fig. 8.19b**). Both St. Louis, Missouri, and Memphis, Tennessee, could be damaged if a major earthquake were to take place in the region today.

TAKE-HOME MESSAGE

Most, but not all, earthquakes happen along plate boundaries. Most catastrophic earthquakes occur at convergent boundaries, or along continental transforms. Events also occur in rifts and collision zones, and along weak faults within plate interiors.

QUICK QUESTION Why are earthquakes in continental crust more dangerous to society than those along mid-ocean ridges?

FIGURE 8.19 New Madrid, Missouri, is a center of intraplate seismic activity.

(a) The earthquakes of 1811–1812 destroyed cabins and disrupted the Mississippi River.

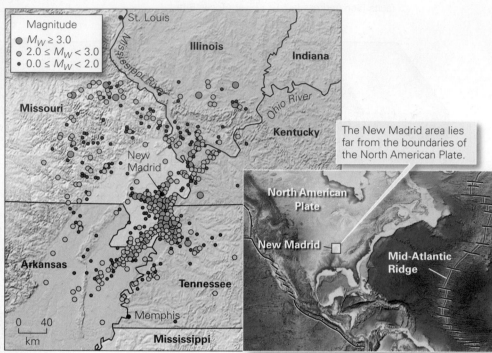

(b) The epicenters of recent small earthquakes in the New Madrid area, as recorded by modern seismic instruments. The region remains active.

8.6 How Do Earthquakes Cause Damage?

On All Saint's Day, November 1, 1755, stresses that had been building due to the northward push of Africa against Europe caused a thrust fault beneath the Atlantic Ocean floor west of Lisbon, Portugal, to slip suddenly. The resulting M_w 9.0 earthquake and its aftermath destroyed Lisbon and led philosophers of the day, such as Voltaire (1694–1778), to question a widely held belief that the ways of the world are always for the best. Ground shaking led to the collapse of buildings in Lisbon, but as is the case for many calamitous earthquakes, shaking represents only part of the story. Below, we examine the many components of earthquake-related destruction.

GROUND SHAKING AND DISPLACEMENT

An earthquake starts suddenly and may last from a few seconds to a few minutes, depending on how long the slip on a fault took, and on how far the earthquake's source lies from the locality where people feel shaking. Different kinds of earthquake waves cause different kinds of ground motion (**Fig. 8.20**). Since waves arrive at different times, and the arrivals of different waves can overlap, ground shaking overall may be quite chaotic. The severity of the shaking at a given location depends on four factors: (1) the magnitude of the earthquake, because larger-magnitude events release more energy; (2) the distance from the focus, because earthquake energy decreases as waves pass through the Earth; (3) the nature of the *substrate* at the location, meaning the character and thickness of materials beneath the ground surface, because earthquake waves tend to be amplified in weaker substrates; and (4) the frequency of the waves, meaning the number of waves that reach a point in a specified interval of time, because frequency affects the way buildings sway.

If you're out in an open field during an earthquake, ground motion alone won't kill you—you may be knocked off your feet and bounced around a bit, but your body won't break.

Did you ever wonder...
if all earthquakes feel the same?

FIGURE 8.20 Types of ground motion during earthquakes. The ground can shake in many ways at once, causing surface structures to move.

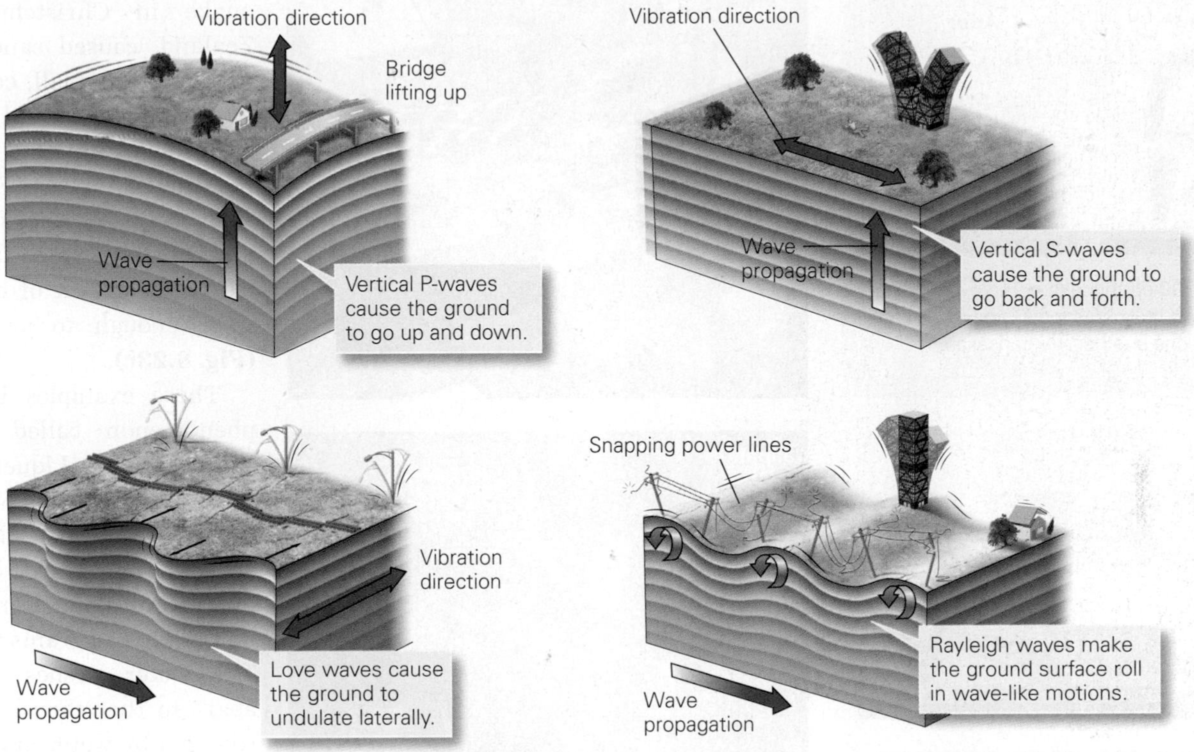

Vibration direction

Bridge lifting up

Wave propagation

Vertical P-waves cause the ground to go up and down.

Vibration direction

Wave propagation

Vertical S-waves cause the ground to go back and forth.

Wave propagation

Vibration direction

Love waves cause the ground to undulate laterally.

Snapping power lines

Wave propagation

Rayleigh waves make the ground surface roll in wave-like motions.

Buildings and bridges aren't so lucky **(Fig. 8.21)**. When earthquake waves pass, they sway, twist back and forth, or lurch up and down, depending on the type of wave motion. As a result, connectors between the frame and facade of a building may separate, causing the facade to crash to the ground. Windows may shatter, and roofs may collapse. Movement may cause floors or bridge decks to rise up and slam down on the columns that support them, thereby crushing the columns. This action can make some buildings collapse so their floors pile on top of one another like pancakes in a stack, and others may crumble into fragments or simply tip over. The majority of earthquake-related deaths and injuries happen when falling debris or collapsing structures crush people. Aftershocks worsen the problem, because they may topple already weakened buildings, trapping rescuers. During earthquakes, roads, rail lines, and pipelines may also buckle and rupture. If a building, fence, road, pipeline, or rail line straddles a fault, slip on the fault can crack the structure and displace one piece relative to the other.

LANDSLIDES

The shaking of an earthquake can cause ground on steep slopes or ground underlain by weak sediment to give way.

This movement results in a **landslide**, during which soil and rock tumble and flow downslope (see Chapter 13). Earthquake-triggered landslides occur commonly along the coast of California, where expensive homes perch on steep cliffs looking out over the Pacific. When the cliffs collapse, the homes tumble to the beach below **(Fig. 8.22a)**. Such events lead to the misperception that "California will someday fall into the sea." Although small portions of the coastline sometimes collapse, the state as a whole remains firmly attached to the continent, despite what Hollywood scriptwriters say. Landslides and slumping are often a major cause of earthquake-related damage **(Fig. 8.22b)**.

Did you ever wonder . . .
if California will fall into the sea?

SEDIMENT LIQUEFACTION

In 1964, an M_W 7.5 earthquake struck Niigata, Japan. A portion of the city had been built on land underlain by a substrate of wet sand. During the ground shaking, foundations of over 15,000 buildings sank into the substrate, causing walls and roofs to crack. Several four-story buildings in a newly built

FIGURE 8.21 Examples of earthquake damage due to vibration.

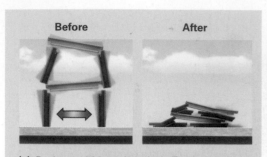

Turkey

Before **After**

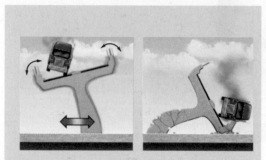

(a) During a 1999 earthquake in Turkey, concrete buildings collapsed when supports gave way and floors piled on one another like pancakes.

Japan

(b) An elevated bridge tipped over during the 1995 earthquake in Kobe, Japan.

California

(c) Concrete bridge supports were crushed during the 1994 Northridge, California, earthquake when the overlying bridge bounced up and slammed back down.

Armenia

(d) A neighborhood of masonry buildings in Armenia collapsed during a 1999 earthquake because the walls broke apart.

apartment complex tipped over **(Fig. 8.23a)**. In 2011, an earthquake in Christchurch, New Zealand, caused sand to erupt and produce small, cone-shaped mounds—called sand volcanoes, sand boils, or sand blows—on the ground surface **(Fig. 8.23b)**. The transfer of sand from underground onto the surface led to formation of depressions large enough to swallow cars **(Fig. 8.23c)**.

These examples illustrate a phenomenon called **sediment liquefaction**. Liquefaction in wet sand happens when shaking causes grains to settle together, for this movement causes the pressure in the water filling the pores between grains to increase. The water pushes the grains apart so that they become surrounded by water and no longer rest against each other, making the sand turn into a slurry called *quicksand,* incapable of supporting weight. A similar weakening phenomenon happens in *quick clay.* The flakes in quick clay are arranged in an unstable structure. When disrupted by ground shaking, what had been a stable, gel-like mass transforms into weak, slippery mud. Weakening due to liquefaction allows material above the liquefied sediment to settle downward. The settling of sedimentary layers down into a liquefied layer can also disrupt bedding, can lead to formation of open fissures of the land surface, and can trigger slumping (see Fig. 8.23b).

The consequences of the 1964 Good Friday earthquake in southern Alaska serve as a case study of these phenomena. This earthquake caused liquefaction beneath the Turnagain Heights neighborhood of Anchorage. The neighborhood was built on a

FIGURE 8.22 Examples of landslide damage triggered by earthquakes.

(a) Shaking triggered a landslide, which caused a steep slope along the coast of California to collapse, carrying part of a home with it.

(b) During the 1964 Alaska earthquake, slumping caused the land to give way beneath parts of Anchorage.

FIGURE 8.23 Examples of liquefaction triggered by earthquakes.

(a) Liquefaction under their foundations caused these apartment buildings in Niigata, Japan, to tip over during a 1964 earthquake.

(c) During the 2011 Christchurch earthquake, liquefied sand spurted out and spread over the pavement. The process produced open space underground, so the pavement collapsed to form a sinkhole.

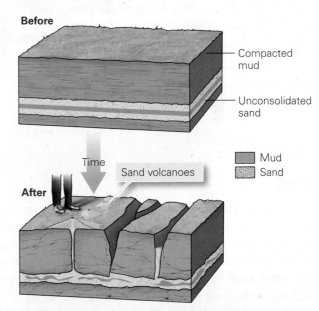

(b) Liquefaction of a sand layer causes the ground to crack and sand volcanoes to erupt.

small terrace of uplifted sediment. The edge of the terrace was a 20-m-high escarpment that dropped down to Cook Inlet, a bay of the Pacific Ocean. As the ground shaking began, a layer of quick clay beneath the development turned into mud, and when this happened, the overlying layers of sediment, along with the houses built on top of them, slid seaward. In the process, the layers broke into separate blocks that tilted, turning the landscape into a chaotic jumble, and resulting in complete destruction of the neighborhood **(Fig. 8.24)**.

FIGURE 8.24 The 1964 Turnagain Heights disaster.

The dark area in this aerial photograph is the slump that was Turnagain Heights.

(a) A landslide carried a neighborhood into the sea.

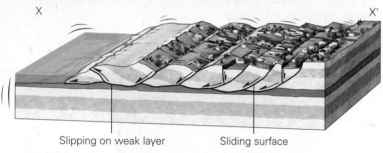

X   X'

Slipping on weak layer   Sliding surface

(b) Slip occurred on a weak layer.

FIRE

The shaking during an earthquake can make lamps, stoves, or candles with open flames tip over, and it may break wires or topple power lines, generating sparks. As a consequence, areas already turned to rubble, and even areas not so badly damaged, may be consumed by fire. Ruptured gas pipelines and oil tanks feed the flames, sending billows of fire erupting skyward **(Fig. 8.25a)**. Firefighters might not even be able to reach the fires, if the doors to the fire station won't open or rubble blocks the streets. Moreover, firefighters may find themselves without water, for ground shaking and landslides damage water lines.

Once a fire starts to spread, it can become an unstoppable inferno. Lisbon was consumed by fire after the 1755 earthquake, and most of the destruction of the 1906 San Francisco earthquake resulted from fire. For three days, the blaze spread through San Francisco until firefighters contained it by blasting a firebreak, but by then, 500 blocks of structures had turned to ash, causing 20 times as much financial loss as the shaking itself. When a large earthquake hit Tokyo in 1923,

fires set by cooking stoves spread quickly through the wood-and-paper buildings, creating an inferno—a *firestorm*—that heated the air above the city **(Fig. 8.25b)**. As hot air rose, cool air rushed in, creating wind gusts of over 100 mph, which stoked the blaze and incinerated 120,000 people.

TSUNAMIS

The azure waters and palm-fringed islands of the Indian Ocean's east coast bordering the Sunda Trench hide one of the most seismically active plate boundaries on the Earth. Along this convergent boundary, the Indian Ocean floor subducts at a rate of about 4 cm per year, leading to episodic

FIGURE 8.25 Fire sometimes follows an earthquake.

(a) Broken gas tanks erupt in fountains of flame after the 2011 Tōhoku, Japan, earthquake.

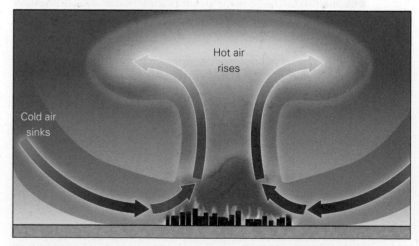

Hot air rises

Cold air sinks

(b) A firestorm develops when cool air rushes in to replace rising hot air above a huge fire. The cool air stokes the blaze, making it larger and hotter.

FIGURE 8.26 On December 26, 2004, a devastating tsunami was triggered by a subduction-related earthquake.

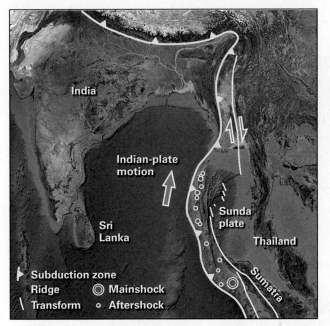

(a) The epicenter of the mainshock lay near Sumatra. The rupture extended from there north, for 1,200 km, as delineated by the distribution of aftershocks.

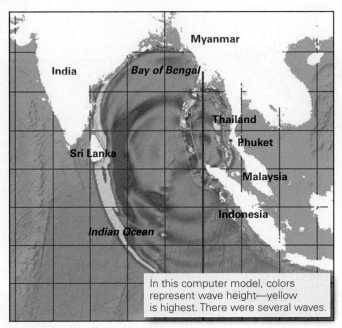

In this computer model, colors represent wave height—yellow is highest. There were several waves.

(b) The 2004 Indian Ocean tsunami hit parts of Indonesia within minutes. It took two hours for the leading wave to reach Sri Lanka and India.

slip on thrust faults. Just before 8:00 A.M. on December 26, 2004, the crust above a 1,300-km-long by 100-km-wide portion of one of these faults lurched westward by as much as 15 m **(Fig. 8.26a)**. The break started at the focus and then propagated north at 2.8 km/s. Rupturing overall took about 9 minutes. This slip triggered a great earthquake (the M_W 9.3 Sumatra earthquake) and pushed the seafloor up by tens of centimeters. The rise of the seafloor, in turn, shoved up the overlying water. Because the area that rose was so broad, the volume of displaced water was immense. As a consequence, a tragedy of an unimaginable extent was about to unfold. Water from above the displaced seafloor began moving outward from above the fault zone, a process that generated a series of broad waves traveling at speeds of about 800 km per hour—almost the speed of a jet plane, striking nearby islands in minutes, and crossing the Indian Ocean in hours **(Fig. 8.26b)**.

Geologists use the term **tsunami** for a wave produced by the sudden displacement of the seafloor **(Fig. 8.27)**. The displacement can be due to an earthquake, a volcanic explosion, or as we'll see in Chapter 13, a submarine landslide. The Japanese word *tsunami* translates literally to harbor wave, an apt name because tsunamis can be particularly damaging to harbor towns. In older literature such waves were called *tidal waves*, because when one arrives, water rises as if a tide were

coming in, even though the waves have nothing to do with daily tidal cycles.

In the case of an earthquake-generated tsunami, the process begins due to the sudden vertical displacement of an area of sea-floor. In the case of a thrust fault, the initial movement pushes water up and produces a broad bulge of the sea surface that may be a few centimeters to a few tens of centimeters high (see Fig. 8.27). Gravity immediately pulls the bulge down, and like the surface of a pond pushed down by a pebble, the surface then rises again and sinks again, multiple times. This up-and-down motion produces a succession of tsunamis that travel sideways, away from the initial site of the bulge. *Near-field tsunamis* travel toward nearby shores, whereas *far-field tsunamis* head out across

FIGURE 8.27 The formation and propagation of a tsunami. After initial uplift of a mound of water, gravity causes the wave to spread out. The waves travel almost as fast as a jet plane.

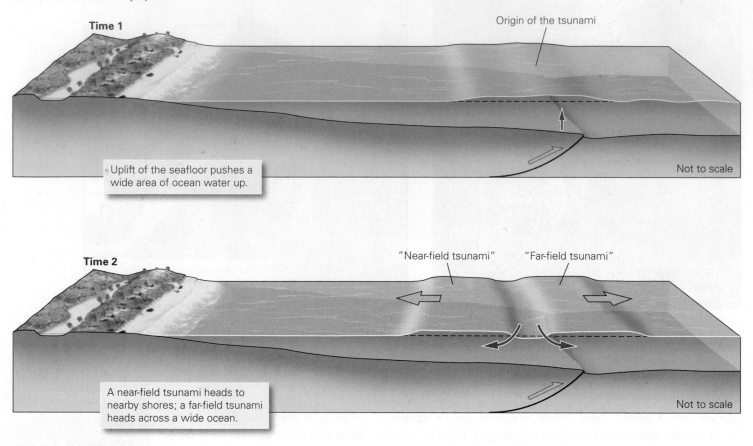

Time 1

Origin of the tsunami

Uplift of the seafloor pushes a wide area of ocean water up.

Not to scale

Time 2

"Near-field tsunami"   "Far-field tsunami"

A near-field tsunami heads to nearby shores; a far-field tsunami heads across a wide ocean.

Not to scale

the ocean. Because waves lose energy as they travel, far-field tsunamis tend to be lower than near-field tsunamis, when they finally reach the coast.

Regardless of their cause, tsunamis are very different from familiar, wind-driven storm waves **(Fig. 8.28)**. Large wind-driven waves can reach heights of 10 to 30 m in the open ocean. But even such monsters have wavelengths of only tens of meters, and thus they contain a relatively small volume of water. In contrast, although a tsunami in deep water may cause a rise in sea level of at most only a few tens of centimeters, which means a ship crossing one wouldn't even notice, a tsunami has a wavelength of 10 to 500 km—so such a wave can involve a huge volume of water. In simpler terms, we can think of a tsunami as being 100 to 1,000 times wider than a wind-driven wave, as measured perpendicular to the wave crest. Because of this difference, a storm wave and a tsunami have very different consequences when they strike the shore.

When any wave approaches the shore, friction between the base of the wave and the seafloor slows the bottom of the wave, so the back of the wave catches up to the front, and the added volume of water builds the wave higher. The top

of the wave may fall over the front of the wave and produce a breaker (see Chapter 15). In the case of a wind-driven wave, the breaker may be tall when it crashes onto the beach, but because the wave doesn't contain much water, it runs out of water and friction slows it to a stop on the beach. Gravity then causes the water to flow seaward, back down the beach. The width of a beach indicates the width of the coast affected by wave action. In the case of a tsunami, the wave is so wide that, as friction slows the wave, it builds into a plateau of water that can be many kilometers wide. If the tsunami is high—some are up to 30 m high—it can cross the beach, and if the land is low-lying, the tsunami can keep moving, eventually submerging a huge area. A smaller tsunami will probably affect only the near-shore area.

Tsunami damage can be catastrophic. For example, when the December 2004 Indian Ocean near-field tsunami struck Banda Aceh, a city at the north end of the island of Sumatra, much of the town and surrounding fields vanished. The waves arrived on a beautiful, cloudless day. First, the sea receded much farther than anyone had ever seen, exposing large areas of reefs that normally remained submerged even at low tide, and people walked out onto the exposed reefs in wonder.

FIGURE 8.28 A comparison of storm waves to a tsunami.

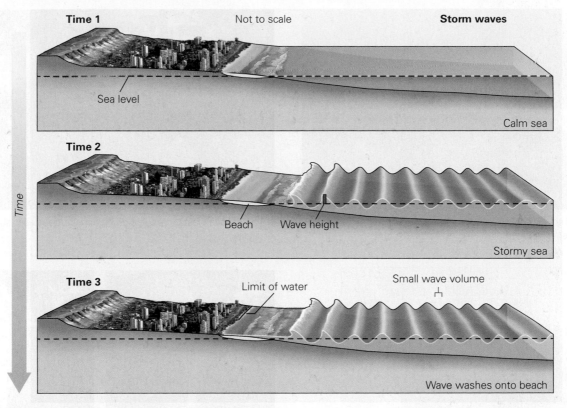

(a) Storm waves can be high, but because they have short wavelengths, they contain relatively little water. Most waves run out of water by the upslope edge of the beach.

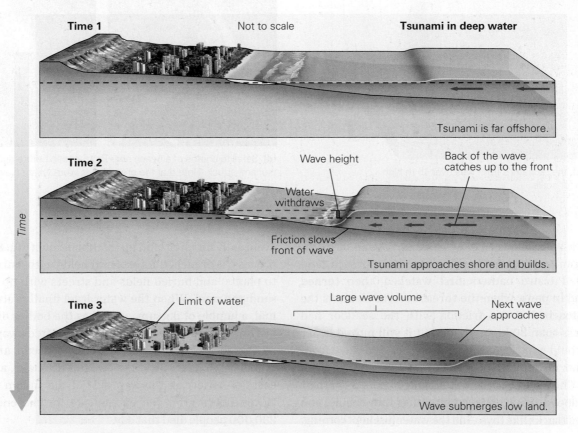

(b) A tsunami is not very high out in the open ocean, but as it approaches the land, friction slows the front of the wave, so the rear catches up and the wave grows. It is so wide that it can cover a wide area of low-lying land.

FIGURE 8.29 The great Indian Ocean tsunami of 2004.

(a) This snapshot shows the wave rushing toward the coast of Sumatra. Recession of water in advance of the wave exposed a reef.

(b) The wave blasts through a grove of palm trees as it strikes the coast of Thailand.

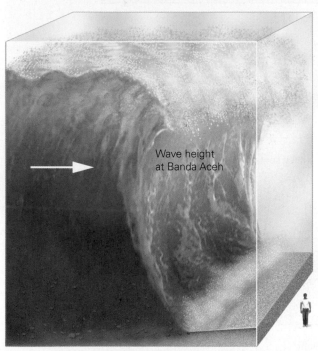

(c) At its highest, the tsunami's front was over 15 m high. Note the person for scale.

(d) Satellite photos of a beach near Banda Aceh before and after the 2004 tsunami struck. Note that the green fields were washed away and the beach vanished.

But then, with a rumble that grew to a roar, a wall of frothing water began to build in the distance and approach land **(Fig. 8.29a)**. Puzzled bathers first watched, then turned and ran inland in panic when the threat became clear. As the tsunami approached shore, friction with the seafloor had slowed it to less than 30 km an hour, but it still moved faster than people could run **(Fig. 8.29b)**. In places, the wave front reached heights of 15 to 30 m as it slammed into Banda Aceh **(Fig. 8.29c)**. The impact of the water ripped boats from their moorings, snapped trees, battered buildings into rubble, and tossed cars and trucks like toys. And the water just kept coming,

eventually flooding land up to 7 km inland **(Fig. 8.29d)**. In the process, it drenched forests and fields with saltwater (deadly to plants) and buried fields and streets with up to a meter of sand and mud. When the water level finally returned to normal, a jumble of flotsam, as well as the bodies of unfortunate victims, were dragged out to sea and drifted away.

The far-field tsunami crossed the ocean and struck Sri Lanka three hours after the 2004 earthquake, and struck the coast of Africa, on the west side of the Indian Ocean, seven to ten hours after the earthquake. In the end, more than 230,000 people died that day.

A tsunami that struck Japan soon after the 2011 Tōhoku earthquake was vividly captured in high-definition video that was seen throughout the world, generating a new level of international awareness. The 10-m-high seawalls that fringe the coast proved to be only a temporary impediment to the advance of the wave, which, in places, was 10 to 30 m high when it reached shore. Racing inland, the wave erased whole towns, submerging airports and fields (see Fig. 8.2b, c). As the wave picked up dirt and debris, it evolved into a viscous slurry, moving with such force that nothing could withstand its impact. The devastation of coastal towns was so complete that they looked as though they had been struck by nuclear bombs **(Fig. 8.30a)**.

But the catastrophe was not over. The wave had also struck a nuclear power plant. Although the plant had withstood ground shaking and had shut down automatically, its radioactive reactor cores still needed to be cooled by water in order to remain safe. The tsunami not only destroyed power lines, cutting the plant off from the electrical grid, but it also drowned the backup diesel generators, so the cooling pumps stopped functioning. Eventually, water surrounding the reactor cores boiled away. Some of the superheated water separated into hydrogen and oxygen gas, which then exploded, thereby breaching the integrity of three of the four reactor buildings and releasing radioactivity into the environment **(Fig. 8.30b)**.

Because tsunamis are so dangerous, predicting their arrival can save thousands of lives. A tsunami warning center in Hawaii keeps track of earthquakes around the Pacific and uses data relayed from tide gauges and seafloor pressure gauges to determine whether a particular earthquake has generated a tsunami. If observers detect a tsunami, they flash warnings to authorities around the Pacific.

FIGURE 8.30 Damage due to the 2011 Tōhoku tsunami.

(a) This Japanese coastal town was completely destroyed by a tsunami that followed the Tōhoku earthquake of March 2011.

Before

The power plant was built next to the shore.

Reactor building 4

After

Reactor building 4

(b) Each cubic building houses a reactor of the Fukushima nuclear power plant. The tsunami washed over the seawalls and inundated the plant to a depth of 14 m, cutting off power to the cooling pumps. Hydrogen explosions destroyed three of the four reactor buildings. The after photo shows reactor buildings 3 and 4 after destruction.

BOX 8.1 CONSIDER THIS...

The 2010 Haiti Catastrophe

On the sunny afternoon of January 12, 2010, at 4:53 P.M., a 70-km-long segment of a large strike-slip fault suddenly slipped by up to 4 m. The motion began at 25 km west-southwest of Port-au-Prince, and 13 km beneath the ground surface, so the shock waves of the resulting M_W 7 earthquake reached the capital city in a matter of seconds, causing the ground to lurch violently over the next 35 seconds.

The ground shaking caused the support columns of poorly constructed buildings to crack and crumble, bringing floors down into gruesome pancake-like stacks **(Fig. Bx8.1a)**. Similarly,

FIGURE Bx8.1 The disastrous January 2010 earthquake in Haiti, and its geologic setting.

(a) Survivors salvage what they can in Port-au-Prince, the capital of Haiti, after the devastating earthquake of January 2010.

brick or block walls broke apart and collapsed, roads buckled, and hillslopes slumped **(Fig. Bx8.1b)**. Beneath the harbor, sediment liquefied, causing wharfs to sink into the sea. When the shaking, which reached an intensity of IX on the Mercalli scale **(Fig. Bx8.1c)**, finally stopped, most of Port-au-Prince had collapsed. As a dense cloud of white dust slowly rose over the rubble, survivors began the frantic scramble to dig out victims, a task made more hazardous by numerous aftershocks, which measured between M_W 4.5 and 6.1. The renewed shaking caused still-standing but weakened structures to collapse on rescuers. Some estimates place the death toll at 230,000, about 2.5% of the country's population.

Why did the earthquake occur? Haiti sits astride the transform boundary along which the North American Plate moves westward at about 2 cm per year, relative to the Caribbean Plate. Therefore, earthquakes in Haiti are inevitable. But the last major earthquakes on this plate boundary happened about 240 years ago, so stress has been building for quite some time.

The impact of an earthquake on society depends not only on earthquake size but also on the nature of the substrate, on the steepness of the slopes, on construction practices, and on the quality of emergency services in the affected area. Much of Port-au-Prince was built on a basin of weak sediment, which amplified ground movements. Sadly, the buildings in Haiti were not designed to withstand ground vibration. In addition, many of the city's neighborhoods perch on steep slopes, which slid downhill during the quake. In the days that followed, local emergency services were overwhelmed, and access to the victims was nearly impossible. An air caravan of aid arrived to help out, but even so, in the months that followed, disease spread. The country will take many years to recover completely.

(b) Ground shaking during the 2010 Haiti earthquake caused most of the houses in this residential neighborhood to collapse.

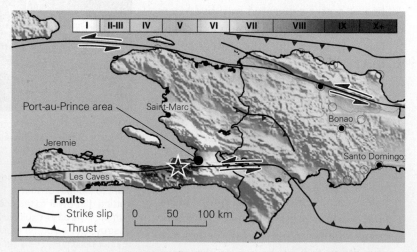

(c) A map showing the intensity of shaking in Haiti on the MMI scale. The star marks the epicenter of the quake.

DISEASE

Once the ground shaking and fires have stopped, disease may still threaten lives in an earthquake-damaged region. Earthquakes cut water and sewer lines, destroying clean-water supplies and exposing the public to bacteria, and they also cut transportation lines, preventing food and medicine from reaching the area. The severity of such problems depends on the ability of emergency services to cope. The lack of sufficient clean water after the 2010 Haiti earthquake contributed to a cholera epidemic later that year **(Box 8.1)**.

TAKE-HOME MESSAGE

Earthquakes cause devastation in many ways. Ground shaking, landslides, sediment liquefaction, and tsunamis can topple buildings and disrupt the land. Fire and disease may follow.

QUICK QUESTION Is ground shaking the major cause of loss of life in all earthquakes?

8.7 Can We Predict the "Big One"?

Can seismologists predict earthquakes? The answer depends on the time frame of the prediction. With their present understanding of the distribution of seismic belts and the frequency at which earthquakes occur, seismologists can make *long-term predictions* (on the time scale of decades to centuries). For example, with some certainty, they can say that a major earthquake will rattle Istanbul during the next 100 years, and that a major earthquake probably won't strike north-central Canada during the next 10 years. But despite extensive research, seismologists cannot make accurate *short-term predictions* (on the time scale of hours to weeks or even decades). Thus, they cannot say, for example, that an earthquake will happen in Montreal next month. In this section, we look at the scientific basis of both long- and short-term predictions and consider the consequences of a prediction. Seismologists refer to studies leading to predictions as *seismic-risk assessment*.

LONG-TERM PREDICTIONS

A long-term prediction estimates the *probability*, or likelihood, that an earthquake will happen during a specified time range. For example, a seismologist may say, "The probability of a major earthquake occurring in the next 50 years in this state is 20%." This implies that there's a 1-in-5 chance that the earthquake will happen before 50 years have passed. Urban planners can use long-term predictions to help create building codes for a region—codes requiring stronger buildings make sense for regions with greater seismic risk. Predictions can help to determine whether to build vulnerable structures such as nuclear power plants, hospitals, or dams in a given region. Seismologists base long-term earthquake predictions on two pieces of information: the identification of seismic belts and the recurrence interval.

The basic premise of long-term earthquake prediction can be stated as follows: a region where many earthquakes have occurred in the past will be likely to experience earthquakes in the future. Seismic belts, regions where many earthquakes happen, are therefore regions of greater seismic risk. This doesn't mean that a disastrous earthquake can't happen far from a seismic belt—they can and do—but the probability is less that an event will happen outside a seismic belt in a given time window. To identify a seismic belt, seismologists produce a map showing the epicenters of earthquakes that have happened during a set period of time (say, 30 years).

The **recurrence interval** is the average time between successive events of a specified size. Since earthquakes do not happen at predictably spaced times, a recurrence interval is not the exact time between successive events. To avoid confusion about the meaning of a recurrence interval, seismologists sometimes use another number, the *annual probability*, to represent seismic risk at a given location. Annual probability equals 1 divided by the recurrence interval. So, if the recurrence interval is 100 years, the annual probability is $1 \div 100 = 1\%$. Note that, since stress builds up over time on a fault, the elastic-rebound theory hints that the annual probability may increase as time passes since the last earthquake.

To determine the recurrence interval for large earthquakes within a given seismic belt, seismologists must note when large earthquakes happened within the belt. For places where the historical record does not provide information far enough back to reveal multiple large events, researchers look for evidence of great earthquakes as preserved in the geologic record. For example, existence of an unweathered fault scarp in an area may indicate that faulting affected the area relatively recently, even if it was before local human habitation. In places where sedimentary strata have accumulated in a basin over a fault, researchers may dig a trench and look for buried layers of sand volcanoes and disrupted bedding in the stratigraphic record. Each layer, whose age can be determined by using radiocarbon dating of plant fragments, records the time of an earthquake **(Fig. 8.31)**. By calculating the average number of years between successive events, seismologists obtain the recurrence interval. Information about a recurrence interval allows seismologists to refine regional maps illustrating seismic risk **(Fig. 8.32)**.

FIGURE 8.31 Evidence of past earthquakes can be used to determine a recurrence interval. The block shows the walls of two trenches cut into the ground at a fault zone.

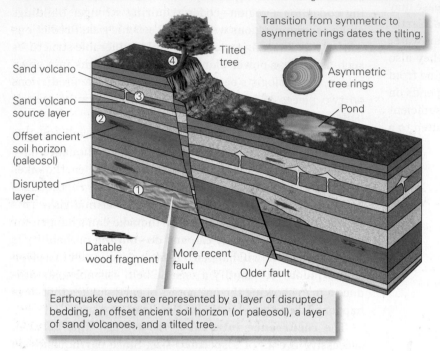

Sand volcano

Sand volcano source layer

Offset ancient soil horizon (paleosol)

Disrupted layer

Tilted tree

Transition from symmetric to asymmetric rings dates the tilting.

Asymmetric tree rings

Pond

Datable wood fragment

More recent fault

Older fault

Earthquake events are represented by a layer of disrupted bedding, an offset ancient soil horizon (or paleosol), a layer of sand volcanoes, and a tilted tree.

sends out emergency signals to nearby populated areas that might be affected. This signal can provide enough warning (several seconds to a minute) for computers to shut down gas pipelines, trains, nuclear reactors, and power lines. The signal can also automatically set off sirens and trigger broadcasts on radio, TV, and cell-phone networks, alerting people to take precautions.

TAKE-HOME MESSAGE

Researchers can determine regions where earthquakes are more likely, but not exactly when and where a given event will occur. Seismic risk is greater where seismicity has happened more frequently in the past and where the recurrence interval of large earthquakes is shorter. Early warning systems can provide seconds of warning by sending out signals that travel faster than seismic waves.

QUICK QUESTION What is the relation between recurrence interval and the annual probability of an earthquake?

SHORT-TERM PREDICTIONS AND WARNINGS

Short-term predictions, specifying that an earthquake will happen on a given date or within a time window of days to years, are not and may never be reliable. Seismologists have considered, and discounted as unreliable, many supposed bases for short-term prediction, but most can be identified only in hindsight. For example, a *swarm* of foreshocks may indicate that rock is beginning to crack in advance of a mainshock, or the surface of the ground may warp slightly prior to an earthquake. But no one knows which swarms are foreshocks or how much warping will take place before an earthquake will happen. Predictions focused on measuring changes in water levels in wells, radon gas in spring water, electrical signals emitted by minerals, or agitation of animals are similarly unsubstantial.

The concept of a short-term prediction should not be confused with the concept of an **earthquake early warning system**, which is based on a real signal and can potentially save lives. An early warning system works as follows: When an earthquake happens, the seismic waves it produces start traveling through the Earth. Seismometers close to the epicenter detect an earthquake before the seismic waves have had time to reach populated areas farther from the epicenter. The instant that seismometers detect the earthquake, a computer sends a signal to a control center, which automatically

8.8 Earthquake Engineering and Zoning

The destruction and loss of human life from an earthquake of a given size vary widely and depend on a number of factors. The most important include the proximity of an epicenter to a population center, the depth of the focus, the style of construction in the epicentral region, whether the earthquake occurred in a region of steep slopes, whether faulting displaced the seafloor, whether building foundations are on solid bedrock or on weak substrate, whether the earthquake happened when people were outside or inside, and whether the government was able to provide emergency services promptly.

For example, a 1988 earthquake in Armenia was not much bigger than the 1971 San Fernando earthquake in southern California, but it caused vastly more deaths (24,000 compared to 65). The difference in death toll reflected differences in the style and quality of construction and the characteristics of the substrate. The unreinforced concrete-slab buildings and masonry houses of Armenia collapsed, whereas the structures in California had, by and large, been erected according to building codes that require structures to withstand stresses caused by earthquakes. Most flexed and twisted but did not fall down and crush people.

FIGURE 8.32 Examples of seismic-hazard maps.

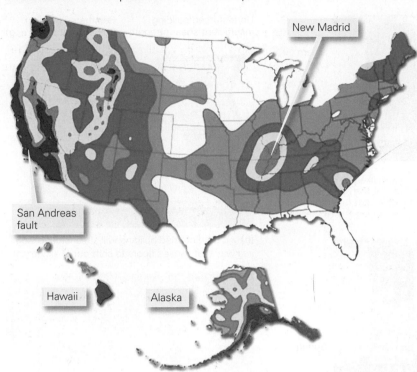

(a) A seismic-hazard map of the United States. Red and pink regions have the greatest probability of experiencing large earthquakes. The San Andreas fault plate boundary is particularly hazardous. Risk in the New Madrid area remains subject to debate.

The terrible 1976 earthquake in T'ang-shan, China, killed over a quarter of a million people, partly because the ground beneath the epicenter had been weakened by coal mining and collapsed, and partly because buildings were poorly constructed. During the 2008 M_W 7.9 Sichuan earthquake that rocked China, 70,000 people died and almost 5 million were left homeless because of building failures. During the 1989 Loma Prieta quake in California, portions of Route 880 in Oakland that were built on a weak landfill collapsed, whereas portions built on bedrock remained standing. Mexico City's 1985 earthquake and a more recent 2017 event were devastating because the city lies over a sedimentary basin whose composition and bowl-like shape focused seismic energy.

Communities can mitigate or diminish the consequences of earthquakes by taking sensible precautions. *Earthquake engineering* (designing buildings that can withstand shaking) and *earthquake zoning* (determining where land is stable and where it is not) can help save lives and property. In regions prone to large earthquakes, buildings and bridges should be constructed to be somewhat flexible so that ground motions can't crack them, but they should have

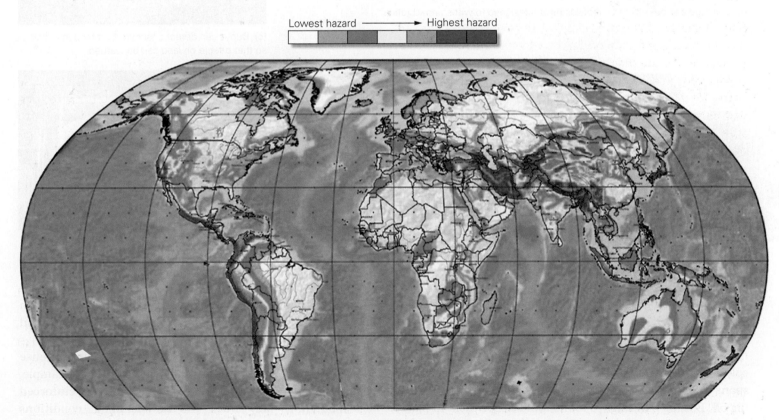

(b) A global seismic-hazard map. The redder regions have the greater probability of experiencing a large earthquake. Probability is high along plate boundaries.

FIGURE 8.33 Preventing damage and injury during an earthquake.

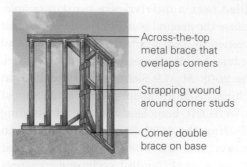

- Across-the-top metal brace that overlaps corners
- Strapping wound around corner studs
- Corner double brace on base

Adding corner struts, braces, and connectors can substantially strengthen a wood-frame house.

Cross-beam

Buildings are less likely to collapse if they are wider at the base and if crossbeams are added for strength.

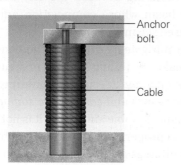

- Anchor bolt
- Cable

Wrapping a bridge's support columns in cable and bolting the span to the columns will prevent the bridge from collapsing so easily.

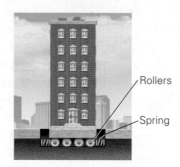

- Rollers
- Spring

Placing buildings on rollers or shock absorbers lessens the severity of the vibrations.

(a) Damage can be prevented if buildings are designed to withstand vibration.

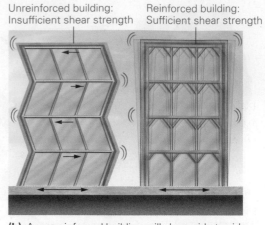

Unreinforced building: Insufficient shear strength   Reinforced building: Sufficient shear strength

(b) An unreinforced building will shear side to side in a way that causes floors to shift out of alignment.

(c) Buoys can detect a tsunami in the open ocean so that people on land can be warned.

(e) If an earthquake strikes, take cover under a sturdy table near a wall.

(d) Signs point out tsunami evacuation routes along the coast of Peru.

sufficient bracing so movements don't become too severe **(Fig. 8.33a, b)**. Also, supports should be strong enough to maintain loads far in excess of the loads caused by static (non-moving) weight. Wrapping steel cables around bridge support columns makes them many times stronger. Bolting bridge spans to the top of a column prevents the spans from bouncing off. Bolting buildings to foundations keeps them in place, and adding diagonal braces to frames minimizes twisting and

shearing. In some cases, shock absorbers in foundations can diminish the amount of energy transferred from the ground to the building.

In regions with significant seismic risk, it makes sense to avoid certain kinds of construction. For example, concrete-block, unreinforced concrete, and unreinforced brick buildings may crack and tumble under conditions in which wood-frame, steel girder, or reinforced concrete buildings remain standing. Traditional heavy, brittle tile roofs can shatter and bury the inhabitants inside, whereas sheet-metal or asphalt-shingle roofs do not. Loose decorative stone and huge open-span roofs also do not fare well when vibrated. Of note, inadequate structures can be made safer by *seismic retrofitting,* the process of strengthening existing buildings in potentially seismically hazardous areas. Examples of retrofitting include adding braces, jacketing support columns, and coating masonry with resins. In some cases, substrates can be strengthened by draining water.

Also, in risk-prone regions, developers should avoid construction on land underlain by weak mud and clay or wet sand that could liquefy. They should not build on top of, on, or at the base of steep escarpments, because the escarpments could fail and tumble downslope in landslides. They should avoid locating communities downstream of dams, which could crack and collapse, causing a flood. Critical buildings (schools, hospitals, fire stations, communications centers, power plants) should never be built over active faults, because fault slip could crack and destroy the buildings. Cities in seismic belts need to prepare emergency plans to deal with disaster, with strategies in place for providing supplies under circumstances where roads may be impassable. And in coastal areas, tsunami warning systems can save lives and property **(Fig. 8.33c, d).**

Finally, individuals should learn to protect themselves during an earthquake. In your home, keep a stock of emergency supplies, bolt bookshelves to walls, strap the water heater in place, install locking latches on cabinets, know how to shut off the gas and electricity, know how to find the exit, have a fire extinguisher handy, and know where to go to find family members. Schools and offices should have earthquake-preparedness drills. When an earthquake strikes, stay away from buildings. If you are trapped inside, crouch beneath a heavy table or solid door frame to be out of the way of falling objects **(Fig. 8.33e).** As long as lithosphere plates continue to move, earthquakes will continue to shake. But we can learn to live with them.

TAKE-HOME MESSAGE

Earthquakes are a fact of life on this dynamic planet. People in regions facing high seismic risk should build on stable ground, avoid unstable slopes, and design construction that can survive shaking. Evacuation planning saves lives after an event.

QUICK QUESTION What factors influence the degree of devastation during an earthquake?

ANOTHER VIEW This apartment building broke apart and tipped over during a major earthquake (magnitude 8.0) that struck Beichuan, Sichuan, China in 2008. Violent earthquakes, which are inevitable on our dynamic planet, can have catastrophic consequences.

Chapter 8 Review

CHAPTER SUMMARY

> Earthquakes are episodes of ground shaking. Most earthquake activity, or seismicity, happens when rock slips during faulting. Rock begins to break at the focus. The point on the ground directly above the focus is the epicenter.

> Active faults are likely to host future movement. Inactive faults ceased being active long ago. Displacement on an active fault that intersects the ground surface may yield a fault scarp.

> During fault formation, rock bends elastically, then cracks. Eventually, cracks link to form a throughgoing fracture on which sliding occurs. When this happens, the elastic rebound generates vibrations (an earthquake).

> Once formed, faults exhibit stick-slip behavior. Most earthquakes happen when stress overcomes friction on a pre-existing fault.

> Earthquake energy travels in the form of seismic waves. Body waves (P and S) pass through the interior of the Earth, whereas surface waves (L and R) pass along the surface of the Earth.

> We can detect earthquake waves by using a seismometer. In principle, a seismometer consists of a weight whose inertia keeps it in place, while the frame around it moves with the Earth's vibrations.

> Seismograms demonstrate that different types of earthquake waves arrive at different times. This difference allows seismologists to pinpoint the epicenter location.

> The Modified Mercalli Intensity scale characterizes earthquake size by perception of ground shaking and of damage.

> Magnitude scales, such as the Richter scale, are based on the amount of ground motion, as indicated by the amplitude of waves traced on a seismogram. The moment magnitude scale takes into account the amplitudes of several seismic waves, the area of slip, and the amount of displacement.

> An M_W 8 earthquake yields about 10 times as much ground motion as an M_W 7 earthquake and releases about 32 times as much energy. The largest earthquake ever measured had a magnitude of M_W 9.5.

> Most earthquakes occur in seismic belts or zones, the majority of which lie along plate boundaries. Intraplate earthquakes happen in the interiors of plates. Different kinds of faulting happen at different kinds of plate boundaries.

> Deep earthquakes only happen in Wadati-Benioff zones.

> Earthquake damage results from ground shaking, landslides, sediment liquefaction, fire, and tsunamis. A large tsunami carries so much water that it can submerge land several kilometers in from the shore.

> Seismologists can identify seismic belts where earthquakes are most likely, and can determine the recurrence intervals for major earthquakes. It may never be possible to pinpoint the exact time and place at which an earthquake will happen.

> Earthquake hazards can be reduced with better construction practices and zoning, and by educating people about what to do during an earthquake.

GUIDE TERMS

aftershock (p. 251)
body wave (p. 255)
compressional wave (p. 255)
displacement (p. 249)
earthquake (p. 247)
earthquake early warning system (p. 280)
earthquake intensity (p. 260)
earthquake magnitude (p. 260)
elastic behavior (p. 250)

elastic-rebound theory (p. 251)
epicenter (p. 255)
fault (p. 247)
fault scarp (p. 249)
focus (p. 254)
foreshock (p. 251)
friction (p. 251)
intraplate earthquake (p. 266)
landslide (p. 269)

mainshock (p. 251)
Modified Mercalli Intensity (MMI) scale (p. 260)
moment magnitude (p. 262)
recurrence interval (p. 279)
Richter scale (p. 261)
sediment liquefaction (p. 270)
seismic belt (p. 263)
seismicity (p. 248)
seismic wave (p. 249)

seismic zone (p. 263)
seismogram (p. 257)
seismologist (p. 248)
seismometer (p. 255)
shear wave (p. 255)
stick-slip behavior (p. 251)
stress (p. 250)
surface wave (p. 255)
travel-time curve (p. 258)
tsunami (p. 273)
Wadati-Benioff zone (p. 265)

GEOTOURS *THIS CHAPTER'S GEOTOURS WORKSHEET (H) FEATURES QUESTIONS AND GOOGLE EARTH SITES ON:*

> Locating earthquakes   > Earthquake activity along a plate boundary   > Intraplate earthquake activity

> Earthquake prediction   > Tsunami devastation

REVIEW QUESTIONS

The letters following each Review Question refer to the corresponding Learning Objective from the Chapter Opener.

1. Compare normal, reverse, and strike-slip faults. **(A)**

2. Describe elastic rebound theory and the concept of stick-slip behavior. **(A)**

3. Identify the focus and epicenter of the earthquake shown in the diagram. **(A)**

4. Explain how the vertical and horizontal components of earthquake motion are detected on a seismometer. **(C)**

5. Explain the differences among the scales used to describe the size of an earthquake. **(E)**

6. How does seismicity on mid-ocean ridges compare with seismicity at convergent or transform boundaries? Do all earthquakes occur at plate boundaries? **(B)**

7. What is a Wadati-Benioff zone? At what depth do the deepest earthquakes occur? **(B)**

8. Describe the types of damage caused by earthquakes. **(F)**

9. What are the four types of seismic waves? Which are body waves, and which are surface waves? Which type is depicted in the diagram? **(D)**

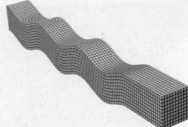

10. What is a tsunami, and why does it form? **(G)**

11. Explain how sediment liquefaction occurs in an earthquake and how it can cause damage. **(F)**

12. How are long-term and short-term earthquake predictions made? What is the basis for determining a recurrence interval, and what does a recurrence interval mean? What is an earthquake early warning system? **(H)**

13. What types of structures are most prone to collapse in an earthquake? What types are most resistant to collapse? **(I)**

ON FURTHER THOUGHT

14. Is seismic risk greater in a town on the west coast of South America or in one on the east coast? Explain your answer. **(B)**

15. The northeast-trending Ramapo fault crops out north of New York City. (You can see the fault on Google Earth by going to Latitude 41°10'21.12" N, Longitude 74°5'12.36" W. The fault trace follows a distinct escarpment.) Where the fault crosses the Hudson River, there is an abrupt bend in the river. A nuclear power plant was built near this bend on sediments deposited by the river. Imagine that you are a geologist with the task of determining the seismic risk of the fault. What evidence of prehistoric seismic activity could you look for? **(B)**

16. On the seismogram of an earthquake recorded at a seismic station in Paris, France, the S-wave arrives 6 minutes after the P-wave. On the seismogram obtained by a station in Mumbai, India, for the same earthquake, the difference between the P-wave and S-wave arrival times is 4 minutes. Which station is closer to the epicenter? From the information provided, can you pinpoint the location of the epicenter? Explain. **(D)**

17. Will the duration of shaking recorded at a seismic station be longer or shorter as the distance between the epicenter and the station increases? **(D)**

ONLINE RESOURCES

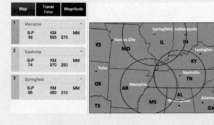

Animations This chapter features animations that simulate and show the effects of tsunamis and an interactive activity on locating an earthquake's epicenter.

Videos This chapter features videos on tsunami awareness and earthquake impacts.

Smartwork5 This chapter features analysis questions on earthquake magnitude, faults, and tsunamis.

Magnetic anomaly
strength (nanoteslas)

+400
+100
+50
0
−50
−100
−400

INTERLUDE D

THE EARTH'S INTERIOR REVISITED: INSIGHTS FROM GEOPHYSICS

By the end of this interlude, you should be able to . . .

A. explain how seismic waves behave as they pass through the Earth's interior.

B. describe how the study of seismic waves can tell us about layering inside the Earth.

C. characterize the Earth's internal layers.

D. provide a model illustrating how the surface elevation of lithosphere generally represents the consequences of Archimedes' principle.

E. explain how gravitational attraction varies with location, and what these variations mean.

F. explain why the Earth's magnetic field exists, and why its strength varies with location.

D.1 **Introduction**

In this interlude, we briefly survey the insight that the study of geophysics can provide to help refine our image of the Earth's interior. The term **geophysics** refers to a subdiscipline of geoscience that focuses on the study of seismic waves, magnetism, gravity, and other physical characteristics of the Earth. Geophysicists use mathematical calculations, measurements with instruments, and computer simulations to carry out their work. (Seismology, the subject of Chapter 8, is one aspect of geophysics.) At the level of this book, we can't delve into the mathematical foundation of geophysics, but we can introduce some of the key concepts that geophysical studies provide.

We begin this interlude by examining how seismic waves interact with layer boundaries, in a general sense, and then discuss how studying seismic waves can define the specific depths of layer boundaries within the Earth. We next turn our attention to the Earth's gravity field and examine why gravitational pull varies with location. We conclude by revisiting the Earth's magnetic field, this time focusing on the question of why the magnetic field exists. Interlude D ties together several subjects covered in Chapters 1, 2, and 8, so we recommend that you read (or reread) relevant parts of those chapters before proceeding.

D.2 **The Basis for Seismic Study of the Interior**

SETTING THE STAGE

As we discussed in Chapter 1, the first clues to what's inside our planet came from measuring the Earth's overall mass and shape. These measurements allowed researchers to conclude, by the end of the 19th century, that the Earth consists of three concentric layers that differ from one another in terms of relative density **(Fig. D.1)**. From the surface down, these are the *crust*, which has a low density; the *mantle*, which has an intermediate density; and the *core*, which has a high density. To go beyond this basic understanding, and to define the specific depths at which layer boundaries occur, researchers searched for a tool that could provide an actual image of the interior. The study of seismic waves provides that tool.

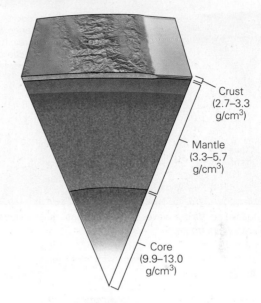

FIGURE D.1 Simplified 19th-century image of the Earth's interior. Geologists realized there were three layers, but they didn't know the depths of the boundaries.

By measuring how fast seismic waves travel through the Earth, and how the waves bend or reflect as they travel, it's possible to define the thicknesses of the main layers, and even to recognize sublayers.

SEISMIC WAVE FRONTS AND TRAVEL TIMES

As described in Chapter 8, the energy released by an earthquake moves through rock in the form of waves, similar to the waves that propagate outward from the impact point of a pebble on the surface of a pond. The boundary between the rock through which a wave has passed and the rock through which it has not yet passed is called a *wave front*. In three dimensions, a wave front initially expands outward from the earthquake focus like a growing bubble. We can represent a succession of wave fronts close to the focus by thinking about a series of concentric spheres or, in cross section, circles **(Fig. D.2a)**. The changing position of an imaginary point on a wave front, as the front moves through rock, is called a **seismic ray**. To represent a seismic ray, we draw a line perpendicular to a wave front—each point on a curving wave front follows a slightly different ray. The time it takes for a wave to travel from the earthquake focus to a seismometer along a given ray is the **travel time** along that ray.

The ability of a seismic wave to travel through a material, as well as the velocity at which it travels, depend on several characteristics of that material. Factors such as *density* (mass per unit volume), *rigidity* (how stiff or resistant to bending a material is), and *compressibility* (how easily a material's volume changes in response to squashing) all affect seismic-wave

FIGURE D.2 The propagation of earthquake waves.

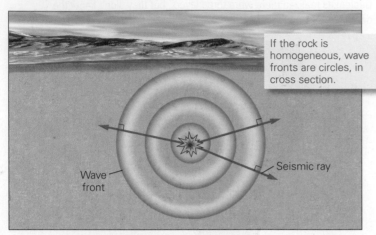

If the rock is homogeneous, wave fronts are circles, in cross section.

Wave front

Seismic ray

(a) An earthquake sends out waves in all directions. Seismic rays are perpendicular to wave fronts.

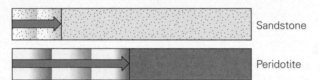

Sandstone

Peridotite

(b) Seismic waves travel at different velocities in different rock types. After a given time, the wave will have traveled farther in peridotite than in sandstone.

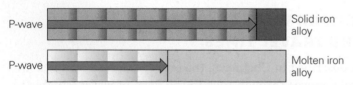

P-wave

Solid iron alloy

P-wave

Molten iron alloy

(c) P-waves travel faster in solid iron alloy than in liquid, such as molten iron alloy.

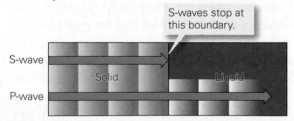

S-waves stop at this boundary.

S-wave

Solid

Liquid

P-wave

(d) Both P-waves and S-waves can travel through a solid, but only P-waves can travel through a liquid.

velocity, and therefore, travel times. Studies of seismic waves reveal the following:

> Seismic waves travel at different velocities in different rock types **(Fig. D.2b)**. For example, P-waves travel at 8 km per second in peridotite (an ultramafic igneous rock) but at only 3.5 km per second in sandstone (a porous sedimentary rock). Therefore, waves accelerate or slow down if they pass from one rock type into another.

> Seismic waves travel more slowly in a liquid than in a solid of the same composition. So seismic waves travel more

slowly in magma than in solid rock, and more slowly in molten iron alloy than in solid iron alloy **(Fig. D.2c)**.

> Both P-waves and S-waves can travel through a solid, but only P-waves can travel through a liquid **(Fig. D.2d)**. That's because one part of a liquid can easily move sideways relative to another part, so the energy of a shear motion, the motion in an S-wave, cannot be transmitted from one part of a liquid to another.

REFLECTION AND REFRACTION OF WAVE ENERGY

Shine a flashlight into a container of water so that the light ray hits the boundary between water and air at an angle. Some of the light bounces off the water surface and heads back up into the air, while some enters the water **(Fig. D.3a)**. The light ray that enters the water bends at the air-water boundary. This means that the angle between the ray and the boundary in the air is different from the angle between the ray and the boundary in the water. Physicists refer to the light ray that bounces off the

FIGURE D.3 Refraction and reflection of waves.

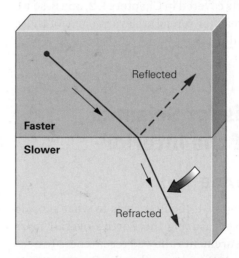

Reflected

Faster

Slower

Refracted

A ray that enters a material through which it travels more slowly bends away from the boundary.

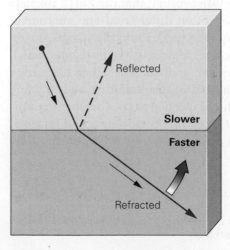

Reflected

Slower

Faster

Refracted

A ray that enters a material through which it travels more rapidly bends toward the boundary.

air-water boundary and heads back into the air as the *reflected light*, and the ray that bends at the boundary is called the *refracted light*. The phenomenon of bouncing is **reflection**, and the phenomenon of bending is **refraction**. Ray reflection and refraction take place at the interface between two materials if the wave travels at different velocities in the two materials.

The angle at which a reflected ray bounces off a boundary is always the same as the angle at which the incoming, or *incident*, ray strikes the surface. The angle by which a refracted ray bends at a boundary, however, depends on the contrast in wave velocity between the two materials in contact at the boundary, and on the angle at which a ray hits the boundary. As a rule, if a ray enters a material through which it will travel more slowly, the ray bends down and away from the interface **(Fig. D.3b)**. (To see why, picture a car driving from a paved surface diagonally onto a sandy beach—the wheel that rolls onto the sand first slows down, relative to the wheel still on the pavement, causing the car to veer toward the sand.) For example, the light ray in Figure D.3b bends down when hitting the air-water boundary because light waves travel more slowly in water. Alternatively, if the ray were to pass from a layer in which it travels slowly into one in which it travels more rapidly, it would bend up and toward the boundary.

D.3 **Seismic Study of the Earth's Interior**

Let's use your knowledge of seismic velocity, refraction, and reflection to see how each of the major boundaries inside the Earth was discovered.

DISCOVERING THE CRUST-MANTLE BOUNDARY

In 1909, Andrija Mohorovičić, a Croatian seismologist, noted that P-waves arriving at seismometer stations less than 200 km from the epicenter traveled at an average speed of 6 km per second, whereas P-waves arriving at seismometers more than 200 km from the epicenter traveled at an average speed of 8 km per second. To explain this observation, he suggested that P-waves reaching nearby seismometers followed a shallow path that kept them entirely within the crust, in which they traveled relatively slowly. However, P-waves reaching distant seismometers had refracted at the crust-mantle boundary, and for part of their route they had passed through the mantle, in which they traveled relatively rapidly **(Fig. D.4a, b)**. Mohorovičić was able to calculate the depth of the crust-mantle boundary from this observation

FIGURE D.4 Discovery of the Moho.

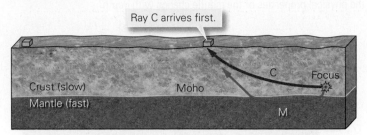

(a) Seismic waves traveling only in the crust reach a nearby seismometer first.

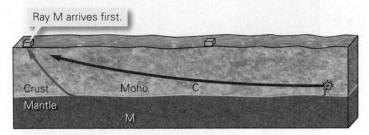

(b) Seismic waves traveling for most of their path in the mantle reach a distant seismometer first.

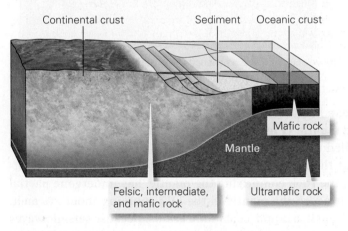

(c) Oceanic crust is thinner than continental crust and has a different composition. The lower part of the continental crust tends to be more mafic than the upper.

and proposed that, beneath continents, it occurred at a depth of about 35 to 40 km. Later studies showed that the depth of the crust-mantle boundary beneath continents varies from 25 to 70 km, and beneath oceans it varies from about 7 to 10 km **(Fig. D.4c)**. As we learned in Chapter 1, the crust-mantle boundary is called the **Moho**, in honor of Mohorovičić.

DEFINING THE STRUCTURE OF THE MANTLE

Seismologists have determined that seismic waves travel at different speeds at different depths in the mantle. Between a depth of about 100 and 200 km in the mantle beneath the

FIGURE D.5 The velocity of P-waves in the mantle changes because the physical properties of the mantle change with depth.

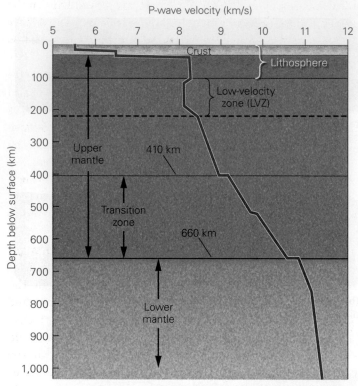

(a) The velocity of P-waves changes with depth in the mantle.

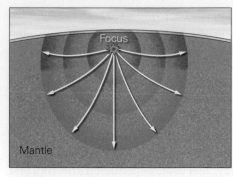

(b) The velocity of seismic waves increases with depth in the mantle, so rays curve and wave fronts are oblong.

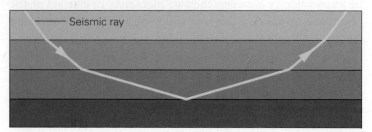

(c) In a stack of discrete layers, rays bend at each boundary. If the velocity is progressively faster in each lower layer, the ray eventually bends back and returns to the surface.

(d) In a material in which the velocity increases gradually with depth, rays curve smoothly and eventually return to the surface.

ocean floor, seismic velocities are slower than in the overlying mantle **(Fig. D.5a)**. This 100- to 200-km-deep layer is called the **low-velocity zone (LVZ)**. Seismologists suggest that the LVZ corresponds to an interval in which peridotite, the rock comprising the mantle, has undergone partial melting, so the mantle in the LVZ contains about 2% melt. This melt, a liquid, coats solid grains. Because seismic waves travel more slowly through liquids than through solids, the coatings of melt slow seismic waves down. In the context of plate tectonics theory, the top of the LVZ delineates the base of the lithosphere beneath oceanic plates—lithospheric mantle lies above this boundary, and asthenospheric mantle lies below. Therefore, the LVZ serves as the weak layer on which oceanic lithosphere plates move. Seismologists do not find a well-developed LVZ beneath continents, meaning that the base of the lithosphere is not delineated by a change in seismic velocity beneath continents.

Below about 200 km, seismic-wave velocities throughout the mantle increase progressively with depth **(Fig. D.5b)**. Because of this change, seismic wave fronts travel faster in the down direction than they do in the up direction or to the sides. As a result, after a seismic wave front has moved some distance from the focus, it becomes elliptical. Seismologists interpret the increase in velocity with depth to mean that mantle peridotite becomes progressively less compressible,

more rigid, and denser with depth. This idea makes sense, considering that the weight of overlying rock increases with depth, and as pressure increases, the atoms making up minerals squeeze together more tightly and are less free to move.

The increase in seismic velocity with depth causes seismic waves to refract, so seismic rays curve in the mantle. To understand the shape of a curved ray, let's represent a portion of the mantle by a series of imaginary layers, each of which has a greater seismic-wave velocity than the layer above **(Fig. D.5c)**. Every time a seismic ray crosses the boundary between adjacent layers, it refracts slightly toward the boundary. After the ray has crossed several layers, it has bent so much that it begins to head back up toward the top of the stack. If we replace the stack of distinct layers with a single layer in which velocity increases with depth at a constant rate, the ray follows a smoothly curving path **(Fig. D.5d)**.

At depths of between 410 km and 660 km in the mantle, seismic velocity increases in a series of abrupt steps (see Fig. D.5a), so in this interval, the stack of layers depicted in Figure D.5c actually provides a somewhat realistic image. Experiments suggest that such **seismic-velocity discontinuities** occur at depths where pressure causes atoms in minerals to rearrange and pack together more tightly. This results in a different mineral with a more compact structure—a process called a *phase change* (see Chapter 7). Where such phase changes take place, the overall characteristics of mantle rock change, and seismic waves travel at a different velocity. Because of these seismic-velocity discontinuities, seismologists subdivide the mantle into the **upper mantle** (above 660 km) and the **lower mantle** (below 660 km). The lower portion of the upper mantle (between 410 and 660 km), in which several small seismic discontinuities have been recognized, is called the **transition zone**.

DISCOVERING THE CORE-MANTLE BOUNDARY

During the first decade of the 20th century, seismologists installed seismometer stations around the world, expecting to be able to record waves produced by a large earthquake anywhere on the Earth. In 1914, one of these seismologists, Beno Gutenberg, noticed that P-waves from an earthquake do not arrive at seismometer stations lying in a band at a distance of between 103° and 143° from the epicenter, as measured along the surface of the Earth. This band is now called the **P-wave shadow zone (Fig. D.6a)**. If the density of the Earth increased gradually with depth all the way to the center, the shadow zone would not exist, because rays passing through the interior would reach every point on the surface. The presence of a shadow zone, therefore, means that deep in the Earth, a major interface exists at which seismic waves abruptly refract downward, implying that the velocity of the waves suddenly decreases. This interface, the **core-mantle boundary**, lies at a depth of about 2,900 km.

To see why the P-wave shadow zone exists, follow the two seismic rays labeled A and B in Figure D.6a. Ray A curves smoothly in the mantle (we are ignoring seismic-velocity discontinuities in the mantle, for simplicity) and passes just above the core-mantle boundary before returning to the surface. It reaches the surface at 103° from the epicenter. In contrast, Ray B penetrates the boundary and refracts down into the core. Ray B then curves through the core and refracts again when it crosses back into the mantle. As a consequence, Ray B intersects the surface at more than 143° from the epicenter.

DISCOVERING THE NATURE OF THE CORE

Based on the study of meteorites thought to be fragments of a large planetesimal's interior, and on density calculations of the Earth, researchers have concluded that the core consists of iron alloy. This alloy contains about 85% iron, 5% nickel, and 10% of lighter elements (probably oxygen, silicon, and sulfur). The downward bending of seismic waves when they pass from the mantle down into the core indicates that seismic velocities, at least in the outer core, are slower than in the mantle. Therefore, even though the core is deeper and denser than the mantle, at least the outer part of the core must be less rigid than the mantle. How can this be?

Seismologists have found that S-waves do not arrive at stations located beyond 103° from the epicenter, defining a band called the **S-wave shadow zone**. This means that S-waves cannot pass through the core at all. If they could, an S-wave headed straight down through the Earth should reach the ground surface on the other side of the planet. S-waves are shear waves, which by their nature can travel only through solids. So, the fact that S-waves do not pass through the core means that the core, or at least part of it, consists of liquid **(Fig. D.6b)**.

At first, seismologists thought that the entire core might be liquid iron alloy. But in 1936, a Danish seismologist, Inge Lehmann, discovered that P-waves passing through the core reflected off a boundary within the core. She proposed that the core is made up of two parts: an *outer core* consisting of liquid iron alloy, and an *inner core* consisting of solid iron alloy. Lehmann's work defined the existence of the inner core but could not locate the depth at which the inner core–outer core interface occurs. This depth was eventually located by measuring the exact time it took for seismic waves to penetrate the Earth, bounce off the inner core–outer core boundary, and return to the surface **(Fig. D.6c)**. The measurements showed that this boundary occurs at a depth of about 5,155 km.

Why does the core have two layers—a liquid outer one and a solid inner one? An examination of **Figure D.6d** provides some insight. This graph shows two curves: the *geotherm* indicates how temperature changes with increasing depth, and the *melting curve* indicates how the temperature at which a material comprising the Earth's interior begins to melt changes with increasing depth. As the graph shows, the geotherm lies to the left of the melting curve through most of the mantle and in the inner core. This means that the temperature in most of the mantle and in the inner core does not become high enough to cause melting, under the very high pressures found in these regions, so these regions are solid. But the geotherm lies to the right of the melting curve in the low-velocity zone of the mantle and in the outer core, so these regions contain molten material.

A MODERN IMAGE OF THE EARTH'S LAYERS

Through painstaking effort, seismologists used data on travel times to develop the whole-Earth *velocity-versus-depth curve*, which can be plotted on a graph with depth on the vertical axis and seismic velocity on the horizontal axis. This graph

FIGURE D.6 Shadow zones and the discovery of the Earth's core. (Note that the circumference of a circle is 360°, so it is 180° from a given point to a locality on the other side of the planet.)

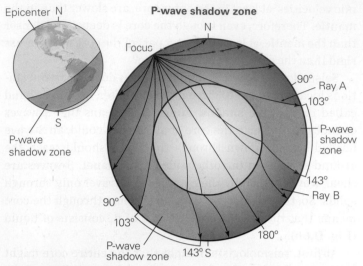

P-wave shadow zone

(a) P-waves do not arrive in the P-wave shadow zone because they refract at the core-mantle boundary.

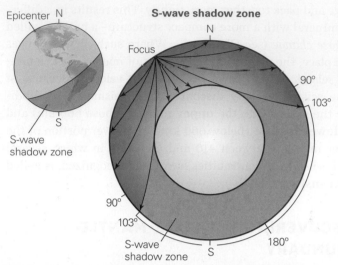

S-wave shadow zone

(b) S-waves do not arrive in the S-wave shadow zone because they cannot pass through the liquid outer core.

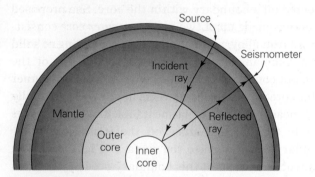

(c) Seismic waves reflect off the inner core–outer core boundary.

shows the average depths at which seismic velocity suddenly changes, and the average amount of change. Depths at which major changes take place define the principal layers and sub-layers of the Earth, down to its center **(Fig. D.7)**. Note that the graph does not show a velocity for S-waves in the outer core, because S-waves cannot travel through molten iron (a liquid).

SEISMIC TOMOGRAPHY

In recent years, seismologists have developed a technique, called **seismic tomography**, to produce three-dimensional images of variation in seismic-wave velocities in the Earth's interior. This technique resembles the method used to produce three-dimensional CAT (or CT) scans of the human body. In seismic tomographic studies, seismologists compare the observed travel time of seismic waves following a specific ray path with the predicted travel time that waves following the same path would have if the average velocity-versus-depth model depicted by Figures D.5 and D.7 were completely correct. Seismologists can distinguish locations within the

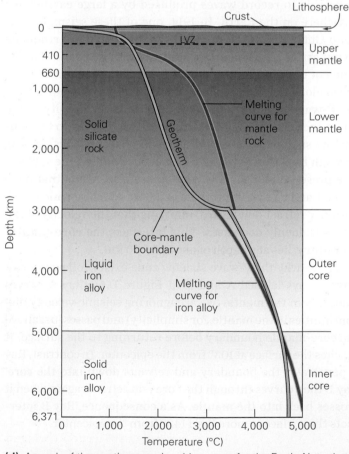

(d) A graph of the geotherm and melting curve for the Earth. Note that the melting temperature is less than the Earth's temperature in the outer core, so the outer core is molten.

Earth where seismic waves travel faster than expected from regions where the waves travel slower than expected.

FIGURE D.7 The velocity-versus-depth profile of the whole Earth.

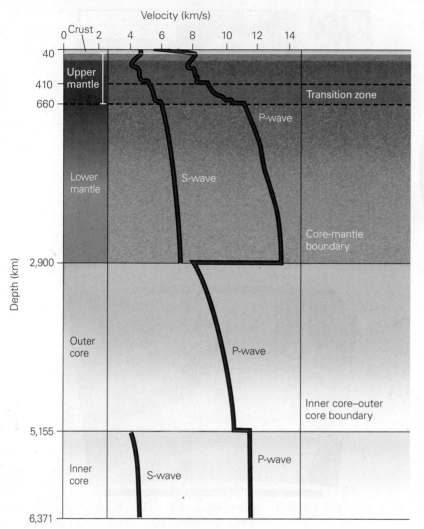

mometers in an array across the United States. Because the seismometers are relatively closely spaced, they record the velocity of seismic waves along a much greater number of ray paths, and therefore can provide an image of the interior with higher resolutions, showing more detail.

SEISMIC-REFLECTION PROFILING

During the past half century, geologists have found that by exploding dynamite, by banging large weights against the Earth's surface, or by releasing bursts of compressed air into the ocean, they can produce artificial seismic waves that propagate down into the Earth and reflect off the boundaries between layers of rock in the crust. By recording the time at which these reflected waves return to the surface, geologists can determine the depth to these boundaries. With this information, they can produce a cross-sectional view of the crust called a **seismic-reflection profile (Fig. D.9)**. This image can define subsurface bedding, stratigraphic formation contacts, folds (bends in layers), and faults. Oil and gas companies use seismic-reflection profiles, despite their high cost, to identify likely places where energy reserves have accumulated underground.

In recent years, computers have become so powerful that geologists can produce three-dimensional seismic-reflection images of the crust. These images provide enough detail that they can even show an individual ribbon of sandstone, representing the channel of an ancient stream, now buried several kilometers below the Earth's surface.

Tomographic studies emphasize that the simple onion-like layered image of the Earth that we've shown so far, in which velocities increase with depth at the same rate everywhere, is an oversimplification. In reality, the velocities of seismic waves vary significantly with location at a given depth. Tomographic studies can display these variations in maps, cross sections, or three-dimensional models **(Fig. D.8)**. Generally, reds on these images indicate slower regions, whereas blues and purples indicate faster regions. Researchers interpret the slower regions to be warmer regions of the mantle, and faster regions to be cooler regions, for as rock gets hotter and softer, it can't transmit seismic waves as rapidly. The occurrence of warm and cold regions is a consequence of convection in the mantle, so seismic tomography has led researchers to picture the inside of the Earth as a dynamic place.

The image of the Earth's interior has become even clearer, in recent years, due to a major research project called **EarthScope**. This project involves placing hundreds of seis-

D.4 **The Earth's Gravity**

The Earth emanates two field forces, gravity and magnetism. Recall that a *field force* is a push or pull that applies across a distance. A field force can cause an object to accelerate (change speed or direction) without ever touching it. In the remainder of this interlude, we see how measurements of variations in the magnitudes of these forces provide important clues to the nature of the Earth's interior. In this section, we discuss gravity, and in the next, we'll turn our attention to magnetism.

THE GEOID

Gravity is an attractive field force that one mass exerts on another. As Isaac Newton showed, the magnitude of gravitational pull depends on the size of the masses and on the distance between them—larger masses exert stronger pulls than do smaller ones, and closer masses produce stronger pulls

FIGURE D.8 Tomographic images of the Earth's interior and their interpretation.

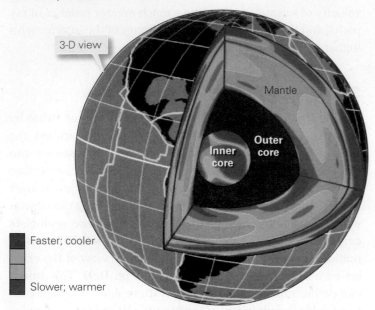

3-D view

Mantle

Inner core

Outer core

Faster; cooler

Slower; warmer

(a) A three-dimensional tomographic image of the Earth.

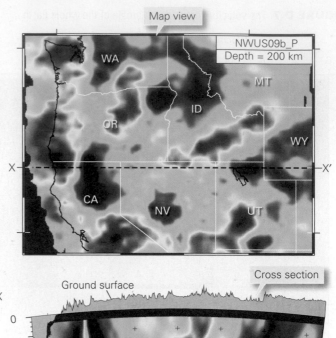

Map view

NWUS09b_P
Depth = 200 km

WA

MT

OR

ID

WY

X —— X'

CA

NV

UT

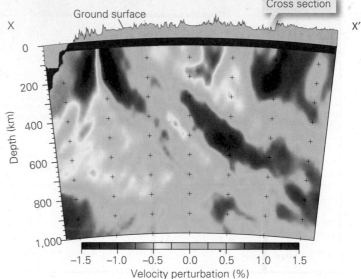

Cross section

Ground surface

X —————— X'

0

Depth (km)

200

400

600

800

1,000

−1.5   −1.0   −0.5   0.0   0.5   1.0   1.5
Velocity perturbation (%)

(b) A tomographic map (at a depth of 200 km) (upper image) and a cross section of the Earth beneath the western United States (lower image), constructed using tomographic data from the EarthScope array.

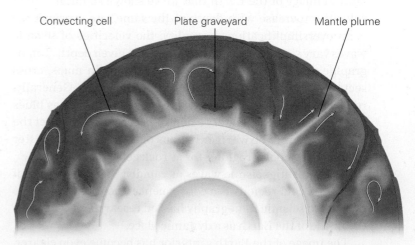

Convecting cell   Plate graveyard   Mantle plume

(c) An artist's rendition of a modern view of the complex and dynamic interior of the Earth. Note the convecting cells, the mantle plumes, and the graveyards of subducted plates.

than do farther ones. When gravity acts on an object but can't make the object move, then the object stores *gravitational potential energy*. For example, a boulder sitting at rest on a hillslope has gravitational potential energy. A surface on which all points have the same gravitational potential energy, such as the surface of standing water on a flat sea, is an **equipotential surface**.

Is the real surface of the Earth an equipotential surface? In other words, is the pull of gravity the same everywhere on the Earth? No. Many different phenomena cause variation in the amount of gravitational pull with location. To start with, our planet is not a perfect sphere, in part because it spins on an axis. This rotation produces centrifugal force, which flattens the Earth. If the Earth were perfectly smooth and homogeneous, this flattening would transform the planet into a *spheroid*—a slice through a spheroid is an ellipse **(Fig. D.10a)**. The imaginary spheroid that serves as the best match to the Earth's overall shape has a radius at the equator of 6,378 km and a radius at the pole of 6,357 km. Geologists refer to this imaginary equipotential surface as the *reference spheroid*.

While the reference spheroid serves as a starting point for describing spatial variations in the Earth's gravitational pull, it doesn't provide a completely accurate picture for many reasons. First, the density of material within the Earth's layers is not uniform throughout the layers. Second, the surface of the Earth is not smooth, for the land surface is higher than the seafloor surface, and both the land and the seafloor have mountains and valleys. Third, rock making up the outermost

FIGURE D.9 Seismic-reflection profiling. The method allows geologists to see layers underground.

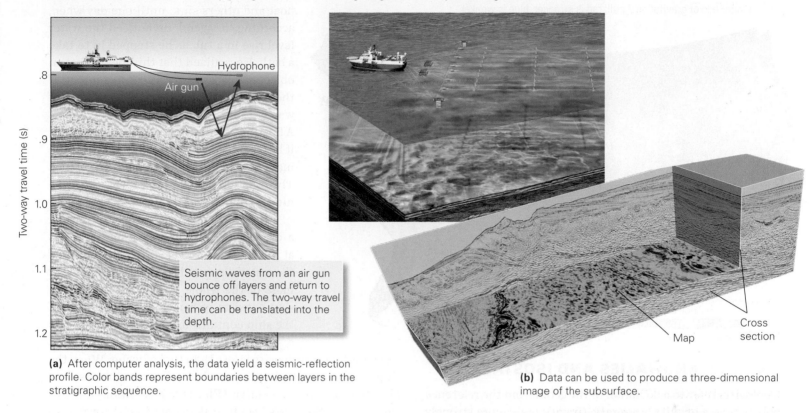

Two-way travel time (s)

Hydrophone

Air gun

Seismic waves from an air gun bounce off layers and return to hydrophones. The two-way travel time can be translated into the depth.

Map

Cross section

(a) After computer analysis, the data yield a seismic-reflection profile. Color bands represent boundaries between layers in the stratigraphic sequence.

(b) Data can be used to produce a three-dimensional image of the subsurface.

layer of the Earth, the lithosphere, is strong enough to hold up heavy loads such as volcanoes or glaciers, or to allow depressions such as trenches to persist. Because of these factors, the Earth at a particular location can exert a greater pull or a lesser pull than the reference spheroid predicts. Therefore, the real equipotential surface for the Earth has bumps and dimples. Geologists refer to this irregular surface, which

provides the best representation of the Earth's equipotential surface, as the **geoid (Fig. D.10b)**. The surface of the Earth's geoid differs from that of the reference spheroid only slightly—the highest point on the geoid lies 85 m above the reference spheroid, and the lowest point lies 107 m below the reference spheroid. But even these small differences can noticeably influence the orbits of satellites or the accuracy of surveys.

FIGURE D.10 Representations of the geoid, the shape of an equipotential surface representing the Earth's gravity. Colors represent elevations of the geoid relative to the reference spheroid.

−100   −50   0   50 meters

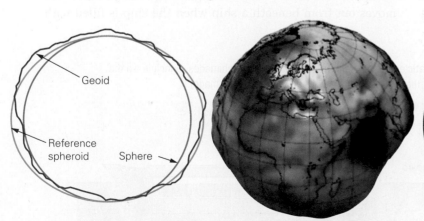

Geoid

Reference spheroid

Sphere

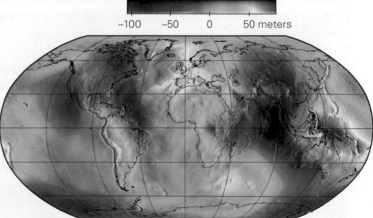

(a) The geoid is a spheroid with bumps and dimples (greatly exaggerated here). A 3-D view emphasizes the lows (blue) and highs (red).

(b) A map of the geoid. Note the deep low in the Indian Ocean and the high in the northeastern Atlantic.

FIGURE D.11 A gravity anomaly map of the United States. Colors represent variations in the magnitude of gravitational pull: red is stronger, blue is weaker.

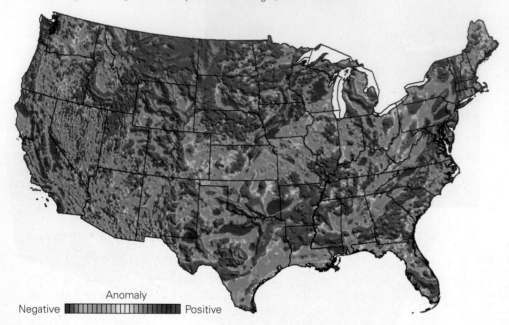

Negative ▮▮▮▮▮▮▮▮▮▮▮▮▮▮▮ Positive
Anomaly

GRAVITY ANOMALIES AND ISOSTASY

Geologists refer to a deviation of the geoid from the reference spheroid as a **gravity anomaly**. Gravity pulls more strongly over a *positive gravity anomaly* and pulls less strongly over a *negative gravity anomaly*. A positive anomaly indicates that extra mass lies below the site, perhaps due to a body of particularly dense rock underground, whereas a negative anomaly means that a deficit of mass lies below the site, perhaps due to the presence of less-dense rock **(Fig. D.11)**. Some gravity anomalies are due to variations in rock composition within the crust, while others are due to variations in temperature in the mantle, for both composition and temperature can affect density.

To better interpret gravity anomalies, we first need to introduce the concept of isostasy. Archimedes (ca. 287–212 B.C.E.), an ancient Greek mathematician, had been puzzling over the question of why some objects float and others sink, until one day when, according to legend, he noticed the water level rise in a tub as he stepped in to take a bath. He realized that this meant that an object displaces water as it sinks into it. If the density of the object exceeds that of water, it sinks, but if it is less dense, it floats. A floating object sinks into water only until it has displaced an amount of water whose mass equals the mass of the whole object. This relationship is now known as **Archimedes' principle**.

According to Archimedes' principle, a cargo ship anchored in a harbor floats at just the right level so that the mass of the water displaced by the ship equals the mass of the whole ship. Even though the ship consists of heavy steel, the inside of the ship contains air. In effect, the ship is a steel-sheathed bubble—it floats because its overall average density is less than that of water. When the ship remains empty, its *freeboard*, the distance that its deck lies above the water, is large **(Fig. D.12)**. Addition of heavy cargo causes the ship's keel to sink deeper into the water, so the deck moves down, and the freeboard decreases. This movement can take place because the water beneath the ship can flow out of the way as the ship settles downward. When the freeboard of the ship becomes just right for a given cargo, we say that the ship is in *isostatic equilibrium*, or that a condition of **isostasy** exists. (The word comes from the Greek *iso*, meaning equal.)

As you learned in Chapter 2, we can think of the Earth's outer layer, the lithosphere, as a relatively rigid shell that "floats" on the underlying, relatively soft asthenosphere. Asthenosphere can flow out of the way of sinking lithosphere or flow in to fill space under rising lithosphere, just as water moves out from beneath a ship when the ship is filled with

FIGURE D.12 A cargo ship illustrates the concept of isostasy. At isostatic equilibrium, the freeboard reflects Archimedes' principle, so the freeboard decreases as the mass of the ship increases. Water flows out of the way as the ship sinks downward (right image).

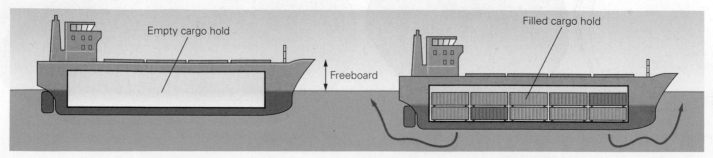

Empty cargo hold

Freeboard

Filled cargo hold

cargo and sinks down, or flows in to fill space when the ship rises as it's unloaded.

The lithosphere approaches a condition of isostatic equilibrium in many places. Where this happens, the elevation of the lithosphere's surface reflects the density and thickness of the lithosphere below. In such locations, no gravity anomaly exists. In order to maintain isostasy, addition of a load, such as growth of an ice sheet or building of a volcano, causes the lithosphere's surface to sink, whereas removal of a load, say by melting of an ice sheet, causes the lithosphere's surface to rise. But, since asthenosphere flows very slowly, isostasy can't be achieved instantly. In addition, because the lithosphere has strength, it can support small, heavy loads in places, or keep relatively low-density materials from rising, without the surface of the lithosphere sinking or rising overall. Places on the Earth where isostasy does not exist are the places where we observe gravity anomalies.

Geophysicists can use the study of gravity anomalies to characterize important structures related to plate boundaries or to delineate areas where a particular rock type is present. For example, a large negative gravity anomaly occurs over trenches at convergent plate boundaries. There, subducting plates are bent downward to a depth greater than where they would be if at isostatic equilibrium. The deep water over a trench has less density than rock, so it does not produce as much gravitational pull. Significant positive gravity anomalies occur over continental rifts filled with thick successions of dense basalt layers, because the dense basalt produces a stronger gravitational pull than does the less-dense surrounding rock.

D.5 The Earth's Magnetic Field, Revisited

As we discussed in Chapters 1 and 2, the Earth produces a magnetic field which, like the field produced by a toy bar magnet, can be represented by a dipole with a north end and a south end, and by field lines that curve through space **(Fig. D.13a)**. We can indicate the dipole by an arrow that points from the north magnetic pole to the south magnetic pole. Here, we discuss why the field exists and why variations occur in the strength of the field that we measure at the Earth's surface.

ORIGIN OF THE EARTH'S MAGNETIC FIELD

Why does the Earth have a magnetic field? The path toward discovering the origin of the Earth's field started in 1926, when researchers realized that our planet's outer core consists of liquid iron alloy. Physicists demonstrated that flow of liquid metal can produce an electric current, so it seemed that the flow of the outer core must be responsible for the magnetic field. But how?

Insight into the generation of the Earth's magnetic field comes from examining how an electric power plant works. In a power plant, the flow of water, wind, or steam spins a coil made of metal (an electrical conductor) around an already magnetic iron bar. The iron bar is a *permanent magnet*, meaning that it always produces a magnetic field, even in the absence of an electric current. The motion of the wire in the bar's magnetic

FIGURE D.13 The Earth's magnetic field may be due to spiral-like convection cells in the outer core.

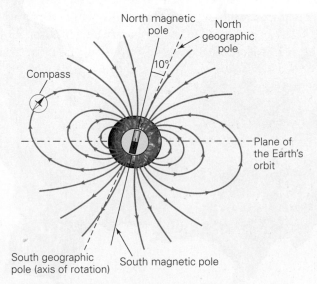

(a) The geometry of the Earth's dipole field.

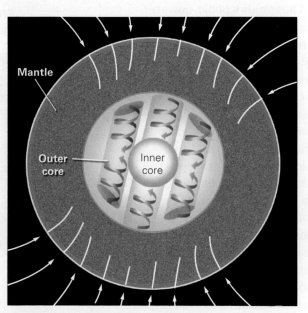

(b) An artist's image of spiral convection cells in the outer core.

field generates an electric current in the wire—we use that current to power our homes. An apparatus that produces electricity by moving a wire around a magnet is called a **dynamo**. In the case of the Earth, flow in the outer core serves the role of the spinning wire coil, making the outer core behave like an electromagnet. But the inner core can't serve the role of a permanent magnet, because it is way too hot—materials can only be permanent magnets at relatively low temperatures. Researchers suggest, therefore, that the Earth is a *self-exciting dynamo*. Evidently, during the Earth's early history, flow in the outer core took place in the presence of a magnetic field. This flow generated an electric current, which in turn generated a magnetic field. (A physics book provides further detail on the relationship between electricity and magnetism.) Continued flow in the presence of the generated magnetic field produced more electric current, which maintained the magnetic field. Once started, the system perpetuates itself, as long as an input of energy can keep the outer core in motion. The input of energy comes from variations in temperature and composition in the outer core that make the outer core convect.

At present, the magnetic poles of the Earth are close to, but not exactly on, the geographic poles. The poles are moving at about 55 km per year, so their position constantly changes. Geophysicists suggest that, when averaged over time, the magnetic poles and geographic poles roughly coincide. What causes this relationship? The key to understanding the orientation of the Earth's magnetic dipole comes from understanding the geometry of the convection cells in the outer core. Researchers suggest that, due to the Earth's rotation, the Coriolis force (see Chapter 15) causes convective cells in the outer core to become spirals that align roughly with the Earth's spin axis **(Fig. D.13b)**. The spirals wobble, however, so at any given time, the overall dipole axis they generate is not exactly parallel to the spin axis. Notably, the spirals apply a force to the inner core, causing it to rotate slightly faster than the rest of the Earth, a phenomenon known as *superrotation*.

In Chapter 2, we noted that the polarity of the Earth's magnetic field—the direction that the dipole points—flips (reverses) every now and then. This has happened about two dozen times during the past 5 million years. Today's *normal polarity* (with the dipole pointing to the south) was established 780,000 years ago. Between 780,000 and 900,000 years ago, the field had *reversed polarity*, with the dipole pointing to the north. Why does the polarity of the Earth's field undergo such reversals every now and then? Computer models suggest that reversals happen because circulation spirals in the outer core are unstable. Over time, they slow and fade away as they interact,

causing the magnetic field to weaken or disappear temporarily. As new spirals become established, the field reappears, sometimes with a different polarity. The details of this process remain a subject of active research.

MAGNETIC ANOMALIES, REVISITED

If the Earth's magnetic field were a perfect dipole, geophysicists could predict the field strength at every point on the planet with a high degree of confidence. However, that isn't the case, because magnetization locked into rocks in the crust can contribute to the field strength measured at a given location, causing the field to be either stronger than or weaker than expected at the location. As we discussed in Chapter 2, the difference between the expected field and the measured field at a locality is called a **magnetic anomaly**. A *positive magnetic anomaly* occurs where the field is stronger than expected, and a *negative magnetic anomaly* occurs where the field is weaker than expected.

We've seen that magnetic anomalies due to the magnetization of seafloor basalt during seafloor spreading take the form of parallel stripes on the seafloor. On continents, in contrast, the pattern of magnetic anomalies is much more irregular, because the distribution of rock types is more complex **(Fig. D.14)**. We see anomalies due to igneous intrusions, lava flows, and concentrations of iron-rich sediments. Rifts, for example, typically stand out as a strong positive magnetic anomaly, because they contain a thick succession of relatively magnetic basalts. Regions with thick accumulations of nonmagnetic sedimentary rocks may appear as regions of negative anomalies.

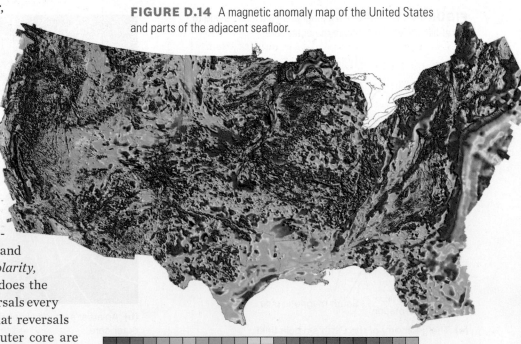

FIGURE D.14 A magnetic anomaly map of the United States and parts of the adjacent seafloor.

←— Negative   Positive —→

Interlude D Review

INTERLUDE SUMMARY

> The depths of the Earth's internal layers, and of divisions within those layers, were discovered through the study of how seismic waves passing through the Earth refract and reflect.

> Seismic waves travel faster through the mantle below the Moho than through the crust above. Seismic-velocity discontinuities define boundaries within the mantle.

> Seismic shadow zones reveal the depth to the Earth's core. S-waves cannot pass through the outer core, indicating that it is a liquid.

> Seismic tomographic studies indicate the Earth's interior layers are not homogeneous.

> Seismic-reflection profiling allows geologists to identify strata and structures in the upper crust.

> The geoid, the representation of the Earth's surface on which the pull of gravity is the same at all locations, is not a perfect sphere. Local gravitational anomalies occur.

> The elevation of the lithosphere's surface regionally reflects isostatic equilibrium.

> The Earth's magnetic dipole exists because of convective circulation in the outer core. Local anomalies may exist where rocks contain magnetic minerals.

GUIDE TERMS

Archimedes' principle (p. 296)	gravity anomaly (p. 296)	Moho (p. 289)	seismic tomography (p. 292)
core-mantle boundary (p. 291)	isostasy (isostatic equilibrium) (p. 296)	P-wave shadow zone (p. 291)	seismic-velocity discontinuity (p. 291)
dynamo (p. 298)	lower mantle (p. 291)	reflection (p. 289)	S-wave shadow zone (p. 291)
EarthScope (p. 293)	low-velocity zone (LVZ) (p. 290)	refraction (p. 289)	transition zone (p. 291)
equipotential surface (p. 294)	magnetic anomaly (p. 298)	seismic ray (p. 287)	travel time (p. 287)
geoid (p. 295)		seismic-reflection profile (p. 293)	upper mantle (p. 291)
geophysics (p. 287)			

REVIEW QUESTIONS

The letters following each Review Question refer to the corresponding Learning Objective from the Chapter Opener.

1. What basic layers of the Earth were recognized before the use of seismic studies? (**B**)

2. Do seismic waves travel at the same velocities in all rock? (**A**)

3. What is the difference between refraction and reflection of seismic waves? (**A**)

4. What observation led to the discovery of the Moho? (**B**)

5. Why do seismic waves bend in the mantle? (**A**)

6. What are seismic-velocity discontinuities, and what do they tell us? (**C**)

7. What are P-wave and S-wave shadow zones, and what do they tell us? Which type does the figure show? (**B**)

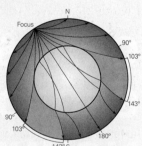

8. Is seismic velocity constant at a given depth? Explain what the answer tells us about the mantle. (**B**)

9. What can we learn from seismic-reflection profiling? (**B**)

10. What is the principle of isostasy? (**D**)

11. Explain what the geoid is. What is a gravity anomaly? (**E**)

12. What causes magnetic anomalies? (**F**)

ONLINE RESOURCES

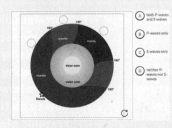

Animations This interlude features interactive animations on isostasy and the effects of different conditions on isostasy.

Smartwork5 This interlude features simulation and analysis questions on isostasy, seismic waves, and interpreting geophysical waves.

CHAPTER 9

CRAGS, CRACKS, AND CRUMPLES: GEOLOGIC STRUCTURES AND MOUNTAIN BUILDING

By the end of this chapter, you should be able to . . .

A. distinguish between mountain belts (orogens) and other regions of continental crust.

B. define what geologists mean by deformation when referring to rock, and explain how and why rocks undergo deformation and develop geologic structures.

C. contrast brittle with plastic deformation, and give examples of each.

D. sketch the various kinds of folds, faults, and foliation that develop during deformation, and interpret these structures.

E. correlate different kinds of mountain belts with specific geologic settings, in the context of plate tectonics.

F. explain why mountain uplift develops, and why topography in mountain belts becomes so rugged.

G. explain how a craton differs from an orogen, and show how to distinguish between basins and domes in cratonic platforms.

> Innumerable peaks, black and sharp, rose grandly into the dark blue sky, their bases set
> in solid white, their sides streaked and splashed with snow, like ocean rocks
> with foam. . . . [Mountains] are nature's poems carved on tables of stone.
>
> JOHN MUIR (American naturalist, 1838–1914)

9.1 Introduction

Geographers call the peak of Mt. Everest "the top of the world," because this mountain, which lies in the Himalayas of south-central Asia, rises higher than any other on the Earth. The cluster of flags on Mt. Everest's summit flap at 8.85 km above sea level, almost the cruising height of modern jets. In 1953, Sir Edmund Hillary, from New Zealand, and Tenzing Norgay, a Nepalese guide, became the first to reach the summit. Since then, thousands of other people have also succeeded, but hundreds have died trying. Mountains appeal to nonclimbers as well, for almost everyone loves a vista of snow-crested peaks. The stark cliffs, clear air, meadows, forests, streams, and glaciers of mountainous landscapes provide a refuge from the mundane.

Geologists have a particular fascination with mountains, because mountains represent one of the most obvious indications of the Earth's dynamic activity. Think about it . . . to make a single mountain, cubic kilometers of rock must rise skyward against the pull of gravity. Furthermore, with the exception of large volcanoes formed over hot spots, mountains do not occur in isolation, but rather they are parts of elongate ranges called **mountain belts** or **orogens** (from the Greek words *oros*, meaning mountain, and *genesis*, meaning formation). Uplift during the growth of an orogen may have moved a million or more cubic kilometers of rock.

A map of the present-day Earth reveals about a dozen major orogens, and numerous smaller ones **(Fig. 9.1)**. Many places were mountainous in the geologic past but are not mountain belts now. **Mountain building**, the process of forming a mountain belt, has happened many times and in many places over the Earth's history. Mountain building not only leads to *uplift*, the vertical rise of the land surface and the rock beneath, but also causes rocks to undergo *deformation*, when they bend, break, or flow. Features produced by deformation are known as *geologic structures*. These include *joints* (cracks), *faults* (fractures on which one

body of rock slides past another), *folds* (bends, curves, or wrinkles of rock layers), and *foliation* (a fabric or layering in rock). Mountain building may also cause metamorphism and igneous activity to take place.

A mountain-building event, or **orogeny**, may last for tens of millions of years. As the land rises, erosion starts to grind it down, producing sediment and sculpting spectacular, jagged landscape. Eventually, erosion may bring the height of a range down to near sea level. But even after the high peaks are gone, a low-lying belt of fractured, contorted, and metamorphosed rock remains. Such crustal scars serve as a permanent monument to what had once been a region of high peaks.

In this chapter, first we learn about the process of deforming rocks, and how to describe and interpret the geologic structures that result from deformation. Then, we turn our attention to the question of why mountain belts form, in the context of plate tectonics theory. Finally, we consider the phenomena that sculpt regions of uplift to form the dramatic terrains that we can see when visiting mountains.

9.2 Rock Deformation in the Earth's Crust

DEFORMATION AND STRAIN

To get a visual sense of what geologists mean by the term *deformation*, let's contrast rock that has not been affected by an orogeny with rock that has been affected. Our undeformed example comes from a road cut in the Great Plains of North America, and our deformed example comes from a cliff in the Alps of Europe.

The road cut exposes nearly horizontal beds of sandstone, shale, and limestone cut by a few joints **(Fig. 9.2a)**. These beds have the same orientation that they had when first deposited. Sand grains in the sandstone of this outcrop are nearly spherical (the same shape as when deposited). Also, the clay flakes in the shale lie roughly parallel to the bedding, due to compaction accompanying burial and lithification. When we say that rock of this outcrop is *undeformed*, we mean that it contains no geologic structures, other than the joints.

◀ (facing page) The surface of this fault displays streak-like slip lineations formed when movement took place on the fault in the past. Faults are one manifestation of the deformation that occurs during mountain building on our own dynamic planet.

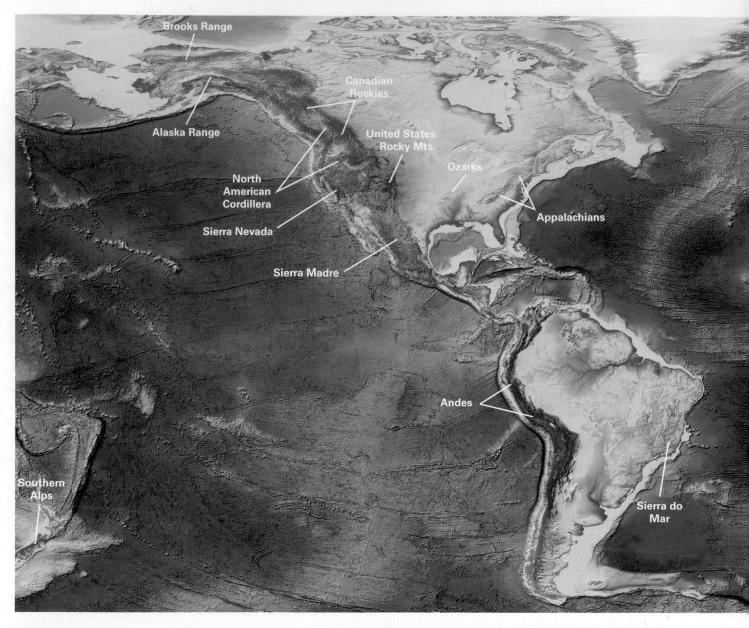

FIGURE 9.1 Digital map of world topography, showing the locations of major mountain ranges.

Deformed rocks of the Alpine cliff look very different (**Fig. 9.2b**). Here we find layers of quartzite, slate, and marble, the metamorphic equivalents of sandstone, shale, and limestone. Here too, joints cut the layers, but in addition, folds contort the layers. On microscopic examination, we may find that the grains in the quartzite aren't spherical, but have been flattened into elliptical shapes. And in the slate, clay flakes no longer align with bedding but rather align in a new orientation to define a new foliation; in this example, the foliation is called *slaty cleavage* (see Chapter 7). Because of changes in grain shapes and alignments, the fabric of this deformed rock no longer looks like the fabric that the rock had before deformation. Finally, if we try tracing the quartzite and slate layers along the cliff face, we find that they terminate abruptly at a fault, bordered by broken-up rock. In our

example, thick layers of marble lie below the fault, indicating that the quartzite and slate moved along the fault from where they first formed to get to their present location.

Study of the types of geologic structures that we see in the Alpine cliff emphasizes that during deformation, rocks can undergo one or more of the following changes (**Fig. 9.3**): (1) a change in location, or *displacement*; (2) a change in orientation, or *rotation*; and (3) a change in shape, or *distortion*. Geologists commonly refer to distortion as **strain.** We distinguish among different kinds of strain according to the nature of the shape change observed. If a layer becomes longer, it has undergone *stretching*, whereas if the layer becomes shorter, it has undergone *shortening* (**Fig. 9.4a–c**). Movement of one part of a rock body past another, so that angles between features in the rock change, results in *shear strain* (**Fig. 9.4d, e**).

FIGURE 9.2 Deformation changes the character and configuration of rocks.

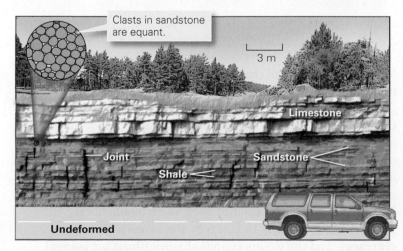

(a) These flat-lying beds along a highway are essentially undeformed. A few joints, formed when overlying rock eroded away, are visible.

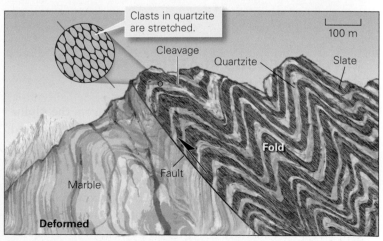

(b) In a mountain belt, deformation may cause layers to undergo folding and faulting. In addition, foliation (such as slaty cleavage) and stretched clasts may develop.

FIGURE 9.3 The components of deformation include displacement, rotation, and distortion.

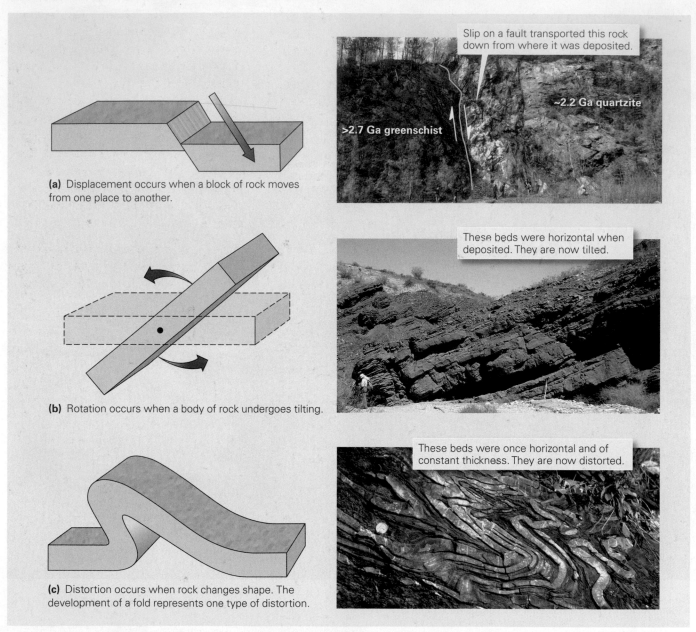

(a) Displacement occurs when a block of rock moves from one place to another.

Slip on a fault transported this rock down from where it was deposited.

~2.2 Ga quartzite

>2.7 Ga greenschist

(b) Rotation occurs when a body of rock undergoes tilting.

These beds were horizontal when deposited. They are now tilted.

(c) Distortion occurs when rock changes shape. The development of a fold represents one type of distortion.

These beds were once horizontal and of constant thickness. They are now distorted.

BRITTLE VERSUS PLASTIC DEFORMATION

Look again at the example of deformed rocks from the Alpine cliff in Figure 9.2b. Forming the joints and the fault involved cracking and breaking rock, a process that geologists refer to as **brittle deformation (Fig. 9.5a, b)**. You've seen this process in daily life if you've ever dropped a plate on a hard floor, so that the plate breaks into pieces. Figure 9.2b also shows a fold that formed when the rock changed shape without breaking, a process that geologists refer to as **plastic deformation**, or *ductile deformation* **(Fig. 9.5c, d)**. You can produce plastic deformation by squeezing a ball of soft dough between a book and a tabletop

or by bending a stick of chewing gum. During plastic deformation, objects change shape without visibly breaking.

What actually happens within mineral grains during these two different kinds of deformation? Recall that the atoms that make up mineral grains in rock are connected by chemical bonds. During brittle deformation, many bonds break and stay broken, leading to the formation of a permanent fracture across which material no longer connects. During plastic deformation in rock some bonds break, but new ones quickly form. In this way, the atoms within grains rearrange, and the grains change shape without permanent cracks forming.

FIGURE 9.4 Different kinds of strain in rock. Strain is a measure of the distortion, or change in shape, that takes place in rock during deformation.

Unstrained

Fossil shell (brachiopod)

(a) An unstrained cube and an unstrained fossil shell (brachiopod).

Stretching (elongation)

(b) Horizontal stretching changes the cube into a horizontal brick and elongates the shell.

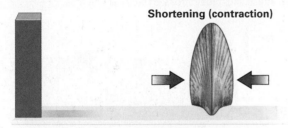

Shortening (contraction)

(c) Horizontal shortening changes the cube into a vertical brick and makes the shell narrower.

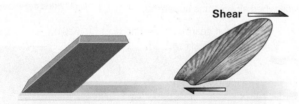

Shear

(d) Shear strain tilts the cube and transforms it into a parallelogram, and changes angular relationships in the shell.

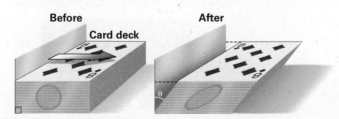

Before   **After**
Card deck

(e) You can produce shear strain by moving a deck of cards so that each card slides a little with respect to the one below.

Why do rocks inside the Earth sometimes deform brittlely and sometimes plastically? The behavior of a rock depends on several factors:

> *Temperature:* Warmer rocks tend to deform plastically, whereas colder rocks tend to deform brittlely, for heat makes materials softer. To picture the difference, compare the behavior of a candle warmed in an oven to one cooled

in a freezer, when you try bending it with your hands—the former easily changes shape, but the latter snaps in two.

> *Pressure:* Under great pressures deep in the Earth, rock behaves more plastically than it does under low pressures near the surface. Pressure effectively prevents rock from separating into fragments as it deforms.

> *Deformation rate:* A sudden change in shape causes brittle deformation, whereas a slow change in shape can cause plastic deformation. For example, if you hit a marble bench with a hammer, it shatters, but if you leave the bench alone for a century, it may gradually sag without breaking.

> *Composition:* Some rock types are softer than others; for example, halite (rock salt) deforms plastically under conditions in which granite deforms brittlely.

Considering that pressure and temperature both increase with depth in the Earth, geologists find that, in typical continental crust, rocks generally behave brittlely above a

FIGURE 9.5 Contrast between brittle and plastic deformation.

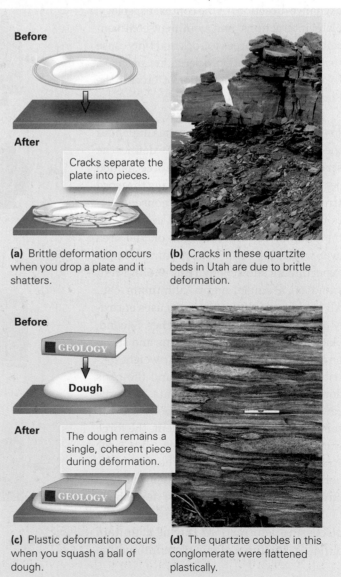

(a) Brittle deformation occurs when you drop a plate and it shatters.

(b) Cracks in these quartzite beds in Utah are due to brittle deformation.

(c) Plastic deformation occurs when you squash a ball of dough.

(d) The quartzite cobbles in this conglomerate were flattened plastically.

depth of about 10 to 15 km and plastically below this depth. The depth at which this change in behavior takes place is called the *brittle-plastic transition*. Earthquakes in continental crust happen mostly above this depth, because generation of earthquakes involves brittle breaking.

FORCE, STRESS, AND THE CAUSES OF DEFORMATION

Up to this point, we've focused on picturing the consequences of deformation. Describing the causes of deformation in the Earth's crust is a bit more challenging. Captions for museum displays about mountain building typically dispense with the issue by using the phrase, "The mountains were caused by forces deep within the Earth." But what does this mean? Isaac Newton stated that a *force* can cause an object to speed up, slow down, or change direction. In the context of geology, plate interactions and continent-continent collisions apply forces to rock and thus cause rock to change location, orientation, or shape. In other words, the application of forces in the Earth indeed causes deformation.

Geologists commonly use the word *stress* instead of *force* when talking about the cause of deformation. We define the **stress** acting on a plane as the force applied per unit area of the plane. By using the word stress, we emphasize that the consequences of applying a force depend not only on the amount of force but also on the area over which the force acts. A pair of simple experiments shows why **(Fig. 9.6a)**. First, stand on a single, empty aluminum can. All of your weight—a force—focuses entirely on the can, and the can crushes. Second, place a board on top of 100 cans and stand on the board. In this case, your weight distributes across all 100 cans, and the cans don't crush. In both experiments, the force caused by the weight of your body was the same, but in the first experiment, the force was applied over a small area, so a large stress developed, whereas in the second experiment, the same force was applied over a large area, so only a small stress developed.

How does the concept of stress apply to discussing deformation? During mountain building, the force of one plate interacting with another acts across the area of contact between the two plates, so the deformation resulting at any specific location actually depends on the stress developed at that location, not on the total force produced by the plate interaction.

Different kinds of stress can exist in rock **(Fig. 9.6b–e)**. **Compression** takes place when a rock is squeezed together,

FIGURE 9.6 There are several kinds of stress.

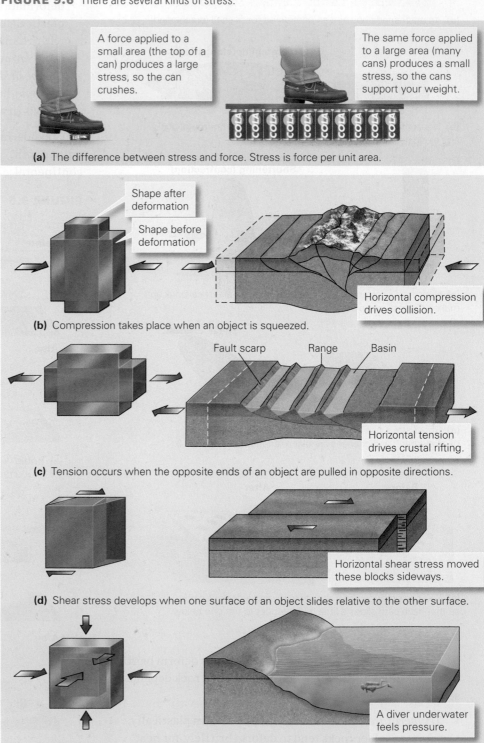

(a) A force applied to a small area (the top of a can) produces a large stress, so the can crushes.

The same force applied to a large area (many cans) produces a small stress, so the cans support your weight.

(a) The difference between stress and force. Stress is force per unit area.

Shape after deformation

Shape before deformation

Horizontal compression drives collision.

(b) Compression takes place when an object is squeezed.

Fault scarp   Range   Basin

Horizontal tension drives crustal rifting.

(c) Tension occurs when the opposite ends of an object are pulled in opposite directions.

Horizontal shear stress moved these blocks sideways.

(d) Shear stress develops when one surface of an object slides relative to the other surface.

A diver underwater feels pressure.

(e) Pressure occurs when an object is subjected to the same stress on all sides.

tension occurs when a rock is pulled apart, and *shear stress* develops when one part of a rock body moves sideways past another. In general, the stress acting in one direction is not the same magnitude as the stress acting in another direction. For example, at a given locality, the horizontal compression in the north-south direction in the crust can be greater than the compression in the east-west or vertical directions. Note that in some special situations, the same amount of compression acts on all sides of an object. We refer to the stress in such situations as **pressure**.

Stress and *strain* have very different meanings to geologists, even though we tend to use the words interchangeably in everyday English. Again, stress refers to the amount of force applied per unit area of a rock, whereas strain refers to a change in shape of a rock. Stress causes strain—compression causes shortening, tension leads to stretching, and shear stress produces shear strain. When stresses are not the same in all directions, the amount of strain can be different in different directions. For example, if compression is greatest in the north-south direction, shortening will be greatest in the north-south direction. Note that pressure can cause an object to become smaller, but will not cause it to change shape. With our knowledge of stress and strain, we can now look at the nature and origin of various classes of geologic structures.

TAKE-HOME MESSAGE

Stress (compression, tension, shear) can cause rocks to change shape, position, and orientation. During brittle deformation, rocks break, whereas during plastic deformation, rocks bend and distort without breaking. Strain is a measure of shape change.

QUICK QUESTION What is the difference between stress and strain to a geologist?

9.3 Brittle Structures

JOINTS AND VEINS

If you look at the photographs of rock outcrops in this book, you'll sometimes notice thin dark lines that cross the rock faces. These lines are the traces of natural cracks, planes along which the rock broke and separated into two pieces during brittle deformation. Geologists refer to such natural cracks as **joints (Fig. 9.7a, b)**. Rock bodies break, but do not slide past each other, on joints. Since joints are roughly planar structures, we define their orientation by their *strike* and *dip*, as described in **Box 9.1**.

Joints develop in response to tensile stress in brittle rock: a rock splits open because it has been pulled slightly apart. Joints may form for a variety of geologic reasons. For example, some joints form when a rock cools and shrinks, because the process makes one part of a rock pull away from the adjacent part. Others develop when rock formerly at depth undergoes a decrease in pressure as overlying rock erodes away, and thus changes shape slightly. Note that the formation of joints in response to cooling or a decrease in pressure can occur even where there has been no mountain building. Some joints, however, form when rock layers bend, and these are associated with mountain building.

If groundwater or hydrothermal fluids seep through cracks in rocks, minerals such as quartz or calcite may precipitate out of the groundwater and fill the cracks. A mineral-filled crack is called a **vein**. Veins typically look like white stripes cutting across a body of rock **(Fig. 9.7c)**.

Geotechnical engineers, people who study the geologic setting of construction sites, pay close attention to jointing when recommending where to put roads, dams, and buildings. Water flows much more easily through joints than it does through solid rock, so it would be a bad investment to situate a water reservoir over rock with lots of joints—the water would leak down into the joints. Also, building a road on a steep cliff composed of jointed rock could be risky, for joint-bounded blocks separate easily from bedrock, and the cliff might collapse.

FAULTS: SURFACES OF SLIP

After the San Francisco earthquake of 1906, geologists found a rupture that had torn through the land surface near the city. Where this rupture crossed orchards, it offset rows of trees, and where it crossed a fence, it broke the fence in two—the western side of the fence moved northward by about 2 m (see Fig. 8.3a). The rupture represents the trace of the San Andreas fault. As we have seen, a **fault** is a fracture on which sliding occurs, and slip events, or *faulting*, can generate earthquakes. Faults, like

BOX 9.1 CONSIDER THIS...

Describing the Orientation of Geologic Structures

When discussing geologic structures, it's important to be able to communicate information about their orientation. For example, does a fault exposed in an outcrop at the edge of town continue beneath the nuclear power plant 3 km to the north, or does it go beneath the hospital 2 km to the east? If we knew the fault's orientation, we might be able to answer this question. To describe the orientation of a geologic structure, geologists picture the structure as a simple geometric shape, and then specify the angles that the shape makes with respect to a horizontal plane (a flat surface parallel to sea level), a vertical plane (a flat surface perpendicular to sea level), and the north direction (a line of longitude).

FIGURE Bx9.1 Specifying the orientation of planar and linear structures.

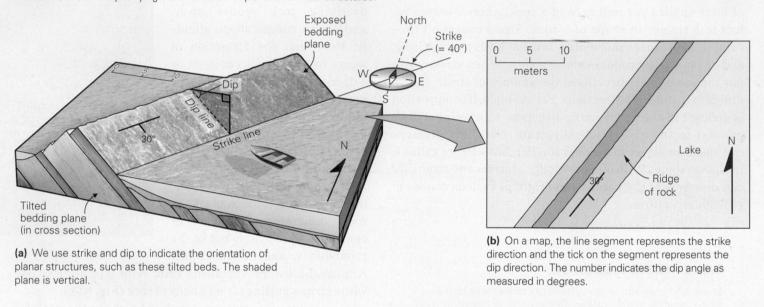

(a) We use strike and dip to indicate the orientation of planar structures, such as these tilted beds. The shaded plane is vertical.

(b) On a map, the line segment represents the strike direction and the tick on the segment represents the dip direction. The number indicates the dip angle as measured in degrees.

FIGURE 9.7 Examples of joints and veins.

(a) Prominent vertical joints cut these red sandstone beds in Arches National Park, Utah.

(b) Vertical joints on a cliff face in shale near Ithaca, New York.

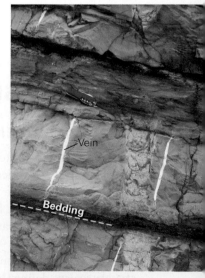

(c) Milky white quartz veins cut across gray limestone beds.

Let's start by considering planar structures such as faults, beds, joints, and foliation. We call these structures *planar* because they resemble a geometric plane. A planar structure's orientation can be specified by its strike and dip. The **strike** is the angle between an imaginary horizontal line (the strike line) on the plane and the direction to true north **(Fig. Bx9.1a, b)**. The **dip** is the angle of the plane's slope or, more precisely, the angle between a horizontal plane and the dip line (an imaginary line parallel to the steepest slope on the structure), as measured in a vertical plane perpendicular to the strike. A horizontal plane has a dip of 0°, and a vertical plane has a dip of 90°. We can measure the strike with a special type of compass **(Fig. Bx9.1c)** and the dip with a clinometer, a type of protractor that uses a bubble to indicate horizontal, and we can represent strike and dip on a geologic map using the symbol shown in Figure Bx9.1b.

A linear structure resembles a geometric line rather than a plane. Examples of linear structures include scratches or grooves on a rock surface or aligned elongate minerals. Geologists specify the orientation of a line by giving its plunge and bearing **(Fig. Bx9.1d)**. The **plunge** is the angle between a line and horizontal in the vertical plane that contains the line. A horizontal line has a plunge of 0°, and a vertical line has a plunge of 90°. The **bearing** is the compass heading of the line, meaning the angle between the projection of the line on the horizontal plane and the direction to true north.

The edge of the compass is parallel to the strike line.

The intersection of the water with the rock is horizontal.

(c) Geologists use a Brunton compass to measure strike and dip.

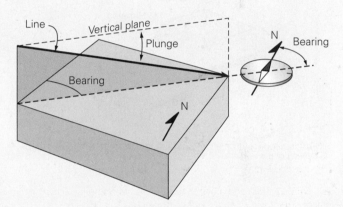

Line — Vertical plane — Plunge — Bearing — N — Bearing — N

(d) To specify the orientation of a line, we use plunge and bearing.

joints, are planar structures, so we can represent their orientation by strike and dip. Geologists refer to the amount of movement that takes place across a fault as the fault's **displacement**.

Faults have formed throughout Earth history. Some are currently *active* in that sliding has been occurring on them in recent geologic time, but most are *inactive*, meaning that sliding on them ceased long ago. Some faults, such as the San Andreas, intersect the ground surface and displace the ground when they move. Others accommodate the sliding of rocks in the crust at depth and remain invisible at the surface unless they are later exposed by erosion.

FAULT CLASSIFICATION

Not all faults result in the same kind of crustal deformation—some faults accommodate shortening, some accommodate

stretching, and some accommodate horizontal shear. Geologists distinguish among different kinds of faults in order to interpret their geologic significance. Fault classification focuses on two characteristics of faults: (1) the *dip* or slope of the fault surface (see Box 9.1)—the dip can be vertical, horizontal, or any angle in between; and (2) the *shear sense* across the fault—by shear sense, we mean the direction that material on one side of the fault moved relative to the material on the other side. With this concept in mind, let's consider the principal kinds of faults, expanding on the discussion presented in Chapter 8.

> *Dip-slip faults:* On a dip-slip fault, movement parallels the dip line, a line going down the slope of the fault surface. Therefore, the *hanging-wall block*, meaning the material above the fault surface, slides up or down relative to the *footwall block*, the material below the fault surface

FIGURE 9.8 The different categories of faults.

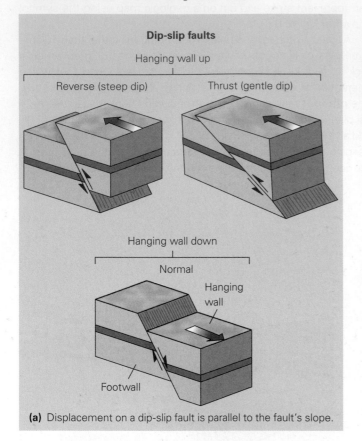

Dip-slip faults

Hanging wall up

Reverse (steep dip)   Thrust (gentle dip)

Hanging wall down

Normal

Hanging wall

Footwall

(a) Displacement on a dip-slip fault is parallel to the fault's slope.

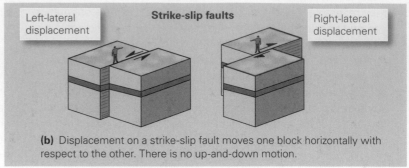

Strike-slip faults

Left-lateral displacement

Right-lateral displacement

(b) Displacement on a strike-slip fault moves one block horizontally with respect to the other. There is no up-and-down motion.

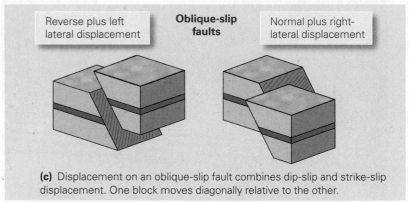

Reverse plus left lateral displacement

Oblique-slip faults

Normal plus right-lateral displacement

(c) Displacement on an oblique-slip fault combines dip-slip and strike-slip displacement. One block moves diagonally relative to the other.

(Fig. 9.8a). If the hanging-wall block slides up, the fault is a **reverse fault**; a reverse fault with a gentle dip (less than 30°) is also known as a **thrust fault**. Reverse faults accommodate shortening of the crust, as happens during continental collision. If the hanging-wall block slides down, the fault is a **normal fault**. Normal faults accommodate stretching of the Earth's crust, as happens during rifting.

› *Strike-slip faults*: A **strike-slip fault** is a fault on which the slip direction is parallel to a horizontal line on the fault surface (the strike line; see Box 9.1). This means that the block on one side of the fault slips sideways, relative to the block on the other side, and there is no up-or-down motion. Most strike-slip faults have a steep to vertical dip. Geologists distinguish between two types of strike-slip faults based on the shear sense as viewed when you are facing the fault and looking across it. If the block on the far side slipped to your left, the fault is a *left-lateral* strike-slip fault, and if the block slipped to the right, the fault is a *right-lateral* strike-slip fault **(Fig. 9.8b)**. Displacement at a transform plate boundary is accommodated by movement on strike-slip faults.

› *Oblique-slip faults*: On an **oblique-slip fault**, sliding occurs diagonally on the fault plane. In effect, an oblique-slip fault is a combination of a strike-slip and a dip-slip fault **(Fig. 9.8c)**.

RECOGNIZING FAULTS

How do you recognize a fault when you see one? In many cases, faults visibly displace distinct layers in rocks, so that layers on one side of a fault don't connect to layers on the other side **(Fig. 9.9a, b)**. If a fault intersects the ground surface, slip on the fault can displace natural landscape features such as stream valleys or glacial moraines **(Fig. 9.9c)**, or human-made features such as highways, fences, or rows of trees in orchards. Displacement on a dip-slip or oblique-slip fault that offsets the ground surface produces a step, known as a **fault scarp (Fig. 9.10a)**.

Faulting under brittle conditions may crush or break rock in a band bordering the fault surface. If this shattered rock consists of visible angular fragments, it's called *fault breccia* **(Fig. 9.10b)**, but if it consists of a fine powder, it's called *fault gouge*. Some fault surfaces have been polished and grooved

FIGURE 9.9 Recognizing fault displacement in the field.

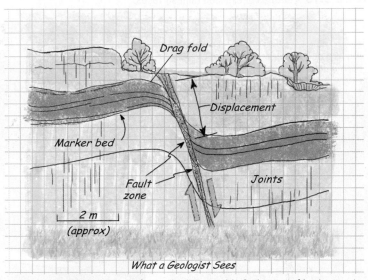

(a) A steep normal fault has displaced a distinctive red bed (a marker bed). Note that the displacement formed a 0.5-m-wide fault zone of broken rock. Drag folds developed adjacent to the fault.

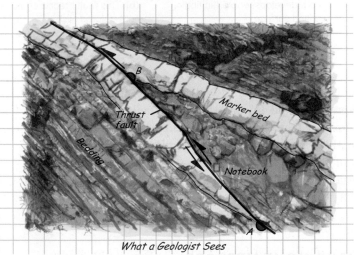

(b) Slip on a thrust fault caused one part of the light-colored marker bed to be shoved over another part, as emphasized in the drawing. Note that the beds are tilted to the right. The distance between A and B (the red dots) is the displacement.

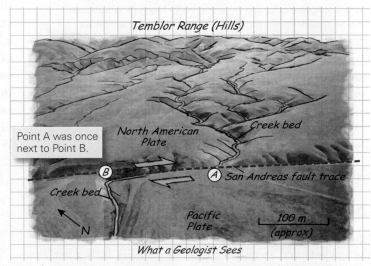

(c) An aerial photograph of a portion of the San Andreas fault. As emphasized by the sketch, the fault offsets a stream channel in a right-lateral sense.

FIGURE 9.10 Features of exposed fault surfaces.

A scarp due to slip on a normal fault in Nevada.

(a) A fault scarp develops where faulting displaces the land surface.

(b) This fault breccia consists of irregular fragments of light-colored rock.

Striation orientation

(c) Slip lineations or striations on the surface of a strike-slip fault may look like grooves or scratches.

by movement on the fault. Polished fault surfaces are called **slickensides**, and linear grooves on fault surfaces are *slip lineations* or fault striations **(Fig. 9.10c)**. Because faults tend to break up and weaken rock, the *fault trace* (the line of intersection between the fault and the ground surface) may preferentially erode to become a linear valley (see inset photo on p. 253).

TAKE-HOME MESSAGE

Brittle structures include joints (cracks), veins (mineral-filled cracks), and faults (fractures on which sliding occurs). Geologists distinguish among different kinds of faults based on the relative displacement across the fault. Faulting can break up rock. Fault surfaces themselves may be polished and grooved due to the slip.

QUICK QUESTION How can you distinguish between a joint and a fault in the field?

9.4 Folds and Foliations

GEOMETRY OF FOLDS

Imagine a carpet lying flat on the floor. Push on one end of the carpet, and it will wrinkle or contort into a series of wave-like curves. Stresses developed during mountain building can similarly cause bedding, foliation, or other planar features in rock to warp or bend. The result—a curve in the shape of a rock layer—is called a **fold** (see Fig. 9.2b).

Not all folds look the same—some look like arches, some look like troughs, and some have other shapes. To describe these shapes, we must first label the parts of a fold **(Fig. 9.11a)**. The **limbs** are the sides of the fold that have less curvature, and the **hinge** refers to a line along which the curvature of the fold is greatest. The **axial surface** is an imaginary plane that contains the hinge lines of successive layers and effectively divides the fold into two halves. With these terms in hand, we can distinguish among the following:

› *Anticlines, synclines, and monoclines:* Folds that have an arch-like shape in which the limbs dip away from the hinge are called **anticlines** (see Fig. 9.11a), whereas folds with a trough-like

FIGURE 9.11 Geometric characteristics of folds.

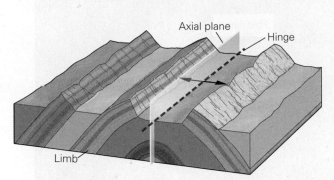

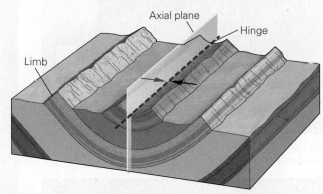

(a) An anticline looks like an arch. The beds dip away from the hinge.

(b) A syncline looks like a trough. The beds dip toward the hinge.

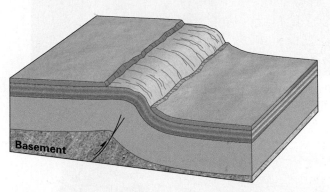

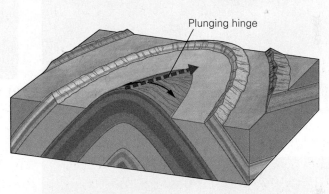

(c) A monocline looks like a stair step and is commonly draped over a fault block.

(d) A plunging anticline has a tilted hinge.

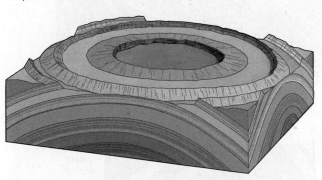

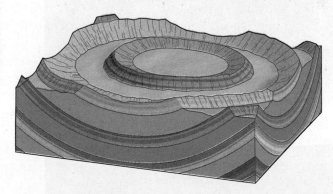

(e) A dome has the shape of an overturned bowl.

(f) A basin has the shape of an upright bowl.

shape in which the limbs dip toward the hinge are called **synclines (Fig. 9.11b)**. Therefore, an anticline resembles an arch, whereas a syncline resembles a trough. Geologists use the term **monocline** for a fold that looks like the shape of a carpet draped over a stair **(Fig. 9.11c)**.

> *Nonplunging and plunging folds:* A fold hinge does not necessarily have to be horizontal. Geologists refer to folds that have a horizontal hinge as *nonplunging folds*, and those that have a plunging (tilted) hinge as *plunging folds* **(Fig. 9.11d)**.

> *Domes and basins:* A fold with the shape of an overturned bowl is called a **dome**, whereas a fold shaped like an upright bowl is called a **basin (Fig. 9.11e, f)**. Domes and basins both display circular patterns that look like bull's-eyes on a geologic map. If domes and basins involve sedimentary strata, note that the oldest layer crops out in the center of a dome, whereas the youngest layer crops out in the center of a basin.

Using these terms, see if you can identify the various folds shown in **Figure 9.12**.

FIGURE 9.12 Examples of folds on outcrops and in the landscape.

(a) This anticline, exposed in a road cut near Kingston, New York, involves beds of Paleozoic limestone.

(b) This syncline, exposed in a road cut in Maryland, involves beds of Paleozoic sandstone and shale.

(d) A train of folds exposed in sea cliffs in eastern Ireland includes anticlines and synclines. The folds affect beds of Paleozoic sandstone and shale.

Fold hinge

Fold limb

(c) This fold, exposed along the coast of Brazil, occurs in Precambrian gneiss.

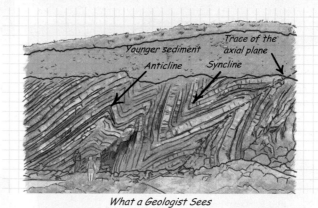

Younger sediment

Trace of the axial plane

Anticline

Syncline

What a Geologist Sees

Aerial view

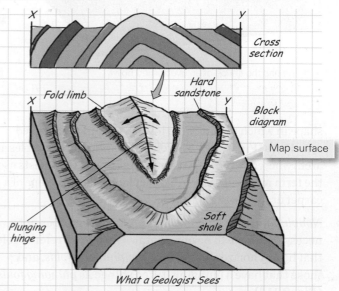

Cross section

X Y

Fold limb

Hard sandstone

Block diagram

X Y

Map surface

Plunging hinge

Soft shale

What a Geologist Sees

(e) The plunging anticline of Sheep Mountain, Wyoming, is easy to see because of the lack of vegetation. Resistant rock layers (sandstone) stand out as ridges, whereas weak rock layers (shale) erode away. A block diagram shows how surface exposures relate to underground structure.

FIGURE 9.13 Fold development in flexural-slip and passive-flow folding.

(a) During the formation of flexural-slip folds, layers maintain constant thickness, so for the fold to form, layers must bend. To accommodate the bending, each bed slips relative to its neighbor.

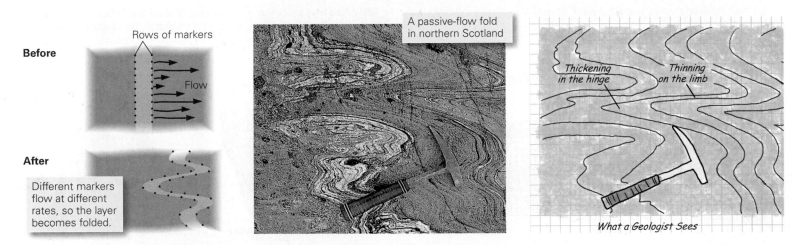

(b) During the formation of passive-flow folds, the rock slowly flows. Different points along a marker line flow at different rates, causing the layer to become folded. Note that the layer's thickness doesn't stay constant during folding. In outcrops, the limb tends to thin and the hinge tends to thicken.

FORMATION OF FOLDS

Folds develop in two principal ways **(Fig. 9.13)**. During formation of *flexural-slip folds,* a stack of layers bends, and slip occurs between the layers to accommodate the bending without producing gaps between layers. The same phenomenon happens when you bend a deck of cards—to accommodate the change in shape, the cards slide with respect to one another. *Passive-flow folds,* in contrast, form when the rock, overall, behaves like weak plastic and slowly flows; these folds develop simply because different parts of the rock body flow at different rates.

Why do folds form? Some layers wrinkle up, or buckle, in response to end-on compression **(Fig. 9.14)**. Other folds develop where shear stress gradually shifts one part of a layer up and over another part. Still others form when new slip on a fault causes a block of basement to move up so that the overlying sedimentary layers must warp. And finally, some result from movement of rock layers up and over step-like bends in an underlying fault, for the layers must curve to conform to the fault's shape as they move.

FOLIATION IN ROCKS

In an undeformed sandstone, the grains of quartz are roughly spherical, and in an undeformed shale, clay flakes press together into the plane of bedding so that shales tend to split parallel to the bedding. During deformation, internal changes take place in a rock that gradually modify the original shape and arrangement of grains. For example,

FIGURE 9.14 Folding is caused by several different processes.

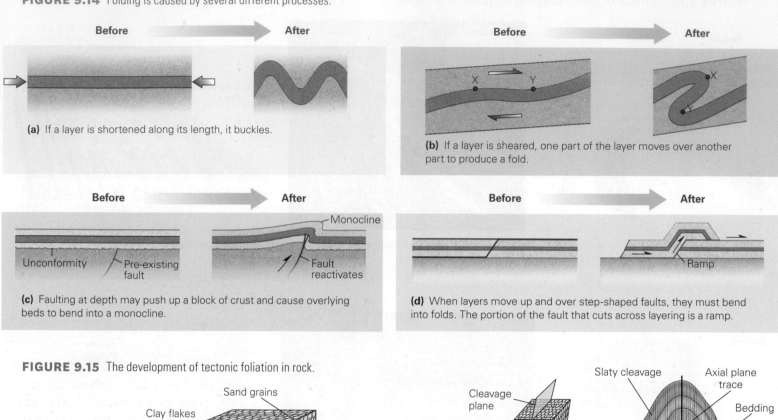

(a) If a layer is shortened along its length, it buckles.

(b) If a layer is sheared, one part of the layer moves over another part to produce a fold.

(c) Faulting at depth may push up a block of crust and cause overlying beds to bend into a monocline.

(d) When layers move up and over step-shaped faults, they must bend into folds. The portion of the fault that cuts across layering is a ramp.

FIGURE 9.15 The development of tectonic foliation in rock.

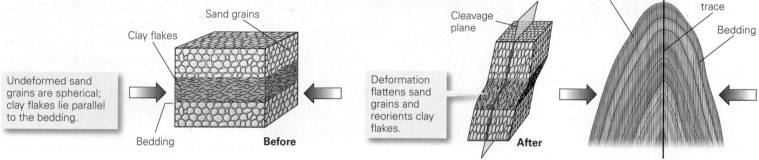

Undeformed sand grains are spherical; clay flakes lie parallel to the bedding.

Deformation flattens sand grains and reorients clay flakes.

Cleavage is parallel to the axial plane.

(a) Compression shortens beds, flattens sand grains, and reorients clay flakes. Clay flakes were originally parallel to bedding, but they became parallel to slaty cleavage during deformation. Folding may accompany cleavage formation.

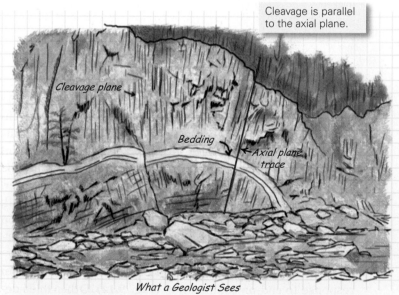

What a Geologist Sees

(b) An example of slaty cleavage developed in Paleozoic strata exposed in a stream cut in New York. Relict bedding is still visible, but note that the rock breaks more easily on the cleavage plane. This slaty cleavage formed in association with folding.

quartz grains may transform into tiny pancakes, and clay flakes may recrystallize and undergo reorientation. Overall, deformation can produce inequant grains that align parallel to one another to form a new layering **(Fig. 9.15a)**. Layering developed when inequant grains align in response to deformation is called **foliation**.

We introduced foliation, such as slaty cleavage, schistosity, and gneissic layering, while discussing the effects of metamorphism in Chapter 7. Here we add to the story by noting that because such foliation forms in response to flattening and shearing in plastically deforming rocks **(Fig. 9.15b)**, the presence of foliation in a rock indicates that the rock developed a strain under metamorphic conditions.

TAKE-HOME MESSAGE

Folds are bends or curves defined by the shape of rock layers. Arch-like folds are anticlines, and trough-like folds are synclines. Deformation can change the shape and orientation of grains, aligning them to produce a planar fabric called foliation.

QUICK QUESTION What mechanisms allow layers to undergo folding?

9.5 Causes of Mountain Building

Before plate tectonics theory became established, geologists were just plain confused about how mountains formed. In the context of the theory, however, the many processes driving mountain building became clear—mountains form primarily in response to convergent-boundary deformation, continental collisions, and rifting. In this section,

Did you ever wonder...
why mountains occur in distinct belts?

we look at these different settings and the types of mountains and geologic structures that develop in each one.

MOUNTAINS RELATED TO SUBDUCTION

As we saw in Chapter 2, subduction at a convergent plate boundary produces a volcanic arc. But that's not the only feature to develop in response to interactions at such boundaries. At some, compressional stress develops and drives crustal shortening in the overriding plate. Such shortening produces a *fold-thrust belt*. In such belts, the sedimentary strata of the upper several kilometers of crust shorten as if being pushed toward the interior of the continent by an imaginary giant bulldozer. Thrust faults transport rock away from the push, and folds develop as strata move up the thrusts. The Andes orogen of western South America displays the rugged topography that can develop in a compressional convergent-margin orogen **(Fig. 9.16)**. A fold-thrust belt has formed along its eastern edge.

MOUNTAINS RELATED TO COLLISION

Once the oceanic lithosphere between two relatively buoyant crustal blocks completely subducts, the blocks themselves collide with each other. The buoyant blocks may be large or small continents, island arcs, or large oceanic plateaus. Continental collision results in the formation of large mountain ranges, such as the present-day Himalayas of Asia, the Alps of Europe, and the Paleozoic Appalachian Mountains of eastern North America **(Geology at a Glance**, pp. 320–321). The final stage in the growth of the Appalachians happened when Africa and North America collided.

During collision, intense compression generates fold-thrust belts on the margins of the orogen **(Fig. 9.17a–c)**. In the interior of the orogen, where one block overrides the edge of the other, high-grade metamorphism occurs, accompanied by formation of passive-flow folds and tectonic foliation. The boundary between blocks that had been separate before the collision is called a **suture**. During collision, *crustal thickening* takes place as the crust shortens horizontally and thickens vertically, and thrust faults place slices of crust on top of one another.

If subduction continues over a long time, many offshore volcanic arcs, oceanic plateaus, and *microcontinents* (small blocks of continental crust), may drift into a convergent margin. Such crustal blocks may collide with and attach to the edge of the overriding plate. Geologists refer to the process of building out a continent by attaching new crustal fragments as **accretion**. The buoyant crustal block is called an *accreted terrane* once it has attached to the overriding plate. Once accretion occurs, the convergent boundary may jump to the seaward side of the accreted terrane so that

FIGURE 9.16 Characteristics of convergent-margin orogens.

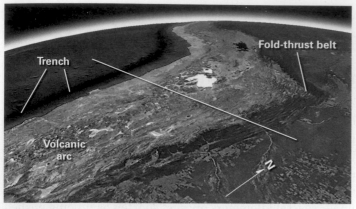

(a) An oblique view of the central Andes, showing the fold-thrust belt on the east side of the orogen and the trench on the west.

(c) Peaks of the Andes in Chile.

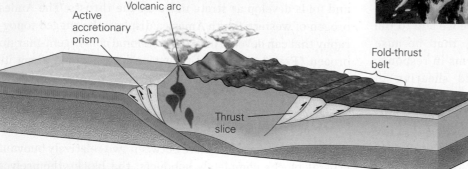

(b) A simplified cross section of the crustal structure beneath the Andes along the white line in (a). Compression uplifts the range and shortens the crust.

subduction can continue and perhaps bring in additional crustal blocks. The process of accretion can add substantial new crust to a convergent-boundary orogen. For example, the western part of the North American Cordillera, the huge (7,000 km long and 600–1,500 km wide) orogen that extends from Alaska through Mexico, consists of accreted terranes that attached to the continent between 250 and 150 million years ago (Ma) **(Fig. 9.17d)**. The belt of accreted crust reaches a width of 500 km, as measured in an east-west direction.

MOUNTAINS RELATED TO CONTINENTAL RIFTING

A continental rift is a place where a continent undergoes stretching, and may eventually split in two. During rifting, tensional stress causes normal faulting in the upper crust **(Fig. 9.18a)**. Movement on normal faults drops down blocks of crust, which typically tilt over as they move. As a result, rifts contain several elongate mountain ranges—the tilted blocks of crust—separated by deep, sediment-filled basins. (Ranges in a rift are sometimes called *fault-block mountains*.) Stretching during rifting thins the lithosphere, allowing hot asthenosphere to rise and undergo decompression melting (see Chapter 4). This process produces magmas that rise to form volcanoes within the rift. Heating may cause uplift of the rift and its borders. Today, the East African Rift clearly shows the configuration of rift-related mountains and volcanoes. In North America, rifting yielded the broad Basin and Range Province of Utah, Nevada, and Arizona **(Fig. 9.18b)**.

MEASURING MOUNTAIN BUILDING IN PROGRESS

Not all mountains are just "old monuments," as John Muir mused. The rumblings of earthquakes and the eruptions of volcanoes in some ranges attest to present-day, continuing movements that geologists can measure through field studies and satellite technology. For example, researchers can determine where coastal areas have been rising

FIGURE 9.17 Characteristics of collisional orogens.

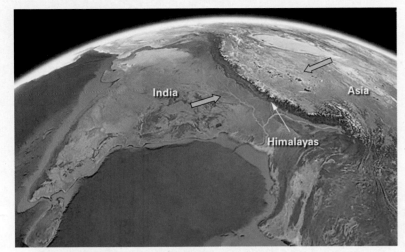

(a) The ongoing collision between India and Asia generated the Himalaya orogen.

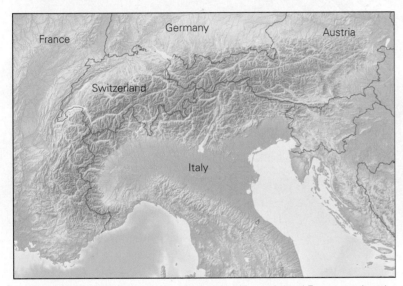

(b) The collision of the Italian Peninsula with continental Europe produced the Alps.

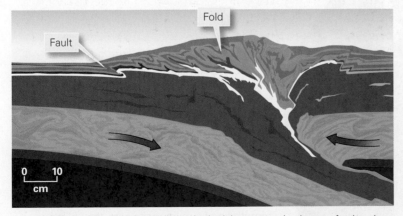

(c) Geologists simulate collision in the laboratory using layers of colored sand. Dragging the left side of the model under the right produces structures and uplift, as shown in this sketch of a model.

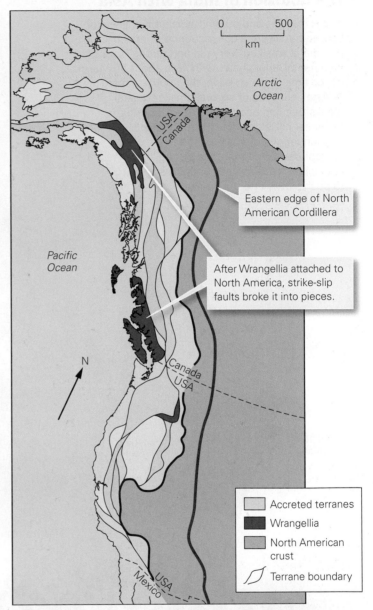

(d) The western portion of the North American Cordillera consists of accreted terranes that attached to the continent during the Mesozoic. During docking, a distinct terrane called Wrangellia (highlighted in red) was sliced into pieces that were displaced by strike-slip faults.

relative to sea level by locating ancient beaches that now lie high above the water. And they can tell where the land surface has risen relative to a river by identifying places where a river has recently carved a new valley. In addition, the **global positioning system (GPS)** provides a means to measure rates of uplift and horizontal shortening in orogens. Research-quality GPS systems can specify locations to within 1 mm. By comparing a location within an orogen to a location outside an orogen over a few years, it is possible to detect crustal motion. Amazingly, we can "see" the

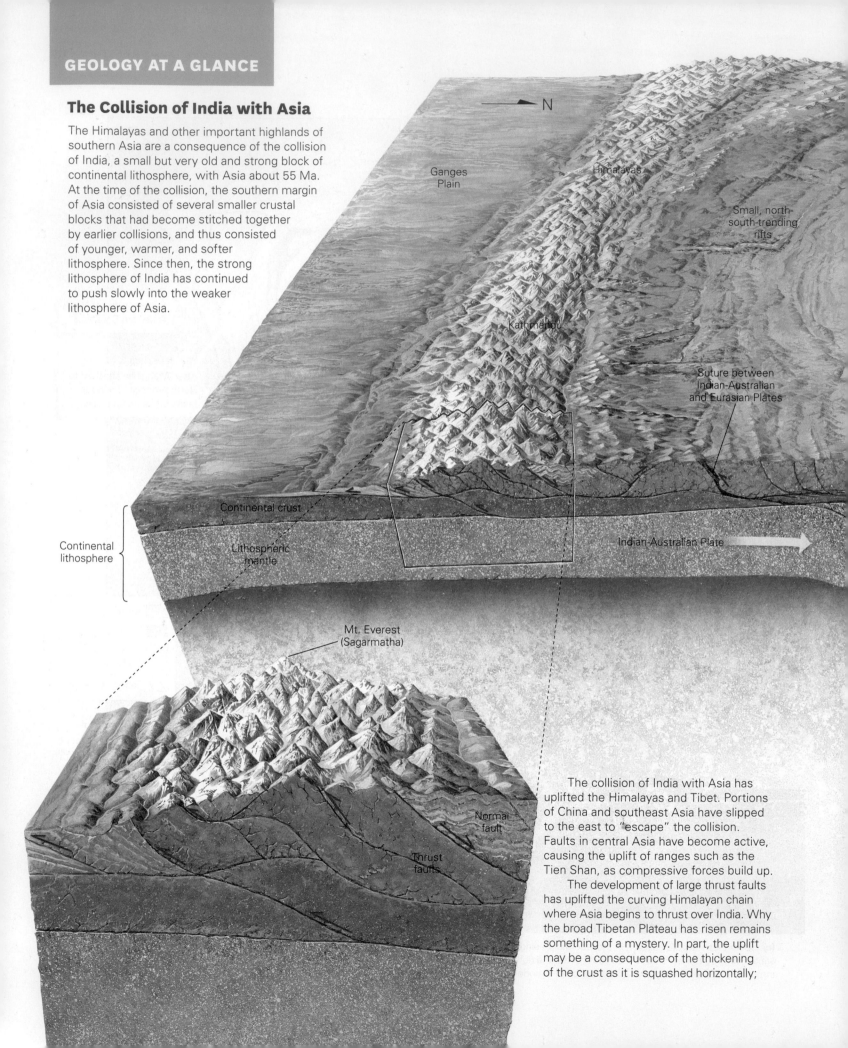

The Collision of India with Asia

The Himalayas and other important highlands of southern Asia are a consequence of the collision of India, a small but very old and strong block of continental lithosphere, with Asia about 55 Ma. At the time of the collision, the southern margin of Asia consisted of several smaller crustal blocks that had become stitched together by earlier collisions, and thus consisted of younger, warmer, and softer lithosphere. Since then, the strong lithosphere of India has continued to push slowly into the weaker lithosphere of Asia.

Ganges Plain

Himalayas

Small, north-south-trending rifts

Kathmandu

Suture between Indian-Australian and Eurasian Plates

Continental crust

Continental lithosphere

Lithospheric mantle

Indian-Australian Plate

Mt. Everest (Sagarmatha)

Normal fault

Thrust faults

The collision of India with Asia has uplifted the Himalayas and Tibet. Portions of China and southeast Asia have slipped to the east to "escape" the collision. Faults in central Asia have become active, causing the uplift of ranges such as the Tien Shan, as compressive forces build up.

The development of large thrust faults has uplifted the curving Himalayan chain where Asia begins to thrust over India. Why the broad Tibetan Plateau has risen remains something of a mystery. In part, the uplift may be a consequence of the thickening of the crust as it is squashed horizontally;

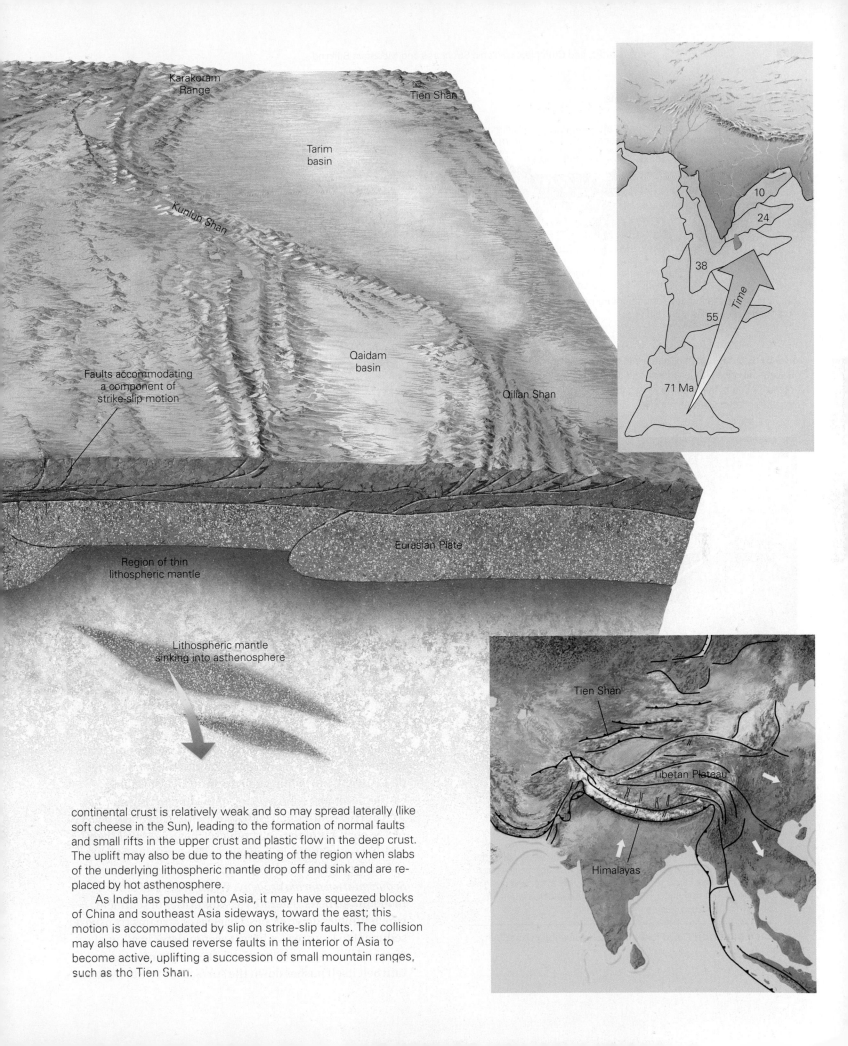

Karakoram
Range

Tien Shan

Tarim
basin

Kunlun Shan

Faults accommodating
a component of
strike-slip motion

Qaidam
basin

Qilian Shan

Eurasian Plate

Region of thin
lithospheric mantle

Lithospheric mantle
sinking into asthenosphere

10

24

38

55

Time

71 Ma

Tien Shan

Tibetan Plateau

Himalayas

continental crust is relatively weak and so may spread laterally (like soft cheese in the Sun), leading to the formation of normal faults and small rifts in the upper crust and plastic flow in the deep crust. The uplift may also be due to the heating of the region when slabs of the underlying lithospheric mantle drop off and sink and are replaced by hot asthenosphere.

As India has pushed into Asia, it may have squeezed blocks of China and southeast Asia sideways, toward the east; this motion is accommodated by slip on strike-slip faults. The collision may also have caused reverse faults in the interior of Asia to become active, uplifting a succession of small mountain ranges, such as the Tien Shan.

FIGURE 9.18 Rift-related mountains.

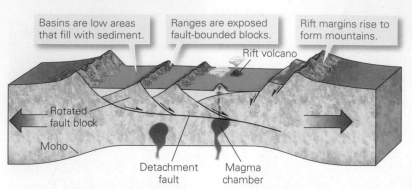

Basins are low areas that fill with sediment.

Ranges are exposed fault-bounded blocks.

Rift margins rise to form mountains.

Rift volcano

Rotated fault block

Moho

Detachment fault

Magma chamber

(a) Rifting leads to the development of numerous narrow mountain ranges. Rift-margin mountains may also form.

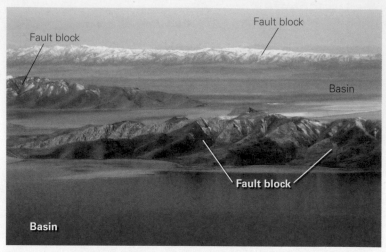

Fault block

Fault block

Basin

Fault block

Basin

(b) The Basin and Range Province of the western United States formed during Cenozoic rifting.

FIGURE 9.19 GPS measurements of shortening in the Andes. The arrows indicate the velocity of locations in the Andes relative to locations in the interior of South America. The white arrow indicates relative plate motion.

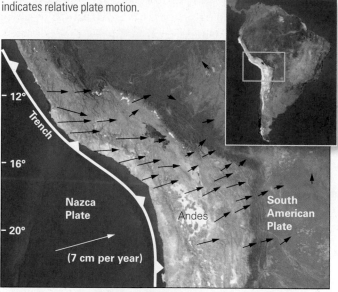

Trench

−12°

−16°

−20°

Nazca Plate

Andes

South American Plate

(7 cm per year)

Andes shorten horizontally at a rate of a couple of centimeters per year **(Fig. 9.19)**, and we can "watch" as mountains along this convergent boundary rise by a couple of millimeters per year.

TAKE-HOME MESSAGE

Mountain belts form in association with convergence, collision, and rifting. During convergent and collisional orogeny, continental crust thickens, large thrust faults and folds form, and regional metamorphism develops. Rifting yields fault-block mountains, separated by narrow basins.

QUICK QUESTION Could fold-thrust belts develop in rifts? Why or why not?

9.6 Other Consequences of Mountain Building

Mountain building not only causes deformation but also leads to conditions that produce new rocks, significant crustal uplift, and distinctive landforms. We've already mentioned these phenomena and have discussed deformation in detail. Now let's look more closely at examples of rock formation, uplift, and topography development.

FORMING ROCKS IN AND NEAR MOUNTAINS

The process of orogeny establishes geologic conditions appropriate for the formation of a great variety of rocks. Examples from all three rock categories can form in mountain belts **(Fig. 9.20a)**:

> *Igneous activity during orogeny:* At convergent boundaries, melting takes place in the mantle above the subducting plate. In rifts, stretching and thinning of lithosphere causes decompression melting of the underlying mantle. And during continental collision, melting may take place where portions of the crust undergo heating. All of these melting regimes produce magma, which rises and freezes to form igneous rocks in or on the overlying mountains.

> *Sedimentation during orogeny:* Weathering and erosion in mountain belts generate vast quantities of sediment, which tumbles down slopes. Glaciers or streams transport sediment to low areas, where it accumulates in alluvial fans or deltas. In some locations, the weight of the mountain belt itself pushes down the surface of the lithosphere

FIGURE 9.20 An example of the various rocks formed during orogeny.

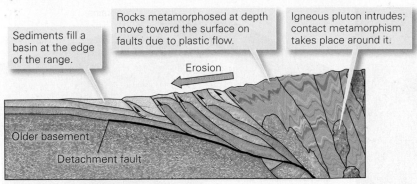

Sediments fill a basin at the edge of the range.

Rocks metamorphosed at depth move toward the surface on faults due to plastic flow.

Igneous pluton intrudes; contact metamorphism takes place around it.

Erosion

Older basement

Detachment fault

(a) In the internal zone of the range, metamorphic rocks and igneous rocks form. At the edge of the range, sedimentary rocks form.

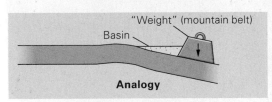

"Weight" (mountain belt)

Basin

Analogy

(b) The sedimentary basin develops because the mountain range acts as a weight that pushes down the surface of the lithosphere.

bordering the mountain belt, producing a deep sedimentary basin **(Fig. 9.20b)**.

> *Metamorphism during orogeny:* Contact metamorphic aureoles (see Chapter 7) form adjacent to igneous intrusions in orogens. Regional metamorphism occurs where mountain building thrusts one part of the crust over another; when this happens, rock of the footwall ends up at great depth and thus can be subjected to high temperature and pressure. Because deformation accompanies this process, the resulting metamorphic rocks contain tectonic foliation.

PROCESSES CAUSING UPLIFT AND PRODUCING MOUNTAINOUS TOPOGRAPHY

Leonardo da Vinci, the great Renaissance artist and scientist, enjoyed walking in the mountains, where he sketched ledges and examined the rocks he found there. To his surprise, he discovered fossil marine shells in limestone beds cropping out a kilometer above sea level. He puzzled over this observation, and finally concluded that the rock containing the fossils had risen from below sea level up to its present elevation. Modern geologists agree with Leonardo and, as we've noted, refer to vertical movement of the Earth's surface from a lower to a higher elevation as *uplift*. What processes can cause uplift? The list is long because mountain building happens in numerous different geologic settings. To understand how uplift processes work, we begin by revisiting the concept of isostasy introduced in Interlude D.

Continental lithosphere, which consists of relatively rigid crust and lithospheric mantle, rests on the softer, but still solid, asthenospheric mantle below. The elevation of the top surface of the lithosphere, over a broad region, represents a balance between buoyancy force pushing lithosphere up and gravitational force pulling the lithosphere down. Therefore, the downward push at a given depth in the asthenosphere is the same everywhere. Geologists refer to the condition that exists when this balance has been achieved as **isostasy**, or *isostatic equilibrium*. Put another way, isostasy exists where the elevation of the Earth's surface reflects the level at which the lithosphere naturally rests.

To picture the concept of isostasy, imagine placing a block of wood into a bathtub full of water. If the block is less dense than water, it floats, with part of the block remaining above the water surface and most of the block submerged below. Now place a denser block of the same thickness next to the first block. The top of the denser block sits lower than that of the less-dense block of the same thickness. Similarly, the top surface of a thicker block sits higher than the top surface of a thinner block of the same density. If you were to add another block of wood on top of one that is already floating, the lower block would sink, adjusting for the additional weight so as to maintain isostasy. If you could measure the pressure in the water at the floor of the tub, it would have the same value beneath all the blocks.

From our bathtub experiment, we can deduce that any phenomenon that changes the thickness or density of a floating block will affect the elevation of the block's surface above the water surface. Similarly, the elevation of the lithosphere's surface depends on the thickness and density of the lithosphere. So, to answer the question of why mountain belts can rise, we must identify geologic processes that can change the thickness or density of layers in the lithosphere. Let's consider some examples of why such changes can take place.

Crustal Shortening and Thickening During collisional orogeny or during certain types of convergent-margin orogeny, horizontal compression causes the crust to shorten horizontally and thicken vertically **(Fig. 9.21a)**. To isostatically compensate for the thickening of the crust, the geologic equivalent to adding another block of low-density wood to the top of a floating block, the base of the crust and underlying lithospheric mantle sink, or subside **(Fig. 9.21b)**. For example, continental crust is about 70 km thick, almost twice its normal thickness, beneath the Himalayas. Because the Himalayas rise about 8 km above sea level, most of the thickened crust extends downward beneath the range just as most of a floating ice cube is underwater. This downward protrusion of crust is called a **crustal root**. We can illustrate this relationship in a bathtub

FIGURE 9.21 The concept of isostasy as applied to collisional mountain ranges.

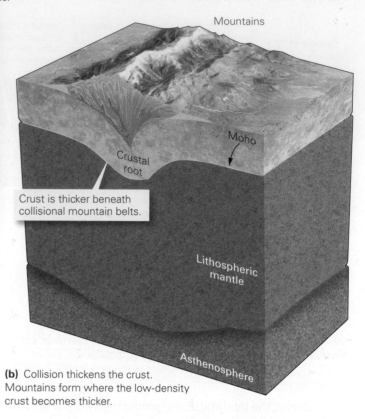

(a) The Himalayas are the world's highest mountains. The crust beneath them is almost twice the normal thickness.

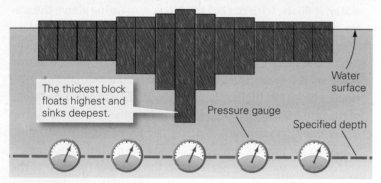

(c) Because of isostasy, blocks of wood floating in water sink to different depths. The pressure at a specified depth below the blocks is constant.

(b) Collision thickens the crust. Mountains form where the low-density crust becomes thicker.

model by lining up a row of floating blocks of different thickness but the same density **(Fig. 9.21c)**.

Adding Igneous Rock to the Crust When lava or pyroclastic debris accumulate on the Earth's surface, a volcano grows and may become a mountain. Growth of mountains associated with igneous activity may also occur because intrusions at depth may add rock to the crust and, therefore, thicken the crust.

Removal of Lithospheric Mantle The weight of the lithospheric mantle (composed of very dense rock) pulls the lithosphere down, just as heavy ballast makes a ship settle deeper into the water. Removal of some or all of the lithospheric mantle from the base of a plate, therefore, causes the surface of the remaining lithosphere to rise to maintain isostasy, even if the thickness of the crustal component remains unchanged **(Fig. 9.22)**. Such removal, a process known as *delamination*, resembles removal of ballast from the hold of a ship—as the weight of the ballast disappears, the deck of the ship rises.

Thinning and Heating of Lithosphere In rifts, the lithosphere undergoes stretching and thinning. As a result, relatively less-dense asthenosphere rises beneath the rift, and the remaining lithosphere heats up. Replacing denser lithospheric mantle with less-dense asthenosphere, together with heating the remaining thinned lithosphere (which causes the lithosphere to expand and become less dense), causes the overall region of the rift, as well as the margin of the rift, to rise in order to maintain isostasy.

WHAT GOES UP MUST COME DOWN

When the land surface rises significantly, for any reason, gravity and the Sun's heat begin to drive a variety of erosive processes. For example, as slopes steepen, landslides cause rock and debris to tumble from higher to lower elevations. When winds blow clouds over the mountains, rain provides water that collects in streams whose flow carries away debris and sculpts valleys and canyons. If temperatures remain cold enough during the year, glaciers grow and flow, carving sharp peaks and deepening valleys. The net result of all these processes is to grind away elevated areas and produce the jagged landscapes that we associate with mountain terrains **(Fig. 9.23)**. It's important to keep in mind that uplift and erosion happen simultaneously in active mountain belts, so for the elevation of a range to increase over time, the rate of uplift must exceed the rate of erosion. If uplift rate becomes less than erosion rate, the elevation of the range decreases.

FIGURE 9.22 Uplift due to delamination of the lithosphere root may happen after collision.

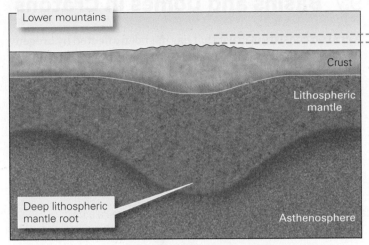

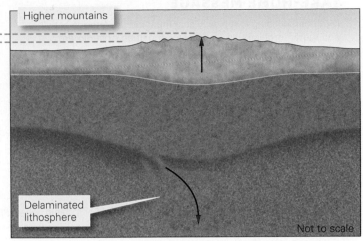

(a) Soon after collision, the crust and lithospheric mantle have thickened. The lithospheric mantle root is like ballast, holding the surface of the crust down.

(b) If the lithospheric mantle root detaches and sinks, a process called delamination, the surface of the lithosphere may rise, like a ship dropping ballast.

The highest point on our planet, the peak of Mt. Everest, lies 8.85 km above sea level. Can the Earth's mountain ranges get significantly higher? Probably not. Mountains as high as Olympus Mons on Mars, which rises 27 km above the plain at its base, couldn't form on the Earth because of the relatively high *geothermal gradient* (the rate of increase in temperature with depth) in the Earth's crust. At the high temperatures that occur at depths of 10 to 30 km, quartz-rich crustal rocks become so weak that it is relatively easy for them to flow plastically. When this flow begins, overlying mountains effectively begin to collapse under their own weight and spread laterally like soft cheese that has been left out in the summer sun. Geologists call this process **orogenic collapse**. During orogenic collapse, the upper crust breaks and a system of normal faults develops to accommodate the horizontal stretching.

The simultaneous activity of uplift, erosion, and in some cases, orogenic collapse ultimately brings rock that was metamorphosed at great depth up to the surface of the Earth. This process of revealing deeper rocks by removal of the overlying crust is called unroofing or **exhumation**.

> **Did you ever wonder...**
> whether mountain belts last forever?

FIGURE 9.23 Manifestations of erosion in mountain ranges. When land rises, water and ice start cutting into it.

(a) Glaciers carved these rugged peaks in Switzerland.

(b) Streams cut valleys into weathered bedrock in Brazil.

TAKE-HOME MESSAGE

Mountain building is typically accompanied by igneous activity and metamorphism, as well as by deposition of sediment. Mountains exist where major uplift takes place. Beneath some belts, the relatively buoyant crust has thickened, so the lithosphere sits higher. Once uplifted, erosion sculpts rugged topography. Because deep crust is warm and weak, mountain belts may eventually collapse under their own weight.

QUICK QUESTION How do rocks metamorphosed at 20 km below the surface eventually become exposed in a mountain range?

9.7 Basins and Domes in Cratons

A **craton** consists of crust that has not been affected by orogeny for at least about the last 1 billion years. As a result, cratons have cooled substantially, and therefore they have become relatively strong and stable. In North America, the craton forms the interior of the continent—during Phanerozoic time, mountain belts (the North American Cordillera on the west, the Appalachians on the east, and the Ouachitas on the south) formed along the margins, not within the craton. Geologists divide cratons into

FIGURE 9.24 North America's craton consists of a shield, where Precambrian rock is exposed, and a cratonic platform, where Paleozoic sedimentary rock covers the Precambrian rock.

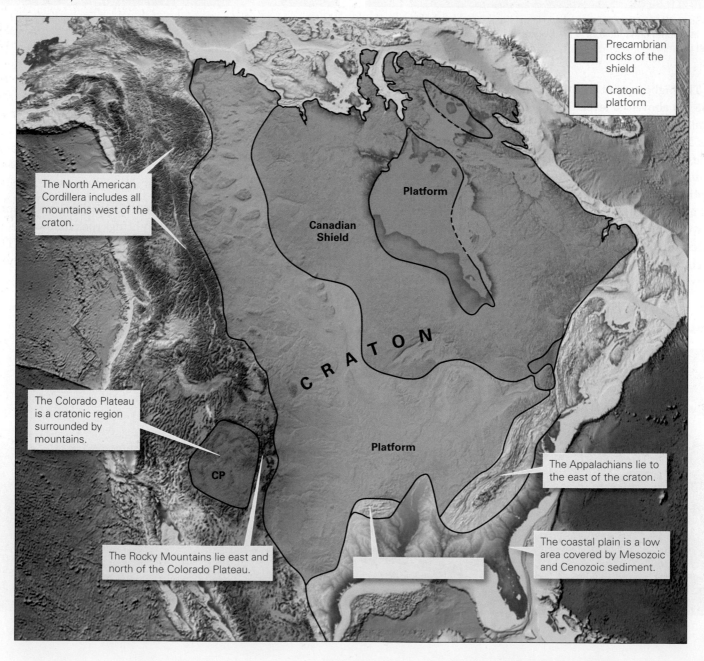

The North American Cordillera includes all mountains west of the craton.

The Colorado Plateau is a cratonic region surrounded by mountains.

The Rocky Mountains lie east and north of the Colorado Plateau.

The Appalachians lie to the east of the craton.

The coastal plain is a low area covered by Mesozoic and Cenozoic sediment.

Precambrian rocks of the shield

Cratonic platform

Platform

Canadian Shield

CRATON

Platform

CP

FIGURE 9.25 Domes and basins of the North American cratonic platform.

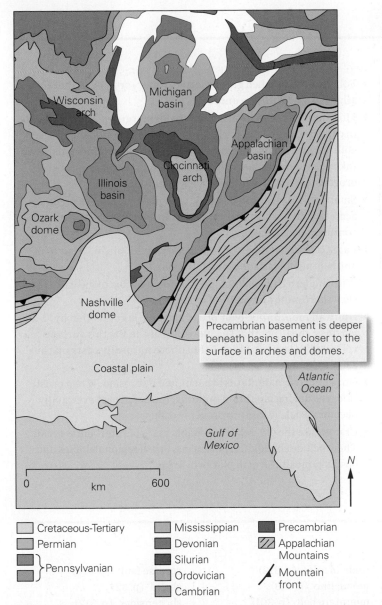

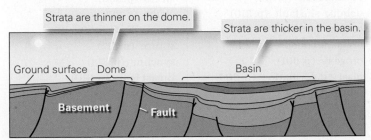

(a) A geologic map of the U.S. mid-continent platform region showing the locations of basins and domes.

Legend:
- Cretaceous-Tertiary
- Permian
- Pennsylvanian
- Mississippian
- Devonian
- Silurian
- Ordovician
- Cambrian
- Precambrian
- Appalachian Mountains
- Mountain front

Labels on map: Wisconsin arch, Michigan basin, Appalachian basin, Cincinnati arch, Illinois basin, Ozark dome, Nashville dome, Coastal plain, Gulf of Mexico, Atlantic Ocean. Scale: 0 — km — 600. N.

Annotation: Precambrian basement is deeper beneath basins and closer to the surface in arches and domes.

(b) A cross section showing how strata thin toward the crest of a dome and thicken toward the center of a basin. The cross section is vertically exaggerated and symbolically represents the Ozark dome and Illinois basin.

Labels on cross section: Strata are thinner on the dome. Strata are thicker in the basin. Ground surface, Dome, Basin, Basement, Fault.

two provinces: **shields**, where Precambrian metamorphic and igneous rocks crop out at the ground surface, and **cratonic platforms**, where a relatively thin layer of Phanerozoic sediment covers the Precambrian rocks **(Fig. 9.24)**.

In shields, we find widespread exposures of intensively deformed metamorphic rocks with abundant examples of flow folds and tectonic foliation. That's because the crust making the cratons was deformed during a succession of orogenies in the Precambrian. These orogens are so old that erosion has worn away the original peaks, in the process exhuming deep crustal rocks.

In cratonic platforms, the pattern of contacts between stratigraphic formations typically defines regional domes and basins **(Fig. 9.25)**. For example, in Missouri, strata arch across a broad uplift, the Ozark dome, whose diameter is 300 km. Individual sedimentary layers thin toward the top of the dome, because less sediment accumulated on the dome than in adjacent basins. Erosion during more recent geologic history has produced the characteristic bull's-eye pattern of a dome, with the oldest rocks (Precambrian granite) exposed near the center. To the east, in the Illinois basin, strata warp downward into a huge bowl that is also about 300 km across. Strata get thicker toward the basin's center, indicating that the floor of the basin was subsiding at the same time sediment was accumulating—there was more room for sediment to accumulate where the basin subsided the most. The basin also has a bull's-eye shape, but here the youngest strata are exposed in the center. Geologists refer to the vertical movements that generate huge, but gentle, mid-continent domes and basins as **epeirogeny** (from the Greek *epeiros*, meaning mainland or continent).

TAKE-HOME MESSAGE

Cratons are portions of continents that consist of very old and relatively stable crust. Parts of cratons may be covered by Paleozoic sedimentary strata. Variations in the dip and in the thickness of these strata define regional basins and domes.

QUICK QUESTION Imagine that coal occurs in a particular stratigraphic unit. Will mines to reach the coal be deeper or shallower in the center of a basin?

Chapter 9 Review

CHAPTER SUMMARY

> Mountains occur in elongate ranges called mountain belts or orogens. An orogen forms during an orogeny, or mountain-building event.

> Mountain building causes rocks to bend, break, shorten, stretch, and shear. Because of such deformation, rocks can change their location, orientation, and shape.

> During brittle deformation, rocks break into pieces. During plastic deformation, rocks change shape without breaking.

> Stress can be compressional, tensional, or shear. Pressure refers to a condition in which a material feels the same amount of compression in all directions.

> Strain refers to the shape change that materials undergo when subjected to a stress. For example, compression can cause shortening, and tension can cause stretching.

> Deformation results in the development of geologic structures.

> Joints are natural cracks in rock, formed in response to tension under brittle conditions. Veins develop when minerals precipitated from water fill cracks.

> Faults are fractures on which there has been shear. Geologists distinguish among normal, reverse, strike-slip, and oblique-slip faults based on the direction of movement.

> Folds are curves in rock layers. Anticlines are arch-like, synclines are trough-like, monoclines resemble the shape of a carpet draped over a stair step, basins are shaped like an upright bowl, and domes are shaped like an overturned bowl.

> Foliation forms when grains flatten, rotate, or grow so that they align parallel with one another.

> Mountain belts can form at convergent margins along continents, in collision zones, and in rifts. At convergent-boundary and collisional orogens, intense deformation and metamorphism can take place. Fold-thrust belts develop along the margins of such orogens. Rifting yields fault-block mountains.

> Small crustal blocks that collide with plate margins become accreted terranes. Accretion results in lateral growth of a continent over time.

> Large collisional mountain ranges are underlain by deep roots and contain folds, faults, and foliations.

> With modern GPS technology, it is now possible to measure the slow shortening and uplift of mountains.

> Uplift in mountains, over broad regions, is controlled by isostasy, meaning that the elevation of the Earth's surface reflects the level at which lithosphere naturally rests on the asthenosphere.

> Once uplifted, mountains are sculpted by erosion. When crust thickens during mountain building, the deep crust eventually becomes weak, leading to orogenic collapse.

> Cratons are the old, relatively stable parts of continental crust. They include shields and platforms. Broad regional domes and basins typically form in platform areas.

GUIDE TERMS

accretion (p. 317)
anticline (p. 312)
axial surface (p. 312)
basin (p. 313)
bearing (p. 309)
brittle deformation (p. 304)
compression (p. 306)
craton (p. 326)
cratonic platform (p. 327)
crustal root (p. 323)
dip (p. 309)
displacement (p. 309)

dome (p. 313)
epeirogeny (p. 327)
exhumation (p. 325)
fault (p. 307)
fault scarp (p. 310)
fold (p. 312)
foliation (p. 317)
global positioning system (GPS) (p. 319)
hinge (p. 312)
isostasy (p. 323)
joint (p. 307)

limb (of fold) (p. 312)
monocline (p. 313)
mountain belt (p. 301)
mountain building (p. 301)
normal fault (p. 310)
oblique-slip fault (p. 310)
orogen (p. 301)
orogenic collapse (p. 325)
orogeny (p. 301)
plastic deformation (p. 304)
plunge (p. 309)
pressure (p. 307)

reverse fault (p. 310)
shield (p. 327)
slickensides (p. 312)
strain (p. 302)
stress (p. 306)
strike (p. 309)
strike-slip fault (p. 310)
suture (p. 317)
syncline (p. 313)
thrust fault (p. 310)
vein (p. 307)

 GEOTOURS *THIS CHAPTER'S GEOTOURS WORKSHEET (I) FEATURES QUESTIONS AND GOOGLE EARTH SITES ON:*

> Calculating strike and dip from flatirons   > Fractures   > Normal, reverse, and strike-slip faults   > Folds, domes, and basins

REVIEW QUESTIONS

The letters following each Review Question refer to the corresponding Learning Objective from the Chapter Opener.

1. What changes do rocks undergo during formation of a mountain belt? **(A, B)**

2. Contrast brittle and plastic deformation. **(C)**

3. What factors determine whether a rock will behave in brittle or plastic fashion? **(C)**

4. Discuss the differences between stress and strain. **(B)**

5. How is a fault different from a joint? **(D)**

6. Compare normal, reverse, strike-slip, and oblique-slip faults. Which type of fault does the figure show? **(D)**

7. How do you recognize faults in the field? **(D)**

8. Describe the differences among an anticline, a syncline, and a monocline. Which is depicted in the figure? **(D)**

9. Discuss the relationship between foliation and deformation. **(D)**

10. Discuss the processes by which mountain belts form in convergent margins, during collisions, and in continental rifts. **(E)**

11. Describe the causes of uplift and the concept of isostasy. **(F)**

12. How are the structures of a craton different from those of an orogenic belt? **(G)**

ON FURTHER THOUGHT

13. Imagine that a geologist sees two outcrops of resistant sandstone, as depicted in the cross-section sketch. The region between the outcrops is covered by soil. A distinctive bed of cross-bedded sandstone occurs in both outcrops, so the geologist correlated the western outcrop (on the left) with the eastern outcrop. The curving lines in the bed indicate the shape of the cross beds. Keeping in mind how cross beds form (see Chapter 6), sketch how the cross-bedded bed connected from one outcrop to the other, before erosion. What geologic structure have you drawn? **(D)**

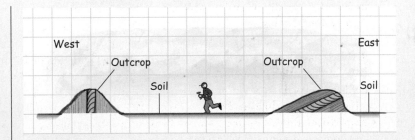

ONLINE RESOURCES

Animations This chapter features animations on types of rock deformation, faults, and folds.

Videos This chapter features a video on continental collision and the formation of mountains.

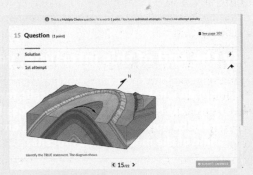

Smartwork5 This chapter features identification exercises on types of faults, folds, and deformations.

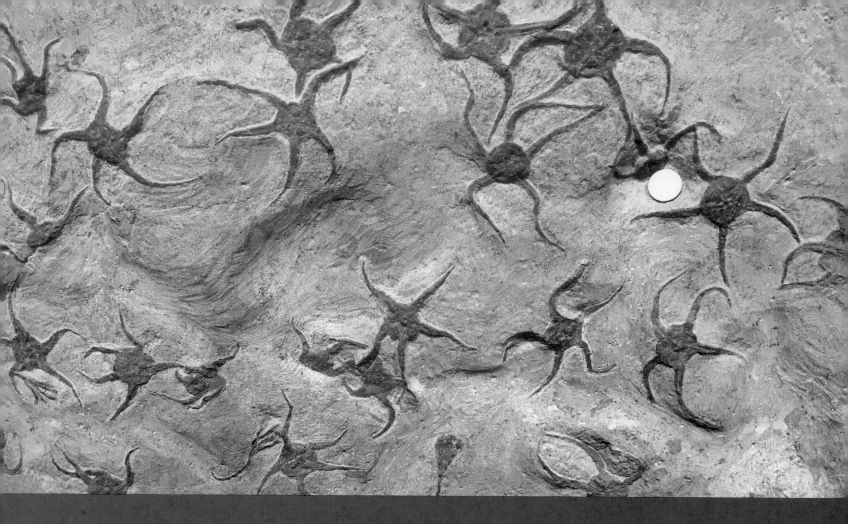

INTERLUDE E

MEMORIES OF PAST LIFE: FOSSILS AND EVOLUTION

By the end of this interlude, you should be able to . . .

A. explain what a fossil is, how fossils form, and why relatively few organisms become fossils.

B. describe how fossils are classified, and recognize some of the more common fossils.

C. describe how the study of fossils contributes to an understanding of life's evolution.

D. discuss the concept of extinction, and list the various phenomena that can cause extinction.

E.1 **The Discovery of Fossils**

If you look at bedding surfaces of sedimentary rocks, you may find shapes that resemble shells, bones, leaves, or footprints (Fig. E.1a, b). The origin of these shapes mystified early scientists. Some thought that the shapes had simply grown underground in place. Today geologists interpret such **fossils**—from the Latin word *fossilis*, which means dug up—to be the remnants or traces of ancient living organisms that were buried with the material from which the rock formed and were preserved after lithification. Surprisingly, this interpretation, though proposed by the Greek historian Herodotus in 450 B.C.E. and revived by Leonardo da Vinci in 1500 C.E., did not become widely accepted until the 1669 publication of a book in which a Danish physician named Nicolas Steno (1638–1686) argued that components of organisms could be incorporated in rock without losing their distinctive shape.

An understanding of fossils increased greatly thanks to the efforts of a British scientist, Robert Hooke (1635–1703), who described the characteristics of many fossil species. (A *species* is a distinct group of organisms capable of breeding.) Hooke eventually realized that most fossils represent *extinct species*, those that lived in the past but no longer survive anywhere today. The concept of extinction remained controversial until Georges Cuvier (1769–1832), a French zoologist, carefully demonstrated that the skeletons and teeth of fossil elephant-like mammoths and mastodons differ from those of any modern elephants. These days, we take for granted that species can go extinct, because we've seen a great number vanish during human history. Before Cuvier, however, most people thought that fossil species must have living relatives hidden somewhere on this planet. Considering that large parts of the Earth remained unexplored, this idea wasn't so far-fetched. But by the end of the 18th century, it became clear that many fossil organisms had no modern-day counterparts anywhere—mastodons and mammoths, for example, were too big to hide.

Gradually, **paleontology**, the study of fossils, ripened into a science. Cuvier and others described and classified fossils using an approach similar to the way biologists classify living organisms. **Paleontologists**, scientists who study fossils, described thousands of specimens and established museum collections to preserve and display them (Fig. E.1c). The value of research on fossils increased when William Smith, a British engineer who surveyed canal construction sites in England during the 1790s, noted that different fossil organisms occur in different layers of strata within a sequence of

◀ (facing page) Fossil brittle stars of Ordovician age on a slab of rock from Morocco. The coin, for scale, is 2.5 cm across.

FIGURE E.1 Fossils, and their collection.

(a) Fossil shells in a 400-Ma sandstone (Ma = million years old or ago).

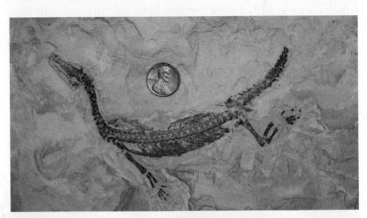

(b) A fossil skeleton in 200-Ma sandstone.

(c) A drawer of fossil specimens in a museum.

sedimentary rocks. In fact, Smith realized that strata contain a predictable succession of fossils, and that a given species can be found only in a specific interval of strata. After this discovery, researchers could use fossils as a basis for determining the age of one sedimentary rock layer relative to another, and for documenting the order in which species appeared and went extinct in the geologic record. In this way, fossils became an indispensable tool for studying geologic history, and for documenting **evolution**, the progressive change over time in characteristics of species that has led to the appearance of new species. In this interlude, we introduce fossils and discuss how their study teaches us about life's evolution. An understanding of fossils serves as essential background for the next two chapters.

E.2 Fossilization

WHAT KINDS OF ROCKS CONTAIN FOSSILS?

Most fossils are found in sedimentary rocks. Fossils sometimes occur in tuff, a rock formed from volcanic ash, but they are not found in other igneous rocks and tend to be destroyed by metamorphism. Fossils can form when organisms die and

become buried, or when their burrows and footprints become buried. The degree of a fossil's preservation reflects the context of burial. For example, rocks formed from sediment deposited under *anoxic* (oxygen-poor) conditions in quiet water (such as lakebeds or lagoons) can preserve particularly fine specimens. In contrast, rocks made from sediment deposited in high-energy environments—where strong currents tumble the remains of organisms and break them up—contain at best only small fragments of fossils mixed with other grains.

FORMING FOSSILS

Paleontologists refer to the process of forming a fossil as **fossilization**. To see how a typical fossil develops in sedimentary rock, let's consider an example **(Fig. E.2)**. Imagine an elderly dinosaur searching for food along the muddy shore of a lake on a scalding summer day in the geologic past. The hungry dinosaur succumbs to the heat and collapses dead into the mud. Soon after, scavengers strip the skeleton of meat

FIGURE E.2 The stages in the fossilization of a dinosaur.

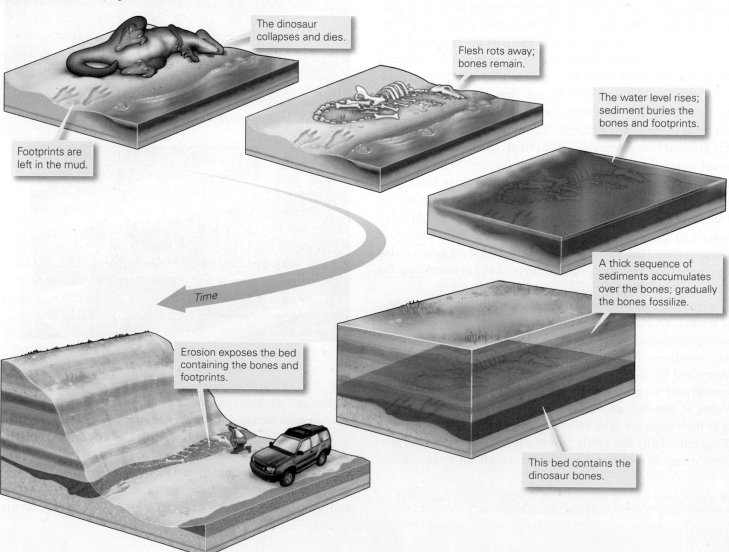

The dinosaur collapses and dies.

Footprints are left in the mud.

Flesh rots away; bones remain.

The water level rises; sediment buries the bones and footprints.

A thick sequence of sediments accumulates over the bones; gradually the bones fossilize.

Time

Erosion exposes the bed containing the bones and footprints.

This bed contains the dinosaur bones.

and scatter the bones. But before the bones have had time to weather away, the river floods and buries the bones, along with the dinosaur's footprints, under a layer of silt. More silt from succeeding floods buries the bones and prints still deeper, so that the bones cannot be reworked by currents or disrupted by burrowing organisms. Later, sea level rises and a thick sequence of marine sediment buries the fluvial sediment. Eventually, the sediment containing the bones and footprints turns to rock (siltstone and shale). The footprints remain outlined by the contact between the siltstone and shale, while the bones reside within the siltstone. Minerals precipitating from groundwater passing through the siltstone gradually replace some of the chemicals constituting the bones, until the bones themselves have become rock-like. The buried bones and footprints are now fossils.

One hundred million years later, uplift and erosion expose the dinosaur's grave. Part of a fossil bone protrudes from a rock outcrop. A lucky paleontologist observes the fragment and starts excavating, gradually uncovering enough of the bones to permit reconstruction of the beast's skeleton. Further digging uncovers the footprints. The dinosaur rises again, this time in a museum. In recent years, bidding wars have made some fossil finds extremely valuable. For example, a skeleton of a *Tyrannosaurus rex*, a 67-million-year-old dinosaur, sold at auction in 1997 for $7.6 million. The specimen, named Sue after its discoverer, now stands in the Field Museum of Chicago.

Similar tales can be told for fossil seashells buried by sediment settling in the sea, for insects trapped in *amber* (hardened tree sap), and for mammoths drowned in the muck of a tar pit. In all cases, fossilization involves the burial and preservation of an organism or the trace (a footprint or burrow) of an organism.

MANY KINDS OF FOSSILS

Perhaps when you think of a fossil, you picture either a dinosaur bone or the imprint of a seashell in rock. In fact, paleontologists distinguish among many different kinds of fossils, according to the specific way in which the organisms were fossilized. Let's look at examples of these categories.

› *Frozen or dried body fossils:* In a few environments, whole bodies of organisms may be preserved. Most of these fossils are fairly young, by geologic standards, in that their ages can be measured in thousands, not millions, of years. Examples include woolly mammoths that became incorporated in the permafrost (permanently frozen ground) of Siberia **(Fig. E.3a)**. In desert climates, organisms can become desiccated (dried out) and can survive for millennia in caves.

› *Body fossils preserved in amber or tar:* Insects landing on the bark of trees may become trapped in the sticky sap or resin the trees produce. This golden syrup envelops the insects and over time hardens into amber, the

semiprecious "stone" used for jewelry. Amber can preserve insects, as well as other delicate organic material such as feathers, for 40 million years or more **(Fig. E.3b)**. Tar similarly acts as a preservative. In isolated regions where oil has seeped to the surface, the more volatile components of the oil evaporate away and bacteria degrade what remains, leaving behind a sticky residue of tar. At the La Brea Tar Pits in Los Angeles, tar accumulated in a swampy area. While grazing, drinking, or hunting at the swamp, animals became mired in the tar and sank into it. Their bones have been remarkably well preserved for over 40,000 years **(Fig. E.3c)**.

› *Preserved or replaced bones, teeth, and shells:* Bones (the internal skeletons of vertebrate animals) and shells (the external skeletons of invertebrate animals) consist of durable minerals, which may survive in rock. Some bone or shell minerals are not stable and may recrystallize. But even when this happens, the shape of the bone or shell can be preserved.

› *Molds and casts of bodies:* When a shell pushes into soft sediment, a **mold** in the shape of the shell forms, and if the inside of the shell then fills with more sediment, that forms a **cast** of the shell **(Fig. E.3d)**. (Similar terms are used by sculptors who fill a mold with molten metal to form a cast of a statue.) Eventually, when the sediment turns into rock, the mold and cast display the shape of the shell and remain in place, even if the shell itself dissolves away. Typically, a mold appears as an indentation into a bed of rock, whereas a cast protrudes from the surface of a bed of rock.

› *Carbonized impressions of bodies:* Impressions are simply flattened molds and casts produced when soft or semi-soft organisms (leaves, insects, shell-less invertebrates, sponges, feathers, jellyfish) are pressed between layers of sediment **(Fig. E.3e)**. Chemical reactions eventually remove the organic material, leaving only a thin film of carbon on the surface of the impression.

› *Permineralized fossils:* **Permineralization** refers to the process by which minerals precipitate in porous material, such as wood or bone, underground. The ions from which the minerals grow come from groundwater solutions that slowly seep into the pores. **Petrified wood**, an example of a permineralized fossil, forms when volcanic ash buries a forest. Groundwater passing through the ash dissolves silica and carries it into the wood, where the silica replaces cell interiors. Eventually, the wood completely transforms into hard chert. (The word *petrified* literally means turned to stone.) During the process, the cell walls of the wood transform into organic films that survive permineralization, so you can see the fine detail of the wood's cell structure and bark in a petrified log **(Fig. E.3f)**. The colors of petrified logs come from impurities such as iron or carbon in the chert.

FIGURE E.3 Examples of different kinds of fossils.

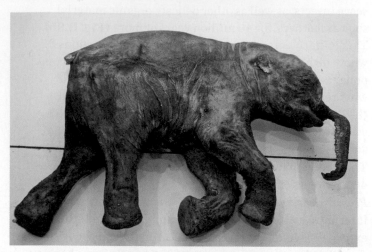

(a) This 1-m-long baby mammoth, found in Siberia, died 37,000 years ago.

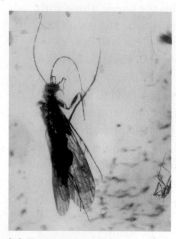

(b) This insect became embedded in amber about 200 million years ago.

(c) A fossil skeleton of a 2-m-high giant ground sloth from the La Brea Tar Pits in California.

(d) Fossil shells, the hard parts of invertebrates.

(e) The carbonized impressions of fern fronds in a shale.

(f) Petrified wood from Arizona. It is so hard that it remains after the rock that surrounded it has eroded away.

> *Trace fossils:* These include footprints, feeding traces, burrows, and dung that organisms leave behind in sediment **(Fig. E.4a)**. Paleontologists refer to fossilized dung as a *coprolite.*

> *Chemical fossils:* Living plants or animals consist of complex organic chemicals. When buried with sediment and subjected to diagenesis, some of these chemicals are destroyed, some remain intact, and some break down to form different, but still distinctive, chemicals. A distinctive chemical derived from an organism and preserved in rock is called a *chemical fossil*, a *molecular fossil*, or a **biomarker.**

Paleontologists also find it useful to distinguish among different fossils on the basis of their size. **Macrofossils** are large enough to be seen with the naked eye. But some rocks and sediments also contain abundant **microfossils**, which can be

seen only with a microscope or even an electron microscope **(Fig. E.4b)**. Microfossils include remnants of plankton, algae, bacteria, and pollen.

FOSSIL PRESERVATION

Not all living organisms become fossils when they die—in fact, only a small percentage do. It takes special circumstances to produce a fossil and for a fossil to survive over geologic time. A number of conditions can affect the degree to which a recognizable fossil can be preserved:

> *Oxygen content of the depositional environment:* A dead squirrel by the side of the road will not become a fossil. As time passes, birds, dogs, or other scavengers may come along and eat the carcass. And if that doesn't happen,

FIGURE E.4 Examples of trace fossils and microfossils.

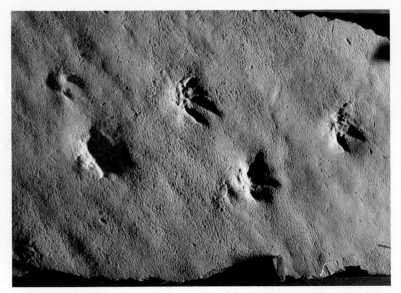

(a) Molds of dinosaur footprints. These traces indent the top of a bed. Each footprint is about 15 cm across.

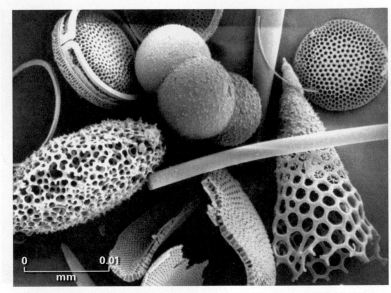

(b) Examples of fossil plankton shells. Because of their size, geologists refer to these fossils as microfossils.

maggots, bacteria, and fungi infest the carcass and gradually decompose it. As the remaining flesh rots in the air, its original organic chemicals can oxidize (react with oxygen), forming new chemicals that wash away in water or escape as gas. Once exposed, the skeleton weathers and turns to dust, so before roadkill can become incorporated in sediment, it has vanished. If, however, a carcass settles into an anoxic environment, oxidation occurs slowly, scavenging organisms are scarce, and bacterial metabolism takes place very slowly. In such environments, the organism will have a chance to be buried and preserved, so the likelihood increases that the organism will become fossilized.

> *Rapid burial:* If an organism dies in a depositional environment where sediment accumulates rapidly, it may be buried before it has time to rot, oxidize, be eaten, or undergo weathering. For example, if a storm suddenly buries an oyster bed with a thick layer of sediment, the oysters die and their shells become part of the sedimentary rock derived from the sediment.

> *The presence of hard parts:* Organisms without durable shells, skeletons, or other such *hard parts* usually won't be fossilized, for under most depositional conditions soft flesh decays long before hard parts do. For this reason, paleontologists have learned more about the fossil record of bivalves (a class of organisms, including clams and oysters, with strong shells) than they have about the fossil record of jellyfish (which have no shells) or spiders (which have very fragile shells).

By carefully studying modern organisms, paleontologists have been able to provide rough estimates of the **preservation potential** of organisms, meaning the likelihood that an organism will be buried and eventually transformed into a fossil. For example, in a typical modern-day shallow-marine environment, such as the mud-and-sand-covered seafloor close to a beach, about 30% of the organisms have sturdy shells and thus a high preservation potential, 40% have fragile shells and a lower preservation potential, and the remaining 30% have no hard parts at all and are not likely to be fossilized except in special circumstances. Out of the 30% with sturdy shells, though, only a small number of organisms happen to die in a depositional setting where they become fossilized. Consequently, fossilization is the exception rather than the rule.

EXTRAORDINARY FOSSILS: A SPECIAL WINDOW TO THE PAST

Although only hard parts survive in most fossilization environments, paleontologists have discovered a few special locations where rock contains fossils of soft parts as well—such fossils are known as **extraordinary fossils (Fig. E.5).** Examples can include tissue, fur, feathers, or their impressions. We've already discussed how extraordinary fossils of insects and feathers can be preserved in amber. Extraordinary fossils have also been found in deposits of tar pits and in fine-grained limestone or shale lithified from sediment that accumulated on the anoxic floors of quiet-water lakes, lagoons, or the deep ocean.

FIGURE E.5 Extraordinary fossils. These examples are particularly well preserved.

(a) A 50-million-year-old mammal fossil was chiseled from oil shale near Messel, Germany. It still contains the remains of skin.

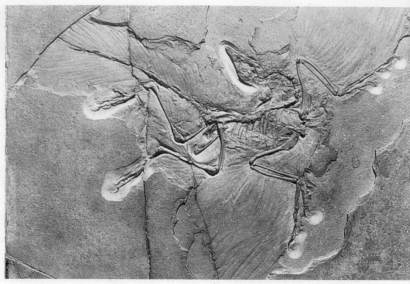

(b) *Archaeopteryx* from the 150-million-year-old Solnhofen Limestone of Germany. The imprints of feathers are clearly visible.

E.3 **Taxonomy and Identification**

ORGANIZING LIFE

The principles of **taxonomy**, the study of how to classify organisms in a systematic way, were first proposed in the 18th century by Carolus Linnaeus, a Swedish biologist. Linnaeus based his classification scheme, which divides organisms into a hierarchy of divisions, or *taxa*, on observable similarities in form and function. Initially, researchers applied taxonomy only to present-day organisms. Cuvier and his contemporaries realized that the same concept applies to fossils.

Modern taxonomists divide all life into three **domains**: Archaea, Bacteria, and Eukarya. (Note that we capitalize these terms only when talking about domain names.) The domains differ from one another based on fundamental characteristics of their DNA. *Archaea* include a vast array of tiny single-celled microorganisms (microbes) that occur not only in the mild environments of oceans, soils, and wetlands, but also in the harsh environments of hot springs, black smokers, salt lakes, very acidic streams, and deep subsurface water. Microbes that can survive in harsh environments are known as *extremophiles*—they do not need light, but rather live off the energy stored in the chemical bonds of minerals. *Bacteria* are also tiny single-celled microbes that inhabit almost all livable environments on the Earth. Although it may be difficult to distinguish cells of bacteria from archaea visually, on a genetic level they differ profoundly. Both archaea and bacteria are *prokaryotes*, meaning that their cells do not contain a nucleus, a distinct envelope containing the cell's DNA.

Taxonomists divide *Eukarya*—the domain of *eukaryotes*, organisms whose cells contain a nucleus—into **kingdoms**. Traditionally, these include: *Protista* (various unicellular and simple multicellular organisms, including algae); *Fungi* (mushrooms and yeast); *Plantae* (such as trees, grasses, mosses, and ferns); and *Animalia* (such as sponges, corals, dinosaurs, birds, and humans). Each kingdom consists of one or more *phyla*. A phylum, in turn, includes several *classes*, a class includes several *orders*, an order includes several *families*, a family includes several *genera*, and a genus includes one or more *species*. Kingdoms, therefore, represent the broadest category of eukaryotic life, and species represent the narrowest. **Table E.1** illustrates how taxonomic classification works for humans.

IDENTIFYING FOSSILS

There's nothing magical about identifying fossils. You can often recognize common fossils in the field simply by examining their **morphology** (form or shape). If the fossil is well preserved and has distinctive features, the process can be straightforward, but if the fossil broke into fragments and parts are missing, or if the fossil shares many characteristics with other fossils, identification can be challenging.

Many fossil organisms resemble modern organisms, so it's relatively easy to figure out how to classify them taxonomically. For example, a fossil clam (class Bivalvia) looks like a clam and not like, say, a snail (class Gastropoda). At the taxonomic levels below class, identification may involve recognizing such details as the number of ridges on the surface of the shell. Not all fossils resemble living organisms, so figuring

TABLE E.1 Taxonomic Classification of Humans

Domain	Eukarya	Organisms composed of cells containing nuclei
Kingdom	Animalia	All animals
Phylum	Chordata	Animals with backbones
Class	Mammalia	Chordates with fur, warm blood, and ability to secrete milk
Order	Primates	Great apes, monkeys, and humans
Family	Hominidae	Great apes (gorillas, chimpanzees, orangutans) and humans
Genus	*Homo*	Humans, as well as Neanderthals and other extinct species
Species	*sapiens*	Modern humans (first appearing about 200,000 years ago)

FIGURE E.6 Common types of invertebrate fossils.

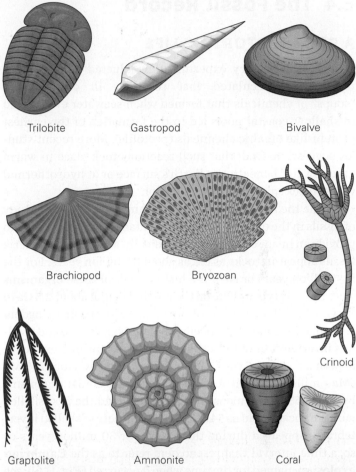

Trilobite   Gastropod   Bivalve

Brachiopod   Bryozoan

Crinoid

Graptolite   Ammonite   Coral

(a) Sketches of representative fossils.

out their taxonomic relation to other organisms may be a challenge, and some interpretations remain controversial.

Figure E.6a, which illustrates common invertebrate fossils, can help you identify many of the fossils you might find in a bed of sedimentary rock. We list the names and notable characteristics of these fossils below:

> *Trilobites:* These arthropods have a segmented shell that is divided lengthwise into three parts.

> *Gastropods (snails):* Most fossil specimens of gastropods have a spiral shell that does not contain internal chambers.

> *Bivalves (clams and oysters):* These invertebrates have a shell that can be divided into two similar halves. The plane of symmetry lies parallel to the plane of the shell.

> *Brachiopods (lamp shells):* The top and bottom parts of lamp shells have different shapes, and the plane of symmetry lies perpendicular to the plane of the shell. Fossils typically have ridges radiating out from the hinge.

> *Bryozoans:* These invertebrates are colonial animals. Their fossils resemble a screen-like grid. Each box in the grid represents the shell of a single animal.

> *Crinoids (sea lilies):* These organisms look like flowers but are actually animals. They have a stalk consisting of numerous circular plates stacked one on top of the other.

> *Graptolites:* These fossils look like tiny saw blades. They are remnants of colonial animals that floated in the sea.

> *Cephalopods:* These squid-like organisms include ammonites, with a spiral shell, and nautiloids, with a straight shell. Their shells contain internal chambers and have ridged surfaces.

> *Corals:* These invertebrates include colonial organisms that form distinctive mounds or columns in tropical reefs. Paleozoic examples include solitary individuals with a cone-like shell.

(b) An artist's reconstruction of a Cambrian ecosystem. The painting is based on Burgess Shale fossils.

E.4 **The Fossil Record**

A BRIEF HISTORY OF LIFE

Based on laboratory experiments conducted in the 1950s, researchers speculated that reactions in concentrated "soups" of chemicals that formed when seawater evaporated in shallow, coastal pools led to the formation of the earliest protein-like organic chemicals (protolife). More recent studies suggest, instead, that such reactions took place in warm groundwater beneath the Earth's surface or at hydrothermal vents (black smokers) on the seafloor.

While the nature of protolife remains a mystery, the study of fossils in the oldest-known sedimentary rocks has started to provide an image of early life. Specifically, archaea and bacteria fossils appear in rocks as old as about 3.7 billion years. For the first billion years or so of life history, cells of these organisms were the only types of life on the Earth. Then, at about 2.5 Ga to 3.8 Ga (Ga = billion years ago), organisms of the protist kingdom first appeared. Early multicellular organisms (fungi, and shell-less invertebrates of the animal kingdom) came into existence at perhaps 1.0 to 1.5 Ga. Green algae appeared by about 750 Ma (Ma = million years ago), and complex multicellular organisms by about 635 Ma. Paleontologists have found the first shelled fauna in rocks as old as 541 Ma. A great variety of shelled invertebrates appeared during the next 20 or 30 million years or so, a time interval that researchers refer to as the **Cambrian explosion**, named for the time when it occurred (**Fig. E.6b**; see Chapter 10). The succession of fossils found in the stratigraphic record indicates that fish, land plants, amphibians, reptiles, and finally birds and mammals appeared in sequence over the next several hundred million years.

Researchers have been working hard to understand the **phylogeny** (evolutionary relationships) among organisms, using both the morphology of organisms and, more recently, the study of genetic material. Ideas about which taxa radiated from which ancestors are shown in a chart called the *tree of life,* or more formally, the **phylogenetic tree** (**Fig. E.7**). Study of the DNA from different organisms has enabled researchers to understand the relationship between molecular processes and evolutionary change.

IS THE FOSSIL RECORD COMPLETE?

By some estimates, more than 250,000 species of fossils have been collected and identified to date, by thousands of paleontologists working on all continents during the past two centuries. These fossils define the framework of life's evolution. But the record is not complete—known fossils cannot account for every intermediate step in the evolution of every organism.

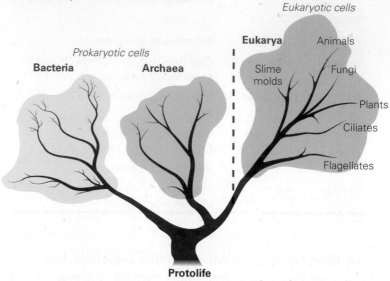

FIGURE E.7 A phylogenetic tree. The tree of life symbolically illustrates relations among different domains of organisms. Archaea and Bacteria are both prokaryotes and don't have nuclei. All other life forms are Eukarya because they have cells with a nucleus.

Over the billions of years that life has existed, 5 billion to 50 billion species of life may have existed. Clearly, known fossils represent at most a tiny percentage of these species. Why is the fossil record so incomplete?

First, despite all the fossil-collecting efforts of the past two centuries, paleontologists have not even come close to sampling every cubic centimeter of sedimentary rock exposed on the Earth. Just as biologists have not yet identified every living species of insect, paleontologists have not yet identified every species of fossil. New species and even genera of fossils continue to be discovered every year.

Second, not all organisms are represented in the rock record, because not all organisms have a high preservation potential. As noted earlier, fossilization occurs only under special conditions, so only a minuscule fraction of the organisms that have lived on the Earth have left a fossil record. We will never be aware of the existence or appearance of the vast numbers of organisms that have not been fossilized.

Finally, as you will learn in Chapter 10, the sequence of sedimentary strata that exists on the Earth does not account for every moment of time at every location since the formation of our planet. Sediments accumulate only in environments where conditions are appropriate for deposition and not for erosion. They do not accumulate, for example, on dry plains or on steep mountain peaks, but they do accumulate in the sea and in the floodplains and deltas of rivers. Because the Earth's climate changes through time and because the sea level rises and falls, certain locations on continents are sometimes sites of deposition and sometimes aren't, and on occasion are sites of erosion. Consequently, strata accumulate only episodically.

E.5 **Evolution and Extinction**

DARWIN'S GRAND IDEA

As a young man in England in the early 19th century, Charles Darwin had been unable to settle on a career but had developed a strong interest in natural history. Therefore, he jumped at the opportunity to serve as a naturalist aboard HMS *Beagle* on an around-the-world surveying cruise. During the five years of the voyage, from 1831 to 1836, Darwin made detailed observations of plants, animals, and geology in the field, and he amassed an immense specimen collection from South America, Australia, and Africa. Just before Darwin departed, a friend gave him a copy of Charles Lyell's 1830 textbook *Principles of Geology*, which argued in favor of James Hutton's proposal that the Earth had a long history and that geologic time extended much further into the past than did human civilization.

A visit to the Galápagos Islands, off the coast of Ecuador, led to a turning point in Darwin's thinking during the voyage. The naturalist was most impressed with the variability of Galápagos finches. He marveled not only at the fact that different varieties of the bird occurred on different islands, but also at how each variety had adapted to utilize a particular food supply. Keeping in mind Lyell's writings about the long duration of Earth history, Darwin proposed that all the different species of finches had evolved from a single initial species that arrived on the Galápagos in the distant past. According to Darwin's hypothesis, when a population of finches became isolated, it gradually developed new traits. Such change could happen because offspring can differ from their parents, and new traits can be transferred to succeeding generations. If enough change accumulates over many generations, the living population ends up being so different from its distant ancestors that the population can be classified as a new species. During the course of evolution, old species vanish and new species appear. The accumulation of changes may eventually yield a population that differs so much from its ancestors that taxonomists consider it to be a new genus—greater differences may result in a new class, an order, or even a new phylum. Such change in a population over a succession of generations, due to the transfer of inheritable characteristics, as we noted earlier, is the process of *evolution*.

Darwin and his contemporary, Alfred Russel Wallace, not only proposed that evolution took place, but also came up with an explanation for why it occurs. They stated that populations of organisms cannot increase in numbers forever, because they are limited by competition for scarce resources in the environment. In nature, only organisms capable of survival can pass on their characteristics to the next generation. In each new generation, some individuals have characteristics that make them more fit, whereas some have characteristics that make them less fit. The fitter organisms are more likely to survive and produce offspring. As a result, the survivors' beneficial characteristics get passed on to the next generation. Darwin referred to the survival of the fittest as **natural selection**, because it occurs on its own in nature. If environmental conditions change, or if competitors enter the environment, species that do not evolve to better adapt to the new conditions eventually die off and become extinct.

Darwin's view of evolution by natural selection has been successfully supported by many observations over time, and has not been definitively disproven by any observation or experiment. Also, it can be used to make testable predictions. Thus, scientists now refer to the concept as the **theory of evolution by natural selection** or just the *theory of evolution* (see Box P.1 in the Prelude for the definition of a theory). The theory of evolution provides a conceptual framework in which to understand paleontology. By studying fossils in sequences of strata, paleontologists are able to observe progressive changes in a species or an assemblage of species through time, and can determine when some species die out and other species appear.

Because of the incompleteness of the fossil record, however, many questions remain as to the rates at which evolution takes place during the course of geologic time. Originally, it was assumed that evolution took place at a constant, slow rate, a concept known as *gradualism*. More recently, researchers have suggested that evolution occurs in fits and starts: it goes very slowly for quite a while and then, during a relatively short period, it takes place very rapidly. This concept is called *punctuated equilibrium*. Factors that could cause sudden accelerations in the rate of evolution include: (1) widespread extinction during which many organisms disappear, leaving ecological niches open for new species to colonize; (2) a relatively rapid change in the Earth's climate that puts stress on organisms—new species that can survive the new climate survive, whereas those that can't become extinct; (3) formation of new environments, as may happen when rifting splits apart a continent and generates a new ocean with new coastlines; and (4) the isolation of a breeding population.

Regardless of the process of evolution, survivability is not the same for different kinds of organisms. Some populations prove to be very durable, in that they survive as an identifiable genus or even species for long intervals of geologic time (tens of millions to over 100 million years). But others appear and then disappear within a relatively short interval of geologic time (less than a few million years).

EXTINCTION: WHEN SPECIES VANISH

Extinction occurs when the last members of a species die, so no parents can pass on their genetic traits to offspring. Some species become extinct as a population evolves into new species, whereas other species just vanish, leaving no hereditary offspring. Paleontologists include the following geologic factors among many phenomena that cause extinction:

> *Global climate change:* At times, the Earth's mean temperature has been significantly colder than today's, whereas at other times it has been much warmer. Because of a changing climate, an individual species may lose its habitat, and if it cannot adapt to the new habitat or migrate to stay with its old one, the species will disappear.

> *Tectonic activity:* Tectonic activity causes vertical movement of the crust over broad regions, changes in seafloor spreading rates, and changes in the amount of volcanism. These phenomena can modify the distribution and area of habitats. Species that cannot adapt die off.

> *Asteroid or comet impacts:* Many geologists have concluded that impacts of large meteorites with the Earth have been catastrophic for life. A large impact would send dust and debris into the atmosphere that could blot out the Sun and plunge the Earth into darkness and cold (see Chapter 19). Such a change, though relatively short lived, could interrupt the food chain.

> *Voluminous volcanic eruptions:* Several times during Earth history, incredible quantities of lava have spilled out onto the surface and vast volumes of ash and gas have spewed into the air. These eruptions, perhaps due to the rise of particularly large mantle plumes, may have altered the climate globally.

> *The appearance of a new predator or competitor:* Extinctions may happen simply because a new predator appears on the scene. Researchers suggest that this phenomenon explains the widespread extinction that occurred during the past 20,000 years, when a vast number of large mammal species vanished from North America. The timing of these extinctions appears to coincide with the appearance of the first humans (fierce predators) on the continent. If a more efficient competitor appears, the competitor steals an ecological niche from the weaker species, whose members can't obtain enough food and thus die off.

Some extinctions happen over long time intervals, when the replacement rate of a population simply becomes lower than the mortality rate. But others happen suddenly, when a cataclysmic event leads to the rapid extermination of many organisms. In 1870, for example, the population of passenger pigeons in North America exceeded 3 billion. Due to widespread hunting by people, the population dropped rapidly during the next two decades, and the last member of the species died in 1914.

Paleontologists have found that the total number of genera of fossils, which represents **biodiversity** (the overall variation of life), is different over time and has abruptly decreased at specific times during Earth history. A worldwide abrupt decrease in the number of fossil genera is called a **mass-extinction event**. At least five major mass-extinction events have happened during the past half-billion years **(Fig. E.8)**. These events define the boundaries between some of the major intervals into which geologists divide time. For example, a major extinction event marks the end of the Cretaceous Period, 66 million years ago. During this event, all dinosaur species (with the exception of their modified descendants, the birds) vanished, along with most marine invertebrate species. A huge extinction event also brought the Permian to a close. Significantly, the rate at which species have been disappearing during the past few centuries has been so rapid that some researchers and writers refer to our present time as the *sixth extinction*.

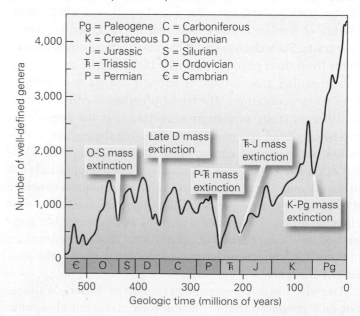

FIGURE E.8 This graph shows how the diversity of life has changed over time. Sudden drops indicate periods when mass extinctions occurred.

Interlude E Review

INTERLUDE SUMMARY

> Fossils form when organisms, or the traces of organisms (burrows or footprints) are buried and preserved in rock. Typically the hard parts (shells or bones) of fossils are better preserved.

> Examples of fossils include molds or casts of shells, bones, or footprints; permineralized wood or bones; carbonized impressions; preserved shells or bones; and bodies preserved in amber, tar, or permafrost. Certain distinctive molecules can serve as molecular fossils or biomarkers.

> The fossil record is very incomplete, because preserved strata do not record all of geologic time, not all organisms have high preservation potential, and paleontologists have only found a tiny fraction of fossil species.

> Fossils provide a record of life's evolution and a basis for geologists to determine the ages of rock layers relative to one another.

> Fossil organisms can be classified using the same taxonomic concepts used to classify modern organisms.

> Darwin's theory of evolution by natural selection states that the fittest organisms survive and pass on their traits to their offspring. Over time, organisms become so different from their distant ancestors that they can be considered to be a new species.

> When the last of a species dies out, the species goes extinct. At several times during the Earth's history, a high percentage of genera went extinct. These events are called mass extinctions, and may be a consequence of catastrophic events, such as meteorite impacts or unusually abundant volcanism.

GUIDE TERMS

biodiversity (p. 340)
biomarker (p. 334)
Cambrian explosion (p. 338)
cast (p. 333)
domain (p. 336)
evolution (p. 332)
extinction (p. 340)

extraordinary fossil (p. 335)
fossil (p. 331)
fossilization (p. 332)
kingdom (p. 336)
macrofossil (p. 334)
mass-extinction event (p. 340)
microfossil (p. 334)

mold (p. 333)
morphology (p. 336)
natural selection (p. 339)
paleontologist (p. 331)
paleontology (p. 331)
permineralization (p. 333)
petrified wood (p. 333)

phylogenetic tree (p. 338)
phylogeny (p. 338)
preservation potential (p. 335)
taxonomy (p. 336)
theory of evolution by natural selection (p. 339)

REVIEW QUESTIONS

The letters following each Review Question refer to the corresponding Learning Objective from the Chapter Opener.

1. What is a fossil? Give examples of various types. (**A**)

2. What are the various processes that can yield fossils? (**A**)

3. What conditions can lead to preservation of an organism as a fossil? (**A**)

4. On what basis do paleontologists classify fossils? What fossil organism does the figure show? (**B**)

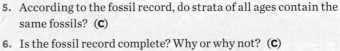

Brachiopod

5. According to the fossil record, do strata of all ages contain the same fossils? (**C**)

6. Is the fossil record complete? Why or why not? (**C**)

7. Describe the basic principles of the theory of evolution by natural selection. (**C**)

8. Explain different ideas characterizing the rate of evolution. (**C**)

9. What phenomena may cause extinction? (**D**)

10. What is a mass-extinction event, and why might such events take place? (**D**)

ONLINE RESOURCES

Animations This interlude features animations on fossil formation and preservation.

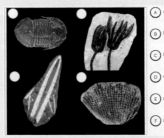

Smartwork5 This interlude features questions about fossils and life over geologic time.

CHAPTER 10

DEEP TIME:
HOW OLD IS OLD?

By the end of this chapter, you should be able to . . .

A. explain the meaning of geologic time and the difference between relative and numerical ages.

B. use geologic principles (uniformitarianism, cross-cutting relations, superposition, fossil succession) to determine relative ages.

C. draw a sketch illustrating how unconformities form and what they represent.

D. explain the basis for correlating stratigraphic formations, and show how correlation led to the development of the geologic column.

E. illustrate the concept of a half-life, and use it to explain how geologists determine the numerical age of rocks.

F. describe the basis for determining dates on the geologic time scale and the age of the Earth.

> If the Eiffel Tower were now representing the world's age, the skin of paint on the pinnacle-knob at its summit would represent man's share of that age; and anybody would perceive that that skin was what the tower was built for. I reckon they would, I dunno.
>
> MARK TWAIN (American writer, 1835–1910)

10.1 Introduction

In May of 1869, a one-armed Civil War veteran named John Wesley Powell set out with a team of nine geologists and scouts to explore previously unmapped expanses of the Grand Canyon, the greatest gorge on the Earth. Though Powell and his companions battled fearsome rapids and the pangs of starvation, most managed to emerge from the mouth of the canyon 3 months later. During their voyage, seemingly insurmountable walls of rock both imprisoned and amazed the explorers and led them to pose important questions about the Earth and its history, questions that even casual tourists at the canyon ponder today: Did the Colorado River sculpt this marvel, and if so, how long did it take? When did the rocks making up the walls of the canyon form? Was there a time before the colorful layers accumulated? Such questions pertain to **geologic time**, the span of time since the Earth's formation.

In this chapter, we first learn the geologic principles that allowed geologists to develop the concept of geologic time, and from these principles we develop a reference frame for describing the *relative ages* of rocks, fossils, structures, and landscapes. This information sets the stage for introducing the *geologic column*, the representation that geologists use to divide geologic time into named intervals. Then, we look at how *isotopic dating* (also known as *radiometric dating*) allows geologists to determine the *numerical age* of rocks, meaning the age of rocks in years, and how numerical ages can be added to the geologic column to produce the *geologic time scale*. With the concept of geologic time in hand, a hike down a trail into the Grand Canyon becomes a trip into what many authors have called *deep time*. The geologic discovery that our planet's history extends billions of years into the past changed humanity's perception of time and the Universe as profoundly as did the astronomical discovery that the limit of deep space extends billions of light-years beyond the edge of our Solar System.

◀ (facing page) This satellite image, available via Google Earth, reveals layer upon layer of strata deposited long ago and now exposed by erosion in the Dome Rock Mountains of Arizona. These layers preserve a partial record of the Earth's very long history.

10.2 The Concept of Geologic Time

SETTING THE STAGE FOR STUDYING THE PAST

Until relatively recently, people in most cultures believed that geologic time began about the same time that human history began, and that our planet has been virtually unchanged since its birth. This view was challenged by James Hutton (1726–1797), a Scottish gentleman farmer and doctor. Hutton lived during the Age of Enlightenment when, motivated by the discovery of physical laws by Sir Isaac Newton, many scientists sought natural rather than supernatural explanations for features of the world around them. While wandering in the highlands of Scotland, where rocks are well exposed, Hutton noted that many features (such as ripple marks and cross beds) found in sedimentary rocks resembled features that he could see forming in modern depositional environments. These observations led Hutton to speculate that rocks and landscapes formed, in general, as a consequence of processes that he could see happening in his own time.

Hutton's idea came to be known as the principle of **uniformitarianism**. According to this principle, physical processes that operate in the modern world also operated in the past, at roughly the same rates, and these processes were responsible for forming geologic features preserved in outcrops. More concisely, the principle can be stated as "the present is the key to the past." Hutton deduced that the development of individual geologic features took a long time, and that not all features formed at the same time. Therefore, the Earth must have a history that includes a succession of slow geologic events. Since no one in recorded history has seen the entire process of sediment first turning into rock and then later rising into mountains, Hutton also realized that the Earth's processes must have been active for a long time before human history began.

Hutton was not a particularly clear writer, and it took years, and the efforts of subsequent geologists, to interpret and clarify the principle of uniformitarianism and to publicize its significance. Once this had been accomplished, geologists around the world began to apply their growing understanding of geologic processes to define and interpret geologic events of the Earth's past.

RELATIVE VERSUS NUMERICAL AGE

Like historians, geologists strive to establish the sequence of events that produced an array of geologic features (such as rocks, structures, and landscapes). When possible, they also try to find the date on which each event took place. We specify the age of one feature with respect to another in a sequence as its **relative age** and the age of a feature given in years as its **numerical age** or its *absolute age*. Geologists learned how to determine relative age long before they could determine numerical age, so we look next at the principles leading to relative-age determination.

> ### TAKE-HOME MESSAGE
>
> The principle of uniformitarianism—the present is the key to the past—implies that the Earth must be very old, for geologic processes happen slowly. Geologists distinguish between relative age (older or younger) and numerical age (how many years ago).
>
> ---
>
> **QUICK QUESTION** What observations led Hutton to propose uniformitarianism?

10.3 Relative Age

GEOLOGIC PRINCIPLES

The British geologist Charles Lyell (1797–1875) popularized a set of formal, usable geologic principles in the first modern textbook of geology (*Principles of Geology*, published between 1830–1833). These principles, described below, continue to provide the basic framework within which geologists read the record of Earth history and determine relative ages.

> *The principle of uniformitarianism:* As we've seen, the principle of uniformitarianism means that physical processes we observe operating today also operated in the past, at roughly comparable rates. Put concisely, "the present is the key to the past" **(Fig. 10.1a)**.

> *The principle of original horizontality:* Sediments on the Earth settle out of fluids in a gravitational field. The surfaces on which sediments accumulate (such as floodplains or the seafloor) are fairly flat. Therefore, layers of sediment, when first deposited, are fairly horizontal **(Fig. 10.1b)**. If sediments collect on a steep slope, they typically slide downslope before lithification, so they will not be preserved as sedimentary rocks. With this in mind, we conclude that folding, tilting, and faulting of sedimentary beds must occur after the beds were deposited.

> *The principle of superposition:* In a sequence of sedimentary rock layers, each layer must be younger than the one below, for a layer of sediment cannot accumulate unless there is already a substrate on which it can collect. Thus, the layer at the bottom of a sequence is the oldest, and the layer at the top is the youngest **(Fig. 10.1c)**.

> *The principle of lateral continuity:* Sediments generally accumulate in continuous sheets within a given region. When you see a sedimentary layer cut by a canyon, you can assume that the layer once spanned the area and was later eroded by the river that formed the canyon **(Fig. 10.1d)**.

> *The principle of cross-cutting relations:* If one geologic feature cuts across another, the feature that has been cut is older. For example, if an igneous dike cuts across a sequence of sedimentary beds, the beds must be older than the dike **(Fig. 10.1e)**. If a fault cuts across and displaces layers of sedimentary rock, then the fault must be younger than the layers. But if a layer of sediment buries a fault, the layer must be younger than the fault.

FIGURE 10.1 Examples of the major geologic principles.

Present-day mudcracks form in clay-rich sediment.

Ancient mudcracks in solid rock

(a) Uniformitarianism: The processes that form cracks in the dried-up mud puddle on the left also formed the mudcracks preserved in the ancient, solid rock on the right. We can see the ancient mudcracks because erosion removed the adjacent bed.

FIGURE 10.1 (*continued*)

Modern sediment, exposed at low tide, on the coast of France.

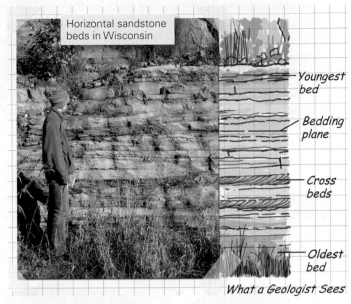

Horizontal sandstone beds in Wisconsin

Youngest bed

Bedding plane

Cross beds

Oldest bed

What a Geologist Sees

(b) Original horizontality: Gravity causes sediments to accumulate in horizontal sheets.

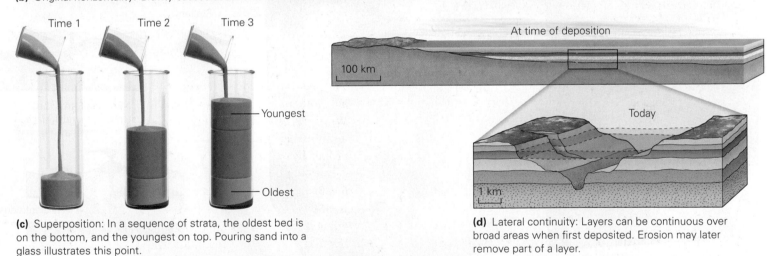

Time 1   Time 2   Time 3

Youngest

Oldest

(c) Superposition: In a sequence of strata, the oldest bed is on the bottom, and the youngest on top. Pouring sand into a glass illustrates this point.

At time of deposition

100 km

Today

1 km

(d) Lateral continuity: Layers can be continuous over broad areas when first deposited. Erosion may later remove part of a layer.

(e) Cross-cutting relations: The dike cuts across the lower three sedimentary beds, so the dike is younger. The top bed buries the dike, so it is younger than the dike.

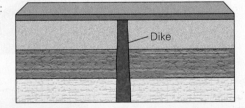

Dike

(f) Baked contacts: The pluton baked the adjacent rock, so the adjacent rock is older.

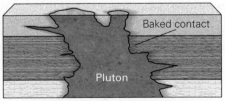

Baked contact

Pluton

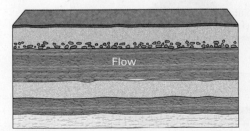

Flow

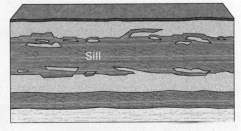

Sill

(g) By the principle of inclusions, the pebbles of basalt in a conglomerate must be older than the conglomerate, and xenoliths of sandstone must be older than the basalt containing them.

> *The principle of baked contacts:* An igneous intrusion "bakes" (metamorphoses) surrounding rocks, so the rock that has been baked must be older than the intrusion **(Fig. 10.1f)**.

> *The principle of inclusions:* A rock containing an *inclusion* (fragment of another rock) must be younger than the inclusion. For example, a conglomerate containing pebbles of basalt is younger than the basalt, and a sill containing fragments of sandstone must be younger than the sandstone **(Fig. 10.1g)**.

Geologists apply geologic principles to determine the relative ages of rocks, structures, and other geologic features at a given location. They then interpret the development of each feature to be the consequence of a specific *geologic event*.

Examples of geologic events include deposition of sedimentary beds, erosion of the land surface, intrusion or extrusion of igneous rocks, deformation (folding or faulting), and episodes of metamorphism. The succession of events in order of relative age that have produced the rock, structure, and landscape of a region is called the *geologic history* of the region.

By applying geologic principles, we can determine relative ages of the features shown in **Figure 10.2a**, and in so doing, we can develop a geologic history of the region. The geologic history depicted in this figure includes several events: deposition of the sedimentary sequence in order from beds 1 to 8; intrusion of the sill; folding of the sedimentary beds and the sill; intrusion of the granite pluton; faulting; intrusion of the dike; and erosion to form the present land surface **(Fig. 10.2b)**.

FIGURE 10.2 Interpreting the geologic history of a region, using geologic principles as a guide.

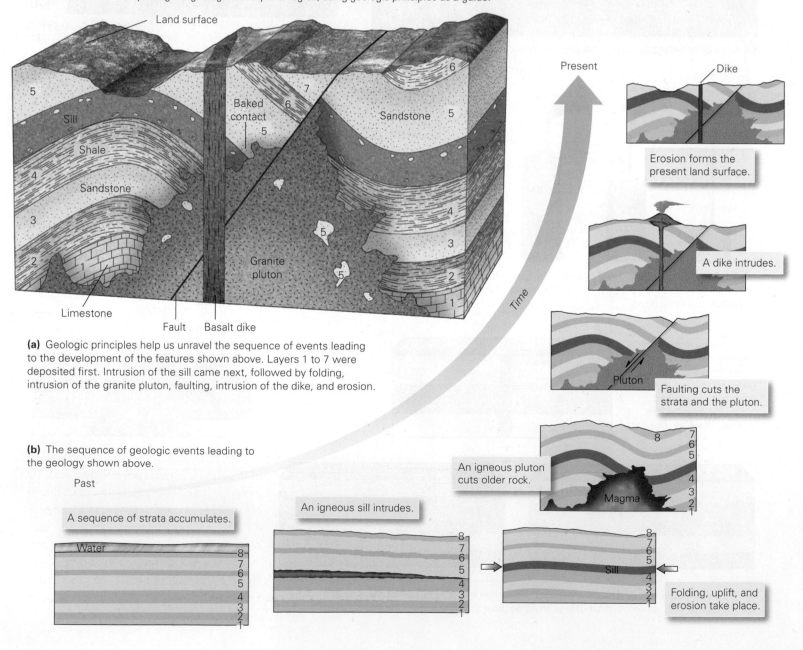

(a) Geologic principles help us unravel the sequence of events leading to the development of the features shown above. Layers 1 to 7 were deposited first. Intrusion of the sill came next, followed by folding, intrusion of the granite pluton, faulting, intrusion of the dike, and erosion.

(b) The sequence of geologic events leading to the geology shown above.

A sequence of strata accumulates.

An igneous sill intrudes.

Folding, uplift, and erosion take place.

An igneous pluton cuts older rock.

Faulting cuts the strata and the pluton.

A dike intrudes.

Erosion forms the present land surface.

FOSSIL SUCCESSION

As Britain entered the industrial revolution in the late 18th and early 19th centuries, new factories demanded coal to fire their steam engines. Companies decided to construct a network of canals to transport coal and iron, and they hired an engineer named William Smith (1769–1839) to survey the excavations. Canal digging provided fresh exposures of bedrock, which previously had been covered by vegetation. Smith learned to recognize distinctive layers of sedimentary rock and to identify the **fossil assemblage** (the group of fossil species) that they contained. He also realized that a particular assemblage can be found only in a limited interval of strata, and not above or below this position. Therefore, once a fossil species disappears at a horizon in a sequence of strata, it never reappears higher in the sequence. Put another way, extinction is forever. Smith's observation has been repeated at millions of locations around the world and has been codified as the *principle of fossil succession*. It provides the geologic underpinning for the theory of evolution (see Interlude E).

To see how this principle works, examine **Figure 10.3**, which depicts a sequence of strata. Bed 1 at the base contains fossil species A, Bed 2 contains fossil species A and B, Bed 3 contains B and C, Bed 4 contains C, and so on. From these data, we can define the *range* of specific fossils in the sequence, meaning the interval in which the fossils occur. Note that the sequence contains a definable succession of fossils (A, B, C, D, E, F), that the range where a particular species occurs may overlap with the range of other species, and that once a species vanishes, it does not reappear higher in the sequence. Some fossil species are widespread but survived for only a relatively short interval of geologic time. Such species are called **index fossils** (or *guide fossils*), because they can be used by geologists to associate the strata with a specific time interval. Because of the principle of fossil succession, we can define the relative ages of strata by looking at fossils. For example, if we find a bed containing fossil A, we can say that the bed is older than a bed containing, say, fossil F. Geologists have now determined the relative ages of over 200,000 fossil species.

FIGURE 10.3 The principle of fossil succession.

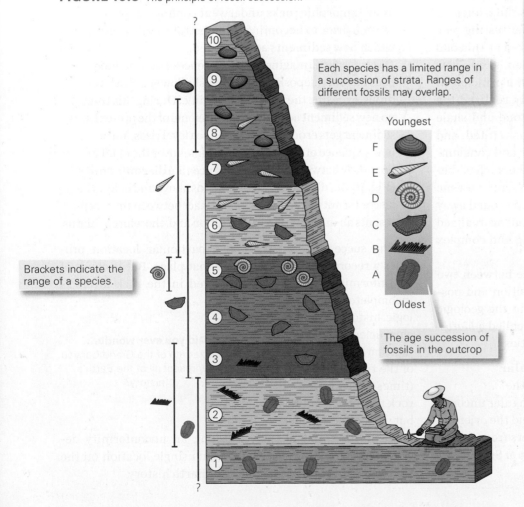

Each species has a limited range in a succession of strata. Ranges of different fossils may overlap.

Youngest

Brackets indicate the range of a species.

The age succession of fossils in the outcrop

Oldest

TAKE-HOME MESSAGE

Geologic principles (including uniformitarianism, superposition, cross-cutting relations, and fossil succession) provide the basis for determining the relative ages of rocks and other geologic features. By working out relative ages, we can reconstruct the geologic history of a region.

QUICK QUESTION If offset on a fault produces a fault scarp, what are the relative ages of the fault and the landscape surface?

10.4 Unconformities: Gaps in the Record

James Hutton used a boat to explore the seaside of Scotland, where shore cliffs provided great exposures of rock, stripped of soil and shrubbery. He was puzzled by an outcrop he found at Siccar Point on the east coast. Here he saw a sequence of strata consisting of red sandstone and conglomerate that rested on a distinctly different sequence of gray sandstone and shale (**Fig. 10.4**). Notably, the beds

FIGURE 10.4 The unconformity at Siccar Point, Scotland.

(a) The layers above were deposited long after the beds below had been tilted.

What a Geologist Sees

(b) The contact between the two units is an unconformity.

of gray sandstone and shale had a nearly vertical dip (slope), whereas the beds of red sandstone and conglomerate had a dip of less than 15°. (See Box 9.1 for the definition of dip.) The gently dipping layers seemed to lie across the truncated ends of the vertical layers, like a handkerchief resting on a row of books. Perhaps as Hutton sat and stared at this odd geometric relationship, the tide came in and deposited a new layer of sand on top of the rocky shore. With the principle of uniformitarianism in mind, Hutton suddenly realized the significance of the outcrop—the gray sandstone and shale sequence had been deposited, turned into rock, tilted, and truncated by erosion before the red sandstone and conglomerate beds had been deposited. Therefore, the surface between the gray and red rock sequences represented a time interval during which the gray strata had been eroded away and new strata had not yet accumulated. Hutton realized that Siccar Point exposed the record of a long and complex saga of geologic history.

Geologists now refer to a boundary surface between two units, which represents a period of nondeposition and possibly erosion, as an **unconformity**. The gap in the geologic record that is reflected in an unconformity is called a *hiatus*. We recognize three main types of unconformities.

> *Angular unconformity:* Rocks below an **angular unconformity** were tilted or folded before the unconformity developed **(Fig. 10.5a)**. An angular unconformity cuts across the underlying layers, and the orientation of layers below an unconformity differs from the orientation of the layers above. (The outcrop at Siccar Point exposes an angular unconformity.)

> *Nonconformity:* At a **nonconformity**, sedimentary rocks overlie generally much older intrusive igneous rocks or metamorphic rocks **(Fig. 10.5b)**. These igneous or metamorphic rocks underwent cooling, uplift, and erosion prior to becoming the substrate, or *basement*, on which new sediments accumulated.

> *Disconformity:* Imagine that a sequence of sedimentary beds has been deposited beneath a shallow sea. Sea level drops, exposing the beds for some time. During this time, no new sediment accumulates, and some of the pre-existing sediment gets eroded away. Later, sea level rises, and a new sequence of sediment accumulates over the old. The boundary between the two sequences is a **disconformity** **(Fig. 10.5c, d)**. Even though the beds above and below the disconformity are parallel, the contact between them represents an interruption in deposition and, therefore, a hiatus.

The succession of strata at a particular location provides a record of Earth history in that place. But because of unconformities, the record preserved in the rock layers is incomplete. It's as if geologic history were being chronicled by a recording that runs only some of the time—when it's on (times of deposition), the rock record accumulates, but when it's off (times of nondeposition and possibly erosion), an unconformity develops. Because of unconformities, no single location on the Earth contains a complete record of Earth history.

Did you ever wonder...
if the strata of the Grand Canyon represent all of the Earth's history?

FIGURE 10.5 The three kinds of unconformities and their formation.

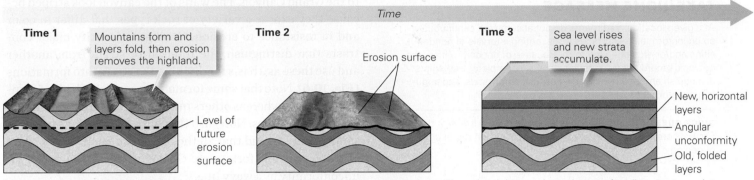

Time

Time 1 — Mountains form and layers fold, then erosion removes the highland.

Level of future erosion surface

Time 2 — Erosion surface

Time 3 — Sea level rises and new strata accumulate.

New, horizontal layers

Angular unconformity

Old, folded layers

(a) An angular unconformity: (1) layers undergo folding; (2) erosion produces a flat surface; (3) sea level rises and new layers of sediment accumulate.

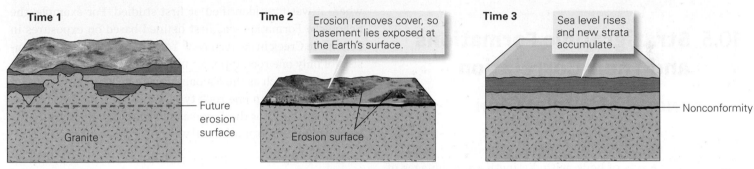

Time 1

Future erosion surface

Granite

Time 2 — Erosion removes cover, so basement lies exposed at the Earth's surface.

Erosion surface

Time 3 — Sea level rises and new strata accumulate.

Nonconformity

(b) A nonconformity: (1) a pluton intrudes; (2) erosion cuts down into the crystalline rock; (3) new sedimentary layers accumulate above the erosion surface.

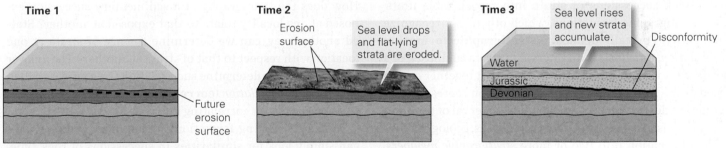

Time 1

Future erosion surface

Time 2 — Erosion surface

Sea level drops and flat-lying strata are eroded.

Time 3 — Sea level rises and new strata accumulate.

Disconformity

Water

Jurassic

Devonian

(c) A disconformity: (1) layers of sediment accumulate; (2) sea level drops and an erosion surface forms; (3) sea level rises and new sedimentary layers accumulate.

Channel (disconformity surface)

Paleosol

(d) This road cut in Utah shows a sand-filled channel cut down into floodplain mud. The mud was exposed between floods, and a soil formed on it. When later buried, all the sediment turned into rock; the channel floor is now a disconformity, and the ancient soil is now a paleosol. Note that the channel cut across the paleosol. The paleosol also represents a disconformity, a time during which deposition did not occur.

TAKE-HOME MESSAGE

At a given location, sediments do not accumulate continuously, so unconformities—surfaces representing intervals of nondeposition and/or erosion—can form. Because of unconformities, the geologic record at any given location is incomplete. Geologists distinguish among nonconformities, disconformities, and angular unconformities.

QUICK QUESTION Is there any one place on the surface of the Earth where the exposed stratigraphic succession represents all of geologic time? Explain.

10.5 Stratigraphic Formations and Their Correlation

THE CONCEPT OF A FORMATION

When William Smith first began to explore the strata exposed along the newly dug canals of England, he realized that distinctive sets of beds, with distinctive assemblages of fossils, crop out at many locations. Geologists now routinely divide thick successions of strata into recognizable units, called stratigraphic formations, which others can recognize and identify. Formally defined, a **stratigraphic formation** (or *formation*, for short) is an interval of strata composed of a specific rock type or group of rock types that together can be traced over a fairly broad region. A formation represents the products of deposition during a definable interval of time and typically consists of many beds. In some cases, geologists subdivide a formation into two or more *stratigraphic members*, or lump several formations together to form a **stratigraphic group** (or simply, a *group*). Geologists commonly use the informal term *stratigraphic unit* (or simply *unit*) for a set of beds, a member, a formation, or a group. The boundary surface between two formations is a type of **geologic contact** (or *contact*, for short). Fault surfaces, unconformities, and the boundary between an igneous intrusion and its wall rock are also types of contacts.

Geologists summarize information about the sequence of sedimentary strata at a location by drawing a **stratigraphic column**, a chart that depicts the order of the units (with the oldest at the bottom and the youngest at the top), the relative thicknesses of formations, and in some examples, the rock types within each unit. Geologists may also represent the relative resistance to erosion of beds or formations by making the side of the column irregular to symbolize the way the units might erode on a cliff face—units that stick out further are more resistant.

Let's see how the concept of a stratigraphic column applies to the Grand Canyon. The walls of the canyon look striped because they expose a variety of rock types that differ in color and in resistance to erosion. Geologists identify major contrasts that distinguish one interval of strata from another and use these as a basis for dividing the strata into formations **(Fig. 10.6)**. Note that some formations consist of beds of a single rock type, whereas others include interlayered beds of two or more rock types. Note also that not all formations have the same thickness, and that all the beds of a region must be part of a stratigraphic formation. Geologists generally depict an unconformity by a wavy line.

Commonly, the name of a formation comes from a locality where it was first identified or first studied. For example, the Schoharie Formation was first defined based on exposures in Schoharie Creek in eastern New York State. If a formation consists of only one rock type, we may incorporate that rock type in the name (such as the Kaibab Limestone), but if a formation contains more than one rock type, we use the word formation in the name (such as the Toroweap Formation). Note that in the formal name of a formation, all words are capitalized.

CORRELATING STRATA

How does the stratigraphy of a sedimentary succession exposed at one locality relate to that exposed at another? Stated another way, can we determine the age of strata at one location with respect to that of strata at another? The answer is yes. Geologists determine such relations by a process called *stratigraphic correlation* (**correlation**, for short), and they use two approaches for correlating intervals of strata.

When correlating formations among nearby regions, we can simply look for similarities in successions of rock type. We call this method *lithologic correlation* **(Fig. 10.7)**. For example, the sequence of strata on the southern rim of the Grand Canyon clearly correlates with the sequence on the northern rim, because they contain the same rock types in the same order. In some cases, a sequence contains a *marker bed*, a unique layer that provides a definitive basis for correlation.

Lithologic correlation doesn't necessarily work over broad areas because the depositional setting and the source of sediments can change from one location to another. Therefore, beds deposited at one location during a given time interval may look quite different from the beds deposited at another location during the same time interval. To correlate rock units over broad areas, we must rely on fossils to define the relative ages of sedimentary units. We call this method *fossil correlation*. If fossils of the same relative age occur at both locations, we can say that the strata at the two locations correlate. The fossils in correlative units are not necessarily the same species—they won't be if the depositional environments

FIGURE 10.6 The stratigraphic column and stratigraphic formation: examples from the Grand Canyon in Arizona.

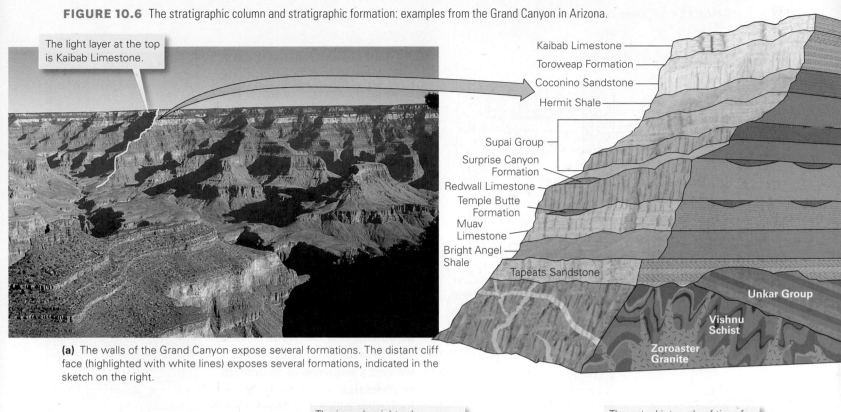

The light layer at the top is Kaibab Limestone.

Kaibab Limestone
Toroweap Formation
Coconino Sandstone
Hermit Shale
Supai Group
Surprise Canyon Formation
Redwall Limestone
Temple Butte Formation
Muav Limestone
Bright Angel Shale
Tapeats Sandstone
Unkar Group
Vishnu Schist
Zoroaster Granite

(a) The walls of the Grand Canyon expose several formations. The distant cliff face (highlighted with white lines) exposes several formations, indicated in the sketch on the right.

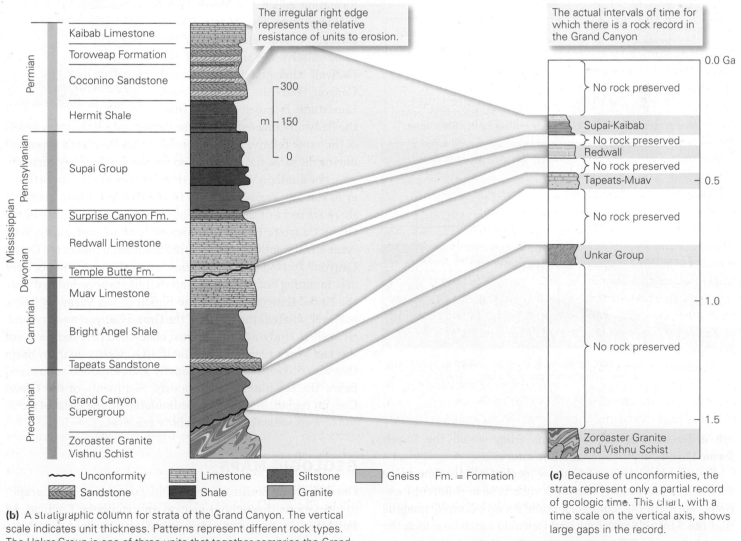

The irregular right edge represents the relative resistance of units to erosion.

The actual intervals of time for which there is a rock record in the Grand Canyon

Permian — Kaibab Limestone, Toroweap Formation, Coconino Sandstone, Hermit Shale

Pennsylvanian — Supai Group

Mississippian — Surprise Canyon Fm., Redwall Limestone

Devonian — Temple Butte Fm.

Cambrian — Muav Limestone, Bright Angel Shale, Tapeats Sandstone

Precambrian — Grand Canyon Supergroup, Zoroaster Granite Vishnu Schist

m — 300, 150, 0

0.0 Ga
No rock preserved
Supai-Kaibab
No rock preserved
Redwall
No rock preserved
Tapeats-Muav — 0.5
No rock preserved
Unkar Group
No rock preserved — 1.0
1.5
Zoroaster Granite and Vishnu Schist

⌇⌇⌇ Unconformity   Limestone   Siltstone   Gneiss   Fm. = Formation
Sandstone   Shale   Granite

(b) A stratigraphic column for strata of the Grand Canyon. The vertical scale indicates unit thickness. Patterns represent different rock types. The Unkar Group is one of three units that together comprise the Grand Canyon Supergroup.

(c) Because of unconformities, the strata represent only a partial record of geologic time. This chart, with a time scale on the vertical axis, shows large gaps in the record.

FIGURE 10.7   The principles of correlation.

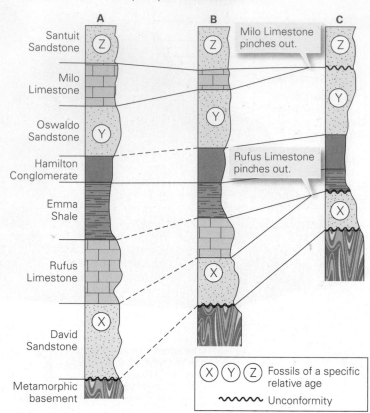

(a) Stratigraphic columns can be correlated by matching rock types (lithologic correlation). The Hamilton Conglomerate is a marker horizon. Because some strata pinch out, column C contains unconformities. Fossil correlation indicates that the youngest beds in C are Santuit Sandstone.

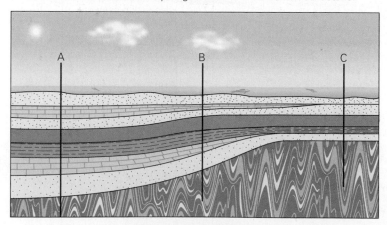

(b) At the time of deposition, locations A, B, and C (which correlate with the columns in part a) were in different parts of a basin. The basin floor was subsiding fastest at A.

are different—but the organisms from which the fossils formed lived during the same time interval.

Fossil correlation can allow geologists to determine whether or not beds of the same rock type at different locations represent the same formation. As an example, imagine that the Santuit Sandstone and Oswaldo Sandstone look the same but are of different ages (see Fig. 10.7a). In location A,

the units are separated by the Milo Limestone, but at location C the units are in direct contact. At first glance, the combination of the two units at location C may look like a single unit. But a sharp-eyed geologist, looking closely at the strata, will see that the fossils have significantly different ages and will depict the contact between the two units by an unconformity.

Let's now apply correlation principles to the challenge of determining the relative ages of formations exposed in the Grand Canyon compared to those exposed in the mountains near Las Vegas, Nevada, 150 km to the west. Near Las Vegas, we find a sequence of sedimentary rocks that includes a limestone formation called the Monte Cristo Limestone. The Monte Cristo Limestone contains fossils of the same relative age as those of the Redwall Limestone of the Grand Canyon, but the Monte Cristo Limestone is much thicker than the Redwall Limestone. Because the formations contain fossils of the same relative age, we conclude that they were deposited during the same time interval, so we can say that they correlate with one another. We also find that not only are the units thicker in the Las Vegas area than in the Grand Canyon area, but there are more of them. Consequently, the contact beneath the Grand Canyon's Redwall Limestone is an unconformity. Why is the stratigraphy of Las Vegas different from that of the Grand Canyon? During part of the time when thick sediments were accumulating near Las Vegas, no sediments accumulated near the Grand Canyon, because the region of Las Vegas lay below sea level, whereas the region of the Grand Canyon was dry land. Geologists studying this contrast concluded that sediments of the Las Vegas region accumulated in a passive-margin basin that was sinking (subsiding) rapidly and remained submerged below the sea almost continuously. Sediments of the Grand Canyon region, in contrast, accumulated on the crust of a craton that episodically emerged above sea level.

GEOLOGIC MAPS

Once William Smith succeeded in correlating stratigraphic formations throughout central and southern England, he faced the challenge of communicating his ideas to others. Since Smith was a surveyor and worked with maps, it

FIGURE 10.8 A geologic map depicts the distribution of rock units and structures.

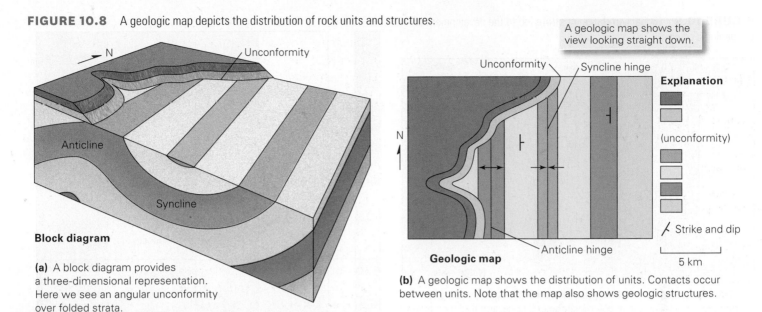

N

Unconformity

Anticline

Syncline

Block diagram

(a) A block diagram provides a three-dimensional representation. Here we see an angular unconformity over folded strata.

A geologic map shows the view looking straight down.

Unconformity   Syncline hinge

Explanation

(unconformity)

N

Anticline hinge

Strike and dip

Geologic map

5 km

(b) A geologic map shows the distribution of units. Contacts occur between units. Note that the map also shows geologic structures.

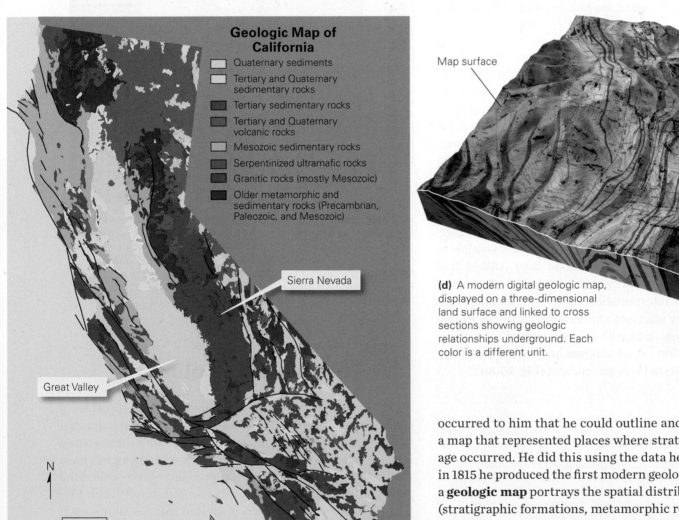

Geologic Map of California

☐ Quaternary sediments

☐ Tertiary and Quaternary sedimentary rocks

■ Tertiary sedimentary rocks

■ Tertiary and Quaternary volcanic rocks

☐ Mesozoic sedimentary rocks

■ Serpentinized ultramafic rocks

■ Granitic rocks (mostly Mesozoic)

■ Older metamorphic and sedimentary rocks (Precambrian, Paleozoic, and Mesozoic)

Sierra Nevada

Great Valley

N

0   100
    km

(c) A geologic map of California indicates that the state is underlain by many different rock units. Granite underlies the Sierra Nevada, and Quaternary sediments underlie the Great Valley. The black lines are fault traces.

Map surface

Cross section

(d) A modern digital geologic map, displayed on a three-dimensional land surface and linked to cross sections showing geologic relationships underground. Each color is a different unit.

occurred to him that he could outline and color the areas on a map that represented places where strata of a given relative age occurred. He did this using the data he had collected, and in 1815 he produced the first modern geologic map. In general, a **geologic map** portrays the spatial distribution of rock units (stratigraphic formations, metamorphic rocks, and extrusive or intrusive igneous rocks), the contacts among them, and the structures that affect them, at the Earth's surface.

The pattern displayed on a geologic map can provide insight into the presence and orientation of geologic structures in the map area (**Fig. 10.8a–c**). Formations on a geologic map

FIGURE 10.9 Global correlation of strata led to the development of the geologic column.

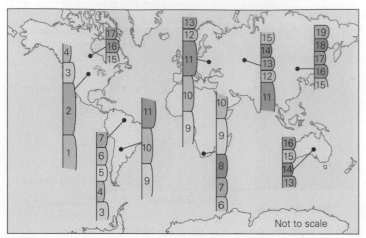

(a) Each of these small columns represents the stratigraphy at a given location. By correlating these columns, geologists determined their relative ages, filled in the gaps in the record, and produced the geologic column.

Eon	Era	Period	Epoch
		Quaternary	Holocene
			Pleistocene
	Cenozoic	Neogene	Pliocene
		Tertiary	Miocene
			Oligocene
		Paleogene	Eocene
			Paleocene
Phanerozoic	Mesozoic	Cretaceous	
		Jurassic	
		Triassic	
	Paleozoic	Permian	
		Carboniferous	Pennsylvanian
			Mississippian
		Devonian	
		Silurian	
		Ordovician	
		Cambrian	
Precambrian	Proterozoic		
	Archean		
	Hadean		

(b) By correlation, the strata from localities around the world were stacked in a chart representing geologic time to create the geologic column. Geologists assigned names to time intervals, but since the column was built without knowledge of numerical ages, it does not depict the duration of these intervals. Subdivisions of eons in the Precambrian are not shown. The Hadean is not shown because rocks do not preserve a record of it.

can be depicted by colors or patterns, contacts on a geologic map by lines, the orientation of layers by strike-and-dip marks (see Box 9.1), and the position of structures by various symbols. With experience, a geologist can use the configuration of contacts and distribution of formations on a map as a basis for identifying folds, faults, unconformities, and other features. For example, repetition of a succession of strata may indicate the presence of a fold. If the oldest strata crop out along the hinge and the strata on opposite limbs of the fold dip away from the hinge, the fold is an anticline. But if the youngest strata crop out along the hinge and the limbs dip toward the hinge, it's a syncline. Using computer technology, it's now possible to plot geologic contacts on *digital elevation maps* (DEMs) that portray the ground surface in three dimensions. *Geologic cross sections* provide an interpretation of geologic relationships underground as they intersect a vertical plane. By combining depictions on a geologic map with depictions on cross sections, a geologist can develop a *block diagram* of a region, representing contacts and units in three dimensions **(Fig. 10.8d)**.

TAKE-HOME MESSAGE

A stratigraphic formation is a recognizable sequence of beds that can be mapped across a broad region. Geologists correlate formations regionally on the basis of rock type and fossil content, and portray the configuration of formations in a region on geologic maps, cross sections, and block diagrams.

QUICK QUESTION What does an angular unconformity look like on a geologic map?

10.6 **The Geologic Column**

As stated earlier, no one locality on the Earth provides a complete record of our planet's history, because stratigraphic columns can contain unconformities. But by correlating rocks at millions of places around the world, geologists have pieced together a composite stratigraphic column, called the **geologic column**, that represents the entirety of Earth history **(Fig. 10.9)**. The column is divided into segments, each of which represents a specific interval of time. The largest subdivisions break Earth history into the Hadean, Archean, Proterozoic, and Phanerozoic **Eons**—the first three together constitute the **Precambrian**. The suffix *–zoic* means life, so

FIGURE 10.10 Life evolution in the context of the geologic column. The Earth formed at the beginning of the Hadean Eon.

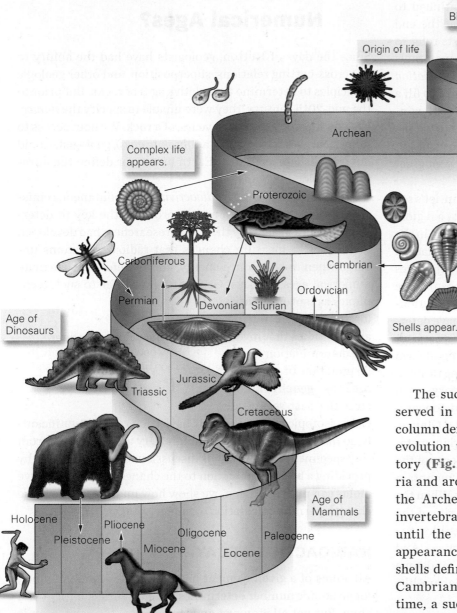

from the Carboniferous Period contain coal). The terminology was not set up in a planned fashion that would make it easy to learn. Instead, names came into use haphazardly between 1760 and 1845, as geologists began to refine their understanding of geologic history and fossil succession. Considering that the divisions were defined before numerical ages could be determined, it's not surprising that the durations of the divisions vary greatly.

The succession of fossils preserved in strata of the geologic column defines the course of life's evolution throughout Earth history **(Fig. 10.10)**. Simple bacteria and archaea appeared during the Archean Eon, but complex invertebrates did not evolve until the late Proterozoic. The appearance of invertebrates with shells defines the Precambrian-Cambrian boundary. At this time, a sudden diversification in life took place, with many new types of organisms appearing over a relatively short interval. Geologists refer to this event as the **Cambrian explosion**. Progressively more complex organisms populated the Earth during the Paleozoic. For example, the first fish appeared in Ordovician seas, land plants started to spread over the continents during the Silurian (prior to this time the land surface was unvegetated), and amphibians appeared during

SEE FOR YOURSELF...

Vermilion Cliffs, Arizona

LATITUDE
36°49'4.81" N

LONGITUDE
111°37'56.59" W

Look obliquely north from 2 km.

Marble Canyon (foreground) is the entry to the Grand Canyon. Outcrops in the distance are the Vermilion Cliffs, exposing reddish-brown sandstone and shale of the Moenkopi Formation. The canyon walls consist of underlying Kaibab Limestone.

Phanerozoic means visible life, and Proterozoic means first life. These names can be a bit misleading, though, because decades after the eons had been named, geologists discovered that the earliest living cells, of bacteria and archaea, actually appeared during the Archean Eon. Eons, in turn, can be subdivided into **eras**. We further divide each era into **periods** and each period into **epochs**.

Where do the names of the periods come from? They refer either to localities where a fairly complete stratigraphic column representing that time interval was first identified (for example, rocks representing the Devonian Period crop out near Devon, England) or to a characteristic of the time (rocks the Devonian. Although the first reptiles appeared during the Pennsylvanian Period, the first dinosaurs did not stomp across the land until the Triassic. Dinosaurs continued to inhabit the Earth until their sudden extinction at the end of the Cretaceous Period. For this reason, geologists refer to the Mesozoic Era as the *Age of Dinosaurs*. Small mammals appeared during the Triassic Period, but the *diversification* (branching into many different species) of mammals to fill a wide range of environments did not happen until the beginning of the Cenozoic Era, so geologists call the Cenozoic the *Age of Mammals*. Birds also appeared during the Mesozoic (at the beginning of the Cretaceous Period) but underwent great diversification in the Cenozoic Era.

To conclude our discussion of the geologic column, let's see how it comes into play when correlating strata across a region. We return to the Colorado Plateau of Arizona and Utah, in the southwestern United States (**Fig. 10.11**). Because of the lack of vegetation in this region, you can easily see bedrock exposures on the walls of cliffs and canyons. Some of these exposures are so beautiful that they have become the centerpieces of national parks. Using correlation techniques, geologists have determined that the oldest sedimentary rocks of the region crop out near the base of the Grand Canyon, whereas the youngest form the cliffs of Cedar Breaks and Bryce Canyon. Walking through these parks is like walking through the Earth's history—each rock layer gives an indication of the climate and topography of the region at a time in the past (**Geology at a Glance**, pp. 358–359). For example, when the Precambrian metamorphic and igneous rocks exposed in the inner gorge of the Grand Canyon first formed, the region was a high mountain range, perhaps as dramatic as the Himalayas today. When the fossiliferous beds of the Kaibab Limestone at the rim of the canyon first developed, the region was a Bahamas-like carbonate reef and platform, bathed in a warm, shallow sea. And when the rocks making up the towering red cliffs of sandstone in Zion Canyon were deposited, the region was a Sahara-like desert, blanketed with huge sand dunes.

TAKE-HOME MESSAGE

Correlation of stratigraphic sequences from around the world led to the production of a chart, the geologic column, that represents the entirety of Earth history. The column, developed using only relative-age relations and fossil correlation, is subdivided into eons, eras, periods, and epochs. We can track life evolution by the succession of fossil assemblages tied to intervals of the geologic column.

QUICK QUESTION What feature of living organisms appeared at the Precambrian-Cambrian boundary?

10.7 How Do We Determine Numerical Ages?

Since the days of Hutton, geologists have had the ability to use cross-cutting relations, superposition, and other geologic principles to determine the relative ages of rocks. But prior to the mid-20th century, they were unable to specify the *numerical age*, meaning the age in years, of a rock. Without access to numerical ages (also known as *absolute ages*), geologists could not establish a timeline of Earth history or define the duration of geologic events.

The 1896 discovery of *radioactivity*, the spontaneous emission of energy by certain atoms, provided the key to determine numerical ages. By the 1950s, researchers had developed techniques to measure changes that radioactive atoms undergo when they produce energy, and therefore, to calculate numerical ages of rocks. Geologists now refer to such techniques as **isotopic dating** (or **radiometric dating**), and to studies that involve the determination and interpretation of numerical ages as **geochronology**. Over the past several decades, isotopic dating methods have been refined significantly, so that now it's possible to obtain very accurate dates from tiny specimens. We begin our discussion of isotopic dating by providing a brief introduction to the changes that radioactive atoms can undergo. Then, we show how understanding these changes permits calculation of numerical ages.

> **Did you ever wonder...**
> how geologists can specify the age of some rocks in years?

RADIOACTIVE DECAY

All atoms of a given element have the same atomic number, for an atomic number determines the elemental identity of an atom. But not all atoms of an element have the same atomic mass. Various versions of an element, which differ from one another because they have different numbers of neutrons in the nucleus, and therefore, they have different atomic masses, are called **isotopes** of the element. (Of note, not all elements have different isotopes, but most do.)

To clarify what we mean by an isotope, let's compare two isotopes of uranium. All uranium atoms have 92 protons, and therefore, have an atomic number of 92. But atoms of the uranium-235 isotope (^{235}U) have an atomic mass of 235, whereas atoms of the uranium-238 isotope (^{238}U) have an atomic mass of 238. Therefore, ^{235}U atoms contain 143 neutrons, whereas ^{238}U atoms contain 146 neutrons. Significantly, some isotopes of an element are *stable*, in that atoms of the isotope

FIGURE 10.11 Correlation of strata among the national parks of Arizona and Utah.

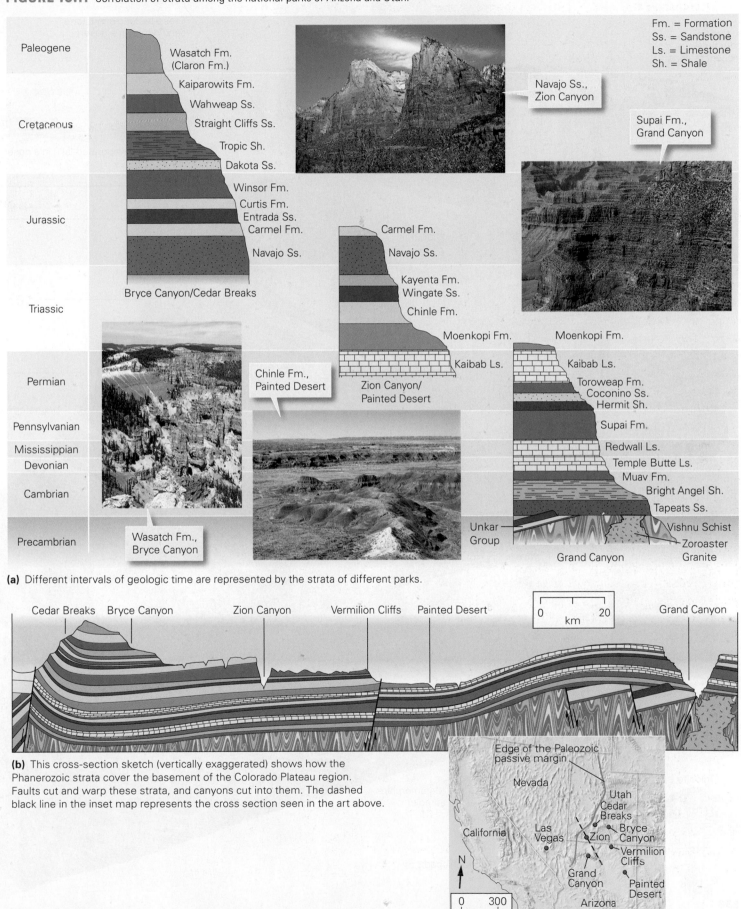

Fm. = Formation
Ss. = Sandstone
Ls. = Limestone
Sh. = Shale

(a) Different intervals of geologic time are represented by the strata of different parks.

(b) This cross-section sketch (vertically exaggerated) shows how the Phanerozoic strata cover the basement of the Colorado Plateau region. Faults cut and warp these strata, and canyons cut into them. The dashed black line in the inset map represents the cross section seen in the art above.

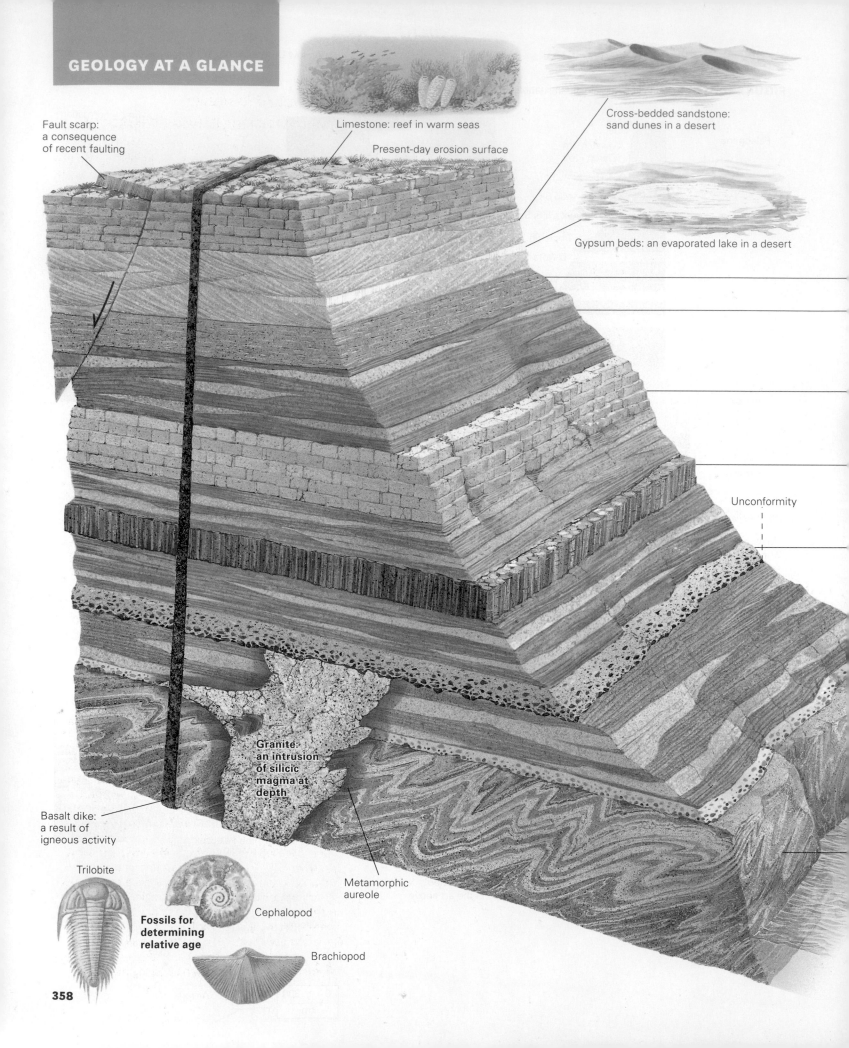

GEOLOGY AT A GLANCE

Limestone: reef in warm seas

Cross-bedded sandstone: sand dunes in a desert

Present-day erosion surface

Gypsum beds: an evaporated lake in a desert

Fault scarp: a consequence of recent faulting

Unconformity

Basalt dike: a result of igneous activity

Granite: an intrusion of silicic magma at depth

Metamorphic aureole

Trilobite

Cephalopod

Fossils for determining relative age

Brachiopod

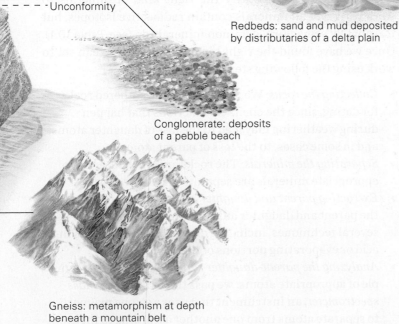

Ignimbrite (welded tuff): an explosive volcanic eruption

Limestone: reef in warm seas

Redbeds: sand and mud deposited in a river channel and bordering floodplain

Basalt lava: flows from a volcano

Conglomerate: debris eroded from a cliff

Unconformity

Redbeds: sand and mud deposited by distributaries of a delta plain

Conglomerate: deposits of a pebble beach

Gneiss: metamorphism at depth beneath a mountain belt

The Record in Rocks: Reconstructing Geologic History

When geologists examine a sequence of rocks exposed on a cliff, they see a record of Earth history that they interpret by applying the basic principles of geology, by searching for fossils, and by using isotopic dating. On this cliff, we see evidence for many geologic events. The layers of sediment (and the sedimentary structures they contain), the igneous intrusions, and the geologic structures tell us about past climates and past tectonic activity.

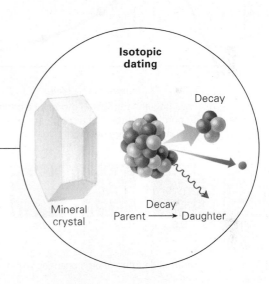

Isotopic dating

Decay

Mineral crystal

Decay
Parent ⟶ Daughter

The insets show the way the region looked in the past, based on the record in the rocks. For example, the presence of gneiss at the base of the canyon indicates that at one time the region was a mountain belt in which deeply buried crust underwent metamorphism and deformation. Unconformities indicate that the region later underwent uplift and erosion. Sedimentary successions record transgressions and regressions of the sea. The land surface portrayed in this painting was sometimes a river floodplain or a delta (indicated by redbeds), sometimes a shallow sea (limestone), and sometimes a desert dune field (cross-bedded sandstone). The presence of igneous rocks indicates that, at times, the region was volcanically active, and the presence of faults indicates it was seismically active. We can gain insight into the age of the sedimentary rocks by studying the fossils they contain, and into the age of the igneous and metamorphic rocks by using isotopic dating methods.

359

FIGURE 10.12 The concept of a half-life, in the context of radioactive decay.

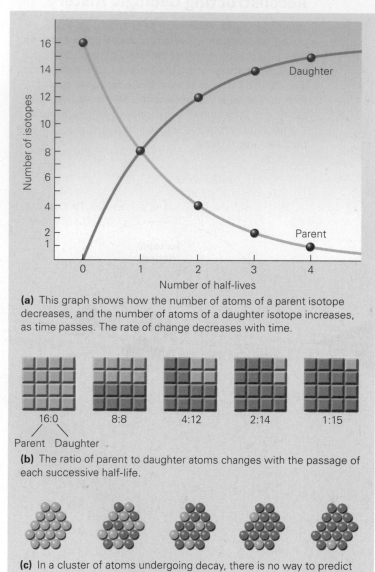

(a) This graph shows how the number of atoms of a parent isotope decreases, and the number of atoms of a daughter isotope increases, as time passes. The rate of change decreases with time.

16:0   8:8   4:12   2:14   1:15

Parent Daughter

(b) The ratio of parent to daughter atoms changes with the passage of each successive half-life.

(c) In a cluster of atoms undergoing decay, there is no way to predict which parent atom will decay next.

Physicists cannot specify how long an individual radioactive atom will survive before it decays, but they can measure how long it takes for half of a group of parent atoms to decay. This time is called the **half-life**. To visualize the concept of a half-life, imagine a crystal containing 16 radioactive parent atoms **(Fig. 10.12)**. (Note that in a real crystal, the number of atoms would be immensely larger.) After one half-life, 8 atoms have decayed, so the crystal now contains 8 parents and 8 daughters. After a second half-life, 4 of the remaining parents have decayed, so the crystal contains 4 parents and 12 daughters. And after a third half-life, 2 more parents have decayed, so the crystal contains 2 parents and 14 daughters. For a given decay reaction, the half-life is a constant, generally measured in years.

ISOTOPIC DATING TECHNIQUE

Since radioactive decay of a population of radioactive atoms proceeds at a known rate, like the tick-tock of a clock, it provides a basis for telling time. In other words, because an element's half-life is a constant, we can calculate the age of a mineral by measuring the ratio of parent to daughter atoms in the mineral.

In practice, how can we obtain an isotopic date? First, we must find an appropriate radioactive isotope to work with. Only a few have long enough half-lives and occur in sufficient abundance in minerals to be useful for isotopic dating. **Table 10.1** lists particularly useful elements. Each radioactive element has its own half-life. Note that *carbon dating*, based on the decay of ^{14}C, is not used for dating rocks because appropriate carbon isotopes occur only in organisms, not minerals, and radioactive ^{14}C has a very short half-life.

Second, we must identify the right kind of minerals to work with. Not all minerals contain radioactive isotopes, but fortunately some fairly common minerals do (see Table 10.1). Once we have found the right kind of minerals, we can set to work using the following steps to obtain a date:

> *Collecting the rocks:* We need to find unweathered rocks for dating, since the chemical reactions that happen during weathering may lead to the loss of daughter atoms, and in some cases, to the loss of parent atoms.
> *Separating the minerals:* The rocks are crushed, and the appropriate minerals are separated from the debris.
> *Extracting parent and daughter atoms:* To separate out the parent and daughter atoms from minerals, we can use several techniques, including dissolving the minerals in acid or evaporating portions of them with a laser.
> *Analyzing the parent-daughter ratio:* Once we have a sample of appropriate atoms, we pass them through a *mass spectrometer*, an instrument that uses a strong magnet to separate atoms from one another according to their

last forever without change. Other isotopes are *unstable*, in that they eventually undergo **radioactive decay**, or *nuclear decay*, a process that releases energy stored in nuclear bonds in the nucleus. Some types of radioactive decay result in the expulsion of fragments (known as alpha particles and beta particles) from the nucleus of an atom. When this happens, the atom's atomic number changes, and as a result, the atom becomes a different element. Physicists refer to an isotope that can undergo radioactive decay as a *radioactive isotope*. An atom of a radioactive isotope that has not yet decayed is a **parent atom**, whereas the product of radioactive decay is a **daughter atom**.

TABLE 10.1 Isotopes Used in the Isotopic Dating of Rock

Parent → Daughter	Half-life (years)	Minerals Containing the Isotopes
$^{147}Sm \rightarrow {}^{143}Nd$	106.0 billion	Garnets, micas
$^{87}Rb \rightarrow {}^{87}Sr$	48.8 billion	Potassium-bearing minerals (mica, feldspar, hornblende)
$^{230}U \rightarrow {}^{206}Pb$	4.5 billion	Uranium-bearing minerals (zircon, uraninite)
$^{40}K \rightarrow {}^{40}Ar$	1.3 billion	Potassium-bearing minerals (mica, feldspar, hornblende)
$^{235}U \rightarrow {}^{207}Pb$	713.0 million	Uranium-bearing minerals (zircon, uraninite)

Sm = samarium, Nd = neodymium, Rb = rubidium, Sr = strontium, U = uranium, Pb = lead, K = potassium, Ar = argon.

respective masses **(Fig. 10.13)**. The instrument can count the number of atoms of specific isotopes separately.

At the end of the laboratory process, we can define the ratio of parent to daughter atoms in a mineral and from this ratio calculate the age of the mineral. Needless to say, the description of the procedure here has been simplified—in reality, obtaining an isotopic date can require complex calculations, and it can be time-consuming and expensive. In some cases, however, new technologies permit extraction of parent and daughter atoms from a mineral grain simply by zapping the grain with a very narrow-diameter laser beam. Heat produced by the beam evaporates atoms from the surface of the grain, leaving behind a tiny pit. Gas containing the evaporated atoms then flows into the mass spectrometer for analysis.

WHAT DOES AN ISOTOPIC DATE MEAN?

At high temperatures, atoms in a crystal lattice vibrate so rapidly that chemical bonds can break and reattach relatively easily. As a consequence, atoms can escape from or move into crystals, so parent-daughter ratios are meaningless. Because isotopic dating is based on the parent-daughter ratio, the isotopic clock starts only when crystals become cool enough for atoms to be locked into the lattice. The temperature below which atoms are no longer free to move is called the **closure temperature** of a mineral. When we specify an isotopic date for a mineral, we are defining the time at which the mineral cooled below its closure temperature.

With the concept of closure temperature in mind, we can interpret the meaning of isotopic dates. In the case of igneous rocks, isotopic dating tells us when a magma or lava cooled to form a solid, cool igneous rock. In the case of metamorphic rocks, an isotopic date tells us when a rock cooled from a metamorphic temperature above the closure temperature to a temperature below. Not all minerals have the same closure temperature, so different minerals in an igneous or metamorphic rock that cools very slowly will yield different dates.

FIGURE 10.13 In an isotopic dating laboratory, samples are analyzed using a mass spectrometer. This instrument measures the ratio of parent to daughter atoms.

In recent years, geologists have developed procedures that allow them to date minerals with a very low closure temperature (as low as 75°C). This method identifies the time at which the rock was exhumed, meaning the time when uplift and erosion brought a rock that was once deep within an orogen up to within a few kilometers of the Earth's surface.

Can we isotopically date a clastic sedimentary rock directly? No. If we date minerals in a sedimentary rock, we determine only when these minerals first crystallized as part of an igneous or metamorphic rock, not the time when the minerals were deposited as sediment, or the time when the sediment lithified to form a sedimentary rock. For example, if we date the feldspar grains contained within a granite pebble in a conglomerate, we're dating the time the granite cooled below feldspar's closure temperature, not the time the pebble was deposited by a stream.

FIGURE 10.14 Ways to determine numerical ages.

(a) Patterns of tree rings are like bar codes. Each ring in this slice of wood represents the growth of one year.

(c) Dust settling during the summer highlights boundaries between snow layers, now turned to ice in this Oregon glacier.

OTHER METHODS OF DETERMINING NUMERICAL AGE

The rate of tree growth depends on the season. During the spring, trees grow rapidly and produce lighter, less-dense wood, but during the rest of the season, trees grow slowly or not at all and produce darker, denser wood. Thus, wood contains recognizable annual growth rings, which provide a basis for determining age. If you've ever wondered about the age of a tree that has just been cut down, look at the stump and count the rings. By correlating clusters of distinctive rings in the older parts of living trees with comparable clusters of rings in dead logs, scientists can extend the tree-ring record back for many thousands of years, allowing them to track climate changes back into prehistory (**Fig. 10.14a, b**).

Seasonal changes also affect rates of such phenomena as shell growth, snow accumulation, clastic sediment deposition, chemical sediment precipitation, and production of organic material. Geologists have learned to use growth rings in shells, as well as rhythmic layering in sediments and in glacial ice (**Fig. 10.14c**), to date events numerically back through recent Earth history.

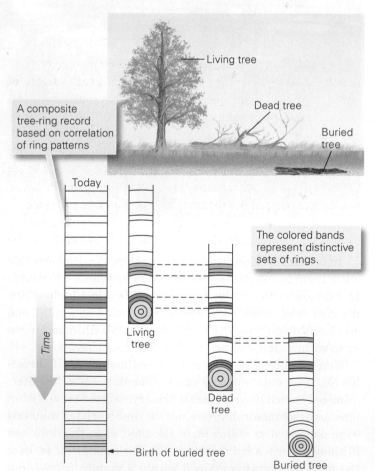

(b) Dendrochronology is based on the correlation of tree rings. Each of the columns in the diagram represents a core drilled out of a tree. Distinctive clusters of closely spaced rings indicate dry seasons. By correlation, researchers extend the climate record back in time before the oldest living tree started to grow.

> ### TAKE-HOME MESSAGE
>
> Isotopic dating specifies numerical ages in years. To obtain an isotopic date, we measure the ratio of radioactive parent atoms to stable daughter atoms in a mineral. A date for a mineral gives the time at which the mineral cooled below a closure temperature.
>
> **QUICK QUESTION** How can we date changes in climate during the past few hundred thousand years?

10.8 Numerical Age and Geologic Time

DATING SEDIMENTARY ROCKS

The mind grows giddy gazing so far back into the abyss of time.
JOHN PLAYFAIR (British geologist, 1747–1819)

We have seen that isotopic dating can be used to date when igneous rocks formed and when metamorphic rocks metamorphosed, but not when sedimentary rocks were deposited. So how do we determine the numerical age of a sedimentary rock? We must answer this question if we want to add numerical ages to the geologic column.

Geologists obtain dates for sedimentary rocks by studying cross-cutting relationships between sedimentary rocks and datable igneous or metamorphic rocks. For example, if we find a sequence of sedimentary strata deposited unconformably on a datable granite, the strata must be younger than the granite **(Fig. 10.15)**. If a datable basalt dike cuts the strata, the strata must be older than the dike. And if a datable volcanic ash buried the strata, then the strata must be older than the ash.

THE GEOLOGIC TIME SCALE

Geologists have searched the world for localities where they can recognize cross-cutting relations between datable igneous rocks and sedimentary rocks, or for localities where layers of datable volcanic rocks are interbedded with sedimentary rocks. By isotopically dating the igneous rocks, researchers have been able to provide numerical ages for the boundaries between all geologic periods. For example, information gathered around the world shows that the Cretaceous Period began about 145 million years ago and ended 66 million years ago. So the Cretaceous sandstone bed in Figure 10.15 was deposited during the middle part of the Cretaceous, not at the beginning or end.

The discovery of new data may cause the numbers defining the boundaries of periods to change. For example, around 1995, new dates on rhyolite ash layers above and below the Cambrian-Precambrian boundary showed that this boundary occurred at 542 million years ago, in contrast to previous, less definitive studies that had placed the boundary at 570 million years ago. More recent dating indicates that the boundary lies at 541 Ma. **Figure 10.16** shows the currently favored numerical ages of periods and eras in the geologic column. This dated column is commonly called the **geologic time scale**.

WHAT IS THE AGE OF THE EARTH?

During the 18th and 19th centuries, before the discovery of isotopic dating, scientists came up with a great variety of clever solutions to the question, "How old is the Earth?" All have since been proved wrong. Lord Kelvin, a 19th-century physicist renowned for his discoveries in thermodynamics, made the most influential scientific estimate of the Earth's age, for his time. Kelvin calculated how long it would take for the Earth to cool down from a temperature as hot as the Sun's, and concluded that our planet is about 20 million years old.

Kelvin's estimate contrasted with those being promoted by followers of Hutton, Lyell, and Darwin, who argued that if the concepts of uniformitarianism and evolution were correct, the Earth must be much older. They argued that physical processes that shape the Earth and form its rocks, as well as the process of natural selection that yields the diversity of species, all take a very long time. Geologists and physicists continued to debate the age issue for many years. The route to a solution didn't appear until 1896, when Henri Becquerel announced the discovery of radioactivity. Geologists immediately realized that if the Earth's interior was producing heat from the decay of radioactive atoms, then the planet has cooled down much more slowly than Kelvin had calculated, and it could be much older. The discovery of radioactivity not only invalidated Kelvin's estimate of the Earth's age, it also led to the development of isotopic dating.

Since the 1950s, geologists have searched for the Earth's oldest rocks. Rock samples from several localities (Wyoming, Canada, Greenland, and China) have yielded dates as old as

FIGURE 10.15 The Cretaceous sandstone bed was deposited on the granite, so it must be younger than 125 Ma. The dike cuts the bed, so the bed must be older than 80 Ma. Thus, the Cretaceous bed was deposited between 125 and 80 Ma. The Paleocene sandstone was unconformably deposited over the dike and lies beneath a 50-million-year-old layer of ash. Therefore, it must have been deposited between 80 and 50 Ma.

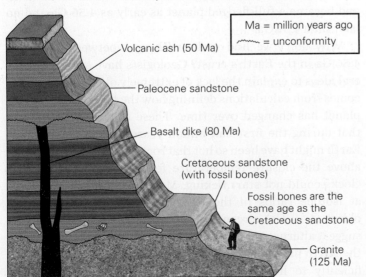

Ma = million years ago
〜〜〜 = unconformity

Volcanic ash (50 Ma)

Paleocene sandstone

Basalt dike (80 Ma)

Cretaceous sandstone (with fossil bones)

Fossil bones are the same age as the Cretaceous sandstone

Granite (125 Ma)

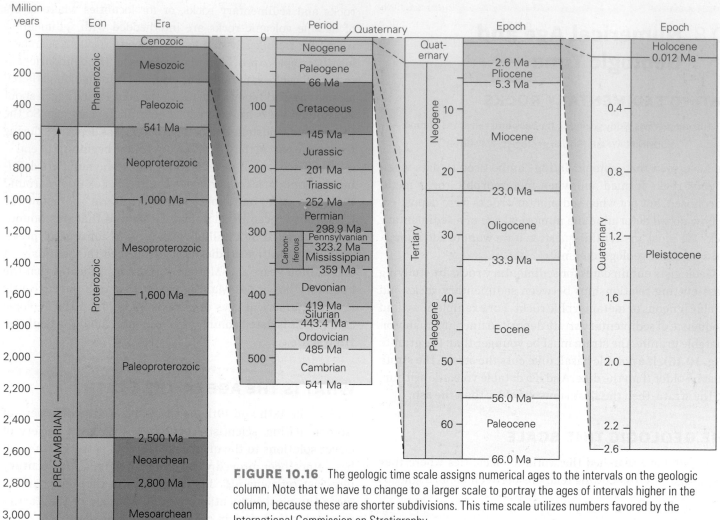

FIGURE 10.16 The geologic time scale assigns numerical ages to the intervals on the geologic column. Note that we have to change to a larger scale to portray the ages of intervals higher in the column, because these are shorter subdivisions. This time scale utilizes numbers favored by the International Commission on Stratigraphy.

planetesimals, those that had separated into a core and mantle, are slightly younger, and these ages are taken to be the same as the age of the Earth itself. Based on these ages, therefore, researchers estimate that the Earth itself differentiated and became a full-fledged planet as early as 4.56 Ga and no later than 4.54 Ga.

Why don't we find rocks with ages between 4.03 and 4.56 Ga in the Earth's crust? Geologists have suggested several ideas to explain the lack of extremely old rocks. One idea comes from calculations defining how the temperature of our planet has changed over time. These calculations indicate that during the first half-billion years of its existence, the Earth might have been so hot that rocks in the crust remained above the closure temperature for minerals, and isotopic clocks could not start ticking. More recent studies, looking at isotope ratios in the oldest (4.40 Ga) zircons, suggest alternatively that the Earth had cooled sufficiently to host oceans of water within only a

4.03 Ga. Individual clastic grains of the mineral zircon have yielded dates of up to 4.40 Ga, indicating that rock as old as 4.40 Ga did once exist. Isotopic dating of Moon rocks yields dates of up to 4.50 Ga, and dates on meteorites have yielded ages as old as 4.57 Ga. Geologists consider these meteorites to be fragments of planetesimals that never differentiated into a mantle and core. Meteorites from the oldest differentiated

Did you ever wonder...
how old the Earth's oldest rock is?

couple of hundred million years of its formation and suggest that intense bombardment of the Earth by meteorites prior to 4.03 Ga destroyed or remelted any crust that existed and vaporized the earliest oceans. As noted earlier, geologists refer to the time interval between the birth of the Earth and the origin of the oldest isotopically dated whole rock as the Hadean Eon, to emphasize that conditions at the surface, at times, resembled literary images of Hades.

PICTURING GEOLOGIC TIME

The number 4.56 billion is so staggeringly large that we can't begin to comprehend it. If you lined up this many pennies in a row, they would make an 87,400-km-long line that would wrap around the Earth's equator more than twice. Notably, at the scale of our penny chain, human history is only about 100 city blocks long.

Another way to grasp the immensity of geologic time is to equate the entire 4.56 billion years to a single calendar year. On this scale, the oldest rocks preserved on the Earth date from early February, and the first archaea appear in the ocean on February 21. The first shelled invertebrates appear on October 25, and the first amphibians crawl out onto land on November 20. On December 7, the continents coalesce into the supercontinent of Pangaea. Birds and the ancestors of mammals appear about December 15, along with the dinosaurs, and the Age of Dinosaurs ends on December 25. The last week of December represents the last 66 million years of Earth history, including the entire Age of Mammals. The first human-like ancestor appears on December 31 at 3 P.M., and our species, *Homo sapiens*, shows up an hour before midnight. The last ice age ends a minute before midnight, and all of recorded human history takes place in the last 30 seconds. To put it another way, human history occupies the last 0.000001% of Earth history. The Earth is so old that there has been more than enough time for the rocks and life forms of the Earth to have formed and evolved.

TAKE-HOME MESSAGE

Numerical dates for sedimentary rocks come from isotopic dating of cross-cutting datable rocks. Such work led to the geologic time scale, which assigns dates to periods. The oldest rock of the Earth's crust is about 4.0 Ga. Dating of meteorites indicates the Earth became a differentiated planet at 4.56 Ga.

QUICK QUESTION Why don't all periods on the geologic time column have the same duration in years?

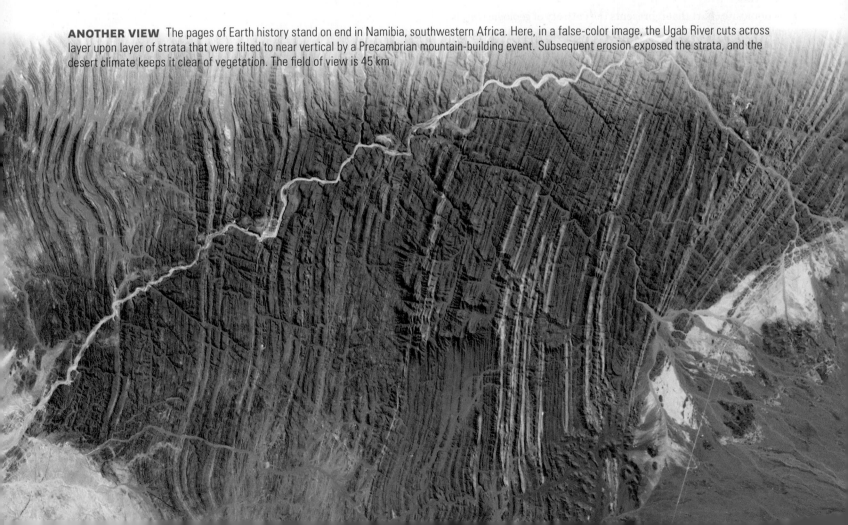

ANOTHER VIEW The pages of Earth history stand on end in Namibia, southwestern Africa. Here, in a false-color image, the Ugab River cuts across layer upon layer of strata that were tilted to near vertical by a Precambrian mountain-building event. Subsequent erosion exposed the strata, and the desert climate keeps it clear of vegetation. The field of view is 45 km.

Chapter 10 Review

CHAPTER SUMMARY

> Geologic time is the time since the Earth's formation.

> Relative age specifies whether one geologic feature is older or younger than another; numerical age provides the age of a geologic rock or feature in years.

> Using such principles as uniformitarianism, original horizontality, superposition, and cross-cutting relations, we can construct the geologic history of a region.

> The principle of fossil succession states that the assemblage of fossils in strata changes from base to top of a sequence.

> Strata are not necessarily deposited continuously at a location. An interval of nondeposition or erosion produces an unconformity. Geologists recognize three kinds: angular unconformity, nonconformity, and disconformity.

> A stratigraphic column shows the succession of strata in a region. A given succession of strata that can be traced over a fairly broad region is called a stratigraphic formation.

> The process of determining the relationship between strata at one location and strata at another is called correlation. A geologic map shows the distribution of formations, as well as of geologic structures.

> A composite chart that represents the entirety of geologic time is called the geologic column. The column's largest subdivisions are eons. Eons are subdivided into eras, eras into periods, and periods into epochs.

> The numerical age of rocks can be determined by isotopic (radiometric) dating. This is because radioactive isotopes decay at a rate characterized by a known half-life. Therefore, the ratio of parent atoms to daughter atoms in a mineral represents the age of the mineral.

> The isotopic age of a mineral specifies the time at which the mineral cooled below a closure temperature. Isotopic dating determines when an igneous rock solidified and when a metamorphic rock cooled. To date sedimentary strata, we must examine cross-cutting relations with dated igneous or metamorphic rock.

> Other methods for dating materials include counting growth rings in trees and seasonal layers in glaciers.

> Isotopic dating indicates that the Earth is 4.56 billion years old. This date represents the time when the planet internally differentiated into a core and mantle. A rock record of the first half billion years of Earth history doesn't exist, for the oldest dated whole rock is about 4 Ga.

GUIDE TERMS

angular unconformity (p. 348)
Cambrian explosion (p. 355)
closure temperature (p. 361)
correlation (p. 350)
daughter atom (p. 360)
disconformity (p. 348)
eon (p. 354)
epoch (p. 355)
era (p. 355)

fossil assemblage (p. 347)
geochronology (p. 356)
geologic column (p. 354)
geologic contact (p. 350)
geologic map (p. 353)
geologic time (p. 343)
geologic time scale (p. 363)
half-life (p. 360)
index fossil (p. 347)

isotope (p. 356)
isotopic dating (p. 356)
nonconformity (p. 348)
numerical age (p. 344)
parent atom (p. 360)
period (p. 355)
Precambrian (p. 354)
radioactive decay (p. 360)
radiometric dating (p. 356)

relative age (p. 344)
stratigraphic column (p. 350)
stratigraphic
   formation (p. 350)
stratigraphic group (p. 350)
unconformity (p. 348)
uniformitarianism (p. 343)

GEOTOURS THIS CHAPTER'S GEOTOURS WORKSHEET (J) FEATURES QUESTIONS AND GOOGLE EARTH SITES ON:

> Relative-age dating and unconformities   > Stratigraphic formations in southern Utah   > Rock layers and monoclines, Circle Cliffs, Utah

REVIEW QUESTIONS

The letters following each Review Question refer to the corresponding Learning Objective from the Chapter Opener.

1. Contrast numerical age with relative age. **(A)**

2. Describe the principles that allow us to determine the relative ages of geologic events. **(B)**

3. How does the principle of fossil succession help determine relative ages? **(B)**

4. How does an unconformity develop? Distinguish among the three kinds of unconformities. Which type is pictured here? **(C)**

5. What is a stratigraphic formation? Describe two different methods of correlating formations. How was correlation used to develop the geologic column? **(D)**

6. What happens during radioactive decay, and what does a half-life indicate? **(E)**

7. How do geologists obtain an isotopic date? What does the age of an igneous rock mean? What does the age of a metamorphic rock mean? **(E)**

8. Why can't we date sedimentary rocks directly? How do we assign numerical ages to intervals on the geologic column to produce a geologic time scale? **(F)**

9. How are growth rings and ice layering useful in determining the ages of geologic events? **(F)**

10. What is the age of the oldest rocks on the Earth? What is the current estimate of the numerical age of the Earth? Why is there a difference? **(F)**

ON FURTHER THOUGHT

11. Examine the photograph and the What a Geologist Sees interpretation of an outcrop in eastern New York State, shown below. Write a brief geologic history that explains the relationships displayed in this outcrop. The strata above the unconformity are Late Silurian (Rondout Formation) and Early Devonian (Helderberg Group), whereas the strata below the unconformity are Middle Ordovician (Austin Glen Formation). **(B, C)**

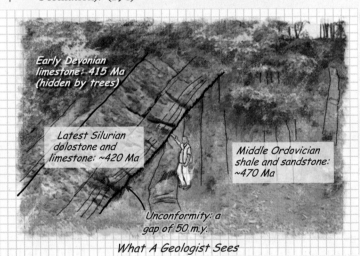

Early Devonian limestone: 415 Ma (hidden by trees)

Latest Silurian dolostone and limestone: ~420 Ma

Middle Ordovician shale and sandstone: ~470 Ma

Unconformity: a gap of 50 m.y.

What A Geologist Sees

ONLINE RESOURCES

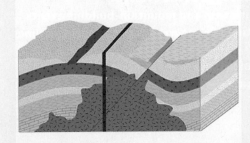

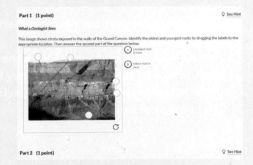

Animations This chapter features a series of animations demonstrating relative age dating and the different types of geologic unconformities.

Videos This chapter features a video on a petrified forest and the clues it reveals to geologic history.

Smartwork5 Questions cover principles for defining relative age, unconformities, and identifying relative ages.

CHAPTER 11

A BIOGRAPHY OF THE EARTH

By the end of this chapter, you should be able to . . .

A. describe the tools geologists use to study the Earth's history.

B. outline changes that our planet underwent during its first 2 billion years.

C. correlate key geologic events with the tectonic conditions and settings in which they formed.

D. provide evidence for Proterozoic glaciations, and explain the concept of snowball Earth.

E. outline major stages in life evolution, and describe how the land surface changed as life evolved.

F. identify supercontinents and mountain-building events (orogenies), and tie their occurrence to the geologic time scale.

G. place the Age of Dinosaurs in the context of Earth history, and indicate the favored hypothesis for why it came to an end.

> The man who should know the true history of the bit of chalk which every carpenter carries about in his breeches pocket, though ignorant of all other history, is likely . . . to have a truer and therefore a better conception of this wonderful universe and of man's relation to it than the most learned student who [has] deep-read the records of humanity [but is] ignorant of those of nature.
>
> THOMAS HENRY HUXLEY (British biologist, 1825–1895), from *On a Piece of Chalk* (1868)

11.1 Introduction

In 1868, a well-known British scientist, Thomas Henry Huxley, presented a public lecture on geology to an audience in Norwich, England. Seeking a way to convey his fascination with the subject to people who had no previous geologic knowledge, he focused his audience's attention on the piece of chalk that he had been writing with (see the epigraph above). And what a tale the chalk has to tell! Chalk, a type of limestone, consists of microscopic marine algae shells and shrimp feces. The piece of chalk that Huxley held came from beds deposited in Cretaceous time—the name *Cretaceous*, in fact, derives from the Latin word for chalk. These beds now form the white cliffs bordering the southeastern coast of England. Geologists in Huxley's day knew of similar chalk beds throughout much of Europe and had discovered that the chalk contains fossils of bizarre swimming reptiles, fish, and invertebrates, species absent in the seas of today. Clearly, when the chalk was deposited, warm seas holding unfamiliar creatures covered some of what is dry land of Europe today. Clues in a humble piece of chalk allowed Huxley to demonstrate to his audience that the geography and inhabitants of the Earth in the past differed markedly from those of today, implying that "the Earth has a history."

In this chapter, we offer a concise geologic biography of our planet, from its birth to the present. We discuss how the oceans and atmosphere first formed, and how continents came into existence and moved about the surface. We also describe mountain-building events, changes in the Earth's climate and sea level through time, and the evolution of life. To streamline discussion, remember that we use the following abbreviations: *Ga* (for billion years ago), *Ma* (for million years ago), and *Ka* (for thousand years ago).

◀ (facing page) The "pancake rocks," exposed along the west coast of New Zealand, consist of limestone formed from tiny shells that settled on the seafloor about 30 million years ago. Much later, tectonic forces caused the rock to be uplifted, and now they are being eroded. Clearly, the Earth has a history!

11.2 Hadean and Archean Time: The Earth's Early Days

THE TIME BEFORE THE ROCK RECORD— THE HADEAN EON

James Hutton, the 18th-century Scottish geologist who was the first to provide convincing evidence that the Earth was very old, could not measure the Earth's age directly, and indeed speculated that there may be "no vestige of a beginning." But, as we discussed in Chapter 10, isotopic dating of meteorites has led geologists to consider 4.56 to 4.54 Ga to be the Earth's birth date. At this time, the planet reached nearly its full size and underwent **differentiation** internally, so that metallic iron alloy sank to the center to form a core, leaving behind a mantle of ultramafic rock. Based on dating Moon rocks, researchers have concluded that the Moon formed at about 4.53 Ga, probably when a sizable protoplanet collided with, and partially merged with, the Earth. The force of the impact blasted material from the Earth's mantle into orbit. Soon after, this debris coalesced into the Moon.

During its infancy, the Earth was vastly different from the green and blue globe we know today. For a time after the Moon formed, the Earth remained so hot that much of its surface was probably a seething pool of magma, occasionally pummeled by meteorites **(Fig. 11.1a)**. But our planet's temperature rapidly decreased. Heat radiated into space from the surface, and the supply of heat from radioactive decay inside diminished as elements with short half-lives rapidly decayed. Calculations suggest that by 4.4 Ga, the Earth's surface had frozen into a skin of solid ultramafic rock. Most likely, this skin underwent rapid recycling, for once it formed, the rock of the skin cooled and became dense enough to sink back into the still very hot, soft mantle. As it sank, it was replaced by new rock at the surface.

During the Earth's early history, rapid *outgassing* also took place, meaning that volatile (gassy) elements or compounds originally incorporated in mantle minerals were released and erupted at the planet's surface, along with lava. These gases accumulated to constitute a toxic atmosphere consisting

FIGURE 11.1 Visualizing the early Earth is a challenge, but by using geologic interpretations, artists have provided useful images.

(a) At a very early stage, during or after differentiation, the surface may have been largely molten. The loss of heat to space would have allowed patches of solid ultramafic crust to form. Meteorite impacts destroyed most of this crust.

(b) When the Earth's surface fell below the boiling point of water, water from the atmosphere rained onto the surface, submerging it beneath early oceans.

mostly of water (H_2O), methane (CH_4), ammonia (NH_3), hydrogen (H_2), nitrogen (N_2), carbon dioxide (CO_2), and sulfur dioxide (SO_2). Some researchers speculate that gases from comets colliding with the Earth contributed additional gases to the early atmosphere.

Recent studies of oxygen isotopes in the oldest known mineral specimens, 4.40-Ga grains of a durable mineral called zircon, found in Western Australia, suggest that some of the H_2O in the young Earth's early atmosphere condensed to form liquid water that rained onto the planet's surface. What might the Earth's surface have looked like at this time? If the hypothesis that liquid water condensed from the early atmosphere is correct, then the surface probably would have been a black, barren, cratered landscape, locally submerged by an acidic sea **(Fig. 11.1b)**. Volcanoes spewing lava would rise here and there. Both land and sea would likely have been obscured by murky, dense (H_2O-, CO_2-, and SO_2-rich) air.

Although the oldest dated minerals are 4.38 Ga, the oldest known whole rock, a sample collected in northwestern Canada, has an age of about 4.03 Ga. Geologists use the name **Hadean Eon** to refer to the mysterious time interval between the birth of Earth at 4.56 Ga and the beginning of the rock record by 4.03 Ga. The name comes from the Greek name Hades, the god of the underworld—during intervals of this eon, the Earth's surface resembled an inferno. Where did pre-4.0-Ga rocks go? They may have sunk back into the mantle where they were recycled, or they may have been pulverized and melted at the end of the Hadean, during a time that astronomers refer to as the *late heavy bombardment*, when the

Earth was pummeled intensely by meteorites. This event took place sometime between 4.1 and 3.8 Ga.

BIRTH OF THE CONTINENTS IN THE ARCHEAN EON

The age of the first whole rocks marks the start of the **Archean Eon** (from the Greek word for beginning). With the advent of the Archean, set at 4.0 Ga, crust was locally cool and stable enough, and meteorite impacts rare enough, for rocks to survive and for isotopic clocks to start ticking. Geologists still debate whether plate tectonic activity was taking place in the early Archean Eon. According to some researchers, the Earth was so hot until sometime between 3.8 Ga and 3.2 Ga that lithospheric mantle could not become cool enough and dense enough to subduct and sink back into the asthenosphere. This would mean that subduction couldn't happen, and without subduction, plates couldn't move. If this view is correct, then most volcanism of the early Earth may have been related to mantle plumes. Other researchers, in contrast, think that plate tectonic–like processes did begin at the end of the Hadean. And still other researchers argue that plate tectonics was operating after 3.2 Ga, and that during the middle of the Archean, there were many rapidly moving small plates, numerous volcanic island arcs, and abundant hot-spot volcanoes.

Significant volumes of new, relatively buoyant rocks formed during the Archean after the Earth's mantle became cool enough that only partial melting could take place. As we

FIGURE 11.2 A model for crust formation during the Archean Eon.

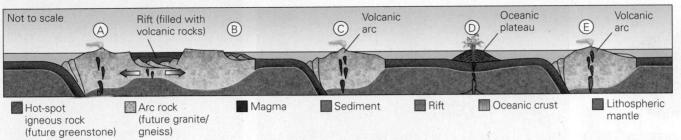

▨ Hot-spot igneous rock (future greenstone)	▦ Arc rock (future granite/ gneiss)	■ Magma	▬ Sediment	▬ Rift	▢ Oceanic crust	▬ Lithospheric mantle

(a) In the Archean, island arcs and hot-spot volcanoes built small blocks of buoyant crust. Rifting of these blocks may have produced flood basalts, and erosion of the blocks produced sediment.

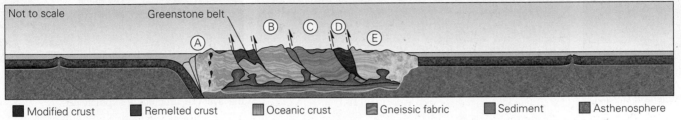

■ Modified crust	■ Remelted crust	▢ Oceanic crust	▨ Gneissic fabric	▬ Sediment	▬ Asthenosphere

(b) Buoyant blocks collided and sutured together, forming protocontinents. Melting at depth produced granite. Eventually, regions of crust cooled, stabilized, and became cratons.

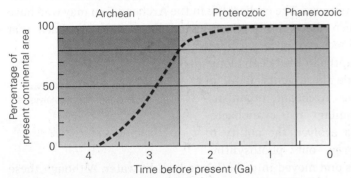

(c) As time progressed, the area of the Earth covered by continental crust increased. Most crust had formed by the beginning of the Proterozoic.

discussed in Chapter 4, magmas formed by partial melting tend to be enriched in silica, relative to their source, so they are more buoyant than their source. Partial melting of the ultramafic mantle, in Archean time, produced mafic magma. This melt extruded or intruded in volcanic arcs at convergent plate boundaries or in hot-spot volcanoes over plumes. Plume-related igneous activity may have produced broad basalt plateaus. When these arcs and plateaus collided with one another, they sutured together, forming larger blocks that were buoyant enough to remain at the Earth's surface. When convergent boundaries developed along the margins of these blocks, and rifts and hot spots developed within the blocks, not only did more basaltic magma form, but in addition, felsic and intermediate magma formed, forming even less-dense rock. As collisions continued, mountain belts formed, and broad regions of crustal rock underwent metamorphism.

Eventually, collisions resulted in the assembly of *protocontinents* (**Fig. 11.2a, b**). Size matters when it comes to the geologic behavior of continents, for the interior of a larger continental block can be isolated from heating by subduction-related igneous activity along its margins. As a result, the crust of the interior region of a protocontinent can slowly cool and strengthen until it becomes a durable and relatively stable crustal block called a **craton**. Between 3.2 and 2.7 Ga, several small Archean cratons came into existence. By the end of the Archean Eon, about 80% of the Earth's continental rock had formed (**Fig. 11.2c**). Notably, much of this rock later passed through stages of the rock cycle one or more times, so only a relative small amount retains Archean isotopic ages. These remnants of Archean crust occur scattered around all of today's continents.

A clear stratigraphic record of marine sediment deposition in crust as old as 3.8 Ga has been found in a few localities, indicating that global oceans existed in the early Archean. Oceans have survived ever since, so for most of the Earth's history, surface temperatures have remained between the boiling and freezing points of water. The accumulation of liquid water in oceans changed the Earth's atmosphere dramatically; prior to ocean formation, H_2O and CO_2 were the dominant gases of the atmosphere. Accumulation of water in the oceans removed most of the H_2O from the atmosphere. And once liquid water existed, most atmospheric CO_2 dissolved into the water. Thus, during the Archean, the atmosphere changed from a foggy mixture of H_2O and CO_2 into a transparent blend whose major component was nitrogen (N_2) gas. Since nitrogen is inert

FIGURE 11.3 Archean life forms.

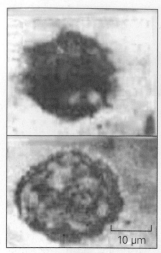

(a) These shapes in 3.2-Ga chert from South America are thought to be fossil bacteria or archaea.

(b) Stromatolites consist of layers of sediment initially stuck to microbes.

(c) This weathered outcrop of 1.85-Ga dolostone near Marquette, Michigan, reveals the layer-like structure of stromatolites.

(it doesn't chemically react with or dissolve in other materials), it remained in the form of a gas.

THE FIRST LIFE

The search for the earliest evidence of life on the Earth continues to make headlines. Sedimentary rocks as old as 3.8 Ga contain biomarkers, chemical signatures of organisms, suggesting that life existed in the early Archean. The oldest shapes that look like fossilized cells, however, occur in 3.5-Ga strata, and the oldest undisputed fossil forms of bacteria and archaea occur in 3.2-Ga strata **(Fig. 11.3a)**. Sedimentary rocks deposited after about 3.2 Ga locally contain **stromatolites**, distinctive layered mounds of sediment formed when cyanobacteria secrete a mucus-like substance to which sediment settling from water sticks **(Fig. 11.3b, c)**. As the mat gets buried, new cyanobacteria colonize the top of the sediment, building a mound upward. Modern examples of stromatolites can be found in shallow, tropical waters.

Did you ever wonder . . .
what the oldest relic of life is?

What specific environment on the Archean Earth served as the cradle of life? Laboratory experiments conducted in the 1950s led many researchers to think that life began in warm pools of surface water, beneath a methane- and ammonia-rich atmosphere streaked by bolts of lightning. More recent researchers suggest instead that submarine hot-water vents, so-called *black smokers*, served as the hosts of the first organisms. These vents emit clouds of ion-charged solutions from which sulfide minerals precipitate and build chimneys. The earliest life in the Archean Eon may well have been thermophilic (heat-loving) bacteria or archaea that dined on pyrite at dark depths in the ocean alongside these vents. Later in the Archean, organisms similar to cyanobacteria evolved the ability to carry out photosynthesis and moved into shallower, well-lit water. Although these organisms produced oxygen, very little accumulated in the atmosphere, for it was either dissolved in the sea or absorbed by weathering reactions with rocks. Oxygen breathers, such as humans, would have suffocated in seconds in the atmosphere of the Archean.

Did you ever wonder...
if the atmosphere has always been breathable?

TAKE-HOME MESSAGE

During the Hadean, which began with the birth of the Earth, the Earth differentiated and the Moon formed. New evidence suggests that liquid water may have appeared as early as 4.4 Ga. But a rock record of the Hadean doesn't exist, because the rock skin of our planet's surface was recycled or was melted and pulverized by meteorite bombardment. During the Archean (4.0–2.5 Ga), the first continental crust formed. As the oceans filled, the atmosphere lost its H_2O and CO_2, but throughout the Archean, it remained devoid of O_2. The oldest definitive fossils of archaea and bacteria first appear in 3.2-Ga strata.

QUICK QUESTION What did continental crust form from?

11.3 The Proterozoic: The Earth in Transition

CONTINUED GROWTH OF CONTINENTS

The **Proterozoic Eon** spans roughly 2 billion years—almost half of the Earth's history—from about 2.5 Ga to the beginning of the Cambrian Period at 541 Ma. During Proterozoic time, the Earth's surface environment changed from an unfamiliar world of small, fast-moving plates, small continents, and an oxygen-free atmosphere, to a more familiar configuration of mostly large, slow-moving plates, large continents, and an oxygenated atmosphere.

New continental crust continued to form during the Proterozoic Eon, though at progressively slower rates, so by the middle of this eon, over 90% of the Earth's continental crustal area had formed. Collisions between Archean cratons, and between cratons and volcanic island arcs or hot-spot volcanoes, gradually produced larger cratons. All cratons that exist today had formed by about 1 Ga (Fig. 11.4), meaning that the crust of cratons ranges from 3.85 Ga to about 1 Ga. Geologists subdivide the basement of North America's craton into distinct, named blocks or provinces, based on the numerical ages of rocks (Fig. 11.5). The *Canadian Shield*, the portion of this craton in which Precambrian rocks are exposed at the Earth's surface, consists of several Archean crust blocks sutured together by Proterozoic orogens. Much of the *cratonic platform* in the United States, a portion of the craton in which a younger sedimentary strata covers Precambrian rock, grew when a series of volcanic island arcs and continental slivers accreted, or attached, to the margin of the Canadian Shield during the time between 1.8 and 1.6 Ga. Granite plutons intruded much of this accreted region, and rhyolite lava and ash flows covered it, between 1.5 and 1.3 Ga.

Successive collisions ultimately brought together most continental crust on the Earth into a single supercontinent, named *Rodinia*, by around 1 Ga. The last major collision during the formation of Rodinia was the *Grenville orogeny*, the products of which crop out in eastern North America. If you look at a popular (though not universally

FIGURE 11.4 Major geologic provinces of the Earth. The black lines indicate the borders of regions underlain by Precambrian crust. Shields are regions where broad areas of Precambrian rocks are exposed.

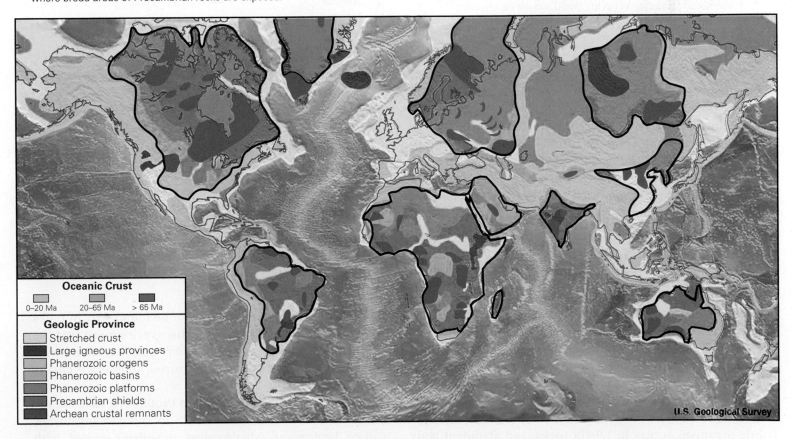

Oceanic Crust
0–20 Ma   20–65 Ma   > 65 Ma

Geologic Province
Stretched crust
Large igneous provinces
Phanerozoic orogens
Phanerozoic basins
Phanerozoic platforms
Precambrian shields
Archean crustal remnants

U.S. Geological Survey

FIGURE 11.5 North America's craton, and the blocks and belts that comprise it.

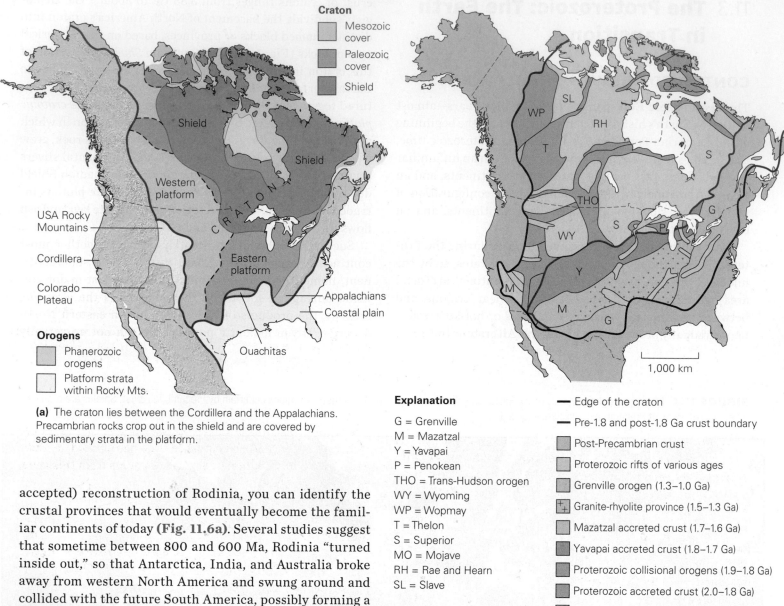

(a) The craton lies between the Cordillera and the Appalachians. Precambrian rocks crop out in the shield and are covered by sedimentary strata in the platform.

Explanation

G = Grenville
M = Mazatzal
Y = Yavapai
P = Penokean
THO = Trans-Hudson orogen
WY = Wyoming
WP = Wopmay
T = Thelon
S = Superior
MO = Mojave
RH = Rae and Hearn
SL = Slave

— Edge of the craton
— Pre-1.8 and post-1.8 Ga crust boundary
Post-Precambrian crust
Proterozoic rifts of various ages
Grenville orogen (1.3–1.0 Ga)
Granite-rhyolite province (1.5–1.3 Ga)
Mazatzal accreted crust (1.7–1.6 Ga)
Yavapai accreted crust (1.8–1.7 Ga)
Proterozoic collisional orogens (1.9–1.8 Ga)
Proterozoic accreted crust (2.0–1.8 Ga)
Archean provinces (>2.5 Ga)

(b) The craton consists of many different blocks, each of which has a name. Most of Canada, and the northwestern United States, had assembled by 1.8 Ga. Belts of younger terranes accreted to the southeastern margin nucleus between 1.8 and 1.1 Ga.

accepted) reconstruction of Rodinia, you can identify the crustal provinces that would eventually become the familiar continents of today (**Fig. 11.6a**). Several studies suggest that sometime between 800 and 600 Ma, Rodinia "turned inside out," so that Antarctica, India, and Australia broke away from western North America and swung around and collided with the future South America, possibly forming a short-lived supercontinent that some geologists refer to as *Pannotia* (**Fig. 11.6b**).

LIFE BECOMES MORE COMPLEX

The map of the Earth clearly changed radically during the Proterozoic. But that's not all that changed—fossil evidence suggests that this eon also saw important steps in the evolution of life. When the Proterozoic began, most life was *prokaryotic*, meaning that it consisted of single-celled organisms (archaea and bacteria) without a nucleus. Though studies of chemical fossils (see Interlude E) hint that *eukaryotic* life, consisting of cells that have nuclei, originated as early as 2.7 Ga, the first possible body fossil of a eukaryotic organism occurs in 1.9-Ga rocks, and abundant body

fossils of eukaryotic organisms can be found only in rocks younger than about 1.5 Ga. Such discoveries mean that the proliferation of eukaryotic life, the foundation from which complex organisms eventually evolved, took place during the Proterozoic.

The last half-billion years of the Proterozoic Eon saw the remarkable transition from simple to complex organisms. Ciliate protozoans (single-celled organisms coated with

FIGURE 11.6 Supercontinents in the late Precambrian.

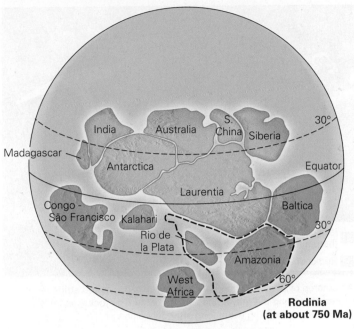

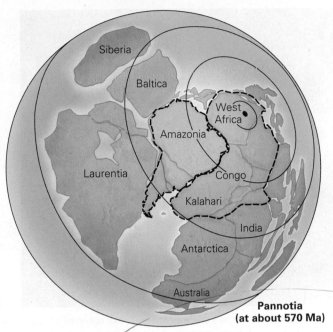

(a) Rodinia formed around 1 Ga and lasted until about 700 Ma. North America and Greenland together comprise Laurentia.

(b) According to one model, by 570 Ma Rodinia had broken apart; continents that once lay to the west of Laurentia ended up to the east of Africa. The resulting supercontinent, Pannotia, broke up soon after it formed.

fibers that give them mobility) appear at about 750 Ma. A great leap in complexity of organisms occurred during the next 150 million years of the eon, for sediments deposited perhaps as early as 620 Ma and certainly by 565 Ma contain several types of multicellular organisms that together constitute the *Ediacaran fauna*. Ediacaran species, named for a region in southern Australia where their fossils were first found, survived into the beginning of the Cambrian before becoming extinct. Their fossil forms suggest that some of these invertebrate organisms resembled jellyfish, while others resembled worms **(Fig. 11.7a)**.

The evolution of life played a key role in the evolution of the Earth's atmosphere. When photosynthetic archaea and bacteria appeared, oxygen began to enter the atmosphere. But it was not until between 2.4 Ga and 1.8 Ga that the concentration of oxygen in the atmosphere began to increase dramatically. This **great oxygenation event** happened when other environments could no longer absorb or dissolve the oxygen produced by organisms, so the oxygen began to accumulate as a gas in air. Because of this change, the oceans became oxidizing environments that could no longer contain large quantities of dissolved iron. Between 2.4 Ga and 1.8 Ga, huge amounts of iron settled out of the ocean to form colorful sedimentary beds known as *banded iron formation (BIF)*. BIF consists of alternating layers of iron-oxide minerals (hematite or magnetite) and jasper (red chert) **(Fig. 11.7b)**.

SNOWBALL EARTH

Radical climate shifts occurred on the Earth at the end of the Proterozoic Eon, and glaciation became widespread, as indicated by the distribution of glacial sediments from this time. Surprisingly, these sediments can be found even in regions that were then located at the equator **(Fig. 11.7c)**. This observation implies that the entire planet was cold enough for glaciers to form at the end of the Proterozoic. Geologists speculate that late Proterozoic glaciers not only covered the land, but that the entire ocean surface froze, resulting in **snowball Earth (Fig. 11.7d)**. The Earth might have remained a snowball forever were it not for volcanic CO_2. The icy sheath covering the oceans prevented atmospheric CO_2 from dissolving in seawater, but it did not prevent volcanic activity from adding CO_2 to the atmosphere. CO_2 is a greenhouse gas, meaning that it traps heat in the atmosphere much as glass panes trap heat in a greenhouse (see Chapter 19), so as the CO_2 concentration increased, the Earth's climate warmed up, and eventually the glaciers melted. Life may have survived snowball Earth conditions only near submarine black smokers and near hot springs. When the ice vanished, life rapidly expanded into

> **Did you ever wonder...**
> if the oceans have ever frozen over entirely?

FIGURE 11.7 Major changes in the Earth System during the Proterozoic Eon.

(a) *Dickinsonia*, a fossil of the Ediacaran fauna. These complex, soft-bodied marine organisms appeared in the late Proterozoic.

Iron oxide minerals (hematite or magnetite)

Red chert (jasper)

(b) An outcrop of BIF in the Iron Ranges of Michigan. The red stripes are jasper (red chert) and the gray stripes are hematite. The rock was folded during a mountain-building event long after deposition.

Strata contain large clasts surrounded by mudstone, for glaciers can carry clasts of all sizes.

Bedding

(c) Layers of Proterozoic glacial till crop out in Africa, indicating that low-latitude landmasses were glaciated during the Proterozoic.

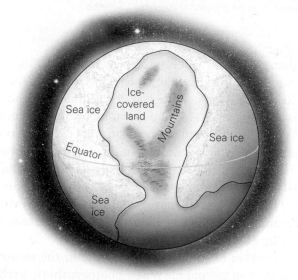

Sea ice

Ice-covered land

Mountains

Sea ice

Equator

Sea ice

(d) This planet may have frozen over completely to form "snowball Earth." Glaciers first grew on land, and eventually the sea surface froze over.

new environments, where new species, such as those of the Ediacaran fauna, evolved.

TRANSITION TO THE PHANEROZOIC EON

As the Proterozoic came to a close, the Earth's climate continued to warm and continents drifted apart—life evolved and diversified to occupy the newly formed environments. Over a relatively short period of time, shells appeared and the fossil record became much more complete. This event

defines the end of the Proterozoic Eon, and therefore of the Precambrian, and the start of the Phanerozoic Eon. Based on the fossil record, geologists recognized the significance of this event long before they could assign it a numerical age (currently 541 Ma).

The **Phanerozoic Eon** consists of three eras: the Paleozoic (Greek for ancient life), the Mesozoic (middle life), and the Cenozoic (recent life). The Mesozoic and Cenozoic each consist of three periods and the Paleozoic of six periods. In the sections that follow, we consider changes in the map of our

FIGURE 11.8 Land and sea in the early Paleozoic Era.

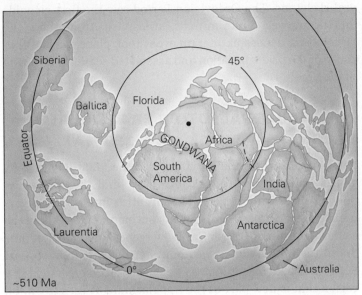

(a) The distribution of continents in the Cambrian Period (510 Ma), as viewed looking down on the South Pole.

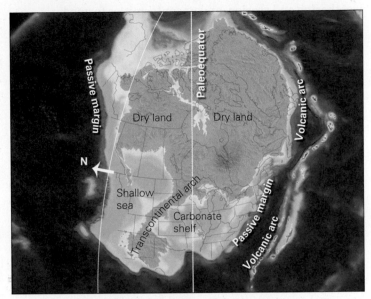

(b) A paleogeographic map of North America shows the regions of dry land and shallow sea in the Late Cambrian Period.

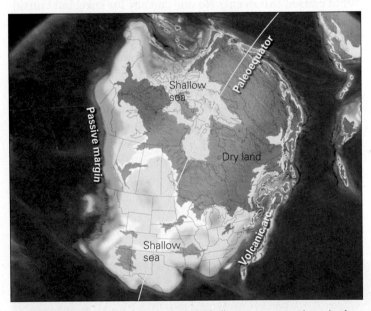

(c) During the Middle Ordovician Period, shallow seas covered much of North America. A volcanic arc formed off the east coast.

planet's surface (its *paleogeography*), as manifested by the distribution of continents, seas, and mountain belts, as well as life evolution during these three eras.

TAKE-HOME MESSAGE

During the Proterozoic (2.5–0.541 Ga), cratons formed and then sutured together to form continents and, eventually, supercontinents, which later rifted apart. Multicellular organisms appeared, and the atmosphere began to accumulate significant oxygen.

QUICK QUESTION What evidence suggests that oxygen began to accumulate in the air during the Proterozoic?

11.4 The Paleozoic Era: Continents Reassemble, and Life Gets Complex

THE EARLY PALEOZOIC ERA (CAMBRIAN–ORDOVICIAN PERIODS, 541–444 MA)

Paleogeography At the beginning of the Paleozoic Era, Pannotia broke up, yielding smaller continents, including **Laurentia** (North America and Greenland), **Gondwana** (South America, Africa, Antarctica, India, and Australia), *Baltica* (Europe), and *Siberia* (**Fig. 11.8a**). New passive-margin basins formed along the edges of these new continents. Sea level rose and fell during this time, and at times, vast areas of continental interiors were flooded with shallow, *epicontinental seas* (**Fig. 11.8b**). The sediment deposited from epicontinental seas forms the strata of cratonic platforms. In the well-lit shallower waters of epicontinental seas, life abounded, so strata of cratonic platforms contain abundant fossils. Sea level, however, did not stay high for the entire early

Paleozoic. When regressions took place, land was exposed to weathering and erosion, and unconformities formed.

The geologically peaceful world of the early Paleozoic in Laurentia ended abruptly in the Middle Ordovician Period when the eastern passive-margin basin of Laurentia collided with a volcanic island arc and other crustal fragments. The resulting collision, called the *Taconic orogeny*, deformed and metamorphosed strata that had been deposited in the continent's passive-margin basin and produced a mountain range in what is now the eastern part of the Appalachians (**Fig. 11.8c**).

Life Evolution Although complex invertebrates existed prior to the Paleozoic, invertebrates with shells didn't appear until the dawn of the Cambrian. Shells may have evolved as a means of protection against predation by conodonts, small, eel-like organisms with hard parts that resemble teeth. The fossil record indicates that soon after the Cambrian began, life began to undergo remarkable diversification, and many new classes of creatures appeared. This event, which paleontologists refer to as the **Cambrian explosion**, took several million years. What caused this event? No one can say for sure, but considering that it occurred roughly at the time a supercontinent broke up, the explanation may involve the production of new ecological niches and the isolation of populations that resulted when smaller continents formed and drifted apart.

By the end of the Cambrian, trilobites were grazing the seafloor, sharing their environment with mollusks, brachiopods, nautiloids, gastropods, graptolites, and echinoderms (**Fig. 11.9**; see Interlude E). A complex food chain arose, which included plankton, bottom feeders, and at the top of the chain, predators. Many of the organisms crawled over or

FIGURE 11.9 A museum diorama illustrates what early Paleozoic marine organisms may have looked like.

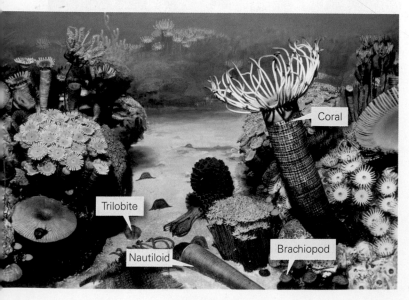

swam around reefs composed of mounds of sponges with mineral skeletons. The Ordovician Period saw the first crinoids and the first vertebrate animals, jawless fish. At the end of the Ordovician, a mass-extinction event took place, perhaps because of the brief glaciation and associated sea-level lowering of the time.

THE MIDDLE PALEOZOIC ERA (SILURIAN–DEVONIAN PERIODS, 444–359 MA)

Paleogeography As the Earth entered the Middle Paleozoic, global climate warmed, sea level rose, and the continents flooded once again. In parts of Laurentia's interior, huge carbonate reef complexes grew in epicontinental seas. Along the eastern margin of Laurentia, convergence and collision produced the *Caledonian orogen* in eastern Greenland, western Scandinavia, and Scotland, as well as the *Acadian orogen* in a region that is now part of the Appalachians (**Fig. 11.10a, b**). Along the western margin of Laurentia, a passive-margin basin continued to subside until the Late Devonian, when an island arc collided. This event caused the *Antler orogeny*, during which the once quiet passive-margin basin became a site of deformation. The Caledonian, Acadian, and Antler orogenies all shed deltas of sediment onto the continents—deposits that formed thick successions of redbeds, such as those visible today in the Catskill Mountains of New York State.

Life Evolution Life on the Earth underwent radical changes during the middle Paleozoic. In the sea, new species of trilobites, gastropods, crinoids, and bivalves replaced species that had disappeared during the mass extinction at the end of the Ordovician. On land, vascular plants with veins (for transporting water and food), woody tissues, and seeds rooted for the first time. The evolution of veins and wood allowed plants to grow much larger, and by the Late Devonian Period the land surface hosted swampy forests with tree-sized relatives of club mosses and ferns. Also at this time, spiders, scorpions, insects, and crustaceans began to exploit both dry-land and freshwater habitats, and jawed vertebrates such as sharks and bony fish began to cruise the oceans. Finally, at the end of the Devonian, the first tetrapods, animals with legs, crawled out onto land and inhaled air with lungs (**Fig. 11.10c**).

THE LATE PALEOZOIC ERA (CARBONIFEROUS–PERMIAN PERIODS, 359–252 MA)

Paleogeography The climate cooled significantly in the late Paleozoic. Seas gradually retreated from the continents, so that during the Carboniferous Period, regions that had hosted the limestone-forming reefs became coastal areas

FIGURE 11.10 Paleogeography and fossils of the Silurian and Devonian Periods.

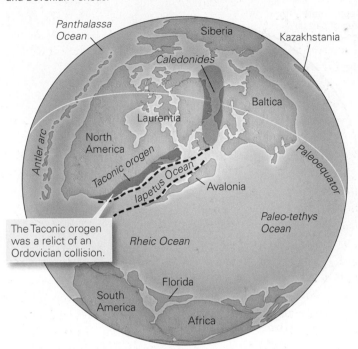

(a) During the Silurian and Devonian Periods, Laurentia collided with Baltica, Avalonia, and South America in succession, as oceans in between were consumed. The Antler arc formed off the west coast.

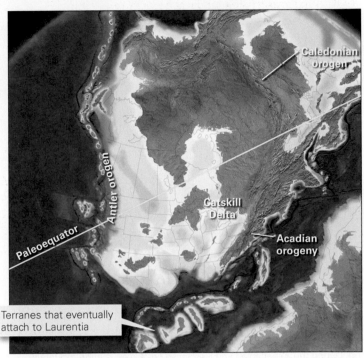

(b) During the Devonian Period, the Acadian orogeny shed sediments into a shallow sea to form the Catskill Delta on the east coast of Laurentia. The Antler orogeny shed sediments eastward.

(c) A Late Devonian fossil skeleton of *Tiktaalik*; this lobe-finned fish was one of the first animals to walk on land.

and river deltas in which sand and mud accumulated. During the Carboniferous, Laurentia lay near the equator and rainfall was heavy, so the coastal plains as well as lowlands inshore became swamps harboring lush vegetation. Plant debris of these swamps transformed into coal after burial. Much of Gondwana and Siberia, in contrast, lay at high latitudes and became ice-covered during the Permian Period.

The late Paleozoic Era also saw a succession of continental collisions. The largest of these took place during Carboniferous and Permian time. This event, the *Alleghanian orogeny*, was the final stage in the development of the Appalachians. It happened when eastern North America rammed into northwestern Africa, and what is now the Gulf Coast region of North America squashed against the northern margin of South America (**Fig. 11.11a**). The resulting supercontinent, **Pangaea**, included most of the Earth's land (**Fig. 11.11b**). The event generated a broad fold-thrust belt whose eroded remnants underlie the valley-and-ridge provinces of the Appalachian and Ouachita Mountains (**Fig. 11.12**). Stresses generated during the Alleghanian orogeny were so great that pre-existing faults in the continental crust clear across North America became active again. The movement reactivated faulting that

produced uplifts, bordered by basins, in the Midwest and in the region of the present-day Rocky Mountains. Geologists refer to the late Paleozoic uplifts of the Rocky Mountain region as the *Ancestral Rockies*.

The assembly of Pangaea involved other collisions around the world as well. Notably, Africa collided with southern Europe to form the *Hercynian orogen*, and a rift or small ocean in Russia closed, leading to the uplift of the Ural Mountains. In addition, parts of China, along with other fragments of Asia, attached to southern Siberia.

FIGURE 11.11 Paleogeography at the end of the Paleozoic Era.

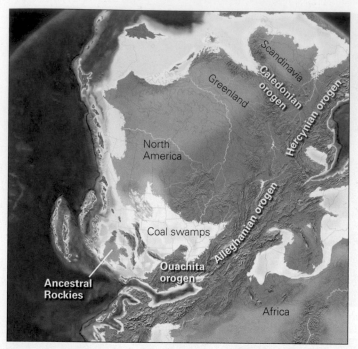

(a) During the Alleghanian and Hercynian orogenies, a huge mountain belt formed. The Caledonian orogeny had formed earlier. Coal swamps bordered interior seas, and the Ancestral Rockies rose.

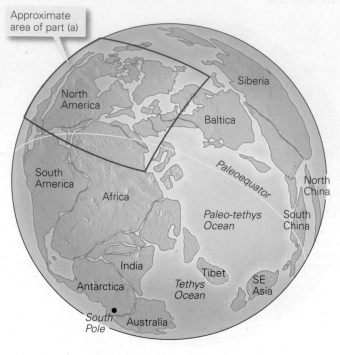

(b) At the end of the Paleozoic, almost all land had combined into a single supercontinent called Pangaea.

Life Evolution The fossil record indicates that during the late Paleozoic, plants and animals continued to evolve toward more familiar forms. In coal swamps, fixed-wing insects including huge dragonflies flew through a tangle of ferns, club mosses, and scouring rushes, and by the end of the Carboniferous, insects such as the cockroach, with foldable wings, appeared **(Fig. 11.13)**. Forests containing gymnosperms ("naked seed" plants, such as conifers) and cycads (trees with a palm-like stalk and fern-like fronds) became widespread in the Permian. Amphibians and, later, reptiles populated the land. The appearance of reptiles marked the evolution of a radically new component in animal reproduction—eggs with a protective shell. Such eggs permitted reptiles to reproduce without returning to the water.

The Permian, the last period of the Paleozoic, came to a close with a major mass-extinction event, during which over 95% of marine species disappeared. Why this event occurred remains a subject of debate. According to one hypothesis, the *Permian mass extinction* occurred as a result of an episode of extraordinary voluminous volcanic activity in the region that is now Siberia. The eruptions could have clouded the atmosphere, acidified the oceans, and disrupted the food chain. Another hypothesis relates the mass extinction to a huge meteorite impact.

TAKE-HOME MESSAGE

Breakup of the late Precambrian supercontinent ushered in the Paleozoic. During the Paleozoic, shallow seas covered continents at times, and life diversified and moved onto land. As the era ended, collisions led to the assembly of Pangaea.

QUICK QUESTION What event, indicated by the fossil record, marks the end of the Paleozoic?

11.5 The Mesozoic Era: When Dinosaurs Ruled

THE EARLY AND MIDDLE MESOZOIC ERA (TRIASSIC–JURASSIC PERIODS, 252–145 MA)

Paleogeography Pangaea, the supercontinent formed at the end of the Paleozoic, lasted for about 100 million years. Then during the Triassic and Early Jurassic Periods, rifting commenced and the supercontinent began to break up. During the Early Jurassic, rifting succeeded in splitting

North America from Europe and Africa. At this time, the Mid-Atlantic Ridge formed and the North Atlantic Ocean started to grow **(Fig. 11.14)**.

The Earth had a warm climate during the Triassic and Early Jurassic. Large areas of North America's interior were nonmarine environments in which red sandstones and

FIGURE 11.12 Features of the Appalachian Mountains in the eastern United States.

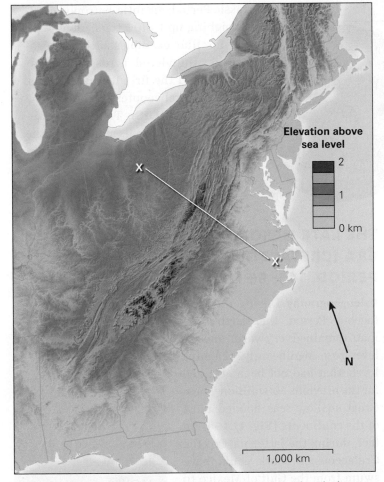

(a) The eroded remnants of the Appalachian orogen stand out in the eastern United States. Line XX' shows the approximate position of the cross section in part (b).

FIGURE 11.13 A museum diorama of a Carboniferous coal swamp includes a giant dragonfly with a wingspan of about 1 m. The inset gives a sense of its size relative to a human.

shales, such as those now exposed in the spectacular cliffs of Zion National Park, were deposited. By the Middle Jurassic, the climate cooled and sea level began to rise. Eventually, a shallow sea submerged much of the Rocky Mountain region of the western United States. Toward the end of the Jurassic, the climate warmed once again.

On the western margin of North America, convergent-margin tectonics became the order of the day. Beginning with Late Permian and continuing through Mesozoic time, subduction generated volcanic island arcs. Over time, as intervening oceanic lithosphere was consumed, several volcanic arcs,

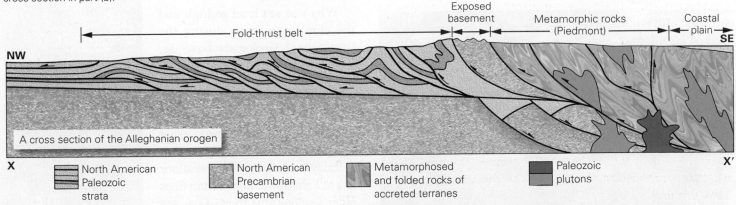

(b) In the fold-thrust belt, strata have been pushed westward and have been folded and faulted. The Blue Ridge exposes a slice of Precambrian basement. Metamorphic and plutonic rocks underlie the Piedmont. The location of this cross section is shown in part (a).

FIGURE 11.14 Pangaea began to break up in the Triassic, and by Jurassic time a narrow North Atlantic Ocean existed.

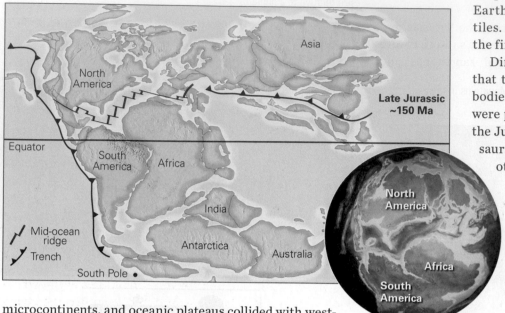

microcontinents, and oceanic plateaus collided with western North America. Accretion (attachment) of these crustal fragments added land to the continent. Because these fragments formed elsewhere, not originally on or adjacent to the continent, geologists call them *exotic terranes*. Near the end of the Jurassic, a major continental volcanic arc, the Sierran arc, developed on the western margin of North America itself— we'll discuss this arc later.

Life Evolution During the early Mesozoic Era, numerous new plant and animal species appeared, filling the ecological niches left vacant by the Late Permian mass extinction. Reptiles swam in the oceans, and new kinds of corals became the

> Did you ever wonder . . .
> when the dinosaurs lived?

FIGURE 11.15 During the Jurassic, giant dinosaurs roamed the land. This painting shows several species.

predominant reef builders. On land, gymnosperms and reptiles diversified, and the Earth saw its first turtles and flying reptiles. And at the end of the Triassic Period, the first true dinosaurs evolved.

Dinosaurs differed from other reptiles in that their legs were positioned under their bodies rather than off to the sides, and they were probably warm-blooded. By the end of the Jurassic Period, gigantic sauropod dinosaurs weighing up to 100 tons, along with other familiar examples such as stegosaurus, thundered across the landscape, and the first feathered birds, such as archaeopteryx, took to the skies (Fig. 11.15). The earliest ancestors of mammals appeared at the end of the Triassic, in the form of small, rat-like creatures.

THE LATE MESOZOIC ERA (CRETACEOUS PERIOD, 145–66 MA)

Paleogeography During the Cretaceous Period, the Earth's climate remained very warm, and sea level rose significantly, reaching levels that had not been attained for the previous 200 million years. Great shallow seas flooded most of the continents (Fig. 11.16a). In fact, during the latter part of the Cretaceous, a shark could have swum from the Gulf of Mexico to the Arctic Ocean, or across much of western Europe.

Why was sea level so high and the climate so warm during the Cretaceous? Geologists have determined that seafloor-spreading rates may have been as much as three times faster during the Cretaceous than they are today. As a result, more of the oceanic crust was younger and warmer than it is today, and since young seafloor lies at a shallower depth than does older seafloor (due to isostasy; see

SEE FOR YOURSELF...

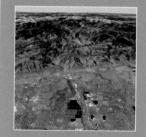

Rocky Mountain Front, Colorado

LATITUDE
39°46'2.32" N

LONGITUDE
105°13'45.35" W

Look obliquely down from 8 km.

We see the steep face of the Rocky Mountains in Colorado. These mountains were uplifted during the Laramide orogeny. During the event, reactivation of large faults thrust Precambrian rocks up and caused overlying Paleozoic strata to fold.

FIGURE 11.16 Cretaceous paleogeography.

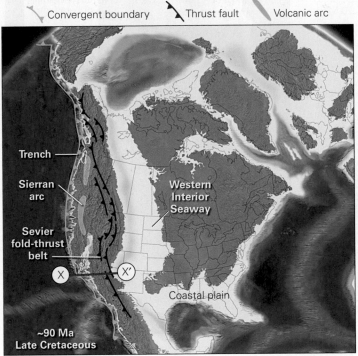

(a) A long seaway flooded the western interior.

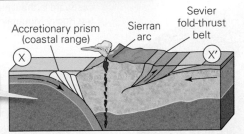

This cross section (X–X') shows the relationship of the Sierran arc to the Sevier fold-thrust belt.

(b) In the Late Cretaceous, a continental volcanic arc formed. A fold-thrust belt formed to the east, as did a transcontinental seaway.

Nevada range. A fold-thrust belt, whose remnants crop out in the Canadian Rockies and in western Wyoming, formed to the the east of the Sierran arc (**Fig. 11.16b**). Geologists refer to the deformation that produced this fold-thrust belt as the *Sevier orogeny*.

The breakup of Pangaea continued through the Cretaceous Period, with the opening of the South Atlantic Ocean and the separation of South America and Africa from Antarctica and Australia. India broke away from Gondwana and headed rapidly northward toward Asia (**Fig. 11.17a, b**). Passive-margin basins developed along the margins of the newly formed oceans—these basins have existed ever since, and have filled with immensely thick accumulations of sediments.

At the end of the Cretaceous, deformation in the western United States swept eastward, perhaps due to a decrease in the angle of subduction or to the collision of an oceanic plateau with the margin of the continent. Slip occurred on large reverse faults in the region of what is now Wyoming, Colorado, eastern Utah, and northern Arizona. This event, which geologists call the *Laramide orogeny*, generated the structure of the present Rocky Mountains in the United States (**Fig. 11.17c**). In contrast to the faults of the Sevier fold-thrust belt, Laramide faults penetrated deep into the Precambrian basement of the continent, and movement on them brought Precambrian rocks up to the surface in *basement-cored uplifts* (**Fig. 11.17d**). As the basement rose, overlying layers of Paleozoic strata warped into large monoclines, folds whose shape resembles the drape of a carpet over a step.

Interlude D), Cretaceous mid-ocean ridges occupied more volume than they do today, and they displaced seawater. Also during the Cretaceous, huge submarine plateaus formed from basalts erupted at hot-spot volcanoes—these plateaus similarly displaced seawater and contributed to sea-level rise. Formation of these plateaus may have been due to mantle upwelling in particularly large plumes, known as **superplumes**. The occurrence of voluminous volcanic activity during the Cretaceous likely released CO_2, a greenhouse gas, whose presence in the atmosphere trapped heat. Rising temperatures caused seawater to expand and polar ice sheets to melt, both of which would make sea level go up even more.

In western North America, the *Sierran arc*, a large continental volcanic arc that initiated at the end of the Jurassic Period, continued to be active. This arc resembled the present-day Andean arc of western South America. Although the volcanoes of the Sierran arc have long since eroded away, we can see their roots in the form of the plutons that now constitute the granitic batholith of the Sierra

Life Evolution   In the seas of the Cretaceous world, modern fish appeared. In contrast with earlier fish, modern fish had short jaws, rounded scales, symmetrical tails, and specialized fins. Huge swimming reptiles and gigantic turtles (with shells up to 4 m across) preyed on the fish. On land, cycads largely vanished, and angiosperms (flowering plants), including hardwood trees, began to compete successfully with conifers for dominance of the forest. Dinosaurs reached their peak of success at this time, inhabiting almost all environments on the Earth. Social herds of grazing dinosaurs roamed the plains, preyed on by the fearsome *Tyrannosaurus rex* (a Cretaceous, not a Jurassic, dinosaur, despite what Hollywood says!). Pterosaurs, with wingspans of up to 11 m, soared overhead, and birds began to diversify. Mammals also diversified and developed larger brains and more specialized teeth, but for the most part, they remained small and rat-like.

Did you ever wonder...
if dinosaurs and humans lived at the same time?

FIGURE 11.17 Paleogeography in Late Cretaceous through Eocene time.

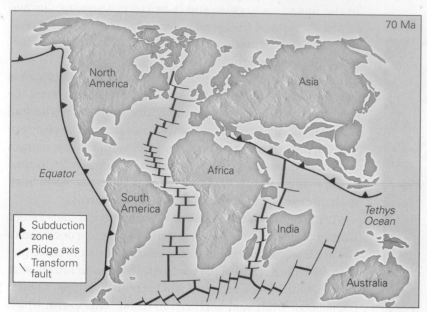

70 Ma

Subduction zone
Ridge axis
Transform fault

(a) By the Late Cretaceous Period, the Atlantic Ocean had formed, and India was moving rapidly northward to eventually collide with Asia.

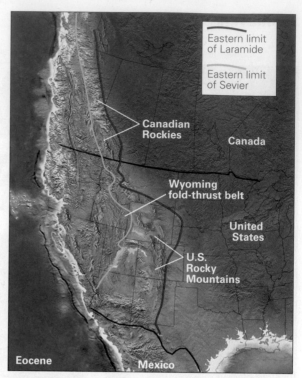

Eastern limit of Laramide
Eastern limit of Sevier

Canadian Rockies
Canada
Wyoming fold-thrust belt
United States
U.S. Rocky Mountains
Eocene
Mexico

(c) During the Laramide orogeny, deformation shifted eastward in the United States, moving from the Sevier belt to the Rocky Mountains, and the style of deformation changed.

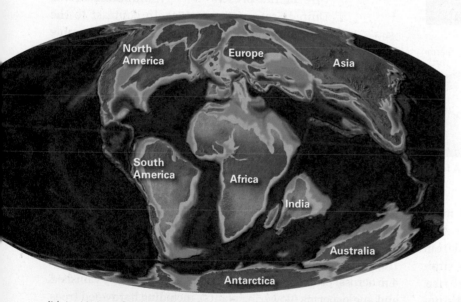

North America
Europe
Asia
South America
Africa
India
Australia
Antarctica

(b) In this Late Cretaceous paleogeographic reconstruction, southern Europe and Asia are beginning to form from a collage of many crustal blocks.

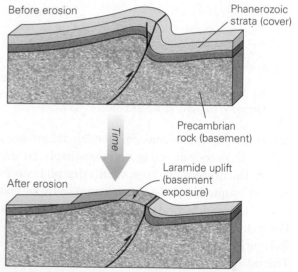

Before erosion
Phanerozoic strata (cover)
Time
Precambrian rock (basement)
After erosion
Laramide uplift (basement exposure)

(d) The Laramide orogeny produced "basement-cored uplifts." In these, faults lifted up blocks of basement, causing the overlying strata to bend into a stair-step-like fold.

The K-Pg Boundary Event Geologists first recognized the K-Pg boundary from 18th-century studies that identified an abrupt global change in fossil assemblages. (K stands for Cretaceous and Pg for Paleogene, the earliest period of the Cenozoic Era.) Until the 1980s, most geologists assumed that this faunal turnover took millions of years. But modern dating techniques indicate that the change happened almost instantaneously and that it signaled a sudden mass extinction of most species on the Earth. The dinosaurs, who ruled the planet for over 150 million years, simply vanished, along with

90% of plankton species and up to 75% of plant species, at 66 Ma. What catastrophe could cause such a sudden and extensive mass extinction? From data collected in the 1970s and 1980s, most geologists have concluded that the Cretaceous Period came to a close, at least in part, as a result of the impact of a 13-km-wide meteorite at the site of the present-day Yucatán Peninsula in Mexico **(Fig. 11.18a)**.

FIGURE 11.18 The Cretaceous-Paleogene (K-Pg) impact. The event caused a mass extinction.

(a) An artist's image of the 13-km-wide object as it hit.

The Earth at 66 Ma

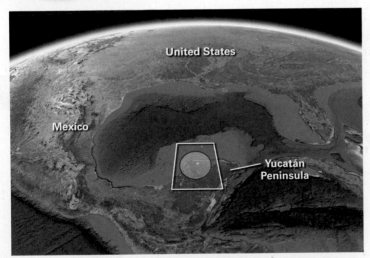

(b) The location of the buried crater today. Gravity anomalies reveal the shape of the crater.

United States

Mexico

Yucatán Peninsula

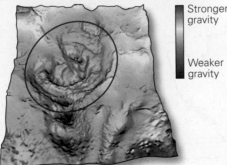

Stronger gravity

Weaker gravity

(c) The crater is now buried, but gravity anomalies outline its shape (see Interlude D).

The discoveries leading up to this conclusion provide fascinating insight into how science works. The story began when Walter Alvarez, a geologist studying strata in Italy, noted that a thin layer of clay interrupted the deposition of deep-sea limestone precisely at the K-Pg boundary. Cretaceous plankton shells constituted the limestone below the clay layer, whereas Cenozoic plankton shells made up the limestone just above the clay. Apparently, for a short interval of time at the K-Pg boundary, all the plankton died, so only clay settled out of the sea. When Alvarez, his father, Luis (a physicist), and other colleagues analyzed the clay, they learned that it contained *iridium*, a rare element found only in meteorites. Soon geologists were finding similar iridium-bearing clay layers at the K-Pg boundary all over the world. Further study showed that the clay layer contained other unusual materials, such as tiny glass spheres (formed from the flash freezing of molten rock), wood ash, and shocked quartz (grains of quartz that had been subjected to intense pressure). Only an immense meteorite impact could explain all these features. The glass spheres formed when melt sprayed into the air from the impact site, the iridium came from fragments of the colliding object, and the shocked quartz grains were produced and scattered by the force of the impact. The wood ash resulted when forests were set ablaze, conceivably because the impact ejected super-heated debris at such high velocity that the debris almost went into orbit and could reach forests worldwide. The impact also generated 2-km-high tsunamis that inundated the shores of continents, and it generated a blast of superheated air.

Researchers suggest that the impact caused unfathomable destruction because the dust, ash, and aerosols that it produced lofted into the atmosphere and transformed the air into a murky haze that reflected incoming sunlight. As a result, photosynthesis became difficult, so all but the hardiest plants died, and winter-like cold might have lasted all year, perhaps for several years. Finally, sulfur-bearing aerosols could have combined with water to produce acid rain, acidifying the ocean and land. These conditions could have broken the food chain and, therefore, could have triggered extinctions.

Geologists suggest that the meteorite responsible for the K-Pg boundary event landed on the northwestern coast of the Yucatán Peninsula, because beneath the reefs and sediments of this tropical realm lies a 100-km-wide by 16-km-deep scar called the *Chicxulub crater* (**Fig. 11.18b, c**). A layer of glass spheres up to 1 m thick occurs at the K-Pg boundary in strata near the site. And radiometric dating indicates that igneous melts in the crater formed at the time of the K-Pg boundary event. The discovery of this event has led geologists to speculate that other such collisions may have punctuated the path of life evolution throughout Earth history and has led to modern-day efforts to track asteroids that pass close to the Earth.

TAKE-HOME MESSAGE

The Mesozoic Era began with the breakup of Pangaea and the formation of the Atlantic Ocean. A convergent boundary formed along the west coast of North America, yielding the Sierran arc and eventually leading to the uplift of the Rocky Mountains. Dinosaurs ruled the planet, and modern forests appeared. A huge meteorite impact marks the K-Pg boundary event—the end of the era—which may have caused the extinction of the dinosaurs.

QUICK QUESTION Did all of today's continents break away from Pangaea at the same time?

11.6 The Cenozoic Era: The Modern World Comes to Be

Paleogeography During the last 66 million years, the map of the Earth has continued to change, gradually producing the configuration of continents and plate boundaries we can see today. The final stages of Pangaea's breakup separated Australia from Antarctica and Greenland from North America. The Atlantic Ocean continued to grow because of seafloor spreading on the Mid-Atlantic Ridge, so the Americas have moved progressively away from Europe and Africa. Meanwhile, the continents that once constituted Gondwana drifted northward as the intervening Tethys Ocean was consumed by subduction (see Fig. 11.17). Collisions of the former Gondwana continents with the southern margins of Europe and Asia resulted in the formation of the largest orogenic belt on the Earth today, the **Alpine-Himalayan chain** (Fig. 11.19). India and a series of intervening volcanic island arcs and microcontinents collided with Asia to form the Himalayas and the Tibetan Plateau to the north, while Africa along with some volcanic island arcs and microcontinents collided with Europe to produce the Alps.

FIGURE 11.19 The two main active continental orogenic systems on the Earth today. The Alpine-Himalayan system formed when Africa, India, and Australia collided with Asia. The Cordilleran and Andean systems reflect the consequences of convergent-boundary tectonism along the eastern Pacific Ocean.

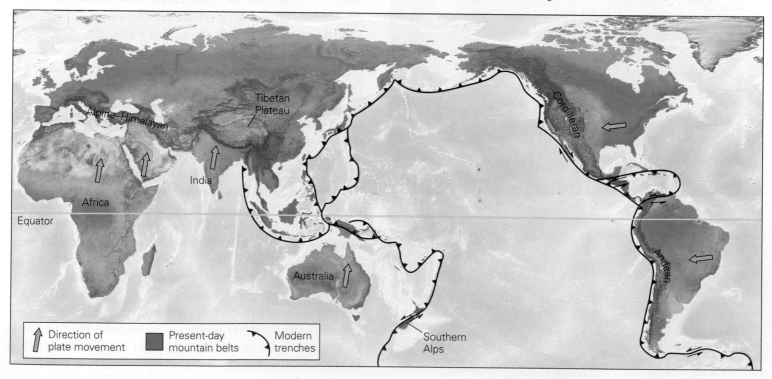

Direction of plate movement — Present-day mountain belts — Modern trenches

FIGURE 11.20 The Basin and Range Province is a rift. The inset shows a cross section along the red line.

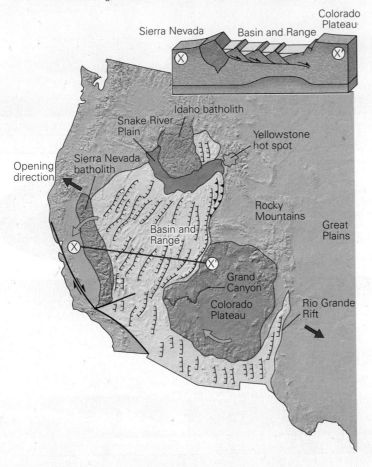

As the Americas moved westward, convergent boundaries evolved along their western margins. In South America, convergent-boundary activity built the Andes, still an active orogen. In North America, convergent-boundary activity continued without interruption until about 40 Ma (the Eocene Epoch). Then, because of the rearrangement of plates off the western shore of North America, a transform boundary gradually replaced the convergent boundary. When this happened, the Laramide orogeny ceased, the *San Andreas fault system* formed along the western coast of the United States, and the Queen Charlotte fault system formed off the coast of Canada. Along these strike-slip faults today, the Pacific Plate moves northward with respect to North America at a rate of about 6 cm per year. In the western United States, convergent-boundary tectonics continues only in Washington, Oregon, and northern California, where subduction of the Juan de Fuca Plate generates the volcanism of the Cascade Range.

As convergent tectonics ceased in the western United States south of the Cascades, the region began to undergo extension in a roughly east-west direction, forming the **Basin and Range Province**, a broad continental rift. It has continued to grow ever since, stretching a portion of the western

United States to twice its original width (**Fig. 11.20**). The Basin and Range Province, as its name suggests, encompasses long, narrow mountain ranges separated from each other by flat, sediment-filled basins. This topography formed when the crust of the region was broken up by normal faults (see Chapter 9), and blocks of crust in the hanging wall of these faults slipped down, tilting in the process. Crests of the tilted blocks form the ranges, and the depressions between them, which rapidly filled with sediment eroded from the ranges, became basins. The Basin and Range Province terminates just north of the *Snake River Plain*. This plain, underlain by products of extrusive volcanism, marks the track of the hot spot that now lies beneath Yellowstone National Park.

Recall that during the Cretaceous Period, the Earth was relatively warm and sea level rose to submerge extensive areas of continents. Climate continued to warm through the early part of the Eocene, but after about 50 Ma, climate rapidly became cooler, and by the early Oligocene Epoch, Antarctic glaciers reappeared for the first time since the Triassic, leading to sea-level fall. In the Early Miocene, warming led to thawing of the Antarctic ice cap, but by the Late Miocene Epoch, colder temperatures led to reglaciation of Antarctica. About 2.5 Ma, the Isthmus of Panama formed, separating the Atlantic completely from the Pacific, changing the configuration of oceanic currents, and perhaps leading to freezing over of the Arctic Ocean.

During the overall cold climate of the past 2.6 million years, continental glaciers have expanded and retreated across northern continents at least 20 times. Geologists therefore refer to this time as the **Pleistocene Ice Age** (**Fig. 11.21**). Each time the glaciers grew, sea level fell so much that the continental shelf became exposed to air, and at times, a *land bridge* formed across the Bering Strait, west of Alaska, providing routes for people from Asia to migrate into North America. A partial land bridge also formed between southeast Asia and Australia, making human migration to Australia easier. Erosion and deposition by the glaciers yielded much of the landscape we see today in northern temperate regions. About 11,000 years ago, the climate warmed, and we entered the *interglacial* time interval we are still experiencing today. The time since the last glaciation is the **Holocene Epoch** (see Chapter 18).

Life Evolution When the skies finally cleared in the wake of the K-Pg boundary catastrophe, plant life recovered, and soon forests of both angiosperms and gymnosperms grew. Grasses, which first appeared in the Cretaceous, spread across the plains in temperate and subtropical climates by the middle of the Cenozoic Era, transforming them into vast grasslands. The dinosaurs, except for their descendants the birds, were gone for good. Mammals rapidly diversified into

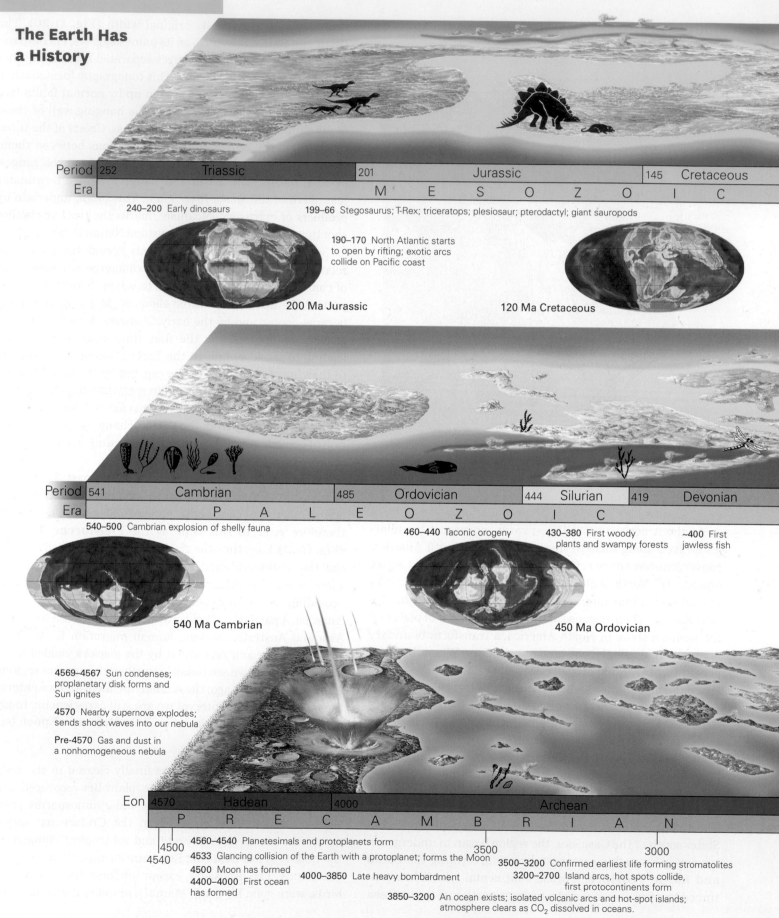

The Earth Has a History

Period	252	Triassic	201	Jurassic	145	Cretaceous
Era			M E S O Z O I C			

240–200 Early dinosaurs

199–66 Stegosaurus; T-Rex; triceratops; plesiosaur; pterodactyl; giant sauropods

190–170 North Atlantic starts to open by rifting; exotic arcs collide on Pacific coast

200 Ma Jurassic

120 Ma Cretaceous

Period	541	Cambrian	485	Ordovician	444	Silurian	419	Devonian
Era			P A L E O Z O I C					

540–500 Cambrian explosion of shelly fauna

460–440 Taconic orogeny

430–380 First woody plants and swampy forests

~400 First jawless fish

540 Ma Cambrian

450 Ma Ordovician

4569–4567 Sun condenses; proplanetary disk forms and Sun ignites

4570 Nearby supernova explodes; sends shock waves into our nebula

Pre-4570 Gas and dust in a nonhomogeneous nebula

Eon	4570	Hadean	4000	Archean
		P R E C A M B R I A N		

4560–4540 Planetesimals and protoplanets form

4533 Glancing collision of the Earth with a protoplanet; forms the Moon

4500 Moon has formed

4400–4000 First ocean has formed

4000–3850 Late heavy bombardment

3500–3200 Confirmed earliest life forming stromatolites

3200–2700 Island arcs, hot spots collide, first protocontinents form

3850–3200 An ocean exists; isolated volcanic arcs and hot-spot islands; atmosphere clears as CO_2 dissolved in oceans.

4500

4540

3500

3000

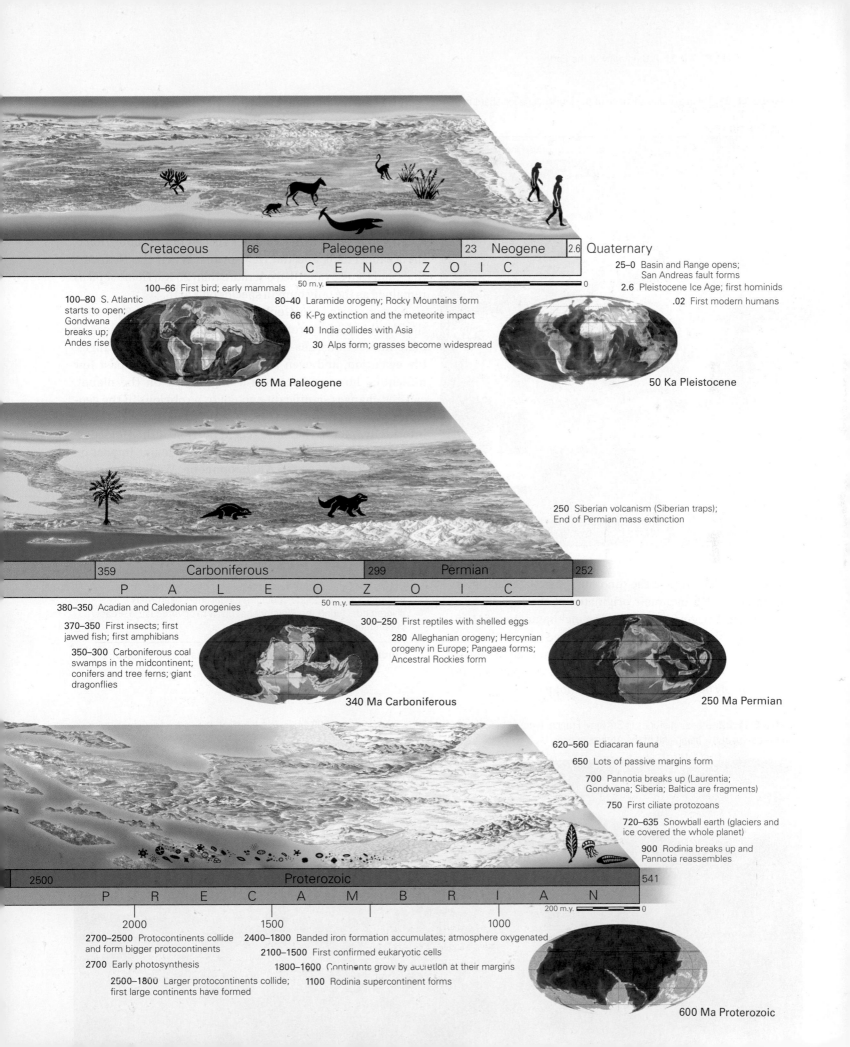

Cretaceous	66	Paleogene	23	Neogene	2.6	Quaternary

C E N O Z O I C

50 m.y. |————————————————| 0

100–66 First bird; early mammals

100–80 S. Atlantic starts to open; Gondwana breaks up; Andes rise

80–40 Laramide orogeny; Rocky Mountains form
66 K-Pg extinction and the meteorite impact
40 India collides with Asia
30 Alps form; grasses become widespread

25–0 Basin and Range opens; San Andreas fault forms
2.6 Pleistocene Ice Age; first hominids
.02 First modern humans

250 Siberian volcanism (Siberian traps); End of Permian mass extinction

65 Ma Paleogene

50 Ka Pleistocene

359	Carboniferous	299	Permian	252

P A L E O Z O I C

50 m.y. |————————————————| 0

380–350 Acadian and Caledonian orogenies

370–350 First insects; first jawed fish; first amphibians

350–300 Carboniferous coal swamps in the midcontinent; conifers and tree ferns; giant dragonflies

300–250 First reptiles with shelled eggs

280 Alleghanian orogeny; Hercynian orogeny in Europe; Pangaea forms; Ancestral Rockies form

340 Ma Carboniferous

250 Ma Permian

620–560 Ediacaran fauna

650 Lots of passive margins form

700 Pannotia breaks up (Laurentia; Gondwana; Siberia; Baltica are fragments)

750 First ciliate protozoans

720–635 Snowball earth (glaciers and ice covered the whole planet)

900 Rodinia breaks up and Pannotia reassembles

2500	Proterozoic	541

P R E C A M B R I A N

2000   1500   1000

200 m.y. |————————————————| 0

2700–2500 Protocontinents collide and form bigger protocontinents

2700 Early photosynthesis

2500–1800 Larger protocontinents collide; first large continents have formed

2400–1800 Banded iron formation accumulates; atmosphere oxygenated

2100–1500 First confirmed eukaryotic cells

1800–1600 Continents grow by accretion at their margins

1100 Rodinia supercontinent forms

600 Ma Proterozoic

FIGURE 11.21 The maximum advance of the Pleistocene ice sheet in North America.

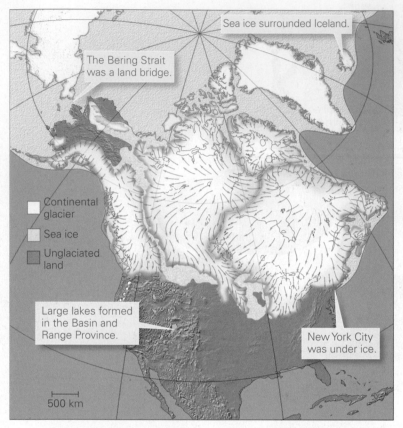

Sea ice surrounded Iceland.

The Bering Strait was a land bridge.

☐ Continental glacier

☐ Sea ice

☐ Unglaciated land

Large lakes formed in the Basin and Range Province.

New York City was under ice.

500 km

It was during the later Cenozoic that our own direct ancestors first appeared. Specifically, ape-like primates diversified in the Miocene Epoch (about 20 Ma), and the first human-like primate evolved at about 4 Ma, followed by the first members of the human genus, *Homo*, at about 2.4 Ma. Fossil evidence, primarily from Africa, indicates that *Homo erectus*, capable of making stone axes, appeared about 1.6 Ma. According to the fossil record, modern people appeared about 200 Ka, initially sharing the planet with two other species of the genus *Homo*, the Neanderthals and the Denisovans. When the Neanderthals and Denisovans died out more than 25 Ka, *Homo sapiens* remained as the only human species on Earth.

As summarized in **Geology at a Glance** (pp. 388–389), the Earth's history reflects the complex consequences of plate interactions, sea-level changes, atmospheric changes, life evolution, and even meteorite impact. In the past few millennia, humans have had a huge effect on the planet, causing changes significant enough to be obvious in the geologic record of the future **(Fig. 11.22)**. In fact, geologists now informally refer to the more recent portion of the Holocene, during which human activities have had a major impact on the Earth System, as the **Anthropocene**. We'll pick up the thread of this story in Chapter 19, where we develop the concept of global change.

a variety of forms to take the dinosaurs' place. In fact, most modern groups of mammals originated at the beginning of the Cenozoic Era, giving this time the nickname *Age of Mammals*. During the latter part of the era, huge mammals (such as mammoths, giant beavers, giant bears, and giant sloths) thrived, but these became extinct during the past 10,000 years, probably because of hunting by humans.

TAKE-HOME MESSAGE

During the Cenozoic, the mountain belts of today rose, and modern plate boundaries became established. Mammals diversified. During the Pleistocene, glaciers covered large areas of continents, and humans appeared.

QUICK QUESTION What interval of time does the Anthropocene refer to?

FIGURE 11.22 Standing stones in Brittany, France, were put in place by prehistoric people over 5,000 years ago. In effect, they are a trace fossil of human activity at the beginning of the Anthropocene.

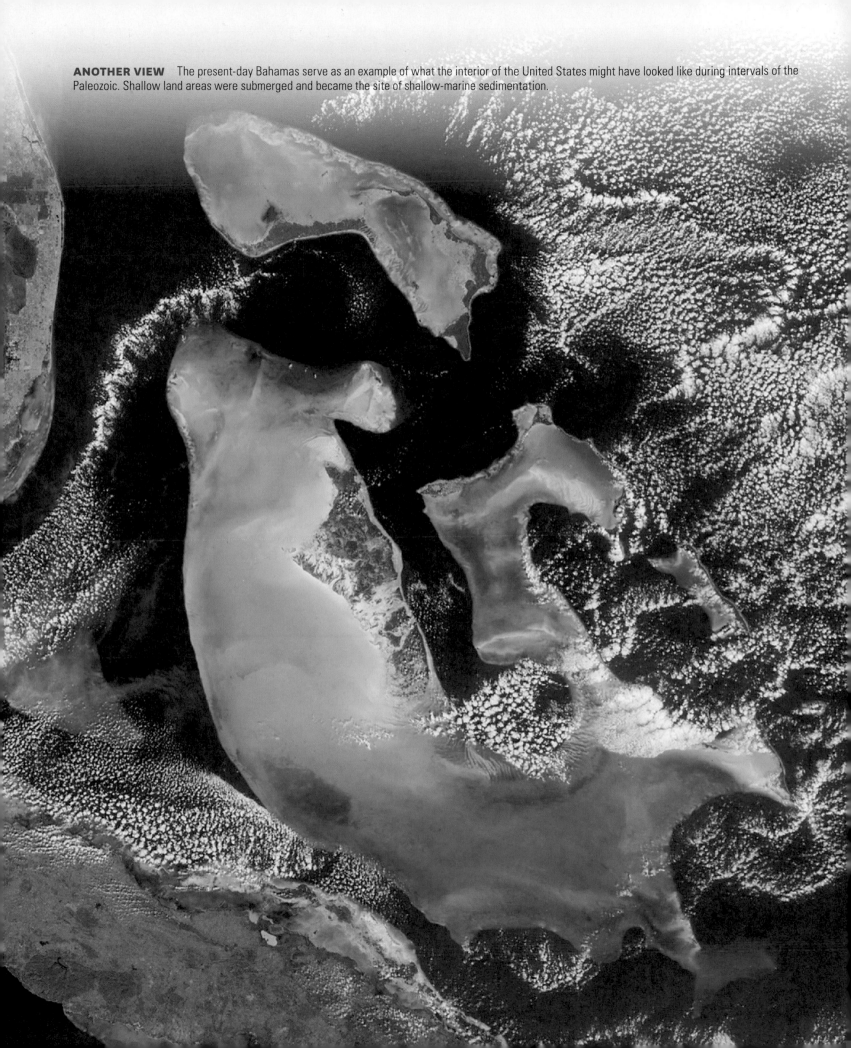

ANOTHER VIEW The present-day Bahamas serve as an example of what the interior of the United States might have looked like during intervals of the Paleozoic. Shallow land areas were submerged and became the site of shallow-marine sedimentation.

Chapter 11 Review

CHAPTER SUMMARY

> The Earth formed about 4.56 billion years ago. At times during the Hadean Eon, the planet was so hot that its surface was a magma ocean.

> The Archean Eon began at 4.0 Ga, about when the oldest rock that remains formed. Continental crust, assembled out of volcanic arcs and hot-spot volcanoes that were too buoyant to subduct, grew during the Archean, and the first life forms—bacteria and archaea—appeared.

> During the Proterozoic Eon, which began at 2.5 Ga, Archean cratons collided and sutured together along orogenic belts, forming large Proterozoic cratons. Photosynthesis added oxygen to the atmosphere.

> By the end of the Proterozoic, complex shell-less marine invertebrates populated the planet. Most continental crust assembled to form a supercontinent called Rodinia at about 1 Ga.

> At the beginning of the Paleozoic Era, rifting yielded several separate continents. During the Paleozoic, sea level rose and fell, yielding sequences of strata in continental interiors. Collisional orogenies produced another supercontinent, Pangaea, at the end of the era.

> Early Paleozoic evolution produced many invertebrates with shells, as well as jawless fish. Land plants and insects appeared in the middle Paleozoic. And, by the end of the era, there were land reptiles and gymnosperm trees.

> In the Mesozoic Era, Pangaea broke apart and the Atlantic Ocean formed. Convergent-boundary tectonics dominated along the western margin of North America. Dinosaurs appeared in Late Triassic time and became prominent land animals through the Mesozoic Era.

> During the Cretaceous Period, sea level was relatively high, and the continents flooded. Angiosperms appeared at this time, along with modern fish. A huge mass-extinction event, which wiped out the dinosaurs, occurred at the end of the Cretaceous Period, probably because of the impact of a large meteorite.

> In the Cenozoic Era, continental fragments of Pangaea collided again. The collision of Africa and India with Asia and Europe formed the Alpine-Himalayan orogen.

> Convergent tectonics has persisted along the margin of South America, producing the Andes, but ceased in North America when the San Andreas fault formed. Rifting in the western United States during the Cenozoic Era produced the Basin and Range Province.

> During the Cenozoic, various kinds of mammals filled niches left vacant by the disappearance of dinosaurs. The human genus, *Homo*, appeared and evolved throughout the radically shifting climate and ice ages of the Pleistocene Epoch.

GUIDE TERMS

Alpine-Himalayan chain (p. 386)
Anthropocene (p. 390)
Archean Eon (p. 370)
Basin and Range Province (p. 387)

Cambrian explosion (p. 378)
craton (p. 371)
differentiation (p. 369)
Gondwana (p. 377)
great oxygenation event (p. 375)

Hadean Eon (p. 370)
Holocene Epoch (p. 387)
Laurentia (p. 377)
Pangaea (p. 379)
Phanerozoic Eon (p. 376)
Pleistocene Ice Age (p. 387)

Proterozoic Eon (p. 373)
snowball Earth (p. 375)
stromatolite (p. 372)
superplume (p. 383)

 GEOTOURS *THIS CHAPTER'S GEOTOURS WORKSHEET (K) FEATURES QUESTIONS AND GOOGLE EARTH SITES ON:*

> Paleogeography of the Earth

REVIEW QUESTIONS

The letters following each Review Question refer to the corresponding Learning Objective from the Chapter Opener.

1. What do geologists date to determine when the Earth formed? **(A)**

2. Why are there no whole rocks on the Earth that yield isotopic dates older than about 4 billion years? **(B)**

3. Describe the formation of the crust, atmosphere, and oceans during the Hadean Eon. **(B)**

4. How did the atmosphere change during the first 2 billion years of Earth history? When did most continental crust form? **(B)**

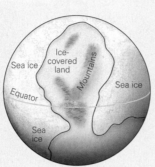

5. What evidence do we have that the Earth's surface froze during the Proterozoic Eon? What term do geologists use for the globe shown here? **(D)**

6. Did supercontinents exist in the Proterozoic? **(F)**

7. How did the Cambrian explosion of life change the nature of the living world? **(E)**

8. What was the Earth's land surface like during the early Paleozoic? **(E)**

9. When in Earth history did vascular plants appear? When did trees appear? **(E)**

10. How did the Alleghanian and Ancestral Rockies orogenies affect North America? **(F)**

11. Describe the plate-tectonic conditions that led to the formation of the Sierran arc and the Sevier fold-thrust belt. What mountain belt in the United States formed during the Laramide Orogeny? **(C)**

12. What life forms appeared during the Mesozoic? **(E)**

13. What may have caused the flooding of the continents during the Cretaceous Period? **(C)**

14. What could have caused the K-Pg mass extinction? **(G)**

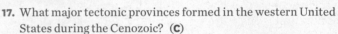
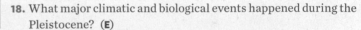

15. What continents formed as a result of the breakup of Pangaea? Which ocean is growing on this paleogeographic map? **(F)**

16. Why did the Himalayas and the Alps form? **(F)**

17. What major tectonic provinces formed in the western United States during the Cenozoic? **(C)**

18. What major climatic and biological events happened during the Pleistocene? **(E)**

ON FURTHER THOUGHT

19. During intervals of the Paleozoic, large areas of continents were submerged by shallow seas. Using Google Earth, tour North America from space. Do any present-day regions within North America consist of continental crust that was submerged by seawater? What about regions off shore? (Hint: Look at the region just east of Florida.) **(E)**

20. Geologists have concluded that 80% to 90% of the Earth's continental crust had formed by 2.5 Ga. But if you look at a geologic map of the world, you find that only about 10% of the Earth's continental crustal surface is labeled "Precambrian." Why? **(B)**

ONLINE RESOURCES

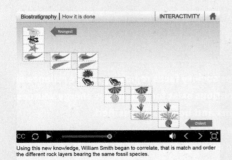

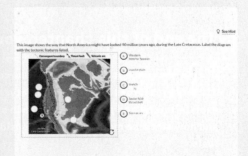

Animations This chapter features interactive animations testing your knowledge of biostratigraphy.

Videos This chapter features a video on how the study of Antarctic rocks reveals clues to the Earth's past.

Smartwork5 This chapter features understanding and ranking exercises about the Earth throughout history, including how the planet's tectonic features have evolved over time.

CHAPTER 12

RICHES IN ROCK: ENERGY AND MINERAL RESOURCES

By the end of this chapter, you should be able to . . .

A. discuss the concept of an energy resource, and list the variety of energy resources available to society.

B. explain how oil and gas differ from each other and from coal, and how all these materials can be considered to be fossil fuels.

C. describe the steps that take place in geologic environments to form fossil fuel reserves.

D. distinguish between conventional and unconventional hydrocarbon reserves, and describe the technologies needed to access them.

E. describe the challenges society faces regarding future reliance on fossil fuels and what options exist for alternative energy sources.

F. explain how coal forms and how it is classified.

G. describe where nuclear fuel comes from.

H. describe the geologic settings in which ore deposits form, and how we find ores.

I. provide examples of nonmetallic resources and their origins.

J. describe the environmental consequences of extracting and using mineral resources, and why reserves of mineral supplies may run out in the coming years.

> All the gold which is under or upon the Earth is not enough
> to give in exchange for virtue.
>
> PLATO (Greek philosopher, ca. 428–347 B.C.E.)

12.1 Introduction

The earliest humans were hunter-gatherers who needed only food and water to survive. But when people discovered that fire could be used for cooking and heating, weapons could facilitate hunting, shelters could make daily life more comfortable, and farming could make food supplies more reliable, their needs expanded. Specifically, people began to require *energy resources* (sources of heat or power) and *mineral resources* (materials from which metals, stone, and chemicals can be derived). Before the 18th century, energy resources were used only to warm homes and to drive pumps or millstones, and mineral resources were used only for constructing buildings or for making weapons and tools. But when industrialization took hold, society's demands increased dramatically, for people began to rely on a huge variety of energy-hungry machines and, eventually, motorized vehicles. It's no surprise that contemporary industrial societies use 100 times more resources per capita than did pre-industrial societies.

Where do people obtain energy and mineral resources? Most come from geologic materials or processes. That's why we discuss resources in a geology textbook. The first half of this chapter focuses on the nature and origin of energy resources. Our discussion begins with a focus on the most widely used energy sources of the past century (oil, gas, coal, and nuclear). Then we address alternative resources that are gradually providing a greater proportion of our needs (hydroelectric, wind, and solar, among others). The second half of the chapter considers both metallic and nonmetallic mineral resources. Knowing how resources form, why they occur where they do, how they can be extracted, and how their extraction and use impacts the environment provides an essential context for facing sustainability challenges of the future. Searching for resources, and dealing with the consequences of their production and use, provides jobs for tens of thousands of geologists worldwide.

◀ (facing page) Heavy equipment scrapes coal from the ground in an Indiana mine. The coal stores energy that came to the Earth from the Sun about 300 million years ago.

12.2 Sources of Energy in the Earth System

Physicists define **energy** as the capacity to do work. In other words, energy causes objects to move or materials to change. Where does the energy in the Earth System originally come from? There are several fundamental sources:

> *Energy directly from the Sun:* Solar energy, resulting from nuclear fusion reactions in the Sun, bathes the Earth's surface. This energy may be converted directly to electricity or used to heat water.

> *Energy involving solar energy and gravity:* Solar radiation heats the air, which becomes buoyant and rises. As this happens, gravity causes cooler air to sink. The resulting air movement—wind—powers sails and windmills. Solar energy also evaporates water, which enters the atmosphere. Wind can blow this moisture over land. When water condenses, forms clouds, and rains on elevated land, it accumulates in streams that flow downhill in response to gravity. This moving water can drive waterwheels that power machines or generators that produce electricity.

> *Energy via photosynthesis:* Algae and green plants absorb solar energy and, through *photosynthesis*, produce sugar. From this sugar, they manufacture complex organic chemicals. Animals eat plants and produce more organic chemicals, as well as waste. For centuries humans have burned **biomass** (plants, wood, manure) to produce energy. When biomass burns, organic chemicals react with oxygen and break apart to produce carbon dioxide, water, and carbon (soot). These reactions release the energy stored in chemical bonds. More recently, chemists have developed processes to convert biomass into flammable liquids, including ethanol.

> *Energy from fossil fuels:* **Fossil fuels**, such as oil, natural gas, and coal, store solar energy that reached the Earth long ago. This solar energy was converted into organic matter; which, when buried, can be preserved in rocks over geologic time.

> *Energy from inorganic chemical reactions:* Release of energy stored in chemical bonds does not always require the use of organic chemicals. The heat of a dynamite explosion or the electricity of hydrogen fuel cells, for example, comes from inorganic chemical reactions.

FIGURE 12.1 The proportion of different sources of energy that people use has changed over time, and the amount of energy used increases almost continuously.

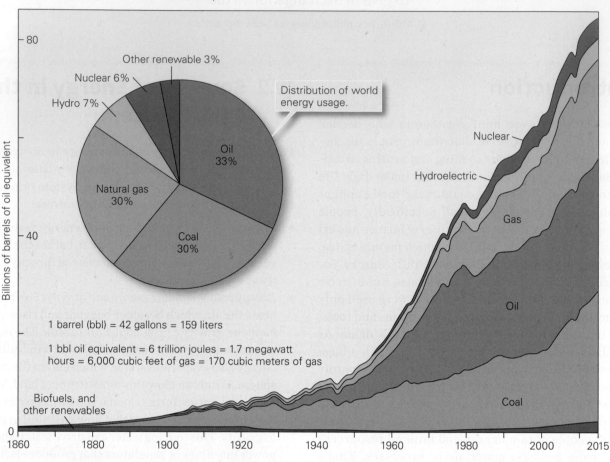

> *Energy from nuclear fission:* Radioactive atoms can split into smaller pieces, a process called *nuclear fission*. During fission, a tiny amount of mass transforms into a large amount of energy, called *nuclear energy*. This type of energy runs nuclear power plants and nuclear submarines.

> *Energy from the Earth's internal energy:* The Earth remains very hot inside—some of this *internal energy* dates from the birth of the planet, while some comes from later radioactive decay. Internal energy heats underground water, which can warm buildings at the surface or, when transformed to steam, can drive electrical generators.

During the course of civilization, people have relied on various sources of energy **(Fig. 12.1)**. Prior to the industrial age, direct burning of wood and other biomass provided most of humanity's energy. But by the second half of the 19th century, deforestation had so depleted this resource, and energy needs had increased so dramatically, that other types of **fuel** (a transportable material that can burn or react to produce energy) came into use. In the next sections, we discuss commonly used energy sources, beginning with fossil fuels, the dominant type in use today.

TAKE-HOME MESSAGE

Energy comes from several sources, such as solar radiation, gravity, chemical reactions, radioactive decay, and the Earth's internal heat. The chemicals in living organisms can be preserved in rocks as fossil fuels.

QUICK QUESTION What kinds of energy sources are available on the Moon?

12.3 Introducing Hydrocarbon Resources

WHAT ARE OIL AND GAS?

Oil and gas are **hydrocarbons**, chain-like or ring-like molecules composed exclusively of carbon and hydrogen atoms. Some hydrocarbons are gaseous and invisible, some resemble a watery liquid, some flow slowly like syrup, and some are solid. The *viscosity* (ability to flow) and the *volatility* (ability

FIGURE 12.2 The formation of oil. The process begins when organic debris settles with sediment. As burial depth increases, heat and pressure transform the sediment into organic shale in which organic matter becomes kerogen. At appropriate temperatures, kerogen becomes oil, which then seeps upward.

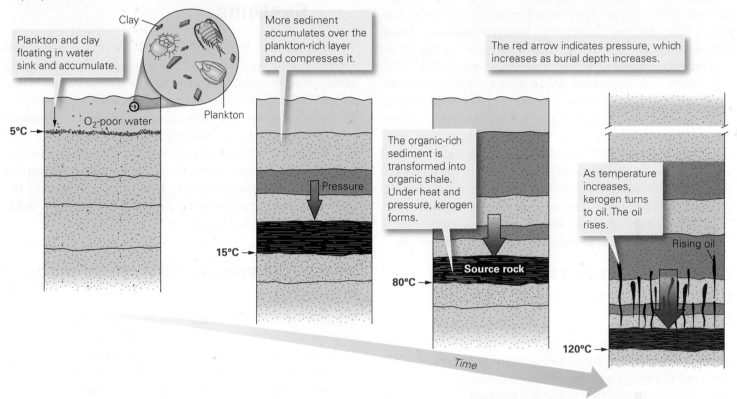

to evaporate) of a hydrocarbon product depend on the size of its molecules. Products made of short chains of molecules tend to be less viscous (they can flow more easily) and more volatile (they evaporate more easily) than products composed of long chains, because the long chains tend to tangle up with each other. Consequently, at room temperature, short-chain molecules occur in gaseous form (such as cooking gas), moderate-length-chain molecules occur in liquid form (such as gasoline and motor oil), and long-chain molecules occur in solid form (tar).

Industrialized societies today rely heavily on hydrocarbons for their energy needs. Why? Oil and gas have a high **energy density**, meaning that they provide a relatively large amount of energy per unit volume. For example, a kilogram of jet fuel provides about 67 times as much energy as a lead-acid battery (car battery) of the same weight. So it's possible to fly an airplane halfway around the world on the jet fuel that it can carry—but the same plane would never get off the ground if it had to be powered by lead-acid batteries.

HYDROCARBON GENERATION IN SOURCE ROCKS

News stories commonly imply, incorrectly, that oil and gas came from the carcasses of dinosaurs. In fact, the chemicals

that make up oil and gas came mostly from *plankton*, very tiny floating organisms. Plankton can include algae, protists, and microscopic animals. When plankton dies, it sinks to the floor of the lake or sea in which it lived. In relatively quiet (nonflowing) water, significant quantities of dead plankton can accumulate, and if the water in the depositional environment does not contain much oxygen, the organic chemicals in the plankton can survive and mix with the clay that also settled from the water. (If the depositional environment contains oxygen, the dead plankton will likely be consumed by other organisms, or will decay, before it can accumulate.) The mixture of clay and organic chemicals becomes an *organic ooze*. Over time, the ooze may become buried by more sediment, and when buried deeply enough, it undergoes lithification to become a black, *organic shale* (*black shale*). Organic shale contains the raw materials from which hydrocarbons form, so geologists refer to it as a **source rock** (**Fig. 12.2**).

At relatively low temperatures, buried organic matter decays, releasing *biogenic natural gas*. As organic shale becomes buried more deeply, it warms, for temperature increases with depth in the Earth. Chemical reactions take place in source

Did you ever wonder...
what the "fossils"
in fossil fuels are?

rocks as they exceed 50°C, slowly transforming the organic material in the shale into a mass of waxy molecules called **kerogen**. Shale with more than 25% to 75% kerogen is called *oil shale*. If oil shale becomes buried even more deeply, warming to temperatures above 90°C, kerogen molecules break into smaller oil and gas molecules, a process known as **hydrocarbon generation**. At temperatures between 150°C and 225°C, oil molecules break down into *thermogenic* (produced by heat) natural gas, and at temperatures over 225°C, organic matter loses all its hydrogen and transforms into graphite (pure carbon). Thus, oil forms only at temperatures between 90°C and 150°C, a range called the **oil window**. *Thermogenic natural gas* can form in the larger *gas window* (90°C to 225°C). A typical value for the *geothermal gradient*, the rate at which temperature increases with depth, is 25°C/km. At this gradient, the oil window occurs at depths of 3.5 to 6.5 km, and the gas window extends down to 9 km.

CONVENTIONAL VERSUS UNCONVENTIONAL HYDROCARBON RESERVES

Oil and gas do not occur in all rocks at all locations. A known supply of oil and gas held underground is a **hydrocarbon reserve**—if the reserve consists predominantly of oil, it's usually called an *oil reserve,* and if it consists predominantly of gas, it's a *gas reserve*. Geologists refer to a region with hydrocarbon reserves underground as an *oil field* or *gas field*, depending on which substance dominates. Until relatively recently, oil companies could economically obtain only those hydrocarbons that could be pumped from the ground easily. We refer to such hydrocarbon reserves as **conventional reserves**. In recent years, however, advances in technology have made it possible to obtain oil and gas from reserves that previously could not be pumped out easily. Such reserves are known as **unconventional reserves**. The ability to use unconventional reserves has led to major changes in the energy industry. We'll discuss these two types of reserves separately.

> ### TAKE-HOME MESSAGE
>
> Oil and gas are hydrocarbons derived from the remains of plankton buried with clay in an organic ooze. Burial transforms ooze into source rock, a black organic shale, and heating underground transforms organic molecules to kerogen. If a rock attains temperatures within the oil or gas window, kerogen transforms into oil or gas. Hydrocarbon reserves are underground accumulations of oil or gas.
>
> **QUICK QUESTION** Could the material that makes up a source rock accumulate in a fast-flowing mountain stream?

12.4 Conventional Hydrocarbon Systems

We've just noted that the hydrocarbons of a conventional reserve can be pumped out of the ground easily. Developing such a reserve requires a specific association of materials, conditions, and time, collectively known as a *conventional hydrocarbon system*. We have discussed the first two components of the system, namely, the formation of a source rock of organic shale and the existence of thermal conditions that transform the kerogen in the shale into smaller hydrocarbon molecules. Let's now look at the remaining components, the migration of oil into a reservoir rock that lies within a configuration of rocks called an oil trap.

RESERVOIR ROCKS AND HYDROCARBON MIGRATION

The clay flakes of an oil shale pack together tightly, so hydrocarbons within the rock are unable to flow easily through the rock. Therefore, an energy company can't simply drill a hole into a source rock and pump out oil or gas—the hydrocarbons won't flow into the well fast enough to make the process cost-efficient. Instead, to obtain oil or gas from a conventional oil or gas field, companies drill into **reservoir rocks**, rocks with high porosity and high permeability. To understand this statement, we need to define two key characteristics of rock or sediment: porosity and permeability.

Porosity refers to the proportion of a rock that consists of open spaces, or **pores**. *Primary porosity* exists if, during lithification, grains didn't fit together perfectly, and cement didn't fill all spaces available. *Secondary porosity* exists if, after lithification, some of the rock dissolved in groundwater or the rock underwent fracturing. Not all rocks have the same porosity. For example, shale typically has porosity of less than 10%. In contrast, a poorly cemented sandstone has porosity of up to 35%, so about a third of a block of sandstone would consist of open space. Note that oil and gas in a reservoir rock reside in the pores, not in underground pools.

Permeability refers to the degree to which pore spaces and cracks connect to one another, either because adjacent pores happen to open into each other or because fracturing broke apart otherwise solid rock. Permeability allows fluids (water, oil, or gas) to flow through the rock. When we say that a good reservoir rock has

Did you ever wonder...
if there are actually pools of oil underground?

high porosity and permeability, we mean that the rock can contain a lot of oil or gas, and that oil or gas can be pumped out of the rock relatively easily.

Reservoir rocks do not contain oil to start with, for the sediments from which reservoir rocks form do not contain organic matter when deposited. So for oil or gas to get into the pores of a reservoir rock, it must first **migrate** (move) upward from the source rock into a reservoir rock, a process that can take thousands to millions of years **(Fig. 12.3)**. What causes hydrocarbon migration? Because oil and gas are less dense than water, they try to rise toward the Earth's surface and get above groundwater, just as salad oil rises above vinegar in a bottle of salad dressing. In other words, buoyancy drives oil and gas upward. (Natural gas, being less dense than oil, eventually rises above oil.) Typically, a conventional hydrocarbon system must have a good *migration pathway*, such as a set of permeable fractures, in order for large volumes of hydrocarbons to move from source to reservoir.

TRAPS AND SEALS

If a migration pathway connects the reservoir rock to the Earth's surface, oil can leak away at an **oil seep**, so none remains underground to extract later. Thus, for a conventional oil or gas reserve to exist, the hydrocarbons must be held underground in the reservoir rock by means of a geologic configuration called a **trap**. An oil or gas trap has two components. First, a **seal rock**, a relatively *impermeable* rock such as shale, salt, or unfractured limestone, must lie above the reservoir rock and stop the hydrocarbons from rising further. Second, the seal and reservoir rock bodies must be arranged in a geometry that holds the hydrocarbons in a restricted area to make pumping economical. Geologists recognize several types of trap geometries—Box 12.1 describes four types.

BIRTH OF THE CONVENTIONAL OIL INDUSTRY

In the United States, during the first half of the 19th century, people obtained "rock oil"

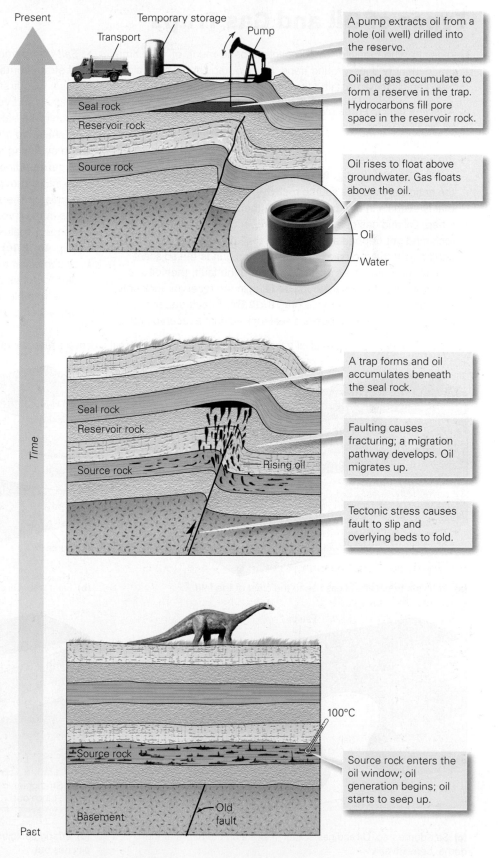

FIGURE 12.3 Initially, oil resides in the source rock. Because it is buoyant relative to groundwater, the oil migrates into the overlying reservoir rock and accumulates in a trap.

Present

Transport · Temporary storage · Pump

Seal rock
Reservoir rock
Source rock

A pump extracts oil from a hole (oil well) drilled into the reservo.

Oil and gas accumulate to form a reserve in the trap. Hydrocarbons fill pore space in the reservoir rock.

Oil rises to float above groundwater. Gas floats above the oil.

Oil
Water

Time

Seal rock
Reservoir rock
Source rock · Rising oil

A trap forms and oil accumulates beneath the seal rock.

Faulting causes fracturing; a migration pathway develops. Oil migrates up.

Tectonic stress causes fault to slip and overlying beds to fold.

100°C

Source rock

Source rock enters the oil window; oil generation begins; oil starts to seep up.

Basement · Old fault

Past

BOX 12.1 CONSIDER THIS...

Types of Oil and Gas Traps

Geologists who work for oil companies spend much of their time trying to identify underground traps. No two traps are exactly alike, but we can classify most into the following four categories.

› *Anticline trap:* In some places, sedimentary beds are not horizontal, as they were when originally deposited, but have been bent by the stresses involved in mountain building. These bends, as we have seen, are called folds. An anticline is a type of fold with an arch-like shape (**Fig. Bx12.1a**; see Chapter 9). If the layers in the anticline include a source rock overlain by a reservoir rock that is overlain in turn by a seal rock, then we have the recipe for a trap. Oil and gas rise from the source rock, enter the reservoir rock, and get trapped at the crest of the anticline.

› *Fault trap:* If the slip on a fault crushes and grinds the adjacent rock to make an impermeable layer along the fault, then oil and gas may migrate upward along bedding in the reservoir rock until they stop at the fault surface (**Fig. Bx12.1b**). A fault trap may also develop if the slip juxtaposes a seal rock against a reservoir rock.

› *Salt-dome trap:* In some sedimentary basins, the sequence of strata contains a thick layer of salt, deposited when the basin was first formed and seawater covering the basin was shallow and very salty. Sandstone, shale, and limestone overlie the salt. The salt layer is not as dense as sandstone or shale, so it is buoyant and tends to rise slowly through the overlying strata. Once the salt starts to rise, the weight of surrounding strata squeezes the salt out of the layer and up into a growing, bulbous **salt dome**. As the dome rises, it bends up the adjacent layers of sedimentary rock. Oil and gas in reservoir rock layers migrate upward until they are trapped against the boundary of the impermeable salt dome (**Fig. Bx12.1c**).

› *Stratigraphic trap:* In a stratigraphic trap, a tilted reservoir rock bed "pinches out" (thins and disappears) up-dip between two impermeable layers. Oil and gas migrating upward along the bed accumulate at the pinch-out (**Fig. Bx12.1d**).

FIGURE Bx12.1 Examples of oil traps. A trap is a configuration of a seal rock over a reservoir rock in a geometry that keeps the oil underground.

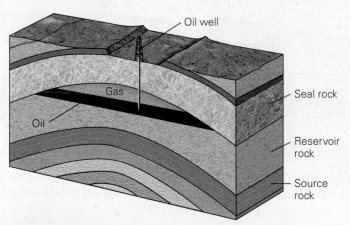

(a) Anticline trap. Oil and gas rise to the crest of the fold.

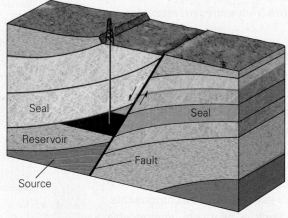

(b) Fault trap. Oil and gas collect in tilted strata adjacent to the fault.

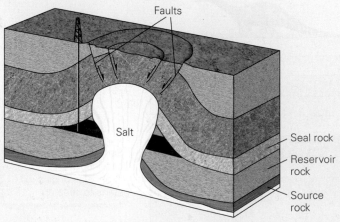

(c) Salt-dome trap. Oil and gas collect in strata on the flanks of the dome, beneath salt.

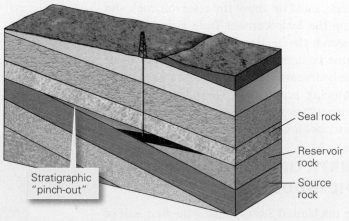

(d) Stratigraphic trap. Oil and gas collect where the reservoir layer pinches out.

(later called petroleum, from the Latin words *petra*, meaning rock, and *oleum*, meaning oil) at seeps and used it to grease wagon axles. But such oil was rare and expensive. In 1854, George Bissell, a New York lawyer, realized that oil might have broader uses, particularly as fuel for lamps, to replace increasingly scarce whale oil. Bissell and a group of investors contracted a colorful character named Edwin Drake to find a way to drill for oil in rocks beneath a hill near Titusville, Pennsylvania, where oily films floated on the water of springs. Using the phony title "Colonel" to add respectability, Drake hired drillers and obtained a steam-powered drill. Work was slow and the investors became discouraged, but the very day that a letter arrived ordering Drake to stop drilling, his drillers found that the hole, which had reached a depth of 21.2 m, had filled with oil. They set up a pump, and on August 27, 1859, for the first time in history, pumped oil out of the ground. No one had given much thought to the question of how to store the oil, so workers dumped it into empty whisky barrels. A *barrel* of oil (bbl) has become a standard unit of measurement: 1 bbl = 42 gallons = 159 liters. (Another unit, a metric ton [tonne] of oil, equals 6.5 barrels.) This first oil well yielded 10 to 35 barrels a day.

Within a few years, thousands of oil wells had been drilled in many places in the United States and in other countries, and by the turn of the 20th century, civilization had begun its addiction to oil. Initially, most oil was used to make kerosene for lamps. Later, when electricity became the primary source for illumination, gasoline derived from oil was the fuel of choice for the newly invented automobile. Oil also fueled electric power plants. In its early years, the oil industry was disorganized. When "wildcatters" discovered a new oil field, there would be a short-lived boom during which the price of oil could drop to pennies a barrel. In the midst of this chaos, John D. Rockefeller established the enormous Standard Oil Company, which monopolized the production, transport, and marketing of oil. In 1911, the Supreme Court divided Standard

Oil into smaller companies, some of which have since recombined. Oil became a global industry governed by the complex interplay of politics, profits, supply, and demand.

THE MODERN SEARCH FOR OIL

Wildcatters discovered the earliest *oil fields* or *gas fields* either by blind luck or by searching for surface seeps. But in the 20th century, when most known seeps had been drilled and blind luck became too risky, oil companies realized that finding new oil fields would require systematic exploration. The modern-day search for oil is a complex, sometimes dangerous, and often exciting procedure with many steps.

Source rocks are always sedimentary, as are most reservoir and seal rocks, so geologists begin exploration by looking for a region with appropriate sedimentary rocks. Then they compile a geologic map of the area, showing the distribution of rock units. From this information, it may be possible to construct a preliminary cross section depicting the geometry of the sedimentary layers underground as they would appear on an imaginary vertical slice through the Earth.

To add detail to the cross section, an exploration company makes a **seismic-reflection profile** of the region **(Fig. 12.4)**. This is done by using either special vibrating trucks or dynamite explosions to send artificial seismic waves into the ground (see Interlude D). The seismic waves reflect off contacts between rock layers, just as sonar waves sent out by a submarine reflect off the bottom of the sea. Reflected seismic waves then return to the ground surface, where small seismometers called *geophones* record their arrival. A computer measures the time between the generation of a seismic wave and its return and from this information defines the depth to the contacts at which waves reflected. With such information, geologists can construct an image of the underground rock layers and identify traps.

FIGURE 12.4 Exploring for oil and gas.

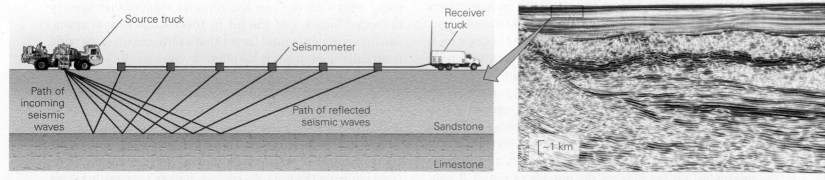

(a) Obtaining a seismic-reflection profile. The source truck sends vibrations downward. These reflect off layer boundaries up to small seismometers (geophones). The travel time indicates the depth to the boundary.

(b) An example of a two-dimensional seismic-reflection profile. The colored bands represent underground sedimentary beds.

FIGURE 12.5 Drilling for oil. The process is challenging.

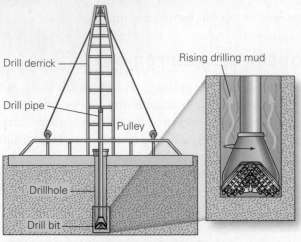

(a) A pipe tipped by a rotating drill bit grinds a hole into the ground. Drilling mud, pumped down through the pipe, cools the bit and flushes up cuttings.

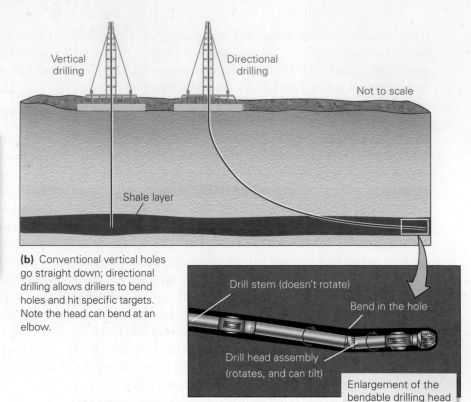

(b) Conventional vertical holes go straight down; directional drilling allows drillers to bend holes and hit specific targets. Note the head can bend at an elbow.

DRILLING AND REFINING

If geologists identify a trap, and if the geologic history of the region indicates the presence of good source rocks, good reservoir rocks, and temperatures within the oil or gas windows, they may recommend drilling. They do not make such recommendations lightly, for drilling a deep well costs millions of dollars. These days, drillers use rotary drills to grind a hole down through rock. A rotary drill consists of a pipe tipped by a rotating bit, a bulb of metal studded with very hard prongs **(Fig. 12.5a)**. Drillers use derricks (towers) to hoist the heavy drill pipe into place so it can be lowered into the hole. On land, derricks have concrete or gravel platforms, but in order to access offshore hydrocarbon reserves that occur in strata beneath the continental shelf, a derrick must be placed on an *offshore-drilling platform*. Such platforms may be built on huge towers rising from the seafloor or on giant submerged pontoons.

During drilling, the bit rotates rapidly. As it does so, it scratches and gouges bedrock, turning the rock into powder and chips. Drillers pump **drilling mud**, a slurry of water mixed with clay and other materials, down the center of the pipe. The mud flows down, past the propeller that rotates the drill bit, and then squirts out of holes at the end of the bit. The extruded mud cools the bit, which otherwise would heat up due to friction as it grinds against rock. Then, the mud flows up the hole on the outside of the drill pipe, carrying *cuttings* (fragments of rock that had been broken up by the drill bit) along with it. Mud also serves another very

important purpose—its weight counters the pressure of the oil and gas in underground reservoir rocks, and thus prevents the hydrocarbons from entering the hole. Were it not for the mud, the natural pressure in the reservoir rock would drive oil and gas into the hole, and if the pressure were great enough, the hydrocarbons would rush up the hole and spurt out of the ground as a gusher or *blowout*. Blowouts are disastrous, because they spill oil onto the land or into the sea and, in some cases, ignite into a deadly inferno.

Traditional drilling employed nonadjustable drill bits and could yield only a vertical hole. Modern drilling, in contrast, uses adjustable drill bits that can be reoriented during drilling, so a drillhole can now curve to become diagonal or even horizontal **(Fig. 12.5b)**. In fact, such **directional drilling** has become so precise that a driller, using a joystick to steer the bit, and sensors that specify the exact location of the bit in three-dimensional space, can hit an underground target that's only centimeters wide from a distance of a few kilometers. Once drilling has been finished, workers "complete" the hole. *Hole completion* involves removing the drill pipe, inserting a *casing* (a metal pipe through which rising hydrocarbons will ultimately flow), and filling the space between the casing and the walls with concrete. The casing and concrete prevent fluids from beds other than the hydrocarbon-rich target beds from entering the hole. To provide access to specific target beds, drillers puncture holes in the casing at the depth of the target bed.

FIGURE 12.6 Pumping, transporting, and refining oil.

The head of the pump goes up and down.

Storage tanks

Well

(a) When drilling is complete, the derrick is replaced by a pump, which sucks oil out of the ground. This design is called a pumpjack.

(b) The Trans-Alaska Pipeline transports oil from fields on the Arctic coast to a tanker port on the southern coast of Alaska.

(c) An oil tanker capable of carrying 2.6 million barrels (enough to supply the entire United States for 7 hours).

After completion, workers remove the derrick and either cap the hole or install a pump. When pumping begins, the completed well becomes a *producing well*. Some pumps resemble a bird pecking for grain—their heads move up and down to pull up oil that has seeped out of pores in the reservoir rock into the well **(Fig. 12.6a)**. You may be surprised to learn that simple pumping brings only about 20% to 30% of the oil from a reservoir rock out of the ground. Oil companies, therefore, commonly use *secondary recovery techniques* to coax out as much as 20% more oil from the reservoir. To do this, a company may drive oil toward a producing well by forcing steam down nearby wells. The steam heats the oil in the ground, making it less viscous, and pushes it along. In some cases, drillers enhance permeability by opening up existing fractures, or by creating artificial fractures in rock adjacent to the hole. They do this by pumping a mixture of water, various chemicals, and sand into a portion of the hole at high pressure. We'll describe this process, called **hydrofracturing** (also known as hydraulic fracturing, hydrofracking, or fracking), in more detail later, in our discussion of how to pump out unconventional reserves.

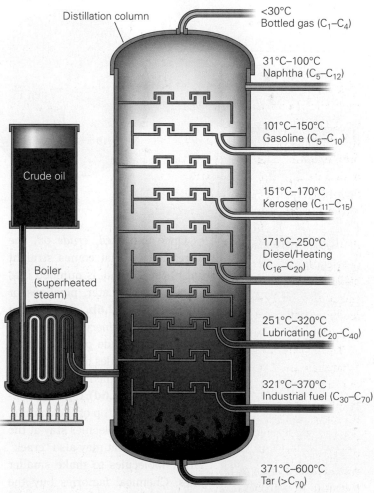

Distillation column

<30°C
Bottled gas (C_1–C_4)

31°C–100°C
Naphtha (C_5–C_{12})

101°C–150°C
Gasoline (C_5–C_{10})

Crude oil

151°C–170°C
Kerosene (C_{11}–C_{15})

171°C–250°C
Diesel/Heating (C_{16}–C_{20})

Boiler (superheated steam)

251°C–320°C
Lubricating (C_{20}–C_{40})

321°C–370°C
Industrial fuel (C_{30}–C_{70})

371°C–600°C
Tar (>C_{70})

(d) A distillation column works by gravity. Heated oil separates into bubbles of lighter hydrocarbons and droplets of heavier ones. Heavier ones sink and light ones rise.

FIGURE 12.7 The distribution of oil reserves around the world.

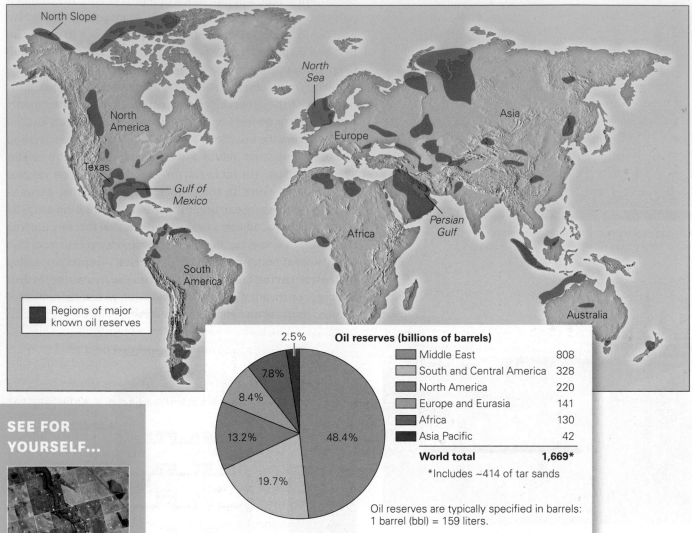

Regions of major known oil reserves

Oil reserves (billions of barrels)

Middle East		808
South and Central America		328
North America		220
Europe and Eurasia		141
Africa		130
Asia Pacific		42
World total		**1,669***

*Includes ~414 of tar sands

Oil reserves are typically specified in barrels:
1 barrel (bbl) = 159 liters.

Oil Field Near Lamesa, Texas

LATITUDE
32°33'18.42" N

LONGITUDE
101°46'55.39" W

Look down from 8 km.

Within a grid of farm fields, 1.6 km wide, small roads lead to patches of dirt. Each patch hosts or hosted a pump for extracting oil. These wells are tapping an oil reserve in Permian sandstone reservoirs.

Once extracted, *crude oil*, the unrefined oil that comes straight out of the ground, is pumped into storage tanks. Later, pipelines or tankers **(Fig. 12.6b, c)** transport crude oil to a *refinery*, where workers separate crude oil into different-sized molecules by heating it in a vertical pipe called a **distillation column (Fig. 12.6d)**. Lighter molecules rise to the top of the column, while heavier molecules stay at the bottom. The heat may also "crack" larger molecules to make smaller ones. Chemical factories buy the largest molecules left at the bottom and transform them into plastics.

Natural gas often occurs in association with oil, or in place of oil, in conventional reserves. This gas must be compressed for transport, so transportation requires high-pressure pipelines or special ships. Because compressing and transporting gas can be very expensive, available quantities must be large enough to make the process economical. Where the costs are too high to process natural gas that comes up with oil, rig operators vent the gas from a pipe and "flare it" (burn it) where it enters the air.

WHERE DO CONVENTIONAL RESERVES OF OIL OCCUR?

Conventional hydrocarbon reserves are not randomly distributed around the Earth **(Fig. 12.7)**. They occur in sedimentary basins along passive continental margins, such as the Gulf Coast of the United States and the Atlantic coasts of Africa and Brazil, as well as in intracratonic and foreland basins of continents (see Chapter 6). By far the largest reserves lie beneath countries bordering the Persian Gulf in the Middle East. In fact, this region has almost 60% of the world's reserves.

Why is there so much oil in the Middle East? Reconstructions of past positions of continents indicate that much of this region was situated in tropical areas, between latitude 20° south and 20° north, during the time between the Jurassic (135 Ma) and the Late Cretaceous (66 Ma). Biological productivity was high, so sediments deposited at that time were rich in organic matter. Thick successions of porous sandstone buried the source rocks of the Middle East, and crustal compression due to the collision between Africa and Asia folded the strata to produce excellent traps.

TAKE-HOME MESSAGE

Conventional hydrocarbon reserves are those in which oil has migrated from a source rock into a porous and permeable reservoir rock, situated within an oil trap, so that the oil or gas reserve can be pumped fairly easily. The first oil was pumped from the ground in 1859. Modern methods for finding, drilling, producing, and refining oil are very complex and expensive.

QUICK QUESTION Where do most of the conventional oil reserves in the world occur today?

12.5 Unconventional Hydrocarbon Reserves

In an *unconventional hydrocarbon reserve*, rock or sediment contains significant quantities of hydrocarbons, but because the rock is impermeable or the hydrocarbons are too viscous to flow, hydrocarbons cannot be extracted simply by drilling and pumping. Extracting these resources, therefore, was not possible until engineers developed new technologies and until the products' selling prices reached a level where extraction was profitable when balanced against the expense of using the new technologies. These conditions were met around 2005, and since then, efforts to extract hydrocarbons from unconventional reserves have grown so rapidly that such reserves now provide a significant portion of the global energy supply. Let's examine the geologic context of three types of unconventional reserves: shale oil and shale gas, tar sands, and oil shale.

SHALE OIL AND SHALE GAS

Huge quantities of oil and natural gas remain in source beds of organic shale, even after the strata have been heated to the oil window or gas window. These resources are called **shale oil** or **shale gas**, respectively. Using traditional drilling and producing methods, it was not cost-effective to extract such source-bed hydrocarbons, for two reasons. First, since the hydrocarbons had not migrated into a trap, a vertical well could not access a significant supply. Second, the source beds are so impermeable that the hydrocarbons would not flow to a pumping well. The development of directional drilling and hydrofracturing radically changed this picture. As a result of directional drilling, a single well can follow a source bed horizontally for many kilometers, and several wells can be drilled from the same platform in different directions into the source bed, so a large volume of reserves can be accessed (**Fig. 12.8a**). Because of hydrofracturing, a rock's permeability can be increased sufficiently to permit the hydrocarbons to flow into the drillhole (**Box 12.2**).

In 2007, Terry Engelder of Pennsylvania State University and other geologists pointed out that previous reports had greatly underestimated the amount of source-bed hydrocarbons available in sedimentary basins that lie near populated areas, so that transport costs would be relatively low (**Fig. 12.8b**). An explosion in exploration and drilling began. In the United States, thousands of wells have been drilled into the Marcellus Shale of Pennsylvania, Ohio, and New York; the Bakken Shale of North Dakota and adjacent regions; and the Barnett Shale of Texas.

Energy companies have spent billions of dollars to lease the rights to obtain hydrocarbons from large areas overlying shale deposits, as a result of which some local landowners became millionaires overnight and thousands obtained jobs. The boom remains controversial, though, because of lingering questions about economic practicality and environmental concerns. Residents worry that chemicals in hydrofracturing fluids may contaminate water supplies and that leaking gas may enter the groundwater or the atmosphere (see Box 12.2). Also, in some cases, extraction brings up large volumes of water, much of which had resided within rock pores for millions of years. This water tends to be too salty to be used for irrigation or drinking, so in order to dispose of it, drillers pump it underground into deep injection wells. Researchers have shown that if injection takes place at very high pressure and at great depth, it may trigger small earthquakes. The water pressure effectively pushes apart fault walls slightly, thereby decreasing frictional resistance to sliding.

TAR SAND (OIL SAND)

In several locations around the world, most notably Alberta (in western Canada) and Venezuela, vast reserves of very viscous, tar-like *heavy oil* exist. This heavy oil, known also as *bitumen*, has the consistency of gooey molasses, and thus

BOX 12.2 CONSIDER THIS...

Hydrofracturing (Fracking)

Hydrofracturing (also known as *hydraulic fracturing*, *hydrofracking*, or *fracking*) is a technique that drillers use to push open existing joints, as well as to generate new cracks in rock at depth. Open joints and cracks provide a permeability pathway through which hydrocarbons can reach a drillhole and be extracted. Hydrofracturing was invented in the 1950s, but it has become headline news in recent years. This method has led to a substantial increase in the amount of unconventional oil and gas that energy companies are extracting in industrialized countries such as the United States, where conventional reserves have been depleted. That's because, while conventional wells can extract hydrocarbons only from reservoir rocks in a trap, hydrofracturing allows extraction of hydrocarbons directly from source beds (black shales).

What is hydrofracturing, how does it work, and what risks are potentially involved in its use? To hydraulically fracture a hole, drillers start by sealing off a length of a well with packers. These inflatable balloons, when filled with high-pressure fluid, press tightly against the wall of the hole. Then the drillers insert a pipe through one of the packers into the sealed-off section of the well and pump *fracking fluid* into this section, under high pressure. When the pressure generated by the fluid within the sealed section becomes great enough, it forces open existing joints that intersect the hole and may cause these joints to lengthen at their tip **(Fig. Bx12.2a)**. The process may also generate new cracks in the rock adjacent to the hole. The area affected by hydrofracturing can extend tens of meters out from the drillhole. Once hydrofracturing concludes, drillers pump out the fracking fluid.

FIGURE Bx12.2 Hydrofracturing in a horizontal shale bed.

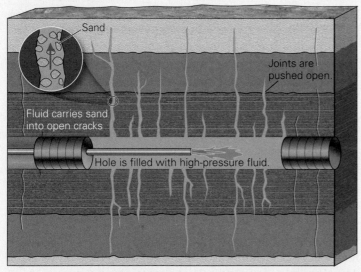

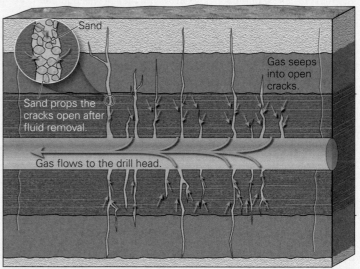

(a) High-pressure fluid is pumped into the segment of hole. The pressure pushes open cracks and forms new ones. Sand injected with the fluid keeps the cracks from closing.

(b) After the packers and the fluid are removed, oil and gas can seep into the pipe and flow to the drill head.

cannot be pumped directly from the ground. It fills the pore spaces of poorly cemented sandstone, constituting up to 12% of the rock volume. Sandstone containing such high concentrations of bitumen is known as **tar sand** or *oil sand*.

Producing usable oil from tar sand is difficult and expensive, but not impossible. It takes about 2 tons of tar sand to produce one barrel of oil. Oil companies mine near-surface deposits in vast open-pit mines, then transport the tar sand to kilns where it undergoes heating. The heat decreases the viscosity of the oil, so that it can be separated from sand. Producers crack the heavy oil molecules to produce smaller, more

usable molecules, and trucks dump the drained sand back into the mine pit. To extract oil from deeper deposits of tar sand, companies drill wells and pump steam or solvents down through injection wells into the sand to liquefy the oil enough so that it can be pumped out of an extraction well.

OIL SHALE

Vast reserves of organic shale have not been subjected to temperatures of the oil window, or if they were, they did not stay within the oil window long enough to complete the

Hydrocarbons then flow along the fractures into the drillhole and up to the surface, where drillers capture it **(Fig. Bx12.2b)**. The volume of fluid used to hydrofracture a section of hole is roughly equivalent to the volume in an Olympic swimming pool. Usually several sections of a hole are subjected to hydrofracturing, so the process needs to be repeated many times in a given drillhole.

What's in fracking fluid? A typical example consists of about 90% water, 9.5% quartz sand, and 0.5% other chemicals. The sand props the holes open once the fluid has been removed—without the sand, the cracks opened by hydrofracturing would close up tightly under the pressure applied by surrounding rock, and thus they could not serve as permeability pathways. The 0.5% portion of fracking fluid that is not water or sand contains many chemicals, including: oils that make the fluid slippery so it can inject farther into the rock; acid, which dissolves cement between grains and increases porosity; detergent, which lowers the surface tension of water so it doesn't stick to grains; guar gum, to make the fluid more viscous so that it can carry more sand; antifreeze, to prevent scale buildup or fluid freezing; and biocides, which prevent the growth of bacteria that could clog pores.

Hydrofracturing requires vast amounts of equipment and materials, and large work crews. At a drilling site, trucks come and go carrying water, sand, and other chemicals, as well as portable pumps and giant mixing vats. Holding tanks or retaining ponds store fluid that has been removed from the ground, once hydrofracturing has finished **(Fig. Bx12.2c)**.

Concerns about hydrofracturing have become the subject of intense public debate and efforts at regulation. The most common concern is that fracking fluid can contaminate drinking water underground. To understand the nature of this risk, it's necessary to understand how groundwater changes with depth. As we'll discuss further in Chapter 16, *groundwater* is water that fills or saturates pores and cracks in rock or sediment underground, beneath a surface called the *water table*—above the water table, pores and cracks contain some air. Typically, groundwater in the upper several hundred to a few thousand meters can be fresh and drinkable, but below that depth, groundwater tends to be saline **(Fig. Bx12.2d)**. If the section of the hole subjected to hydrofracturing lies deeper than the saline boundary, fluids from that section probably won't mix with drinkable groundwater, for they are denser than groundwater and thus are not buoyant. Leakage from the portion of the vertical hole above the saline boundary, however, can be problematic, so it is important that this portion of the hole be sealed thoroughly before fracking fluid is pumped in. Leakage at the surface, from tanks, holding ponds, or transporting trucks, is also of concern. To avoid contamination, handling the fluids at the surface must be very carefully monitored.

(c) A drilling site. The trucks and the holding pond are used during hydrofracturing. Many holes can be drilled from this site, like spokes of a wheel.

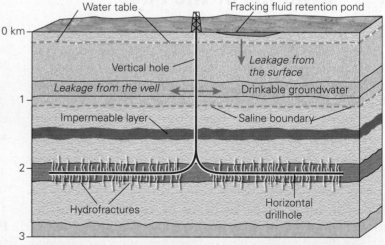

(d) Potential for the contamination of groundwater by hydrofracturing.

transformation of kerogen to oil. Such rock still contains a high proportion of kerogen. Shale with at least 25% to 75% kerogen is called **oil shale**. (Note that *oil shale*, a rock, is different from *shale oil*, oil extracted from source beds; strata containing the latter have been through the oil window.) Lumps of oil shale can be burned directly. In general, however, energy companies produce liquid oil from oil shale. The process involves heating the oil shale in a kiln to a temperature of 500°C, which causes the shale to decompose and the kerogen to transform into liquid hydrocarbons and gas. As is the case with tar sand, producing oil from oil shale is possible but very expensive.

TAKE-HOME MESSAGE

In unconventional reserves, the rock is too impermeable for hydrocarbons to be pumped directly, or the hydrocarbons are too viscous to flow. Directional drilling and hydrofracturing now allow extraction of shale oil and shale gas directly from impermeable source beds. Obtaining hydrocarbons from tar sand or oil shale requires input of heat.

QUICK QUESTION Why does directional drilling allow economical extraction of oil or gas from source beds, even where the source beds do not lie in a trap?

FIGURE 12.8 Unconventional shale oil and shale gas fields in North America.

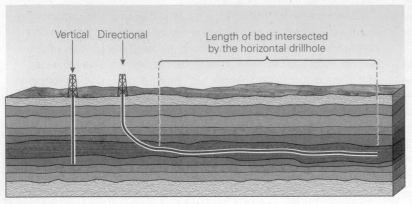

(a) Horizontal directional drilling can allow a very large area of a source bed to be accessed. The length of the hole must be hydrofractured, to provide permeability.

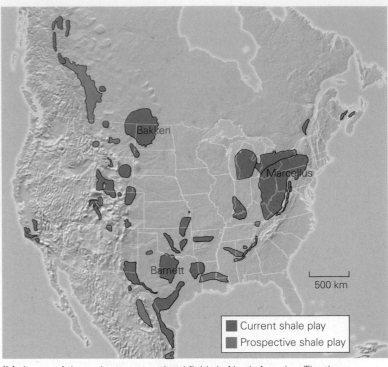

(b) A map of the major unconventional fields in North America. The three largest are labeled. A "play" is a region being drilled.

12.6 Coal: Energy from the Swamps of the Past

Coal is a black, brittle, sedimentary rock that burns. It consists of elemental carbon mixed with minor amounts of organic chemicals, quartz, and clay. Like oil and gas, coal is a fossil fuel because it stores solar energy that reached the Earth long ago. But coal does not have the same composition or origin as oil or gas. In coal, carbon atoms have bonded to form large, complicated molecules called coal *macerals*, whereas oil and gas consist of hydrocarbons. And in contrast to oil and gas, coal forms from vascular plant material (wood, stems, leaves), not plankton. Coal occurs in beds, called *coal seams*, that may be centimeters to meters thick and may be traceable over broad regions.

Significant coal deposits could not form until vascular land plants appeared on the Earth in the late Silurian Period, about 420 million years ago. The most extensive deposits of coal in the world occur in Carboniferous-age strata (359 to 299 Ma) **(Fig. 12.9a)**. In fact, geologists coined the name *Carboniferous* because strata representing this interval of the geologic column contain so much coal. Not all coal reserves, however, are Carboniferous. Notably, during the Cretaceous (145 to 66 Ma), thick layers of organic matter that later became coal accumulated in Wyoming and adjacent states.

THE FORMATION OF COAL

How do the remains of plants transform into coal? The process begins when vegetation of an ancient swamp dies, falls down, and becomes buried in an oxygen-poor environment, such as stagnant water. In this way, the organic matter can become incorporated in a sedimentary sequence before it gets eaten or undergoes complete decay. Compaction and partial decay of the vegetation transforms it into **peat**, which consists of up to 50% to 60% carbon and can itself be dried and then burned as fuel.

To transform peat into coal, the peat must be buried by 4 to 10 km of overlying sediment **(Fig. 12.9b)**. Such deep burial can happen only in sedimentary basins, places where subsidence (sinking) of the surface of the lithosphere takes place, providing the necessary space for thick successions of sediment to accumulate. Notably, when a transgression takes place, due to a relative rise in sea level, marine sediments may be deposited over the coal. The weight of overlying sediment compacts the peat and squeezes out any remaining water. Because temperature increases downward in the Earth, deeply buried peat gradually heats up. Heat accelerates chemical reactions that gradually destroy plant fiber and release molecules of hydrogen, ammonia, methane, and sulfur dioxide as gases. These gases seep out of the reacting peat layer, leaving behind a residue concentrated with carbon. Once the proportion of carbon in the residue exceeds about 60%, the deposit formally becomes coal. With additional burial and higher temperatures, chemical reactions continue, yielding progressively higher concentrations of carbon.

FIGURE 12.9 The formation of coal. Coal forms when plant debris becomes deeply buried.

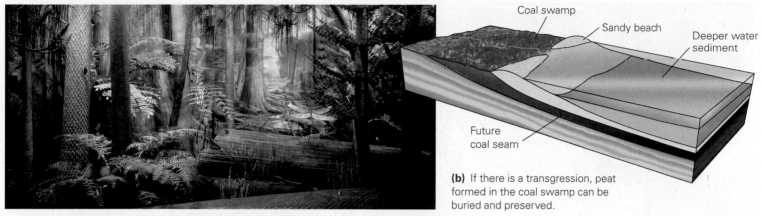

(a) A museum diorama depicting a Carboniferous coal swamp.

(b) If there is a transgression, peat formed in the coal swamp can be buried and preserved.

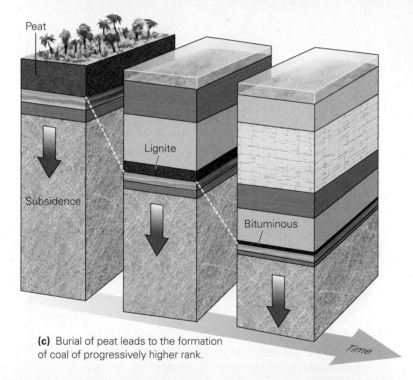

(c) Burial of peat leads to the formation of coal of progressively higher rank.

(d) An example of coal beds interlayered with beds of sandstone and shale.

THE CLASSIFICATION OF COAL

Geologists classify coal based on carbon concentration. Shallower burial transforms peat into a soft, dark-brown coal called *lignite*. At greater depths and higher temperatures (about 100°C–200°C), lignite, in turn, becomes dull, black *bituminous coal*. At still greater depths and even higher temperatures (about 200°C–300°C), bituminous coal transforms into shiny, black *anthracite*. The progressive transformation of peat to anthracite coal, as an organic layer undergoes more burial and becomes warmer, reflects the completeness of chemical reactions that remove water and other chemicals, to leave carbon behind **(Fig. 12.9c)**. Specifically, typical lignite contains about 60% carbon, bituminous about 70%, and anthracite about 90%. As the carbon content of coal increases, we say the **coal rank** increases.

Notably, the formation of anthracite coal requires high temperatures that develop only on the borders of mountain belts. In such locations, mountain-building processes push thick sheets of rock up along thrust faults and over the coal-bearing sediment, burying it to depths of 8 to 10 km. In addition, groundwater that passes through the deeply buried coal may flow along a curving path that first takes the water deep into the subsurface, where it heats up. When this groundwater flows back up into the coal, it brings heat with it, and this heat further warms the coal.

FINDING, MINING, AND USING COAL

Because the vegetation that eventually becomes coal was initially deposited in a sequence of sediment, coal seams are interlayered with other sedimentary rocks **(Fig. 12.9d)**. To find coal, geologists search for sequences of strata that were deposited in tropical to semitropical, shallow-marine to terrestrial environments—the environments in which a swamp could

FIGURE 12.10   The distribution of coal reserves.

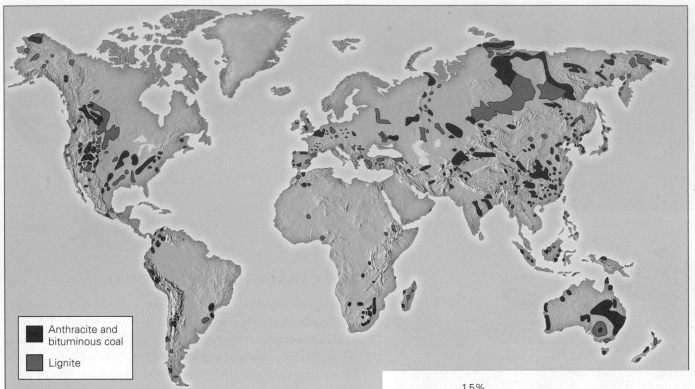

Anthracite and
bituminous coal

Lignite

(a) A map showing global distribution of coal reserves. Most coal accumulated in continental interior basins.

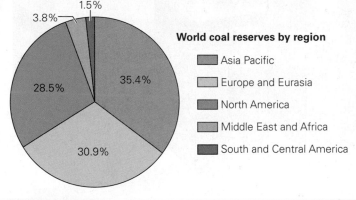

(b) A graph illustrating the distribution of coal reserve quantities, by region.

exist. The sedimentary strata of continents contain huge quantities of discovered coal, or **coal reserves** (Fig. 12.10).

The way in which companies mine coal depends on the depth of the coal seam. If the coal seam lies within about 100 m of the ground surface, *strip mining* proves to be the most economical method. In strip mines, miners use a giant shovel called a *dragline* to scrape off soil and layers of sedimentary rock above the coal seam (Fig. 12.11a, b). Draglines are so big that the shovel could swallow a two-car garage without a trace. Once the dragline has dug an open pit or trench to expose the seam, miners use smaller power shovels to dig out the coal and dump it into trucks or onto a conveyor belt. Before modern environmental awareness took hold, strip mining left huge scars on the landscape. Without topsoil, the rubble and exposed rock of the mining operation remained barren of vegetation. In many contemporary mines, however, the dragline operator separates out the rock that was once above the coal, as well as the soil that once covered bedrock. Then, after the coal has been scraped out, the operator fills the pit or trench with the rock, covers the rock back up with the saved soil, and then plants grass or trees on the soil. The overall process of restoring the landscape of what was once a mine is called *mine reclamation*.

In hilly areas, mining companies may use a practice called *mountaintop removal*, in which they blast off the top

of mountains and dump the debris into adjacent valleys. This practice disrupts the landscape permanently—the mountaintop cannot be replaced, and the debris in valleys disrupts drainage.

Deep coal can be obtained only by *underground mining*. To develop an underground mine, miners dig a shaft down to the depth of the coal seam and then create a maze of tunnels, using huge grinding machines that chew their way into the coal (Fig. 12.11c, d). Underground coal mining can be very dangerous, not only because the sedimentary rocks forming the roof of the mine are weak and can collapse, but because grinding rocks produces vast amounts of coal dust. If miners inhale such dust, they can develop potentially fatal black-lung disease.

FIGURE 12.11 Coal can be mined in strip mines or in underground mines.

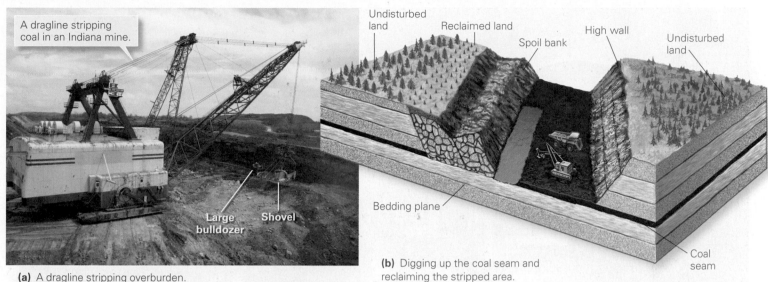

A dragline stripping coal in an Indiana mine.

Large bulldozer   Shovel

(a) A dragline stripping overburden.

Undisturbed land   Reclaimed land   Spoil bank   High wall   Undisturbed land

Bedding plane

Coal seam

(b) Digging up the coal seam and reclaiming the stripped area.

(c) Piles of recently mined coal.

(d) Underground coal mining.

In addition, methane gas released by chemical reactions in coal can accumulate in a mine, along with coal dust, and a small spark can trigger a deadly mine explosion.

COAL GASIFICATION

Solid coal can be transformed into various gases, a process called **coal gasification**, which involves the following steps. First, workers place pulverized coal in a large container called a reaction vessel. Then, they send a mixture of steam and oxygen through the coal at high pressure. As a result, the coal heats to a high temperature but does not ignite. Under these conditions, chemical reactions break down and oxidize the molecules in coal to produce carbon monoxide (CO), hydrogen (H_2), H_2O, and CO_2. The CO gas and H_2 gas can then be combined to produce hydrocarbons. During the process of

coal gasification, contaminants such as ash, sulfur, and mercury concentrate at the bottom of the reaction vessel, from where they can be removed before the gases produced by coal gasification are burned.

UNDERGROUND COAL-BED FIRES

Coal will burn not only in furnaces but also in surface and subsurface mines, as long as the fire has access to oxygen. Coal mining of the past two centuries has exposed much more coal to the air and has provided many more opportunities for fires to begin. Once started, a *coal-bed fire* progresses underground, sucking in oxygen from joints that connect to the ground surface. Such fires may be very difficult or impossible to extinguish because there is no easy way to reach them. Some coal-bed fires have made the land above uninhabitable, for they produce toxic

FIGURE 12.12   Producing electricity at a nuclear power plant.

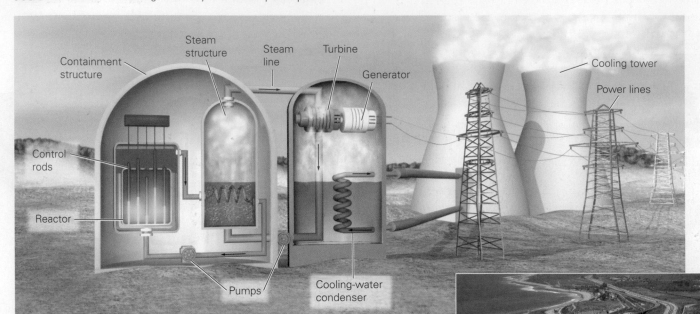

(a) In a nuclear power plant, a reactor heats water, which produces high-pressure steam. The steam drives a turbine that in turn drives a generator to produce electricity. A condenser transforms the steam back into water.

(b) This nuclear power plant in California has two reactors, each in its own containment building.

fumes that rise through joints and pores to the surface, and may lead to the collapse of the land. For these reasons, a coal-bed fire that began in 1962, near the town of Centralia, Pennsylvania, caused the town to be abandoned. Today, over 100 major coal-bed fires are burning in northern China. Recent estimates suggest that 200 million tons of coal burn underground in China every year, an amount equal to approximately 20% of the annual national production of coal in China.

TAKE-HOME MESSAGE

Coal forms from accumulations of plant material over time in anoxic coal swamps. When buried deeply, this organic material loses water and undergoes chemical reactions that concentrate carbon. Higher-rank coal contains more carbon. Coal occurs as beds, called seams, in sedimentary successions and must be mined underground or in open pits.

QUICK QUESTION How can coal be turned into a gas?

12.7 Nuclear Power

HOW DOES A NUCLEAR POWER PLANT WORK?

So far, we have looked at fuels (oil, gas, and coal) that release energy when they undergo *chemical burning*. During such burning, a chemical reaction takes place between the fuel and oxygen, releasing the potential energy stored in the chemical bonds of the fuel. The energy that drives a nuclear power plant comes from a totally different process—*nuclear fission*—which involves breaking the nuclear bonds that hold protons and neutrons together in the nucleus, to split a large atom into smaller pieces. During this process, a small amount of mass transforms into a large amount of thermal and electromagnetic energy, as Einstein recognized. In fact, nuclear fuel has an energy density that is many times that of gasoline.

Nuclear power plants were first built to produce electricity during the 1950s. A **nuclear reactor**, the heart of the plant, commonly lies within a *containment building*, a dome-like structure made of reinforced concrete (**Fig. 12.12**). The reactor itself contains nuclear fuel, consisting of pellets of concentrated uranium oxide or a comparable radioactive material, packed into metal tubes called *fuel rods*. Fission occurs when a neutron strikes a radioactive atom, causing it to split and release more neutrons. For example, radioactive uranium-235 splits into barium-141 and krypton-92, plus three neutrons. The neutrons released during fission of one atom will then strike three other atoms, thereby triggering more fission in a self-perpetuating process called a **chain reaction** that produces lots of heat. Pipes carry water close to the heat-generating fuel rods, and the heat transforms the water into high-pressure steam. The pipes then carry this steam to a turbine, where it rotates fan blades. The rotation, in turn, drives a dynamo that generates electricity. Eventually, the steam goes into cooling towers, where it condenses back into water that can be reused in the plant or returned to the environment.

THE GEOLOGY OF URANIUM

Where does uranium come from? The Earth's radioactive elements, including uranium, probably formed during the explosion of a supernova before the existence of the Solar System. Uranium atoms from this explosion became part of the nebula out of which the Earth grew and thus were incorporated into the planet. They gradually rose into the upper crust, carried along with granitic magma. Although most granite contains uranium, a typical granite contains insufficient quantities of uranium to be worth mining. Geologic processes, however, can produce local concentrations of uranium. For example, hot water circulating through a pluton after intrusion may dissolve uranium and carry it to another location where it precipitates in veins as the mineral pitchblende (UO_2). Uranium may be further concentrated once a pluton, with its associated uranium-rich veins, weathers and erodes at the ground surface. When sand derived from a weathered pluton washes down a stream, the current may carry away relatively low-density grains of feldspar and quartz, leaving behind a concentration of high-density uranium-rich grains. Uranium deposits may also form where groundwater percolates through uranium-rich sedimentary rocks—the uranium dissolves in the water and moves with the water to another location, where it precipitates out of solution and fills the pores of the host sedimentary rock.

You can't just put unprocessed uranium from a mine into a reactor and expect it to produce enough heat to transform water into steam. That's because ^{235}U, the radioactive isotope of uranium that serves as the most common fuel for nuclear power plants, accounts for only about 0.7% of

naturally occurring uranium—most natural uranium consists of ^{238}U. Thus, to make a fuel suitable for use in a power plant, the ^{235}U concentration in a mass of natural uranium must be increased by a factor of 2 or 3, an expensive process called *uranium enrichment*.

CHALLENGES OF USING NUCLEAR POWER

Maintaining safety at conventional nuclear power plants requires hard work. Operators must constantly cool the nuclear fuel with circulating water, and the rate of nuclear fission must be regulated by the insertion of *control rods*, which absorb neutrons and thus decrease the number of collisions between neutrons and radioactive atoms. If not properly controlled, the fuel could become so hot that it would melt. Such a **meltdown** might cause a steam explosion that could breach the containment building and scatter radioactive debris into the air. In some cases, the water gets so hot that it splits into hydrogen and oxygen gases that can recombine explosively. Note that a meltdown is not the same as an atomic bomb explosion. Why? The fuel of a nuclear power plant has not been enriched sufficiently for it to become explosive. Specifically, reactor-grade uranium only contains about 3% to 4% ^{235}U, whereas weapons-grade uranium must contain about 90% ^{235}U. A relatively low concentration of ^{235}U means it's impossible for enough fission reactions to happen in a short enough time to make reactor-grade fuel explode.

Over the past 60 years, three significant accidents have occurred at nuclear power plants. The first occurred in 1979 at the Three Mile Island plant in Pennsylvania, when a stuck valve allowed cooling water to escape, causing the reactor to overheat. Eventually, some radioactive water leaked out into the environment. Fortunately, contamination due to the leak was limited. The next, and much more serious, nuclear accident occurred at the power plant in Chernobyl, Ukraine, in April 1986, while engineers were conducting a test. The fuel became too hot, triggering a hydrogen explosion that ruptured the roof of the containment building and spread fragments of the reactor and its fuel around the plant grounds. Within 6 weeks, 20 people had died from radiation sickness. In addition, some of the radioactive material entered the atmosphere and dispersed over eastern Europe and Scandinavia, but no one yet knows whether this fallout

Did you ever wonder...
if a nuclear power plant could explode like an atomic bomb?

affected the health of exposed populations. To prevent additional radiation hazard, the reactor has been entombed in concrete, and the surrounding region has remained closed. In 2011, an accident second only to Chernobyl, in terms of the amount of radiation released, occurred at the Fukushima power plant along the east coast of northern Japan. This disaster occurred when the catastrophic tsunami that followed the magnitude 9.0 Tōhoku earthquake knocked out the power lines providing electricity to the pumps that circulated cooling water. The tsunami also swamped the plant's backup diesel generators (see Chapter 8). As a result, some of the reactors overheated, and they suffered partial meltdowns and hydrogen gas explosions. The long-term health effects of this event are not known.

One of the biggest challenges to the nuclear power industry pertains to the storage of **nuclear waste**, the radioactive material produced in a nuclear plant. It includes spent fuel, which contains radioactive elements, as well as water and equipment that have come in contact with radioactive materials. Radioactive elements emit gamma rays and X-rays, types of radiation that can damage living organisms and cause cancer. Some radioactive waste decays quickly (in decades to centuries), but some remains dangerous for thousands of years or more. Nuclear waste cannot just be stashed in a warehouse or buried in a town landfill. Some of the waste stays hot enough that it needs to be cooled with water. Even cooler waste, which doesn't need to be submerged in water, has the potential to leak radioactive elements into municipal water supplies or nearby lakes or streams. Ideally, waste should be sealed in containers that will last for thousands of years (the time needed for the short-lived radioactive atoms to undergo decay) and stored in a place where it will not come in contact with the environment. Finding an appropriate place is not easy, and so far, experts disagree about the best way to dispose of nuclear waste. For several years the U.S. government explored the option of storing waste in tunnels under Yucca Mountain, in the Nevada desert. But the site has not been used, and most nuclear waste remains on the sites of the power plants that produced it.

TAKE-HOME MESSAGE

Controlled fission in reactors produces nuclear power. The fuel consists of enriched uranium, or other radioactive elements, obtained by mining. Reactors can produce large amounts of energy, but if not properly monitored, they can run the risk of meltdown. They also yield radioactive waste that is challenging to store.

QUICK QUESTION What is the difference between a meltdown in a reactor and the explosion of an atomic bomb?

12.8 Other Energy Sources

GEOTHERMAL ENERGY

As the name suggests, *geothermal energy* comes from heat in the Earth's crust. We can distinguish between two forms: *high-temperature geothermal energy*, which is used for producing heat and electricity at a commercial scale, and low-temperature geothermal energy (also known as *ambient geothermal energy*), which can warm or cool the water used by an individual household.

High-temperature geothermal energy exists because the crust becomes progressively hotter with increasing depth, at a rate defined by the *geothermal gradient*. In active volcanic areas, the increase produces a temperature of 100°C, the boiling point of water, only several hundred meters from the Earth's surface. Therefore, homeowners in volcanic areas can pump hot groundwater through pipes to heat houses or spas directly, and power companies can use geothermal energy to produce electricity. Geothermal electrical power generation can be accomplished in a few different ways. If the groundwater is hot enough, it turns to steam when pumped up and out of the ground. This happens because the state of water (gas vs. liquid) depends both on temperature and pressure—decompression of very hot groundwater, as happens when it is pumped from depth up to the Earth's surface, causes it to change state from liquid to gas. The steam can drive turbines, which, in turn, drive electrical dynamos **(Fig. 12.13)**. In Iceland and New Zealand, countries with active volcanic areas, geothermal energy provides a substantial portion of electricity needs, but on a global basis, accessibility to geothermally produced electricity is limited.

Ambient geothermal energy takes advantage of the fact that below a depth of a few meters, ground temperature remains nearly constant all year, about

FIGURE 12.13 Geothermal power plants utilize hot groundwater. In some cases, the water is so hot that it becomes steam when it rises and undergoes decompression. Most geothermal plants are in volcanic areas.

Steam turns turbines to generate electricity.

Geothermal well

Rain

Hot groundwater

Hot water rises in a well and turns to steam.

Cold groundwater

Magma

Steam-filled fracture

Natural heat warms groundwater.

A geothermal power plant in New Zealand

equal to the average annual temperature of the air above. By running a home's water supply through an array of pipes buried in the ground, at a depth below 5 m, the water can be warmed in the winter or cooled during the summer before running through the home's electrical or gas-powered cooling or heating system. This process decreases the amount of energy needed to run furnaces or air conditioners, and thus lowers energy costs.

BIOFUELS

In recent years, a portion of our energy needs have been met by **biofuels**, liquid hydrocarbons produced from crops. The most commonly used biofuel is *ethanol*, a type of alcohol, sometimes used as a substitute for gasoline in car engines. The process of producing ethanol from corn includes the following steps. First, producers grind corn into a fine powder, mix it with water, and cook it to produce a mash of starch. They then add an enzyme to the mash, which converts the starch to sugar. The sugar, when mixed with yeast, ferments to produce ethanol, water, and CO_2. Finally, the ethanol is distilled to concentrate the ethanol. Ethanol can also be produced directly from sugar extracted from sugarcane, without fermentation, so sugarcane ethanol is less expensive than corn alcohol.

Researchers are working to develop processes that yield ethanol and other hydrocarbons from other sources that are not used as food crops, so that biofuel production doesn't raise food costs. Examples of alternate biofuel sources include cellulose from perennial grasses, hydrocarbons from algae (which naturally produces fatty organic chemicals), and *biodiesel* (a fuel produced by chemical modification of fats and vegetable oils).

HYDROELECTRIC AND WIND POWER

For millennia, people have used flowing water and air to produce energy. Many towns were established next to rivers, where streams could rotate the waterwheels that powered mills and factories. And in agricultural areas, farmers used windmills to pump water for irrigation. In the past century, engineers have begun to employ the same basic technology to drive generators that produce electricity.

In a modern hydroelectric power plant, water flows from a higher elevation to a lower elevation. The flowing water turns turbine blades placed in a pipe, and the turbine drives an electrical generator. Most hydroelectric plants rely on water from a reservoir held back by a dam. The largest of these, the Three Gorges Dam, blocks the Yangtze River in China **(Fig. 12.14a)**. Dam construction increases the available potential energy of the water because the water level in a filled reservoir is higher than the level of the valley floor that the dam spans. Hydroelectric energy is clean, in the sense that its production does not release chemical or radioactive pollutants, and it is renewable, in that its production does not consume limited resources.

FIGURE 12.14 Alternative energy sources.

(a) Gravity causes water held back by the Three Gorges Dam to flow through turbines and generate electricity.

(b) A wind farm in southwestern England. The towers are about 50 m high.

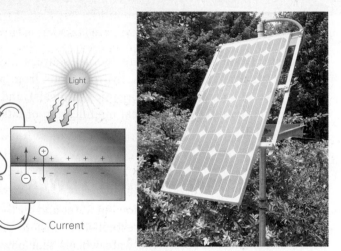

(c) A photovoltaic cell produces electricity directly from solar radiation.

Also, reservoirs may have the added benefit of providing flood control, irrigation water, and recreational opportunities. But damming a river can negatively affect the environment. For example, it may flood a spectacular canyon, eliminate exciting rapids, destroy ecosystems, or submerge towns. Further, reservoirs trap sediment and nutrients, thus disrupting the supply of these materials to downstream floodplains or deltas, a process that may adversely affect agriculture.

When you think of wind energy, you may picture a classic Dutch windmill driving a water pump. Modern efforts to harness the wind are on a much larger scale. To produce wind-generated electricity, engineers identify regions, either onshore or just offshore, with steady breezes. These regions host wind farms made of numerous towers, each with a wind turbine, a giant fan blade that turns even in a gentle breeze **(Fig. 12.14b)**. Some towers can be 100 m tall, with fan blades over 40 m long, and large wind farms can host hundreds of towers. Wind energy produces no pollutants, but as with any type of energy source, it has some drawbacks. Cluttering the horizon with towers may spoil a beautiful view, and the constant loud hum of the towers can disturb nearby residents. The towers may also be a hazard to migrating birds, and offshore towers may interfere with marine life.

SOLAR ENERGY

The Sun drenches the Earth with energy in quantities that dwarf the amounts stored in fossil fuels. Were it possible to harness this energy directly, humanity would have a reliable and totally clean solution for powering modern technology. But using solar energy is not quite so simple because converting light into heat remains fairly inefficient. Let's consider two options for producing solar energy.

A *solar collector* is made up of mirrors that focus light striking a broad area into a smaller area. On a small scale, such devices can be used for cooking; on a large scale, they can produce steam to drive turbines. **Photovoltaic cells** (solar cells), in contrast, convert light energy directly into electricity **(Fig. 12.14c)**. Most photovoltaic cells consist of two wafers of silicon pressed together. One wafer also contains atoms of arsenic, and the other wafer includes atoms of boron. When light strikes the cell, arsenic atoms release electrons that flow over to the boron atoms. If a wire loop connects the back side of one wafer to the back side of the other, this phenomenon produces an electrical current. The production costs of solar cells have decreased substantially in recent years, so their use has increased exponentially. Between 2012 and 2018, global production of electricity by photovoltaic cells more than tripled.

TAKE-HOME MESSAGE

A variety of alternative energy resources are now under development. Biomass can be transformed into burnable alcohol and biodiesel. Geothermal energy utilizes groundwater that has been warmed by heat from the Earth's interior. Flowing water, flowing air, and solar panels can also produce energy.

QUICK QUESTION Is it possible to use ambient geothermal energy in nonvolcanic regions? Why?

12.9 Energy Choices, Energy Problems

THE AGE OF OIL?

Energy usage in industrialized countries grew with dizzying speed through the mid-20th century, and during this time people came to rely increasingly on oil. Oil remains the single largest source of energy globally, accounting for about 33% of global energy consumption.

During the latter half of the 20th century, conventional oil supplies within the borders of industrialized countries could no longer match the demand, and these countries began to import more oil than they produced themselves. In 1973, a complex tangle of politics and war led the Organization of Petroleum Exporting Countries (OPEC) to limit its oil exports. In the United States, fear of an oil shortage turned to panic, and motorists lined up at gas stations, in many cases waiting for hours to fill their tanks. The price of oil rose dramatically, and newspaper headlines proclaimed, "Energy Crisis!" Governments in industrialized countries instituted new rules to encourage oil conservation. During the past few decades, the oil price has bounced up and down. After rising to about $100/bbl around 1980, oil dropped to about $18/bbl in 1999. Prices surged again, reaching a peak of $147/bbl in 2008, only to collapse below $50/bbl a year

later. Since then, it's varied between $40/bbl and $120/bbl **(Fig. 12.15a)**. Oil price changes affect both the price of gasoline at the pump and the number of jobs in the energy sector. Such changes also affect international trade. When the price decreases below around $60/bbl, many unconventional oil and gas fields in industrialized countries become unprofitable, so energy companies choose to import oil from countries that host large conventional reserves. The large fluctuations in oil price emphasize that oil is a commodity whose value reflects market rules of supply and demand.

Will a day come when the supply of conventional oil drops, not for political or economic reasons, but because there is no more oil to produce? To address such a question, we must keep in mind the distinction between renewable and nonrenewable energy resources. We can call a particular resource *renewable* if nature can replace it within a short time relative to a human life span (in months or, at most, decades), whereas a resource is *nonrenewable* if nature takes a very long time (hundreds to millions of years) to replenish it. Oil is a nonrenewable resource in that the rate at which humans consume it far exceeds the rate at which nature replenishes it, so we will inevitably run out of oil. The question is, when?

Historians in the future may refer to our time as the **Oil Age** because so much of our economy depends on oil. How long will the Oil Age last? A reliable answer to this question is hard to come by, because there is not total agreement on the numbers that go into the calculation, especially as the use of unconventional reserves increases, so estimates vary widely. Geologists estimate that we've already used a substantial proportion of our conventional reserves, but that about 1,250 billion barrels of proven conventional oil reserves remain. Proven reserves have been documented and are still in the ground. Optimistically, an additional 2,000 billion barrels of conventional oil supplies may exist. Thus, the world conceivably holds as much as 3,250 billion barrels of conventional oil. Presently, humans consume oil at a rate of about 35 billion barrels per year. At this rate, conventional oil supplies may last no more than about 100 years.

Did you ever wonder...
how much longer the world's oil supply will last?

Some researchers argue that the beginning of the end of the Oil Age has begun, because the rate of consumption now exceeds the rate of discovery, and in many regions the rate of production has already started to decrease. The

FIGURE 12.15 Price, discovery, and consumption of oil.

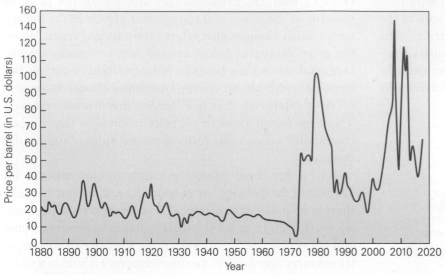

(a) The price of oil held fairly steady for almost 100 years. Starting in 1970, it has risen and fallen dramatically.

M. King Hubbert correctly predicted the peak of production in the United States.

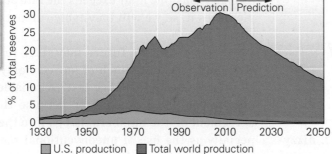

(b) A plot of production versus time suggests that some time in the early 21st century, the production of oil globally will start to decrease. The high point of production is called Hubbert's peak.

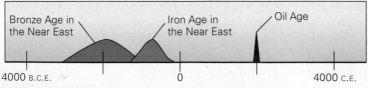

(c) The Oil Age is a relatively short period on the timeline of human history.

peak of production for a given reserve is called **Hubbert's peak**, named for the geologist who first emphasized that the production of nonrenewable resources eventually must decline **(Fig. 12.15b)**. Hubbert's peak for conventional hydrocarbons in the United States appears to have been passed in the 1970s. Some researchers argue that the global peak may occur before 2020. Conservation measures, such as improving the gas mileage of cars and increasing the amount of insulation in buildings, could stretch out supplies and make them last decades longer.

Of course, the picture of oil supplies changes significantly if unconventional reserves are included in estimates. The huge increase in production of unconventional reserves in the United States has changed the slope of the curve showing production related to time. According to current estimates, up to 7,200 billion barrels of unconventional reserves may exist. Even the combination of conventional and accessible unconventional oil reserves means that at current rates of consumption, supplies can last for only another 300 years, so the Oil Age could last a total of about 400 years, a very short blip on a timeline of human history **(Fig. 12.15c)**.

ENVIRONMENTAL ISSUES OF FOSSIL FUEL USE

Environmental concerns about energy resources begin right at the source. Oil drilling requires substantial equipment, the use of which can damage the land. And as demonstrated by the Deepwater Horizon disaster, oil drilling can lead to tragic loss of life and catastrophic marine oil spills **(Box 12.3)**. Transport of oil can also be problematic. Oil spills from pipelines or trucks sink into the subsurface and contaminate groundwater, and oil spills from ships and tankers create slicks that spread over the sea surface and foul the shoreline **(Fig. 12.16)**.

Coal and mining not only scar the land but can lead to the production of *acid mine runoff*, a dilute solution of sulfuric acid that forms when sulfur-bearing minerals such as pyrite (FeS_2) in mines react with rainwater. The runoff enters streams and kills fish and plants. Collapse of underground coal mines may cause the ground surface to sink.

Numerous air-pollution issues also arise from the burning of fossil fuels, which sends soot, carbon monoxide, sulfur dioxide, nitrous oxide, and unburned hydrocarbons into the air. Coal, for example, commonly contains sulfur, primarily in the form of pyrite, which enters the air as sulfur dioxide (SO_2) when coal is burned. This gas combines with rainwater to form dilute sulfuric acid (H_2SO_4), or **acid rain**.

BOX 12.3 CONSIDER THIS...

Offshore Drilling and the *Deepwater Horizon* Disaster

A substantial proportion of the world's oil reserves reside in the sedimentary basins that underlie the continental shelves of passive continental margins. To access such reserves, oil companies must build offshore drilling platforms. In water less than 600 m deep, companies position fixed platforms on towers resting on the seafloor. In deeper water, semi-submersible platforms float on huge submerged pontoons. By drilling from these platforms, oil companies can now access fields lying beneath 3 km of water. North America's largest offshore fields occur in the passive-margin basin that fringes the coast of the Gulf of Mexico. More than 3,500 platforms operate in the Gulf at present, together yielding up to 1.7 million bbl/day.

During both onshore and offshore exploration, drillers worry about the possibility of a **blowout**. A blowout happens when the pressure within a hydrocarbon reserve penetrated by a well exceeds the pressure that drillers had planned for, causing the hydrocarbons (oil or gas) to rush up the well in an uncontrolled manner and burst out of the well at the surface in an oil gusher or gas plume. Blowouts are rare because, although fluids below the ground are under great pressure due to the weight of overlying material, engineers fill the hole with drilling mud, which has a density greater than that of clear water. The weight of drilling mud can counter the pressure of underground hydrocarbons and hold the fluids underground. But if drillers encounter a bed in which pressures are unexpectedly high, or if they remove the mud before the walls of the well have been sealed with a casing (a pipe, cemented in place by concrete), a blowout may happen.

A catastrophic blowout occurred on April 20, 2010, when drillers on the *Deepwater Horizon*, a huge semi-submersible platform, were finishing a 5.5-km-long hole in 1.5-km-deep water south of Louisiana. Due to a series of errors, the casing was not sufficiently strong when workers began to replace the drilling mud with clear water. Thus the high-pressure, gassy oil in the reservoir strata that the well had accessed rushed up the drillhole. A backup safety device called a blowout preventer, which should have clamped the pipe shut, failed, so the gassy oil reached the platform and sprayed 100 m into the sky. Sparks from electronic gear triggered an explosion, and the platform became a fountain of flame and smoke that killed 11 workers. An armada of fireboats could not douse the conflagration **(Fig. Bx12.3a)**, and after 36 hours, the still-burning platform tipped over and sank.

Robot submersibles sent to the seafloor to investigate found that the twisted mess of bent and ruptured pipes at the wellhead was billowing oil and gas. On the order of 50,000 to 62,000 bbl/day of hydrocarbons entered the Gulf's water from the well. Stopping this underwater gusher proved to be an immense challenge, and initial efforts to block the well, or to put a containment dome over the wellhead, failed. It was not until July 15 that the flow was finally stopped, and not until September 19 that a new relief well intersected the blown well and provided a conduit to pump concrete down to block the original well permanently. All told, about 4.2 million bbl of hydrocarbons contaminated the Gulf from the *Deepwater Horizon* blowout **(Fig. Bx12.3b)**. The spill was devastating to wetlands, wildlife, and the fishing and tourism industries.

FIGURE Bx12.3 The *Deepwater Horizon* disaster (2010).

(a) Fireboats dousing the burning rig before the rig sank.

(b) A satellite image of the oil slick in the Gulf of Mexico, southeast of the Mississippi Delta.

FIGURE 12.16 Marine oil spills. These can come from drilling rigs or from tankers.

(a) An oil tanker leaking oil on the sea surface.

(b) Oil spills can contaminate the shore and can be very difficult to clean.

For this reason, many countries now regulate the amount of sulfur that coal can contain when it is burned. Even if pollutants can be decreased, burning fossil fuels still releases carbon dioxide (CO_2) into the atmosphere. As we discuss in Chapter 19, CO_2 is a *greenhouse gas,* so a change in the amount of CO_2 in the atmosphere can affect climate. Because of concern about CO_2 production, people are pursuing energy conservation efforts, improving technologies that use alternatives to fossil fuels, and developing techniques for *carbon capture and sequestration (CCS).* CCS involves three steps. First, CO_2 is removed from the effluents sent up smokestacks at power plants or factories; second, CO_2 gas is transformed into liquid; and third, CO_2 is pumped down wells into reservoir rocks deep underground.

ENERGY OPTIONS FOR THE FUTURE

In the early 21st century, hydrocarbons continue to be the dominant energy source for human society (**Geology at a Glance**, pp. 426–427). Of course, technologies already exist to substitute coal for oil or natural gas in power plants, and the planet still hosts immense coal reserves—by some estimates, there are about 9,800 billion tons of *recoverable coal* (coal that can be accessed by mining), which contains as much energy as 4,700 billion barrels of oil. Large-scale substitution of coal for oil seems unlikely, though, because other energy sources have become cheaper, and because many people consider the environmental

consequences of mining and burning coal to be unacceptable. Vast supplies of uranium, the fuel of nuclear power plants, remain untapped. Further, nuclear engineers have designed new generations of power plants that produce far less waste and cannot have a meltdown. But many people view nuclear plants with concern because of issues pertaining to radiation, accidents, terrorism, and waste storage, and these concerns have slowed down the industry. Nevertheless, China has recently decided to replace coal-burning plants with nuclear ones in order to combat its air pollution problem. The growth of geothermal energy also seems limited.

Because of the potential problems that might result from relying more on coal, hydroelectric, or nuclear energy, researchers have been increasingly exploring clean energy options such as solar power, whose cost has decreased exponentially in the past few decades, and wind power. However, because sunlight reaching the Earth changes during the day and with the amount of overcast, and because wind speed varies, solar power and wind power can't provide the steady supply of energy that the present-day *energy grid* (the interconnected network of power lines and transformers that distribute energy around the country) requires. This problem might someday be overcome with "smart grids," which can accommodate rapid fluctuations in energy input. Clearly, society will be facing difficult choices in the not-so-distant future about where to obtain energy, and we will need to invest in the research required to discover new alternatives.

FIGURE 12.17 Gold occurs as native metal within quartz veins. The quartz breaks up to form sand, leaving nuggets of gold.

Gold bracelets on display in a Kuwaiti marketplace

12.10 **Mineral Resources**

In January of 1848, James Marshall and a crew of workers were putting the finishing touches on a new sawmill in the foothills of the Sierra Nevada. They had been deepening the channel that would guide a stream to the water wheel that would drive the mill. As Marshall examined the excavation, he noticed a glimmer of metal in the gravel that littered its floor. He picked up a sample and banged it between two rocks, knowing that if the sample crushed into fragments, it was only pyrite (fool's gold), but if it flattened without breaking, it was real gold. Marshall looked up and said to his crew, "I have found it . . . gold!" Word soon spread, and gold fever spread across the country. As a result, 1849 brought 40,000 prospectors to California. These "forty-niners" had abandoned their friends and relatives on the gamble that they could strike it rich. Marshall himself, sadly, did not, and died penniless.

Gold is but one of many **mineral resources**, meaning minerals extracted from the Earth's crust for use by people. Without these resources, industrialized societies could not function. Geologists divide mineral resources into two categories: *metallic mineral resources* (materials containing gold, copper, aluminum, iron, or other metals) and *nonmetallic mineral resources* (building stone, gravel, sand, gypsum, phosphate, salt) used in construction or for chemical production. Below, we look at the nature of mineral resources, the geologic phenomena responsible for their formation, and the methods that people use to obtain them. We conclude by considering the sustainability of mineral reserves in the future.

METALS AND THEIR DISCOVERY

A **metal** is an opaque, shiny, smooth solid that can conduct electricity and is *malleable*, meaning it can be bent, drawn into wire, or hammered into thin sheets. Metals look and behave quite differently from wood, plastic, meat, or rock because the atoms that make up metals are held together by *metallic bonds*, so electrons can flow from atom to atom fairly easily and atoms can, in effect, slide past each other without breaking chemical bonds.

The first metals that people used—copper, silver, and gold—can occur in rock in the form of *native metal*. Native metal consists only of metal atoms, and thus looks and behaves like metal. Gold nuggets, for example, are chunks of native gold that have eroded free of bedrock **(Fig. 12.17)**. Over the ages, people have collected nuggets of native metal from streambeds and pounded them together with stone hammers to make arrowheads, scrapers, and later, coins and jewelry. But if we had to rely solely on native metals as our source of metal, we would have access to only a tiny fraction of our current metal supply. Most of the metal atoms we use today originated as ions bonded to nonmetallic elements in a great variety of minerals that themselves look nothing like metal. To obtain these metals, their atoms have to be extracted from rock by *smelting*. Smelting involves heating ore and reacting it with chemicals to yield metal and a solid, nonmetallic residue called *slag*.

WHAT IS AN ORE?

The minerals from which metals can be extracted are called **ore minerals**, or *economic minerals*. Ore minerals contain metal in high concentrations and in a form that can be easily extracted. Galena (PbS), for example, is about 50% lead, so we consider it to be an ore mineral of lead **(Fig. 12.18a)**. We obtain most of our iron from hematite and magnetite. Copper comes from a variety of minerals, none of which look like copper **(Fig. 12.18b)**. Geologists have identified a great variety of ore minerals. Many ore minerals are sulfides, in which the metal occurs in combination with sulfur (S); or oxides, in which the metal occurs in combination with oxygen (O). A few are carbonates.

FIGURE 12.18 Examples of ore minerals.

(a) This lead ore, from Missouri, consists of galena (PbS) crystals that grew in dolostone.

(b) Most copper comes from ore minerals that look nothing like metallic copper. This rock has a coating of malachite, a copper carbonate mineral.

Ore is a rock containing a relatively high concentration of ore minerals or native metals. The weight percent of a useful metal in an ore determines the **grade** of the ore—the higher the weight percent, the higher the grade. For particularly valuable metals, the grade of ore can be quite low for the ore to be worth mining. For example, most copper ore used today has a grade of 0.4% to 1.0%. Mining copper of such a grade can be profitable, because copper metal sells for about $7.50 per kilogram. Iron, in contrast, sells for only about $0.55 per kilogram, so iron ore must have a grade of between 30% and 60% to be profitable. (By comparison, the grade of iron in common granite is about 2%.) Geologists refer to a significant accumulation of ore as an **ore deposit**.

HOW DO ORE DEPOSITS FORM?

Geologists distinguish among several different types of ore deposits based on the process by which the deposit formed.

Magmatic Deposits When a magma intrusion cools and starts to solidify, sulfide ore minerals may grow preferentially in distinct lenses or bands. These concentrations are identified as *magmatic deposits*, to emphasize that they form directly by crystallization in a magma. Typically, such deposits consist almost exclusively of sulfide ore minerals, so they represent a type of *massive sulfide deposit* (Fig. 12.19a).

Hydrothermal Deposits The heat from an igneous intrusion can cause groundwater to start convecting through the intrusion and the surrounding wall rocks. This water heats up as it passes through the intrusion and becomes a hot, hydrothermal fluid that can dissolve metal ions. When the fluid enters a region of lower pressure, lower temperature, different acidity,

or different availability of oxygen, the metal ions bond to other elements and precipitate as ore minerals in fractures and pores. The resulting concentration of ore is called a *hydrothermal deposit* (Fig. 12.19b).

In recent decades, geologists have discovered that hydrothermal activity also occurs when seawater enters the crust and convects through the crust in the vicinity of submarine volcanoes along mid-ocean ridges. As the fluids flow through the crust, they dissolve metals, and when the fluids erupt from vents along the ridge axis, they contain high concentrations of dissolved metal and sulfur. The instant the hydrothermal solutions come in contact with cold seawater, the dissolved components precipitate as tiny crystals of metal-sulfide minerals (Fig. 12.19c). This phenomenon makes the erupting water look like a black cloud, so the vents are called *black smokers* (see Chapter 2). Minerals in the cloud sink and form a pile of ore minerals around the vent. Since the ore minerals typically are sulfides, the resulting hydrothermal deposits constitute another type of massive sulfide deposit.

Secondary-Enrichment Deposits Sometimes groundwater passes through ore-bearing rock long after the rock first formed. This groundwater leaches (dissolves) some of the metal in the ore and carries the ions away. When the groundwater eventually flows into a different chemical environment (for instance, one containing a different concentration of oxygen or acid), it precipitates new ore minerals, commonly in concentrations exceeding that of the original deposit. A new ore deposit formed from metals that were dissolved and carried away from a pre-existing ore deposit is called a *secondary-enrichment deposit*. Some of these deposits contain spectacularly beautiful copper-bearing carbonate minerals, such as azurite and malachite (see Fig. 12.18b).

FIGURE 12.19 Various processes that form ore deposits.

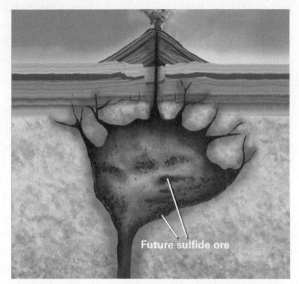

(a) Massive sulfide deposits can form when sulfide ore minerals form concentrations in a magma chamber.

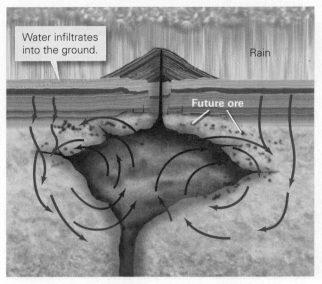

(b) Hydrothermal deposits form when water circulating around and through magma dissolves and redistributes metals (arrows indicate flowing water).

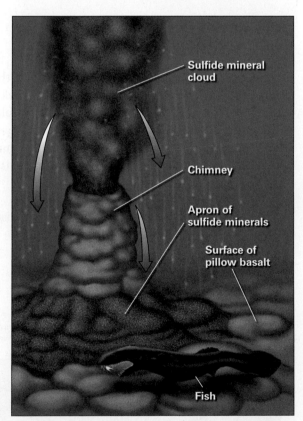

(c) Massive sulfide deposits also form when ore minerals precipitate around hydrothermal vents (black smokers) along a mid-ocean ridge.

MVT Deposits Rain falling along one margin of a large sedimentary basin may sink into the subsurface and then flow as groundwater along a curving path that takes it first down into rocks in the basin's lower section, several kilometers below the surface, and then eventually back up to the distant margin of the basin. At the base of the basin, temperatures can become high enough that the water dissolves metals. When the water returns to the surface and enters cooler rock, these metals precipitate in ore minerals. Ore deposits formed in this way, typically containing lead- and zinc-bearing minerals, appear in dolomite beds of the Mississippi Valley region of the United States, and thus have come to be known as *Mississippi Valley–type* (*MVT*) ore deposits.

Did you ever wonder...
where the iron in the steel bodies of cars comes from?

Sedimentary Deposits of Metals Some ore minerals accumulate in sedimentary depositional environments under special circumstances. For example, between about 2.5 billion and 1.8 billion years ago, the oxygen concentration in the atmosphere began to increase, and more oxygen began to dissolve in seawater. This change affected the chemistry of seawater such that it could no longer hold large quantities of iron in solution. The iron precipitated as iron oxide minerals that settled as sediment on the seafloor. As we learned in Chapter 11, the resulting iron-rich sedimentary deposits are known as **banded iron formation (BIF)** **(Fig. 12.20)** because after lithification they consist of alternating layers of gray iron oxide (magnetite or hematite) and red layers of jasper (iron-bearing chert). Microbes may have participated in the precipitation process.

FIGURE 12.20 Precambrian BIF from northern Michigan consists of hematite interbedded with jasper.

Magnetite or hematite layer

Jasper layer

The chemistry of seawater in some parts of the ocean today leads to the deposition of manganese oxide minerals on the seafloor. These minerals grow into lumpy accumulations known as **manganese nodules**. Mining companies have begun to explore technologies for vacuuming up these nodules.

Residual Mineral Deposits Recall from Interlude B that as rainwater sinks into the Earth, it leaches certain elements and leaves behind others, as part of the process of forming soil. In rainy, tropical environments, the residue left behind in soils after leaching includes concentrations of iron and aluminum oxide, which do not dissolve easily in the downward percolating rainwater. Locally, these metals become so concentrated that the soil itself becomes an ore deposit. We refer to such deposits as *residual mineral deposits*. Most of the aluminum ore mined today comes from *bauxite*, a residual mineral deposit created by the extreme leaching of rocks (such as granite) containing aluminum-bearing minerals.

Placer Deposits Ore deposits may develop when rocks containing native metals erode, producing a mixture of sand grains and metal flakes or nuggets that can be concentrated by streams, because the moving water carries away lighter mineral grains but can't move the heavy metal grains (gold) so easily. Concentrations of metal grains in stream sediments are a type of *placer deposit* (Fig. 12.21). Panning further concentrates gold flakes or nuggets, for swirling water in a pan causes the lighter sand grains to wash away, leaving the gold behind.

WHERE ARE ORE DEPOSITS FOUND?

The Inca Empire of 15th-century Peru boasted elaborate cities and temples, decorated with fantastic masks, jewelry, and sculptures made of gold. Then, around 1532, Spanish conquistadors arrived. Within 6 years, the Inca Empire had vanished, and Spanish ships were transporting Inca treasure back to Spain. Why did the Incas possess so much gold? Or, to ask the broader question, what geologic factors control the distribution of ore? Once again, we can find the answer by considering the consequences of plate tectonics.

Several of the ore-deposit types we've mentioned occur in association with igneous rocks. As we learned in Chapter 4, igneous activity does not happen randomly around the Earth, but rather concentrates along convergent plate boundaries (specifically, in the overriding plate of a subduction zone), along divergent plate boundaries (mid-ocean ridges), along continental rifts, or at hot spots. Therefore, magmatic and hydrothermal deposits (and secondary-enrichment deposits derived from them) occur in these geologic settings. Placer deposits are typically found in the sediments eroded from magmatic or hydrothermal deposits. The Inca gold formed in the Andes along a convergent boundary.

ORE-MINERAL EXPLORATION AND PRODUCTION

Imagine an old prospector of days past clanking through the desert with a worn-out donkey. To find bedrock ore, such prospectors would eye hillsides for a "show," visible evidence on the ground surface of ore below. What does a show look like? It may be an outcropping of milky-white quartz veins, or brightly colored stains due to oxidation (rusting) of ore

FIGURE 12.21 Placer deposits form where erosion produces clasts of native metals. Sorting by the stream concentrates the metals.

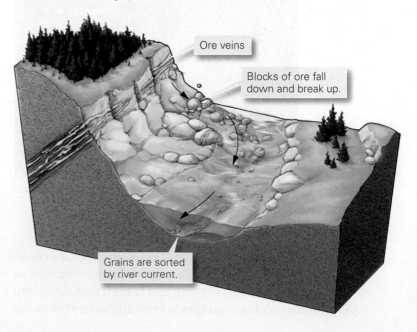

Ore veins

Blocks of ore fall down and break up.

Grains are sorted by river current.

FIGURE 12.22 Mining ore deposits.

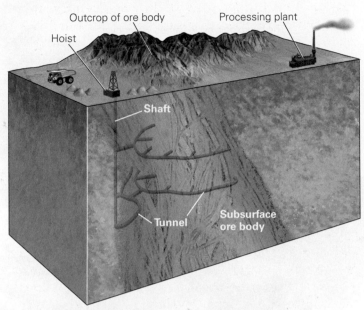

(a) The three-dimensional shape of an ore body underground. Shafts and tunnels access the body.

(b) A tunnel in an old mine.

(c) Open-pit mining utilizes ore that lies fairly close to the ground surface. Terracing the mine walls helps to stabilize them.

minerals. On finding a possible ore, a prospector would take samples back to town for an *assay*, a test to determine how much extractable metal the samples contain. If the assay indicated a significant concentration of metal, the prospector might "stake a claim" by literally marking off an area of ground with stakes.

These days, large mining companies employ professional geologists to survey ore-bearing regions systematically. The geologists focus their studies on rocks that developed in settings appropriate for ore formation. Once such a region has been identified, geologists may measure gravity or magnetic anomalies (see Interlude D), because ore minerals tend to be denser and more magnetic than average rocks, to define the limits of the ore body. They may also sample rocks and soils to test for metal content, and may even analyze plants in the area to detect traces of metals, for plants absorb metals through their roots. Once geologists have identified a possible ore deposit, they drill holes to sample subsurface rock, to help determine the ore deposit's shape, extent, and grade.

If calculations indicate that mining an ore deposit will yield a profit, and if environmental concerns can be addressed properly, a company builds a mine. Mines can be below or above ground, depending on how close the ore deposit lies to the surface. Underground mines require excavating shafts and tunnels deep underground **(Fig. 12.22a, b)**. To develop an *open-pit mine* **(Fig. 12.22c)**, workers first drill a series of holes into the solid bedrock and then fill the holes with high explosives. They must space the holes carefully and must set off the charges in a precise sequence, so that the bedrock shatters into appropriate-sized blocks for handling. When the dust settles, large front-end loaders dump the ore into giant ore trucks, which can carry as much as 200 tons of ore in a single load. (In comparison, a loaded cement mixer at a construction site weighs about 70 tons.)

At both underground and open-pit mines, ore must be separated from waste rock next to the mine. The trucks transport waste rock or *tailings* (rock that doesn't contain ore) to

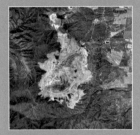

Power from the Earth

Water, Wind, and Tides

The hydrologic cycle carries water over land. Water flows back toward the sea.

Forming and Mining Coal

Plants in coastal swamps and forests die, become buried, and transform into coal.

Coal at shallow depths can be accessed by strip mining.

Forming and Finding Oil

Plankton, algae, and clay sink through quiet water to settle on the floor of a lake or sea. Eventually, this organic sediment becomes buried deeply and becomes a source rock. Chemical reactions yield oil, which percolates upward.

Tectonic processes form oil traps. Oil accumulates in reservoir rock within the trap; a seal rock keeps the oil underground.

Regardless of whether an oil reserve is under land or under the seafloor, modern drilling technology can reach it and pump it.

Exploration for oil uses seismic-reflection profiling, which can reveal the configuration of layers underground.

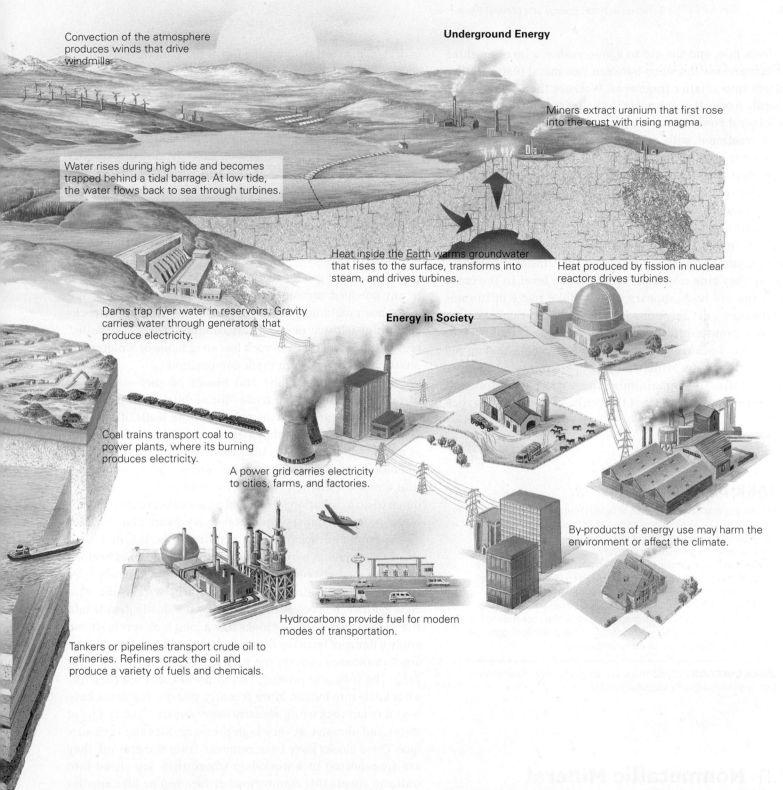

Convection of the atmosphere produces winds that drive windmills.

Underground Energy

Miners extract uranium that first rose into the crust with rising magma.

Water rises during high tide and becomes trapped behind a tidal barrage. At low tide, the water flows back to sea through turbines.

Heat inside the Earth warms groundwater that rises to the surface, transforms into steam, and drives turbines.

Heat produced by fission in nuclear reactors drives turbines.

Dams trap river water in reservoirs. Gravity carries water through generators that produce electricity.

Energy in Society

Coal trains transport coal to power plants, where its burning produces electricity.

A power grid carries electricity to cities, farms, and factories.

By-products of energy use may harm the environment or affect the climate.

Hydrocarbons provide fuel for modern modes of transportation.

Tankers or pipelines transport crude oil to refineries. Refiners crack the oil and produce a variety of fuels and chemicals.

Modern society, for better or worse, uses vast amounts of energy to produce heat, to drive modes of transportation, and to produce electricity. This energy comes either from geologic materials stored in the Earth or from geologic processes happening at our planet's surface. For example, oil and gas fill the pores of reservoir rocks at depth below the surface, coal occurs in sedimentary beds, and uranium concentrates in ore deposits. A hydroelectric power plant taps into the hydrologic cycle, windmills operate because of atmospheric convection, and geothermal energy comes from hot groundwater.

Ultimately, the energy in the sources just listed comes from the Sun, from gravity, from the Earth's internal heat, and from nuclear reactions. Oil, gas, and coal are fossil fuels because the energy they store first came to the Earth as sunlight, long ago.

As energy usage increases, easily obtainable energy resources dwindle, the environment can be degraded, and the composition of the atmosphere changes. The pattern of energy use that forms the backbone of society today may have to change radically in the not-so-distant future if we wish to avoid a decline in living standards.

a tailings pile, and the ore to a *jaw crusher*, a large machine that compresses the stone between two metal plates until it shatters into smaller fragments. Workers then separate ore minerals from other minerals in the fragments and send the ore-mineral concentrate to a processing plant, where smelting or treatment with acidic solutions can separate metal atoms from other atoms. Eventually, workers melt the metal and then pour it into molds to make ingots (brick-shaped blocks) for transport to a manufacturing facility.

If the ore deposit lies too far below the ground surface for it to be accessed by an open pit, miners must dig an *underground mine*. To do so, they either bore a tunnel into the side of a mountain (the entrance to the tunnel is called an *adit*) or they sink a vertical *shaft*. At the level in the crust where the ore body appears, they build a maze of tunnels into the ore by drilling holes into the rock and then blasting. The rock removed in this process must be carried back to the surface. Rock columns between the tunnels hold up the ceiling of the mine. Miners face danger from mine collapse and rockfalls. To help minimize these risks, operators bolt steel beams and nets against the ceiling, to hold falling rock in place.

TAKE-HOME MESSAGE

Metals are malleable materials in which metallic bonds hold atoms together. While some occur as native metals, most occur in other minerals and have to be extracted by smelting. An ore is a rock that contains a relatively high concentration of a useful metal, and an ore deposit is an accumulation of ore. Ore deposits form for a variety of reasons. For example, some crystallize from melts, some precipitate from hydrothermal fluids, and some collect as sediment. To find ore deposits, geologists search for ore shows, make maps of gravity and rock magnetism, and obtain drill cores. Mining takes place either in open-pit mines or in shafts and tunnels underground.

QUICK QUESTION Why do many ore deposits occur in association with convergent plate boundaries?

12.11 Nonmetallic Mineral Resources

So far, this chapter has focused on resources that contain metal. Society uses many nonmetallic mineral resources, also known as *industrial minerals*, as well. From the ground, we get the stone used to make roadbeds and buildings, the chemicals for fertilizers, the gypsum in drywall, the salt filling salt shakers, and the sand used to make glass—the list is endless. This section looks at a few of these materials and describes their sources.

DIMENSION STONE

The Parthenon, a colossal stone temple that is rimmed by 46 carved columns, has stood atop a hill overlooking the city of Athens for almost 2,500 years. No wonder—*stone*, rock being used for practical purposes, outlasts nearly all other construction materials. We use stone to make facades, roofs, curbs, steps, countertops, and floors. We value stone for its visual appeal as well as its durability. The names that architects give to various types of stone may differ from the formal rock names that geologists use. For example, architects refer to any polished carbonate rock as "marble," whether or not it has been metamorphosed. Likewise, any crystalline rocks containing feldspar or quartz are commonly called "granite," regardless of whether the rock has an igneous or a metamorphic texture, or a felsic or mafic composition.

To obtain intact slabs and blocks of rock—known as **dimension stone** in the trade—for architectural purposes, workers must carefully cut rock out of the walls of quarries (**Fig. 12.23a, b**). (Note that a *quarry* provides stone, whereas a *mine* supplies ore.) Traditionally, quarry operators split the blocks off of bedrock by hammering a series of wedges into a crack, causing the crack to grow until it propagates down to an exfoliation joint (a natural, subhorizontal crack). More recently, quarry operators have been able to slice blocks from bedrock by using wireline saws, thermal lances, or water jets. A *wireline saw* consists of a loop of braided wire moving between two pulleys. Operators spill abrasive (sand or garnet grains) and water onto the wire as the wire rubs against the rock surface, so the wire slowly grinds into the rock. A *thermal lance* looks like a long blowtorch—it can apply a flame of burning diesel fuel, stoked by high-pressure air, to a focused spot on the rock surface. The intense heat turns the rock into powder, and thus cuts into the rock like a hot knife into butter. More recently, quarry operators have begun to cut rock using *abrasive water jets* that direct a jet of water and abrasive at very high pressure onto the rock surface. Once blocks have been removed from the ground, they are transported to a workshop where they are sliced into uniform sheets (for countertops or facades) or into smaller blocks for gravestones, monuments, and curbs.

CRUSHED STONE AND CONCRETE

Crushed stone forms the substrate of highways and railroads and serves as the raw material for manufacturing cement, concrete, mortar, and asphalt. In crushed-stone quarries (**Fig. 12.23c**), operators use explosives to break up bedrock into rubble that

FIGURE 12.23   Stone production in quarries.

(a) An active quarrying operation in Missouri that produces large blocks of cut dimension stone.

(b) Sheets of cut dimension stone being measured for cutting to become a kitchen countertop.

(c) A large crushed-stone quarry in Silurian limestone of Illinois. Drillers are working on the shelf in the distance.

(d) The long tube is a rotating kiln. Rock fed in at one end is heated to a high temperature so that cement comes out the other end.

> **Did you ever wonder...**
> how concrete differs from rock?

then goes by truck to a jaw crusher.

Most of the buildings constructed in the past two centuries consist of bricks attached to each other by mortar, or of walls and floors made of *concrete* that has been spread into a layer or poured into a form. Both mortar and concrete start out as a slurry, but when allowed to set, they harden into a hard, rock-like substance. The slurry from which mortar and concrete form consists of *aggregate* (sand and/or gravel) mixed with water and **cement**. Before mixing, the cement in mortar and concrete is a powder made of lime (CaO), quartz (SiO_2), aluminum oxide (Al_2O_3), and iron oxide (Fe_2O_3). Typically, lime accounts for 66% of cement and silica for 25%. When cement powder mixes with water, the chemicals in it dissolve. Mortar and concrete set (solidify) when dissolved chemicals recombine to produce a complex assemblage of new mineral crystals. These minerals bind together the grains of aggregate, much as quartz or calcite binds together grains of quartz in a sandstone.

The ancient Romans were among the first people to use cement. They made it from a mixture of volcanic ash and limestone. In the 18th and early 19th centuries, cement was produced by heating specific types of limestone (which happened to contain calcite, clay, and quartz in the correct proportions) in a kiln up to a temperature of about 1,450°C. The heating releases CO_2 gas and produces "clinker," chunks consisting of lime and other oxide compounds **(Fig. 12.23d)**. Manufacturers crushed the clinker into cement powder and packed it in bags for transport. Limestone with the exact composition necessary to make good cement, however, is fairly rare.

Most cement used today is **Portland cement**, produced by mechanically mixing limestone, sandstone, and shale in just the right proportions, then heating it in a kiln to yield a powder with the correct chemical makeup to mix into cement. Isaac Johnson, an English engineer, came up with the recipe for Portland cement in 1844, and chose the name because he thought that concrete made from this material resembled rock exposed in the town of Portland, England.

NONMETALLIC MINERALS FOR HOMES AND FARMS

We use an astounding variety of nonmetallic geologic resources in daily life. Consider the materials in a typical house or apartment. The foundation generally consists of concrete, derived from limestone mixed with gravel. The bricks in many exterior walls originated as clay, formed from the chemical weathering of silicate rocks and perhaps dug from the floodplain of a stream. To make *bricks*, workers mold wet clay into blocks and then bake it to drive out water and cause metamorphic reactions that recrystallize the clay. The glass used to glaze windows consists largely of silica, formed by melting and then freezing quartz sand from a beach deposit or a sandstone formation. Quartz may also produce the silica in solar cells. Gypsum board (drywall), used for many interior walls, is made from a slurry of water and the mineral gypsum, sandwiched between sheets of paper. Gypsum ($CaSO_4 \cdot 2H_2O$) occurs in evaporite strata precipitated from seawater or saline lake water. Evaporites also provide other useful minerals, such as halite, and serve as the source for lithium, a key element in computer batteries.

TAKE-HOME MESSAGE

Society uses a great variety of nonmetallic geologic materials. These include dimension stone, crushed stone, cement (made from roasted limestone), evaporites (including gypsum), and clay (to make bricks).

QUICK QUESTION What is the difference between natural cement and Portland cement?

12.12 Global Mineral Needs

HOW LONG WILL RESOURCES LAST?

Each person in the United States uses about 600 kg of metallic resources, about 9,400 kg of nonmetallic resources, and about 7,600 kg of energy resources each year **(Table 12.1)**. Summing these numbers means that each person uses about 17,000 kg

(= 17 metric tons) of energy and mineral resources each year. To obtain this material, workers must mine, quarry, or pump about 20 billion metric tons of Earth materials. (By comparison, the Mississippi River carries about 0.2 billion metric tons of sediment in a given year.)

Mineral resources, like oil and coal, are nonrenewable resources. Once mined, an ore deposit or a limestone hill disappears forever. Natural geologic processes do not happen fast enough to replace the deposits as quickly as we use them. Geologists have calculated reserves for various mineral deposits just as they have for oil. Reserve estimates depend on price, because as price goes up, the grade of ore that can be mined economically goes down. A comparison of reserve estimates to consumption rates suggests that supplies of some metals may run out in only decades to centuries **(Table 12.2)**. But these estimates may

TABLE 12.1 Yearly Per Capita Usage of Geologic Materials (USA)

Material	Weight Used (kg)
Stone	4,100
Sand and gravel	3,860
Petroleum	3,050
Coal	2,650
Natural gas	1,900
Iron and steel	550
Cement	360
Clay	220
Salt	200
Phosphate	140
Aluminum	25
Copper	10
Lead	6
Zinc	5

1 kg = 2.205 pounds; 1,000 kg = 1 metric ton

TABLE 12.2 Expected Lifetimes of Currently Known Ore Reserves (in Years)

Metal	World Reserves	U.S. Reserves
Iron	120	40
Aluminum	330	< 2
Copper	65	40
Lead	50	25
Zinc	30	25
Gold	30	20
Platinum	45	< 1
Nickel	75	< 1
Cobalt	50	< 1
Manganese	70	0
Chromium	75	0

FIGURE 12.24 Environmental consequences of producing metallic mineral resources.

Much of the color comes from bacteria and archaea living in the water.

Smelter

Tailings pile

(a) The orange color in this mine runoff is due to iron and sulfide in the water.

(b) Acidic smelter smoke killed off vegetation near Sudbury, Ontario, in the 1970s. A large tailings pile can be seen in the distance.

change as prices rise, if geologists discover new reserves, or if engineers develop new technologies for mining and processing ore. Further, increased efforts at conservation and recycling can cause a dramatic decrease in rates of consumption and thereby stretch the lifetime of existing reserves.

Ore deposits do not occur everywhere because their formation requires special geologic conditions. As a result, some countries possess vast supplies, whereas others have none. In fact, no single country owns all the mineral resources it needs, so nations must trade with each other to maintain supplies, and global politics inevitably affects prices. Countries are particularly concerned about access to the **strategic minerals** that are essential for the security of a nation. These include minerals used for the production of specialized types of steel, and those used for electronics. Many wars and treaties have their roots in competition for mineral reserves.

MINING AND THE ENVIRONMENT

Mining leaves a big footprint in the Earth System. Some of the gaping holes that open-pit mining creates in the landscape have become so big that astronauts can see them from space. Both open-pit and underground mining yield immense quantities of waste rock, which miners dump in tailings piles. Some of these piles grow into artificial hills as much as 200 m high and many kilometers long. Because they lack soil, tailings piles tend to remain unvegetated for a long time. Mining also exposes ore-bearing rock to the atmosphere, and since many ore minerals are sulfides, they react with rainwater to produce *acid mine runoff*, which can severely damage vegetation downstream **(Fig. 12.24a)**. In some cases, mining companies douse tailings with toxic solutions to leach out more metals, and these solutions sometimes escape into the environment. Ore processing also tends to yield pollutants, which, if they enter the atmosphere, can impact vegetation and animal life downwind **(Fig. 12.24b)**.

To help counter some of the problems caused by mining, reclamation efforts have begun in some areas, with the goal of isolating exposed mine surfaces and tailings piles from the environment. And, in recent years, new regulations have been established that require smoke from processing plants to be "scrubbed" before it enters the atmosphere. Finally, engineers are developing new technologies that allow metals to be processed in ways that produce less pollutants in the first place.

TAKE-HOME MESSAGE

People use a vast quantity of mineral resources during the course of a lifetime. Mineral resources are nonrenewable, so minerals have limited reserves, and because reserves are not distributed uniformly around the planet, all supplies are not accessible to all consumers. Thus, the trade in economic minerals is politically charged. Also, mineral extraction and utilization has significant environmental consequences that are challenging to address.

QUICK QUESTION If one country restricts the export of a strategic mineral, would that change the minimum concentration of the mineral necessary to make an ore deposit elsewhere worth mining? Why?

Chapter 12 Review

CHAPTER SUMMARY

> Energy resources come in a variety of forms: energy directly from the Sun; energy from tides, flowing water, or wind; energy from chemical reactions; energy from nuclear fission; and energy from the Earth's internal heat.

> Oil and gas are hydrocarbons formed from the organic remains of plankton, which settle out and become incorporated in black organic shale. Chemical reactions at elevated temperatures convert the organic matter to kerogen and then to oil.

> To form a conventional oil reserve, oil must migrate from a source rock into a reservoir rock. The subsurface configuration of strata that holds oil is called an oil trap.

> Substantial volumes of hydrocarbons also exist in unconventional reserves. These include shale gas and shale oil, tar sand, and oil shale. Extraction of unconventional reserves has increased dramatically in recent years due to the development of directional drilling and hydrofracturing.

> For coal to form, abundant plant debris must be deposited in an oxygen-poor environment. Compaction changes the debris into peat that, when buried deeply and heated, transforms into coal.

> Geologists distinguish among three ranks of coal, based on the amount of carbon the coal contains. Coal occurs in beds and can be mined by either strip mining or underground mining.

> Nuclear power plants generate energy by using the heat released from the fission of uranium.

> Nuclear reactors must be carefully controlled to avoid overheating or meltdown. The disposal of radioactive nuclear waste can create environmental challenges.

> Geothermal energy uses the Earth's internal heat to transform groundwater into steam that drives turbines; hydroelectric power uses the potential energy of water; and solar energy uses solar cells to convert sunlight to electricity.

> We now live in the Oil Age, but oil supplies may last for only another century, and the production and use of energy resources have many environmental consequences.

> Industrial societies use many types of minerals, all of which must be extracted from the upper crust.

> Metals come from ore. An ore is a rock containing native metals or ore minerals in sufficient quantities to be worth mining.

> Magmatic ore deposits form when ore minerals grow during solidification of melt. In hydrothermal deposits, ore minerals precipitate from hot-water solutions. Secondary-enrichment deposits and MVT deposits precipitate from groundwater. Sedimentary deposits settle out of water, residual mineral deposits are the result of soil formation, and placer deposits develop when native metal grains accumulate in sediment.

> Nonmetallic resources include dimension stone, crushed stone, clay, sand, and many other materials. A large proportion of materials in your home have a geologic ancestry.

> Mineral resources are nonrenewable. Many, including strategic minerals, are now or may soon be in short supply.

GUIDE TERMS

acid rain (p. 418)
banded iron formation BIF (p. 423)
biofuel (p. 415)
biomass (p. 395)
blowout (p. 419)
cement (p. 429)
chain reaction (p. 413)
coal gasification (p. 411)
coal (p. 408)
coal rank (p. 409)
coal reserve (p. 410)
conventional reserve (p. 398)
dimension stone (p. 428)
directional drilling (p. 402)

distillation column (p. 404)
drilling mud (p. 402)
energy density (p. 397)
energy (p. 395)
fossil fuel (p. 395)
fuel (p. 396)
grade (p. 422)
Hubbert's peak (p. 418)
hydrocarbon generation (p. 398)
hydrocarbon (p. 396)
hydrocarbon reserve (p. 398)
hydrofracturing (p. 403)
kerogen (p. 398)
manganese nodule (p. 424)

meltdown (p. 413)
metal (p. 421)
migrate (p. 399)
mineral resource (p. 421)
nuclear reactor (p. 412)
nuclear waste (p. 414)
Oil Age (p. 417)
oil seep (p. 399)
oil shale (p. 407)
oil window (p. 398)
ore deposit (p. 422)
ore mineral (p. 421)
ore (p. 422)
peat (p. 408)
permeability (p. 398)

photovoltaic cell (p. 416)
pore (p. 398)
porosity (p. 398)
Portland cement (p. 430)
reservoir rock (p. 398)
salt dome (p. 400)
seal rock (p. 399)
seismic-reflection profile (p. 401)
shale gas (p. 405)
shale oil (p. 405)
source rock (p. 397)
strategic mineral (p. 431)
tar sand p. 406)
trap (p. 399)
unconventional reserve (p. 398)

 GEOTOURS *THIS CHAPTER'S GEOTOURS WORKSHEET (L) FEATURES QUESTIONS AND GOOGLE EARTH SITES ON:*

> Hydrocarbon reserves   > Coal resources   > Other energy resources

REVIEW QUESTIONS

The letters following each Review Question refer to the corresponding Learning Objective from the Chapter Opener.

1. What are the fundamental sources of energy? **(A)**

2. Why do we refer to oil, gas, and coal as fossil fuels? **(B)**

3. What is the source of the organic material in oil, and how is it transformed into oil? **(C)**

4. What is the oil window, and what happens to oil at temperatures higher than the oil window? **(C)**

5. Explain how a conventional oil or gas reserve forms. **(D)**

6. Explain the difference between a conventional and unconventional oil reserve. Which type does the diagram show? **(D)**

7. Where is most of the world's oil found? At present rates of consumption, how long will oil supplies last? **(E)**

8. How is coal formed, and in what class of rocks is coal considered to be? **(F)**

9. What determines the rank of coal? **(F)**

10. How is electricity generated by a nuclear power plant? **(G)**

11. Where do uranium deposits form in the Earth's crust? **(G)**

12. What are some of the drawbacks of nuclear energy? **(G)**

13. Define geothermal energy, and describe the ways that it can be used. Can geothermal energy be used to generate energy economically everywhere? **(E)**

14. What is the difference between renewable and nonrenewable resources? **(E)**

15. What is the likely future of hydrocarbon production and use in the 21st century? **(E)**

16. Describe various kinds of economic mineral deposits and how they form. **(H)**

17. What procedures are used to locate and mine mineral resources today? **(H)**

18. Which type of ore deposit does the figure show? **(H)**

19. What are the ingredients of concrete and brick, and where do they come from? **(I)**

20. Will the supply of mineral resources run out? Can a country survive without importing minerals? **(J)**

21. What are some environmental hazards of mining? **(J)**

ON FURTHER THOUGHT

22. Do you think it would make sense for an energy company to drill for oil in a locality where beds of anthracite occur in the stratigraphic sequence? Explain your answer. **(D)**

23. An ore deposit at a location in Arizona has the following characteristics: One portion of the ore deposit is an intrusive igneous rock in which tiny grains of copper sulfide minerals are dispersed among the other minerals of the rock. Another nearby portion of the ore deposit consists of limestone in which malachite fills cavities and pores in the rock. What types of ores are these? Describe the geologic history that led to the formation of these deposits. **(H)**

ONLINE RESOURCES

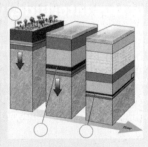

Videos This chapter features videos on the basic process of hydraulic fracturing and on oil sands and plumes.

Smartwork5 This chapter features questions that check students' understanding of energy technologies and the formation of coal and oil.

AN INTRODUCTION TO LANDSCAPES AND THE HYDROLOGIC CYCLE

By the end of this interlude, you should be able to . . .

A. explain what a landscape is and what questions geologists ask about landscapes.

B. describe the difference between uplift and subsidence, and describe the forces that drive them.

C. contrast internal and external energy in the Earth System.

D. differentiate between erosional and depositional landscapes.

E. discuss the reservoirs and exchange processes of the hydrologic cycle.

Nothing that is can pause or stay;
The moon will wax, the moon will wane,
The mist and cloud will turn to rain,
The rain to mist and cloud again,
To-morrow be to-day.

HENRY WADSWORTH LONGFELLOW (American poet, 1807–1882)

F.1 Introduction

The Earth's surface is a place of both endless variety and intricate detail. Observe the height of its mountains, the expanse of its seas, the desolation of its deserts, and you may be inspired, frightened, or calmed. It's no wonder that artists and writers across the ages have sought inspiration from the **landscape**—the character and shape of the land surface in a region—for landscapes engage the full range of human emotion (**Fig. F.1**). Geologists, like artists and writers, savor the impression of a dramatic landscape. But on seeing one, they can't help but ask: How did it come to be, and how will it change in the future?

The subject of landscape development and evolution, and of the **landforms** (individual shapes such as mesas, valleys, cliffs, and dunes) that constitute it, dominate many of the remaining chapters in this book. This interlude, a general introduction to our planet's surface and near-surface realms, characterizes the variety of landscapes, the processes that shape landforms, and the many interrelationships among climate, life, water, rock, and sediment in the Earth System. We will see that water, in both its solid and liquid states, plays a key role in landscape evolution, so we include a focus on the *hydrologic cycle*, the pathway that water molecules follow as they move from ocean to air to land and back to ocean.

F.2 Shaping the Earth's Surface

If the Earth's surface were totally flat, the great diversity of the landscapes that embellish our vistas would not exist. But the surface isn't flat, because a variety of geologic processes cause portions of the surface to move up or down relative to adjacent regions. We refer to the upward movement of the Earth's surface as **uplift** and the sinking or downward movement of the Earth's surface as **subsidence**.

◀ (facing page) We can see key stages of the hydrologic cycle at Rainbow Falls, near Hilo, Hawaii. Water evaporates from the ocean surrounding Hawaii, forming clouds that rain on the land. Some of the water sinks into soil and gets absorbed by plants, and some flows down streams to end up back in the sea.

The occurrence of uplift and subsidence generates **relief**, the elevation difference between two points separated by a specified horizontal distance on a map (**Box F.1**). Wherever relief develops on land, various components of the Earth System kick into action to modify and shape the land surface. Rock at or near the ground surface weathers, fractures, and weakens. On slopes, weakened or loose material becomes susceptible to **downslope movement**, gravity-driven tumbling or sliding from higher elevations to lower ones. Moving water, ice, and air cause **erosion**, the grinding away and removal of the *substrate* (material at or just below the ground surface). And where moving fluids slow down, **deposition**, the accumulation of transported sediment, takes place. Downslope movement, erosion, and deposition redistribute rock and sediment on the Earth, ultimately stripping it from higher areas and collecting it in low areas.

How rapidly do vertical movements and erosion of the Earth's surface take place? The ground can rise or sink by as much as 3 m during a single major earthquake. But, averaged over geologic time, rates of uplift and subsidence range between 0.01 and 10 mm per year (**Fig. F.2a**). Similarly, a single flood, storm, or landslide can carve out several meters of substrate in an hour or less (**Fig. F.2b**), and deposition during a single event can produce a layer of debris several meters thick in a matter of minutes to days. But, averaged over time, erosional and depositional rates also vary between 0.01 and 10 mm per year. Although these average rates seem small, a change in surface elevation of just 0.5 mm (the thickness of a fingernail) per year can yield a net change of 5 km in just 10 million years. Uplift can build a towering mountain range, and erosion can whittle one down to near sea level—it just takes time!

The energy that drives landscape evolution comes from three sources: *internal energy*, the heat within the Earth, which drives the plate motions and mantle plumes that ultimately cause vertical displacements of the surface; *external energy* that comes to the Earth from the Sun, warming the atmosphere and ocean; and *gravitational energy*, which pulls material down slopes at the surface and causes uplift or subsidence necessary to maintain isostasy. Gravity, working together with external energy, also causes *convection*

FIGURE F.1 Examples of the great variety of landscapes on the Earth.

(a) Rounded "sugarloaf" mountains surround Rio de Janeiro, Brazil.

(b) Glaciated peaks of the Alps, France.

(c) Buttes of sandstone, Monument Valley, Arizona.

(d) Cliffs rise from the forest in the Blue Mountains, Australia.

FIGURE F.2 The processes of uplift, subsidence, erosion, and deposition can be slow or rapid.

Uplifted terrace

New terrace forming

(a) Uplifted beach terraces form where the coast is rising relative to sea level. Present-day wave erosion is forming a new terrace and cutting a cliff on the edge of the old one.

(b) So much erosion can take place during a single hurricane that houses built along the beach become undermined. This example happened as a result of "Superstorm Sandy" in 2012.

Topographic Maps and Profiles

We can distinguish one landform from another by its shape—for example, as you will see in succeeding chapters, a river-carved valley does not look like a glacially carved valley. Landform shapes are manifested by variations in elevation within a region. Geologists use the term *topography* to refer to such variations. How can we convey information about topography—a three-dimensional feature—on a

two-dimensional sheet of paper? Geologists and cartographers do this by means of a *topographic map*, which uses contour lines to represent elevation **(Fig. BxF.1a, b)**. A *contour line* represents an imaginary line on the land surface along which all points have the same elevation. You can picture a contour line as the intersection between the land surface and an imaginary horizontal plane. A topographic map that displays many contour lines represents a region with slopes, whereas a map with very few contour lines represents a plain.

The elevation difference between two adjacent contour lines on a topographic map is the *contour interval*. For a given topographic map, the contour interval stays constant, so spacing between contour lines represents slope steepness. Specifically, closely spaced contour lines represent a steep slope, whereas widely spaced contour lines represent a gentle slope. Computer technology can help make slopes on a map stand out by adding shading, to give the effects of shadows cast when the sun lies low in the sky—maps that display slopes this way are called *shaded-relief maps*. Radar beamed from satellites can produce highly detailed digital data on elevation variations on the Earth's surface. The resulting data sets, called *digital elevation models* (*DEMs*), can be analyzed with a computer to produce a shaded and colored image that gives the impression of three dimensions.

Geologists represent variations in elevation along a given traverse by means of a **topographic profile**, the trace of the ground surface as it would appear on a vertical plane that sliced into the ground **(Fig. BxF.1c)**. If we add a representation of geologic features under the ground surface, then we have a **geologic cross section**. In some cases, researchers can gain insight into subsurface geology simply by looking at the shape of a landform **(Fig. BxF.1d)**. For example, a steep cliff in a region of dipping sedimentary strata may indicate the presence of a *resistant layer* (one that stands up to erosion), whereas low areas may be underlain by a *nonresistant layer* (one that erodes easily).

FIGURE BxF.1 Topographic maps and profiles.

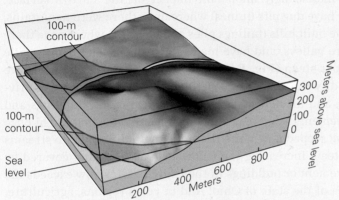

(a) A contour line is the intersection of a horizontal plane with the land surface. This block diagram shows the map area.

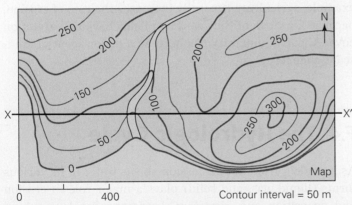

(b) A topographic map depicts the shape of the land surface through the use of contour lines. The difference in elevation between two adjacent lines is the contour interval.

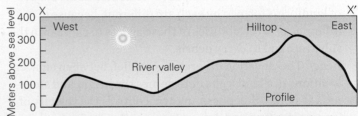

(c) A topographic profile (along section line X–X') shows the shape of the land surface as seen in a vertical slice.

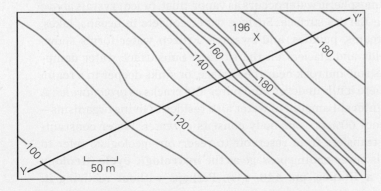

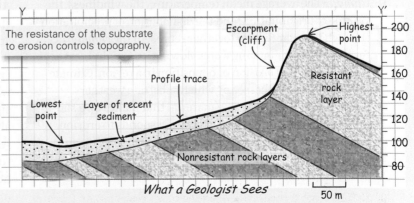

(d) This topographic map (left) shows a distinct cliff. A geologic cross section (right) depicts a geologist's interpretation of the subsurface along Y–Y'. The cliff is the edge of a resistant rock layer.

in the atmosphere and oceans, as manifested by currents and winds. Considering that all these energy sources are always operating, we can think of landscape evolution as a battle between tectonic processes (such as collision, convergence, and rifting), which move land up, and erosional processes, which tear it down. If, in a particular region, the rate of uplift exceeds the rate of erosion, the land surface rises. If, however, the rate of erosion exceeds the rate of uplift or the rate of subsidence exceeds the rate of deposition, the land surface sinks. Without uplift, relief could never develop, and without erosion, high areas would have lasted for the entirety of Earth history.

F.3 Factors Controlling Landscape Development

Imagine traveling across a continent. On your journey, you may cross plains, swamps, hills, valleys, mesas, and mountains. Some of these features are **erosional landforms**, in that they result from the breakdown and removal of rock or sediment when moving water, ice, or air carve into the substrate. Of these three **agents of erosion**, water has the greatest effect on a global basis. Other features are **depositional landforms**, in that they result from the deposition of sediment where the medium carrying the sediment evaporates, slows down, or melts. The specific landforms that develop at a given locality, and that together make up the landscape, reflect several factors.

> *Eroding or transporting agent:* Moving water, ice, and wind all cause erosion and transport sediment. But the shapes of landforms formed by each are different, because of differences in the abilities of these agents to carve into the substrate and to carry debris.

> *Relief:* The elevation difference, or relief, between adjacent locations in a landscape determines the height and steepness of slopes. Steepness, in turn, controls the velocity of ice or water flow and determines whether rock or soil stays in place or tumbles downslope.

> *Climate:* The average mean temperature, the windiness, the volume of precipitation, and the distribution of precipitation through the year (which together represent the *climate*) determine whether moving water, flowing ice, or wind serves as the main agent of erosion or deposition in a region. Climate also affects the processes by which substrate weathers.

> *Substrate composition:* The material that makes up a substrate determines how the substrate responds to erosion. For example, resistant rocks can stand up to form steep cliffs, whereas nonresistant sediment collapses to generate gentle slopes.

> *Life activity:* Some living organisms weaken the substrate (by burrowing, wedging, or digesting), while others hold it together (by binding it with roots). Therefore, the character of life at a location can be affected by life activity.

> *Time:* Landscapes evolve over time in response to continued erosion or deposition. For instance, a gully that has just started to form in response to the flow of a stream does not look the same as the deep canyon that develops after the same stream has existed for a long time.

Although water, wind, and ice are responsible for the development of most landscapes, human activities have had an increasingly important impact on the Earth's surface. We have dug pits (mines) where once there were mountains, have built hills (tailings piles and landfills) where once there were valleys, and have made steep slopes gentle and gentle slopes steep (Fig. F.3). By constructing concrete walls, we modify the shapes of coastlines, change the courses of rivers, and fill new lakes (reservoirs). In cities, buildings and pavements seal the ground and cause water that might once have seeped down into the soil to spill directly into streams instead, increasing their flow. The area of land covered by pavement or buildings in the United States now exceeds the area of the state of Ohio! And in rural regions, agriculture, grazing, water use, and deforestation substantially alter the rates at which natural erosion and deposition take place. For example, agriculture generally increases the rate of erosion, because for much of the year farm fields have no vegetation cover. Clearly, humanity has become a major agent of landscape change.

F.4 The Hydrologic Cycle

As is evident from our discussion above, water in its various forms (liquid, gas, and solid) plays a major role in erosion and deposition and, therefore, landscape development on the Earth's surface. Our planet's water occupies several distinct reservoirs that constitute the **hydrosphere (Table F.1)**. Atmospheric water occurs as vapor, mist, or ice crystals above the Earth's surface. Surface water collects in oceans, lakes, streams, puddles, and swamps. Frozen water forms snowfields and glaciers on the surface. Subsurface water dampens soil and rock near the surface, or sinks deeper to a realm where it fills underground pores and cracks as *groundwater*. A significant amount of water also resides in living organisms—about 60% of your body consists of water. Water constantly transfers from reservoir to reservoir—geologists refer to this never-ending passage as the **hydrologic cycle (Geology at a Glance**, pp. 440–441). Perhaps without realizing it,

FIGURE F.3 Human influence on a geologic scale.

(a) The pyramids of Egypt are human-made hills that rise above the desert sands. They have lasted for thousands of years.

(b) To construct highways through high ridges, workers effectively carve deep valleys. This example borders a highway near Denver.

(c) This stone dam holds back a reservoir in Colorado. Think about how long it would take a glacier to pile up so much sediment.

TABLE F.1 Major Water Reservoirs of the Earth

H_2O Reservoir	Volume (km³)	% of Total Water	% of Fresh Water
Oceans and seas	1,338,000,000	96.5	—
Glaciers, ice caps, snow	24,064,000	2.05	68.7
Saline groundwater	12,870,000	0.76	—
Fresh groundwater	10,500,000	0.94	30.1
Permafrost	300,000	0.022	0.86
Freshwater lakes	91,000	0.007	0.26
Salt lakes	85,400	0.006	—
Soil moisture	16,500	0.001	0.05
Atmosphere	12,900	0.001	0.04
Swamps	11,470	0.0008	0.03
Rivers and streams	2,120	0.0002	0.006
Living organisms	1,120	0.0001	0.003

The Hydrologic Cycle

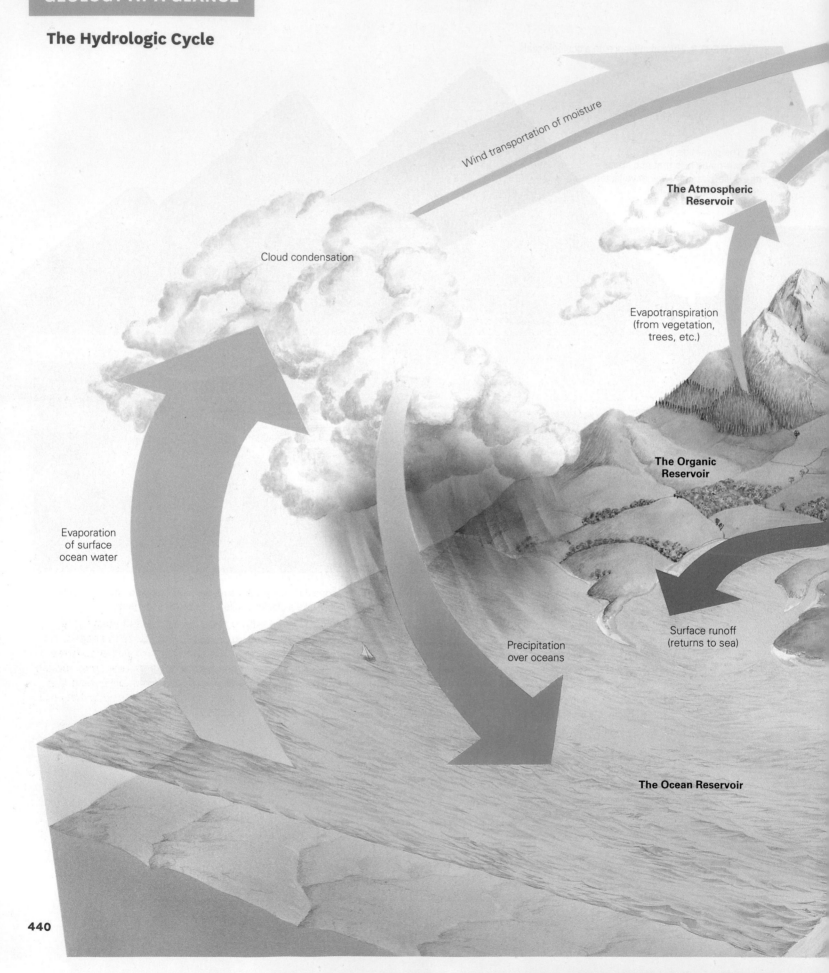

Wind transportation of moisture

Cloud condensation

The Atmospheric Reservoir

Evapotranspiration (from vegetation, trees, etc.)

The Organic Reservoir

Evaporation of surface ocean water

Precipitation over oceans

Surface runoff (returns to sea)

The Ocean Reservoir

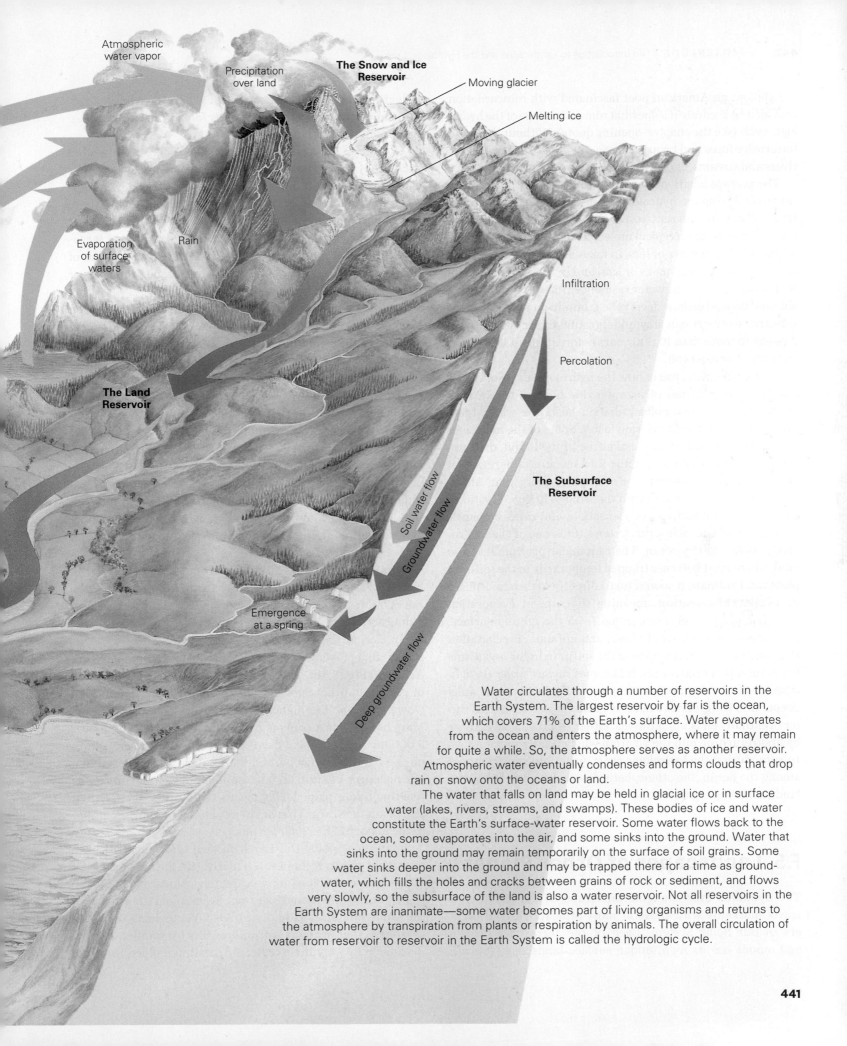

Atmospheric
water vapor

Precipitation
over land

**The Snow and Ice
Reservoir**

Moving glacier

Melting ice

Evaporation
of surface
waters

Rain

Infiltration

Percolation

**The Land
Reservoir**

Soil water flow

Groundwater flow

**The Subsurface
Reservoir**

Emergence
at a spring

Deep groundwater flow

Water circulates through a number of reservoirs in the
Earth System. The largest reservoir by far is the ocean,
which covers 71% of the Earth's surface. Water evaporates
from the ocean and enters the atmosphere, where it may remain
for quite a while. So, the atmosphere serves as another reservoir.
Atmospheric water eventually condenses and forms clouds that drop
rain or snow onto the oceans or land.
  The water that falls on land may be held in glacial ice or in surface
water (lakes, rivers, streams, and swamps). These bodies of ice and water
constitute the Earth's surface-water reservoir. Some water flows back to the
ocean, some evaporates into the air, and some sinks into the ground. Water that
sinks into the ground may remain temporarily on the surface of soil grains. Some
water sinks deeper into the ground and may be trapped there for a time as groundwater, which fills the holes and cracks between grains of rock or sediment, and flows
very slowly, so the subsurface of the land is also a water reservoir. Not all reservoirs in the
Earth System are inanimate—some water becomes part of living organisms and returns to
the atmosphere by transpiration from plants or respiration by animals. The overall circulation of
water from reservoir to reservoir in the Earth System is called the hydrologic cycle.

Longfellow, an American poet fascinated with reincarnation, provided an accurate if somewhat romantic image of the hydrologic cycle (see the chapter-opening quote). Without this cycle, the erosive force and transporting activity of running water in rivers and streams, or of flowing ice in glaciers, would not exist.

The average length of time that water stays in a particular reservoir during the hydrologic cycle is called the **residence time**. Water in different reservoirs has different residence times. For example, a typical molecule of water remains in the oceans for 4,000 years or less, in lakes and ponds for 10 years or less, in rivers for 2 weeks or less, and in the atmosphere for 10 days or less. Groundwater residence times are highly variable and depend on how deeply the groundwater flows into the subsurface—water can stay underground for anywhere from 2 weeks to more than 10,000 years before it inevitably moves on to another reservoir.

To get a clearer sense of how the hydrologic cycle operates, let's follow the journey of seawater that has just reached the surface of the ocean. Solar radiation heats the water, making the water molecules vibrate faster. Some of the molecules evaporate—they effectively shake free of the liquid, drift upward in a gaseous state, and mix into the atmosphere. About 30% of the total ocean volume evaporates every year. Atmospheric water vapor moves with the wind to higher elevations, where it cools, undergoes condensation, and rains or snows. About 76% of this water *precipitates* (falls out of the air) directly back into the ocean. The remainder precipitates onto land, where most becomes trapped temporarily in the soil or in plants and animals. It soon returns directly to the atmosphere by **evapotranspiration,** meaning the sum of evaporation from bodies of water, evaporation from the ground surface, and release of water from plants and animals. Precipitation that does not become trapped in the soil or in living organisms has several potential routes. It becomes *surface water* (held in lakes, rivers, or swamps), it becomes ice in glaciers, or it sinks deeper into the ground to become groundwater. All of this water eventually ends up in the sea, either by flowing directly or by first returning into the air from which it later precipitates. In sum, during the hydrologic cycle, water constantly moves among the ocean, the atmosphere, reservoirs on or below the land, and living organisms.

F.5 Landscapes of Other Planets

The dynamic, ever-changing landscape of the Earth contrasts markedly with those of most other terrestrial planets in our Solar System. Each of the terrestrial planets and moons has its own unique surface landscape features, reflecting the interplay between the object's particular tectonic and erosional processes. Let's look at a few examples: the Moon, Mars, and Venus.

Our Moon has a static, pockmarked landscape generated exclusively by meteorite impacts and volcanic activity. Because no plate motions occur on the Moon, no new mountains form. Because no atmosphere or ocean exists, there is no hydrologic cycle and no erosion from rivers, glaciers, or winds. Therefore, the lunar surface has remained largely unchanged for most of its history **(Fig. F.4a)**. Some of the craters that you may see with a telescope formed over 3 billion years ago.

Landscapes on Mars differ from those of the Moon because Mars does have an atmosphere, though one much less dense than that of the Earth. Martian winds generate huge dust storms, some of which obscure nearly the entire surface of the planet for months at a time. The two sets of landscapes also differ because Mars once had surface water, and the Moon did not **(Box F.2)**. Therefore, the Martian surface consists of four kinds of materials: volcanic flows and deposits (primarily of basalt), debris from impacts, windblown sediment, and water-laid sediment. There is even evidence that soil-forming processes affected surface materials and hints that small amounts of liquid water exist temporarily on Mars at present. Martian winds not only deposit sediment, they also slowly erode impact craters and polish surface rocks.

Landscapes on Mars also differ from those on the Earth, because Mars does not have plate tectonics. So, unlike the Earth, Mars has no mountain belts or volcanic arcs. However, plume activity occurred in the planet's distant past. A plume likely caused the uplift of a 9-km-high bulge (the *Tharsis Ridge*) that covers an area comparable to that of North America **(Fig. F.4b)**. Thermal activity also led to the eruption of gargantuan hot-spot volcanoes, such as the 22-km-high *Olympus Mons*, the highest mountain in the Solar System. Mars boasts a vast canyon, the *Valles Marineris*, a gash over 3,000 km long and 8 km deep, larger than any known feature on the Earth, or anywhere. Because Mars has no vegetation and no longer has rain, its surface does not weather and erode like that of our planet. This means that, unlike the Earth, it still bears the scars of impact by swarms of meteorites earlier in the history of the Solar System.

Venus resembles the Earth in size and may still have operating mantle plumes. Virtually the entire surface of Venus was resurfaced by volcanic eruptions about 300 to 1,600 Ma, so the planet's surface is younger than those of the Moon and Mars **(Fig. F.4c)**. Also, Venus has a dense atmosphere that protects it from impacts by smaller objects. Because relatively little cratering has taken place since the

FIGURE F.4 Landscapes of other planets and moons.

(a) A close-up of the lunar landscape, with a lunar rover and an astronaut for scale.

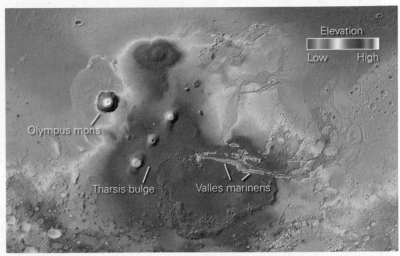

(b) A DEM shows major topographic features on Mars. Colors depict elevation variations.

(c) A DEM of Venus, based on radar data. The colors depict elevation variations.

(d) A photo by the *New Horizons* space probe shows the icy plains, mountains, and craters of Pluto.

resurfacing event, volcanic and tectonic features dominate the landscape of Venus. Satellites using radar have revealed a variety of volcanic constructions on Venus, such as shield volcanoes, lava flows, and calderas. Rifting on Venus produced faults, some of which occur in association with volcanic features. Liquid water cannot survive the scalding temperatures of Venus's surface, so no hydrologic cycle operates there and no life exists. Because of the density of the atmosphere, winds are too slow to cause much erosion or deposition. New images of Pluto's surface became available when the *New Horizons* space probe flew past the icy dwarf planet in 2015. These images reveal icy plains (cracked into polygons) and rugged mountains, in addition to many meteorite craters **(Fig. F.4d)**.

Water on Mars?

In 1877, an Italian astronomer named Giovanni Schiaparelli studied the surface of Mars with a telescope and announced that long, straight *canali* crisscrossed the planet's surface. *Canali* might have been translated into the English word *channel*, but perhaps because of the recent construction of the Suez Canal, newspapers of the day translated the word into the English *canal*, with the implication that the features had been constructed by intelligent beings. An eminent American astronomer began to study the "canals" and suggested that they had been built to carry water from polar ice caps to Martian deserts.

Late-20th-century satellite mapping of Mars showed that the so-called canals do not exist—they were simply optical illusions. No lakes, oceans, rainstorms, or flowing rivers exist on the surface of Mars today. The atmosphere of Mars has such low density, and exerts so little pressure on the planet's surface, that slight amounts of liquid water that may be locally released at the surface quickly evaporate. So, Mars has no hydrologic cycle like the Earth's. But crucial questions remain: Did Mars ever have significant amounts of running water or standing water in the past? If so, where is the water now? The question of the presence of water lies at the heart of an even more basic question: Given that even the simplest life forms, as far as we know, require water to exist, is there, or was there ever, life on Mars?

The case for the existence of liquid water in the past on Mars has become overwhelming. Much of the evidence comes from comparing landforms on the planet's surface with landforms of known origin on the Earth. High-resolution images of Mars reveal landforms that look very much like those formed on the Earth in response to flowing water. Examples include networks of river channels resembling those on the Earth **(Fig. BxF.2)**, scour features, deep gullies, and streamlined deposits of sediment.

Studies by the *Odyssey* satellite in 2003, and by Mars rovers (*Spirit* and *Opportunity*) that landed on the planet in 2004, added intriguing new data to the debate. *Odyssey* detected hints that hydrogen, an element in water, exists beneath the surface of the planet over broad regions, and the Mars rovers have documented the existence of hematite and gypsum, minerals that precipitate from water. The rovers have also found sediment layers that appear to have been deposited in water. The *Phoenix* lander, in 2008, confirmed the existence of water ice by digging into the surface to expose some. And the *Mars Reconnaissance Orbiter* has identified features that likely formed by hydrothermal activity. Researchers speculate that Mars was much wetter in its past, perhaps billions of years ago. But since the atmosphere became less dense, almost all water now lies hidden underground or trapped in polar ice caps.

FIGURE BxF.2 Evidence for water on Mars earlier in the planet's history.

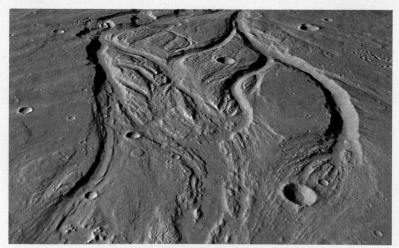

(a) An oblique view showing landforms on Mars resembling channels cut by rivers on the Earth.

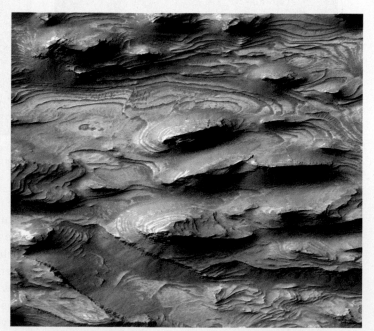

(b) Layers of strata, which appear to have been deposited in water, in Chasma Canyon, Mars. Note the elongate islands in the channel, resembling islands that have been shaped by rivers on the Earth.

Interlude F Review

INTERLUDE SUMMARY

> The character and shape of the land surface in a region is a landscape. Individual shapes are landforms. Topographic maps, shaded-relief maps, and DEMs can portray the shape of landscapes.

> Land can undergo uplift or subsidence to yield relief. Debris formed by weathering, of uplifted land moves downslope and collects in lower areas.

> The energy driving landscape development comes from three sources: the Earth's internal energy, gravitational energy, and energy radiating from the Sun.

> The nature of a landscape depends on climate, time, relief, slope angles, elevation, the activity of organisms, substrate composition, and the rate of uplift and subsidence.

> Water is the dominant agent of erosion on the Earth. Driven by gravity and by energy from the Sun, water moves among various reservoirs (such as the ocean, the atmosphere, the land surface, the subsurface, and life) during the hydrologic cycle.

> Landscapes on other planets differ markedly from those on the Earth.

GUIDE TERMS

agent of erosion (p. 438)
deposition (p. 435)
depositional landform (p. 438)
downslope movement (p. 435)
erosion (p. 435)

erosional landform (p. 438)
evapotranspiration (p. 442)
geologic cross section (p. 437)
hydrologic cycle (p. 438)
hydrosphere (p. 438)

landform (p. 435)
landscape (p. 435)
relief (p. 435)
residence time (p. 442)
subsidence (p. 435)

topographic profile (p. 437)
uplift (p. 435)

REVIEW QUESTIONS

The letters following each Review Question refer to the corresponding Learning Objective from the Chapter Opener.

1. What is the difference between uplift and subsidence? **(B)**

2. Why do landscapes on the Earth change over geologic time, while they remain static on the Moon? **(C)**

3. What is topography, and how can we portray it on a sheet of paper? What does the image show? **(A)**

4. What are the principal agents of erosion on the Earth? **(B)**

5. What factors affect the character of erosional or depositional landforms that develop in a region? **(D)**

6. Explain the steps in the hydrologic cycle. **(E)**

7. How do landscapes of other planets differ from those of the Earth? **(C)**

ONLINE RESOURCES

Videos This interlude features videos on the Earth's water cycle and evapotranspiration.

Topographic maps represent the shape of the surface of the land by using contour lines to show points of equal elevation. Enter the approximate elevation value of the red dot shown on the map.

Smartwork5 This interlude includes questions on features and processes that change the landscape and on the water cycle.

CHAPTER 13

UNSAFE GROUND: LANDSLIDES AND OTHER MASS MOVEMENTS

By the end of this chapter, you should be able to . . .

A. identify different types of mass movements, and explain the differences among them.

B. sketch a model illustrating the forces acting on the material of a slope, and discuss factors that determine whether a slope is stable or unstable.

C. describe the evidence that major mass movements have taken place under the sea, and that some of these movements triggered tsunamis.

D. highlight events that may trigger mass movements, and explain why some regions are particularly susceptible to mass movement events.

E. evaluate hazards related to mass movements, and discuss how such hazards can, in some cases, be prevented.

13.1 Introduction

It was Sunday, May 31, 1970, a market day, and thousands of people had crammed into the Andean town of Yungay, Peru, to shop. Suddenly they felt the jolt of an earthquake, strong enough to topple some masonry houses. But worse was yet to come. Vibrations from the earthquake broke an 800-m-wide ice slab off the end of a glacier at the top of Nevado Huascarán, a nearby 6.6-km-high mountain peak. As the ice tumbled downward for a distance of more than 3.7 km, it disintegrated and became a chaotic avalanche of chunks traveling at speeds of more than 300 km per hour. Near the base of the mountain, most of the avalanche channeled into a valley and thickened into a churning cloud, as high as a 10-story building, that ripped up rocks and soil along the way. Frictional heating transformed the ice into water, which mixed with rock and dust to produce 50 million m³ of a muddy slurry viscous enough to carry boulders larger than houses. This mass, sometimes floating on a compressed air cushion that allowed it to pass without disturbing the grass below, traveled 14.5 km in less than 4 minutes.

On rounding a curve near the mouth of the valley, part of the mass shot up the sides and flew over the ridge between the valley and Yungay. As the town's inhabitants and visitors stumbled out of earthquake-damaged buildings, they heard a deafening roar and looked up to see a wall of debris burst above the nearby ridge. The debris engulfed the town and buried it. Only the top of a church and a few palm trees remained visible to show where Yungay once stood (Fig. 13.1). Today, the site is a grassy meadow, spotted with crosses left by relatives mourning the 18,000 people entombed below.

People often assume that the ground beneath them is *terra firma*, a solid foundation on which they can build their lives. But the catastrophe at Yungay says otherwise. Some areas of the Earth's surface are unstable and might start moving downslope if disturbed. Geologists refer to the downslope transport of rock, **regolith** (soil and loose sediment or debris), snow, and ice as **mass movement**, or *mass wasting*. Like earthquakes, volcanic eruptions, storms, and floods, mass movements are a type of **natural hazard**, meaning a natural feature of the environment that can cause damage to environments and societies. Unfortunately, mass movements have

◀ (facing page) During a rockfall in 2014, a huge boulder rolled right through a farmhouse next to a vineyard in Italy. Clearly, uplifted land doesn't last forever, as gravity inevitably pulls it downslope. Fortunately, in this case there were no injuries.

become more of a threat every year, because as the world's population has grown, cities have expanded into areas of unsafe ground. Mass movement also plays a critical role in the rock cycle, as the first step in the production and transportation of sediment and, therefore, in the development of landscapes.

In this chapter, we look at the types, causes, and consequences of mass movement, and consider the precautions society can take to protect people and property from its dangers. You might want to keep this information in mind when selecting a site for your home or when voting on land-use propositions that could affect your community.

FIGURE 13.1 The May 1970 Yungay landslide disaster in Peru.

(a) Before the landslide, the town of Yungay perched on a hill near the ice-covered mountain Nevado Huascarán.

(b) The landslide completely buried the town beneath debris. A landslide scar is visible on the mountain in the distance.

13.2 **Types of Mass Movement**

Most people refer to any mass movement of rock or regolith down a slope as a **landslide**. Geologists and engineers, however, find it useful to distinguish among different kinds of mass movements based on four features: (1) the type of material involved (rock or regolith); (2) the velocity of movement (slow, intermediate, or fast); (3) the character of the moving mass (coherent or chaotic; wet or dry); and (4) the environment in which the movement takes place (subaerial or submarine). In this section, we first examine mass movements that occur on land, roughly in order from slow to very fast. Then we briefly introduce submarine mass movements.

FIGURE 13.2 The process and consequences of slow mass movements (creep and solifluction).

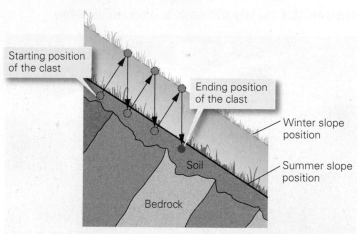

(a) Creep due to freezing and thawing: The clast rises perpendicular to the ground during freezing and sinks vertically during thawing. After 3 years, it migrates to the position shown.

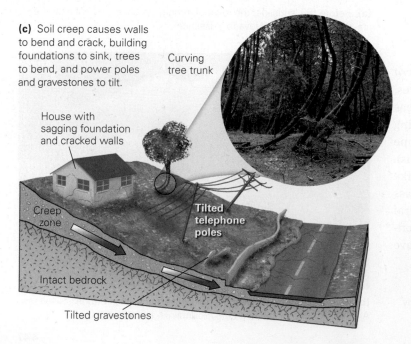

(c) Soil creep causes walls to bend and crack, building foundations to sink, trees to bend, and power poles and gravestones to tilt.

CREEP AND SOLIFLUCTION

Creep refers to the slow, gradual downslope movement of regolith on a slope. Creep happens when regolith alternately expands and contracts in response to freezing and thawing, wetting and drying, or warming and cooling. To see how the process of creep works, let's focus on the consequences of seasonal freezing and thawing. In the winter, when water freezes, the regolith expands, and particles move outward, perpendicular to the slope. During the spring thaw, water becomes liquid again, and gravity makes the particles sink vertically and, therefore, migrate downslope slightly **(Fig. 13.2a, b)**. You can't see creep by staring at a hillslope because it occurs so slowly, but over a period of years, creep causes trees, fences, gravestones, walls, and foundations built

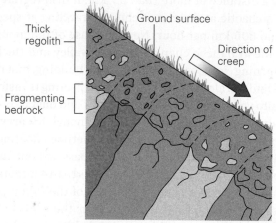

(b) As rock layers weather and break up, the resulting debris creeps downslope.

(d) Solifluction on a hillslope in the tundra.

on a hillside to tilt downslope **(Fig. 13.2c)**. Notably, trees that continue to grow after they have been tilted display a pronounced curvature at their base.

In arctic or high-elevation regions, regolith freezes solid to great depth, producing permanently frozen ground, or *permafrost*. In the brief summer thaw, only the uppermost 1 to 3 m of the permafrost thaws. Since meltwater cannot sink into permafrost, the melted layer becomes soggy and weak and flows slowly downslope in overlapping sheets. Geologists refer to this kind of creep as **solifluction (Fig. 13.2d)**.

SLUMPS

During the summer of 2011, after weeks of drenching rains, a 1.5-km-wide portion of a slope began to move down and out into the floor of Keene Valley, New York. The mass moved at only centimeters to tens of centimeters per day, but even at this slow rate, the accumulated displacement destroyed several expensive homes. The boundary between the moving mass and the unmoving land upslope developed into a 5-m-high escarpment.

Geologists refer to a relatively slow-moving mass movement event such as the Keene Valley event, during which moving rock or regolith does not disintegrate but rather stays somewhat coherent, as a **slump**, and they refer to the moving mass itself as a *slump block* **(Fig. 13.3)**. A slump block slides on a **failure surface**. Some failure surfaces are planar, but commonly they curve and resemble a spoon lying concave side up. The exposed upslope edge of a failure surface forms a *head scarp*, a new cliff face, and the downslope end of a

FIGURE 13.3 The process of slumping on a hillslope. Note the scarps that form at the head of the slump.

(a) A head scarp on a hillslope.

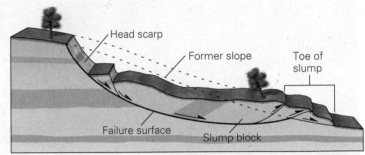

(b) Cross section of a slump.

(c) Slumping dumped sediment into this river in Costa Rica.

(d) A slump beginning to form along a highway in Utah.

slump block is the block's *toe*. On a hillslope, the toe can move up and over the pre-existing land surface to form a curving ridge, but along sea coasts or river banks, the toe ends up in the water, where it eventually washes away. In some cases, the upslope and downslope portions of a slump block break into a series of discrete slices, each separated from its neighbor by a sliding surface. Slumps come in all sizes, from only a few meters across to tens of kilometers across. In general, they move at speeds ranging from millimeters per day to meters per minute.

MUDFLOWS, DEBRIS FLOWS, AND LAHARS

Rio de Janeiro, a coastal city in Brazil, originally occupied only the flatlands bordering beautiful crescent beaches that had formed between steep rock hills. But in recent decades, the city's population has increased so much that, in many places, densely populated communities of makeshift shacks have grown up on steep slopes. These communities, with inadequate storm drains, were built on the thick regolith that resulted from long-term weathering of bedrock in Brazil's tropical climate. Particularly heavy rains can saturate the regolith, transforming it into a viscous slurry of mud, resembling wet concrete, that flows downslope. When this happens, whole communities can disappear overnight, replaced by a muddle of mud and debris. At the bases of the cliffs, the flowing mud may knock over and bury buildings of all sizes **(Fig. 13.4a)**. In unpopulated areas of Brazil, such mass movements can rip away forests **(Fig. 13.4b)**. Geologists refer to a moving slurry of mud as a **mudflow** or

FIGURE 13.4 Examples of mudflows and lahars.

(a) This 2011 mudslide in Nova Friburgo, Brazil, destroyed a high-rise building at the base of the hill.

(b) Mudslides in 2011 stripped away forests on hillslopes in Brazil.

(c) A recent debris flow in Utah. Note the chaotic mixture of rock chunks and mud.

Lahar

(d) A lahar that rushed down the side of Mt. St. Helens, Washington.

mudslide, and a slurry consisting of a mixture of mud and larger, pebble- to boulder-sized fragments as a **debris flow** or *debris slide* (**Fig. 13.4c**).

The speed at which a mudflow or debris flow moves depends on the slope angle and on the water content. Drier mudflows have higher viscosity than do wetter mudflows, so they move more slowly, and mudflows on gentler slopes move more slowly than do those on steeper slopes. Because mudflows and debris flows have greater viscosity than clear water does, they can carry large rock chunks as well as houses and cars. They typically follow channels downslope, and at the base of the slope they may spread out into a broad lobe (**Box 13.1**).

Particularly devastating mudflows spill down the river valleys bordering volcanoes. These mudflows, known as **lahars**, consist of a mixture of volcanic ash and water from the snow and ice that melts in a volcano's heat or from heavy rains (**Fig. 13.4d**; see Chapter 5). A very destructive lahar occurred on November 13, 1985, in the Andes Mountains in Colombia. That night, an eruption melted a volcano's thick snowcap, producing hot water that mixed with ash. A scalding lahar rushed down river valleys and swept over the nearby town of Armero while most inhabitants were asleep. Of the 25,000 residents, 20,000 perished.

ROCKSLIDES AND DEBRIS SLIDES

In the early 1960s, engineers built a huge new dam across a river on the northern side of Monte Toc, in the Italian Alps, to create a reservoir for generating electricity. This dam, the Vaiont Dam, was an engineering marvel, a concrete wall rising 260 m (as high as an 85-story skyscraper) above the valley floor (**Fig. 13.5a**). Unfortunately, the dam's builders did not recognize the hazard posed by nearby Monte Toc. Limestone beds interlayered with weak shale beds underlie the side of Monte Toc facing the reservoir. These beds dip parallel to the surface of the mountain and curve under the reservoir (**Fig. 13.5b**). As the reservoir filled, water seeped into the beds, which weakened them. As a result, the flank of

the mountain started to crack, and therefore, to shake. Local residents began to call Monte Toc *la montagna che cammina* (the mountain that walks).

After several days of rain, Monte Toc began to rumble so much that on October 9, 1963, engineers lowered the water level in the reservoir. They thought that, at worst, the wet ground might slump a little into the reservoir, with minor consequences. So no one ordered the evacuation of the town of Longarone, a few kilometers down the valley below the dam. Unfortunately, the engineers underestimated the problem. At 10:30 that evening, a huge chunk of Monte Toc— 600 million tons of rock—detached from the mountain and slid downslope into the reservoir. Some debris rocketed up the opposite wall of the valley to a height of 260 m above the original reservoir level. The displaced water of the reservoir spilled over the top of the dam and rushed down into the valley below. When the flood had passed, nothing of Longarone or its 1,500 inhabitants remained. Though the dam itself still stands, it holds back only debris and has never provided any electricity.

A sudden movement of rock and debris down a nonvertical slope can be called a **rockslide** if the mass consists only of rock or a **debris slide** if it consists mostly of regolith. Once a slide has taken place, it leaves a scar on the slope and forms a debris pile at the base of the slope. Slides happen when bedrock or regolith detaches from a slope, slips rapidly (at speeds of up to 300 km/h) downhill on a failure surface, and breaks up into a chaotic jumble. Rockslides and debris slides sometimes have enough momentum to climb the opposite side of the valley into which they fell. Slides, like slumps, come at a variety of scales.

AVALANCHES

In the winter of 1999, an unusual weather system passed over the Austrian Alps. First it snowed. Then the temperature warmed and the snow began to melt. But then the weather turned cold again, and the melted snow froze into a hard, icy crust. This cold snap ushered in a blizzard that blanketed the ice crust with tens

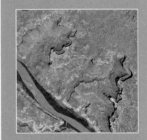

BOX 13.1 CONSIDER THIS...

What Goes Up Must Come Down

Along the coast of California, waves slowly erode the land and produce low, flat areas called *wave-cut benches* (see Chapter 15). As this happens, tectonic motions slowly raise the land surface. When uplifted, these benches form small plateaus, or *terraces*. One such terrace lies at an elevation of 180 m above sea level, about 500 m east of the present-day beach at La Conchita; the west face of this terrace is a cliff-like bluff **(Fig. Bx13.1a)**. Repeated slip along the San Andreas plate boundary has broken up the bedrock of the area, and fragments have weathered substantially, so the substrate of the terrace and bluff consists of weak clay and debris.

Relatively little vegetation covers the bluff or the terrace above. Rain that falls on the face of the bluff drains away quickly via a network of small, temporary streams. But the water falling on the terrace infiltrates into the ground, sinks down, and saturates clay and debris on the terrace and its bluff, turning the clay into very weak mud. When this happens, the weight of surface material causes the bluff to give way, and a mass of mud and debris flows downslope at rates of up to 10 m per second.

If the region of La Conchita were uninhabited, such mass wasting would just be part of the natural process of landscape evolution—gravity brings down land that had been raised by tectonic activity. But when downslope movements take place in La Conchita, they can make headlines, because on the modern wave-cut bench between the shore and the base of the bluff, developers built a community housing 350 people. In 1995, a flow of mud and debris overwhelmed 9 houses at the base of the bluff. An even more devastating flow happened in 2005, burying 13 houses and damaging 23, and killing 10 people **(Fig. Bx13.1b, c)**. These events have sent a clear message about the importance of reading the landscape before planning construction.

FIGURE Bx13.1 The 2005 La Conchita mudslide along the coast of California.

(a) A housing development was built in a narrow strip between the beach and steep cliffs.

(b) During heavy rains, the slope gave way and heavy mud flowed down, burying houses and taking several lives.

(c) Rescuers at the toe of the mudslide.

FIGURE 13.5 The Vaiont Dam disaster—a catastrophic landslide that displaced the water in a reservoir with rock debris.

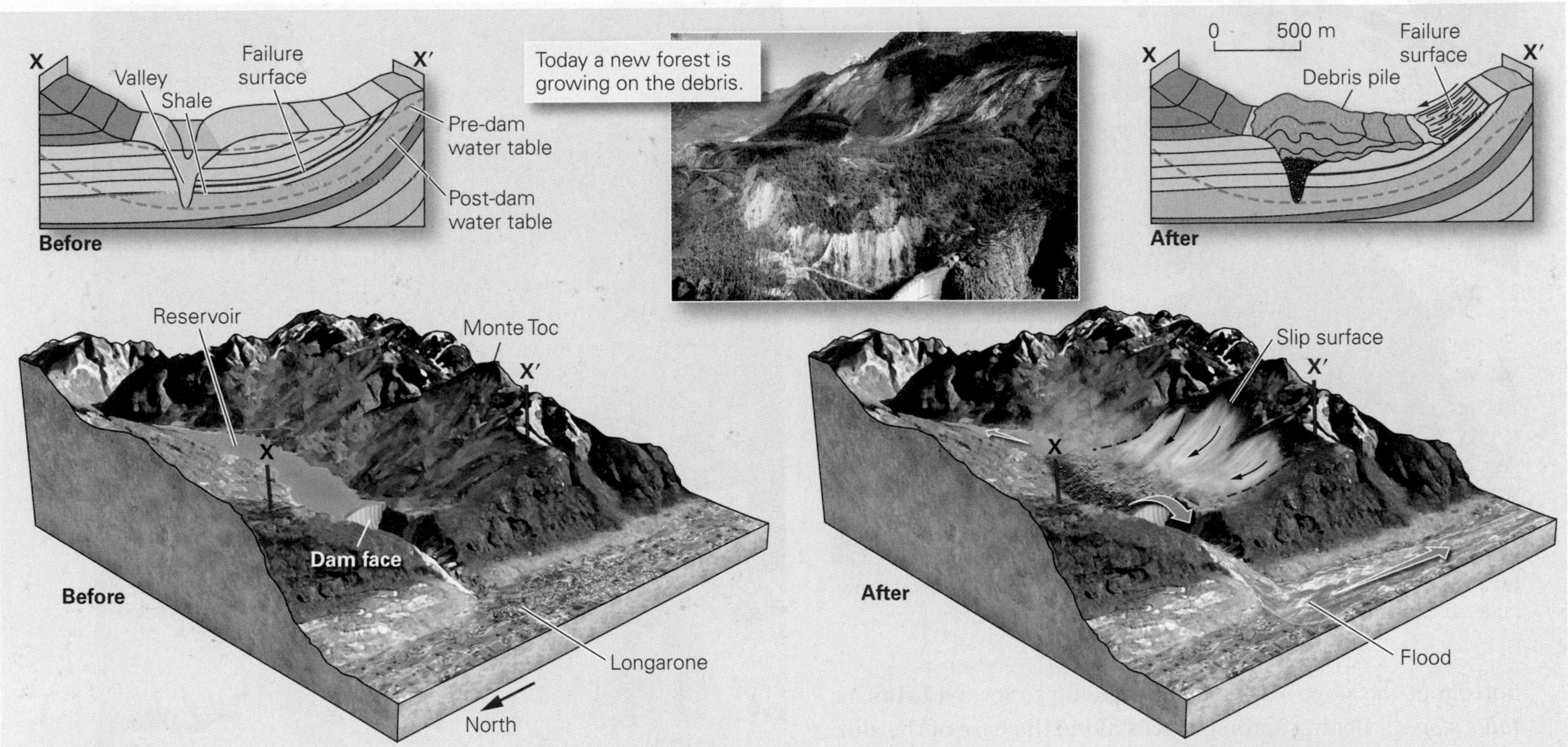

Today a new forest is growing on the debris.

(a) Before the landslide, the north flank of Monte Toc was forested. When the reservoir filled, the slope became unstable. A shale bed a few hundred meters below the ground surface became a failure surface.

(b) Thirty-three million cubic meters of debris slid and displaced water in the reservoir. The water surged over the dam and swept away a village in the valley below.

of centimeters of new snow. With the frozen snow layer underneath acting as a failure surface, 200,000 tons of new snow began to slide down the mountain. As it accelerated, the mass transformed into a **snow avalanche**, a chaotic jumble of snow surging downslope. At the bottom of the slope, the avalanche overran a ski resort, crushing and carrying away buildings, cars, and trees, and killing more than 30 people. It took searchers and their specially trained dogs many days to find buried survivors and victims under the 5- to 20-m-thick pile of snow that the avalanche deposited (Fig. 13.6a).

What triggers snow avalanches? Some happen when a *cornice*, a large drift of snow that builds up on the lee side of a windy mountain summit, suddenly gives way and falls onto slopes below, where it knocks free additional snow. Others happen when a broad slab of snow on a moderate slope detaches from its substrate along an icy failure surface. As we described above, the failure surface is a frozen crust on the surface of older snow. It may also be a layer of rounded grains, consisting of snow that either partly melted or sublimated, and lost its spiky protrusions. Wet avalanches, formed at warmer temperatures, move as a slurry of solid and liquid water, whereas dry avalanches, formed when the snow

mass contains no liquid water, tumble as a cloud of powder (Fig. 13.6b).

Not all avalanches consist only of snow and ice. In some cases, they form when the material in a rock or debris fall mixes with air and forms a turbulent cloud that races downslope at high velocity. Geologists sometimes refer to such mass movements as *rock avalanches* or *debris avalanches*.

ROCKFALLS AND DEBRIS FALLS

Rockfalls and **debris falls**, as their names suggest, occur when a mass free-falls from a cliff for part of its journey (Fig. 13.7). Commonly, rockfalls happen when a body of rock separates from a cliff face along a *joint*, a natural crack in rock across which a block of rock no longer connects to bedrock (see Chapter 9). Most rockfalls involve only a few blocks, but some falls dislodge immense quantities of rock. Friction and collision with other rocks may bring some blocks to a halt before they reach the

Did you ever wonder... why highway engineers erect "falling rock" signs?

FIGURE 13.6 Examples of avalanches.

(a) Aftermath of a 1999 avalanche in the Austrian Alps. Masses of snow buried and destroyed several homes.

bottom of the slope—these blocks pile up to form a **talus**, or *talus slope*, a sloping apron of rocks along the base of the cliff. Large, fast rockfalls push the air in front of them, creating a short blast of hurricane-like wind. For example, the wind in front of a 1996 rockfall in Yosemite National Park flattened over 2,000 trees.

Small rockfalls happen fairly frequently along steep highway road cuts, leading to the posting of falling-rock zone signs. Such rockfalls commonly take place soon after construction because blasting and excavation leave loose rocks on the slope above the road. But rockfalls may continue long

(b) A dry-snow avalanche in Alaska is a turbulent cloud.

FIGURE 13.7 Examples of rockfalls.

(a) Successive rockfalls have littered the area below this sandstone cliff with boulders. Note the talus at the base of the cliff.

(b) This rockfall buried the forest bordering a lake in the Uinta Mountains, Utah. Fresh rock exposed by the rockfall has a lighter color.

after construction, for mechanical and chemical weathering weakens slopes over time (see Interlude B). In temperate climates, rockfalls are particularly common in the spring, after winter ice, which may cause frost wedging, melts.

SUBMARINE MASS MOVEMENTS

So far, we've focused on mass movements that occur subaerially, for these are the ones we can see easily and that affect us the most. But mass movements also happen underwater. Geologists distinguish three types of submarine mass movements, or submarine landslides, according to whether the

FIGURE 13.8 Examples of huge submarine slumps and debris flows.

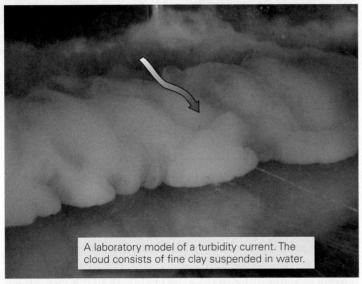

A laboratory model of a turbidity current. The cloud consists of fine clay suspended in water.

(a) A turbidity current is a cloud of sediment suspended in water; it flows near the seafloor because it is denser than clear water.

mass remains coherent or disintegrates as it moves. In **submarine slumps**, semicoherent blocks slip downslope on weak horizons. In some cases, the layers constituting the blocks become contorted as they move, like a tablecloth that slides off a table. In *submarine debris flows*, the moving mass breaks apart to form a slurry containing larger clasts (pebbles to boulders) suspended in a mud matrix. And in **turbidity currents**, sediment disperses in water to yield a turbulent cloud of suspended sediment that rushes downslope as a submarine avalanche or density current **(Fig. 13.8a)**.

In recent years, geologists have used satellites as well as sonar to map out the extent of submarine landslides. The shapes of slumps and landslide deposits stand out on the resulting high-resolution seafloor maps **(Fig. 13.8b)**. Geologists have found that submarine slopes bordering both hot-spot volcanoes and active plate boundaries are scalloped by many immense slumps and debris flows, because tectonic activity frequently jars these areas with earthquakes that set masses of material in motion. Debris from countless events over millions of years has substantially modified the flanks of the Hawaiian Islands **(Fig. 13.8c)**. Some of this debris has moved more than 150 km away from the islands, leaving huge head scarps along the edges of the islands. Significantly, passive-margin coasts are not immune to slumping, and immense slumps have been mapped along the coasts of the Atlantic Ocean.

Since a submarine slump can develop fairly quickly, and since its movement can displace a large area of the seafloor, it can trigger a tsunami. A huge slump called the Storegga Slide occurred west of Norway, along the Atlantic passive margin, thousands of years ago. The area affected by the slump is

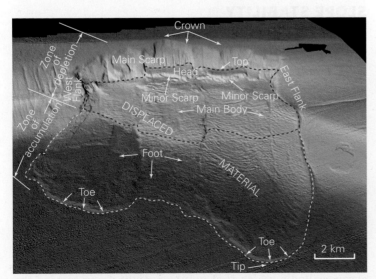

(b) A digital bathymetric map of a slump along the coast of California. The parts of the slump are labeled.

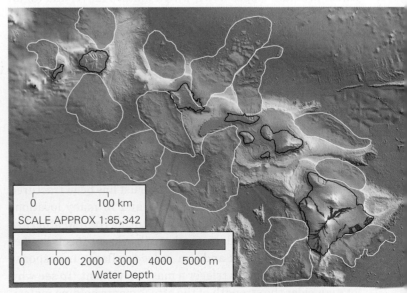

(c) A bathymetric map of the area around Hawaii shows several huge slumps, shaded in tan.

about 100 km wide, and the slump debris traveled underwater for over 600 km. Tsunamis generated by the Storegga Slide may have wiped out Stone-Age villages all around the coast of the North Sea.

TAKE-HOME MESSAGE

Mass movements differ from one another based on speed and character. Creep, slumping, and solifluction are slow. Mudflows and debris flows move faster, and avalanches and rockfalls move the fastest. Mass movements occur on land and underwater. A large submarine mass movement may generate a tsunami.

QUICK QUESTION In what way is a snow avalanche like a turbidity current?

13.3 Why Do Mass Movements Occur?

We've seen that mass movements travel at a range of different velocities (**Geology at a Glance**, pp. 458–459). These range from slow (creep) to faster (slumps, mudflows and debris flows, and rockslides and debris slides) to fastest (snow avalanches, and rock and debris falls). The velocity depends on the steepness of the slope and the water or air content of the mass. For any movement to take place, the stage must be set by the following phenomena: (1) fracturing and weathering of the substrate, which weakens the substrate so it cannot hold up against the pull of gravity; (2) the development of relief, which provides slopes down which masses move; and (3) an event that sets mass in motion. Let's look at these phenomena more closely.

WEAKENING THE SUBSTRATE: FRAGMENTATION AND WEATHERING

If the Earth's surface were covered by completely unfractured and unweathered rock, mass movements would be of little concern, for intact fresh rock has great strength and could form stalwart mountain faces that would not tumble. But, in reality, the rock of the Earth's upper crust has been fractured by jointing and faulting **(Fig. 13.9)**, and in many locations the surface has a cover of regolith resulting from weathering. Regolith and fractured or weathered rock are much weaker than intact fresh rock and can indeed collapse in response to gravitational pull to trigger a mass movement. To see why, let's consider the strength of the attachments holding materials together.

FIGURE 13.9 Jointing broke up this thick sandstone bed along a cliff in Utah. Blocks of sandstone break free along joints and tumble downslope.

A mass of intact bedrock is relatively strong because the chemical bonds within its interlocking grains, or within the cements between grains, can't be broken easily. A mass of loose rocks or of regolith, in contrast, is relatively weak because the grains are held together only by friction, electrostatic attraction, or the surface tension of water. All of these forces combined are weaker than chemical bonds holding together the atoms in the minerals of intact rock. To picture this contrast, think about how much easier it is to flatten a sand castle (whose strength comes primarily from the surface tension of water films on the sand grains) than it is to flatten a sculpture of a castle carved from rock.

SLOPE STABILITY

Mass movements do not take place on all slopes, and even on slopes where such movements are possible, they occur only occasionally. Geologists distinguish between *stable slopes*, on which sliding is unlikely, and *unstable slopes*, on which sliding will likely happen. When material starts moving on an unstable slope, we say that **slope failure** has occurred. Whether a slope fails depends on the balance between two forces—the *downslope force*, caused by gravity, and the *resistance force*, which inhibits sliding. If the downslope force exceeds the resistance force, the slope fails and mass movement results.

Let's examine slope failure more closely by imagining a block sitting on a slope. We can represent the force of gravitational attraction between this block and the Earth by an arrow, called a *vector*, that points straight down toward the Earth's center of gravity. This vector can be separated into

FIGURE 13.10 Forces that trigger downslope movement.

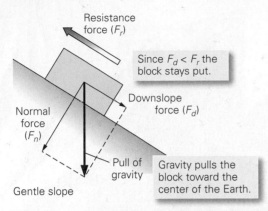

Since $F_d < F_r$ the block stays put.

Resistance force (F_r)

Downslope force (F_d)

Normal force (F_n)

Pull of gravity

Gravity pulls the block toward the center of the Earth.

Gentle slope

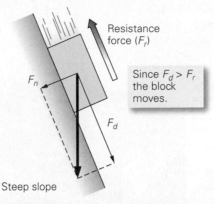

Resistance force (F_r)

F_n

F_d

Since $F_d > F_r$ the block moves.

Steep slope

(a) Gravity can be divided into a normal force and a downslope force. If the resistance force, caused by friction, exceeds the downslope force, the block does not move.

(b) If the slope angle increases, the downslope force due to gravity increases. If the downslope force becomes greater than the resistance force, the block starts to move.

In many locations, the resistance force can be less than expected because a weak surface exists at some depth below ground level. If downslope movement begins on the weak surface, we can say that the weak surface has become a **failure surface**. Geologists recognize several different kinds of weak surfaces that may become failure surfaces **(Fig. 13.12)**. Examples include wet clay layers; wet, unconsolidated sand layers; joints; weak bedding planes (shale beds and evaporite beds are particularly weak); and metamorphic foliation planes.

two components—the downslope force component lies parallel to the slope, whereas the normal force component points perpendicular to the slope. In this representation, we can symbolize the resistance force by an arrow pointing uphill. If the downslope force becomes larger than the resistance force, then the block moves; otherwise, it stays in place **(Fig. 13.10)**.

What produces a resistance force? As we saw above, chemical bonds in mineral crystals or cement hold intact rock in place, friction holds an unattached block in place, electrical charges and friction hold dry regolith in place, and surface tension holds wet regolith in place. Because of the resistance force, granular debris tends to pile up to produce the steepest slope it can without collapsing. The angle of this slope, known as the **angle of repose**, typically has a value of between 30° and 45° for most dry, unconsolidated materials (such as dry sand). The angle depends partly on the shape and size of grains, which determine the amount of friction across grain boundaries. For example, steeper angles of repose tend to form on slopes composed of irregularly shaped grains **(Fig. 13.11)**.

FIGURE 13.12 Different kinds of weak surfaces can become failure surfaces.

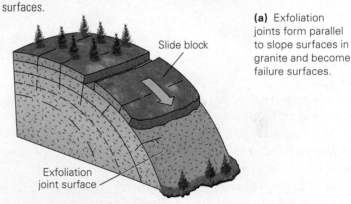

Slide block

Exfoliation joint surface

(a) Exfoliation joints form parallel to slope surfaces in granite and become failure surfaces.

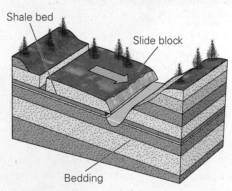

Shale bed

Slide block

Bedding

(b) In sedimentary rocks, bedding planes (particularly in weak shale) become failure surfaces.

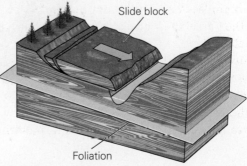

Slide block

Foliation

(c) In metamorphic rock, foliation planes (particularly in mica-rich schist) become failure surfaces.

FIGURE 13.11 The angle of repose is the steepest slope that a pile of unconsolidated sediment can have and remain stable. The angle depends on the shape and size of grains.

30°   45°

Well-rounded sand has a small angle.

Irregularly shaped gravel has a large angle.

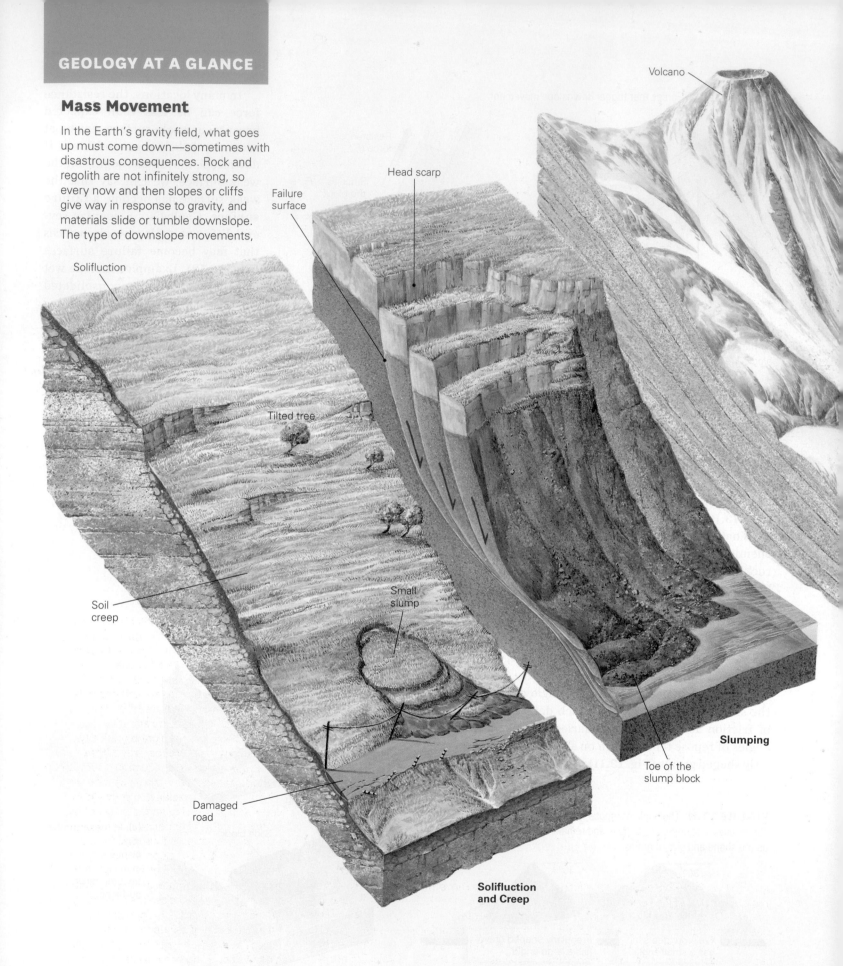

Mass Movement

In the Earth's gravity field, what goes up must come down—sometimes with disastrous consequences. Rock and regolith are not infinitely strong, so every now and then slopes or cliffs give way in response to gravity, and materials slide or tumble downslope. The type of downslope movements,

Solifluction

Failure surface

Head scarp

Volcano

Tilted tree

Soil creep

Small slump

Damaged road

Solifluction and Creep

Slumping

Toe of the slump block

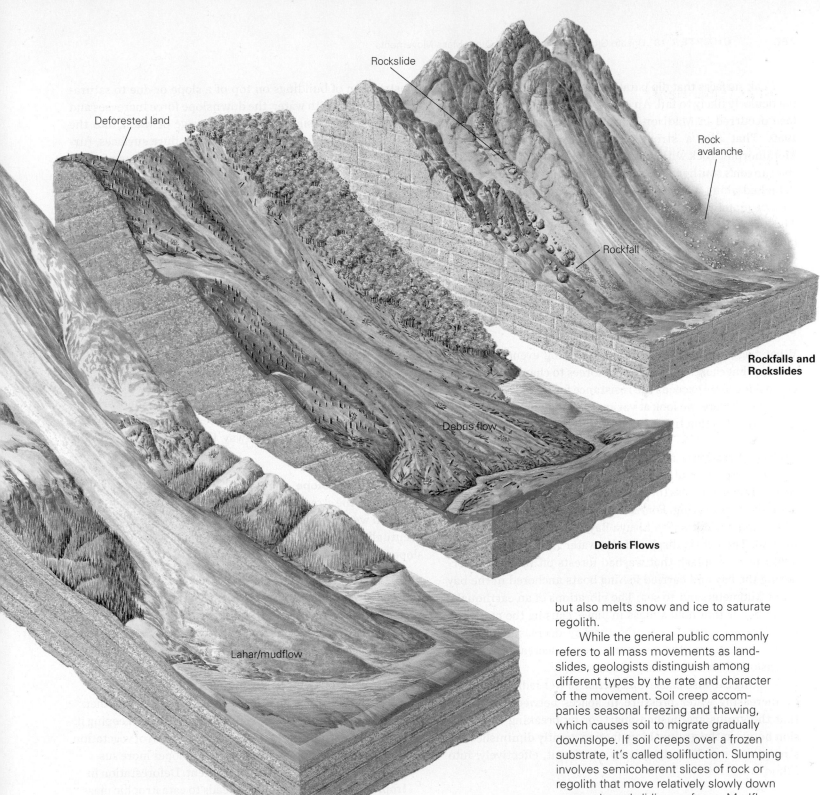

Deforested land

Rockslide

Rock avalanche

Rockfall

Rockfalls and Rockslides

Debris flow

Debris Flows

Lahar/mudflow

Lahars and Mudflows

called mass movement, or mass wasting, that takes place at a given location reflects the composition of the slope, the steepness of the slope, and the climate. Stronger rocks can hold up steep cliffs, but with time, rocks break free along joints and tumble or slide down weak surfaces.

Mass movements may be dry or water-saturated. Episodes of mass movement may be triggered by an over-steepened slope, a heavy rainfall that saturates the slope, an earthquake that shakes debris free, or a volcanic erup-tion, which not only shakes the ground but also melts snow and ice to saturate regolith.

While the general public commonly refers to all mass movements as land-slides, geologists distinguish among different types by the rate and character of the movement. Soil creep accom-panies seasonal freezing and thawing, which causes soil to migrate gradually downslope. If soil creeps over a frozen substrate, it's called solifluction. Slumping involves semicoherent slices of rock or regolith that move relatively slowly down spoon-shaped sliding surfaces. Mudflows and debris flows happen where regolith has become saturated with water and moves rapidly downslope as a slurry. A lahar is a mudflow that forms on the flank of a volcano when ash mixes with water. Steep, rocky cliffs may suddenly give way in rockfalls. If the rock breaks up into a cloud of debris that rushes downslope at high velocity, it becomes a rock avalanche. In snow avalanches, the debris consists of snow.

Weak surfaces that dip parallel to the land-surface slope are particularly likely to fail. An example of failure on such a surface occurred in Madison Canyon, Montana, on August 17, 1959. That day, a strong earthquake jarred the region. Metamorphic rock with weak foliation formed the bedrock of the canyon's southern wall. When the ground vibrated, rock detached along a foliation plane and tumbled downslope. Unfortunately, 28 campers lay sleeping on the valley floor. They were probably awakened by the hurricane-like winds blasting in front of the moving mass, but seconds later they were buried under 45 m of rubble.

FINGERS ON THE TRIGGER: WHAT CAUSES SLOPE FAILURE?

What triggers an individual mass-wasting event? In other words, what causes the balance of forces to change so that the downslope force exceeds the resistance force, and a slope suddenly fails? Here, we look at various phenomena—natural and human-made—that trigger slope failure.

Shocks, Vibrations, and Liquefaction Earthquake tremors, storms, the passing of large trucks, or blasting in construction sites may cause a mass that had been on the verge of moving to actually start moving. For example, an earthquake-triggered slide dumped debris into Lituya Bay, in southeastern Alaska, in 1958. The debris displaced the water in the bay, creating a 300-m-high splash that washed forests off the slopes bordering the bay and carried fishing boats anchored in the bay many kilometers out to sea. The vibrations of an earthquake break bonds that hold a mass in place, causing the mass and the slope to separate slightly, thereby decreasing friction. As a consequence, the resistance force decreases, and the downslope force sets the mass in motion.

Shaking can also cause *liquefaction* of wet sediment, either by increasing water pressure in spaces between grains so that the grains are pushed apart, or by breaking the cohesion between the grains. Liquefaction greatly diminishes the strength of a sediment layer by turning it, effectively, into a fluid.

Changing Slope Loads, Steepness, and Support As we have seen, the stability of a slope at a given time depends on the balance between the downslope force and the resistance force. Factors that change one or the other of these forces can lead to failure. Examples include changes in slope loads, failure-surface strength, slope steepness, and the support provided by material at the base of the slope.

Slope loads change when the weight of the material above a potential failure plane changes. If the load increases due to construction of buildings on top of a slope or due to saturation of regolith with water, the downslope force increases and may exceed the resistance force. Seepage of water into the ground may also weaken underground failure surfaces, further decreasing resistance force. An example of such failure triggered the largest observed landslide in U.S. history, the Gros Ventre Slide, which took place in 1925 on the flank of Sheep Mountain, near Jackson Hole, Wyoming **(Fig. 13.13)**. Almost 40 million m^3 of rock, as well as the overlying soil and forest, detached from the side of the mountain and slid 600 m downslope, yielding a debris flow that filled a valley and formed a 75-m-high natural dam across the Gros Ventre River.

Removing support at the base of a slope due to river or wave erosion or to construction efforts plays a major role in triggering many slope failures. In effect, the material at the base of a slope acts like a dam holding back the material farther up the slope. In some cases, erosion by a river or by waves eats into the base of a cliff and produces an overhang. When such *undercutting* has occurred, rock making up the overhang eventually breaks away from the slope and falls **(Fig. 13.14)**.

Changing the Slope Strength The stability of a slope depends on the strength of the material constituting it. If the material weakens with time, the slope becomes weaker and eventually collapses. Three factors influence the strength of slopes:

> *Weathering:* With time, chemical weathering produces weaker minerals, and physical weathering breaks rocks apart. As a result, a formerly intact rock composed of strong minerals is transformed into a weaker rock or into regolith.

> *Vegetation cover:* In the case of slopes underlain by regolith, vegetation tends to strengthen the slope because the roots hold otherwise unconsolidated grains together. Also, plants absorb water from the ground, thus keeping it from turning into slippery mud. The removal of vegetation therefore has the net result of making slopes more susceptible to downslope mass movement. Deforestation in tropical rainforests, similarly, leads to catastrophic mass wasting of the forest's substrate.

> *Water content:* Water affects materials comprising slopes in many ways. Surface tension in the film of water on grain surfaces may help hold regolith together. But if the water content increases, water pressure may push grains apart so that regolith liquefies and can begin to flow. Water infiltration may make weak surfaces underground more slippery or may push surfaces apart and decrease friction.

FIGURE 13.13 Stages leading to the 1925 Gros Ventre Slide in Wyoming.

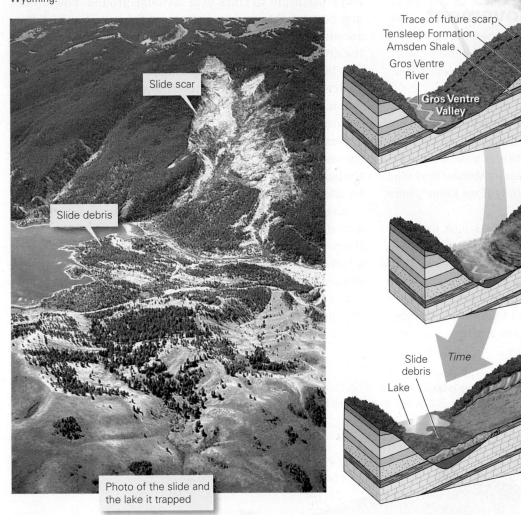

Rain

Trace of future scarp
Tensleep Formation
Amsden Shale
Gros Ventre River

Gros Ventre Valley

At depth, the weak Amsden Shale was a potential failure surface because it lay parallel to the slope.

Rain weakened the Amsden and made the Tensleep heavier. Downslope force caused a mass of rock to start moving.

Time

Slide debris
Lake
Scarp

The debris filled the valley, blocking a stream and forming Slide Lake. The scarp remained on the hillslope.

Slide scar

Slide debris

Photo of the slide and the lake it trapped

FIGURE 13.14 Undercutting and collapse of a sea cliff.

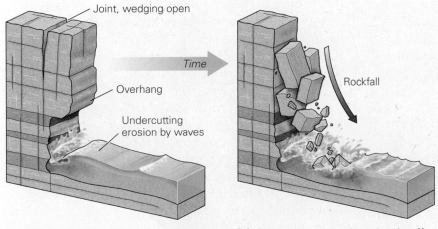

Joint, wedging open

Time

Overhang

Undercutting erosion by waves

Rockfall

(a) Undercutting by waves removes the support beneath an overhang.

(b) Eventually, the overhang breaks off along joints, and a rockfall takes place.

TAKE-HOME MESSAGE

Weathering and fragmentation weaken slope materials and make them more susceptible to mass movement. Failure occurs when the downslope force exceeds resistance force due to shocks, changing slope angles and strength, changing water content, or changing slope support.

QUICK QUESTION If the slope of a sand pile is less than the angle of repose, will the pile fail?

13.4 How Can We Protect Against Mass-Movement Disasters?

IDENTIFYING REGIONS AT RISK

Clearly, landslides, mudflows, and slumps are natural hazards we cannot ignore. Too many people live in regions where mass movements have the potential to kill people and destroy property. In many cases, the best solution is avoidance: don't build, live, or work in an area where mass movement may take place. But avoidance will be possible only if we know where the hazards are greatest.

To pinpoint dangerous regions, geologists look for landforms known to result from mass movements, because where these movements have happened in the past, they might happen again in the future. Features such as slump head scarps, swaths of forest in which trees have been tilted, piles of loose debris at the base of hills, and hummocky (bumpy) land surfaces all indicate recent mass wasting. In some cases, geologists may also be able to detect regions that are beginning to move (**Fig. 13.15**). For example, roads, buildings, and pipes begin to crack over unstable ground. Power lines may be too tight or too loose because the poles to which they are attached move together or apart. Visible cracks form on the ground at the potential head of a slump, and the ground may bulge up at the toe of the slump. In some cases, subsurface cracks may drain the water from an area and kill off vegetation, whereas in other areas land may sink and form a swamp. Slow movements cause trees to develop pronounced curves at their base. More recently, new, extremely precise surveying technologies have permitted geologists to detect the beginnings of mass movements that may not yet have visibly affected the land surface.

Even if there is no evidence of recent movement, a danger may still exist—just because a steep slope hasn't collapsed in the recent past doesn't mean it won't collapse in the future. In recent years, geologists have begun to identify potential hazards by using computer programs

Did you ever wonder . . . why parts of the California coast slip into the sea?

FIGURE 13.15 Surface features warn that a large slump is beginning to develop. Cracks that appear at the head scarp may drain water and kill trees. Power-line poles tilt and the lines become tight. Fences, roads, and houses on the slump begin to crack.

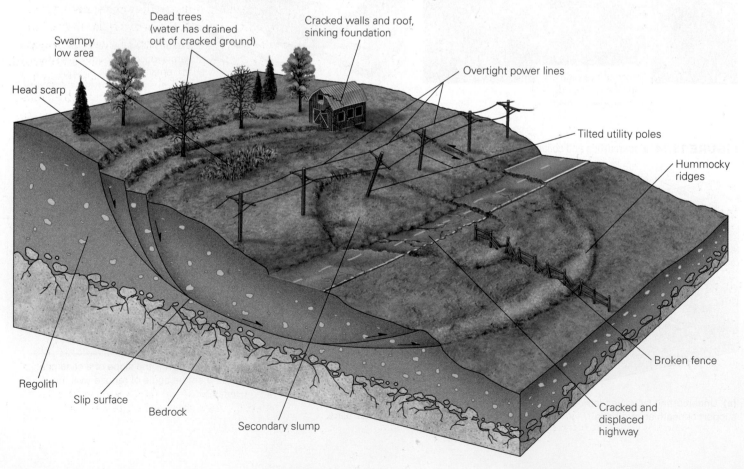

to evaluate factors that trigger mass wasting. These factors include the following: slope steepness; strength of substrate; degree of water saturation; orientation of weak surfaces relative to the slope; vegetation cover; potential for heavy rains; susceptibility to undercutting; and likelihood of earthquakes. From such hazard-assessment studies, geologists compile *landslide-potential maps,* which rank regions according to the likelihood that a mass movement will occur **(Fig. 13.16).**

PREVENTING MASS MOVEMENTS

In areas where a hazard exists, people can take effective steps to remedy the problem and stabilize the slope **(Fig. 13.17).**

FIGURE 13.16 A landslide-potential map of the western United States.

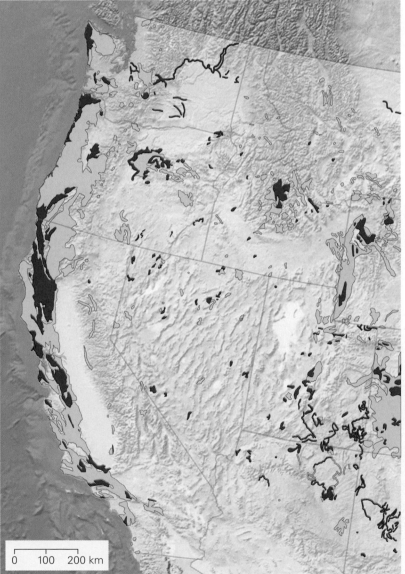

0   100   200 km

High (More than 15% of area involved)   Moderate (1.5–15% of area involved)   Low (Less than 1.5% of area involved)

› *Revegetation:* Stability in deforested areas will be greatly enhanced if landowners replant the region with vegetation that sends down deep roots and binds regolith together.

› *Regrading:* An oversteepened slope can be regraded or terraced so that it does not exceed the angle of repose.

› *Reducing subsurface water:* Because water weakens material beneath a slope and adds weight to the slope, an unstable situation may be remedied either by improving drainage so that water does not enter or remain in the subsurface in the first place, or by pumping water out of the ground.

› *Preventing undercutting:* In places where a river undercuts a cliff face, engineers can divert the river. Similarly, along coastal regions they may build an offshore breakwater or pile **riprap** (loose boulders or concrete) along the beach to absorb wave energy before it strikes the cliff face.

› *Building safety structures:* In some cases, engineers can build a structure that stabilizes a potentially unstable slope or protects a region downslope from debris if a mass movement does occur. For example, civil engineers can build retaining walls or bolt loose slabs of rock to more coherent masses in the substrate in order to stabilize highway embankments. The danger from rockfalls can be decreased by covering a road cut with chain link fencing or by spraying road cuts with concrete. Highways at the base of an avalanche-prone slope can be covered by an avalanche shed, whose roof keeps debris off the road.

› *Controlled blasting of unstable slopes:* When it is clear that unstable ground or snow threatens a particular region, the best solution may be to blast the unstable ground or snow loose at a time when its movement can do no harm. Ski resorts routinely blast cornices at the top of slopes to diminish avalanche hazards.

Clearly, common sense and well-designed precautions can save lives and property.

TAKE-HOME MESSAGE

Various features of the landscape may help geologists to identify unstable slopes and estimate risk. Systematic study allows production of landslide-potential maps. Engineers can use a variety of techniques to protect localities from mass movements.

QUICK QUESTION Why is mass movement of major concern during production of a large road cut?

FIGURE 13.17 A variety of remedial steps can stabilize unstable ground.

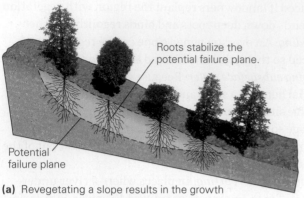

Roots stabilize the
potential failure plane.

Potential
failure plane

(a) Revegetating a slope results in the growth
of roots that can hold a slope together.

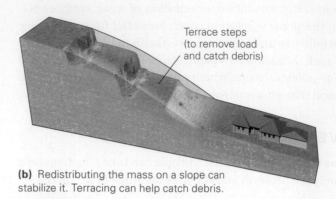

Terrace steps
(to remove load
and catch debris)

(b) Redistributing the mass on a slope can
stabilize it. Terracing can help catch debris.

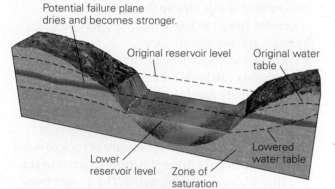

Potential failure plane
dries and becomes stronger.

Original reservoir level

Original water
table

Lower
reservoir level

Zone of
saturation

Lowered
water table

(c) Lowering the level of the water table can
strengthen a potential failure surface.

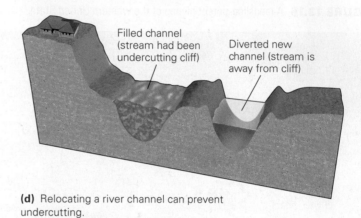

Filled channel
(stream had been
undercutting cliff)

Diverted new
channel (stream is
away from cliff)

(d) Relocating a river channel can prevent
undercutting.

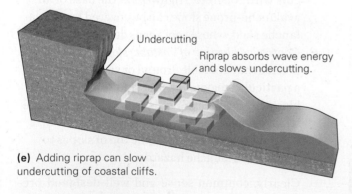

Undercutting

Riprap absorbs wave energy
and slows undercutting.

(e) Adding riprap can slow
undercutting of coastal cliffs.

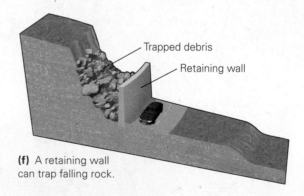

Trapped debris

Retaining wall

(f) A retaining wall
can trap falling rock.

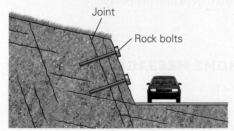

Joint

Rock bolts

(g) Bolting or screening a cliff face can hold
loose rocks in place.

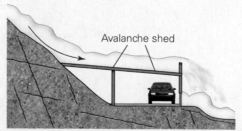

Avalanche shed

(h) An avalanche shed diverts debris or snow
over a roadway.

ANOTHER VIEW A 1991 landslide in Switzerland dumped a huge apron of giant boulders and other rock debris into a valley. Such landslides play a major role in the erosion of mountains.

Chapter 13 Review

CHAPTER SUMMARY

› Mass movement, or mass wasting, is the downslope movement of rock or regolith under the influence of gravity. This process plays an important role in the shaping of landforms and serves as the first step in the transport of sediment.

› Although in everyday language people refer to most mass movements as landslides, geologists distinguish among different types of mass movements based on such factors as the composition of the moving materials and the rate of movement.

› Slow mass movement, caused by the freezing and thawing or wetting and drying of regolith, is called creep. In places where slopes are underlain with permafrost, solifluction causes a melted layer of regolith to flow downslope. During slumping, a semicoherent mass of material moves down a spoon-shaped failure surface. Mudflows and debris flows occur where regolith has become saturated with water and moves downslope as a slurry.

› Rock and debris slides move very rapidly down a slope; the rock or debris breaks apart and tumbles. During avalanches, snow or debris mixes with air and moves downslope as a turbulent cloud. And in a debris fall or rockfall, the material free-falls down a vertical cliff.

› Mass movements of various types can occur beneath the sea and can trigger tsunamis.

› Intact, fresh rock is too strong to undergo mass movement. So, for mass movement to be possible, rock must be weakened by fracturing (joint formation) or weathering.

› Unstable slopes start to move when the downslope force exceeds the resistance force that holds material in place. The angle of repose refers to the steepest angle at which a slope of unconsolidated material can remain without collapsing.

› Downslope movement can be triggered by shocks and vibrations, a change in the steepness of a slope, removal of support from the base of the slope, a change in the strength of a slope, deforestation, weathering, or heavy rain.

› Geologists produce landslide-potential maps to identify areas susceptible to mass movement. Engineers can help detect incipient mass movements and can help prevent mass movements by using a variety of techniques to stabilize slopes.

GUIDE TERMS

angle of repose (p. 457)
creep (p. 448)
debris fall (p. 453)
debris flow (p. 451)
debris slide (p. 451)
failure surface (pp. 449, 457)

lahar (p. 451)
landslide (p. 448)
mass movement (p. 447)
mudflow (p. 450)
natural hazard (p. 447)
regolith (p. 447)

riprap (p. 463)
rockfall (p. 453)
rockslide (p. 451)
slope failure (p. 456)
slump (p. 449)
snow avalanche (p. 453)

solifluction (p. 449)
submarine slump (p. 455)
talus (p. 454)
turbidity current (p. 455)

GEOTOURS THIS CHAPTER'S GEOTOURS WORKSHEET (M) FEATURES QUESTIONS AND GOOGLE EARTH SITES ON:

› Portuguese Bend Slide, CA
› La Conchita Mudslide, CA
› Gros Ventre Slide, WY

› Vaiont Dam Rockslide, Italy
› Frank Rockslide, Alberta, Canada

› Velarde Slumps, NM
› Landslide Remediation, Japan

REVIEW QUESTIONS

The letters following each Review Question refer to the corresponding Learning Objective from the Chapter Opener.

1. What factors do geologists use to distinguish among various types of mass movements? (**A**)

2. Identify the key differences among a slump, a debris flow, a lahar, an avalanche, a rockslide, and a rockfall. Which type of mass wasting is depicted here? Identify the parts of the sketch. (**A**)

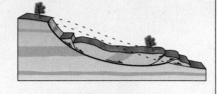

3. Explain the process of creep, and discuss how it differs from solifluction. (**A**)

4. Distinguish among different types of submarine mass movements. Which of these types can trigger a major tsunami, and why? (**C**)

5. Why is intact bedrock stronger than fractured bedrock? Why is it stronger than regolith? (**B**)

6. Explain the difference between a stable and an unstable slope. What factors determine the angle of repose of a material? What features may serve as failure surfaces? (**B**)

7. Discuss the variety of phenomena that can cause a stable slope to become so unstable that it fails. (**D**)

8. How can ground shaking cause fairly solid layers of sand or mud to become weak slurries capable of flowing? (**D**)

9. Discuss the role of vegetation in slope stability. Why can fires and deforestation lead to slope failure? (**B**)

10. Identify the various factors that make the coast of southern California susceptible to mass movements. (**D**)

11. What factors do geologists take into account when producing a landslide-potential map, and how can they detect the beginning of mass movement in an area? (**E**)

12. What steps can people take to avoid landslide disasters? What is the purpose of the rock bolts shown in the sketch? (**E**)

ON FURTHER THOUGHT

13. Imagine that you have been asked by the World Bank to determine whether it makes sense to build a dam across a steep-sided, east-west-trending valley in a small central Asian nation. Initial investigation shows that the rock of the valley floor consists of schist containing a strong foliation that dips south. Outcrop studies reveal that abundant fractures occur in the schist along the valley floor; the surfaces of most fractures are coated with slickensides. Moderate earthquakes have rattled the region. Explain the hazards and what might happen if the reservoir behind the dam were filled. Would you support the bank's decision to finance dam construction? (**B, E**)

ONLINE RESOURCES

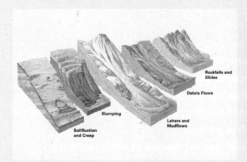

Videos This chapter features videos on developing and detecting slumps, and landslide processes.

Smartwork5 This chapter includes questions on features and processes that change the landscape, and on mass wasting.

CHAPTER 14

STREAMS AND FLOODS: THE GEOLOGY OF RUNNING WATER

By the end of this chapter, you should be able to . . .

A. explain how streams and drainage networks form and evolve.

B. describe the processes that lead to erosion and deposition by streams.

C. characterize the changes that take place along the length of a stream from its headwaters to its mouth.

D. sketch the evolution of meanders in a stream, and explain the concept of a floodplain.

E. distinguish between slow-onset floods and flash floods, and describe the conditions that lead to each.

F. describe the various methods that engineers may use to protect areas from flooding, and interpret statements concerning flood frequency.

G. characterize environmental issues associated with streams.

14.1 Introduction

Wind swooping northward across warm oceans can pick up vast quantities of water vapor. It can then transport that water as moist air into a continent's interior. On a September day in 2013, such a mass of warm, moist air collided with a mass of cold air stalled along the eastern edge of the Rocky Mountains in Colorado. The warm air rose over the cold air, the moisture in it condensed, and drenching rains began to fall. In Boulder County, 430 mm of rain fell over a period of a few days—normally, the county receives 525 mm over the course of a whole year! Some of the water soaked into the ground, but most ended up in the numerous streams that flow from the mountains toward the plains. One of these streams, the Big Thompson River, temporarily carried 30 times more water than normal. The roaring torrent washed away houses and trees, scoured away roads (**Fig. 14.1a**), destroyed rail lines, inundated farmland, broke pipelines, overwhelmed sewage treatment plants, and undermined bridges. During the same year, flooding in Alberta had the same effect in the eastern Canadian Rockies (**Fig. 14.1b**). Where this water spilled out onto the plains, it submerged portions of Calgary (**Fig. 14.1c**).

The people of Colorado and Alberta experienced the immense power of *running water*, surface water that flows down sloping land in response to the pull of gravity. Geologists use the term **stream** for any body of running water that flows in a **channel**, an elongate depression or trough. (In everyday English, we tend to refer to large streams as *rivers* and medium-sized ones as *creeks*.) The edges of a channel are the stream's *banks*, the floor of the channel is the *streambed*, and a defined length of the channel is a *reach*. Water in a stream flows from *upstream* reaches, closer to the source or **headwaters** of the stream, to *downstream* reaches, closer to the end or **mouth** of the stream. Streams drain water from the landscape and carry it into lakes or to the sea, much as culverts drain water from a parking lot. In the process, streams relentlessly erode the landscape and transport sediment to sites of deposition. Generally, a stream stays within the

◀ (facing page) A torrent of water rushes down the floor of a steep-sided canyon in the Andes of Peru. The silt and mud suspended in the water give it a brown color. During a major flood, this stream can move the boulders that litter its bed, and it would become a threat to the people who have built along its banks.

FIGURE 14.1 Examples of flooding damage illustrate the power of running water.

(a) A stream roaring down a Colorado canyon in 2013 undercut and carried away part of a highway.

(b) Heavy rains in the Canadian Rockies caused extensive flooding in Alberta, in 2013.

(c) The 2013 flood inundated parts of Calgary, including the stadium for the city's Stampede (rodeo).

confines of its channel, but when the supply of water entering a stream exceeds the channel's capacity, water spills out and covers the surrounding land, thereby causing a **flood** like the ones that devastated parts of Colorado and Alberta.

The Earth is the only planet in the Solar System that currently hosts flowing streams of liquid water. Streams are of great importance to human society, not only because of how they modify the landscape, especially during floods, but also because they provide avenues for travel and commerce, nutrients and sediment for agriculture, water for irrigation and consumption, and energy to produce electricity and drive factories. In this chapter, we examine how streams operate in the Earth System. First, we learn about the origin of running water and about how the water draining a region organizes into an array of connected streams. Then we look at the process of stream erosion and deposition and at the landscapes that form as a consequence of these processes. Finally, we focus on the nature and consequences of flooding and see how people can mitigate flooding risk.

14.2 Draining the Land

FORMING STREAMS AND DRAINAGE NETWORKS

Where does the water in a stream come from? To answer this question, we first need to revisit the hydrologic cycle (see Interlude F). Water enters the hydrologic cycle by evaporating from the Earth's surface and rising into the atmosphere. After a relatively short residence time, atmospheric water

condenses and falls back to the Earth's surface as rain or snow that accumulates in various reservoirs of the Earth System. Some rain or snow remains on the land as surface water (in puddles, swamps, lakes, snowfields, and glaciers), some flows downslope as a thin film called **sheetwash**, and some sinks into the ground, where it either becomes trapped in soil (as *soil moisture*) or descends below the water table to become *groundwater*. (As we discuss further in Chapter 16, the *water table* defines the level below which groundwater fills all the pores and cracks in subsurface rock or sediment. Above the water table, air partially or entirely fills the pores and cracks.) Streams receive water spilling from the outlets of ponds and lakes, seeping out of the base of melting snowfields or glaciers, and from sheetwash trickling down slopes **(Fig. 14.2)**—you can see this input, because it happens at the Earth's surface. Significant inputs to a stream may come from underground. Specifically, the pressure exerted by the weight of new rainfall can squeeze existing soil moisture out of the ground back to the surface and, if the floor of the channel lies below the water table, groundwater can seep out of the channel walls or streambed into the channel. Geologists use the term **runoff** for all water flowing on the surface of the Earth—runoff includes sheetwash plus the water in streams.

How does a stream channel form in the first place? The process of channel initiation begins when sheetwash starts flowing downslope due to gravity. Like any flowing fluid, sheetwash erodes its *substrate*, the material it flows over. The efficiency of such erosion depends on the velocity of the flow—faster flows erode more rapidly. In nature, the ground is not perfectly planar, not all substrate has the same resistance to erosion, and the amount of vegetation that covers and protects the ground varies with location. Consequently,

FIGURE 14.2 Excess surface water (runoff) comes from rain, melting ice or snow, and groundwater springs. On flat ground, water accumulates in puddles or swamps, but on slopes it flows downslope in streams.

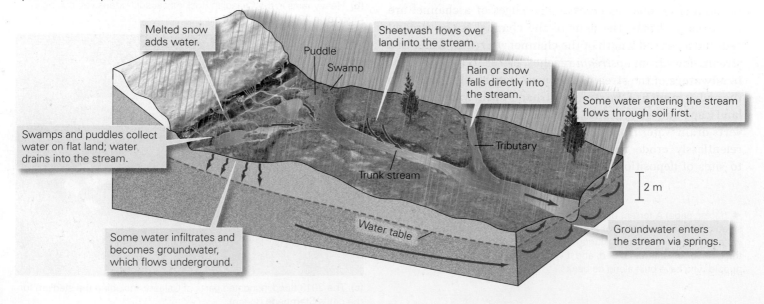

Melted snow adds water.

Puddle

Swamp

Sheetwash flows over land into the stream.

Rain or snow falls directly into the stream.

Some water entering the stream flows through soil first.

Swamps and puddles collect water on flat land; water drains into the stream.

Tributary

Trunk stream

2 m

Some water infiltrates and becomes groundwater, which flows underground.

Water table

Groundwater enters the stream via springs.

FIGURE 14.3 An example of headward erosion. The main stream flows in a deep valley. Side streams are cutting into the bordering plateau.

~1 km

Lower land

Higher plateau

Headward erosion

the velocity of sheetwash also varies with location. Where the flow happens to be a bit faster, or the substrate a little weaker, erosion carves a channel. Since the channel sits lower than the surrounding ground, sheetwash in adjacent areas starts to head toward it. Over time, the extra flow deepens the channel relative to its surroundings, a process called **downcutting**, and a stream forms.

As flow increases and time passes, a stream channel begins to lengthen at its head (origin). This process, called **headward erosion (Fig. 14.3)**, occurs for two reasons. First, it happens when the surface flow converging at the entrance

to a channel has sufficient erosive power to downcut. Second, it happens where groundwater seeps out of the ground and enters the head of the stream channel. Such seepage, called *groundwater sapping*, gradually weakens and undermines the soil or rock just upstream of the channel's head until the material collapses into the channel. Each collapse makes the channel longer. The collapsed debris eventually washes away during a flood.

As downcutting deepens the main channel, the surrounding land surfaces start to slope toward the channel. New side channels, or **tributaries**, begin to form, and these flow into the main channel. Eventually, an array of linked streams evolves, with the smaller tributaries flowing into a *trunk stream*. The array of interconnecting streams together constitute the **drainage network**. Like transportation networks of roads, drainage networks of streams reach into all corners of a region, providing conduits that can remove runoff.

The configuration of tributaries and trunk streams defines the map pattern of a drainage network. This pattern depends on the shape of the landscape and the composition of the substrate. Geologists recognize several types of networks on the basis of the network's map pattern **(Fig. 14.4)**.

> *Dendritic:* When rivers flow over a fairly uniform substrate, they develop a *dendritic network* that looks like the pattern of branches connecting to the trunk of a deciduous tree.

> *Radial:* Drainage networks forming on the surface of a cone-shaped mountain flow outward from the mountain peak, like spokes on a wheel. Such a pattern defines a *radial network*.

FIGURE 14.4 Different types of drainage networks.

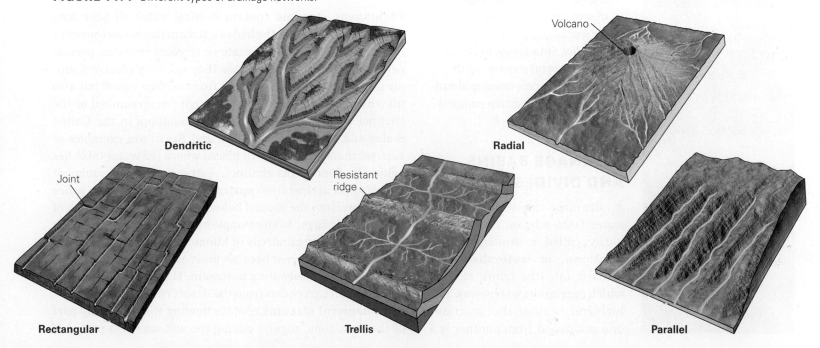

Dendritic

Joint

Rectangular

Resistant ridge

Trellis

Volcano

Radial

Parallel

FIGURE 14.5 Drainage divides and basins.

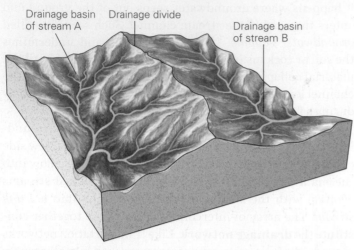

(a) A drainage divide is a relatively high ridge that separates two drainage basins.

> *Rectangular:* In places where a rectangular grid of fractures (vertical joints) breaks up the ground, channels follow pre-existing fractures and streams join each other at right angles, creating a *rectangular network.*
>
> *Trellis:* Where a drainage network develops across a landscape of parallel valleys and ridges, major tributaries flow down a valley and join a trunk stream that cuts across the ridges. The resulting map pattern resembles a garden trellis, so such an arrangement of streams is called a *trellis network.*
>
> *Parallel:* On a steep, uniform slope, several streams with parallel courses develop simultaneously. The group comprises a *parallel network.*

DRAINAGE BASINS AND DIVIDES

A drainage network collects water from a broad region, variously called a *drainage basin, catchment,* or **watershed,** and feeds it into the trunk stream, which carries the water away. The highland, or ridge, that separates one watershed from another is a

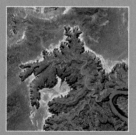

(b) The major drainage basins of North America.

drainage divide (Fig. 14.5). A *continental divide* separates drainage that flows into one ocean from drainage that flows into another. For example, if you straddle the continental divide where it runs along the crest of the Rocky Mountains in the western United States, and you pour a cup of water out of each hand, the water in one hand flows to the Atlantic, and the water in the other flows to the Pacific.

STREAMS THAT LAST, STREAMS THAT DON'T

Permanent streams contain flowing water all year long (Fig. 14.6a, b). Where the bed of a stream lies below the water table, as is typical in temperate or tropical climates, permanent streams can exist because they not only receive a supply of water from upstream and from surface runoff but also fill with groundwater seeping through the streambed or the channel walls. Rivers such as the Mississippi in the United States and the Amazon in Peru and Brazil are examples of such permanent streams. In places where the water table lies below the floor of the channel, a stream can be permanent only if water arrives from upstream more quickly than water seeps down into the ground below. The downstream reach of the Colorado River is an example of such a stream. Although this river crosses hundreds of kilometers of the dry Sonoran Desert, it flows all year because most of its water comes from the river's wet headwaters upstream. Hardly any water that enters the stream comes from the desert runoff.

Ephemeral streams contain flowing water for only part of the year. Some survive during the wet season, a period of

FIGURE 14.6 The contrast between permanent and ephemeral streams.

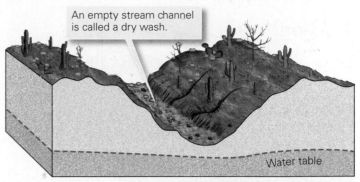

An empty stream channel is called a dry wash.

(a) The channel of a permanent stream in a temperate climate lies below the water table. Springs add water from below, so the stream contains water even between rains.

(c) The channel of an ephemeral stream lies above the water table, so the stream flows only when water enters the stream faster than it can infiltrate into the ground.

(b) An example of a permanent stream in the Wind River Mountains, Wyoming.

(d) An example of a dry wash in the Buckskin Mountains, Arizona.

many months, but dry up during the dry season. Streams whose watersheds lie entirely within desert regions tend to be ephemeral. Such streams may flow for only tens of minutes, or for a few hours, following a heavy rain **(Fig. 14.6c, d)**. The dry bed of an ephemeral stream is called a *dry wash*, an *arroyo*, or a *wadi*.

TAKE-HOME MESSAGE

Stream channels form by downcutting and lengthen by headward erosion. They carry water from sheetwash, lakes, springs, and melting snow or ice. Drainage networks carry water from a watershed to the sea and have a variety of geometries. Permanent streams flow all year, while ephemeral streams flow only intermittently. Stream discharge depends on such factors as watershed area and climate. Water velocity varies across a stream, and tends to be turbulent.

QUICK QUESTION Why do streams develop distinct channels?

14.3 The Work of Running Water

DESCRIBING FLOW IN STREAMS

Imagine two streams—a larger one in which water flows slowly, and a smaller one in which water flows rapidly. Which stream carries more water? The answer isn't obvious. To answer this question completely, geologists must calculate the streams' discharge. Technically speaking, we define **discharge** as the volume of water passing through a cross section of the stream in a given time. We can calculate stream discharge by using a simple formula: $D = A_c \times v_A$. In this formula, A_c is the area of the stream, as measured in an imaginary plane perpendicular to the stream flow, and v_A is the average velocity at which water moves in the downstream direction. For example, if a stream has a cross-sectional area of 100 m², and the water in the stream flows at an average velocity of 0.2 m per second, then discharge $D = 100$ m² $\times$ 0.2 m per

FIGURE 14.7 Flow velocity and turbulence in streams.

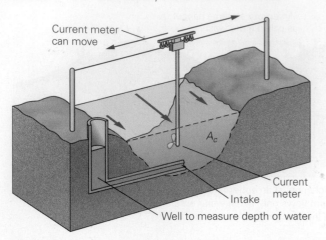

(a) At a stream-gauging station, geologists measure the cross-sectional area (A_c, the area within the dashed line), the depth, and the average velocity of the stream. Note that water flows more slowly near the banks.

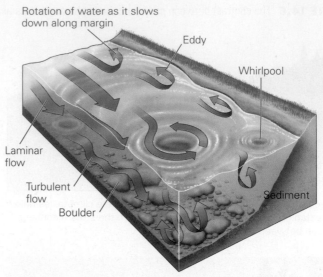

(b) Water in a stream doesn't usually follow a straight path. It swirls and twists, producing turbulence.

second = 20 m³ per second. Note that the discharge is specified in units of volume per second. Stream discharge can be determined at a *stream-gauging station*, where instruments measure the velocity and depth of the water at several points across the stream **(Fig. 14.7a)**.

The average velocity of stream water can be difficult to calculate because the water doesn't all travel at the same velocity, due to friction and turbulence. Specifically, friction along the sides and floor of the stream slows the flow, so water near the channel walls or the streambed moves more slowly than water in the middle of the flow (see Fig. 14.7a). *Turbulence*—the twisting, swirling motion of a fluid—can produce eddies in which water curves and flows upstream or circles in place **(Fig. 14.7b)**. Turbulence develops because the shearing motion of one water volume against its neighbor causes the neighbor to spin, and because obstacles such as boulders deflect water flow.

A stream's average discharge reflects the size of its drainage basin and the climate. The Amazon River drains a huge rainforest and has the largest average discharge in the world—about 200,000 m³ per second, or 15% of the total amount of runoff on the Earth. In contrast, the "mighty" Mississippi River's discharge is only 17,000 m³ per second. The discharge of a given stream varies along its length. For example, discharge in a temperate region increases in the downstream direction, because each tributary that enters the stream adds more water. In contrast, the discharge in an arid region decreases downstream, because progressively more water seeps into the ground or evaporates. Human activity can affect discharge—for example, if people divert the river's water for irrigation, the river's discharge decreases. Finally, the discharge at a given location can vary with time. For example, in a temperate climate, a stream's discharge

during the spring may be double or triple the amount during a dry summer, and a flood may increase the discharge to more than a hundred times normal.

HOW DO STREAMS ERODE?

The energy that makes running water move comes from gravity. As water flows downslope from a higher to a lower elevation, the gravitational potential energy stored in water transforms into kinetic energy, the energy of motion. About 3% of this kinetic energy goes into the work of eroding the walls and beds of stream channels. Running water causes erosion in four ways:

> *Scouring:* Running water can remove and carry away loose fragments of sediment, a process called **scouring**.
> *Breaking and lifting:* In some cases, the push of flowing water can break chunks of solid rock off the channel bed or walls. In addition, the flow of a current over a clast can cause the clast to rise, or lift off the substrate.
> *Abrasion:* Clean water has little erosive effect, but sediment-laden water acts like sandpaper and grinds or rasps away at the channel bed and walls, a process called *abrasion*. In places where turbulence produces long-lived tornado-shaped whirlpools, abrasion by sand or gravel carves a bowl-shaped depression, called a **pothole**, into the bed of the stream **(Fig. 14.8a, b)**.
> *Dissolution:* Running water dissolves soluble minerals, and it carries the minerals away in solution.

The efficiency of erosion depends on the velocity and volume of water and on its sediment content. A large volume of fast-moving, turbulent, sandy water causes much more erosion than does a trickle of quiet, clear water. Therefore, most erosion takes place during floods, when a stream carries a large volume of fast-moving, sediment-laden water.

FIGURE 14.8 Erosion and transportation in streams.

(a) A pothole in the bed of a stream near Ithaca, New York.

(b) This slot canyon in Arizona formed when many potholes were linked together.

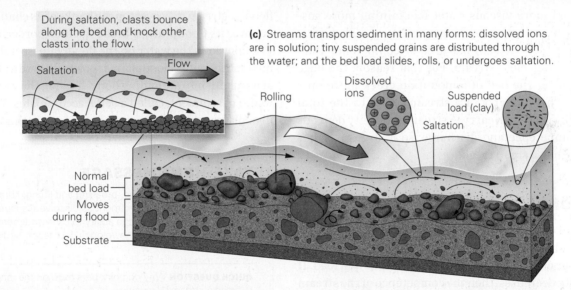

During saltation, clasts bounce along the bed and knock other clasts into the flow.

Saltation   Flow

(c) Streams transport sediment in many forms: dissolved ions are in solution; tiny suspended grains are distributed through the water; and the bed load slides, rolls, or undergoes saltation.

Rolling

Dissolved ions

Suspended load (clay)

Saltation

Normal bed load

Moves during flood

Substrate

HOW DO STREAMS TRANSPORT SEDIMENT?

The Mississippi River received the nickname Big Muddy for a reason—its water can become chocolate brown because of all the clay and silt it carries. Geologists refer to the total volume of sediment carried by a stream as its *sediment load*. The sediment load consists of three components **(Fig. 14.8c):**

> *Dissolved load:* Running water dissolves soluble minerals in the sediment or rock that it flows over, and groundwater seeping into a stream brings dissolved minerals with it. The ions of these dissolved minerals constitute a stream's **dissolved load**.

> *Suspended load:* The **suspended load** of a stream usually consists of tiny solid grains (silt or clay) that swirl along with the water without settling to the streambed.

> *Bed load:* The **bed load** of a stream consists of large particles (such as sand, pebbles, or cobbles) that bounce or

roll along the streambed. Bed-load movement commonly involves **saltation**, in which a multitude of grains bounce along in the direction of flow, within a zone that extends upward from the surface of the streambed for a distance of several centimeters to several tens of centimeters. Each saltating grain in this zone follows a curved trajectory up through the water and then back down to the bed. When it strikes the bed, it knocks other grains upward, and by doing so, supplies more grains to the saltation zone.

When describing a stream's ability to carry sediment, geologists distinguish between competence and capacity. The **competence** of a stream refers to the maximum particle size it carries—a stream with high competence can carry large particles, whereas one with low competence can carry only small particles. Competence depends on both the speed and the viscosity of the water. Faster water applies more force

FIGURE 14.9 Sediment, carried and deposited by streams. The clast size depends on stream velocity.

(a) Gravel in the bed of a mountain stream in Denali National Park, Alaska. The large clasts were carried during floods. This stream has high competence.

(b) Point bars of mud deposited along a gentle, slowly moving stream in Brazil. This stream has low competence.

to particles, and more viscous water (containing more suspended sediment) can buoy up heavier particles. Therefore, a fast-moving, muddy stream has greater competence than does a slow-moving, clear stream. For that reason, the huge boulders that litter the bed of a mountain creek move only during floods. The **capacity** of a stream refers to the total quantity of sediment it can carry. A stream's capacity depends on both its competence and discharge, so a large river has more capacity than a small creek.

DEPOSITIONAL PROCESSES

A raging torrent of water can carry coarse and fine sediment—finer clasts rush along with the water as suspended load, whereas coarser clasts may bounce and tumble as bed load. If the flow velocity decreases, then the competence of the stream decreases and sediment settles out. The size of the clasts that settle at a particular locality depends on the decrease in flow velocity at the locality. For example, if the stream slows by a small amount, only large clasts settle **(Fig. 14.9a)**; if the stream slows by a greater amount, medium-sized clasts settle; and if the stream slows almost to a standstill, the fine grains settle. Because of this process of *sediment sorting*, stream deposits tend to be segregated by size—gravel accumulates in one location and mud and silt in another.

Geologists refer to sediments transported by a stream as *fluvial deposits* (from the Latin *fluvius*, meaning river) or **alluvium**. Alluvium may accumulate along the streambed in elongate mounds, called **bars**. Some stream channels follow broad curves, which, as we'll see, are called *meanders*. Water slows along the inner edge of a meander, so crescent-shaped **point bars** develop along the inner edge **(Fig. 14.9b)**. During

floods, a stream may overtop the banks of its channel and spread out over its **floodplain**, a broad flat area bordering the stream. Friction slows the water on the floodplain, so a sheet of silt and mud settles out as floodplain deposits. Where a stream empties at its mouth into a standing body of water, the water slows and a wedge of sediment, called a *delta*, accumulates. We will discuss floodplains and deltas in more detail later in this chapter.

> **TAKE-HOME MESSAGE**
>
> Streams erode by scouring, breaking and lifting, abrasion, and dissolution. They carry sediment as dissolved, suspended, or bed loads. Competence, the ability to carry sediment, depends on flow velocity. Where the velocity decreases, sediment settles out.
>
> **QUICK QUESTION** Why do point bars form on the inner edge of a curve in a stream?

14.4 **Streams in the Landscape**

HOW DO STREAMS CHANGE ALONG THEIR LENGTH?

In 1803, under President Thomas Jefferson's leadership, the United States bought the Louisiana Territory, a vast tract of land encompassing the western half of the Mississippi drainage basin. At the time, the geography of the territory was a mystery. To fill the blank on the map, Jefferson asked Meriwether Lewis and William Clark to lead a voyage of exploration across the Louisiana Territory to the Pacific.

FIGURE 14.10 Change in the character of a stream along its longitudinal profile.

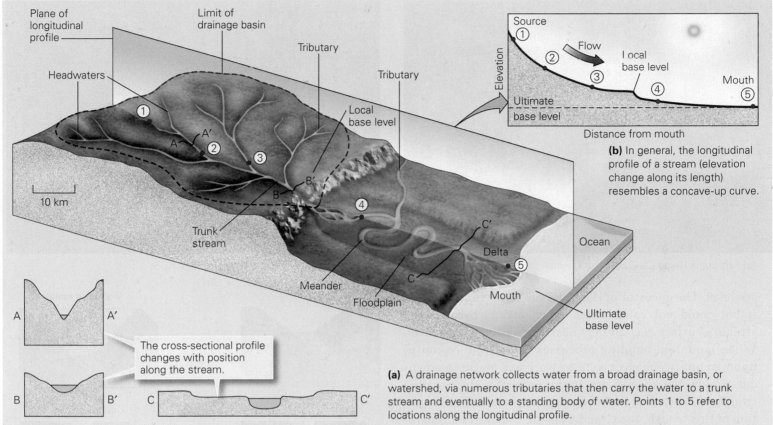

(b) In general, the longitudinal profile of a stream (elevation change along its length) resembles a concave-up curve.

The cross-sectional profile changes with position along the stream.

(a) A drainage network collects water from a broad drainage basin, or watershed, via numerous tributaries that then carry the water to a trunk stream and eventually to a standing body of water. Points 1 to 5 refer to locations along the longitudinal profile.

Lewis and Clark, along with a crew of 40, began their expedition at the mouth of the Missouri River, where it joins the Mississippi. At this juncture, the Missouri is a wide, languid stream of muddy water. The group found that along the Missouri's downstream reach, where the river's channel is deep and the water smooth, progress was easy. But the farther upstream they went, the more difficult their voyage became, for the **stream gradient**, the slope of the stream channel, became progressively steeper, and the stream's discharge decreased. When Lewis and Clark reached the site of what is now Bismarck, North Dakota, they had to abandon their original boats and haul smaller vessels up *rapids*, where turbulent water plunges over a steep, bouldery bed, and occasionally they had to carry their boats around *waterfalls*, where water free-falls through the air. Eventually, they abandoned these boats as well and trudged along the stream valley on foot or on horseback, struggling up steep gradients until they reached the continental divide.

If Lewis and Clark had been able to plot a graph showing their elevation above sea level relative to their distance from the river's mouth, they would have found that the **longitudinal profile** of the Missouri, a cross-sectional image showing the variation in the river's elevation along its length, is roughly a concave-up curve **(Fig. 14.10)**. A stream with this typical concave-up profile has a steep gradient near its headwaters, but flows over an almost horizontal plain near its mouth. As a result, such streams typically cut deep valleys or canyons near their headwaters, but not near their mouths.

The lowest elevation that a stream channel's surface can reach at a locality is the **base level** of the stream. A *local base level* occurs at a location upstream of a stream's mouth, and the *ultimate base level*, the lowest possible elevation along the stream's longitudinal profile, lies at sea level. Why? The surface of a stream cannot be lower than the body of water that it flows into, for if it were, the stream would have to flow upslope. In a drainage network, the surface of a larger stream acts as the base level for the tributaries that flow into it. Similarly, lakes and reservoirs can act as local base levels along a stream, as can a ledge of resistant rock, for the stream level cannot drop below the ledge until the ledge erodes away.

VALLEYS AND CANYONS

Millions of years ago, the region now known as the Colorado Plateau (which covers parts of Arizona, Utah, Colorado, and New Mexico) was a plain whose surface lay close to

FIGURE 14.11 The formation of valleys and canyons.

(a) Part of the Grand Canyon, Arizona, a stair-step canyon.

(b) V-shaped valleys in the Andes of Peru.

sea level. The ancestor of the Colorado River flowed across it but could not cause much downcutting, because the stream's surface was already near the ultimate base level. When mountain-building processes caused the region to undergo uplift, the river went to work and began to downcut. Today, the river flows at the base of a steep-walled trough, 1.6 km below the surface of the plateau. The formation of this trough, the Grand Canyon, illustrates a general phenomenon—if a river flows at an elevation high above its local base level, it can carve a trough that is much deeper than the channel itself. A trough bordered by steep slopes is a **canyon**, whereas one bordered by gentler slopes is a **valley** (Fig. 14.11a, b).

Whether stream erosion produces a valley or a canyon depends on the rate at which downcutting occurs relative to the rate at which mass movement causes the walls on either side of the stream to collapse (**Fig. 14.11c**). In places where a stream downcuts through its substrate faster than the walls of the stream collapse or slump, erosion produces a steep-walled canyon. Such canyons typically form in hard rock, which can hold up cliffs for a long time (see Fig. 14.11a). In places where the walls collapse as fast as the stream downcuts, landslides and slumps gradually cause the slope of the walls to approach the angle of repose (see Chapter 13). When this happens, the stream channel lies at the floor of a valley whose cross-sectional shape resembles the letter V. Not surprisingly, geologists refer to this landform as a *V-shaped valley*. Where the walls of the stream consist of alternating layers of hard and soft rock, the walls develop a stair-step shape such as that of the Grand Canyon.

In places where active downcutting takes place, the valley floor remains relatively clear of sediment, for the stream—especially when it floods—carries away sediment that has

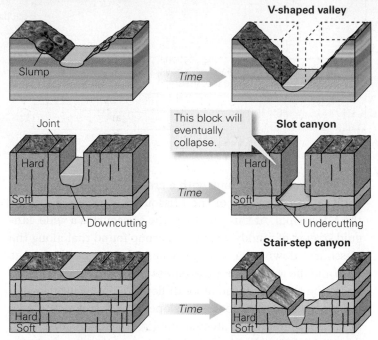

(c) The shape of a canyon or valley depends on the relative resistance of walls to erosion, and the rate of downcutting relative to the rate of mass movement.

fallen or slumped into the channel from the valley or canyon walls. But if the stream's base level rises, its discharge decreases, or its sediment load increases, the valley floor may fill with sediment, creating an *alluvium-filled valley*. The surface of the alluvium becomes a broad floodplain. If the stream's base level later drops again or the discharge increases, the stream will start to cut down into its own alluvium, a process that generates **stream terraces** bordering the present floodplain (**Fig. 14.12**).

FIGURE 14.12 A rise in base level or a decrease in discharge can cause a valley to fill with alluvium. If base level drops or discharge increases, the stream will cut down into the alluvium, leaving terraces.

FIGURE 14.13 Examples of rapids and waterfalls.

(a) These rapids in the Grand Canyon formed when a flood from a side canyon dumped debris into the channel of the Colorado River.

RAPIDS AND WATERFALLS

When Lewis and Clark forged a path up the Missouri River, they came to reaches that could not be navigated by boat because of **rapids**, particularly turbulent water with a rough surface **(Fig. 14.13a)**. Rapids form where water flows over steps or large clasts in the streambed, where the channel abruptly narrows, or where its gradient abruptly changes. The turbulence in rapids produces eddies, waves, and whirlpools that roil and churn the water surface and produce *whitewater*, a mixture of bubbles and water. Modern-day whitewater rafters thrill to the unpredictable movement of rapids.

A **waterfall** forms where the gradient of a stream becomes so steep that some or all of the water free-falls above the streambed **(Fig. 14.13b)**. The energy of falling water may scour a depression, called a *plunge pool*, at the base of the waterfall. Typically, waterfalls form where a river flows over a resistant ledge of rock. Though a waterfall may appear to be a permanent feature of the landscape, all waterfalls eventually disappear because headward erosion slowly eats back the resistant ledge. We can see a classic example of headward erosion at Niagara Falls, where water flowing from

(b) Iguaçu Falls, at the Brazil-Argentina border, spills across layers of basalt. The basalt acts as a resistant ledge.

Lake Erie to Lake Ontario drops over a 55-m-high ledge of resistant Silurian dolostone that overlies weak shale. Over time, turbulence in the plunge pool at the base of the waterfall erodes the shale, causing undercutting of the dolostone. Eventually, the overhang of dolostone becomes unstable, breaks along joints, and collapses, with the result that the location of the waterfall migrates upstream **(Fig. 14.14)**.

Did you ever wonder . . .
why a waterfall forms and whether it will always be there?

FIGURE 14.14 The formation of Niagara Falls, at the border between Ontario, Canada, and New York State. The falls tumble over the Lockport Dolomite, a relatively strong rock layer.

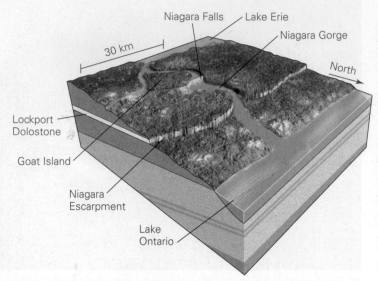

(a) Niagara Falls formed where the outlet of Lake Erie flowed over the Niagara escarpment.

(b) Horseshoe Falls, on the Canadian side of the border.

ALLUVIAL FANS AND BRAIDED STREAMS

Where a fast-moving ephemeral stream abruptly emerges from a mountain canyon onto an open plain, water that had been confined to a narrow channel spreads over a broader surface, slows, and deposits a lens of sediment. Eventually, the accumulation of sediment decreases the gradient at the mouth of the outlet that it came from, so the next time the stream flows, it follows a different, steeper path, to one side of the previous lens. Each new flow event deposits its load in a remaining lower area. Over time, therefore, deposition episodes build a broad, gently sloping, wedge-shaped apron of sediment called an **alluvial fan** (Fig. 14.15a).

In some localities, streams carry abundant coarse sediment during floods, but they cannot carry this sediment during normal flow. As a result, during normal flow, the sediment settles out and chokes the channel. Because the gravelly sediment can't stick together, the stream cannot cut a single deep channel with steep banks—the channel walls simply collapse. As a consequence, the stream divides into numerous strands weaving back and forth between elongate bars of gravel and sand. The result is a **braided stream**—the name emphasizes that the strands entwine like hair in a braid (Fig. 14.15b).

FIGURE 14.15 Examples of depositional landforms produced from stream sediment.

(a) An alluvial fan in Death Valley, California, consists of sand, gravel, and debris flows. The curving black line is a road.

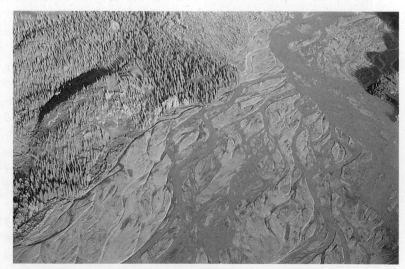

(b) A braided stream, carrying meltwater from a glacier near Denali, Alaska, deposits elongate bars of gravel.

FIGURE 14.16 The character and evolution of meandering streams and floodplains.

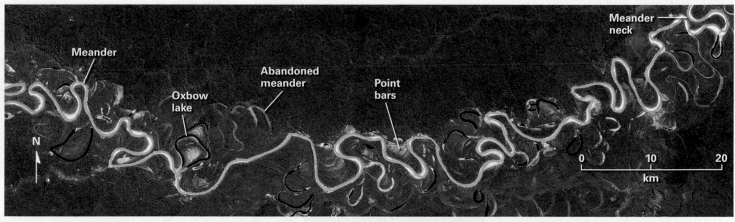

(a) A meandering stream in Brazil, as viewed from space. Note the oxbows, cutoffs, and abandoned meanders.

Time

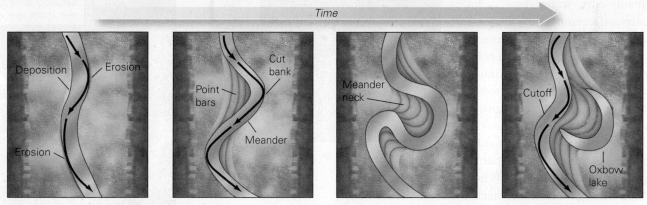

(b) Meanders evolve because erosion occurs faster on the outer edge of a curve, and deposition takes place on the inner curve. Eventually, a cutoff isolates an oxbow lake.

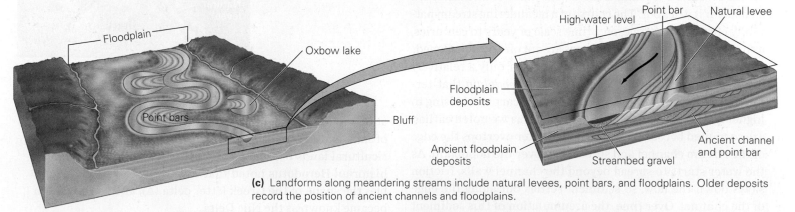

(c) Landforms along meandering streams include natural levees, point bars, and floodplains. Older deposits record the position of ancient channels and floodplains.

MEANDERING STREAMS AND THEIR FLOODPLAINS

A riverboat cruising along the lower reaches of the Mississippi River cannot travel in a straight line, for the river channel winds back and forth in a series of snake-like curves, each of which is called a **meander (Fig. 14.16a).** In fact, the boat may have to go 500 km along the river channel to travel 100 km as the crow flies. Geologists refer to a stream with many meanders as a **meandering stream**—typically, meandering streams exist where the stream's gradient is very gentle.

(d) The flat land of this floodplain hosts farm fields.

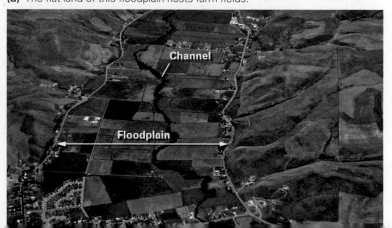

How do meanders evolve? Even if a stream starts out with a straight channel, natural variations in water depth will cause the fastest-moving part of the current, called the *thalweg*, to swing back and forth. The water erodes the side of the stream more effectively where it flows faster, so it begins to cut away faster on the outer arc of the curve. Thus, each curve begins to migrate sideways and grow more pronounced until it becomes a meander **(Fig. 14.16b)**. On the outside edge of a meander, erosion continues to eat away at the channel wall, forming a *cut bank*. On the inside edge, water slows down and the stream's competence decreases, so sediment accumulates, forming a crescent-shaped wedge called a *point bar*, as we noted earlier. With continued erosion, a meander may curve through more than 180°, so that the cut bank at the meander's entrance approaches the cut bank at its end, leaving a *meander neck*, a narrow isthmus of land separating the portions of the meander. When erosion eats through a meander neck, a straight reach called a *cutoff* develops. The meander that has been cut off becomes an **oxbow lake** if it remains filled with water or an *abandoned meander* if it dries out. The course of a meandering stream naturally changes over time, on a time scale of years to centuries, as new meanders grow and old ones are cut off and abandoned.

Most meandering stream channels cover only a relatively small portion of a broad, gently sloping *floodplain* that terminates at its sides along a *bluff*, a small escarpment rising to higher land **(Fig. 14.16c, d)**. Floodplains, as we noted earlier, are so named because during a flood, water overtops the edge of the stream channel and spreads out over the floodplain. As the water starts to spread beyond the channel walls, friction slows down the flow, so sediment settles out along the edge of the channel. Over time, the accumulation of this sediment produces a pair of low ridges, called **natural levees**, on either side of the stream. Natural levees may grow so large that the floor of the stream's channel may actually become higher than the surface of the adjacent floodplain.

DELTAS: DEPOSITION AT THE MOUTH OF A STREAM

Along most of its length, only a narrow floodplain—covered by irrigated farm fields—borders the Nile River in Egypt. But

FIGURE 14.17 Delta shape varies depending on current activity, waves, and vegetation.

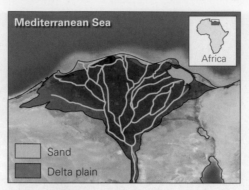

(a) The Nile Delta is a Δ-shaped delta.

(b) The Niger Delta is an arc-like delta.

(c) The Mississippi Delta is a bird's-foot delta.

at its mouth, the trunk stream of the Nile divides into a fan of smaller streams, called **distributaries**, and the area of agricultural lands broadens into a triangular patch. The Greek historian Herodotus noted that this triangular patch resembles the shape of the Greek letter delta (Δ), and so the region became known as the Nile Delta.

Geologists adopted the name **delta** for a wedge of sediment formed where a flowing stream enters standing water. At this intersection, the stream's flow slows, and therefore, sediment settles out. Deltas come in many shapes. Those with the classic Δ shape form where waves or offshore currents redistribute sediment entering the standing water. Where currents are stronger, less sediment accumulates so the edge of the delta is straight, and where currents are weaker, more sediment accumulates so the edge of the delta arcs outward **(Fig. 14.17a, b)**. Deltas that form where the strength of the river current

FIGURE 14.18 A map showing ancient lobes of the Mississippi Delta. A major flood could divert water from the Mississippi into the channel of the Atchafalaya.

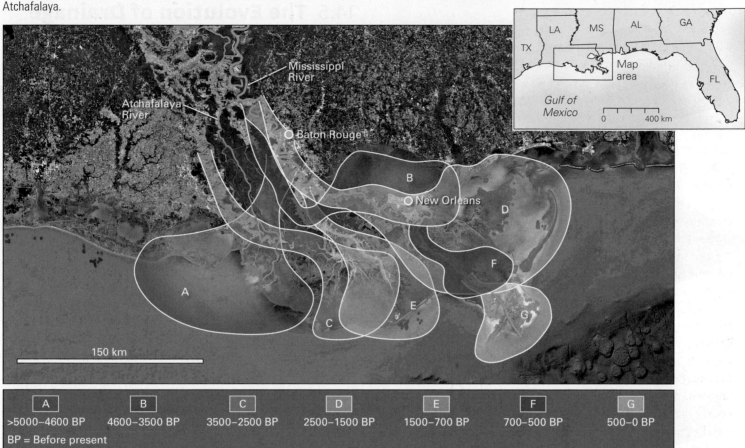

A	B	C	D	E	F	G
>5000–4600 BP	4600–3500 BP	3500–2500 BP	2500–1500 BP	1500–700 BP	700–500 BP	500–0 BP

BP = Before present

exceeds that of ocean currents are called *bird's-foot deltas*, because they resemble the scrawny toes of a bird—they develop where several distributaries extend far out into relatively calm water **(Fig. 14.17c)**.

Large deltas may consist of several distinct parts, or *lobes*, each of which formed during a different time interval. A stream builds a lobe by following a given course a period of time. As the lobe grows seaward, the overall gradient of the stream as it crosses the delta decreases. When the gradient becomes too gentle for the stream to flow, the river overtops a natural levee upstream and begins to follow a new course, with a steeper gradient. Geologists refer to such events as *avulsions*. The existence of several distinct lobes in the Mississippi Delta indicates that avulsions have happened several times during the past 9,000 years **(Fig. 14.18)**. New Orleans, built along the Mississippi, may eventually lose its riverfront, for a break in a levee upstream of the city could divert the Mississippi into the Atchafalaya River channel.

With time, the sediment of a large delta compacts, and the weight of the delta pushes down the crust below. As a consequence, the surface of a delta slowly sinks. Distributaries can

provide sediment that fills the resulting space so that the delta's surface remains at or just above sea level, forming a broad, flat area called a *delta plain*. If people build up artificial levees to keep the river within its channel, sediment gets carried directly to the seaward edge of the delta and the delta's interior "starves" (does not receive sediment). When this happens, the delta's surface drops below sea level.

TAKE-HOME MESSAGE

Stream erosion and deposition yield distinct landforms. Streams have steeper regional gradients in their headwaters and carve canyons or valleys. Gradient decreases toward a stream's mouth—the mouth can't be lower than a base level. Streams choked with sediment become braided, whereas those following snake-like paths are meandering. Alluvial fans build out at the mouth of canyons in deserts; deltas build out into standing water.

QUICK QUESTION What factors cause the formation of rapids and waterfalls?

FIGURE 14.19 Fluvial landscapes evolve over time.

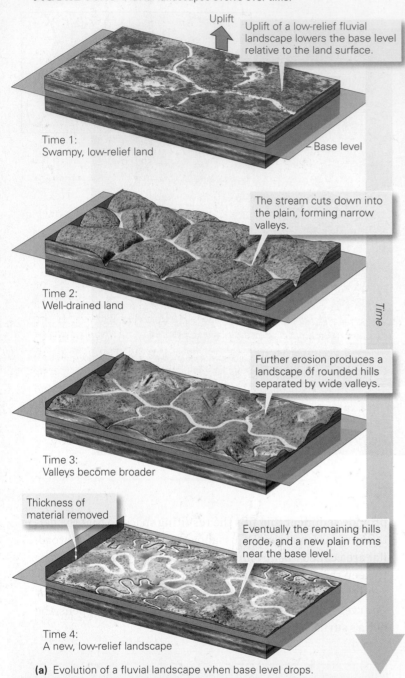

Uplift

Uplift of a low-relief fluvial landscape lowers the base level relative to the land surface.

Time 1:
Swampy, low-relief land

Base level

The stream cuts down into the plain, forming narrow valleys.

Time 2:
Well-drained land

Time

Further erosion produces a landscape of rounded hills separated by wide valleys.

Time 3:
Valleys become broader

Thickness of material removed

Eventually the remaining hills erode, and a new plain forms near the base level.

Time 4:
A new, low-relief landscape

(a) Evolution of a fluvial landscape when base level drops.

(b) The "goosenecks" of the San Juan River, Utah, are incised meanders.

14.5 The Evolution of Drainage

BEVELING TOPOGRAPHY

Imagine a place where continental collision uplifts a region **(Fig. 14.19a)**. At first, in the highlands, rivers have steep gradients, flow over many rapids and waterfalls, and cut deep valleys. Over time, the landscape evolves. When erosion rates exceed uplift rates, the rugged mountains become low, rounded hills, and the once-deep, narrow valleys broaden into wide floodplains with gentler gradients. As more time passes, erosion bevels even the low hills, and the region becomes nearly planar again, lying at an elevation close to that of a stream's base level. (Some geologists have referred to the resulting landscape as a *peneplain*, from the Latin *paene*, which means almost.)

Although this model makes intuitive sense, it is an oversimplification. Plate tectonics can uplift the land again, or global sea-level rise or fall can change the base level, so in reality peneplains rarely develop. **Stream rejuvenation** occurs when a stream starts to downcut into a land surface whose elevation had previously been close to the stream's base level. Rejuvenation can be triggered by a drop in base level, as happens when sea level falls; an uplift event that causes the land to rise relative to the base level; or an increase in stream discharge that makes the stream more able to erode and transport sediment. As we've seen, rejuvenation can lead to formation of stream terraces in alluvium. In cases where rejuvenation causes a stream to erode deeply into bedrock, a new canyon or valley will develop. If the rejuvenated stream had a meandering course, downcutting produces *incised meanders* **(Fig. 14.19b)**.

STREAM PIRACY AND DRAINAGE REVERSAL

Stream piracy sounds like pretty violent stuff. In reality, it's just a natural process that happens when headward erosion by one stream causes the stream to intersect the course of another stream. When this happens, the pirate stream "captures" the water of the stream that it has intersected, so that the water of the captured stream starts to flow down the pirate stream. The channel of the captured stream, downstream of the point of capture, therefore dries up **(Fig. 14.20)**. In some cases, stream capture causes a *water gap* (a stream-carved notch through a ridge) to become a dry *wind gap*. In 1775, Daniel Boone, the legendary pioneer, led settlers through the Cumberland Gap, a wind gap in the Appalachians, to new homesteads in western Kentucky.

The pattern of stream flow in an area can also be altered, over time, on a continental scale. For example, when South

FIGURE 14.20 The concept of stream piracy.

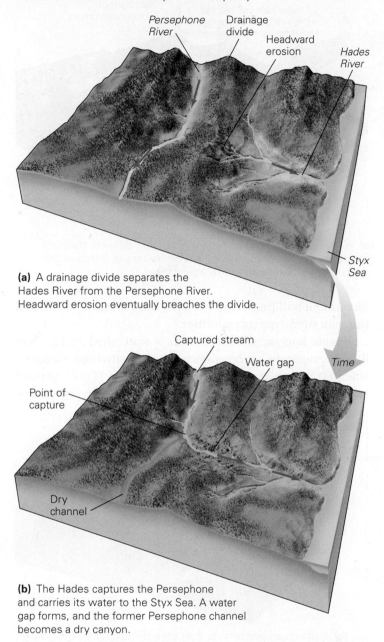

Persephone River — Drainage divide — Headward erosion — Hades River — Styx Sea

(a) A drainage divide separates the Hades River from the Persephone River. Headward erosion eventually breaches the divide.

Captured stream — Water gap — Time
Point of capture
Dry channel

(b) The Hades captures the Persephone and carries its water to the Styx Sea. A water gap forms, and the former Persephone channel becomes a dry canyon.

America and Africa were adjacent to each other in Pangaea, a highland existed between what would become two continents, and the main drainage network of northern South America flowed westward. Later, when South America rifted away from Africa, a convergent boundary developed along the western margin of the South American Plate, causing the Andes Mountains to rise. The uplift of the Andes caused a **drainage reversal**, in that the overall slope direction of the drainage network became the opposite of what it once had been. As a consequence, westward flow became impossible and the eastward-flowing Amazon drainage network developed.

SUPERPOSED AND ANTECEDENT STREAMS

The structure and topography of the landscape do not always appear to control the path, or course, of a stream. For example, consider a stream that carves a deep canyon straight across a strong mountain ridge—why didn't the stream find a way around the ridge? We distinguish two types of streams that cut across resistant topographic highs.

Imagine a region in which a drainage network initially forms on a layer of flat, nonresistant strata that unconformably overlies folded strata. Initially, the streams carve valleys into the flat strata. When they eventually erode down through the unconformity and start to downcut into the folded strata, they may maintain their earlier course, ignoring the structure of the folded strata. Geologists refer to streams whose pre-existing geometry has been laid down on underlying rock structure as *superposed streams* (**Fig. 14.21**).

In some cases, tectonic activity, such as subduction or collision, causes localized uplift, so a mountain range rises beneath the course of an already established stream. If the stream downcuts as fast as the range rises, it can maintain its course and will cut right across the range. Geologists call such streams *antecedent streams* (from the Greek *ante*, meaning before) to emphasize that they existed before the range uplifted. Note that if the range rises faster than the stream can downcut, the new highlands divert the stream's course, and the *diverted stream* starts to flow parallel to the range face (**Fig. 14.22**).

TAKE-HOME MESSAGE

Stream-carved landscapes evolve over time as gradients diminish and ridges between valleys erode away. Superposed streams attain their shape before cutting down into rock structure, whereas antecedent streams cut while the land beneath them uplifts.

QUICK QUESTION What can cause a drainage-reversal event to take place?

14.6 **Raging Waters**

THE INEVITABLE CATASTROPHE

Up until now, this chapter has focused on the process of drainage formation and evolution and on the variety of landscape features formed by streams (**Geology at a Glance**, pp. 490–491). Now we turn our attention to the havoc that streams can cause when flooding takes place. A flood occurs when the volume of water flowing down a stream exceeds the volume of the stream channel, so water spills out of the

FIGURE 14.21 Formation of superposed drainage.

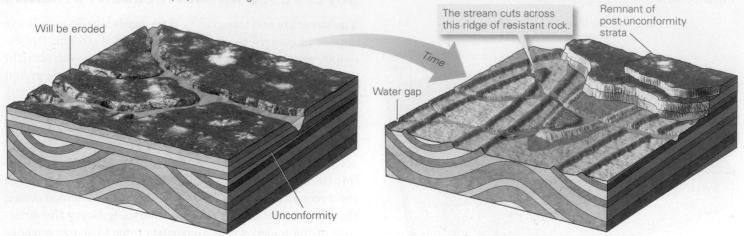

(a) A superposed stream establishes its geometry while flowing over a uniform substrate above an unconformity.

(b) When erosion exposes underlying rock with a different structure, the stream is superposed on the structure. As a result, it cuts across resistant ridges instead of flowing around them.

normal channel and either spreads out over a floodplain or delta plain, or rises above the lip of the normal channel and fills a canyon to a greater depth than normal. We refer to the water level just before flooding begins as the *flood stage*, and the maximum level of water above the flood stage as the *flood crest*. Floods can be catastrophic—they can strip land of

forests and buildings, they can bury land in clay and silt, and they can submerge communities.

Floods happen if water enters a watershed faster than it can evaporate or be absorbed. This situation develops when (1) very heavy downpours take place; (2) the ground is already saturated with water, with no room to absorb any

FIGURE 14.22 Development of antecedent and diverted streams.

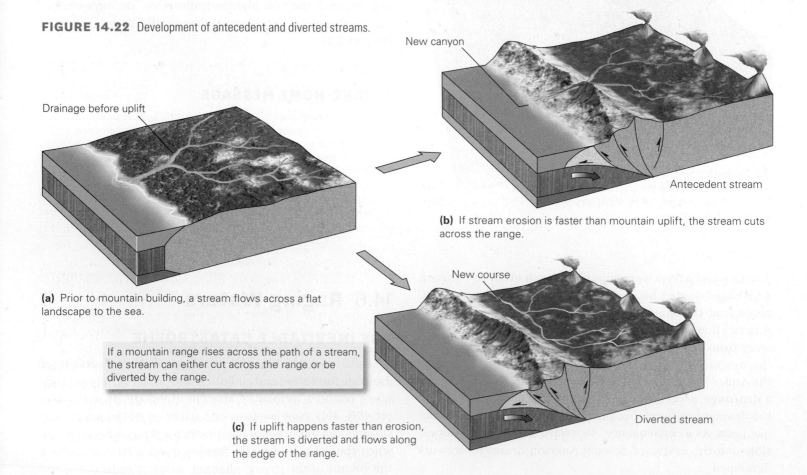

(a) Prior to mountain building, a stream flows across a flat landscape to the sea.

If a mountain range rises across the path of a stream, the stream can either cut across the range or be diverted by the range.

(b) If stream erosion is faster than mountain uplift, the stream cuts across the range.

(c) If uplift happens faster than erosion, the stream is diverted and flows along the edge of the range.

FIGURE 14.23 Examples of slow-onset floods, in a floodplain.

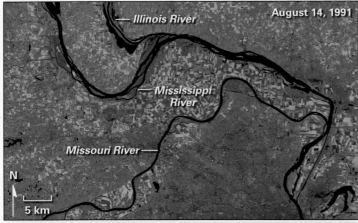

Illinois River

Mississippi River

Missouri River

N

5 km

August 14, 1991

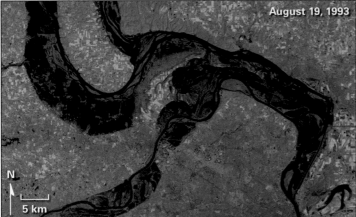

N

5 km

August 19, 1993

(a) Satellite images show how the rivers in the midwestern United States can cover their floodplains. The top image shows the rivers at relatively normal water levels for comparison.

(b) Flooding along the Missouri River near Cedar City, Missouri, in 1993.

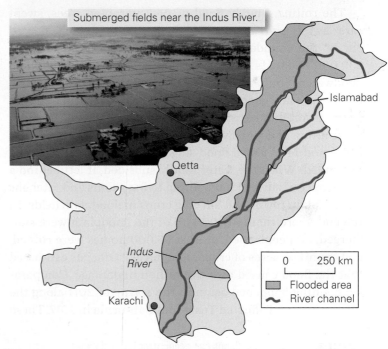

Submerged fields near the Indus River.

Islamabad

Qetta

Indus River

Karachi

0	250 km

Flooded area
River channel

(c) Flooding in Pakistan during 2010 covered 62,000 km².

more rainfall; (3) heavy snows from the previous winter melt rapidly in response to a sudden hot spell or heavy rain; or (4) a dam holding back a lake or reservoir, or a levee holding back a river or canal, suddenly collapses and releases the water that it held back. Geologists find it convenient to divide floods into two general categories: slow-onset floods and flash floods. Let's consider these in turn.

SLOW-ONSET FLOODS

When you read a news story about flooding that takes days to develop and lasts for days or even weeks, you're reading about a **slow-onset flood (Fig. 14.23a)**. Such floods can happen (1) in response to rapid melting of winter snow and ice, supplemented by spring rains; (2) during the sustained rains of a distinct wet season, such as the *monsoon season* of tropical regions when winds blow moist air from the oceans over the land for weeks on end; or (3) when a system of storms remains stationary over a broad region for a long time. In all these circumstances, the ground becomes saturated, so meltwater or rainfall becomes runoff that fills stream channels beyond their capacity.

Slow-onset floods that happen during specific times of the year are also known as **seasonal floods**. If the flood spreads out over a stream's floodplain, it's also a *floodplain flood*, whereas if the water spreads out over a delta plain, it's also a *delta-plain flood*. Along many rivers, slow-onset flooding plays an important role in the sustainability of ecosystems and societies. Seasonal floods along the Nile River traditionally played a major role in replenishing nutrients to agricultural areas of the Nile's flood plain and delta.

Since slow-onset floods take time—hours or days—to develop, authorities have time to evacuate potential victims and organize efforts to protect property. But because preparation or evacuation doesn't always take place, and because so many people live on floodplains and delta plains, these floods can cause a staggering loss of life and property.

For example, slow-onset flooding happens frequently in the Mississippi Valley of the central United States, and in some years, the floods are disastrous. The region's 1993 flood can serve as a case study of such an event. High-altitude (10- to 15-km-high) winds drifted southward and for weeks formed an invisible wall that trapped warm, moist air from the Gulf of Mexico over the central United States. As this air rose to higher elevations and cooled, the water it held condensed and fell as rain, rain, and more rain. In fact, almost a whole year's supply of rain fell during the spring of 1993, and some regions received 400% more than usual. Eventually, the water in the Missouri and Mississippi Rivers rose above the height of their levees and spread out over the floodplain **(Fig. 14.23b)**. By July, parts of nine states were underwater.

The roiling, muddy flood of 1993 uprooted trees, swept cars away, and even unearthed coffins (which floated out of the inundated graveyards). In Des Moines, Iowa, 250,000 residents lost their supply of drinking water when floodwaters contaminated the municipal water supply with raw sewage and chemical fertilizers. Rowboats replaced cars as the favored mode of transportation in towns where only the rooftops remained visible. In St. Louis, Missouri, the river crested 14 m above flood stage. For 79 days, the flooding continued. When the water finally subsided, it left behind a thick layer of silt and mud, filling living rooms and kitchens in floodplain towns and burying crops in floodplain fields. In the end, more than 40,000 km² of the floodplain were submerged, 50 people died, at least 55,000 homes were ruined, and countless acres of crops were buried. Officials estimated that the flood caused over $12 billion in damage. Comparable flooding happened again in the spring of 2011 along the Mississippi, and affected Texas and California in 2017. These west coast floods were associated with *atmospheric rivers*, streams of very moist air from the Pacific that can drop torrential rains when they reach land.

An earlier example of disastrous slow-onset flooding occurred across the globe in 1931, when heavy snows fell in the upstream portion of the watershed for China's Yangtze River. When spring came, not only did the snow begin to melt, but rains drenched much of the watershed for many weeks. And to make matters even worse, summer cyclones swept in from the Pacific, dumping torrents of water over the already saturated landscape. The Yangzte and other rivers of the area overtopped their banks, broke through levees, and spread out over the countryside, eventually inundating vast areas of a densely populated region, and the floodwaters didn't subside for months. No one knows the death toll due to drowning—estimates range between 0.5 and 2 million. Millions more succumbed to disease or starvation during or after the flood, and tens of millions more people lost property and livelihood.

During the 1990 monsoon season in Bangladesh, rain fell almost continuously for weeks. Consequently, the Ganges Delta plain became inundated, and the resulting flood killed 100,000 people. Seasonal floods struck Indonesia in 2007, displacing almost half a million people, nearly half of whom became sick from contact with the filthy water and mud that submerged 60% of the capital and hundreds of square kilometers of farmland.

One of the most devastating floods of recent times began in July 2010, when a seasonal flood fed by intense monsoonal rains submerged floodplains of the Indus River drainage system in Pakistan **(Fig. 14.23c)**. On some days, it rained up to 40 cm in 24 hours! The floods put almost 62,000 km² of the country underwater and severely impacted the lives of over

FIGURE 14.24 Flash floods can occur after torrential rains.

(a) A flash flood in a desert region of Israel has washed over a highway, forcing the evacuation of truckers.

(b) During the 1976 Big Thompson River flash flood, this house was carried off its foundation and dropped on a bridge.

BOX 14.1 CONSIDER THIS...

The Johnstown Flood of 1889

By the 1880s, Johnstown, built along the Conemaugh River in scenic western Pennsylvania, had become a significant industrial town. Recognizing the attraction of the surrounding hills as a summer retreat, speculators built a mud-and-gravel dam across the river, upstream of Johnstown, to trap a pleasant reservoir of cool water. A group of industrialists and bankers bought the reservoir and established the exclusive South Fork Hunting and Fishing Club, a cluster of lavish 15-room "cottages" on the shore. Unfortunately, the dam had been poorly designed, and debris blocked its spillway (the passageway designed to carry surplus water around the dam), setting the stage for a monumental tragedy. On May 31, 1889, torrential rain drenched Pennsylvania, and the reservoir filled until water began to flow over the dam. Despite frantic attempts to strengthen the dam, the soggy structure abruptly collapsed, and the reservoir emptied into the Conemaugh River Valley. A 20-m-high wall of water roared downstream and slammed into Johnstown, transforming bridges and buildings into twisted wreckage **(Fig. Bx14.1)**. When the water subsided, 2,300 people were dead, and Johnstown became the focus of national sympathy. The recently founded Red Cross set to work building dormitories, and citizens nationwide donated everything from clothes

to beds. Nevertheless, it took years for the town to recover, and many residents simply picked up and left.

FIGURE Bx14.1 During the 1889 Johnstown flood, raging waters could tumble large houses.

20 million people (12% of the population)—many people lost all they owned. Crops growing in the fertile floodplain floated away or rotted, and over 200,000 cattle drowned. Due to the destruction of clean-water supplies and of road and rail networks, survivors were stranded for weeks or more, and sadly, disease spread despite relief efforts by organizations from around the world.

FLASH FLOODS

Events during which floodwaters rise so fast that it may be impossible to escape from the path of the water are called **flash floods (Fig. 14.24a)**. They happen during unusually intense rainfall or as a result of a dam collapse or levee failure **(Box 14.1)**. During a flash flood, a canyon or valley may fill to a level many meters above normal in a matter of minutes. Flash floods can be particularly unexpected in arid or semi-arid climates, where isolated thundershowers suddenly fill the channel of an otherwise dry wash. Such a flood may carry water into areas downstream that had not received even a drop of rain.

The historic Big Thompson River flood of July 31, 1976, illustrates the power of a flash flood. This river, which drains the Front Range of the Rocky Mountains, normally carries clear water in a boulder-littered channel. The channel of the river does not occupy the whole valley floor, so a road follows the course of the river, and in places, vacation cabins and campgrounds dot the land between the river and the road. On July 31, towering thunderheads built up over the Front Ranges, and rain began to pour in quantities that even old-timers couldn't recall. In a little over an hour, 19 cm of rain drenched the watershed of the Big Thompson River, so the river's discharge grew to more than four times the maximum recorded during the previous century, and the water level rose by several meters.

Turbulent currents swirled down the canyon at up to 8 m per second and churned up so much sand and mud that the once clear river became a viscous slurry. Landslides of rock and soil tumbled down the steep slopes bordering the river and fed the torrent with even more sediment. The raging water undercut foundations and washed the houses and bridges away **(Fig. 14.24b)**. Parts of the road were swept away, and other parts were buried by debris. Boulders that had stood as landmarks for generations bounced along in the torrent like beach balls, striking and shattering other rocks along the way. Cars bobbed downstream until they wrapped like foil around obstacles. When the flood subsided, the canyon had changed forever, and 144 people had lost their lives.

The Changing Landscape along a Stream

Streams, or rivers, drain the landscape of surface runoff. Typically, an array of connected streams, called a drainage network, develops, consisting of a trunk stream into which numerous tributaries flow. The land drained constitutes the network's watershed. A stream starts from a source, or headwaters. Some headwaters are in the mountains, perhaps collecting water from rainfall or from melting ice and snow. In the mountains, streams carve deep, V-shaped valleys and tend to have steep gradients. For part of its course, a stream may flow over a steep, bouldery bed, forming rapids, and it may drop off an escarpment, creating a waterfall. Streams gradually erode landscapes and carry away debris, so after a while, if there is no renewed uplift, mountains evolve into gentle hills. Over time, streams can bevel once-rugged mountain ranges into nearly flat plains.

Developing drainage networks

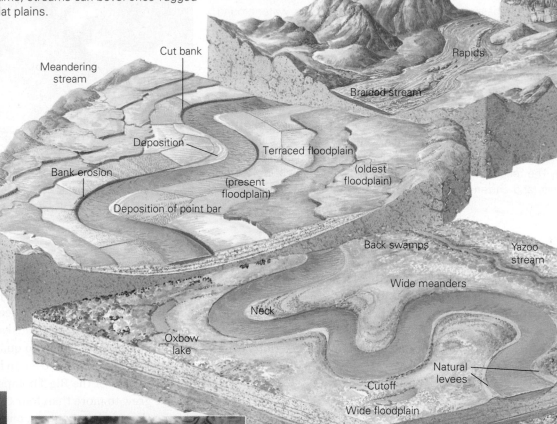

Transportation along the channel

Rapids

Braided stream

Meandering stream

Cut bank

Deposition

Bank erosion

Terraced floodplain

(present floodplain)

(oldest floodplain)

Deposition of point bar

Back swamps

Yazoo stream

Wide meanders

Neck

Oxbow lake

Natural levees

Cutoff

Wide floodplain

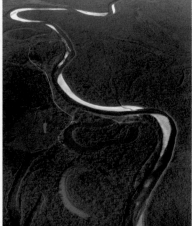

Point bars forming on inner curves.

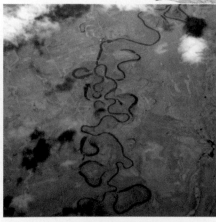

Meanders, abandoned meanders, and cutoffs.

A small delta in a mountain lake.

Headward erosion

Glaciers

Valleys with
high relief

Melting
ice

Lake

Dendritic
drainage

Rapids

Waterfall

**Collection of water
in watershed**

Streams contribute to carving mountains.

Waterfall in Hawaii spilling over a basalt ledge.

Farther along its length, the stream emerges from the mountains. If it is choked with sediment, it may split into numerous entwined channels separated from one another by gravel bars, creating a braided stream. Where a stream that has not been choked by sediment flows over flat ground, it becomes a meandering stream, winding back and forth in snake-like curves called meanders. The current flows faster on the outer edge of a curve, so erosion takes place there, whereas the current flows more slowly on the inner edge, where it drops sediment. Because of erosion and deposition, a meandering stream changes shape over time. Occasionally a meander may be cut off, leaving a curving lake called an oxbow lake. A broad floodplain, covered with water only during floods, may develop on either side of the stream. Natural levees build up between the channel and the floodplain from sediment dropped as a flooding river starts to spill out of its channel. Eventually, a stream reaches a standing body of water and slows down, and the sediment it carries is deposited to form a delta. On a delta, the trunk stream divides into many smaller channels called distributaries.

**Deposition
at mouth**

Delta

Distributaries

Natural levees

Swamps and
marsh

Tidal flats

Bar

Banks

Not all flash floods occur in regions of high relief. For example, in May 2015, intense thunderstorms drenched parts of Texas. Because the region had been suffering from a drought, vegetation had withered and wasn't able to absorb as much water as usual, so the proportion of rain that became runoff was greater than usual. In successive 24-hour periods, about 30 cm of rain fell, much of which entered the drainage network. Some streams rose by up to 12 m, breaking previous records. Low-lying areas, including interstate highways, disappeared beneath the water, and thousands of homes were destroyed.

LIVING WITH FLOODS

Flood Control Mark Twain once wrote of the Mississippi that we "cannot tame that lawless stream, cannot curb it or confine it, cannot say to it, 'go here or go there,' and make it obey." Was Twain right? Since ancient times, people have attempted to control courses of rivers in order to prevent undesired flooding. In the 20th century, flood-control efforts intensified as the population living along rivers increased. For example, since the passage of the 1927 Mississippi River Flood Control Act (after a disastrous flood took place that year), the U.S. Army Corps of Engineers has labored to control the Mississippi. First, engineers built about 300 dams along the river's tributaries so that excess runoff could be stored in the reservoirs and later be released slowly. Second, they constructed **artificial levees** (elongate mounts of sand and clay) and concrete *flood walls* along the banks of the river to increase the channel's volume **(Fig. 14.25a, b)**. Artificial levees and flood walls isolate a discrete area of the floodplain.

Although the Corps' strategy worked for floods up to a certain size, it was insufficient to handle the 1993 and 2011 floods. The river rose until it spilled over the tops of some levees and undermined others. *Levee undermining* occurs when rising water levels increase the water pressure on the river side of the levee, forcing water through sand under the levee **(Fig. 14.25c, d)**. In susceptible areas, water begins to spurt out of the ground on the dry side of the levee, thereby washing away the levee's support. The levee finally becomes so weak that it collapses and water fills in the area behind it. In some cases, the Corps of Engineers will intentionally dynamite levees along a relatively unpopulated reach of the river upstream of a vulnerable town, to divert the flow out onto a portion of the floodplain where the water will do less damage and prevent the floodwaters from overtopping levees close to the town.

Because traditional flood-prevention efforts don't always work, engineers have explored alternatives. For example, restoration of wetland areas along rivers can diminish the volume of water in a channel, for wetlands can absorb significant quantities of floodwater. Also, where appropriate, planners may prohibit construction within designated land areas adjacent to the channel, so that floodwater can fill these areas without causing expensive damage. The existence of such areas, known as *floodways*, effectively increases the volume of water that the river can carry, thereby preventing the water level from rising too high.

Evaluating Flooding Hazard When making decisions about investing in flood-control measures, mortgages, or insurance, planners need a basis for defining the hazard or risk posed by flooding. If floodwaters submerge a locality every year, a bank officer would be ill advised to approve a loan that would promote building there, and insurance, even if available, would be prohibitively expensive. But if floodwaters submerge the locality only very rarely, then the loan or insurance may be worth the risk. Geologists characterize the risk of flooding in two ways. The **annual probability** of flooding indicates the likelihood that a flood of a given size or larger will happen at a specified locality during any given year. For example, if we say that a flood of a given size has an annual probability of 1%, then we mean there is a 1 in 100 chance that a flood of at least this size will happen in any given year. The **recurrence interval** of a flood of a given size is defined as the average number of years between successive floods of at least this size. To understand the concept of a recurrence interval, imagine that a researcher determines that successive floods of a given size happened 96, 32, 200, 14, and 158 years ago. The average of these numbers is 100, so we can refer to a flood of this size as a *100-year flood*. Note that annual probability and recurrence interval are related: Annual probability = 1 ÷ recurrence interval. For example, the annual probability of a 100-year flood is 1/100, which can also be written as 0.01 or 1%. Unfortunately, many people misunderstand the meaning of recurrence interval, and think that they will not face flooding hazard for another century if they buy a home within an area just after a 100-year flood has occurred. Note from our example that such confidence is misplaced. Two 100-year floods can occur in consecutive years, or even in the same year. Alternatively, the interval between such floods could be decades or centuries apart.

The recurrence interval or annual probability for a flood along a particular river reflects the size of a flood. For example, the discharge of a 100-year flood (annual probability of 1%) is larger than that of a 2-year flood (annual probability of 50%),

> **Did you ever wonder . . .**
> what newscasters mean by a "100-year flood?"

FIGURE 14.25 Holding back rivers to prevent floods.

(a) Artificial levees built to protect the downtown of Galena, Illinois, which is built along a tributary of the Mississippi.

High-water marks of past floods.

(b) These floodwalls can be closed to protect Cape Girardeau, Missouri, from Mississippi River floods.

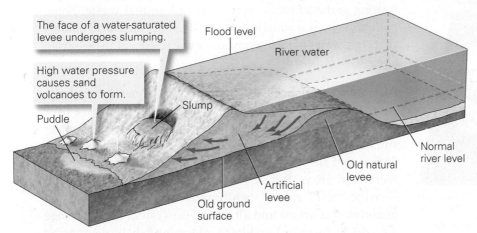

The face of a water-saturated levee undergoes slumping.

High water pressure causes sand volcanoes to form.

Puddle

Slump

Flood level

River water

Old natural levee

Normal river level

Artificial levee

Old ground surface

(c) Levees can be undermined if the river floods.

(d) Breaching of a levee lets water spill into the floodplain.

because the former happens less frequently **(Fig. 14.26a)**. To define this relationship, geologists construct graphs that plot flood discharge on the vertical axis against recurrence interval on the horizontal axis **(Fig. 14.26b)**.

Knowing the discharge during a flood of a specified annual probability, and knowing the shape of the river channel and the elevation of the land bordering the river, geologists can predict the extent of land that will be submerged by such a flood. Using such data, they produce **flood-hazard maps**. In the United States, the Federal Emergency Management Agency (FEMA) produces maps that show the 1% annual probability (100-year) flood area and the 0.2% annual probability (500-year) flood risk zones **(Fig. 14.26c)**. It is always important to keep in mind when thinking about flooding

hazard that the term *floodplain* means what it says—it's an area that will eventually flood.

TAKE-HOME MESSAGE

Seasonal floods submerge floodplains and delta plains at certain times of the year. Flash floods are sudden and short-lived. We can specify the probability that a certain size of flood will happen in a given year, but flood-control efforts meet with mixed success. Statements about flooding hazard are, unfortunately, often misinterpreted.

QUICK QUESTION How do geologists produce a flood-hazard map?

FIGURE 14.26 The conceptual relationship between flood size and probability.

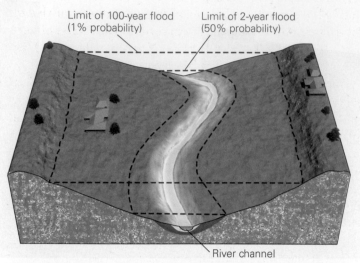

(a) A 100-year flood covers a larger area than a 2-year flood and occurs less frequently.

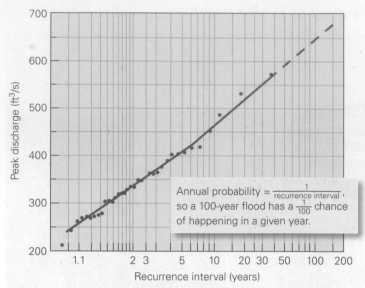

Annual probability = $\frac{1}{\text{recurrence interval}}$, so a 100-year flood has a $\frac{1}{100}$ chance of happening in a given year.

(b) A flood-frequency graph shows the relationship between the recurrence interval and the peak discharge for an idealized river.

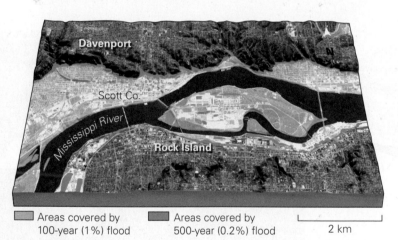

☐ Areas covered by 100-year (1%) flood   ☐ Areas covered by 500-year (0.2%) flood   2 km

(c) A flood-hazard map shows areas likely to be flooded. Here, near Rock Island, Illinois, even large floods are confined to the floodplain.

14.7 Human Impact on Rivers

Streams can serve as avenues for transportation and can be sources of food, irrigation water, drinking water, power, recreation, and (unfortunately) waste disposal. Further, their floodplains provide particularly fertile soil for fields. Considering the vast resources that rivers provide, it's no coincidence that ancient cultures developed in river valleys and on floodplains. Nevertheless, over time, humans have increasingly tended to abuse or overuse the Earth's rivers. Here we note four pressing environmental issues affecting rivers.

Pollution The capacity of some rivers to carry pollutants has long been exceeded, transforming them into deadly cesspools. Pollutants include raw sewage and storm drainage from urban areas, spilled oil, toxic chemicals from industrial sites, floating garbage, excess fertilizer, and animal waste. Some pollutants poison aquatic life directly, some feed algal blooms that strip water of its oxygen, and some settle out to be buried along with sediments.

Dam Construction In 1950, there were about 5,000 large dams (over 15 m high) worldwide, but today there are over 57,000. Damming rivers has both positive and negative results. Reservoirs provide irrigation water and hydroelectric power, and they trap some floodwaters and create popular recreation areas. But in some locations their construction destroys "wild rivers" (whitewater streams in hilly and mountainous areas) and alters the ecosystem of a drainage network by forming barriers to migrating fish, by decreasing the nutrient supply to organisms downstream, and by removing the source of sediment and nutrients for the floodplain and delta.

Overuse of Water Because of growing populations, our thirst for river water continues to increase, but the supply of water does not. The use of water has grown especially in response to the "green revolution" of the 1960s (an intense effort to increase food crops worldwide), during which huge new tracts of land came under irrigation. Today, 65% of the water taken from rivers is used for agriculture, 25% for industry, and 9% for drinking. In some places, human activity consumes the entire volume of a river's water, and as a result the channel contains little more than a saline trickle, if that, at its mouth. For example, except during unusually wet years, the Colorado River's channel contains almost no water where it crosses the Mexican border; pipes and canals carry the water instead to Las Vegas, Phoenix, and Los Angeles **(Fig. 14.27)**.

FIGURE 14.27 The Central Arizona Project canal shunts water from the Colorado River to Phoenix.

forests into parking lots, roads, and buildings, a layer of impermeable concrete and asphalt prevents rainfall from infiltrating the ground, the amount of living biomass available to absorb water decreases, and storm sewers divert water directly to streams. Unfortunately, the changes in volume and lag time can lead to flash flooding.

Agriculture can have a similarly profound effect on streams. Though covered with green during the growing season, fields of many crops actually host relatively little vegetation because farmers keep the space between rows of the crop plants free of weeds. Further, if the crop consists of annual plants, such as corn or soybeans, the fields remain vegetation-free during the winter. As a consequence, the runoff from farm fields can be greater than that from forests or from naturally vegetated fields, and that runoff can carry significant amounts of soil into streams, thus increasing the sediment load that streams carry.

Urbanization and Agriculture When it rains in a naturally vegetated region, much of the water that falls from the sky either soaks into the ground or gets absorbed by plants. Some of that soil moisture and groundwater eventually seeps into a nearby stream, but the remainder flows elsewhere underground. As a result, the amount of water that reaches nearby streams after a rainstorm is less than the total amount of precipitation, and a significant *lag time* occurs between when the rain falls and when the stream's discharge increases. Urbanization changes this picture. When developers transform fields and

TAKE-HOME MESSAGE

Society depends on streams for water supplies, irrigation, energy, and transport, but the growth of populations can negatively affect streams. Diversion of flow may decrease stream discharge to a trickle, dam construction can change flow, pollution can foul the water, and urbanization or runoff can change discharge and sediment load.

QUICK QUESTION How can urbanization change the discharge of a stream?

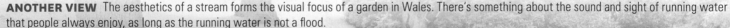

ANOTHER VIEW The aesthetics of a stream forms the visual focus of a garden in Wales. There's something about the sound and sight of running water that people always enjoy, as long as the running water is not a flood.

Chapter 14 Review

CHAPTER SUMMARY

> Streams are bodies of water that flow down channels and drain the land surface. They grow by downcutting and headward erosion. Streams carry water out of a drainage basin; a drainage divide separates two adjacent basins. Drainage networks consist of many tributaries that flow into a trunk stream.

> Permanent streams exist where the water table lies above the bed of the channel or when large amounts of water enter the channel from upstream. Where the water table lies below the channel bed, streams are ephemeral.

> The discharge of a stream is the total volume of water passing a point along the bank in a second. Most streams are turbulent, meaning that their water swirls in complex patterns.

> Streams erode the landscape by scouring, lifting, abrading, and dissolving. The resulting sediment provides dissolved loads, suspended loads, and bed loads. The total quantity of sediment carried by a stream is its capacity. Capacity differs from competence, the maximum particle size a stream can carry. When stream water slows, it deposits alluvium.

> The longitudinal profile of a stream is concave up, meaning that a stream has a steeper gradient at its headwaters than near its mouth. The base level limits the depth of downcutting.

> Streams cut valleys or canyons, depending on the rate of downcutting relative to the rate at which the slopes on either side of the stream undergo mass wasting. Where a stream flows down steep gradients and has a bed littered with large rocks, rapids develop, and where a stream plunges off a vertical face, a waterfall forms.

> A meandering stream wanders back and forth across a floodplain. It erodes its outer bank and deposits a point bar on the inner bank. Eventually, a meander may be cut off and turn into an oxbow lake. Natural levees form on either side of the river channel.

> Braided streams form where a stream that carried abundant sediment during floods slow, so the flowing water divides among many entwined channels.

> Where streams or rivers flow into standing water, they deposit deltas. Different deltas have different shapes.

> Fluvial erosion can bevel landscapes to a nearly flat plain. If the base level drops or the land surface rises, stream rejuvenation takes place. The headward erosion of one stream may capture the flow of another.

> If an increase in rainfall or spring melting causes more water to enter a stream than the channel can hold, a flood results. Some floods are seasonal and submerge floodplains or delta plains. Flash floods happen very rapidly. Officials try to prevent floods by building reservoirs and levees.

> Rivers are becoming a vanishing resource because of pollution, damming, and overuse of water.

GUIDE TERMS

alluvial fan (p. 480)
alluvium (p. 476)
annual probability (p. 492)
artificial levee (p. 492)
bar (p. 476)
base level (p. 477)
bed load (p. 475)
braided stream (p. 480)
canyon (p. 478)
capacity (p. 476)
channel (p. 469)
competence (p. 475)
delta (p. 482)
discharge (p. 473)

dissolved load (p. 475)
distributary (p. 482)
downcutting (p. 471)
drainage divide (p. 472)
drainage network (p. 471)
drainage reversal (p. 485)
ephemeral stream (p. 472)
flash flood (p. 489)
flood (p. 470)
flood-hazard map (p. 493)
floodplain (p. 476)
headward erosion (p. 471)
headwaters (p. 469)
longitudinal profile (p. 477)

meander (p. 481)
meandering stream (p. 481)
mouth (p. 469)
natural levee (p. 482)
oxbow lake (p. 482)
permanent stream (p. 472)
point bar (p. 476)
pothole (p. 474)
rapids (p. 479)
recurrence interval (p. 492)
runoff (p. 470)
saltation (p. 475)
scouring (p. 474)
seasonal flood (p. 487)

sheetwash (p. 470)
slow-onset flood (p. 487)
stream (p. 469)
stream gradient (p. 477)
stream piracy (p. 484)
stream rejuvenation (p. 484)
stream terrace (p. 478)
suspended load (p. 475)
tributary (p. 471)
valley (p. 478)
waterfall (p. 479)
watershed (p. 472)

GEOTOURS *THIS CHAPTER'S GEOTOURS WORKSHEET (N) FEATURES QUESTIONS AND GOOGLE EARTH SITES ON:*

> Headward erosion
> Stream patterns

> Meandering stream features
> Stream piracy

> Discordant streams
> Catastrophic flooding

REVIEW QUESTIONS

The letters following each Review Question refer to the corresponding Learning Objective from the Chapter Opener.

1. What role do streams serve during the hydrologic cycle? **(A)**

2. Describe the four different types of drainage networks. Which type of drainage network does the diagram show? **(A)**

3. Distinguish between permanent and ephemeral streams and explain why the difference exists. **(A)**

4. Why does the velocity of a stream vary with location in a given reach of the stream? **(C)**

5. Describe how streams erode the Earth's surface. **(B)**

6. What are three components of sediment load in a stream, and how do competence and capacity differ? **(B)**

7. Describe how the character of a drainage network changes along its length. **(C)**

8. What is the difference between a local base level and the ultimate base level of a stream? **(C)**

9. Why do some streams become braided? **(D)**

10. Describe how meanders form, develop, are cut off, and then are abandoned. What is a floodplain? Identify the landforms shown in the diagram. **(D)**

11. Describe how deltas grow and develop. How do they differ from alluvial fans? **(D)**

12. How does a stream-eroded landscape evolve over time? **(B)**

13. What is stream piracy? What causes a drainage reversal? **(A)**

14. How are superposed and antecedent drainages similar, and how do they differ? **(A)**

15. What is the difference between a seasonal flood and a flash flood? **(E)**

16. How do people try to protect regions from flood damage? **(F)**

17. What is the recurrence interval of a flood, and how is it related to the annual probability? **(F)**

18. How have humans abused and overused the resource of running water? **(G)**

ON FURTHER THOUGHT

19. Records indicate that flood crests for a given amount of discharge along the Mississippi River have been getting higher since 1927, when a system of levees began to block off portions of the floodplain. Why? **(F)**

20. The Ganges River carries an immense amount of sediment load, which has been building a huge delta in the Bay of Bengal. Look at the region using an atlas or Google Earth, think about the nature of the watershed supplying water to the drainage network that feeds the Ganges, and explain why this river carries so much sediment. **(B)**

ONLINE RESOURCES

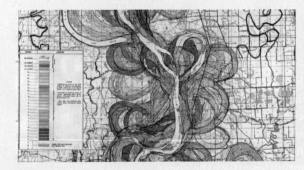

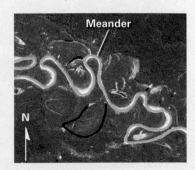

Videos This chapter includes videos on river meanders, aquifer systems, and flash floods.

Smartwork5 This chapter features visual identification exercises on meandering stream systems, erosion, and lake formation.

CHAPTER 15

RESTLESS REALM: OCEANS AND COASTS

By the end of this chapter, you should be able to . . .

A. describe how ocean depth, temperature, and salinity correlate.

B. explain how tectonic processes produce bathymetric features of the seafloor.

C. describe the nature and causes of surface currents and deep currents.

D. relate the behavior of tides to the forces that cause tides.

E. discuss how waves form, and how they change as they approach the shore.

F. differentiate among the great variety of different coastal landforms, and explain how they develop and evolve.

G. explain how changes in sea level, wave erosion, and human activities affect the coast.

> The three great elemental sounds in nature are the sound of rain, the sound of wind in a primeval wood, and the sound of the outer ocean on a beach.
>
> HENRY BESTON (American naturalist, 1888–1968)

15.1 Introduction

When seen from space, the Earth glows blue, for the ocean covers most of its surface (Fig. 15.1). The ocean incubates life, tempers the Earth's climate, and spawns its storms. It also serves as a vast reservoir for water and chemicals that cycle into the atmosphere, the crust, and life, and serves as a resting place for sediment washed off the land. In this chapter, we begin our study of the oceans by addressing the question of why distinct ocean basins exist in the first place. Then, we consider the nature of seawater and its movements. Finally, we focus on landforms that develop along the *coast,* the region where the land meets the sea, and we learn about some of the natural hazards that challenge people who live in coastal areas.

15.2 Landscapes Beneath the Sea

If the surface of the lithosphere were completely smooth, ocean water would surround the Earth as a 2.5-km-deep layer. But the surface of the lithosphere is not smooth; rather, it displays vertical relief of almost 20 km, as measured from the deepest point in the ocean to the highest point on land (Fig. 15.2a). Dry land, on average, rises less than 1 km above sea level, and it covers about 30% of the Earth's surface. In contrast, the seafloor, which accounts for the remaining 70% of the Earth's surface, has an average elevation of 3.7 km below sea level. Because of this difference in elevation, we can

FIGURE 15.1 The oceans of the world. The Pacific is the largest, covering almost half the planet. The Arctic region is an ocean covered by a thin coating of ice, whereas the Antarctic region is a continent.

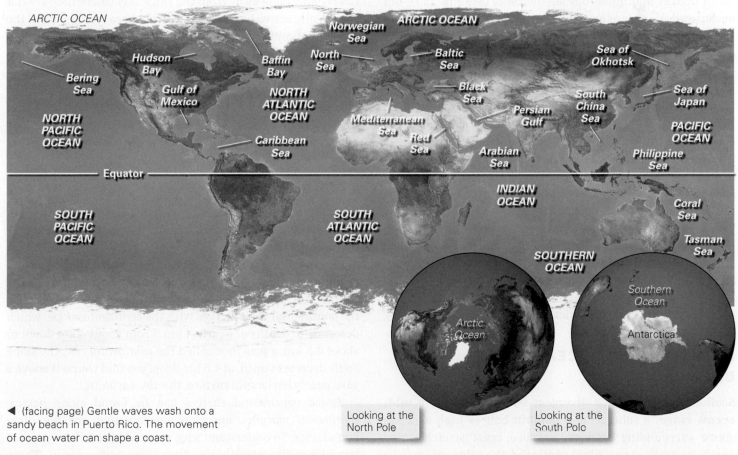

◀ (facing page) Gentle waves wash onto a sandy beach in Puerto Rico. The movement of ocean water can shape a coast.

FIGURE 15.2 Contrasts between continental lithosphere and oceanic lithosphere.

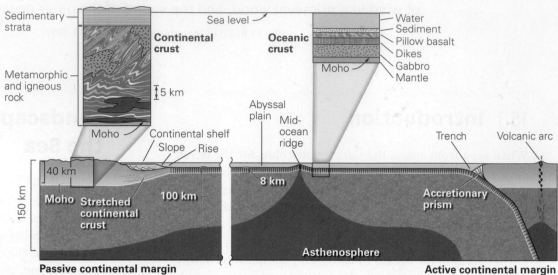

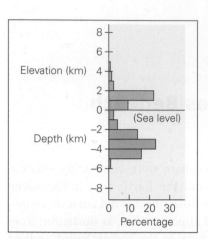

(a) Percentages of the Earth's surface at different elevations above or below sea level.

(b) The crustal portion of the continental lithosphere differs markedly from that of the oceanic lithosphere.

distinguish between distinct *ocean basins*, the low-lying lithosphere surface beneath the sea, and continents, the large land masses between the ocean basins **(Fig. 15.2b)**. The elevation difference between the surfaces of oceanic and continental lithosphere is a manifestation of isostasy (see Interlude D). On the map of the present-day Earth, geographers recognize seven oceans and numerous smaller seas (see Fig. 15.1). But because all the oceans are interconnected, the boundaries between them are, geologically, somewhat arbitrary, and water exchanges among all the oceans.

Have you ever wondered what the ocean floor would look like if all the water evaporated? To gain insight into the seafloor's **bathymetry** (variation in depth), researchers, until the mid-20th century, had to drop a *plumb line*, a weight at the end of a cable, down to the seafloor. Measurements took a long time to obtain using this tedious technique. When sonar became available, about the time of World War II, it became possible to make continuous recordings of depth as a ship crossed the sea. Today, satellites survey the ocean and translate very precise measurements of changes in the pull of gravity into measurements of water-depth variation. Taken together, such studies indicate that ocean basins include several *bathymetric provinces*, distinguished from one another by water depth. Let's take a look at these provinces to explain their characteristics in terms of plate tectonics theory **(Fig. 15.3a)**.

BATHYMETRY OF OCEAN PLATE BOUNDARIES

Seafloor spreading at a divergent boundary yields a **mid-ocean ridge**, a submarine mountain belt as high as 2 km above surrounding seafloor. Because crust stretches and breaks as seafloor spreading continues, the ridge axis may be

bordered by steep escarpments produced by normal faulting. *Oceanic transform faults*, strike-slip faults along which one plate shears sideways past another, typically link segments of mid-ocean ridges (see Chapter 2). Transform faults are delineated by *fracture zones*, narrow belts of steep escarpments and broken-up rock. These fracture zones can be traced into the oceanic plate away from the ridge axis where they are not seismically active but are still distinct because they form the boundary between lithosphere of different ages.

Subduction at convergent boundaries yields a **trench**, a deep, elongate trough at the boundary between the subducting plate and the accretionary prism. Most trenches reach depths of between 6 and 8 km. The deepest, the Mariana Trench of the western Pacific, attains a depth of 11 km. Some trenches border continents, whereas others border island arcs.

CONTINENTAL MARGINS

Imagine you're in a submersible cruising just above the floor of the western half of the North Atlantic. If you start at the shoreline of North America and head east, you will cross the 200- to 500-km-wide **continental shelf**, a relatively shallow portion of the ocean that fringes the continent. Water depth over the continental shelf does not exceed 500 m. At its eastern edge, the continental shelf merges with the *continental slope*, which descends to depths of nearly 4 km. From about 4 km down to about 4.5 km, a province called the *continental rise*, the slope angle decreases until, at 4.5 km deep, you find yourself above a vast, nearly horizontal surface, the *abyssal plain*.

Broad continental shelves can be found along *passive continental margins*, meaning margins that are not plate boundaries. To understand why, we need to recall the process by which sedimentary basins form along such margins. These

FIGURE 15.3 Bathymetric features of the seafloor. The maps are produced by computer using measurements from satellites or submersibles.

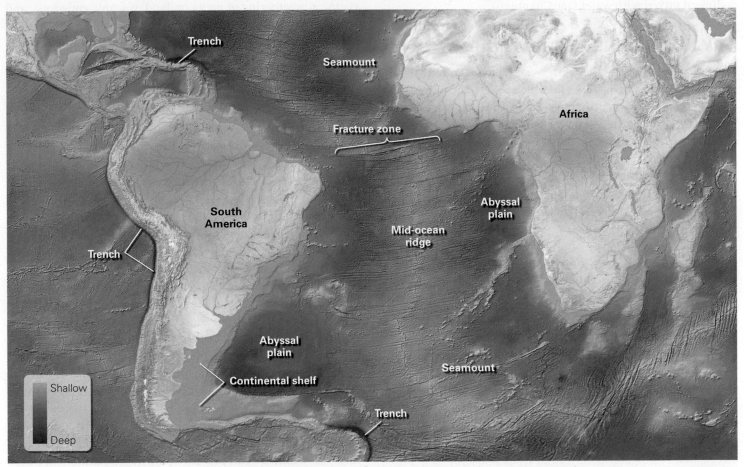

(a) Passive margins appear on both sides of the Atlantic and in the Gulf of Mexico. Active margins border the Caribbean Sea and the western coast of South America.

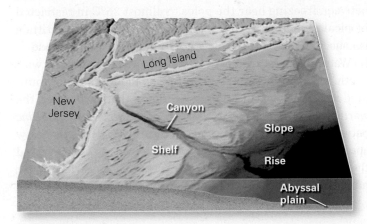

(b) A submarine canyon along the coast of New Jersey. Turbidity currents in submarine canyons carry sediment to the abyssal plain.

passive-margin basins initiate when the rifting that produced the margin ceases and a new mid-ocean ridge forms. At this time, the stretched lithosphere of the now-inactive rift cools, becomes denser, and subsides to maintain isostasy. This subsidence produces a depression that fills with sediments washed in by rivers draining the continent, as well as with sediments

consisting of the shells of marine organisms. Subsidence continues for tens of millions of years, so the succession of strata can achieve a thickness of 15 to 20 km. Accumulation of sediment keeps pace with subsidence, so the surface of the sedimentary succession lies between zero and a few hundred meters of sea level. This surface is the continental shelf.

Active continental margins exist where a plate boundary—either a convergent boundary or a transform boundary—coincides with the edge of a continent. Such margins tend to host relatively narrow continental shelves. For example, at localities where a convergent boundary defines an active continental margin, a trench lies offshore, and the continental shelf forms on the surface of the accretionary prism.

At many locations, relatively narrow and deep valleys called **submarine canyons** downcut into continental shelves and slopes (Fig. 15.3b). Some submarine canyons start offshore of major rivers that had cut into the continental shelf when sea level was low and the shelf was exposed. But river erosion cannot explain the great depth of these canyons—some slice almost 1,000 m down into the continental margin, far deeper than the maximum sea-level change. Researchers have determined that much of the erosion in submarine canyons results from the flow of turbidity currents, submarine avalanches of sediment mixed with water (see Chapter 6).

ABYSSAL PLAINS AND SEAMOUNTS

As oceanic crust ages and moves away from the axis of the mid-ocean ridge, two changes take place. First, the lithosphere cools, and as it does so, its surface sinks. Second, a blanket of deep-marine sediment, known as *pelagic sediment*, gradually accumulates and covers the basalt of the oceanic crust. This sediment consists mostly of microscopic plankton shells and fine flakes of clay, which slowly fall like snow from the ocean water and settle on the seafloor. As the ocean crust moves away from the ridge axis and gets progressively older, sediment thickness increases. Seafloor that has aged more than about 80 million years no longer sinks, and after being buried by a thick layer of pelagic sediment, it tends to be deep, flat, and smooth. Broad areas of old, flat seafloor are called **abyssal plains**.

In dozens of locations around the world, hot-spot eruptions on oceanic lithosphere have produced volcanoes. Active oceanic hot-spot volcanoes that rise above sea level, as well as the remnants of extinct ones, are *oceanic islands*. Those whose peaks lie below sea level become **seamounts**. Particularly voluminous eruptions that produced broader buildups of basalt are known as *submarine plateaus*, the largest of which are 1,200 km across.

TAKE-HOME MESSAGE

Ocean basins exist because oceanic and continental lithosphere differ in thickness and density. Seafloor bathymetric features (ridges, trenches, and fracture zones) reflect plate-tectonic processes. Abyssal plains overlie old oceanic lithosphere, and continental shelves overlie passive-margin basins.

QUICK QUESTION How do submarine canyons form?

15.3 Ocean Water Characteristics

According to Archimedes' principle, a buoyant object sinks only until it displaces a mass of water equal to the mass of the object. Therefore, you float higher in a denser liquid than you do in a less-dense liquid. If you've ever had a chance to swim in the ocean, you may have noticed that you float much more easily in ocean water than you do in freshwater. That's because ocean water contains an average of 3.5% dissolved salt, whereas typical freshwater contains less than 0.02% salt. Dissolved ions fit between water molecules without changing the volume of the water, so adding salt to water increases the water's density.

The ions of dissolved salt in the sea come from three sources: chemical weathering of rocks on land, volcanic emissions, and seafloor hydrothermal vents.

The ocean contains so much salt that if all the water suddenly evaporated, a 60-m-thick layer of salt would coat the ocean floor. Though we call this layer "salt," we don't mean that the layer would consist entirely of the mineral halite, common table salt (NaCl). In fact, this layer would consist of only about 75% halite—the remainder would include gypsum ($CaSO_4 \cdot H_2O$), anhydrite ($CaSO_4$), and several other trace salts.

Oceanographers, researchers who study the oceans, refer to the concentration of salt in water as **salinity**. Although ocean salinity averages 3.5%, a compilation of measurements from around the world reveals that salinity varies with location, ranging from about 1.0% to about 4.1%. Why? Salinity reflects the balance between the addition of freshwater, by rivers or rain, and the removal of water molecules, by evaporation. When seawater evaporates, salt stays behind. Salinity also depends on water temperature, for warmer water can hold more salt in solution than can cold water. Consequently, oceans are saltier along the coasts of hot deserts and are less salty near the mouths of large, cold rivers.

When RMS *Titanic* sank after striking an iceberg in the North Atlantic, most of the unlucky passengers and crew who jumped or fell into the sea died within minutes, because the seawater temperature at the site of the tragedy approached freezing, and cold water removes heat from a body very rapidly. Yet swimmers can play for hours in the Caribbean, where sea-surface temperatures reach 29°C. Though the average global sea-surface temperature hovers around 17°C, it ranges between freezing near the poles to almost 36°C in restricted tropical seas. Average temperatures correlate with latitude because the intensity of solar radiation varies with latitude.

Water temperature in the ocean also varies markedly with depth, for solar energy can penetrate no more than a few hundred meters. As warm water is less dense than cold water, the warmed water remains near the surface. Typically, a *thermocline* between warm water above and significantly colder water below appears at a depth of about 300 m in the tropics. A pronounced thermocline does not develop in polar seas since surface water has nearly the same temperature as deep water.

TAKE-HOME MESSAGE

Water in the ocean contains dissolved ions that, if precipitated, would form salts, such as halite and gypsum. Salinity averages 3.5%, but it varies with location. Ocean temperature also varies with location; it's warmer in low latitudes.

QUICK QUESTION What is a thermocline, and why does it form?

15.4 **Tides and Wave Action**

TIDES

A ship captain hoping to sail out of a port must pay close attention to the **tide**, the generally twice-daily rise and fall of sea level. At high tide, the harbor's water may be deep enough that the ship can float over obstacles easily, but at low tide, the water may be so shallow that the ship could run aground. Tides develop in response to the *tide-generating force*, which in simple terms develops in part due to the gravitational attraction of the Sun and Moon, and in part due to centrifugal force produced as the Earth-Moon system revolves around its center of mass. The tide-generating force results in two *tidal bulges* in the oceans on opposite sides of our planet (**Fig. 15.4a**). The larger of these occurs on the side of the Earth closer to the Moon, because the force due to the pull of the Moon's gravity is stronger than that of other contributors to the tide-generating force. When a location lies under a tidal

bulge, it experiences a high tide, and when it underlies a depression between two bulges, it experiences a low tide.

If the Earth's surface were smooth and completely submerged beneath the ocean, each point on the surface would experience two high tides and two low tides per day, as the Earth spins relative to the tidal bulges. But the shape of the seafloor and the irregularity of the coastline cause the specific timing and magnitude of tides to vary from place to place. For example, the **tidal range**, the elevation difference between sea level at high tide and low tide, can measure from less than a meter up to several meters at different points along a coast (**Fig. 15.4b**). The largest tidal range on Earth, 16.8 m, affects the Bay of Fundy on the Atlantic coast of Canada.

The manifestation of tides along a shore depends not only on the tidal range, but also on the slope of the shore. Along a vertical cliff, the *intertidal zone* (the region of shore submerged at high tide and exposed at low tide) appears as a ring of stained, barnacle- and seaweed-encrusted rock. Where the coast has a gentle slope, the shoreline moves substantially inland during a rising tide and seaward during a falling tide—in some places, the shoreline at high tide lies a few kilometers inland of the shoreline at low tide. In such locations, vast muddy *tidal flats* may be exposed during low tide (**Fig. 15.4c**).

WAVE ACTION

Wind-driven waves make the ocean surface a restless, ever-changing vista. An ocean **wave** is a periodic undulation of the water surface—the low part of the wave is its *trough*, and the high part is its *crest*. The distance between adjacent crests (or adjacent troughs) is the *wavelength*, and the elevation difference between the trough and crest is the *wave height*. Larger waves move horizontally along the surface of

FIGURE 15.4 Ocean tides and their manifestation.

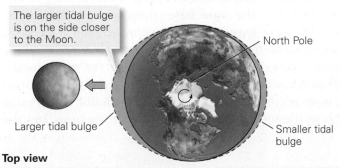

Top view

(a) Tides develop as the Earth spins relative to the two tidal bulges.

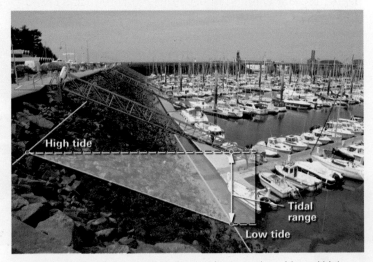

(b) The tidal range is the vertical distance between low tide and high tide. This photo of a French harbor was taken at low tide.

(c) A broad tidal flat is exposed at low tide around Mont-Saint-Michel, on the coast of France.

the ocean at speeds of 30 to 50 km per hour. This movement gives the visual impression that water itself is moving horizontally—in fact, it's not. As a wave passes, the molecules of water affected by the wave follow a somewhat circular path. The diameter of the circle is largest for molecules at the surface, where it equals the wave height of the wave. With increasing depth, though, the diameter of the circle decreases until, at a depth equal to about half the wavelength, no wave movement at all takes place **(Fig. 15.5a)**. Submarines traveling below this **wave base** cruise through smooth water, while ships toss about above. The movement of water in a wave is not a perfect circle, so after about 20 waves have passed, water affected by the wave will have moved a horizontal distance of about one wavelength.

FIGURE 15.5 Ocean waves build in response to the shear of wind blowing over the water surface.

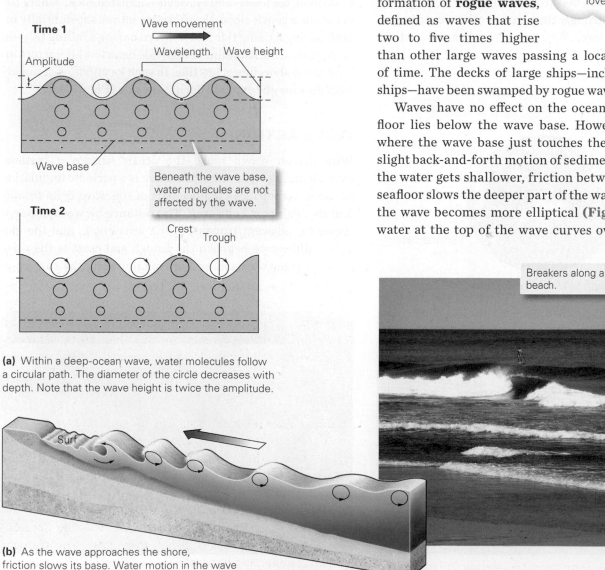

Time 1

Wave movement

Amplitude

Wavelength   Wave height

Wave base

Beneath the wave base, water molecules are not affected by the wave.

Time 2

Crest

Trough

(a) Within a deep-ocean wave, water molecules follow a circular path. The diameter of the circle decreases with depth. Note that the wave height is twice the amplitude.

Surf

(b) As the wave approaches the shore, friction slows its base. Water motion in the wave becomes more elliptical, and the wave becomes a breaker.

Ocean waves develop because of shear between air molecules in the wind and water molecules at the surface of the sea. The character of waves in the open ocean depends on the wind's strength (how fast the air moves) and *fetch* (the distance over which it blows). When the wind first begins to blow, it produces *ripples*, small waves with sharp crests, in the water surface. With continued blowing over a long fetch, *swells*, larger waves with heights of 2 to 10 m and wavelengths of 40 to 500 m, form. Swells may travel for thousands of kilometers across the ocean, well beyond the region where they formed. Particularly large waves develop during storms. For example, hurricanes can generate wave heights of up to 35 m. Wave interference, as well as the interaction of wind-driven waves with strong currents, or the focusing of waves due to the shape of the coastline or seafloor, can lead to the formation of **rogue waves**, defined as waves that rise two to five times higher than other large waves passing a location during a period of time. The decks of large ships—including famous cruise ships—have been swamped by rogue waves in the open ocean.

> **Did you ever wonder . . .**
> why the breakers that surfers love form near the shore?

Waves have no effect on the ocean floor, as long as the floor lies below the wave base. However, near the shore, where the wave base just touches the seafloor, it causes a slight back-and-forth motion of sediment. Closer to shore, as the water gets shallower, friction between the wave and the seafloor slows the deeper part of the wave, and the motion in the wave becomes more elliptical **(Fig. 15.5b)**. Eventually, water at the top of the wave curves over the base, and the

Breakers along a beach.

wave becomes a *breaker*, ready for surfers to ride. Breakers crash onto the shore in the surf zone, sending a surge of water up the beach. This upward surge, or **swash**, continues until friction brings motion to a halt. Then gravity draws the water back down the beach as **backwash**.

Waves may make a large angle with the **shore** (the edge of the land) as they come in, but they bend as they approach, so that when they reach the shore, wave crests generally make an angle of less than 5° with the shoreline **(Fig. 15.6)**. To understand this phenomenon, called **wave refraction**, imagine a wave approaching so that its crest initially makes an angle of 45° with the shore. The end of the wave closer to the shore touches bottom first and slows down because of friction, whereas the end farther offshore continues to move at its original velocity, swinging the whole wave around so that it becomes more parallel with the shore.

Although wave refraction decreases the angle at which waves move close to shore, it does not necessarily eliminate the angle. Where waves arrive at the shore obliquely, water in the nearshore region has a component of motion that aligns roughly parallel to the shore. This **longshore current** causes swimmers floating in the water just offshore to drift gradually in a direction parallel to the shore.

Waves pile water up on the shore incessantly. As excess water moves back to the sea, it may localize into a strong seaward flow, called a *rip current*, perpendicular to the shore. Rip currents cause many people to drown every year, because they can carry unsuspecting swimmers far away from land. To escape from a rip current, swim calmly parallel to the shore.

> ### TAKE-HOME MESSAGE
>
> The tide-generating force produces two tidal bulges beneath which the Earth spins, so locations generally experience the twice-daily rise and fall of tides, which alternately submerge and expose the intertidal zone. Wind also causes waves to form. Within a wave, water moves approximately in a circle—the amount of motion decreases with depth. Near the shore, water piles up into breakers, and the orientation of waves changes due to wave refraction.
>
> **QUICK QUESTION** Why do longshore currents develop?

FIGURE 15.6 Wave refraction and its consequences along the shore.

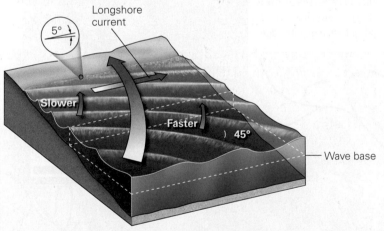

(a) Wave refraction occurs when waves approach the shore at an angle. If the wave reaches the beach at an angle, it causes a longshore current and beach drift of sand.

(b) An example of waves refracting along a beach.

15.5 Currents: Rivers in the Sea

Since first setting sail on the open ocean, people have known that ocean water moves in somewhat continuous, recognizable flows, called **currents**, at velocities of up to several kilometers per hour, so sailors sometimes refer to them as rivers in the sea. But, unlike rivers on land, currents are not confined by physical channels. Rather, oceanographers define their boundaries by detecting changes in velocity. Currents cause water to circulate around an ocean and among oceans at two levels: *surface currents* affect the upper hundred meters of water, whereas *deep currents* affect deeper levels.

SURFACE CURRENTS

Surface currents trace complicated pathways across the world's oceans **(Fig. 15.7)**. They result from interaction between the sea surface and the wind, for as moving air molecules shear across the surface of the water, friction between air and water drags the water along with it. If the Earth did not spin, currents would flow in the same direction as the wind. The Earth's rotation, however, generates the **Coriolis effect (Box 15.1)**, a phenomenon that causes surface currents in the northern hemisphere to veer toward the right and surface currents in the southern hemisphere to veer toward

BOX 15.1 CONSIDER THIS...

The Coriolis Effect and the Generation of Gyres

Imagine that you fire a cannon due north from a point on the Earth's surface in the northern hemisphere. If the Earth were standing still, the cannonball would head due north. But the Earth doesn't sit still; rather, it rotates on its axis. Because of this rotation, a point at the equator zips by at about 1,665 km per hour, relative to a fixed point sitting in space above the equator. On the surface of a spinning object, the path of a moving object undergoes a deflection. In the northern hemisphere, an object initially heading north, like the cannonball we described above, or a volume of water in a current or a volume of air in the wind, deflects to the east **(Fig. Bx15.1a)**. Similarly, an object initially heading east deflects to the south, an object heading south deflects to the west, and an object heading west deflects to the north. In fact, if there were no friction or air resistance to slow a moving object

down, and if it were able to travel a long enough distance without falling back to the Earth's surface, it would follow a clockwise circle in the northern hemisphere **(Fig. Bx15.1b)**. We would see a mirror image of this motion in the southern hemisphere, so that an object would follow a counterclockwise circle. The diameter of these *inertial circles* depends on the velocity and latitude of the object. Researchers refer to the deflection of a moving object on the surface of a spinning object as the **Coriolis effect**, named for the French physicist who first explained it. The apparent force that causes the deflection is known as the *Coriolis force*.

We'll have to leave a more detailed physical explanation of the Coriolis force on the Earth's surface for a more advanced book, for such an explanation requires delving into the relative contributions of centrifugal force and gravitational force acting on a moving

FIGURE 15.7 A simplified map showing the major surface currents of the world's oceans. The arrows indicate flow direction, and colors indicate temperature (red = warm; bule = cold). The inset shows a detail of the Gulf Stream, which carries warm water to the northeast. Note that eddies form along its edge.

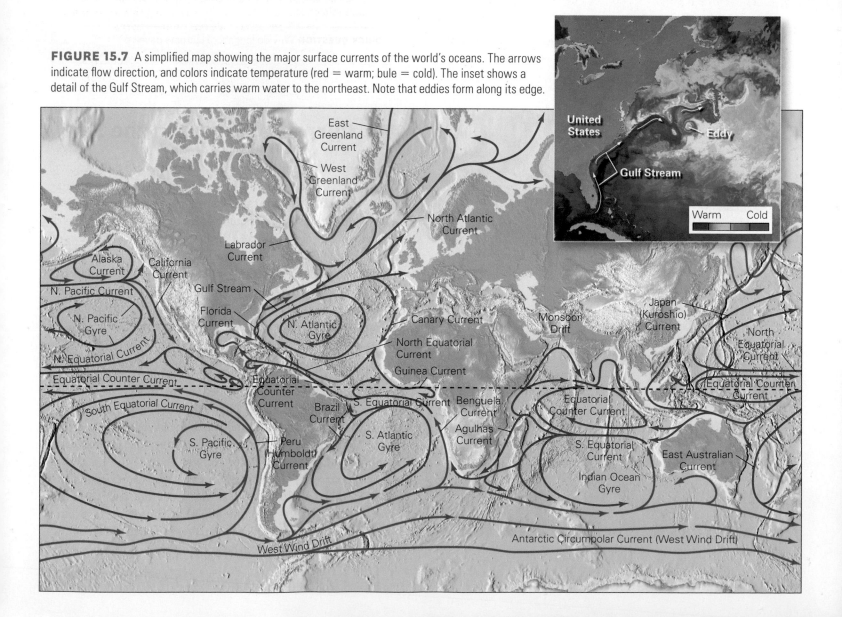

FIGURE Bx15.1 The Coriolis effect occurs because the velocity of a point at the equator, in the direction of the Earth's spin, is greater than that of a point near the pole.

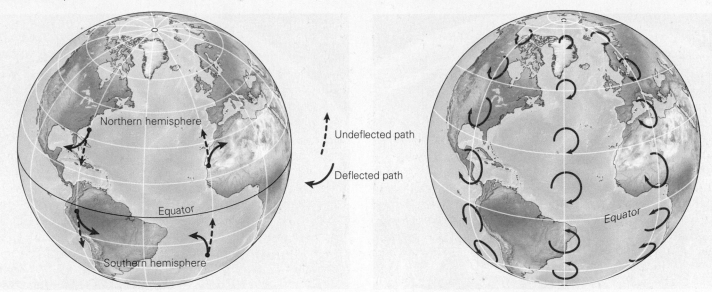

(a) In the northern hemisphere, the Coriolis effect causes moving objects to veer right. In the southern hemisphere, they veer left.

(b) If there were no friction, a moving object would follow an inertial circle due to the Coriolis effect. The diameter of the circle depends on the object's latitude and velocity.

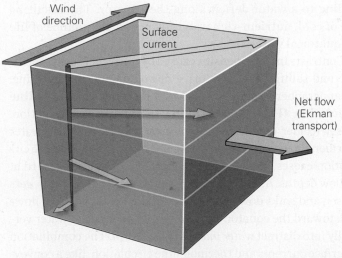

(c) Adding together the flow variations with depth leads to a total flow at right angles to the wind direction. This flow is Ekman transport.

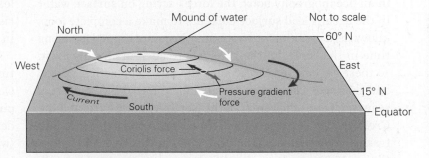

(d) Due to Ekman transport (indicated by white arrows), a mound of water builds in the center of an ocean. The resulting pressure-gradient force balances the Coriolis force and causes oceanic-scale gyres to flow around the mound.

object. Here, we focus instead on the profound influence that the Coriolis effect has on the development of the large gyres that circulate water around an ocean.

The process of gyre formation starts when the wind blows, and moving air molecules exert a drag on the surface of the water, causing water molecules at the surface to move. In the northern hemisphere, the Coriolis effect causes these molecules to deflect to the right. Molecules at the water surface exert drag on water molecules in the layer just below the surface. Due to the Coriolis effect, the molecules of this slightly deeper layer deflect even further to the right. Molecules in this deeper layer, in turn, exert drag on molecules in the next layer below, and these molecules deflect even further to the right. Deflection increases progressively downwards. Due to this pattern, calculations show that a volume of

near-surface ocean water moves, overall, almost at right angles to the direction that the wind above flows. (Oceanographers refer to this pattern of movement as *Ekman transport*; **Fig. Bx15.1c**).

Given the slow velocity of ocean currents, inertial circles due to the Coriolis effect are rather small (tens to hundreds of kilometers in diameter). In contrast, the gyres that circulate around the oceans are thousands of kilometers across, so they are not inertial circles. Gyres form because the deflection of surface water causes it to build into a subtle mound, only 1 to 3 m high, but as much as 3,000 km across, in the middle region of the ocean. Because of gravity, the uplift of the ocean surface in this mound produces an outward-directed force called the *pressure-gradient force*. To picture this force, imagine a mound of maple syrup on a plate; gravity causes the syrup to flow outward. When the outward force equals the inward-directed Coriolis force, a situation known as *geostrophic balance* exists, causing ocean water overall to follow a roughly circular path—a gyre—around the subtle mound of water **(Fig. Bx15.1d)**. Needless to say, understanding the details of currents in the ocean is complex.

FIGURE 15.8 Animations of ocean currents, based on data collected over a 2-year period, show the eddies that develop where currents shear against the margins of the continents or squeeze between land masses.

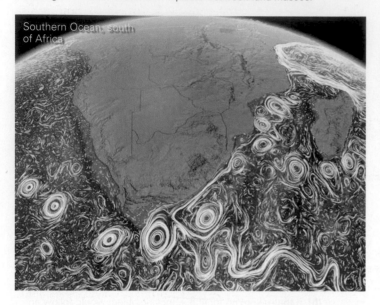

Southern Ocean, south of Africa

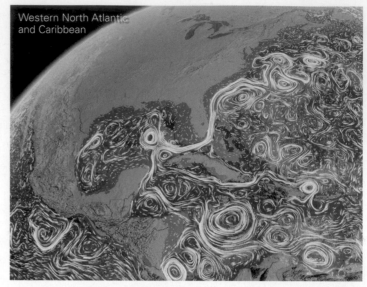

Western North Atlantic and Caribbean

the left of the average wind direction. For reasons discussed in an oceanography book, the forces acting on surface water in the oceans lead surface currents to make a complete loop, known as a *gyre*. Surface water may become trapped for a long time in the center of the gyre, where currents hardly exist, so these regions tend to accumulate floating plastics, sludge, and seaweed. The *Sargasso Sea*, named for a kind of floating seaweed, lies at the center of the North Atlantic gyre, and the *Great Pacific Garbage Patch*, a mass of floating plastic and trash, accumulated at the center of the North Pacific gyre. Figure 15.7 provides a simplified map of currents. In detail, interactions between currents and coastlines, or between currents and neighboring still water, or between currents flowing in different directions or at different rates, produce *eddies*, in which water circulates in small loops, or local map-view undulations **(Fig. 15.8)**.

UPWELLING AND DOWNWELLING

Surface water exchanges with deeper water in the ocean at a number of locations. Specifically, in *downwelling zones*, surface water sinks, and in *upwelling zones*, deeper water rises. Such vertical motion can have a number of causes. For example, where winds cause surface water to flow shoreward, an oversupply of water develops along the coast, so surface water must sink to make room. And where winds cause surface water to flow away from the shore, a deficit of water develops along the coast, so deeper water must rise to fill the gap. Upwelling also occurs near the equator, where winds blow steadily from east to west, because the Coriolis effect deflects surface currents to the right in the northern

hemisphere and to the left in the southern hemisphere, leading to a water deficit along the equator. The resulting rise of cool, nutrient-rich water fosters an abundance of life in equatorial waters.

Contrasts in water density, caused by differences in temperature and salinity, can also drive upwelling and downwelling. Oceanographers refer to this movement as **thermohaline circulation (Fig. 15.9a)**. During thermohaline circulation, denser (colder or saltier) water sinks, while less-dense water (warmer or less salty) rises. On a global scale, thermohaline circulation causes warm, equatorial water to move northward at shallow depths. As this water enters polar latitudes it cools, gets denser, and sinks down to the floor of the ocean, where it flows back toward the equator. This process divides ocean water vertically into distinct *water masses* **(Fig. 15.9b)**. The combination of surface currents and thermohaline circulation, like a conveyor belt, moves water among the various ocean basins. Because water carries heat, this circulation affects global climate.

TAKE-HOME MESSAGE

The friction between wind and water at the ocean surface generates surface currents. Currents do not flow parallel to the wind, however, because of the Coriolis effect. Overall, currents carry water in a big loop, a gyre, around an ocean. Smaller eddies form along the gyre. Water can also undergo upwelling and downwelling.

QUICK QUESTION What is thermohaline circulation, and why does it form?

FIGURE 15.9 Global-scale upwelling and downwelling of ocean water.

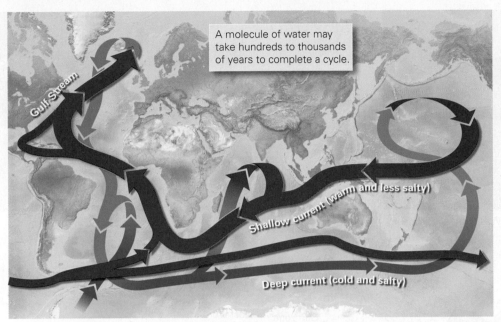

(a) Thermohaline circulation results in a global-scale conveyor belt that circulates water throughout the entire ocean system. The ocean mixes entirely in a 1,500-year period.

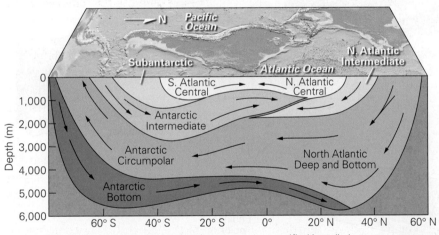

(b) Due to variations in density, the oceans are stratified into distinct water masses.

15.6 Where Land Meets Sea: Coastal Landforms

Tourists along the Amalfi Coast of Italy thrill to the sound of waves crashing on rocky shores, but in the Virgin Islands, sunbathers can find seemingly endless white sand beaches. Along the Mississippi delta, vast swamps border the sea, whereas large, dome-like mountains rise directly from the sea in Rio de Janeiro, Brazil, and a 100-m-high vertical cliff marks the boundary between the Nullarbor Plain of southern Australia and the Great Southern Ocean. As these examples illustrate, *coasts*, the belts of land bordering the sea, vary dramatically in terms of topography and associated landforms. Below, we introduce some of the components of coastal landscapes, and explain how they can change over time.

BEACHES

For millions of vacationers, the ideal holiday includes a trip to a **beach**, a gently sloping fringe of sediment along the shore. Some beaches consist entirely of sand, others are made up of pebbles or cobbles, and still others have a mixture of sand, pebbles, and cobbles (**Fig. 15.10a, b**). The character of sediment on beaches is no accident—waves pick up and stir beach sediment, and over time they winnow out silt and clay grains. These finer grains get carried to quieter water, where they settle. Cobble beaches exist primarily where new blocks fall from nearby cliffs onto the beach, or where fast mountain streams spill their load of coarse sediment into the sea at their outlet. Waves smash them against one another with enough force to round their edges, and may even shatter them into smaller pieces.

Sand, once formed, will not break down into smaller clasts even in a storm wave, for sand grains can't strike each other with enough force to cause fracturing. The composition of sand varies from beach to beach, because different sands come from different sources. Sand derived from the weathering and erosion of felsic to intermediate rocks consists mainly of quartz, for other minerals in these rocks chemically weather to form clay, which washes away. Beaches made from eroded limestone or broken-up coral reefs and shells consist of carbonate sand. And beaches derived from the erosion of basalt boast black sand, made of tiny basalt grains.

A **beach profile**, a cross section drawn perpendicular to the shore, illustrates the shape of a beach (**Fig. 15.10c**). Starting from the sea and moving landward, a beach consists of several zones: a *foreshore zone* or *intertidal zone*, across which the tide rises and falls; the *beach face*, a steeper, concave-up part of the foreshore zone,

Did you ever wonder . . . why beautiful sandy beaches don't form along all coasts?

FIGURE 15.10 Characteristics of beaches, barrier islands, and tidal flats.

(a) A gravel beach along the Olympic Peninsula, Washington. The clasts were derived from erosion of adjacent cliffs.

(b) A sand beach in Puerto Rico. Wave action has carried away finer sediment.

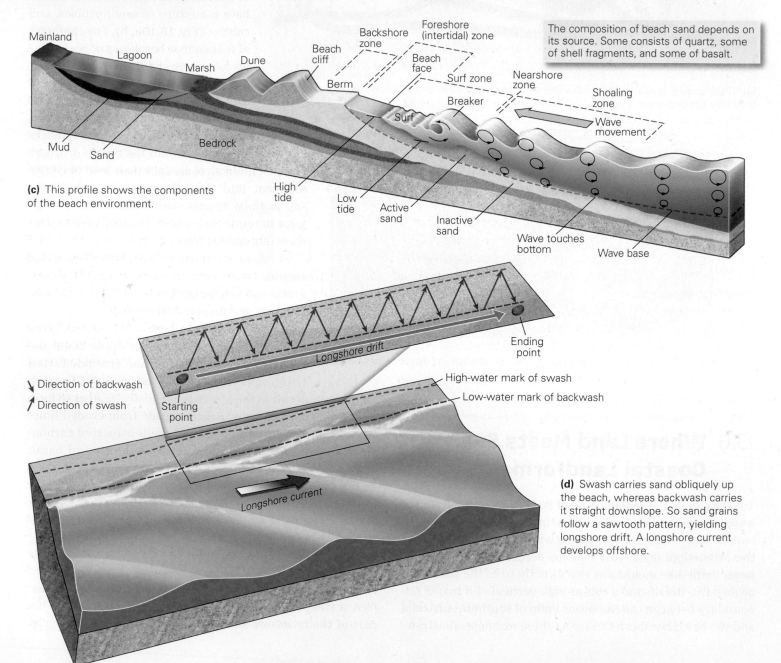

The composition of beach sand depends on its source. Some consists of quartz, some of shell fragments, and some of basalt.

(c) This profile shows the components of the beach environment.

Mainland · Lagoon · Marsh · Dune · Beach cliff · Berm · Backshore zone · Foreshore (intertidal) zone · Beach face · Surf zone · Nearshore zone · Shoaling zone · Breaker · Surf · Wave movement · Mud · Sand · Bedrock · High tide · Low tide · Active sand · Inactive sand · Wave touches bottom · Wave base

↘ Direction of backwash
↗ Direction of swash

Longshore drift · Ending point · Starting point · High-water mark of swash · Low-water mark of backwash · Longshore current

(d) Swash carries sand obliquely up the beach, whereas backwash carries it straight downslope. So sand grains follow a sawtooth pattern, yielding longshore drift. A longshore current develops offshore.

FIGURE 15.11 Barrier islands and tidal flats.

(a) A satellite image showing barrier islands off the coast of North Carolina.

(b) At low tide, boats rest in the mud of a tidal flat on the coast of Brittany, France.

where the swash of the waves actively scours the sand; and the *backshore zone*, which extends from a small step cut by high-tide swash to the front of dunes or cliffs that may lie farther inshore. The backshore zone includes one or more *berms*, horizontal to landward-sloping terraces that receive sediment only during a storm.

When breakers crash onto the shore, they send a surge of water up the beach. This upslope flow, or swash, continues until friction, together with the downslope pull of gravity, brings the water's motion to a halt. Then gravity takes over and draws the water back down the beach as backwash. If the waves' crests trend parallel to the beach face, beach sediment simply moves up and down the slope of the beach face along with the swash and backwash, respectively. Geologists refer to the sediment that moves in response to waves on a calm to moderately windy day as *active sand*; a layer of *inactive sand* lies beneath the active sand. When storms build very large waves, even some of the normally inactive sand will start moving.

Where waves reach the shore at an angle, the active sand moves in a sawtooth pattern that results in the gradual net transport of sediment parallel to the beach, a phenomenon known as **longshore drift**. (Note that longshore drift on the beach moves in the same direction as the *longshore current* just offshore.) The sawtooth pattern happens because the swash of a wave moves perpendicular to the wave crest, so an oblique wave carries sediment diagonally up the beach, but the backwash flows straight down the beach slope due to gravity **(Fig. 15.10d)**. Because of longshore drift, geologists sometimes refer to beaches as "rivers of sand" to emphasize that beach sand moves along the coast over time—it's not a fixed substrate. In fact, longshore drift can transport sand hundreds of kilometers along a coast in a matter of centuries.

Where the coastline indents landward, longshore drift effectively stretches beaches out into open water to create a **sand spit**. Some sand spits grow across the opening of a bay, forming a *baymouth bar*. The scouring action of waves can pile sand up in a narrow ridge away from the shore called an *offshore bar*, which parallels the shoreline. In regions with an abundant sand supply, offshore bars rise above the mean high-water level and become *barrier islands* **(Fig. 15.11a)**. The water between a barrier island and the mainland becomes a quiet-water *lagoon*, a body of shallow seawater separated from the open ocean.

Though developers have covered some barrier islands with expensive resorts, in the time frame of centuries to millennia, barrier islands are temporary features. Wind and waves pick up sand from the ocean side of the barrier island and drop it on the lagoon side, causing the island to migrate landward, and longshore drift gradually transports the sand of the barrier island, thereby modifying the island's shape.

Tidal flats, as noted earlier, are regions of mud and silt exposed or nearly exposed at low tide but totally submerged at high tide. They develop in regions protected from strong wave action **(Fig. 15.11b)**. Tidal flats may be found along the margins of lagoons or on shores protected by barrier islands.

ROCKY COASTS

Many a ship has met its end smashed and splintered in the spray and thunderous surf of a rocky coast, where bedrock cliffs rise directly from the sea. Rocky coasts may feel the full force of ocean breakers. Water pressure generated during the impact of a breaker can pick up boulders and smash them together until they shatter, and it can squeeze air into cracks, creating enough force to split off fragments. Further, because of its turbulence, the water hitting a cliff face carries suspended sand and thus can abrade the cliff. The combined effects of shattering, wedging, and abrading, together called *wave erosion*, can gradually undercut a cliff face and make a *wave-cut notch* (**Fig. 15.12a, b**). Undercutting continues until the overhang becomes unstable, breaks away at a joint, and falls to produce a pile of rubble at the base of the cliff that waves immediately attack and break up. As wave erosion cuts away at a rocky coast, the position of the cliff face gradually migrates inland. Such cliff retreat may leave behind a *wave-cut platform* that becomes visible at low tide (**Fig. 15.12c**).

Many rocky coasts are irregular, with *headlands* protruding into the sea and *embayments* set back from the sea. Wave refraction focuses wave energy on headlands, which accelerates their erosion, whereas it disperses energy in embayments and allows them to be sites of deposition (**Fig. 15.12d**). In some cases, a headland erodes in stages (**Fig. 15.12e**). As waves curve, they attack the sides of a headland and slowly erode through it to produce a *sea arch* connected to the

FIGURE 15.12 Erosion landforms of rocky shorelines.

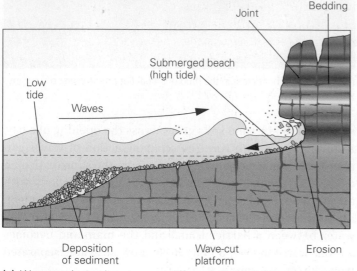

(a) Wave erosion undercuts a sea cliff, producing a notch and a platform.

(b) A wave-cut notch.

(c) A wave-cut platform at the foot of the cliffs at Étretat, France.

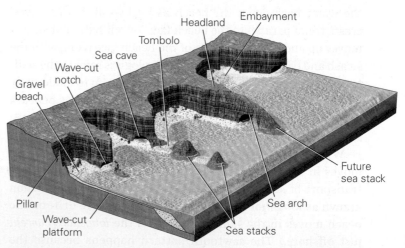

(d) Landforms of a rocky shore. Beaches collect in embayments, whereas erosion concentrates at headlands.

(e) Coastal erosion along Australia's southern coast produced a sea arch (left). Eventually, the bridge will collapse and only sea stacks will remain (right). These sea stacks are among several that together are known locally as the Twelve Apostles.

FIGURE 15.13 The Chesapeake Bay estuary formed when the sea flooded river valleys. The region is sinking relative to other coast areas because it overlies a buried meteor crater.

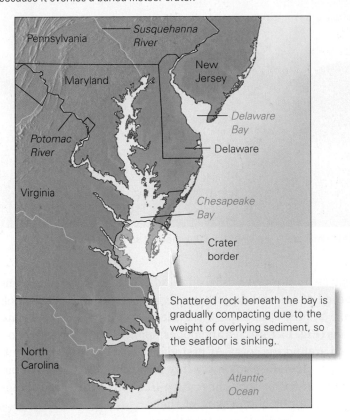

Shattered rock beneath the bay is gradually compacting due to the weight of overlying sediment, so the seafloor is sinking.

a relatively narrow body of water in which seawater and river water interact. If an estuary fills the mouth of a dendritic drainage network, the estuary will have a complex coastline with many finger-like inlets **(Fig. 15.13)**. Oceanic and fluvial waters interact in two ways within an estuary. In *quiet estuaries*, protected from wave action or river turbulence, the water becomes stratified, with denser oceanic saltwater flowing upstream as a wedge beneath less-dense fluvial freshwater. In *turbulent estuaries*, oceanic and fluvial water mix to produce nutrient-rich brackish water with a salinity between that of oceans and rivers. Estuaries are complex ecosystems inhabited by unique species of shrimp, clams, oysters, worms, and fish that can tolerate large changes in salinity.

During the last ice age, glaciers carved deep valleys in coastal mountain ranges. When the ice age came to a close, the glaciers melted away, leaving deep, steep-sided valleys. Water stored in the glaciers, along with the water within the vast ice sheets that covered continents during the ice age, returned to the sea and caused sea level to rise. The rising sea filled the deep, glacially carved valleys, yielding **coastal fjords**, relatively narrow and long finger-like bays. Because of their deep-blue water and steep walls of glacially polished rock, they are distinctly beautiful **(Fig. 15.14)**.

SEE FOR YOURSELF...

Fjords of Norway

LATITUDE
60°53'56.09" N

LONGITUDE
5°12'31.84" E

Look down from 80 km.

During the last ice age, glaciers carved deep, steep-sided valleys into the mountains of western Norway. When the glaciers melted and sea level rose, the valleys filled with water, forming fjords.

mainland by a narrow bridge. When the arch eventually collapses, an isolated *sea stack* remains offshore.

ESTUARIES AND FJORDS

Along some coastlines, a relative rise in sea level causes the sea to flood a river valley where it enters the coast. As a result, the near-coast reach of the river becomes an **estuary**,

FIGURE 15.14 Fjord landscapes form where relative sea-level rise drowns glacially carved valleys.

Fjord

Glacial valleys have a U shape, so fjords have steep sides.

A fjord in Norway

FIGURE 15.15 Examples of coastal wetlands.

(a) A salt marsh along the coast of Cape Cod.

(b) A mangrove growing along the shore of southern Florida.

ORGANIC COASTS

Coasts along which living organisms control landforms are called **organic coasts**. The nature of an organic coast depends on the type of organisms that live there, which, in turn, depends on climate.

A **coastal wetland** is a vegetated, flat-lying stretch of coast that floods at high tide but does not feel the impact of strong waves. In temperate climates, coastal wetlands include *swamps* (wetlands dominated by trees), *marshes* (wetlands dominated by grasses; **Fig. 15.15a**), and *bogs* (wetlands dominated by moss and shrubs). Despite their relatively small area, when compared with the oceans as a whole, wetlands account for 10% to 30% of marine organic productivity.

In tropical or subtropical climates (between 30° N and 30° S), *mangrove swamps* thrive in wetlands (**Fig. 15.15b**). Mangrove tree roots can filter salt out of water, so the trees have evolved to survive in either freshwater or saltwater. Some mangrove species form a broad network of roots above the water surface, making the plant look like an octopus standing on its tentacles. Some trees send up small protrusions from roots that rise above the water and allow the plant to breathe.

Along the azure coasts of some tropical regions, visitors can swim through a different kind of organic coast, one composed of colorful growths of living coral. Some corals look like brains, others like elk antlers, still others like delicate fans (**Fig. 15.16a**). Sea anemones, sponges, and clams grow on and around the coral. At first glance coral looks like a plant, but it's actually a colony of tiny invertebrates. An individual coral animal, or polyp, has a tube-like body with a head of tentacles. Corals obtain part of their livelihood by filtering nutrients out of seawater; the remainder comes from algae that live on the corals' tissue. Corals have a symbiotic (mutually beneficial) relationship with the algae in that the algae photosynthesize and provide nutrients and oxygen to the corals, while the corals provide carbon dioxide and nutrients to the algae.

Coral polyps secrete calcite shells, which gradually build into a mound of solid limestone whose top surface can extend from just below the low-tide level down to a depth of about 60 m. At any given time, only the surface of the mound is alive—the mound's interior consists of shells from previous generations of now-dead coral. The realm of shallow water underlain by coral mounds, associated organisms, and debris composed of loose shells and reef fragments comprises a **coral reef**. Reefs absorb wave energy and thus serve as a living

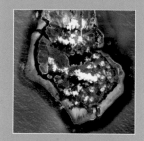

FIGURE 15.16 The character and evolution of coral reefs.

A reef near Honolulu, Hawaii

(a) The surface of this Hawaiian coral reef protects the shore from wave erosion. The underwater view emphasizes that a reef hosts a great variety of organisms. A close-up shows individual coral polyps.

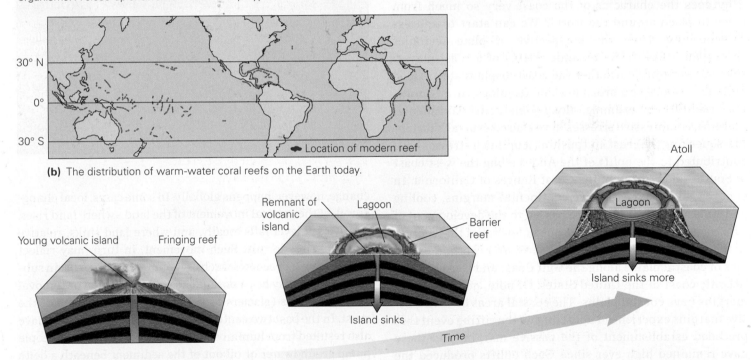

30° N

0°

30° S

← Location of modern reef

(b) The distribution of warm-water coral reefs on the Earth today.

Young volcanic island   Fringing reef

Remnant of volcanic island   Lagoon   Barrier reef

Island sinks

Atoll   Lagoon

Island sinks more

Time

(c) The progressive evolution of a reef surrounding an oceanic island.

buffer zone that protects coasts from erosion. To thrive, coral reefs need clear, well-lit, warm (18°C to 30°C) water with normal oceanic salinity, as can exist along coasts at latitudes of 0° to 30° **(Fig. 15.16b)**.

When Charles Darwin was a young naturalist on HMS *Beagle*, he noticed that some islands of the tropics had *fringing reefs*, reefs that extend right up to the shore **(Fig. 15.16c)**, whereas others were surrounded by a *barrier reef* that lay offshore, separated from the island by a quiet lagoon. In yet other locations, a roughly circular coral reef, called an *atoll*,

surrounded a lagoon without an island in the middle. Darwin correctly surmised that a fringing reef evolves into a barrier reef, which in turn evolves into an atoll, as the island undergoes subsidence. As we've seen, this happens both because the surface of the oceanic lithosphere sinks beneath the island as the lithosphere ages, and because the island erodes and slumps away. Eventually, the reef itself sinks below sea level, for the rate of coral growth can't keep up with the rate of subsidence. When this happens, the resulting seamount has a flat top—geologists refer to a flat-topped seamount as a **guyot**.

15.7 Causes of Coastal Variability

PLATE-TECTONIC SETTING

Why does the character of the coast vary so much from place to place around the world? We can start to address this question from the perspective of plate tectonics theory. Specifically, the tectonic setting of a coast plays a role in determining whether the coast displays steep-sided mountain slopes or a broad lowland (**Geology at a Glance**, pp. 518–519). For example, along some active continental margins, compressive stresses have caused crustal shortening, squeezing the crust and pushing it up. Such stresses have contributed to the uplift of the Andes along the west coast of South America and of the Coast Ranges of California. In contrast, along some passive continental margins, cooling and sinking of the lithosphere has led to the development of a broad **coastal plain**, a landscape of low relief that merges at the shore with the continental shelf. You can find examples of coastal plains along the Gulf Coast and southeastern Atlantic coast of the United States. Of note, not all passive margins have coastal plains. The coastal areas of some passive margins experienced uplift during the rifting event that preceded establishment of the passive margin, and these have remained high ever since. Such uplifts produced the present-day highlands bordering the Red Sea, southeastern Brazil, southern African coasts, and eastern Australia.

RELATIVE SEA-LEVEL CHANGES

Sea level, relative to the land surface, changes during geologic time. Some sea-level changes affect the ocean worldwide. Such *eustatic sea-level changes* may reflect global variations in seafloor-spreading rates, for the volume of a ridge depends on spreading rate (see Chapter 2). Changes may also reflect growth or shrinkage of ice-age glaciers, which store water on land and keep it out of the sea (**Fig. 15.17**). Not all sea-level

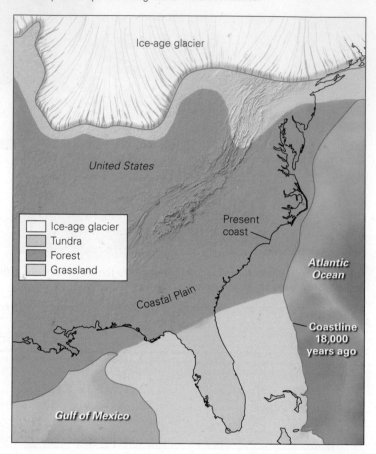

FIGURE 15.17 Because of sea-level drop during the ice age, there was more dry land exposed along eastern North America.

change, however, happens globally. In some cases, local changes reflect the vertical movement of the land—where land rises, relative sea level falls locally, and where land sinks, relative sea level rises locally. Such movement, in turn, may reflect plate-tectonic processes, such as mountain uplift or basin subsidence, or it may be a consequence of the addition or removal of huge weights (glaciers or volcanoes) on the surface of the crust. In the past two centuries, local changes in sea level have also resulted from human activity. For example, where people pump groundwater or oil out of the sediment beneath a delta plain, the sediment packs together more tightly, and thus takes up less volume, so the land surface sinks.

Geologists refer to coasts where the land is currently rising or rose in the past, relative to sea level, as *emergent coasts*. At emergent coasts, steep slopes typically border the shore, and in some locations, a series of step-like terraces form (**Fig. 15.18a**). These terraces reflect uplift of the coast as well as episodic changes in relative sea level. Coasts where the land has been sinking relative to sea level are *submergent coasts* (**Fig. 15.18b**). At such places, landforms include estuaries and fjords that develop when the rising sea floods coastal valleys.

FIGURE 15.18 Features of emergent coastlines (relative sea level is falling) and submergent coastlines (relative sea level is rising).

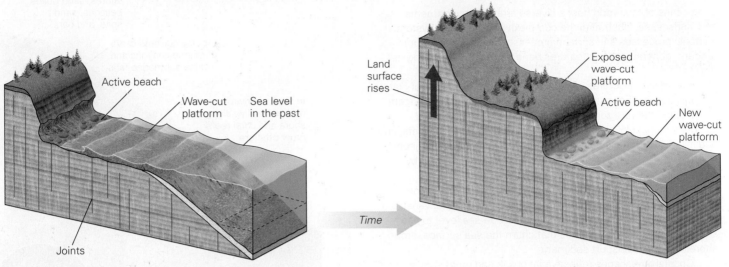

(a) *Emergent coasts:* Wave erosion produces a wave-cut platform along an emergent coast. As the land rises, the platform becomes a terrace, and a new wave-cut platform forms.

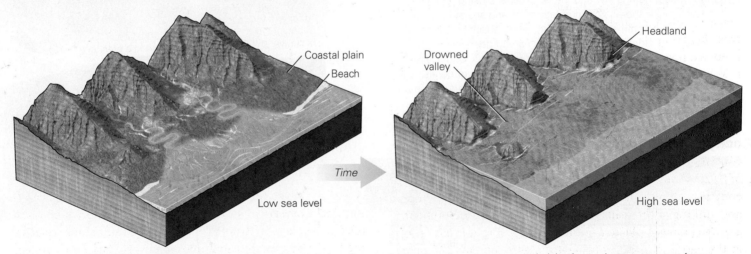

(b) *Submergent coasts:* A coast before sea level rises. Rivers drain valleys and deposit sediment on a coastal plain. As a submergent coast forms, sea level rises and floods the valleys, and waves erode the headlands.

SEDIMENT SUPPLY AND CLIMATE

The type and amount of sediment supplied to a coast also affects its character. At an *erosional coast*, waves wash sediment away faster than it can be supplied, so the coast becomes rocky, whereas at an *accretionary coast*, more sediment arrives than washes away, so sediment builds outward into a broad beach or even a coastal plain. Whether a local stretch of coastline becomes erosional or accretionary depends on sediment supply and climate. Specifically, for given climate conditions, a coast may become erosional if the sediment supply decreases, but it may become accretionary if the sediment supply increases. And for a given sediment supply, a coast can become erosional if subjected to frequent storms but accretionary if it enjoys calm weather. Climate

also influences rates of erosion by affecting the productivity of organisms that can protect the coast from wave erosion.

TAKE-HOME MESSAGE

Many aspects of the Earth System influence the character of a given coastline. Examples include tectonic activity, sediment supply, and climate. At an emergent coast, land rises relative to sea level, whereas at a submergent coast, it sinks. Relative sea level at a location can rise or fall over time. Sediment supply, relative to erosion rates, also affects the character of a coast by determining whether beaches build outward or erode away.

QUICK QUESTION What landforms develop at submergent coasts?

Oceans and Coasts

The oceans of the world host a diverse array of environments and landscapes, illustrating the complexity of the Earth System. Tectonic processes and surface processes, working alone or in tandem, generate unique features beneath the sea and along its coasts.

The major structures of the ocean floor are the result of plate-tectonic activity. For example, mid-ocean ridges define divergent boundaries, fracture zones form along transform faults, and trenches mark subduction zones. Oceanic islands, seamounts, and oceanic plateaus build above hot spots. Along passive continental margins, broad continental shelves develop, locally incised by submarine canyons.

Within the ocean, currents circulate water due to friction between wind and the water surface as well as regional variations in water salinity and temperature. Tides cause sea level to rise and fall, and wind builds waves that churn the sea surface, erode shorelines, and transport sediment.

Coastal landscapes reflect variations in sediment supply, relative sea-level rise or fall, and climate. For example, where the supply of sediment is low and the landscape is rising relative to sea level, rocky shores with dramatic cliffs and sea stacks may evolve. Where sediment is abundant, sandy beaches and sand spits develop. Regions where glaciers carved deep valleys now feature spectacular fjords. Protected coastal areas, especially those in warm climates, host grasses, mangroves, or corals. Corals may contribute to growth of broad reefs along the shore. Human activities can significantly affect coastal landscapes.

Rocky Coast Evolution

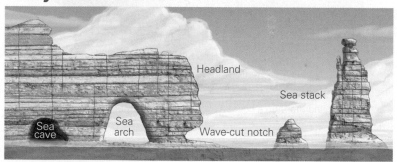

As waves erode rocky coasts, bedrock breaks away at joints and along bedding planes. As a result, several distinct landforms evolve over time.

Water Masses

The oceans are stratified into distinct layers. The coldest, densest layer (Antarctic bottom water) sinks along the coast of Antarctica and flows along the bottom. Winds generate surface currents in the upper layer.

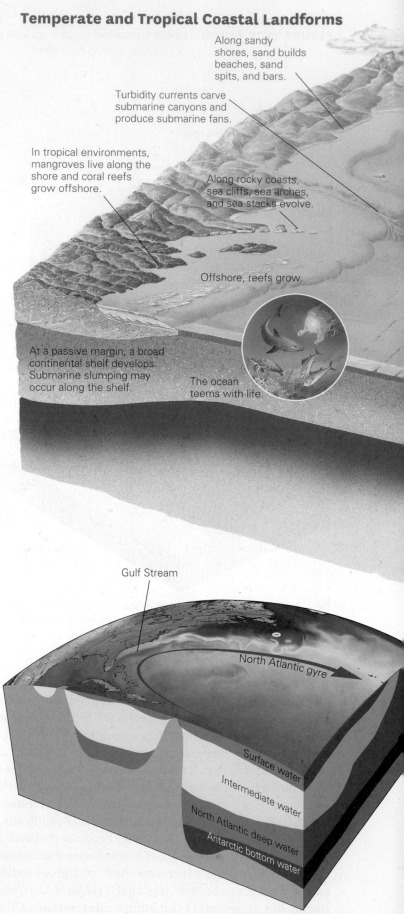

Temperate and Tropical Coastal Landforms

Along sandy shores, sand builds beaches, sand spits, and bars.

Turbidity currents carve submarine canyons and produce submarine fans.

In tropical environments, mangroves live along the shore and coral reefs grow offshore.

Along rocky coasts, sea cliffs, sea arches, and sea stacks evolve.

Offshore, reefs grow.

At a passive margin, a broad continental shelf develops. Submarine slumping may occur along the shelf.

The ocean teems with life.

Gulf Stream

North Atlantic gyre

Surface water

Intermediate water

North Atlantic deep water

Antarctic bottom water

Bathymetry of the Seafloor

Polar and Mountainous Coastal Landforms

Tidewater glaciers produce icebergs.

At high latitudes, fjords form when the rising sea floods glacially carved valleys

A river transports sediment to a delta.

Hot spots build chains of oceanic islands. Only the youngest island of the chain is active.

Mid-ocean ridge

Abyssal plain

Seamounts and guyots are relicts of hot spots.

At a convergent boundary, a trench bordered by an accretionary prism develops.

Trench

At a divergent boundary, a mid-ocean ridge rises. Transform faults, marked by fracture zones, link segments of the ridge.

Volcanic arcs form along active-margin coasts.

The wind forms ocean waves. As a wave passes, water moves in a circular pattern.

A lagoon, in which mud accumulates, may be trapped behind the dunes.

Waves and Beaches

Near the shore, the top of the wave breaks over the base of the wave. Swash carries sand up the beach, and backwash carries sand back.

Beach

Sand may pile into dunes.

15.8 Coastal Problems and Solutions

CONTEMPORARY SEA-LEVEL CHANGES

People tend to view a shoreline as permanent. But in fact, shorelines are ephemeral geologic features. On a time scale of hundreds to thousands of years, a shoreline moves inland or seaward depending on whether relative sea level rises or falls or whether sediment supply increases or decreases. In places where sea level is rising today, shoreline towns will eventually be submerged. For example, if present rates of sea-level rise continue along the east coast of the United States, major coastal cities such as Washington, New York, Miami, and Philadelphia may be underwater within the next millennium **(Fig. 15.19)**.

BEACH DESTRUCTION—BEACH PROTECTION?

The loss of river sediment due to installation of a dam upstream, a change in the shape of a shoreline, a change in wave height, or the destruction of coastal vegetation, can all alter the balance between sediment accumulation and sediment removal on a beach. **Beach erosion**, the net decrease in sediment along a beach and the consequent decrease in beach width,

happens when sediment removal rates exceed sediment accumulation rates. Storms can accelerate beach erosion. In only a few hours, a storm can radically alter a landscape that had taken centuries or millennia to form. The backwash of large storm waves sweeps vast quantities of sand seaward, leaving the beach a skeleton of its former self. In addition, surf submerges barrier islands and shifts them toward the lagoon. Storms can rip out mangrove swamps and salt marshes and break up coral reefs, destroying the organic buffer that can protect a coast, and leaving it vulnerable to erosion for years to come. Because people build along shorelines, beach erosion can be very costly, undermining shoreline buildings and causing them to collapse into the sea. In some cases, *storm surge*, the very high water levels generated when storm winds push water toward the shore, can float buildings off their foundations **(Fig. 15.20)**.

The value of a luxury resort on a coast may depend on the accessibility to a scenic beach, and a harbor can't function if its mouth gets blocked by sediment. In an attempt to prevent or delay such problems, property owners often construct artificial barriers to alter the natural movement of sand along the coast. Unfortunately, such constructions can bring undesirable results. For example, beachfront property owners may build *groins*, concrete or stone walls protruding perpendicular to the shore, to prevent longshore drift from removing sand **(Fig. 15.21a)**. Sand accumulates on the updrift side of the groin, forming a long triangular wedge, but sand erodes away on the downdrift side. Needless to say, the property owner on the downdrift side doesn't appreciate this change. Harbor engineers may build a pair of walls called *jetties* to protect the entrance to a harbor **(Fig. 15.21b)**. But jetties erected at the mouth of a river channel effectively extend the river into deeper water and thus may lead to the deposition of an offshore sandbar. Engineers may also build an offshore wall called

FIGURE 15.19 Areas that may be submerged if sea level rises by about 1.0 to 1.5 m.

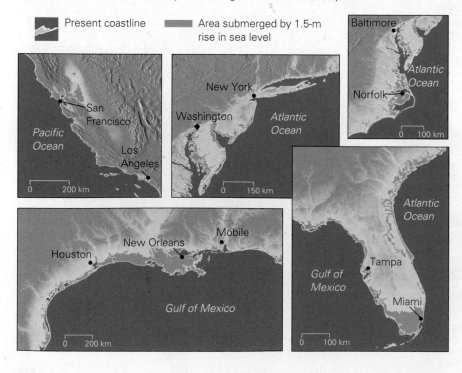

Present coastline | Area submerged by 1.5-m rise in sea level

a *breakwater*, parallel or at an angle to the beach, to prevent the full force of waves from reaching a harbor. With time, however, sand builds up in the lee of the breakwater and the beach grows seaward, clogging the harbor **(Fig. 15.21c)**. To protect expensive shoreline construction, people build *seawalls* out of riprap (large stone or concrete blocks) or reinforced concrete on the landward side of the beach **(Fig. 15.21d)**, but during a storm, these can be undermined.

In some places, people have given up trying to decrease the rate of beach erosion and instead have worked to increase the rate of sediment supply. To do this, they pump sand from farther offshore or bring sand from elsewhere to replenish a beach. This procedure, called *beach nourishment*, can be hugely expensive, and at best it provides only a temporary fix, for the backwash and beach drift that removed the sand in the first place continue unabated as long as the wind blows and the waves break.

COASTAL POLLUTION, AND THE DESTRUCTION OF WETLANDS AND REEFS

Coastal regions, both onshore and offshore, host fragile ecosystems that can be negatively impacted by the by-products of modern society. For example, beaches may accumulate

garbage that is left by tourists, dumped by illegal haulers, washed in from offshore, or transported from up the coast by longshore drift. Over the years, oil spills from ships that flush their bilges, tankers that foundered in stormy seas, or offshore seeps and leaks have contaminated shorelines with tar at many locations around the world.

The influx of nutrients into coastal waters, from sewage and from agricultural runoff, can produce dead zones along coasts. A *dead zone*, a region in which water contains so little oxygen that fish and other organisms within it die, forms when the concentration of nutrients rises enough to stimulate an algal bloom. When the algae dies and decomposes, it can deplete oxygen. One of the world's largest dead zones occurs in the northern Gulf of Mexico, where the Mississippi River brings in fertilizer residue from the fields that occupy its watershed.

Coastal wetlands and coral reefs are particularly susceptible to pollution, and many have been destroyed in recent decades. Their loss increases a coast's vulnerability to erosion, and because these areas provide spawning grounds for marine organisms, their loss disrupts the global food chain. The statistics of wetland and reef destruction worldwide are frightening. Ecologists estimate that between 20% and 70% of wetlands have already been destroyed, and along some coasts, 90% of reefs have died.

Destruction of wetlands and reefs happens for many reasons. Wetlands have been filled or drained to be converted to farmland, housing developments, resorts, or garbage dumps. Reefs have been destroyed by boat anchors, dredging, the activities of divers, dynamite explosions intended to kill fish, and quarrying operations in order to obtain construction

FIGURE 15.20 Examples of beach erosion.

September 9, 2008

September 15, 2008

(a) When Hurricane Ike hit Galveston, Texas, in September 2008, many beachfront homes were washed away.

(b) Wave erosion has completely removed the beach and has started to erode a beach cliff along the coast of Cape Cod.

FIGURE 15.21 Techniques used to preserve beaches.

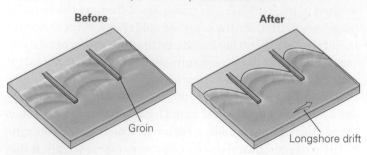

(a) The construction of groins may produce a sawtooth beach.

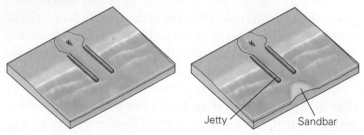

(b) Jetties extend a river farther into the sea but may cause a sandbar to form at the end of the channel.

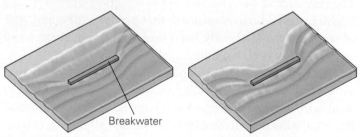

(c) A beach may grow seaward behind a breakwater.

(d) Riprap will slow erosion of a parking lot along a California beach.

materials. Chemicals and particulates entering coastal water from urban, industrial, and agricultural areas can cause havoc in wetlands and reefs, for these materials cloud water or trigger algal blooms, killing filter-feeding organisms. Toxic chemicals in such runoff can also poison plankton and burrowing organisms and, therefore, other organisms progressively up the food chain.

Global climate change also impacts the health of organic coasts (see Chapter 19). For example, transformation of once-vegetated regions into deserts means that the amount of dust carried by winds from the land to the sea has increased. This dust can interfere with coral respiration and can infect coral with dangerous viruses. A global increase in seawater temperature may be contributing to *reef bleaching*, the loss of coral color due to the death of the algae that live in coral polyps. Reef bleaching and other reef diseases have transformed once-living landscapes hosting diverse species of colorful coral into mounds and piles of whitish rubble. Since 2016, most of Australia's Great Barrier Reef has been subjected to record-high water temperatures that rapidly increased an ongoing problem of severe bleaching. Bleaching has already destroyed major portions of the reefs in the Caribbean. In fact, about 30% to 50% of the world's reefs have already died, and by some estimates, hardly any will remain by 2050.

HURRICANES—A COASTAL CALAMITY

What Is a Hurricane? Global-scale convection of the atmosphere, influenced by the Coriolis effect, causes currents of warm air to flow steadily from east to west in tropical latitudes **(Fig. 15.22a)**. As the air flows over the ocean, it absorbs moisture. Because air becomes less dense as it gets warmer, tropical air eventually begins to rise like a balloon. As the air rises, it cools, and the water vapor it contains condenses to form clouds (mists of very tiny water droplets). If the air contains sufficient moisture, the clouds grow into a cluster of large thunderstorms, which consolidate to form a single, very large storm. Because of the Coriolis effect (see Box 15.1), this large storm starts to rotate, and if it remains over warm ocean water, as can happen in late summer and early fall, it grows. The storm gets the energy to continue growing from the evaporation of warm water, a process that releases heat. This heat keeps the moist air within the storm buoyant enough to rise up to the stratosphere. Eventually, the storm evolves into a spiral of intense winds called a *tropical storm*. If a tropical storm becomes powerful enough, it becomes a **tropical cyclone**—a huge rotating storm that forms in tropical latitudes, in which winds exceed 119 km per hour. It resembles a giant counterclockwise spiral of clouds, 300 to 1,500 km wide, when viewed from space **(Fig. 15.22b)**. Such a storm is called a *hurricane* in the Atlantic and eastern Pacific, a *typhoon* in the western Pacific, and simply a *cyclone* around Australia and in the Indian Ocean.

FIGURE 15.22 The structure and distribution of hurricanes.

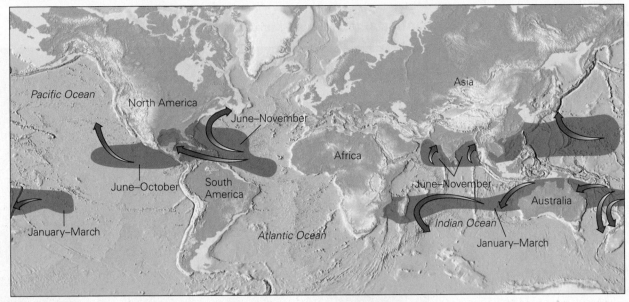

(a) Hurricanes form only in certain regions of the ocean, where water temperatures are high.

(b) Hurricane Katrina approaches the Gulf Coast of the United States in 2005, as seen from space.

(c) In this cutaway drawing, we can see the rain bands, the eye, and the eye wall of a hurricane. Dry air descends in the eye.

Atlantic hurricanes generally form in the ocean to the east of the Caribbean Sea, though some form in the Caribbean itself. At first they drift westward at speeds of up to 60 km per hour, but they may eventually turn north and head into the North Atlantic or into the interior of North America, where they die when they run out of a supply of warm water. Weather researchers classify the strength of hurricanes using the *Saffir-Simpson scale*, which runs from 1 to 5. On the Saffir-Simpson scale, a Category 5 hurricane has sustained winds of 250 km per hour or greater.

A typical hurricane consists of several spiral arms extending inward to a central zone of relative calm known as the *hurricane eye* **(Fig. 15.22c)**. A rotating vertical cylinder of clouds, the *eye wall*, surrounds the eye. As winds spiral toward the eye, they accelerate, so they are fastest at the eye wall. Typically, hurricane-force winds occur within a zone that is only 15% to 35% as wide as the whole storm. On the side of the eye where winds blow in the same direction as the whole storm is moving, the ground speed of winds is greatest, because the storm's overall speed adds to the rotational motion.

Consequences of Hurricanes Hurricanes pose extreme danger in the open ocean, because their winds cause huge waves

BOX 15.2 CONSIDER THIS...

Hurricane Katrina

FIGURE Bx15.2 Hurricane Katrina and the disaster it caused.

Tropical Storm Katrina came into existence over the Bahamas in August, 2005, and headed west. Just before landfall in southeastern Florida, winds strengthened and the storm became Hurricane Katrina. This hurricane sliced across the southern tip of Florida, causing several deaths and millions of dollars in damage. It then entered the Gulf of Mexico and passed directly over an eddy of summer-heated water from the Caribbean that had entered the Gulf of Mexico. This water, which reached a temperature of 32°C, stoked the storm, injecting it with a burst of energy sufficient for the storm to morph into a Category 5 monster whose swath of hurricane-force winds reached a width of 325 km **(Fig. Bx15.2a)**. When it entered the central Gulf of Mexico, Katrina turned north and began to bear down on the Louisiana-Mississippi coast **(Fig. Bx15.2b, c)**. The eye of the storm passed just east of New Orleans, and then

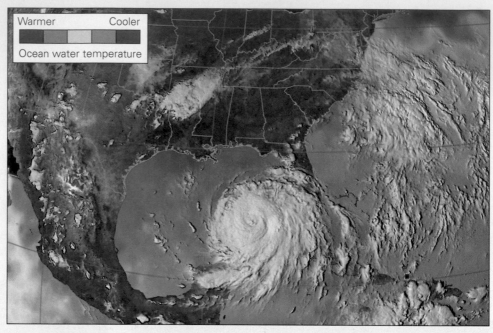

(a) A satellite photo of Hurricane Katrina over the warm water of the Gulf of Mexico.

across the coast of Mississippi. Storm surges broke records, in places rising 7.5 m above sea level, and they washed coastal communities off the map along a broad swath of the Gulf Coast **(Fig. Bx15.2d, e)**.

The winds of Hurricane Katrina ripped off roofs, toppled trees, smashed windows, and triggered the collapse of weaker buildings. Residents of New Orleans gave a sigh of relief when the storm had passed, because wind damage in New Orleans was not catastrophic. But the worst was yet to come. When the winds blew the storm surge into Lake Pontchartrain, the lake's water level rose beyond most expectations, and the weight of elevated water pressed against the network of artificial levees and flood walls that had been built to protect New Orleans, much of which lies below sea level **(Fig. Bx15.2f)**. Hours after the hurricane eye had passed, the high

water of Lake Pontchartrain found a weakness along the floodwall bordering a drainage canal and pushed out a section. Breaks eventually formed in other locations as well. So, a day after the hurricane was over, New Orleans began to flood. As the water level climbed the walls of houses, brick by brick, residents fled to higher levels of their homes, finally to their roofs. Water spread

across the city until the bowl of New Orleans filled to the same level as Lake Pontchartrain, submerging 80% of the city **(Fig. Bx15.2g)**. Floodwaters washed some houses away and filled others with debris **(Fig. Bx15.2h)**. The disaster took on national significance, as the city's trapped inhabitants sweltered without adequate food, drinking water, or shelter. Lacking communications, medical care, and police protection, the city remained in chaos for days. By the time outside relief reached the city, many people had died and parts of New Orleans, a cultural landmark and major port, had become uninhabitable. It has taken years for the city to rebuild.

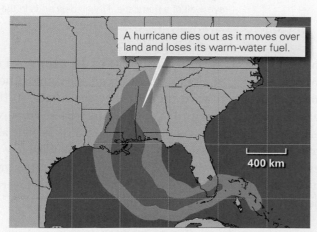

(b) A wind-swath map of Hurricane Katrina. Red areas represent hurricane winds; orange areas represent tropical-storm winds.

A hurricane dies out as it moves over land and loses its warm-water fuel.

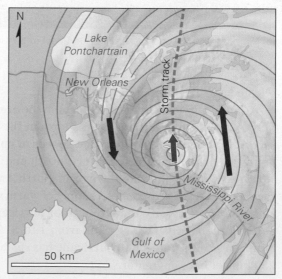

(c) The track of the storm lay just east of New Orleans. Because the storm was moving northward overall (red arrow) and spinning counterclockwise, winds (blue arrows) to the east of the eye were much stronger than those to the west of the eye.

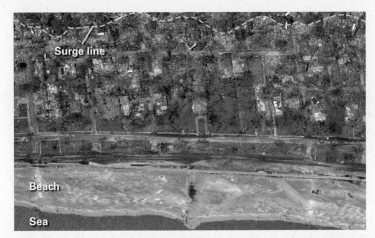

(d) Surge from the storm destroyed homes along the Alabama coast.

(e) Officials survey the storm damage.

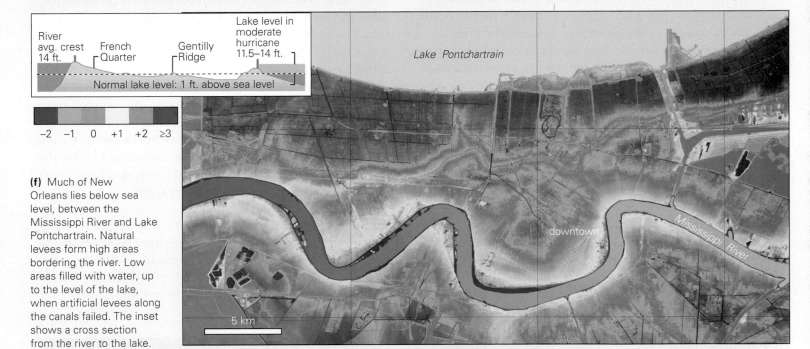

(f) Much of New Orleans lies below sea level, between the Mississippi River and Lake Pontchartrain. Natural levees form high areas bordering the river. Low areas filled with water, up to the level of the lake, when artificial levees along the canals failed. The inset shows a cross section from the river to the lake.

River avg. crest 14 ft. | French Quarter | Gentilly Ridge | Lake level in moderate hurricane 11.5–14 ft.

Normal lake level: 1 ft. above sea level

−2 −1 0 +1 +2 ≥3

Lake Pontchartrain

downtown

Mississippi River

5 km

(g) Water flowing across the levees bordering the 17th Street Canal after the hurricane had passed.

(h) Damage due to flooding in a New Orleans home.

to build, and these waves have led to the foundering of countless ships. They also cause havoc in coastal regions, and even inland. Coastal damage happens for several reasons:

> *Wind:* Winds of weaker hurricanes tear off branches and smash windows. Stronger hurricanes uproot trees, rip off roofs, and collapse walls.
> *Waves:* Winds shearing across the sea surface during a hurricane generate huge waves. In the open ocean, these waves can capsize ships. Near the shore, waves batter and erode beaches, rip boats from moorings, and destroy coastal property.
> *Storm surge:* The sustained winds of a hurricane not only generate waves, but they push water into a mound where sea level becomes significantly higher than normal, over an area of 60 to 80 km in diameter. Beneath the hurricane, the mound rises even higher, because such storms cause the local air pressure to drop significantly, and when the downward push of the air decreases, the sea surface rises. When it reaches the coast, this bulge of water, known as **storm surge**, swamps the land. If the bulge hits the land at high tide, the sea surface will be especially high and will affect a broader area.
> *Rain, stream flooding, and landslides:* Rain drenches the Earth's surface beneath a hurricane. In places, half a meter or more of rain falls in a single day, so streams flood. Saturation of the land with water can trigger landslides.
> *Disruption of social structure:* When the storm passes, the hazard is not over. By disrupting transportation and communication networks, breaking water mains, and washing away sewage-treatment plants, hurricane damage produces severe obstacles to search and rescue, and can lead to the spread of disease, fire, and looting.

Nearly all hurricanes that reach the coast cause death and destruction, but some are truly catastrophic. Storm surge from a 1970 cyclone making landfall on the low-lying delta lands of Bangladesh led to an estimated 500,000 deaths. In 1992, Hurricane Andrew leveled extensive areas of southern Florida, causing over $30 billion in damage and leaving 250,000 people homeless. Hurricane Katrina, in 2005, destroyed much of New Orleans (**Box 15.2**), and Hurricane Sandy in 2012 devastated coastal regions of the eastern United States. Sandy was an immense storm, reaching a diameter of 1,800 km at its peak. Storm surge associated with Sandy flattened countless communities and even submerged tunnels and subways in New York City. The most powerful storm on record, Typhoon Haiyan, struck the Philippines in 2013—during its peak, winds gusted to 378 km per hour.

The 2017 hurricane season was particularly disastrous. During only a little more than a month, Hurricane Harvey stalled over Texas and caused historic floods; Hurricane Irma became one of the most powerful Atlantic hurricanes on record, destroying communities on Caribbean islands and raking the west coast of Florida; and Hurricane Maria made a direct hit on Puerto Rico, ruining the island's electrical grid.

The Hurricanes of 2017 In terms of hurricane destruction, 2017 was a year to remember: 10 hurricanes raged across the western Atlantic! Of these, three hurricanes—Harvey, Irma,

FIGURE 15.23 Houston's streets turned into rivers during Hurricane Harvey.

and Maria—reached Category 5 status, and together they caused close to $300 billion of damage.

Hurricane Harvey dumped historic amounts of rain on portions of Texas, where the storm, after drawing in vast amounts of moisture from the Gulf of Mexico, stalled. Large areas received more than a meter of rainfall over the course of 4 days, and in places, rainfall exceeded 1.5 m, more than had ever been recorded. In the Houston area, much of the land lies at low elevation, so floodwaters damaged or destroyed tens of thousands of homes and businesses and caused major oil refineries to shut down. Private boats plied the submerged streets and highways to rescue stranded inhabitants (Fig. 15.23).

Before Harvey had dissipated, Irma was already cutting its path of destruction across the Caribbean. The strength of its peak winds (295 km per hour) made it the strongest hurricane ever in the Atlantic. Irma slammed into islands of the eastern Caribbean, and on Barbuda and Saint Martin it severely damaged or destroyed about 95% of the buildings. Maria then skirted the north coast of Cuba, cutting a swath of damage across the Florida Keys, before turning north and damaging the entire west coast of Florida.

Maria followed a similar path to that of Irma, so some of the islands devastated by Irma were raked again by the winds of Maria. But while Irma's eye passed just to the north of Puerto Rico, Maria's eye made a direct hit on the island before turning north and heading into the Atlantic. The storm was a catastrophe for Puerto Rico, where the winds wiped out the electrical grid. In parts of the island, people remained without power for many months, and food and clean water became scarce.

TAKE-HOME MESSAGE

Coastal landscapes can be impacted by human activities and climate change. Beaches erode, and pollution destroys wetlands. People try to protect beaches by constructing barriers or replenishing sand, but this can lead to other problems. Hurricanes can be particularly destructive hazards along coasts.

QUICK QUESTION What problems can be caused by construction of groins or seawalls?

ANOTHER VIEW Coasts have always been a hazard for shipping. In the days before GPS, the bright beam of a lighthouse served to warn ships that a coastline, or offshore bar, was nearby. This example, near Portland, Maine, was first built in 1791.

Chapter 15 **Review**

CHAPTER SUMMARY

> The landscape of the seafloor depends on the character of the underlying crust. Wide continental shelves form over passive-margin basins. Continental shelves may be cut locally by submarine canyons. Abyssal plains develop on old, cool oceanic lithosphere. Seamounts form above hot spots.

> The salinity, temperature, and density of seawater vary with location and depth.

> Water in the oceans circulates in currents. Surface currents are driven by the wind and are deflected by the Coriolis effect. Large gyres carry water completely around the margins of an ocean. The vertical upwelling and downwelling of water produce deep currents. Some of this movement is thermohaline circulation.

> Tides—the daily rise and fall of sea level—are caused by a tide-generating force. The largest contribution to this force comes from the gravitational pull of the Moon.

> Waves are caused by friction where the wind shears across the surface of the ocean. Water particles follow a circular motion in a vertical plane as a wave passes. Waves refract (bend) when they approach the shore because of frictional drag with the seafloor.

> Sand on beaches moves with the swash and backwash of waves. If there is a longshore current, the sand gradually moves along the beach and may build spits.

> At rocky coasts, waves grind away at rocks, yielding such features as wave-cut platforms and sea stacks. Some shores are wetlands, where marshes or mangrove swamps grow. Coral reefs grow along coasts in warm, clear water.

> The differences in coasts reflect their tectonic setting, whether sea level is rising or falling, sediment supply, and climate.

> To protect beach property, people build groins, jetties, break-waters, and seawalls.

> Human activities have led to the pollution of coasts. Reef bleaching has become dangerously widespread, and dead zones have formed along some coasts.

> Hurricanes produce winds of between 119 and 380 km per hour. The force of the winds, along with accompanying storm surge and heavy rains, can destroy coastal areas.

GUIDE TERMS

abyssal plain (p. 502)
backwash (p. 505)
bathymetry (p. 500)
beach (p. 509)
beach erosion (p. 520)
beach profile (p. 509)
coastal fjord (p. 513)
coastal plain (p. 516)
coastal wetland (p. 514)

continental shelf (p. 500)
coral reef (p. 514)
Coriolis effect (pp. 505, 506)
current (p. 505)
estuary (p. 513)
guyot (p. 515)
longshore current (p. 505)
longshore drift (p. 511)
mid-ocean ridge (p. 500)

organic coast (p. 514)
rogue wave (p. 504)
salinity (p. 502)
sand spit (p. 511)
seamount (p. 502)
shore (p. 505)
storm surge (p. 526)
submarine canyon (p. 501)
swash (p. 505)

thermohaline circulation
  (p. 508)
tidal range (p. 503)
tide (p. 503)
trench (p. 500)
tropical cyclone (p. 522)
wave (p. 503)
wave base (p. 504)
wave refraction (p. 505)

 GEOTOURS *THIS CHAPTER'S GEOTOURS WORKSHEET (O) FEATURES QUESTIONS AND GOOGLE EARTH SITES ON:*

> Seafloor bathymetry   > Coral reefs   > Barrier islands and spits   > Beach preservation   > Coastlines and sea-level rise

REVIEW QUESTIONS

The letters following each Review Question refer to the corresponding Learning Objective from the Chapter Opener.

1. How much of the Earth's surface is covered by oceans? (**B**)

2. How does the lithosphere beneath a continent differ from that beneath an abyssal plain? (**B**)

3. How do the shelf and slope of an active continental margin differ from those of a passive margin? Why do passive-margin basins exist? (**B**)

4. Where does the salt in the ocean come from? How do the salinity and temperature in the ocean vary? (**A**)

5. What factors control the direction of surface currents in the ocean? Explain thermohaline circulation. (**C**)

6. What causes the tides? Why does tidal range vary? (**D**)

7. In which direction does the longshore current flow in this image? (**E**)

8. Describe the motion of water molecules in a wave. How does wave refraction cause longshore currents? (**E**)

9. Describe the components of a beach profile. How does beach sand migrate as a result of longshore drift? (**F**)

10. Describe how rocky coasts evolve. What features are visible in this image? (**F**)

11. What is an estuary? What is the difference between an estuary and a fjord? (**F**)

12. Discuss the different types of coastal wetlands. What is a coral reef, and how does the reef surrounding an oceanic island change with time? (**F**)

13. How do plate tectonics, sea-level changes, sediment supply, and climate change affect the local shape of a coastline? Explain the difference between an emergent and a submergent coast. (**G**)

14. In what ways do people try to modify or stabilize coasts? How do the actions of people threaten coastal areas? (**G**)

15. What is a hurricane and how can hurricanes affect coasts? (**G**)

ON FURTHER THOUGHT

16. In 1789, the crew of HMS *Bounty* mutinied. Near Tonga, the crew forced the ship's commanding officer, Lieutenant Bligh, along with those crewmen who remained loyal to Bligh, into a rowboat and set them adrift in the Pacific Ocean. The castaways, amazingly, survived, and 47 days later, they landed at Timor (near Sumatra), 6,700 km to the west. Why did they end up where they did? (**E**)

ONLINE RESOURCES

Animations This chapter features animations on living on the coasts and ocean wave motion and refraction.

Videos This chapter includes videos on the science of currents, harnessing wave energy to power coastal communities, how sea-level rise impacts coasts, and the seafloor of central California.

Smartwork5 This chapter features questions on ocean composition and zones, currents and waves, coastal profiles, and sea-level rise.

CHAPTER 16

A HIDDEN RESERVE: GROUNDWATER

By the end of this chapter, you should be able to . . .

A. explain what groundwater is, where it resides in the Earth, and how its composition and flow varies.

B. characterize porosity and permeability.

C. describe the difference between aquifers and aquitards and the nature of the water table.

D. differentiate among the various types of wells and springs that provide access to groundwater.

E. produce a sketch showing how hot springs originate.

F. explain how groundwater supplies can be damaged or depleted and how to address these problems.

G. describe how caves and karst landscapes originate and evolve.

> When the rain falls and enters the earth, when a pearl drops into the depth of the sea, you can dive in the sea and find the pearl, you can dig in the earth and find the water.

> MEI YAO-CH'EN (Chinese poet, 1002–1060)

16.1 Introduction

Imagine Mae Rose Owens's surprise when, on May 8, 1981, she looked out her window and discovered that a large sycamore tree in the backyard of her Winter Park, Florida, home had suddenly disappeared. It wasn't a particularly windy day, so the tree hadn't blown over—it had just vanished! When Owens went outside to investigate, she found that more than the tree had disappeared. Her whole backyard had become a deep, gaping pit. The pit continued to grow for a few days until finally it swallowed Owens's house and six other buildings, as well as the municipal swimming pool, part of a road, and several expensive Porsches in a car dealer's lot (**Fig. 16.1a**).

What had happened in Winter Park? The bedrock beneath the town consists of limestone. **Groundwater**, the liquid

FIGURE 16.1 Development of sinkholes in central Florida.

(a) The Winter Park sinkhole, as seen from a helicopter.

Drained pool

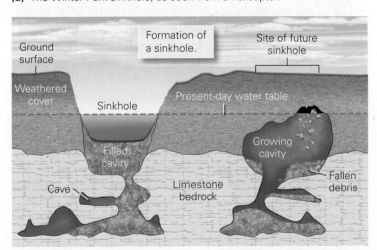

Formation of a sinkhole.

Ground surface

Weathered cover

Sinkhole

Present-day water table

Site of future sinkhole

Filled cavity

Growing cavity

Cave

Limestone bedrock

Fallen debris

(b) As overburden slowly washes into underlying caves, a cavity forms. When the roof of this cavity collapses, a sinkhole forms (not to scale).

◀ (facing page) This large cave in Vietnam (note the people, for scale) formed due to dissolution in groundwater long ago. It now lies well above the water table and has filled with air. Calcite has precipitated from water dripping into the cave to form sculpture-like speleothems.

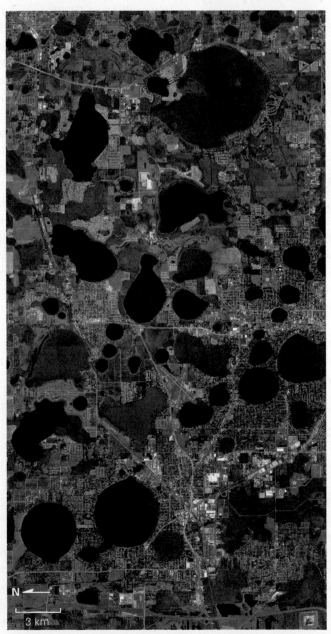

N

3 km

(c) An aerial view of Florida sinkholes that have become lakes.

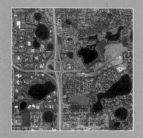
water that resides within sediment or rock under the surface of the Earth, had gradually dissolved the limestone over time, carving open rooms, or *caverns*, underground. On May 8, the roof of a cavern underneath Owens's backyard began to collapse. Soil and rock from up above sank into the opening, and a circular depression called a **sinkhole** developed **(Fig. 16.1b)**. The sycamore tree and the rest of the neighborhood simply dropped down into the sinkhole. It would have taken a huge effort to fill in the sinkhole with gravel or sand, so the community allowed it to fill with water, and now it's a circular lake. Similar lakes occur throughout central Florida **(Fig. 16.1c)** and elsewhere worldwide.

The Winter Park sinkhole serves as a dramatic reminder that significant quantities of water reside underground. But while we can easily see the Earth's surface water (in lakes, rivers, streams, marshes, glaciers, and oceans) and some of its atmospheric water (in clouds and rain), groundwater lies hidden beneath the surface. In this chapter, we examine where this subsurface water comes from, how it flows, how it interacts with rock and sediment, and how, in special places, it erupts in forceful plumes of steam. We also explore the underground landscapes that can be carved by groundwater. It accounts for about two-thirds of the Earth's freshwater resources used by homes, agriculture, and industry.

16.2 Where Does Groundwater Reside?

Water moves among various reservoirs during the hydrologic cycle (see Interlude F). The groundwater reservoir consists of the open spaces within sediment or rock underground. How does water get into this reservoir? Some gets buried with sediment grains during deposition and gets trapped when the rock lithifies. Some bubbles out of magma that has intruded the crust. But most water precipitates from the sky, as rain or snow, and falls on the land. If it doesn't evaporate directly

back into the atmosphere, get trapped in glaciers, or run off in a stream, the water sinks or percolates downward into the ground, a process called *infiltration*. In effect, the upper part of the crust behaves like a giant sponge that can soak up water that falls from the sky. Let's look more closely at the interconnected openings that allow infiltration.

POROSITY: OPEN SPACE IN ROCK AND SEDIMENT

Contrary to popular belief, only a small proportion of groundwater occurs in underground lakes or streams. Most groundwater resides in relatively small open spaces between grains of sediment, between grains of seemingly solid rock, or within cracks of various sizes. Geologists use the term **pore** for any open space within a volume of sediment or within a body of rock, and they use the term **porosity** for the total amount of open space within a material. We specify porosity as a percentage. For example, if we say that a block of rock has 20% porosity, we mean that 20% of the block consists of open space.

We can distinguish between two basic kinds of porosity: primary and secondary. *Primary porosity* refers to the open space that remains in a sediment after deposition, or in a rock after its formation **(Fig. 16.2a, b)**. Primary porosity exists because grains don't fit together perfectly during deposition, because grains don't grow to fill all space during crystallization, or because open spaces do not completely fill with cement. *Secondary porosity* refers to new pore space produced in a rock subsequent to the rock's lithification. For example, secondary porosity can develop when a rock cracks and opposing walls of the cracks do not fit together perfectly **(Fig. 16.2c)**, or when fluids passing through a rock dissolve grains or cements, thereby producing open cavities. Different Earth materials have different porosities. For example, a poorly cemented sandstone may have a porosity of up to 30%, whereas a granite generally has a porosity of less than 1%.

PERMEABILITY: THE EASE OF FLOW

The **permeability** of a material refers to the ease with which fluids can pass through the material. Therefore, for permeability to exist, passageways or conduits must link pores to each other. These conduits may be spaces between grains, or they may be cracks.

To develop an intuitive image of permeability, take a glass jar and fill it with rounded pebbles. You can see the air-filled pores between the pebbles, which exist because the pebbles don't fit together perfectly, and many of the pores connect to one another. If you pour water into the jar, the water can trickle between grains down to the bottom of the jar, where

FIGURE 16.2 The concepts of porosity and permeability.

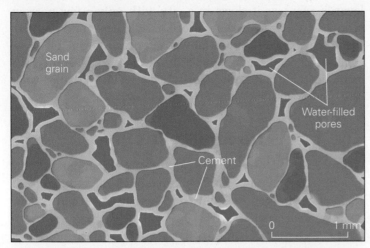

(a) Isolated pores in a sandstone occur in the spaces between grains. Water or air can fill pores.

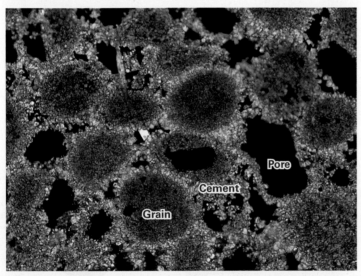

(b) This photo of a sedimentary rock, as seen through a microscope, shows grains, cement, and pores. The field of view is about 3 mm.

(c) Limestone outcrop on the coast of Ireland contains abundant fractures that provide secondary porosity.

Water follows a tortuous path as it flows from pore to pore.

Glass beaker

Air-filled pore

Solid pebble

Screen

Water flows freely

(d) Gravel contains pore space because clasts don't fit together tightly. The connection of pores produces permeability, so water can flow through gravel.

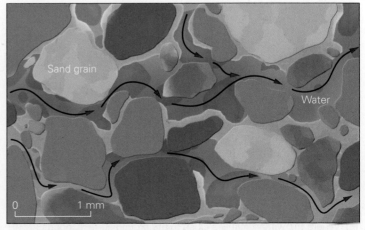

(e) Even at a microscopic scale, if passages connect pores, then water can flow through, and the rock has permeability.

it displaces air and fills the pores **(Fig. 16.2d)**. Water flows easily through a *permeable material*, whereas it flows slowly or not at all through an *impermeable material* **(Fig. 16.2e)**.

Informally, we can describe materials as having "high permeability," "low permeability," or "no permeability" to convey the relative ease of fluid movement. The permeability of a material depends on several factors:

› *Number of available conduits:* As the number of available conduits increases, permeability increases.
› *Size of the conduits:* Water can travel faster through wider conduits than through narrower ones.
› *Straightness of the conduits:* Water can travel faster through straight conduits than through crooked ones.

The factors that control permeability in rock or sediment resemble those that control traffic flow through a city. Traffic can pass quickly in a city with many straight, multilane boulevards, whereas it slows in a city with only a few narrow, crooked streets. Note that porosity and permeability are not the same feature—a material whose pores are isolated from each other can have high porosity but low permeability. For example, shale may have porosity comparable to that of sandstone, but it has much lower permeability.

AQUIFERS AND AQUITARDS

Hydrogeologists, researchers who focus on understanding groundwater, distinguish between **aquifers**, sediments or rocks with high permeability and porosity, and **aquitards**, sediments or rocks with lower permeability, regardless of porosity. Note that an aquitard slows, or retards, the motion of groundwater **(Fig. 16.3a)**.

FIGURE 16.3 Water underground—aquifers, aquitards, and the water table.

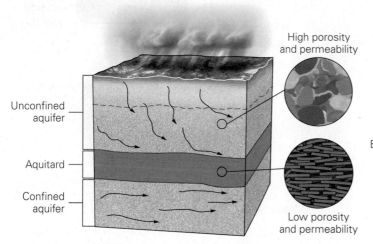

(a) An aquifer is a high-porosity, high-permeability rock. Some aquifers are unconfined, and some are confined.

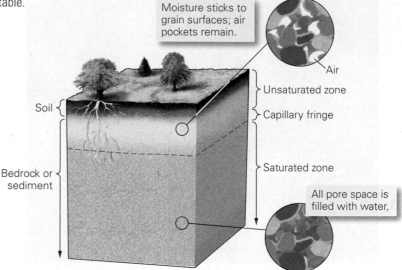

(b) The water table is the top of the groundwater reservoir in the subsurface. It separates the unsaturated zone above from the saturated zone below. A capillary fringe forms at the boundary.

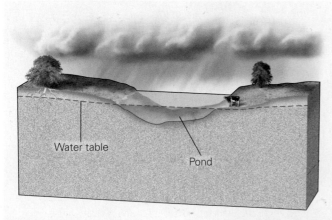

(c) Where the water table lies close to the ground surface, ponds remain filled—the water table is the surface of the pond.

(d) In dry regions, the water table sinks deep below the surface. Water that collects temporarily in low areas sinks into the subsurface.

Geologists distinguish between unconfined and confined aquifers based on whether or not an aquitard separates the aquifer from the ground surface. An *unconfined aquifer* reaches up to the Earth's surface, so that water can infiltrate into it directly from the surface. In contrast, a *confined aquifer* is isolated from the Earth's surface by an aquitard, so water can't infiltrate directly.

THE WATER TABLE

Infiltrating water moves downward from the Earth's surface into the subsurface by percolating along cracks and conduits that connect pores. Nearer the ground surface, in the *unsaturated zone*, this water only partially fills pores, so air remains in some of the open space (Fig. 16.3b). In most places, the top of the unsaturated zone consists of soil, so the water in this part can also be called **soil moisture**. Soil moisture may evaporate back into the atmosphere, seep into streams, or get drawn up by plant roots and transpired back into the atmosphere before it even moves below the soil. Deeper down, in the *saturated zone*, water completely fills, or saturates, pore space. In a strict sense, geologists use the term *groundwater* only for subsurface water in the saturated zone. The **water table** specifies the horizon that separates the unsaturated zone above from the saturated zone below. Typically, surface tension, the electrostatic attraction of water molecules to each other and to mineral surfaces, causes some water to seep up from the water table (just as water rises in a thin straw). This water fills pores in a thin layer, known as the **capillary fringe**, just above the water table.

The depth of the water table below the ground surface varies greatly with location. In temperate or tropical regions, where it rains fairly frequently and water often seeps into the ground, the water table lies within a few meters of the Earth's surface. In such regions, a channel or depression whose floor lies below the water table will fill with liquid water. Consequently, the surface of a permanent stream, lake, or marsh effectively defines the level of the water table in temperate or tropical regions (Fig. 16.3c). Rainfall rates affect the water table depth in a given locality—the water table drops during the dry season and rises during the wet season. Therefore, streams or ponds that contain water during the wet season may dry up during the dry season, when the water in them infiltrates into the ground below. In arid regions, where it rarely rains, the water table typically lies tens to hundreds of meters below the ground surface (Fig. 16.3d).

SHAPE OF THE WATER TABLE

In hilly regions, the water table is not a planar surface. Rather, its shape mimics, in a subdued way, the shape of the overlying land surface (Fig. 16.4a). This means that the water table

is higher beneath hills than beneath valleys. But the *relief* of the water table, meaning the vertical distance between the highest and lowest elevations, tends to be less than that of the overlying land, so the water table tends to be smoother than the ground surface.

At first thought, it may seem surprising that the shape of the water table reflects ground-surface shape. After all, when you pour a bucket of water into a pond, the surface of the pond immediately adjusts so it will remain horizontal. The elevation of the water table varies because groundwater moves so slowly through rock and sediment that it cannot quickly assume a horizontal surface. So when rain falls on a hill and water infiltrates down to the water table, the water table rises a little. When it doesn't rain, the water table sinks a little, but this movement takes place so slowly that when rain falls again, the water table rises before it has had time to sink very far.

In some locations, lens-shaped layers of impermeable rock (such as shale) may lie within a thick aquifer. A mound of groundwater accumulates above such aquitard lenses, producing a **perched water table**, a local water table that lies above the regional water table (Fig. 16.4b).

FIGURE 16.4 Factors that influence the position of the water table.

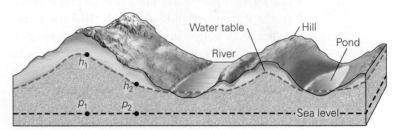

(a) The shape of a water table beneath hilly topography. Point h_1 on the water table is higher than Point h_2, relative to a reference elevation (sea level). The pressure at p_1, is, therefore, more than the pressure at p_2.

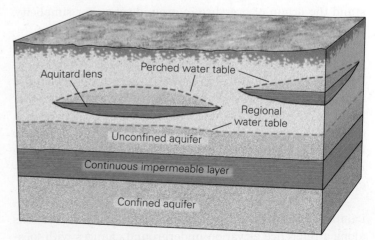

(b) A perched water table occurs where a mound of groundwater becomes trapped above a localized aquitard that lies above the regional water table.

TAKE-HOME MESSAGE

Most underground water fills pores and cracks in rock or sediment. Porosity refers to the total amount of open space within a material, whereas permeability indicates the degree to which pores connect. Aquifers have high porosity and permeability, while aquitards don't. Above the water table, water only partially fills pores, but below the water table, water completely saturates pores. The water table tends to be higher beneath hills and lower beneath valleys.

QUICK QUESTION Why do poorly cemented sandstones make good aquifers?

16.3 Groundwater Flow

What happens to groundwater over time? Does it just sit, unmoving, like the water in a stagnant puddle, or does it flow and eventually find its way back to the surface? Countless measurements confirm that groundwater indeed flows and, in some cases, moves great distances underground. Let's examine factors that drive groundwater flow.

In the unsaturated zone—the region between the ground surface and the water table—water percolates straight down in response to the downward pull of gravity, like the water passing through a drip coffee maker. But in the saturated zone—the region below the water table—water flow tends to be more complex, for water moves in response to variations in pressure. Pressure can cause groundwater to flow sideways or even upward. To picture the influence of pressure, step on a water-filled balloon (the pressure caused by your foot drives the water sideways) or watch water spray from a fountain (a pump generates pressure that pushes water up). So, to understand the nature of groundwater flow, we must first understand the origin of pressure in groundwater. For simplicity, we'll consider only the case of an unconfined aquifer.

Let's start by imagining a point in groundwater at a location where the water table over a broad area is horizontal. Pressure in groundwater at a specific point underground is caused only by the weight of the overlying water from that point up to the water table—the weight of overlying rock does not contribute to the pressure exerted on groundwater because the contact points between mineral grains bear the rock's weight. In this case, pressure acting on water at a specific depth below the water table is the same everywhere.

Now, imagine a location where the water table is not horizontal, as shown in Figure 16.4a. The pressure at locations along a reference elevation underground changes with location. For example, the pressure at location p_1, which lies below a hill, is greater than the pressure at location p_2, which lies below a valley, even though both p_1 and p_2 lie at the same

reference elevation—sea level—because more water is above p_1 than above p_2. The pressure within the water provides energy that can cause groundwater to flow. Hydrogeologists represent the energy available to drive the flow of groundwater, at a given location, by the **hydraulic head**. To measure the hydraulic head at a point in an aquifer, you can drill a vertical hole down to the point and insert a pipe in the hole. The height above a reference elevation to which water rises in the pipe represents the hydraulic head: water rises higher in the pipe where the head is higher. As a rule, groundwater flows from regions where it has higher hydraulic head to regions where it has lower hydraulic head. This statement generally implies that groundwater flows from locations where the water table is higher to locations where the water table is lower.

Hydrogeologists have calculated how hydraulic head changes with location underground, and have determined how water flows in response to these changes. The results indicate that groundwater flows along concave-up curving paths **(Fig. 16.5a)**. The curved paths eventually take

FIGURE 16.5 The flow of groundwater.

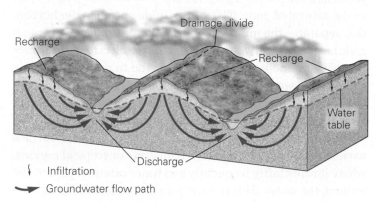

(a) Groundwater flows from recharge areas to discharge areas. Typically, the flow follows curving paths.

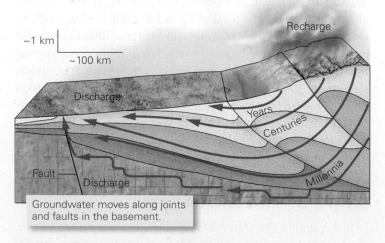

(b) The large hydraulic head resulting from uplift of a mountain belt may drive groundwater hundreds of kilometers across regional sedimentary basins. Deeper flow paths take longer.

BOX 16.1 CONSIDER THIS...

Darcy's Law for Groundwater Flow

In 1856, a French engineer named Henry Darcy carried out a series of experiments designed to characterize factors that control the velocity at which groundwater flows between two locations, 1 and 2, each of which has a different hydraulic head, h_1 and h_2, respectively **(Fig. Bx16.1)**. Darcy represented the velocity of flow by a quantity called the discharge (Q), meaning the volume of water passing through an imaginary vertical plane perpendicular to the groundwater's flow path in a given time. He found that the discharge depends on several factors:

> *Hydraulic gradient* (i): This quantity represents the change in hydraulic head per unit of distance (j) between two locations, as measured along the flow path. Hydraulic gradient can be represented by the equation: $i = (h_1 - h_2) \div j$. Typically, the slope of the water table is so small that the path length is almost the same as the horizontal distance between two points, so the hydraulic gradient roughly equals the slope of the water table.

> *Area* (A): This quantity represents the cross-sectional area, in square meters, of the imaginary plane through which the groundwater passes.

> *Hydraulic conductivity* (K): This quantity represents the ease with which a fluid can flow through a material. It depends primarily on the permeability of the material, but also on other factors such as the viscosity of the fluid.

The relationship that Darcy discovered is now known as **Darcy's law**. A simplified version can be written as

$$Q = KAi.$$

In words, this equation means that as the hydraulic gradient increases, discharge increases, and that as hydraulic conductivity increases, discharge increases. Put in simpler terms, the flow rate of groundwater increases as the permeability increases, and as the slope of the water table gets steeper.

FIGURE Bx16.1 The level to which water rises in a drillhole is the hydraulic head (h). The hydraulic gradient (i) is the difference in head divided by the length of the flow path (j).

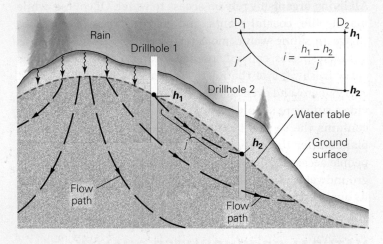

groundwater from regions where the water table is high (under a hill) to regions where the water table is low (below a valley), but because of flow-path shape, some groundwater first descends along the first part of its path and then flows back up along the final part of its path **(Fig. 16.5b)**. A place where water enters the ground and flows down is a **recharge area**, whereas a place where groundwater flows back up to the surface and may seep out at the surface is a **discharge area**.

Water flowing in a steep river channel can reach speeds of up to 30 km per hour, and water flowing in an ocean current moves at up to 3 km per hour. In contrast, groundwater moves at less than a snail's pace, generally from about 4 to 500 m per year. Groundwater moves much more slowly than surface water, for two reasons. First, groundwater moves by percolating through a complex, crooked network of tiny conduits, so it must travel a much greater distance than

it would if it could follow a straight path. Second, friction between groundwater and conduit walls slows down the water flow.

Simplistically, the velocity of groundwater flow depends on the slope of the water table and on the permeability of the material through which the groundwater moves. Groundwater flows faster through high-permeability materials than it does through low-permeability materials, and it flows faster in regions where the water table has a steep slope than it does in regions where the water table has a gentle slope. For example, groundwater flows relatively slowly through a low-permeability aquifer under the Great Plains, but it flows relatively quickly through a high-permeability aquifer under a steep hillslope. The details of calculating groundwater flow rate are a bit more complicated. Hydrogeologists use Darcy's law to calculate the rate for a given location **(Box 16.1)**.

16.4 Tapping Groundwater Supplies

All living organisms rely on access to water. Of course, while marine life, coastal plants, and certain types of microbes thrive in saline water, most terrestrial organisms, such as our own species, can only use freshwater. Lakes and streams carry freshwater at the Earth's surface, but they don't exist everywhere, and even where they do occur, they may not provide a sufficient supply of water for local needs. Groundwater contains the second-largest reservoir of freshwater on the planet, after the glaciers of Antarctica and Greenland. Can people access this groundwater reservoir? Fortunately, yes—groundwater can be obtained from springs and wells, though its quality varies with location.

NATURAL GROUNDWATER QUALITY

Much of the world's groundwater is crystal clear, pure enough to drink right out of the ground. Rocks and sediment serve as natural filters capable of removing suspended solids—these solids get trapped in tiny pores or stick to the surfaces of clay flakes. It's no wonder that commercial distribution of bottled groundwater (labeled as "spring water") has become a major business worldwide.

Dissolved chemicals, however, may make some natural groundwater unusable. For example, groundwater that has been underground for a long time (in some cases, since the deposition of the strata containing it) or has passed through salt-containing strata, may be too saline to be suitable for irrigation or for drinking. Because of its density, this groundwater tends to be deeper, so a boundary between fresh groundwater above and saline groundwater below may exist in the subsurface. Groundwater that has passed through limestone or dolostone contains dissolved calcium (Ca^{2+}) and magnesium (Mg^{2+}) ions. Such water, also known as *hard water*, can be a problem because carbonate minerals precipitate from it to form *scale* that clogs pipes, and soap won't develop a lather in hard water. Groundwater that has passed through iron-bearing rocks may contain dissolved iron oxide that precipitates to form rusty stains. Some groundwater contains dissolved hydrogen sulfide (H_2S), a poisonous gas with a rotten-egg smell that forms due to the metabolic processes of certain underground bacteria. This corrosive gas comes out of solution when groundwater rises to the surface and the pressure in it decreases. Groundwater that occurs above hydrocarbon reserves may contain dissolved methane, which also comes out of solution near the Earth's surface, to form flammable bubbles. In recent years, concern has grown about arsenic, a highly toxic chemical that enters groundwater when arsenic-bearing minerals dissolve in groundwater.

SPRINGS

Many towns grew up next to **springs**, natural outlets from which groundwater flows or seeps onto the Earth's surface. People can use springs as a source of freshwater for drinking or irrigation, without having to go to the expense of drilling or digging. Some springs spill groundwater onto dry land; others bubble up through the bed of a stream or lake. Springs occur in a variety of locations, including:

> where the ground surface intersects the water table (**Fig. 16.6a**); such springs may add water to lakes or streams.
> where flowing groundwater reaches a steep, impermeable barrier, and pressure pushes the groundwater up to the Earth's surface along the barrier (**Fig. 16.6b**).
> where a perched water table intersects the surface of a hill (**Fig. 16.6c**).
> where a network of interconnected fractures intersects the water table and provides a conduit for groundwater to reach the ground surface (**Fig. 16.6d**).
> where downward-percolating water runs into an aquitard and migrates along the top surface of the aquitard to a hillslope (**Fig. 16.6e**).
> where the ground surface intersects a natural fracture (joint) that taps a confined aquifer in which the pressure is sufficient to drive the water to the surface; such an occurrence is an **artesian spring** (**Fig. 16.6f**).

Did you ever wonder . . .
where bottled spring water comes from?

Because of the water that springs supply, lush and more diverse vegetation grows around a spring. Where a spring occurs in a desert, an **oasis**—an area where plants

FIGURE 16.6 Geologic settings in which springs form.

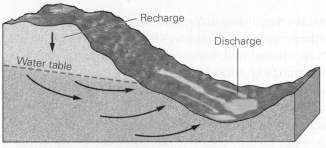

(a) Groundwater reaches the ground surface in a discharge area.

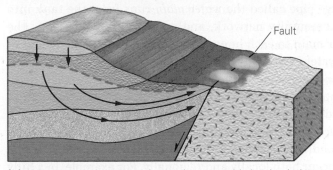

(b) Where groundwater reaches an impermeable barrier, it rises.

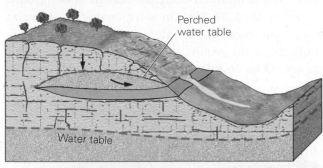

(c) Groundwater seeps out where a perched water table intersects a slope.

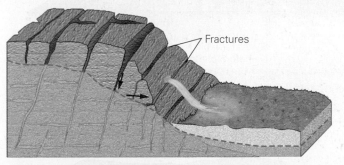

(d) A network of interconnected fractures channels water to the surface of a hill.

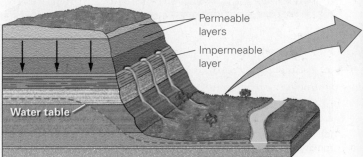

(e) Groundwater seeps out of a cliff face at the top of a relatively impermeable layer.

flourish in an otherwise bone-dry landscape—stands out as a patch of green against the browns and tans of sand and rock.

WELLS

Springs don't occur everywhere, so thousands of years ago, people learned to dig **wells**, holes or pits that provide access to groundwater. Geologists distinguish between *ordinary wells* and *artesian wells*, based on whether water in the well rises under its own pressure.

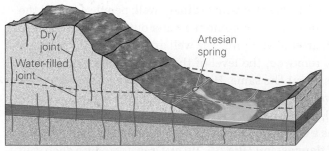

(f) In cases where water under pressure lies below an aquitard, a crack may provide a pathway for an artesian spring to form.

A spring on the wall of the Grand Canyon

In an **ordinary well**, the base of the well penetrates an aquifer below the water table **(Fig. 16.7a)**. Water from the pore space in the aquifer seeps into the well and fills it to the level of the water table. Note that drilling into an aquitard, or into any material that lies above the water table, will not supply water, and thus yields a *dry well*. Some ordinary wells are seasonal and function only during the rainy season, when the water table rises above the base of the well. During the dry season, the water table lies below the base of seasonal wells, so the well becomes dry.

To obtain water from an ordinary well, people can pull the water up in a bucket or pump the water out. As long as the rate at which groundwater fills the well exceeds the rate at which water is removed, the level of the water table near the well remains about the same. However, if people pump water out of the well too fast, then the water table sinks down around the well, a process called *drawdown*, so that the water table becomes a downward-pointing, cone-shaped surface called a **cone of depression (Fig. 16.7b, c)**. Drawdown by a deep well may cause nearby shallower wells to run dry.

An **artesian well**, named for the province of Artois in France, penetrates a confined aquifer in which water has enough pressure to rise on its own to a level above the top surface of the aquifer. If this level lies below the ground surface, the well is a *nonflowing artesian well*, whereas if the level lies above the ground surface, the well is a *flowing artesian well*. At a flowing artesian well, water actively spills or fountains out of the ground **(Fig. 16.8a)**.

Water rises in an artesian well for the same reason that it comes out of your home faucet. To see why, let's examine the configuration of a city water supply **(Fig. 16.8b)**. The city pumps water into an elevated tank, so that the water has a significant hydraulic head relative to the surrounding areas. A large pipe called the *water main* runs from the tank into the community network, and a vertical pipe connects the water main to each house. Pressure caused by the height of the water in the tank pushes the water up each vertical pipe and into home plumbing. If some pipes along the water main did not end at a faucet in a house, the water in these pipes would rise until it reached an imaginary plane, the *potentiometric surface*, that lies above the ground.

Conditions leading to the development of an artesian system occur both locally and regionally. For example, in a hilly area, the local potentiometric surface may lie above ground level in valleys **(Fig. 16.8c)**. On a regional scale, artesian systems develop where water enters a tilted, confined aquifer that intersects the ground in the hills of a high-elevation recharge area **(Fig. 16.8d)**. The confined groundwater flows

FIGURE 16.7 Pumping groundwater from a normal well can affect the water table.

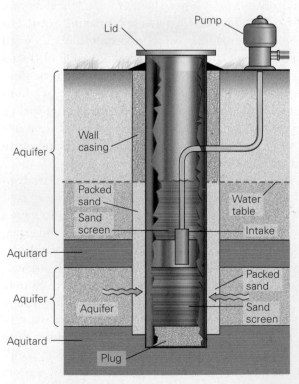

(a) A modern ordinary well sucks up water with an electric pump. The packed sand filters the water.

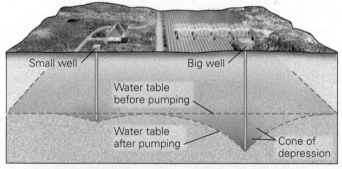

(b) Pumping forms a cone of depression in the water table.

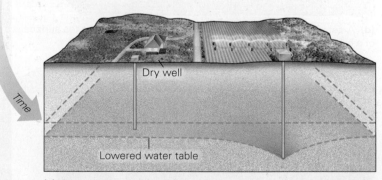

(c) Pumping by the big well may be enough to make the small well run dry.

FIGURE 16.8 Artesian wells, where water rises from the aquifer without pumping.

(a) A flowing artesian well in Wisconsin; the water rises from underground in the corrugated pipe without the need of pumping, and it continuously flows out of the pipe.

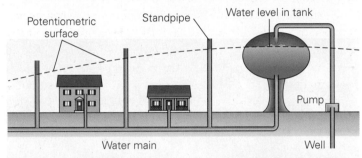

(b) The configuration of a city water supply. Water rises in vertical pipes up to the level of the potentiometric surface.

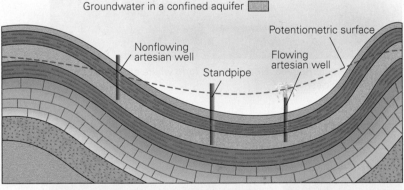

(c) The configuration of a local artesian system. Water rises above the confined aquifer because the water is under pressure.

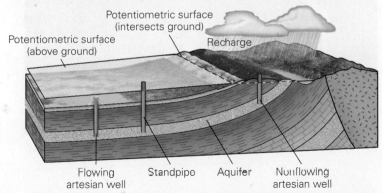

(d) The configuration of a regional artesian system.

down to the adjacent plains, which lie at a lower elevation. The potentiometric surface to which the water would rise were it not confined lies above this aquifer. Pressure in the confined aquifer pushes water up an artesian well or spring. Where the potentiometric surface lies underground, the well will be nonflowing, but where the surface lies above the ground, the well will be flowing.

> **TAKE-HOME MESSAGE**
>
> Groundwater can be obtained at wells (built by people) and springs (natural outlets). In ordinary wells, water must be lifted to the surface, but in artesian wells and springs, it rises on its own. Pumping groundwater can cause a cone of depression, and may lower the surrounding water table. Groundwater can react with rock and sediment and can become mineralized and, in some cases, toxic.
>
> **QUICK QUESTION** Why do some springs emerge along the side of a hill?

16.5 **Hot Springs and Geysers**

Around 60 c.e., the Romans occupying Britain built a large public bathhouse in what is now the city of Bath, England. Soldiers and officials would spend hours relaxing in elaborate lead-lined pools. When the Romans left Britain, in the 5th century, the baths were abandoned, and eventually they fell to ruin. In the early 19th century, during the reign of Queen Victoria, the baths were resurrected and remain a tourist destination to this day. Why build a spa in Bath? Water at a temperature of 46°C bubbles out of a natural spring, a hot spring, there. In a general sense, a **hot spring** (or *thermal spring*) is a conduit from which groundwater ranging in temperature from about 30°C to 104°C flows.

Why do hot springs exist? They can be found in two geologic settings. First, they occur where groundwater, as it slowly flows from recharge area to discharge area, follows a flow path that travels many kilometers down into the crust, where bedrock, naturally warm due to the geothermal gradient, heats the groundwater. The curving path of flow eventually carries the groundwater back up to the surface. Second, hot springs develop in **geothermal regions**, places where igneous activity increases the geothermal gradient substantially, so that hot rock lies close to the ground surface **(Fig. 16.9a)**. In geothermal regions, even shallow groundwater can become very hot. Since it has to rise only a short distance before returning to the surface at a discharge area, it comes out of the ground while still at a high temperature.

FIGURE 16.9 Geothermal waters and examples of their manifestation in the landscape.

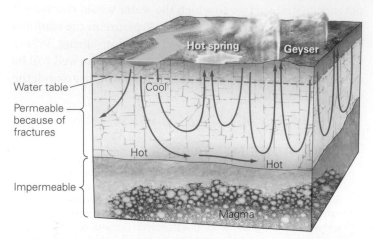

(a) Geothermal springs form where very hot rock in a region that hosts igneous activity lies relatively close to the Earth's surface and heats groundwater.

(b) Hot springs in Iceland, warmed by magma below, attract tourists from around the world.

(c) Colorful bacteria- and archaea-laden pools in Yellowstone National Park, Wyoming.

(d) Mudpots near Rotorua, New Zealand, form where steam bubbles up through volcanic ash.

(e) Terraces of travertine minerals precipitated at Pamukkale, Turkey.

(f) The Old Faithful geyser in Yellowstone National Park erupts predictably.

Water becomes a more effective solvent when hot, so hot groundwater dissolves minerals from rock that it passes through. Consequently, hot springs tend to emit mineralized water. Some people believe that immersion in such hot mineral water can cure ailments. Although this claim carries no scientific proof, hot springs certainly feel relaxing (Fig. 16.9b). In nature, the mineral content and temperature of hot springs can incubate life, and as a result, natural pools of geothermal water may be brightly colored—the gaudy greens, blues, and oranges of these pools come from *thermophilic* (heat-loving) bacteria and archaea that thrive in hot water and eat the sulfur-containing minerals dissolved in groundwater (Fig. 16.9c).

Numerous distinctive geologic features form in geothermal regions as a result of the eruption of hot water. In places where hot water rises into soils rich in volcanic ash and clay, a viscous slurry forms and fills goopy *mud pots* (Fig. 16.9d). Bubbles of steam rising through the mud cause it to splatter about. Where geothermal waters spill out of natural springs and then cool, dissolved minerals in the water precipitate, forming colorful mounds or terraces of travertine and other chemical sedimentary rocks (Fig. 16.9e).

Under special circumstances, a fountain of steam and boiling hot water erupts every now and then from a vent in a geothermal region (Fig. 16.9f). An episodic fountain of hot, steamy water is a **geyser**—the name comes from the Icelandic *geysir*, a word that means gusher. To understand why a geyser erupts, we first need to picture the underground plumbing beneath one. A geyser overlies a network of irregular fractures in very hot rock; one of these fractures connects to the ground surface. Groundwater sinks and fills these fractures and absorbs heat from the rock. Because the boiling point of water—the temperature at which it vaporizes—increases with growing pressure, hot groundwater at depth can remain in liquid form even if its temperature has become greater than the boiling point of water at the Earth's surface (100°C). When such "superheated" groundwater begins to rise through a conduit toward the surface, the pressure in the groundwater decreases until eventually some of the water boils and transforms into steam. The pressure generated by the expansion of this steam drives water upward in the fracture so that it spills out of the conduit at the ground surface. When this spill happens, the pressure deeper in the conduit exerted by the weight of overlying water suddenly decreases. This sudden drop in pressure causes the superheated water at depth to "flash" into steam instantly, and this steam quickly rises, ejecting all the water and steam above it from the conduit in a geyser eruption. Once the conduit empties, the eruption ceases, and the conduit fills once again with water that gradually heats up, starting the eruptive cycle all over again.

> ## TAKE-HOME MESSAGE
>
> Hot springs form either in geothermal regions, heated by magma below, or in localities where discharged groundwater rises from several kilometers down. The water emitted by hot springs tends to be mineralized and can feed microbes in colorful pools. Precipitation of chemical sedimentary rock from this water can build terraces. Geysers develop under special circumstances where sudden boiling of superheated groundwater forcefully ejects water and steam from a vent.
>
> **QUICK QUESTION** Why do geysers generally erupt episodically instead of continuously?

16.6 Groundwater Problems

Since prehistoric times, groundwater has been an important resource that people have relied on for drinking, irrigation, and industry. Dependence on groundwater has increased dramatically in the past century, because of population growth, the human preference to live in warmer climates, and the need to increase production of crops that require lots of water. (It takes about 20 liters of water to grow just one head of broccoli.) Although groundwater accounts for about 95% of the liquid freshwater on the planet, accessible groundwater cannot always be replenished quickly enough, and this leads to shortages. Pollution has also become a major concern; it may be invisible to us but may ruin a water supply for generations to come. In this section, we'll take a look at problems associated with the use of groundwater supplies.

GROUNDWATER DEPLETION

Is groundwater a renewable resource? In a time frame of 10,000 years, the answer is yes, for if we were to stop using groundwater, the natural hydrologic cycle would eventually resupply depleted reserves. But in a time frame of 10 to 1,000 years—the span of a human lifetime or a civilization—groundwater in many regions may behave like a *nonrenewable resource*. By pumping water out of the ground at a rate faster than nature replaces it, people effectively "mine" the groundwater supply, which can cause *groundwater depletion*. In fact, in portions of the desert southwest region of the United States, supplies of young groundwater have already been exhausted, and deep wells are now extracting 20,000-year-old groundwater. Some ancient water has been in rock so long that it has become too mineralized to be usable. A number of problems accompany groundwater depletion.

FIGURE 16.10 Effects of human modification of the water table.

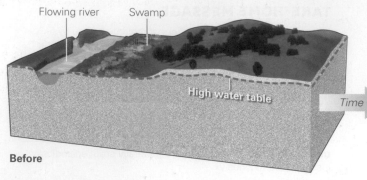

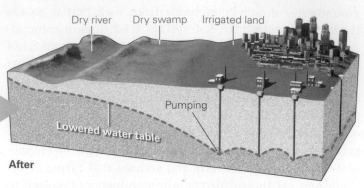

(a) Before humans start pumping groundwater, the water table is high. A swamp and permanent stream exist.

(b) Pumping for consumers in a nearby city causes the water table to sink, so the swamp dries up.

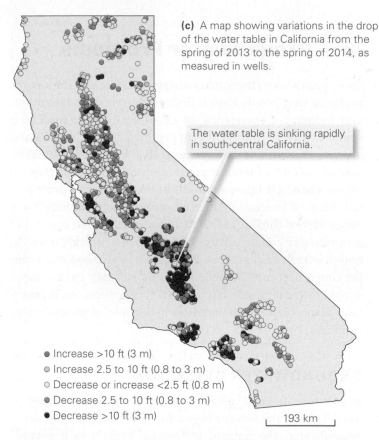

(c) A map showing variations in the drop of the water table in California from the spring of 2013 to the spring of 2014, as measured in wells.

The water table is sinking rapidly in south-central California.

- ● Increase >10 ft (3 m)
- ● Increase 2.5 to 10 ft (0.8 to 3 m)
- ○ Decrease or increase <2.5 ft (0.8 m)
- ● Decrease 2.5 to 10 ft (0.8 to 3 m)
- ● Decrease >10 ft (3 m)

193 km

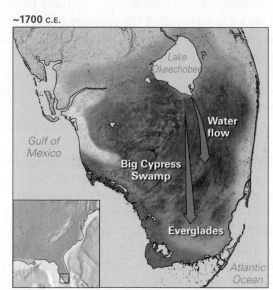

(d) The Florida Everglades before the advent of urban growth and intensive agriculture.

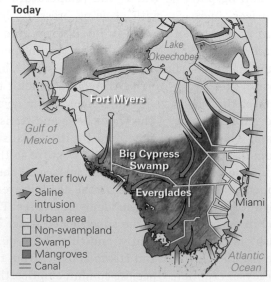

(e) Channelization and urbanization have removed water from recharge areas, disrupting groundwater flow paths.

Lowering the Water Table When we extract groundwater at a rate faster than it can be resupplied by nature, the water table gradually becomes lower over a broad region. As a consequence, existing wells, springs, and rivers dry up **(Fig. 16.10a, b)**, and to continue tapping into the water supply, we must drill progressively deeper. In northern India, water demands of a rapidly growing population have caused the water table to drop at astounding rates, in some places as much as 6 m per year. Due to an intense drought in California, from 2011 through 2016, the water table within the agricultural district of the state's central valley locally dropped by tens of meters **(Fig. 16.10c)**. Without enough rain or snow in the recharge area, water didn't infiltrate to the water table. Needless to say, this water-table drop was headline news and a major political issue worldwide.

Notably, the water table can also drop when people divert surface water from the recharge area. Such a problem has developed in the Everglades of southern Florida. In this huge swamp, before the expansion of Miami and the development of agriculture, the water table lay at the ground surface.

Diversion of water from the Everglades' recharge area into canals has significantly lowered the water table, causing parts of the Everglades to dry up **(Fig. 16.10d, e)**.

Reversing the Flow Direction of Groundwater The cone of depression that develops around a well produces a local slope to the water table. The resulting change in the hydraulic gradient may locally reverse the flow direction **(Fig. 16.11a, b)**.

FIGURE 16.11 Some causes of groundwater problems.

Before

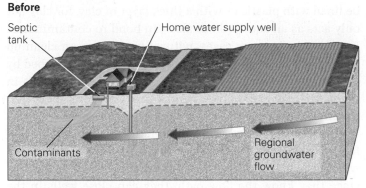

(a) Before pumping, effluent from a septic tank drifts with the regional groundwater flow, and the home well pumps clean water.

After

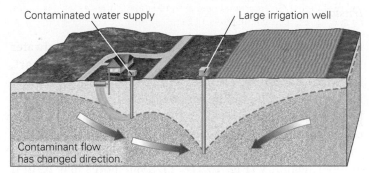

(b) After pumping by a nearby irrigation well, effluent flows into the home well in response to the new local slope of the water table.

Before

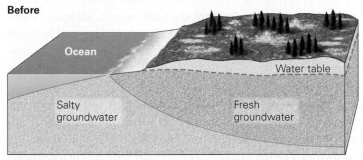

(c) Before pumping, fresh groundwater forms a lens below the ground.

After

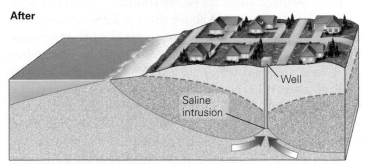

(d) If the freshwater is pumped too fast, saltwater from below is sucked up into the well. This is saltwater intrusion.

Before

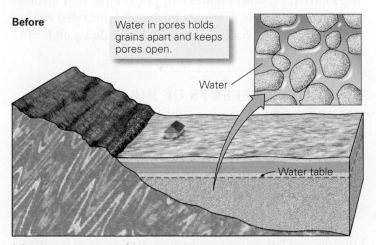

(e) When intensive irrigation removes groundwater, pore space in an aquifer collapses.

After

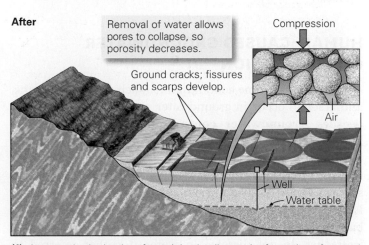

(f) As a result, the land surface sinks, leading to the formation of ground fissures and causing houses to crack.

Such reversals can allow contamination seeping out of a septic tank to head toward the well.

Saline Intrusion In many regions, fresh groundwater lies in a layer above saline (salty) water. Along coasts, the saline water entered the aquifer from the adjacent ocean **(Fig. 16.11c, d)**. Because freshwater is less dense than saline water, it floats above the saline water. If people pump water out of a well too quickly, the boundary between the saline groundwater and the fresh groundwater rises. And if this boundary rises above the base of the well, then the well will start to yield useless saline water. Geologists refer to this phenomenon as *saline intrusion*.

Pore Collapse and Land Subsidence When groundwater fills the pore space underground, it holds the grains apart, for water cannot be compressed. The extraction of water from a pore eliminates the support holding the grains apart, because the air that replaces the water is a gas, and gases can be compressed. As a result, the grains pack more closely together. Such *pore collapse* decreases a rock's porosity and permeability **(Fig. 16.11e, f)**. Some pores may refill if the water table rises again, but some pore collapse may be permanent.

Pore collapse also decreases the volume of an aquifer, with the result that the ground above sinks. Such *land subsidence* may cause fissures to develop at the surface and may cause the ground to tilt. Buildings constructed over regions undergoing land subsidence may themselves tilt or their foundations may crack. In the San Joaquin Valley of California, the land surface subsided by 9 m between 1925 and 1975, because water was removed to irrigate farm fields. It sank by another 0.5 m during the 2011 to 2016 drought. Land subsidence associated with groundwater removal in the Phoenix, Arizona, area has led to the development of open fissures at the land surface.

HUMAN-CAUSED GROUNDWATER CONTAMINATION

As we've noted, some contaminants (such as salts, methane, and arsenic) occur in groundwater naturally. But in recent decades, **groundwater contamination** has increased as human activities have introduced contaminants of many kinds into aquifers **(Fig. 16.12a)**. These contaminants include agricultural waste (pesticides, fertilizers, and animal sewage); industrial waste (both organic and inorganic chemicals); effluent from landfills and septic tanks (including bacteria and viruses); petroleum products and other chemicals that do not dissolve in water (together referred to as nonaqueous-phase liquids, or NAPL); radioactive waste (from weapons manufacture, power plants, and hospitals); and acids leached from sulfide minerals in coal and metal mines. The cloud of contaminated groundwater that moves away from the source of contamination is called a **contaminant plume (Fig. 16.12b)**.

The best way to avoid such groundwater contamination, or groundwater pollution, is to prevent waste products from entering groundwater in the first place. This can be done by sealing contaminants in containers and locating them above impermeable bedrock so that they are isolated from aquifers. If such a site is not available, the storage area should be lined with plastic or with a thick layer of clay, for clay not only acts as an aquitard but also can bond to contaminants. Fortunately, in some cases, natural processes can clean up groundwater contamination. Chemicals may be absorbed by clay, oxygen in the water may oxidize them, and bacteria in the water may metabolize them, thereby turning them into harmless substances.

At locations where contaminants have entered an aquifer, *environmental engineers* can drill test wells to determine which way and how fast the contaminant plume is flowing. Once they know the flow path, they can close wells in the path to prevent consumption of contaminated water. Engineers may also attempt to clean the groundwater by drilling a series of *extraction wells* to pump it out of the ground. If the contaminated water does not rise fast enough, steam or water can be pumped down injection wells into the ground beneath the contaminant plume. Such injection can push the contaminants up into extraction wells, where they can be pumped out **(Fig. 16.12c)**.

More recently, scientists have begun exploring techniques to stop plumes from migrating by burying or injecting certain chemicals in the path of the contaminated groundwater—the contaminants react with these chemicals to form solids that can't flow. And in some locations, engineers have tried **bioremediation**, a technique that involves injecting oxygen and nutrients into a contaminated aquifer to foster growth of bacteria that can break down molecules of contaminants.

UNWANTED EFFECTS OF RISING WATER TABLES

We've seen the negative consequences of sinking water tables, but what happens when the water table rises? Is that necessarily good? Sometimes, but not always. If the water table rises above the level of a house's basement, water seeps through the foundation and floods the basement floor. Catastrophic damage occurs when a rising water table weakens the base of a hillslope or a failure surface underground triggers landslides and slumps.

FIGURE 16.12 Contamination plumes in groundwater.

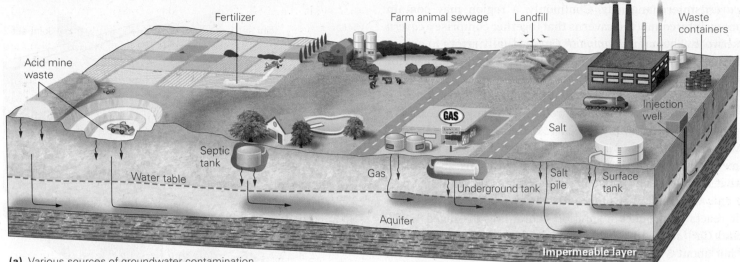

(a) Various sources of groundwater contamination.

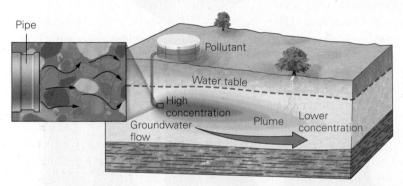

(b) A contaminant plume as seen in cross section. The darker the color, the greater the concentration of contaminant.

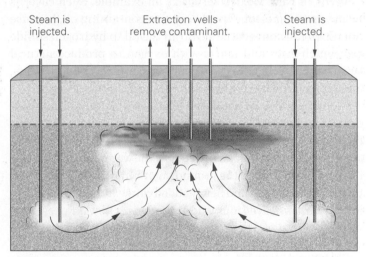

(c) Steam injected beneath the contaminant drives the contaminated water upward in the aquifer, where pumping wells remove it.

TAKE-HOME MESSAGE

Groundwater usage can cause problems. Too much pumping lowers the water table, causing land subsidence and/or salt-water intrusion. Due to population growth and drought, loss of groundwater can have severe effects on society. Contamination can ruin a groundwater supply.

QUICK QUESTION Can contaminants be removed from groundwater?

16.7 Caves and Karst

THE DEVELOPMENT OF CAVES

In 1799, as legend has it, a hunter by the name of Houchins was tracking a bear through the woods of Kentucky when the bear suddenly disappeared on a hillslope. Baffled, Houchins plunged through the brambles trying to sight his prey. Suddenly he felt a draft of surprisingly cool air flowing down the slope from uphill. Now curious, he climbed up the hill and found a dark portal into the hillslope beneath a ledge of rocks. Bear tracks were all around—was the creature inside? Houchins returned later with a lantern and cautiously stepped into the passageway. After walking a short distance, he found himself in a large, underground room. Houchins was the first person of European descent to enter Mammoth Cave.

In a general sense, a **cave** is any underground open space, most or all of which does not receive direct sunlight. A cave may grow due to the dissolution of rock or to the removal of rock by rockfalls or erosion. Many caves open up along a cliff face, whereas others lie completely underground. Some writers use the word **cavern** for caves formed mostly by dissolution or for particularly large caves. In colloquial

English, however, people tend to use the words *cave* and *cavern* interchangeably. Commonly, a region may contain many interconnected caverns that together comprise a *cavern network*. The name *Mammoth Cave* actually refers to an immense cavern network—you would have to walk (or crawl) for 630 km to explore all its known components.

Most large cavern networks develop in limestone bedrock because limestone dissolves relatively easily in corrosive groundwater. Generally, the corrosive component in groundwater is dilute carbonic acid (H_2CO_3), which forms when water absorbs carbon dioxide (CO_2) from materials, such as soil, that it has passed through as it percolates down. When carbonic acid comes in contact with calcite ($CaCO_3$) in limestone, it reacts to produce a solution containing dissolved HCO_3^- and Ca^{2+} ions. In recent years, geologists have discovered that about 5% of limestone caverns around the world form due to reactions with sulfuric-acid-bearing water—Carlsbad Caverns in New Mexico serves as an example. Such caverns form where limestone overlies strata containing oil, because microbes can convert the sulfur in the oil to hydrogen sulfide gas, which rises and reacts with oxygen to produce natural sulfuric acid. When this acid comes in contact with limestone, it reacts to produce gypsum and CO_2 gas.

Geologists debate about the depth at which limestone cavern networks form. Some limestone dissolves above the water table. However, it appears that most cavern formation takes place in limestone that lies just below the water table, for in this interval the acidity of the groundwater remains high, the mixture of groundwater and newly added rainwater has not yet saturated with dissolved ions, and groundwater flow is fastest. The association between cavern formation and the water table helps explain why openings in a cavern network align at the same level.

> **Did you ever wonder . . .**
> why huge underground caverns form?

THE CHARACTER OF CAVE NETWORKS

As we have noted, caverns in limestone usually occur as part of a network. These networks include *rooms*, or *chambers*, which are large, open spaces, sometimes with cathedral-like ceilings, and also tunnel- or slot-shaped *passages* (**Geology at a Glance**, pp. 550–551). Some chambers may host underground lakes, and some passages may serve as conduits for underground streams. The shape of a cave network reflects variations in permeability and in the composition of the rock from which the caves formed. Chambers develop where the limestone dissolves more easily and where groundwater flow is fastest. Passages in cave networks typically follow pre-existing joints, which provide openings along which groundwater can flow faster (**Fig. 16.13a**). Because joints

FIGURE 16.13 Development of karst, dripstone, and flowstone.

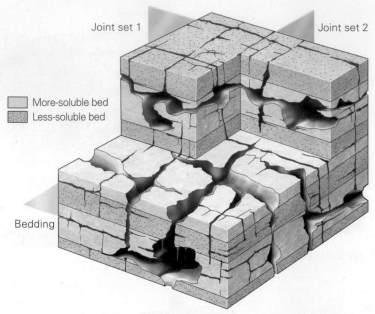

(a) Joints act as conduits for water in cave networks. Chambers and passages follow joints and preferentially form in more-soluble beds.

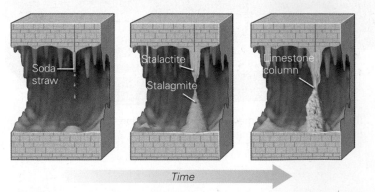

(b) The evolution of a soda straw stalactite into a limestone column.

(c) Flowstone on the wall of a cave in Vietnam.

commonly occur in orthogonal systems (two sets of joints oriented at right angles to each other; see Chapter 9), passages may form a grid.

THE FORMATION OF SPELEOTHEMS

When the water table drops below the level of a chamber, the chamber becomes an open space filled with air. In places where downward-percolating groundwater containing dissolved calcite emerges from the ceiling or walls of the cave, the surface of the cave gradually changes. As this water emerges into the air, it evaporates slightly and releases some dissolved carbon dioxide. As a result, calcite precipitates out of the water, producing dripstone, which can accumulate to form *travertine* and various intricately shaped formations called **speleothems**.

Cave explorers (*spelunkers*) and geologists have developed a detailed nomenclature for different kinds of speleothems **(Fig. 16.13b)**. Where water drips from the ceiling of the cave, calcite initially precipitates around the outside of the drip, forming a delicate, hollow tube called a *soda straw*. But eventually, the soda straw fills up, and water migrates down the margin of the cone to form a more massive, solid icicle-like cone called a **stalactite**, which grows downward from the ceiling. Where the drips hit the floor, the resulting precipitate builds an upward-pointing cone called a **stalagmite**, which rises from the ground. If the process of dripstone formation continues long enough, stalagmites merge with overlying stalactites to produce travertine *columns*. In some cases, groundwater flows along the surface of a wall and precipitates to produce drape-like sheets of travertine called *flowstone* **(Fig. 16.13c)**.

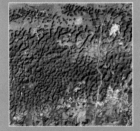

SEE FOR YOURSELF...

Karst Landscape, Puerto Rico

LATITUDE
18°23'53.04" N

LONGITUDE
66°25'49.43" W

Look down from 7 km.

Each of the rounded depressions in this view is a sinkhole. Ridges of limestone separate adjacent sinkholes.

KARST LANDSCAPES

Limestone bedrock underlies most of the Kras Plateau in Slovenia, along the east coast of the Adriatic Sea. The name *kras*, meaning rocky ground, is appropriate for a region with abundant rock exposures **(Fig. 16.14a)**. Geologists refer to such regions, where surface landforms develop when limestone bedrock dissolves both at the surface and in underlying cave networks, as **karst landscapes**—from the Germanized version of *kras*.

Karst landscapes typically display a number of distinct landforms. Where surface streams intersect cracks (joints) or holes that link to caverns or passageways below, the water cascades downward into the subsurface and vanishes **(Fig. 16.14b)**. Such **disappearing streams** may flow through passageways underground and later emerge from a cave entrance downstream. As we noted earlier, where the ground collapses into an underground cavern below, a *sinkhole* develops (see Fig. 16.1). In cases where the ground

FIGURE 16.14 Features of karst landscapes.

(a) Karst terrains typically have a rough, rocky surface. Here, we see sinkholes of the Kras Plateau.

(b) A small disappearing stream in the Hudson Valley region of New York; the water is dropping into a subsurface cave.

Karst Landscapes

Limestone is soluble in acidic water. Much of the water that falls to the ground as rain or seeps through the ground as groundwater is acidic, so in regions of the Earth where bedrock consists of limestone, we find signs of dissolution. Networks of underground openings that develop by dissolution are called caves or caverns. Some parts of these may be large, open chambers, whereas others are long, narrow passages. Underground lakes and streams may cover the floor of caverns. A cavern's location depends on the orientation of bedding and joints, for these features localize the flow of groundwater.

Limestone pavement, Ireland

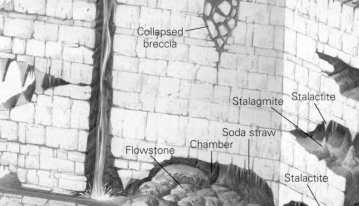

Disappearing stream

Sinkhole

Collapsed breccia

Dissolved joint

Stalagmite Stalactite

Soda straw

Flowstone Chamber

Stalactite

Limestone column

Underground stream

Underground pool

Passage

Emerging spring

Sinkholes

Underground pool, Mexico

Natural Bridge, Virginia

Spelunker crawling in a cave

In many locations, groundwater drips from the ceiling of a cavern or flows along its walls. As the water evaporates and loses its acidity, calcite precipitates. Over time, this calcite builds into speleothems, such as stalactites, stalagmites, columns, and flowstone. Distinctive landscapes, called karst landscapes, develop at the Earth's surface over limestone bedrock that has undergone dissolution.

Soda-straw stalactites, Utah

5 cm

FIGURE 16.15 Tower karst forms a spectacular landscape in southern China.

The landscape is treeless today, a consequence of industrialization policies in the 1950s.

Chinese artists painted scrolls depicting forested towers of karst.

FIGURE 16.16 The progressive formation of caves and a karst landscape.

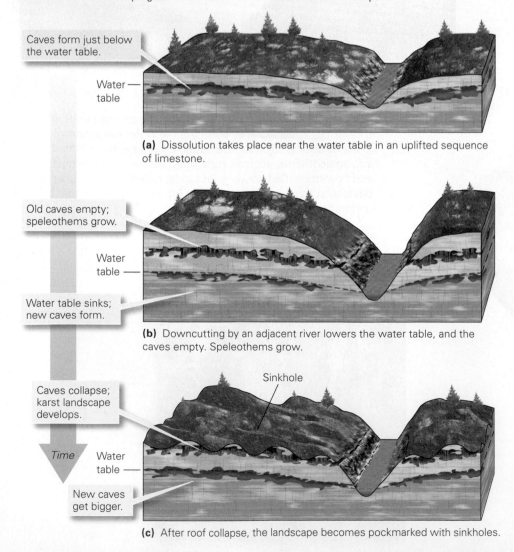

Caves form just below the water table.

Water table

(a) Dissolution takes place near the water table in an uplifted sequence of limestone.

Old caves empty; speleothems grow.

Water table

Water table sinks; new caves form.

(b) Downcutting by an adjacent river lowers the water table, and the caves empty. Speleothems grow.

Caves collapse; karst landscape develops.

Sinkhole

Time

Water table

New caves get bigger.

(c) After roof collapse, the landscape becomes pockmarked with sinkholes.

collapses over a long, joint-controlled passage, sinkholes may be elongate and canyon-like. As most of the cavern network collapses, former walls between caverns are steep-sided ridges, and remnants of cavern roofs can stand as *natural bridges*. Over time, the walls erode, leaving only jagged, isolated limestone *spires*—a karst landscape dominated by such spires is called *tower karst*. The surreal collection of pinnacles constituting the tower karst landscape in the Guilin region of China has inspired generations of artists, who portray them in scroll paintings **(Fig. 16.15)**. Karst landscapes evolve over time **(Fig. 16.16)**. In fact, while karst is forming by the collapse of a cave network closer to the surface, a new cavern network may be developing at depth.

LIFE IN CAVES

Despite their lack of light, caves are not sterile, lifeless environments. Caves that are open to the air provide a refuge for bats as well as for various insects and spiders. Similarly, fish and crustaceans enter caves where streams flow in or out. Species living in caves have evolved some unusual characteristics. For example, cave fish lose their pigment and in some cases their eyes. Explorers have discovered caves in Mexico in which warm, mineral-rich groundwater currently flows. Colonies of bacteria metabolize sulfur-containing minerals in this water and produce thick mats of living ooze in the complete darkness of the cave. Long gobs of this bacteria slowly drip from the ceiling. Because of the mucus-like texture of these drips, they have come to be known as *snottites*.

TAKE-HOME MESSAGE

Reaction with natural acids dissolves limestone underground to form caverns. Most dissolution takes place near the water table. If the water table sinks, dripping of water in caves can produce speleothems. Collapse of a cavern network produces karst terrain.

QUICK QUESTION Can organisms live in the pitch black of caves? If so, how?

ANOTHER VIEW Swamps are swamps because the land surface lies just below the water table. The bald cypress trees of this swamp in southern Illinois have adapted to life in soggy ground. During dry seasons, the water table sinks below the ground surface and the land dries out.

Chapter 16 Review

CHAPTER SUMMARY

> During the hydrologic cycle, water infiltrates the ground and fills the pores and cracks in rock and sediment. This subsurface water is called groundwater.

> The amount of open space in rock or sediment is its porosity, and the ease with which liquid can flow through is its permeability.

> Aquifers are relatively permeable materials, and aquitards are relatively impermeable materials. Groundwater can be obtained from aquifers.

> The water table defines the surface in the ground above which pores contain mostly air and below which pores are filled with water. The shape of a water table resembles the shape of the overlying land surface.

> Groundwater flows wherever a hydraulic gradient exists, and it moves from recharge areas to discharge areas. The velocity of flow depends on permeability and the hydraulic gradient.

> In a general sense, the rate of groundwater flow is faster where the water table has a steeper slope, and where it flows through higher permeability material.

> Groundwater contains dissolved ions. Hard water contains a relatively high concentration of ions; scale in pipes forms when these ions precipitate.

> At a spring, groundwater exits the ground on its own. Springs form for many reasons.

> Groundwater can be extracted in wells. An ordinary well simply penetrates below the water table, so the water level in the well is the water table. In an artesian well, water rises under its own pressure.

> Hot springs release hot water to the Earth's surface. This water may have been heated by flowing down deeply before rising, on its path from a zone of recharge to a zone of discharge. Or, it may have been heated by igneous activity.

> Geysers erupt when decompression causes groundwater to flash into steam.

> Groundwater is a precious resource used for municipal water supplies, industry, and agriculture. In recent years, some regions have lost their groundwater supply because of overuse or contamination.

> Pumping water out of a well too fast causes drawdown, yielding a cone of depression. A drop in the water table across a region can cause pore collapse and land subsidence. Too much pumping can also cause saltwater intrusion.

> When limestone dissolves just below the water table, underground caves develop. Soluble beds and joints determine the location and orientation of caves.

> If the water table drops, caves empty out. Limestone precipitates out of water dripping from cave roofs and produces speleothems.

> Regions where abundant caves have collapsed to form sinkholes are called karst landscapes. These terrains can contain sinkholes, natural bridges, and disappearing streams.

GUIDE TERMS

aquifer (p. 534)
aquitard (p. 534)
artesian spring (p. 538)
artesian well (p. 540)
bioremediation (p. 546)
capillary fringe (p. 535)
cave (p. 547)
cavern (p. 547)
cone of depression (p. 540)

contaminant plume (p. 546)
Darcy's law (p. 537)
disappearing stream (p. 549)
discharge area (p. 537)
geothermal region (p. 541)
geyser (p. 543)
groundwater contamination (p. 546)
groundwater (p. 531)

hot spring (p. 541)
hydraulic head (p. 536)
karst landscape (p. 549)
oasis (p. 538)
ordinary well (p. 540)
perched water table (p. 535)
permeability (p. 532)
pore (p. 532)
porosity (p. 532)

recharge area (p. 537)
sinkhole (p. 532)
soil moisture (p. 535)
speleothem (p. 549)
spring (p. 538)
stalactite (p. 549)
stalagmite (p. 549)
water table (p. 535)
well (p. 539)

GEOTOURS *THIS CHAPTER'S GEOTOURS WORKSHEET (P) FEATURES QUESTIONS AND GOOGLE EARTH SITES ON:*

> Groundwater reserves and irrigation in the Arabian Desert
> Hot springs in Yellowstone National Park

> Surface and groundwater flow in the Everglades
> Karst features around the world

REVIEW QUESTIONS

The letters following each Review Question refer to the corresponding Learning Objective from the Chapter Opener.

1. What is groundwater, and where does it reside on the Earth? **(A)**

2. How do porosity and permeability differ? Give examples of substances with high porosity but low permeability. **(B)**

3. What is a water table, and what factors affect its level? What factors affect the flow direction of the water below the water table? **(C)**

4. In the figure, label the confined and unconfined aquifer, and the aquitard. **(C)**

5. How does the rate of groundwater flow compare with that of moving ocean water or river currents? **(A)**

6. What does Darcy's law tell us about rates of discharge in groundwater? **(A)**

7. How does the chemical composition of groundwater change over time? Why is "hard water" hard, and saline water saline? **(A)**

8. How does excessive pumping affect the local water table? **(F)**

9. Why do natural springs form? **(D)**

10. How is an artesian well different from an ordinary well? **(D)**

11. Explain why hot springs form and what makes a geyser erupt. **(E)**

12. Is groundwater a renewable or a nonrenewable resource? Explain your answer. **(F)**

13. Describe some of the ways in which human activities adversely affect the water table. What impact has water pumping on a large scale had on the nearby environment in this diagram? **(F)**

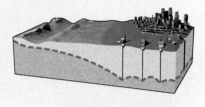

14. What are some sources of groundwater contamination? How can it be prevented? **(F)**

15. Describe the process leading to the formation of caves and the speleothems within caves. **(G)**

16. Describe the various features of a karst landscape, and explain how such landscapes evolve. **(G)**

ON FURTHER THOUGHT

17. The population of Desert Paradise (DP; a fictitious town) has been doubling every 7 years. Most of the new inhabitants are "snowbirds," people escaping the cold winters of northerly latitudes. There are no permanent streams or lakes anywhere near DP. In fact, the only standing water occurs in the ponds of the many golf courses that have been built recently. The water in these ponds needs to be replenished almost constantly, or the ponds will dry up. The golf courses and yards of the suburban-style developments of DP all have lawns of green grass. DP has been growing on a flat, sediment-filled basin between two small mountain ranges. Much of the water supply of DP comes from wells. What do you predict will happen to the water table of the area in coming years, and how might the land surface change as a consequence? Might you suggest a policy to the residents of DP that could slow the process of change? **(F)**

ONLINE RESOURCES

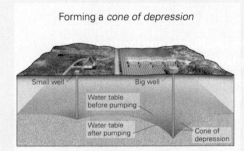

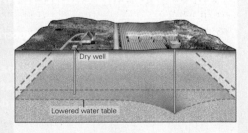

Videos This chapter includes videos on groundwater removal, groundwater as a source to streams, and the recent drought in California.

Smartwork5 This chapter's questions include visual labeling exercises on water tables, wells, and porosity and permeability.

CHAPTER 17

DRY REGIONS:
THE GEOLOGY OF DESERTS

By the end of this chapter, you should be able to . . .

A. define the term desert and characterize factors that cause regions to have deserts.

B. explain how weathering and erosional processes in deserts differ from those in temperate lands.

C. describe the processes that lead to deposition of sediment in deserts.

D. describe distinctive landforms and landscapes of deserts, and explain how they form.

E. discuss how human activity may transform vegetated regions into deserts.

> The bare hills are cut out with sharp gorges, and over their stone skeletons scanty
> earth clings. . . . A white light beat down, dispelling the last trace of shadow, and
> above hung the burnished shield of hard, pitiless sky.
>
> CLARENCE KING (1842–1901), first director of the U.S. Geological Survey, describing a desert

17.1 Introduction

For generations, nomadic traders have saddled camels to traverse the Sahara Desert in northern Africa **(Fig. 17.1a)**. The Sahara, the world's largest desert, receives so little rainfall that it has hardly any surface water or vegetation. So camels must be able to walk for up to three weeks without drinking or eating. They can survive these journeys because they sweat relatively little, they have the ability to metabolize their own body fat to produce new water, and they can withstand severe dehydration.

The survival challenges faced by a camel emphasize that deserts are lands of extremes—extreme dryness, heat, cold, and, in some places, beauty. Desert vistas include everything from sand seas to sagebrush plains, cactus-covered hills to endless stony pavements. Although less populated than other regions of the Earth, deserts cover about 25% of the land surface, and thus make up an important component of the Earth System **(Fig. 17.1b)**. In this chapter, we take a look at deserts and their landscapes. We learn why deserts occur where they do and how erosion and deposition shape their surface. We conclude by exploring life in the desert and by examining the problem of desertification, the gradual transformation of temperate lands into desert.

17.2 The Nature and Locations of Deserts

WHAT IS A DESERT?

Formally defined, a **desert** is a region that is so *arid* (dry) that it supports vegetation on no more than 15% of its surface. In general, desert conditions exist where less than 25 cm of rain falls per year, on average. Because of the lack of water, deserts contain no permanent streams, except for those that bring water in from temperate regions elsewhere.

FIGURE 17.1 Deserts and their hardy inhabitants.

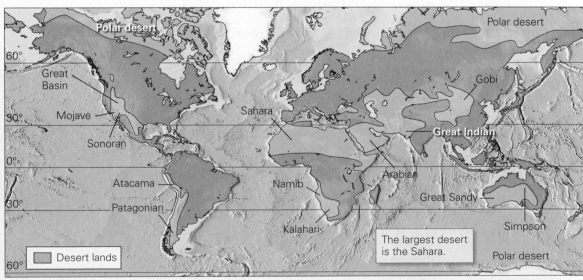

(a) Camels can survive the harsh conditions of the Arabian desert.

(b) The global distribution of deserts. Arid regions cover 25% of the land surface.

◀ (facing page) In arid landscapes such as this one in central Utah, vegetation covers only a small part of the landscape, streambeds stay dry for most of the year, and erosion produces steep cliffs and debris-covered slopes.

Note that the definition of a desert depends on a region's aridity, not on its temperature. Geologists distinguish between *cold deserts*, where temperatures generally stay below about 20°C all year, and *hot deserts*, where daytime temperatures exceed 35°C in the summer. Cold deserts exist at high latitudes, where the Sun's rays strike the Earth obliquely and don't provide much energy, and at high elevations, where thin air can't hold much heat. Hot deserts develop at low latitudes, where the Sun's rays strike the land at a high angle, and at low elevations, where relatively dense air can hold a lot of heat. The hottest recorded temperatures on the Earth—58°C in Libya and 57°C in Death Valley, California—occur in low-latitude, low-elevation deserts.

> **Did you ever wonder . . .**
> how hot it can become in a desert?

TYPES OF DESERTS

Each desert on the Earth has unique landscapes and vegetation that distinguish it from others. Geologists group deserts into five classes, based on the setting in which the desert forms **(Fig. 17.2)**.

> *Subtropical deserts:* The world's largest deserts (such as the Sahara, Arabian, Kalahari, and Australian) form because of the global-scale pattern of air circulation in the atmosphere. At the equator, hot, moisture-laden air rises to great heights. As this air rises, it expands and cools, so the water it contains condenses and falls in downpours that feed the lushness of the equatorial rainforest. The now-dry air high in the troposphere spreads laterally north or south. When this air reaches latitudes of 20° to 30°, a region called the *subtropics*, it has become cold and dense enough to sink. Because the air is dry, no clouds form, and intense solar radiation strikes the Earth's surface all day. The sinking, dry air becomes denser and warmer, soaking up any moisture present. In a *subtropical desert,* swept by this hot air as it heads back to the equator, evaporation rates greatly exceed rainfall rates, so the land becomes parched.

> *Rain-shadow deserts:* As moist air flows from the sea toward a coastal mountain range, the air must rise **(Fig. 17.3)**. The rising air expands and cools, so the water it contains condenses and falls as rain on the seaward flank of the mountains, where it drenches a coastal rainforest. When this air finally reaches the inland side of the mountains, it has lost all its moisture and can no longer provide rain. As a consequence, a *rain shadow* forms, and the land beneath becomes a *rain-shadow desert*. An example lies east of the Cascade Mountains in the state of Washington.

> *Coastal deserts formed near cold ocean currents:* Cold ocean water cools the overlying air by absorbing heat. This reduces the capacity of the air to hold moisture. For example, the cold Humboldt Current, which carries

FIGURE 17.2 Subtropical deserts form because the air that convectively flows downward in the subtropics warms and absorbs water as it sinks.

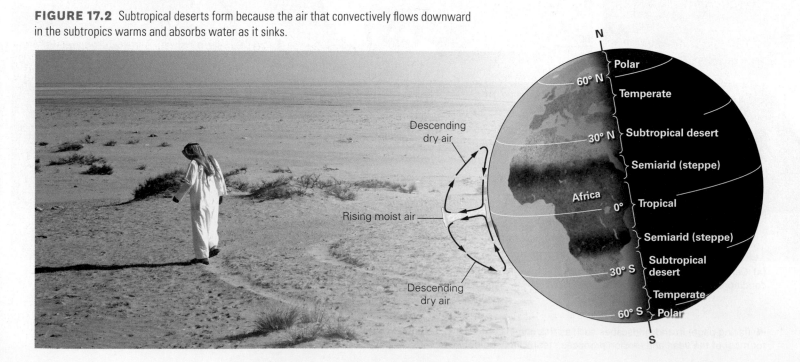

FIGURE 17.3 The formation of a rain-shadow desert. Moist air rises and drops rain on the coastal side of a mountain range. By the time the air has crossed the mountains, it is dry.

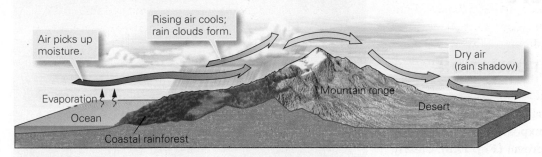

> *Polar deserts:* So little precipitation falls in the Earth's polar regions (north of the Arctic Circle and south of the Antarctic Circle) that these areas are, in fact, arid. Polar regions stay dry for two reasons: First, the general pattern of global-scale air circulation causes the air flowing over these regions to be dry. Second, cold air holds relatively little moisture.

water northward from Antarctica to the western coast of South America, cools the low-elevation air that blows east over the coast. Because this low-elevation air is cooler than the air at higher elevations, it cannot rise buoyantly to high elevations. As a result, conditions that lead to formation of rain clouds do not develop. (Rain clouds only form when moisture-laden air can rise to high elevation where conditions cause the water to condense.) Without rain, western regions of Chile and Peru host coastal deserts, including the Atacama Desert **(Fig. 17.4)**. Astonishingly, portions of the Atacama received no rain at all between 1570 and 1971.

> *Continental-interior deserts:* As an air mass moves across a continent, it progressively loses moisture by dropping rain, even in the absence of a coastal mountain range. By the time an air mass reaches the interior of a broad continent, it has become so dry that the land beneath becomes arid. The largest example of such a *continental-interior desert* is the Gobi, in central Asia.

Different regions of the land surface have become deserts at different times in the Earth's history, because plate movements change the latitude of landmasses, the position of landmasses relative to the coast, and the proximity of landmasses to a mountain range. Because of plate tectonics, some regions that were deserts in the past are temperate or tropical regions now, and vice versa.

TAKE-HOME MESSAGE

Deserts are so arid that they host only very sparse vegetation. They occur in several settings: subtropical dry climates, rain shadows, coasts bordered by cold currents, continental interiors, and polar regions.

QUICK QUESTION Why does the world's largest desert, the Sahara, exist?

FIGURE 17.4 The formation of a coastal desert.

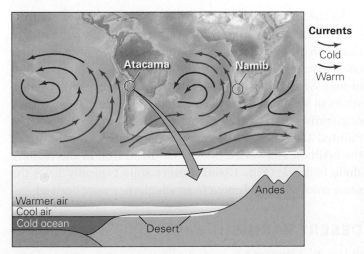

(a) Cold ocean currents cool air along the coast. This air is too dense to rise and produce rain clouds.

(b) The Atacama Desert is one of the driest places on the Earth.

17.3 Producing Desert Landscapes

WEATHERING AND SOIL FORMATION IN DESERTS

If you stand in the Mojave Desert of California and look around, you'll see barren cliffs exposing fractured rock, and slopes or plains littered with gravel **(Fig. 17.5)**. Clearly, *physical weathering* happens in deserts—over time, blocks break off from escarpments at joints and tumble downslope, perhaps shattering into smaller pieces as they collide with other blocks on the way down. Blocks of rock in a desert can remain unchanged for a long time, but they don't last forever. In fact, recent studies suggest that the daily alternation from midday heat to midnight cold might generate sufficient stress to gradually break small blocks apart.

Chemical weathering also takes place in deserts, though it happens very slowly. Moisture from dew and occasional rain allows oxidation, hydrolysis, and dissolution reactions to destroy cements and to transform silicate minerals into clay. These processes cause rocks to disaggregate into pebble- or sand-sized debris.

Does soil develop in a desert? Yes. In places where regolith stays put long enough, unique desert soils slowly develop. After a rare downpour, infiltration of water leaches soluble ions and transports fine clay downward (see Interlude B). But such infiltration events happen infrequently, so the leached ions don't flush away entirely. The ions precipitate underground to form new mineral cement that can bind regolith into a solid, rock-like material known as *caliche* or *calcrete*.

FIGURE 17.5 Dark gravel, formed by physical weathering of rock exposed on this cliff in the Mojave Desert, litters the gentle slope at the cliff's base.

FIGURE 17.6 Soil formation and chemical weathering in deserts.

(a) The red hues of the Painted Desert in Arizona are due to the oxidation of iron in the rock.

(b) By chipping away desert varnish to reveal the lighter rock beneath, Native Americans created art and symbols.

In the absence of organic content, bedrock color controls land-surface color in deserts. Variations in the amounts of iron or in the extent of iron oxidation can result in spectacular color banding. For example, well-oxidized strata are dark red or maroon, strata with moderate iron are light red or orange, strata in which iron has been chemically reduced are whitish or greenish, and strata without much iron are tan or gray. The Painted Desert of northern Arizona earned its name from the brilliant and varied hues of oxidized iron in the region's shale bedrock **(Fig. 17.6a)**. Desert soils typically have the same color as the sediment from which they were derived.

DESERT VARNISH

Shiny **desert varnish**, a dark, rusty brown coating of iron oxide, manganese oxide, and clay, locally coats the surface of rocks in deserts. Desert varnish may form when wind-borne dust sticks to the surface of rock. Microbes in the dust extract metal ions and convert them into oxides. Desert varnish doesn't form in humid climates because frequent rain washes dust away before it has time to be converted into varnish.

Desert varnish takes a long time to form. By measuring the thickness of a desert varnish layer, geologists can provide an estimate of how long a rock has been exposed at the ground surface. Over the centuries, people have used desert-varnished rock as a medium for art, chipping away the varnish to reveal underlying lighter-colored rock to create figures or symbols known as **petroglyphs (Fig. 17.6b)**.

WATER EROSION

Although rain rarely falls in deserts, when it does come, it can radically alter a landscape in a matter of minutes. In the absence of plant cover, rainfall, sheetwash, and stream flow are all extremely effective agents of erosion. It may seem surprising, but more erosion in deserts is due to running water than to wind **(Fig. 17.7a)**.

Water erosion begins with the impact of raindrops, which eject sediment from the ground into the air. On a hill, the ejected sediment lands downslope. The ground quickly becomes saturated with water during a heavy rain, so water starts flowing across the surface, carrying the loose sediment with it. Within minutes after a heavy downpour begins, dry stream channels fill with a turbulent mixture of water and sediment, which rushes downstream as a *flash flood*. Because it is fast and relatively viscous (owing to a load of suspended sediment), water in flash floods can cause intense erosion. Streams in deserts tend to be *ephemeral*, meaning that when the supply of rainwater stops, the water sinks into the gravel

of the streambed and disappears (see Chapter 14). The dry channels of ephemeral streams in desert regions are called *dry washes* or *arroyos* in southwestern North America and *wadis* in the Middle East and North Africa **(Fig. 17.7b)**.

WIND EROSION

In temperate and humid regions, plant cover protects the ground surface from the wind, but in deserts, the wind has direct access to the ground. Wind, like flowing water, can carry sediment both as suspended load and as surface load.

Suspended load, fine-grained sediment such as dust and silt held aloft due to air flow, can be carried so high into the atmosphere (up to several kilometers above the Earth's surface) and so far downwind (tens to thousands of kilometers) that it may move completely out of its source region.

Strong winds can also drive a **surface load**, sediment that undergoes **saltation**, the process of rolling or bouncing along the ground. This begins when turbulence caused by a wind shearing along the ground surface lifts sand grains **(Fig. 17.8a)**. The grains move downwind, following an asymmetric, arch-like trajectory. Eventually, they return to the ground, where they strike other sand grains, causing the new grains to bounce up and drift or roll downwind. The collisions cause sand grains to become rounded and frosted. Saltating grains generally rise no more than 0.5 m, but where they bounce on bedrock, grains may rise as much as 2 m.

A particularly strong wind can generate a dramatic *dust storm*, known as a *haboob* in the Middle East, that can be 100 km long and 1.5 km high. A dust storm, as it approaches, looks like a roiling, opaque wave or cloud. Dust storms can be very hazardous, because they decrease visibility and foul machinery. When a particularly large dust storm engulfed

FIGURE 17.7 Evidence of erosion by running water in deserts.

(a) These hills in the desert near Las Vegas, Nevada, are bone dry, but their shape indicates erosion by water. Note the numerous stream channels.

(b) Gravel and sand are left behind on the floor of a dry wash after a flash flood in Death Valley. Erosion by the water is cutting a channel.

FIGURE 17.8 Sediment transport by wind in deserts.

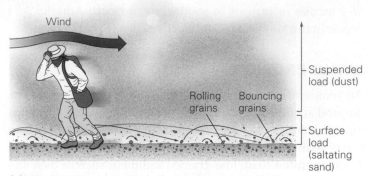

(a) Wind transports desert sediment as suspended load and surface load.

(b) A huge dust storm approaching Phoenix, Arizona.

Phoenix, Arizona, in 2011, the airport had to shut down **(Fig. 17.8b)**.

Because most sand consists of quartz, a hard mineral, sand can not only strip the paint off a car, but it can also abrade rock surfaces in the desert and, over long periods, can carve smooth faces, or *facets*, on pebbles, cobbles, and boulders. If a rock rolls or tips relative to the prevailing wind direction after it has been faceted on one side, or if the wind shifts direction, a new facet with a different orientation forms, and the two facets join at a sharp edge. Geologists refer to rocks whose surface has been faceted by the wind as **ventifacts (Fig. 17.9)**.

FIGURE 17.9 The progressive development of a ventifact.

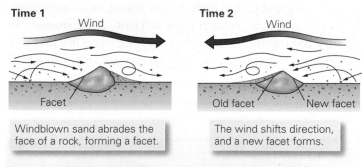

Time 1 Wind   **Time 2** Wind

Facet   Old facet   New facet

Windblown sand abrades the face of a rock, forming a facet.

The wind shifts direction, and a new facet forms.

(a) Ventifacts form when windblown sediment erodes the surface of a rock.

2 cm

(b) An example of a multifaceted ventifact from the Dry Valleys of Antarctica.

Wind abrasion also gradually polishes the surfaces of desert-varnished outcrops into a reflective sheen.

In places where strong winds carrying abrasive sediments blow frequently, **yardangs** (wind-carved elongate ridges or mounds) may develop. Typically, yardangs are much longer than they are wide, and their long axes align with the prevailing wind direction. Mound-like yardangs tend to be asymmetric, with a steeper wall facing the wind, and a gentler slope downwind. If the material being carved into a yardang consists of a layer of stronger rock over weaker rock, or if saltating sand can only rise part way up the side of the eroding rock, the yardang evolves into an elongate knob seemingly perched on a narrower column or ridge.

The size of clasts that wind can carry depends on the wind velocity, so wind does an effective job of sorting sediment. Dust-sized particles loft skyward and sand-sized particles bounce along the ground, while pebbles and larger grains remain behind. Over time, in regions where the substrate consists of soft sediment, wind picks up and removes so much sediment that the land surface becomes lower, a process known as **deflation**. Shrubs can stabilize a small patch of sediment with their roots, so after deflation a forlorn shrub rising from its residual pedestal of soil stands isolated above a lowered ground surface **(Fig. 17.10a)**. In some places, wind carries away so much fine sediment that pebbles and cobbles become concentrated at the ground surface **(Fig. 17.10b)**. Geologists refer to an accumulation of coarser sediment left behind when fine-grained sediment blows away as a **lag deposit**. If the shape of the land surface twists the wind into a turbulent vortex, wind erosion can scour a bowl-like depression called a *blowout*.

FIGURE 17.10 Erosion of sediment by desert winds.

(a) Deflation has removed the sediment between these shrubs in Death Valley, California.

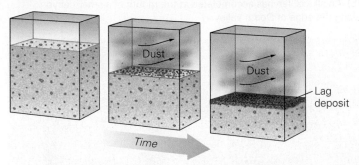

(b) A lag deposit develops when wind blows away finer sediment, leaving behind a layer of coarser grains.

TAKE-HOME MESSAGE

Physical weathering dominates in deserts, but chemical weathering can occur and can contribute to formation of thin soils. Desert varnish may develop to coat rock surfaces. Water causes most erosion and sediment transport but rarely flows, so stream channels are usually dry washes. Wind can transport sediment, generating dust storms and causing saltation. Saltating grains can carve and polish rock.

QUICK QUESTION How is weathering in a desert similar to or different from that of a temperate region?

17.4 Deposition in Deserts

We've seen that erosion relentlessly eats away at bedrock and sediment in deserts. Where does the debris go? In this section, we examine some desert settings in which sediment accumulates.

TALUS SLOPES

Rock that has undergone physical weathering on cliffs eventually tumbles down the cliff and accumulates in an apron-like pile at the cliff's base. Geologists refer to the pile of angular rock debris at the base of a cliff as a **talus**, and to the surface of this pile, which typically represents the angle of repose, as a *talus slope* **(Fig. 17.11a)**. In practice, geologists may also use the word talus for the debris itself, and the phrase *talus pile* for the whole apron. Hikers tend to use the word *scree* for unstable talus that moves when you step on it. While talus occurs commonly in deserts, it can also be found in other climatic realms, as long as new debris falls on the pile faster than the debris already there can weather, erode away, or become soil covered. In desert climates, the rocks of a talus gradually become coated with desert varnish.

ALLUVIAL FANS

Flash floods can carry sediment downstream in a steep-walled canyon. When the turbulent, sediment-laden water flows out onto a plain at the mouth of a canyon, it spreads out over a broader surface with a gentler slope. When this happens, the water slows, and its sediment load drops out and builds into an **alluvial fan**, a conical wedge of sediment that builds outward from the canyon mouth **(Fig. 17.11b)**. Once the fan's overall conical shape has been established, water from a given flash flood tends to flow in a *braided stream* down the steepest part of the fan, meaning that the stream consists of many smaller channels divided from each other by elongate gravel bars. Deposition from a stream builds out sediment at the base of the fan, so the gradient of the stream gradually decreases. Eventually, water emerging from the mouth of the canyon abandons the low-gradient slope and finds a different, steeper pathway down the fan, abandoning the old braided stream and forming a new one. This process repeats itself, so that successive streams follow different paths. Averaged over time, however, the input of sediment at all locations on the fan is about the same, so the fan maintains its overall conical

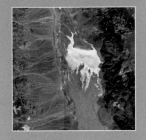

FIGURE 17.11 Production and transportation of debris and sediment in deserts.

(a) This talus apron along the base of a desert cliff formed from rocks that broke off and tumbled down the cliff.

(b) An alluvial fan has accumulated at the mouth of a small canyon along the edge of Death Valley.

(c) A bajada accumulating at the base of a mountain range in the Mojave Desert. Note that a bajada consists of overlapping fans.

geometry. Over time, alluvial fans emerging from adjacent valleys may merge and overlap along the front of a mountain range, producing an elongate wedge of sediment called a *bajada* (Fig. 17.11c).

PLAYAS AND SALT LAKES

Water from larger flash floods may flow to the center of an alluvium-filled basin. If the supply of water is relatively small, the water quickly sinks into the basin's permeable alluvium and does not become a standing body of water. During a particularly large storm or an unusually wet spring, however, a temporary lake may develop over the low part of a basin. During drier times, such desert lakes evaporate and disappear, leaving behind a dry, flat, exposed lake bed known as a **playa** (Fig. 17.12). A crust of clay and various salts (halite,

gypsum, borax, and other minerals) accumulates on the surface of a playa.

Where sufficient water flows into a desert basin, a permanent lake forms. If the basin has no outlet, the lake becomes a salt lake because, although the lake's water escapes by evaporation in the desert sun, its salt stays in place. The Great Salt Lake in Utah serves as an example. Even though the streams feeding the lake are fresh enough to drink, their water contains trace amounts of dissolved ions. Because the lake has no outlet, these ions have become concentrated in the lake over time, making it even saltier than the ocean.

WIND DEPOSITS

As mentioned earlier, wind carries two kinds of sediment loads: a suspended load of dust-sized particles and a surface load of

FIGURE 17.12 Playas form where a shallow, salty lake dries up.

An oblique air photo

Playa

Bajada

Range

(a) This playa in California formed at the base of a bajada.

A close-up of salt crystals

(b) White salt crystals encrust the floor of a playa in Death Valley.

sand. Much of the dust and silt lofts out of the desert and accumulates elsewhere. Sand, however, cannot travel far and accumulates within the desert in elongate mounds called **dunes**, ranging in size from less than a meter to over 300 m high. In favorable locations, dunes accumulate to form vast sand seas. We'll look at dunes in more detail in the following section.

among *hamada* (barren, rocky highlands), *reg* (broad, stony plains), and *erg* (vast sand seas, containing large dunes). In this section, we'll see how the erosional and depositional processes described above can produce such contrasting landscapes. (**Geology at a Glance**, pp. 566–567)

TAKE-HOME MESSAGE

Sediment carried by water and wind in deserts accumulates in a variety of landforms. Alluvial fans form at the outlets of canyons, playas form where water temporarily collects in basins, and dunes form where large amounts of sand are available.

QUICK QUESTION How do bajadas develop?

17.5 Desert Landscapes

The popular media commonly portray deserts as endless seas of sand piled into dunes, which hide the occasional palm-studded oasis. In reality, immense sand seas are only one type of desert landscape. Deserts can host plains covered by loose rock blocks of various sizes, steep highlands bordered by barren cliffs, as well as vast areas buried by drifting sand. Nomads living in the Sahara emphasized these differences by distinguishing

Did you ever wonder . . .
whether all deserts are completely covered by sand?

ROCKY CLIFFS AND MESAS

In hilly or mountainous desert regions, the lack of soil exposes rocky ridges and cliffs. Where bedrock consists of unfoliated rock, such as granite, joint-bounded blocks tend to become rounded, for weathering happens faster at corners and edges than it does at flat faces. Where bedrock consists of horizontal strata, cliff faces tend to form along large, vertical joints. Cliffs are only temporary landscape features, for when rockfalls happen at a cliff face, the position of the cliff effectively steps back. Since rock breaks away from the cliff along a joint that is parallel to the cliff face, the cliff overall retains roughly the same shape as its position migrates. Note that this process, called **cliff retreat**, occurs

SEE FOR YOURSELF...

Uluru, Australia

LATITUDE
25°20'47.64" S

LONGITUDE
131°2'28.96" E

Look down from 5 km.

The red sandstone of Uluru (Ayers Rock), a bornhardt, rises above the stony plains of the central Australian desert. The NW-trending diagonal lines are traces of vertical bedding in the sandstone.

The Desert Realm

The desert of the Basin and Range Province in Utah, Nevada, and Arizona formed due to Cenozoic rifting. It consists of alternating basins (grabens or half-grabens) separated by narrow ranges (tilted fault blocks). The Sierra Nevada, underlain largely by granite, borders the western edge of the province, while the Colorado Plateau, underlain by flat-lying sedimentary strata, borders the eastern edge. Because of the great variety of elevations and rock types, the region hosts several types of desert landscapes.

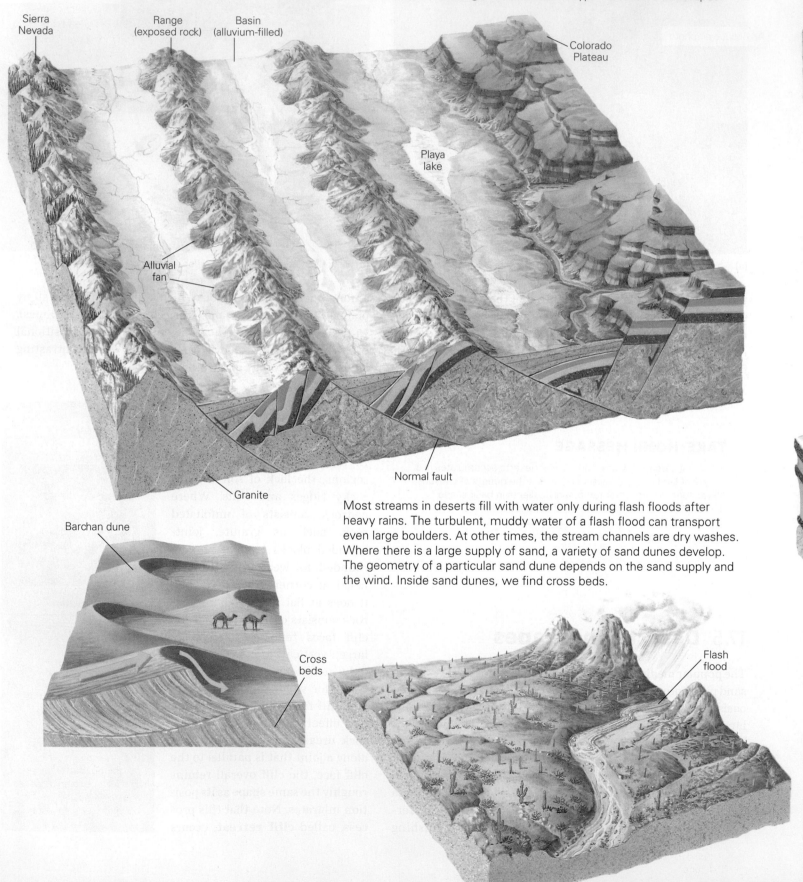

Sierra Nevada

Range (exposed rock)

Basin (alluvium-filled)

Colorado Plateau

Playa lake

Alluvial fan

Granite

Normal fault

Barchan dune

Cross beds

Most streams in deserts fill with water only during flash floods after heavy rains. The turbulent, muddy water of a flash flood can transport even large boulders. At other times, the stream channels are dry washes. Where there is a large supply of sand, a variety of sand dunes develop. The geometry of a particular sand dune depends on the sand supply and the wind. Inside sand dunes, we find cross beds.

Flash flood

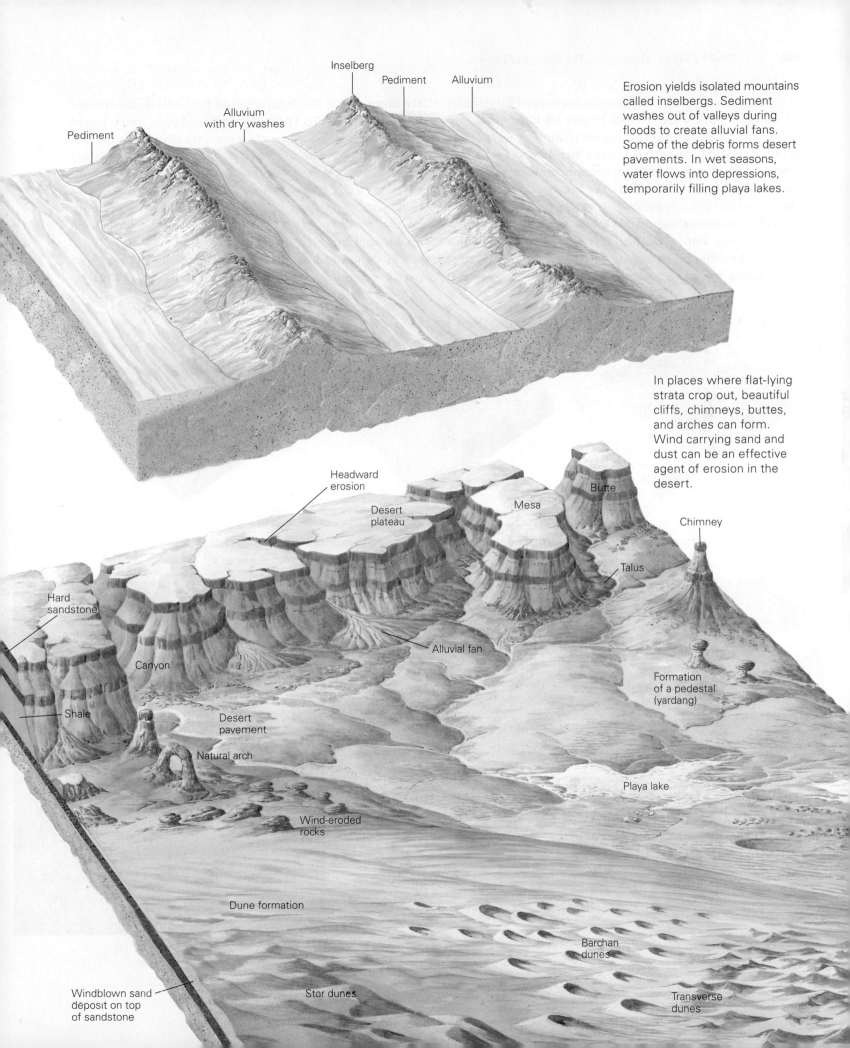

Pediment

Alluvium
with dry washes

Inselberg

Pediment

Alluvium

Erosion yields isolated mountains
called inselbergs. Sediment
washes out of valleys during
floods to create alluvial fans.
Some of the debris forms desert
pavements. In wet seasons,
water flows into depressions,
temporarily filling playa lakes.

In places where flat-lying
strata crop out, beautiful
cliffs, chimneys, buttes,
and arches can form.
Wind carrying sand and
dust can be an effective
agent of erosion in the
desert.

Headward
erosion

Desert
plateau

Mesa

Butte

Chimney

Hard
sandstone

Canyon

Talus

Alluvial fan

Formation
of a pedestal
(yardang)

Shale

Desert
pavement

Natural arch

Playa lake

Wind-eroded
rocks

Dune formation

Barchan
dunes

Windblown sand
deposit on top
of sandstone

Star dunes

Transverse
dunes

in fits and starts—a cliff may remain unchanged for decades or centuries until a slab of rock breaks off to crumble into rubble downslope. Cliffs exposing alternating layers of strata with contrasting strength develop a step-like shape—thick, strong layers of sandstone or limestone become vertical cliffs, and weak layers of thin-bedded shale become rubble-covered slopes. A talus typically accumulates at the base of cliffs.

With continued erosion and cliff retreat, a plateau of rock slowly evolves into a cluster of isolated flat-topped hills. These go by different names, depending on their size—large hills (with a top surface area of several square kilometers)

are **mesas**, from the Spanish word for table; medium-sized hills are **buttes** (Fig. 17.13a, b); and small ones, whose height greatly exceeds their top surface area, are **chimneys**. Erosion of sedimentary strata consisting of alternating resistant and nonresistant layers forms a distinctive type of chimney, known as a *hoodoo*, whose width varies irregularly with height—more resistant layers are wider, and less resistant layers are narrower **(Fig. 17.13c)**. *Natural arches* form when preferential erosion along joints leaves narrow walls or "fins" of rock—when the lower part of a wall erodes before the upper part does, the upper part remains as an arch **(Fig. 17.13d)**.

FIGURE 17.13 Mesas and buttes form in deserts as cliffs retreat over time.

Time (increasing amount of erosion)

(a) Because of cliff retreat, a once-continuous layer of rock evolves into a series of isolated remnants. If the bedding is horizontal, the resulting landforms have flat tops.

(b) Buttes and mesas tower above the floor of Monument Valley, Arizona.

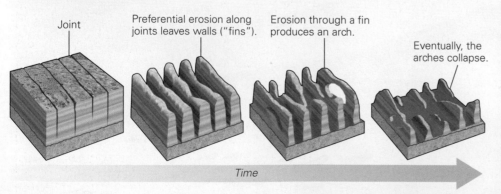

Time

(d) Natural arches form when erosion occurs preferentially along joints, to produce wall-like fins of rock. When the lower part of a fin erodes, an arch remains.

(c) Erosion produced hoodoos, chimney-like columns of rock, in Bryce Canyon, Utah.

FIGURE 17.14 Examples of erosional landscapes in deserts.

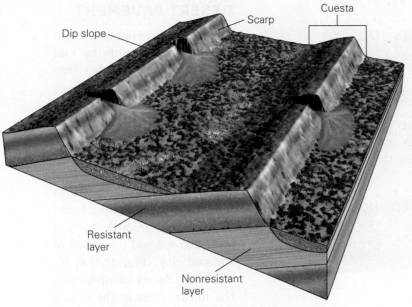

(a) Asymmetric ridges called cuestas develop where strata in a region are not horizontal.

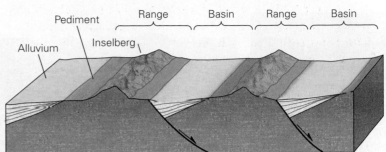

(b) In the Basin and Range Province of the southwestern United States, tilted fault-block ranges evolve into inselbergs, bordered by sediment-filled basins.

In places where bedding dips at an angle, an asymmetric ridge called a *cuesta* develops. A joint-controlled cliff forms the steep front side of a cuesta, and the tilted top surface of a resistant bed forms the gradual slope on the backside **(Fig. 17.14a)**. If the bedding dip is steep to near vertical, a narrow symmetrical ridge, called a *hogback*, forms.

Over time, erosion may grind away all sides of a desert mountain range so that it becomes an *inselberg* (from the German, meaning island rock). In some cases, alluvium-filled basins surround an inselberg and cover bedrock right up to the base of the inselberg's slopes **(Fig. 17.14b)**. Elsewhere, gently sloping bedrock platforms called *pediments* surround inselbergs. Pediments are carved by sediment-laden sheet-wash during heavy rains.

Depending on the rock type or the orientation of stratification in the rock, and on rates of erosion, inselbergs may be sharp-crested, plateau-like, or loaf-shaped (meaning that they have steep sides and a rounded crest). Inselbergs with a loaf geometry, as exemplified by Uluru (Ayers Rock) in central Australia **(Fig. 17.15)**, are also known as *bornhardts*.

FIGURE 17.15 Uluru (Ayers Rock) in central Australia is an inselberg. The red sandstone beds comprising it are the eroded remnant of the vertical limb of a large syncline. Alluvium buried the surrounding bedrock.

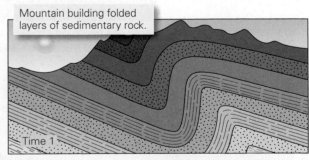

Mountain building folded layers of sedimentary rock.

Time 1

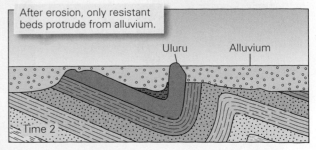

After erosion, only resistant beds protrude from alluvium.

Uluru   Alluvium

Time 2

FIGURE 17.16 Desert pavement and a hypothesis for how it forms by building up a soil from below.

(a) A well-developed desert pavement in the Sonoran Desert, Arizona. The inset shows a close-up of the pavement.

(b) Students standing at the edge of a trench cut into desert pavement. Soil lies between the pavement and the underlying alluvium. The inset shows the top surface of the pavement (lens cap diameter = 8 cm).

STONY PLAINS AND DESERT PAVEMENT

The coarse sediment eroded from desert mountains and ridges ultimately washes into lowland plains to produce gravel-covered surfaces called stony plains. Portions of these stony plains evolve into **desert pavements**, surfaces that resemble a tile mosaic in that they consist of separate stones that fit together tightly, forming a fairly smooth surface layer above a soil composed of silt and clay **(Fig. 17.16a, b)**. Typically, desert varnish coats the top surfaces of the stones forming desert pavement. Geologists have proposed several explanations for the origin of desert pavements. Traditionally, pavements were thought to be lag deposits, formed when wind blows away the fine sediment between clasts, so that the clasts settle down and fit together. Recently, researchers have suggested instead that pavements form when windblown dust slowly sifts down onto the stones and then washes down through the spaces between them. In this model, the pavement is "born at the surface," meaning that

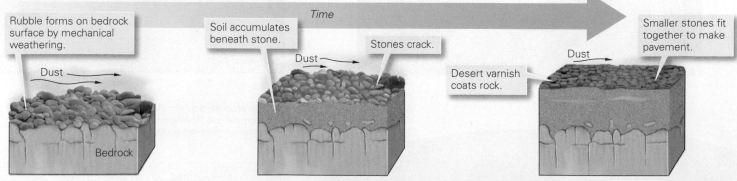

(c) Desert pavement forms in stages. First, loose pebbles and cobbles collect at the surface. Dust settles among the stones and builds up a soil layer below. The stones eventually crack into smaller pieces and settle to form a mosaic-like pavement.

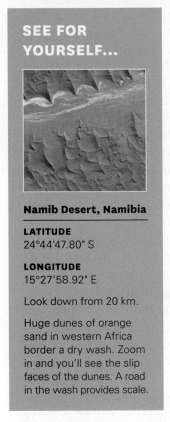
the stones forming the pavement have been progressively lifted up as soil builds up beneath them **(Fig. 17.16c)**. Over time, the rocks at the surface crack, perhaps due to differential heating by the Sun over time. During downpours, sheetwash washes away fine sediment between fragments, and when soils dry and shrink between storms, the clasts settle together, locking into a stable, jigsaw-puzzle-like arrangement.

Desert pavements are remarkably durable and can last for hundreds or thousands of years when left alone. But like many features of the desert, they can be disrupted in a moment by human activity. For example, people driving vehicles across the pavement indent and crack its surface, making it susceptible to erosion. In parts of Arizona, vast desert pavements have become parking lots for campers who migrate to the desert in motor homes for the winter season.

SEAS OF SAND: THE NATURE OF DUNES

A *sand dune* is a pile of sand deposited by wind or by flowing water. Dunes in deserts, of course, are wind-blown deposits. They may start to form where sand becomes trapped on the windward side of an obstacle, such as a rock or a shrub. Gradually, the sand builds downwind into the lee of the obstacle.

Dunes display a variety of shapes and sizes, depending on the character of the wind and the sand supply **(Fig. 17.17a)**. Where sand is relatively scarce and the wind blows steadily in one direction, beautiful crescents called *barchan dunes* develop, with the tips of the crescents pointing downwind. If the wind shifts direction frequently, a group of crescents pointing in different directions overlap one another, yielding a *star dune*. Where enough sand accumulates to bury the ground surface completely, and moderate winds blow, sand piles into simple, wave-like shapes called *transverse dunes*; the crests of transverse dunes lie perpendicular to the wind direction. When strong winds cause a blowout in the middle of a vegetated transverse dune, the sand from the blowout gets deposited downwind, while the portions of the dune on either side of the blowout remain anchored by vegetation. The result is the formation of a *parabolic dune*, whose ends point in the upwind direction. Finally, with abundant sand and a strong,

FIGURE 17.17 The types of sand dunes and the cross beds within them.

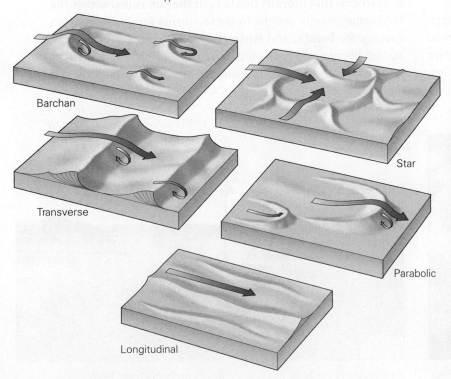

Barchan

Transverse

Star

Parabolic

Longitudinal

(a) The various types of sand dunes.

(b) A sand dune with surface ripples.

(c) Cross bedding inside a dune.

steady wind, the sand streams into *longitudinal dunes*, whose axes lie parallel to the wind direction.

As we've noted, sand saltates up the windward side of the dune, blows over the crest of the dune, and then settles on the steeper, lee face of the dune. The slope of this face attains the angle of repose, the slope angle of a freestanding pile of sand. As sand collects on this surface, it eventually becomes unstable and slides down the slope, so geologists refer to the lee side of a dune as the *slip face*. As more and more sand accumulates on the slip face, the crest of the dune migrates downwind, and former slip faces become preserved inside the dune. In cross section, these slip faces appear as cross beds **(Fig. 17.17b, c)**.

TAKE-HOME MESSAGE

Over time, landscapes evolve in deserts. As rocks collapse along cliffs, the position of a cliff face can migrate, and if bedrock consists of horizontal layers, buttes and mesas form. Erosion of tilted layers yields cuestas. Desert landscapes are variable. Some include vast stony plains, some of which evolve into desert pavements. In sandy areas, many types of dunes will grow.

QUICK QUESTION What factors control the shape, dimensions, and orientation of sand dunes?

17.6 Desert Problems

Natural droughts, overpopulation, overgrazing, careless agricultural practices, and diversion of water supplies can transform a semiarid grassland into a desert landscape in a matter of years to decades. The consequences of such **desertification**, for example, have devastated parts of the Sahel, the belt of semiarid grassland that fringes the southern margin of the Sahara **(Fig. 17.18a)**. In the past, the Sahel provided sufficient vegetation to support a small population of nomadic people and animals. But during the second half of the 20th century, large numbers of people migrated into the Sahel to escape overcrowding in central Africa. The immigrants began farming and maintained large herds of cattle and goats. As a consequence, plowing and overgrazing removed soil-preserving grass and caused the soil to dry out. In addition, trampling animal hooves compacted the ground so it could no longer soak up water. A series of natural droughts has hit the region, bringing catastrophe **(Fig. 17.18b)**. Wind erosion stripped off the remaining topsoil. Without vegetation, the air grew drier, the semiarid grassland of the Sahel became desert, and its inhabitants endured a terrible famine.

Desertification can also happen in industrialized nations. People in the western Great Plains of the United States and Canada suffered from desertification of an agricultural area beginning in 1933, the fourth year of the Great Depression. Banks had failed, workers had no jobs, the stock market had crashed, and hardship burdened all. No one needed yet another disaster—but that year, even nature turned hostile. All through the fall, so little rain fell in the plains, from Texas and Oklahoma up to south-central Canada, that the region's grasslands and croplands browned and withered, and the topsoil turned to powdery dust. Without vegetation to protect it, the topsoil could be stripped away by strong winds. Immense dust storms that literally blotted out the Sun rolled across the landscape. People caught in these storms found themselves gasping for breath, and when the dust finally settled, it had buried houses and roads under drifts, and had dirtied every nook and cranny. What had once been rich farmland turned

FIGURE 17.18 Desertification has been happening in the Sahel.

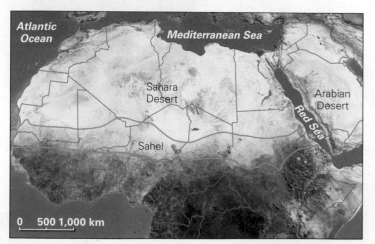

(a) The Sahel is the semiarid land along the southern edge of the Sahara. Large parts have undergone desertification.

(b) Drought in the Sahel has brought deadly consequences. Here residents seek water from a dwindling well.

FIGURE 17.19 In this satellite image, a huge dust cloud that originated in the Sahara blows across the Atlantic.

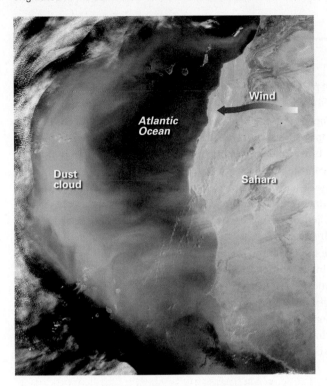

into a wasteland that soon acquired a nickname, the Dust Bowl. It stayed devastated for years.

Why did the fertile soils of Oklahoma and Texas suddenly dry up? The causes were complex—some natural, and some human-induced. Typically, the region has a semiarid climate in which only thin soil develops. But the plains were settled in the 1880s and 1890s, which were unusually wet years. Not realizing its true character, far more people moved into the region than it could sustain, and the land was farmed too intensively. Plowing destroyed the fragile grassland root systems that held the thin soil in place. So when the drought of the 1930s came, it brought catastrophe.

Desertification has an additional dangerous side effect. The dust in dust storms doesn't consist of only clay and silt, but can include salts, toxic chemicals, microbes, and fungi, and these materials can be carried great distances. For example, recent studies suggest that dust blown off the Sahara traverses the Atlantic and settles onto the Caribbean Sea **(Fig. 17.19)**. Geologists are concerned that this dust may infect corals with disease or in some other way inhibit their life processes. So windblown dust may contribute to the destruction of coral reefs.

TAKE-HOME MESSAGE

In lands bordering deserts, droughts and population pressures have led to desertification, transforming vegetated land into desert. As more regions turn into deserts, more wind-transported dust enters the atmosphere, and this dust can include not only fine grains of minerals, but also toxins and pathogens. These damaging materials can be carried across oceans.

QUICK QUESTION What factors transformed Oklahoma and Texas into the Dust Bowl during the 1930s?

ANOTHER VIEW The machinery of an abandoned farm in South Dakota lies buried in dust, as illustrated by this 1936 photograph.

Chapter 17 Review

CHAPTER SUMMARY

> Deserts generally receive less than 25 cm of rain per year. Vegetation covers less than 15% of their surface.

> Subtropical deserts form between latitudes of 20° and 30°, rain-shadow deserts are found on the inland side of mountain ranges, coastal deserts are located on the land adjacent to cold ocean currents, continental-interior deserts exist in landlocked regions far from the ocean, and polar deserts form at high latitudes.

> Physical weathering in deserts produces rocky debris. Chemical weathering happens slowly; it produces ions that precipitate as new minerals just below the surface to form caliche.

> Water causes significant erosion in deserts, mostly during heavy downpours. Flash floods carry large quantities of sediments down ephemeral streams. When rain stops, streams dry up, leaving steep-sided dry washes.

> Wind picks up dust and silt as suspended load, and it causes sand to saltate. Where wind blows away finer sediment, a lag deposit remains. Windblown sediment abrades the ground, creating a variety of features such as ventifacts.

> Desert pavements are mosaics of varnished stones armoring the surface of the ground.

> Talus aprons form when rock fragments accumulate at the base of a slope. Alluvial fans form at a mountain front where water in ephemeral streams deposits sediment on a plain. When temporary desert lakes dry up, they leave playas.

> In some desert landscapes, erosion causes cliff retreat, eventually resulting in the formation of mesas, buttes, cuestas, and inselbergs. Pediments of nearly flat or gently sloping bedrock surround some inselbergs.

> Where sand is abundant, the wind builds it into dunes. Common types include barchan, star, transverse, parabolic, and longitudinal dunes.

> Changing climates and land abuse may cause desertification, the transformation of semiarid land into deserts. Windblown dust, sometimes carrying microbes and toxins, may waft from deserts across oceans.

GUIDE TERMS

alluvial fan (p. 563)	desert (p. 557)	lag deposit (p. 562)	surface load (p. 561)
butte (p. 568)	desertification (p. 572)	mesa (p. 568)	suspended load (p. 561)
chimney (p. 568)	desert pavement (p. 570)	petroglyph (p. 561)	talus (p. 563)
cliff retreat (p. 565)	desert varnish (p. 560)	playa (p. 564)	ventifact (p. 562)
deflation (p. 562)	dune (p. 565)	saltation (p. 561)	yardang (p. 562)

***GEO**TOURS THIS CHAPTER'S GEOTOUR'S WORKSHEET (Q) FEATURES QUESTIONS AND GOOGLE EARTH SITES ON:*

> Deserts and desert features from around the world   > Water use in arid regions

REVIEW QUESTIONS

The letters following each Review Question refer to the corresponding Learning Objective from the Chapter Opener.

1. What factors determine whether a region can be classified as a desert? (**A**)

2. Explain several settings that can cause deserts to form. What type  of desert is depicted in the figure? Explain how it formed. (**A**)

3. Have today's deserts always been deserts? (**A**)

4. How do weathering processes in deserts differ from those in temperate or humid climates? (**B**)

5. Describe how water modifies the landscape of a desert. Be sure to discuss both erosional and depositional landforms. (**B**)

6. Explain the ways in which desert winds transport sediment. (**B**)

7. Explain how the following features form: (a) desert varnish, (b) desert pavement, (c) ventifacts, and (d) yardangs. (**B**)

8. Describe the process of formation of alluvial fans, pediments, bajadas, and playas. (**C**)

9. Describe the process of cliff (scarp) retreat and the landforms that result from it. (**D**)

10. What are the types of sand dunes? What factors determine which type of dune develops at a particular location? Which type is pictured here? (**D**)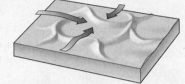

11. What is the process of desertification, and what causes it? How can desertification in Africa affect the Caribbean? (**E**)

ON FURTHER THOUGHT

12. Death Valley, California, lies to the east of a high mountain range, and its floor lies below sea level. During the summer, Death Valley becomes very hot and dry. Explain why it has such weather. (**A**)

13. You are working for an international nongovernmental organization and have been charged with the task of providing recommendations to an African nation that wishes to slow or halt the process of desertification within its borders. What are your recommendations? (**E**)

14. The Namib Desert lies to the north and west of the Kalahari Desert, in southern Africa. The two deserts formed for different reasons. Explain this statement. (**A**)

ONLINE RESOURCES

Videos This chapter includes videos on the evolution of deserts and the transport of Saharan dust to the Amazon.

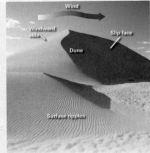

Smartwork5 This chapter includes labeling exercises about desert landscapes and the different types of deserts.

CHAPTER 18

AMAZING ICE: GLACIERS AND ICE AGES

By the end of this chapter, you should be able to . . .

A. describe how glacial ice forms and flows, and how to categorize various kinds of glaciers.

B. explain why glaciers advance and retreat, and how their flow modifies the landscape.

C. recognize sedimentary deposits and associated landforms left by glaciers.

D. describe evidence showing that glaciers covered large areas of continents during the Pleistocene Ice Age, and that there have been four or five earlier ice ages during Earth history.

E. suggest why ice ages happen, and why glaciations during an ice age occur periodically.

> I seemed to vow to myself that some day I would go to the region of ice and snow
> and go on and on till I came to one of the poles of the earth.
>
> ERNEST SHACKLETON (British polar explorer, 1874–1922)

18.1 Introduction

There's nothing like a good mystery, and one of the most puzzling in the annals of geology came to light in northern Europe early in the 19th century. When farmers of the region prepared their land for spring planting, they occasionally broke their plows by running them into hidden boulders buried randomly through otherwise fine-grained sediment. Many of these boulders did not consist of local bedrock, but rather came from outcrops hundreds of kilometers away. Because the boulders had apparently traveled so far, they came to be known as **erratics** (from the Latin *errare*, to wander).

The mystery of the wandering boulders became a subject of great interest to early-19th-century geologists. They realized that such deposits of extremely *unsorted sediment* (containing a variety of clast sizes) could not be examples of typical stream alluvium, for running water sorts clasts by size. Most attributed the deposits to a vast flood that they imagined had been powerful enough to spread a slurry of boulders, sand, and mud across the continent. In 1837, however, a young Swiss geologist named Louis Agassiz proposed a radically different interpretation. Agassiz often hiked among *glaciers* in the Alps near his home. He observed that these slowly flowing masses of solid ice could carry enormous boulders, as well as sand and mud, for ice is strong enough to support the weight of rock. Furthermore, since ice does not sort sediment as it flows, glaciers leave behind unsorted sediment when they melt. On the basis of these observations, Agassiz proposed that the mysterious erratics of Europe were left by vast glaciers that had once spread over much of the continent **(Fig. 18.1)**. In other words, Europe had once been in the grip of an *ice age*, a time when glaciers covered areas that presently have temperate climates.

Agassiz's radical proposal initially faced intense criticism, but by the late 1850s geologists everywhere agreed that, indeed, Europe once had Arctic-like climates. Later in life, Agassiz traveled to the United States and identified many glacial features in North America's landscape, proving that the last ice age did not affect Europe alone. Glaciers cover only about 10% of the land on Earth today, but during the last ice age, which ended less than 12,000 years ago, as much as 30%

◀ (facing page) Glaciers imperceptibly flow down valleys in the mountains of southern Alaska. The rugged topography of many high mountain ranges has been sculpted by glaciers.

FIGURE 18.1 Agassiz envisoned that extensive areas of the northern hemisphere were once covered by vast ice sheets comparable to the one covering Antarctica today.

of continental land surface was covered by an *ice sheet*, a layer of ice hundreds to thousands of kilometers across and as much as a few kilometers thick.

The work of Louis Agassiz brought the subject of glaciers and ice ages into the realm of geologic study and ultimately led people to recognize that major climate changes do happen during the Earth's history. In this chapter, after describing the nature of ice, we see how glaciers form, why they move, and how they modify landscapes by erosion and deposition. We also discuss the evidence for ice ages, with a particular focus on the most recent one, known as the *Pleistocene Ice Age*, whose impact on the landscape can still be seen today. The chapter concludes by considering hypotheses that geologists have put forth to explain why ice ages happen.

18.2 Ice and the Nature of Glaciers

WHAT IS ICE?

Ice consists of solid water, formed when liquid water cools below its freezing point. In effect, ice crystals are minerals—they are naturally occurring solids with a definite chemical composition (H_2O) and a regular crystal structure. Ice crystals have a hexagonal shape, so snowflakes grow into six-pointed stars **(Fig. 18.2a)**. We can consider natural occurrences of ice

FIGURE 18.2 The nature of ice and the formation of glaciers. Snow falls like sediment, and it metamorphoses to ice when buried.

A boundary between layers

(a) The hexagonal shape of snowflakes. No two are alike.

(b) Layers of snow accumulate. They recrystallize to become ice.

The layers in the photo at left are part of this glacier in the Alps.

The wall of a tunnel bored into a glacier

Loose snow
(90% air)

Granular snow
(50% air)

Firn
(25% air)

Fine-grained ice
(<20% air, in bubbles)

Coarse-grained ice
(<20% air, in bubbles)

10,000 years
(250 m)

130,000 years
(2,000 m)

Not to scale

(c) As revealed by a microscope, glacial ice has coarse grains and contains air bubbles.

(d) Snow compacts and melts to form firn, which recrystallizes into ice. Crystal size increases with depth. The numbers indicate the age and depth of the ice for an average glacier.

as comparable to various types of rock. A layer of fresh snow, like a layer of silt, forms when snow (clasts of ice) settle out of the air. A layer of snow that has been compacted, so that the grains stick together, resembles a layer of sedimentary rock **(Fig. 18.2b)**. Ice that appears on the surface of a pond, like an igneous rock, forms when molten ice (liquid water) solidifies. Glacial ice, like a metamorphic rock, develops when a mass of pre-existing ice *recrystallizes* in the solid state so that it develops a new texture **(Fig. 18.2c)**.

HOW A GLACIER FORMS

In order for a glacier to form, the local climate must be cold enough that winter snow does not melt entirely away. Snowfall must occur at such a rate that a large amount of snow can accumulate even during the summer; the surface on which snow accumulates must have a gentle slope so that snow falling on it does not slide away in avalanches; and accumulated snow must be protected from the wind so that it doesn't blow away. These conditions can be met in polar regions and in mountains. Glaciers develop in polar regions, even though these areas are relatively arid, because temperatures remain so cold that any snow that falls can remain. Glaciers develop in mountains, even at low latitudes, because temperature decreases with elevation, so at high elevations, the mean temperature remains cold enough for ice and snow to last all year. Because the temperature of a region depends on latitude, the elevation at which mountain glaciers grow, and the elevation down to which they can flow, depends on latitude. In the Earth's present-day climate, glaciers at the equator do not descend lower than an elevation of 5 km, but glaciers formed at latitudes between 60° and 90° can flow down to sea level.

The transformation from snow to glacial ice takes place progressively, as younger snow buries older snow. Freshly fallen snow consists of delicate hexagonal crystals with sharp points. The crystals do not fit together tightly, so fresh snow contains up to 90% air. With time, the points of the snowflakes become blunt because they either melt (turn into a liquid) or **sublimate** (evaporate directly into vapor). When the flakes become blunt, the snow packs more tightly. As snow becomes buried, the weight of the overlying snow increases the pressure. Under conditions of increased pressure, ice grains slide relative to each other to fill pore spaces, and grains melt slightly at their contact points, forming liquid water that refreezes in pore space. As a result, deeply buried snow transforms into a packed granular material called **firn**, which contains only about 25% air **(Fig. 18.2d)**. As pressure increases still more, firn recrystallizes into a solid mass of new, coarser interlocking ice crystals. Such glacial ice, which may still contain up to 20% air trapped in

bubbles, tends to absorb red light and therefore has a bluish color. The transformation of fresh snow to blue glacial ice can take as little as tens of years in regions with abundant snowfall or as long as thousands of years in regions with little snowfall.

CATEGORIES OF GLACIERS

Formally defined, a **glacier** is a stream or sheet of recrystallized ice that stays mostly frozen all year long and flows under the influence of gravity. Today, glaciers highlight coastal and mountain scenery in Alaska, the Cordillera of western North America, the Alps of Europe, the Southern Alps of New Zealand, the Himalayas of Asia, and the Andes of South America, and they cover most of Greenland and Antarctica. Geologists distinguish between two main categories of glaciers on the Earth, mountain glaciers and continental glaciers, based on their relation to topography and on their overall surface area.

Mountain glaciers (also known as *alpine glaciers*) exist in or adjacent to mountainous regions if temperatures remain cold enough. Overall, they flow from higher elevations to lower elevations **(Fig. 18.3a)**. Mountain glaciers include *cirque glaciers*, which fill bowl-shaped depressions, or *cirques*, on the flank of a mountain; *valley glaciers*, rivers of ice that flow down valleys; *ice caps*, mounds of ice that submerge peaks and ridges at the crest of a mountain range; and *piedmont glaciers*, fans or lobes of ice that form where a valley glacier emerges from a valley and spreads out into the adjacent plain **(Fig. 18.3b–d)**. Mountain glaciers range in size from a few hundred meters to a few hundred kilometers long.

Continental glaciers are vast ice sheets that spread over thousands of square kilometers of continental crust. Today, they exist only in Antarctica and Greenland **(Fig. 18.4)**. These ice sheets rest mostly on solid ground. But locally, liquid water accumulates at the bases of continental glaciers. The largest of these subglacial lakes, Lake Vostok in Antarctica, has an area of 5,400 km^2. Continental glaciers flow outward from their thickest point, where ice may be as thick as 3.5 km, and thin toward their margins, where ice may be only a few hundred meters thick. Space exploration reveals that the Earth is not alone in hosting polar ice sheets—Mars has them too **(Box 18.1)**.

HOW DO GLACIERS MOVE?

Glacial ice is solid, yet glaciers flow. To understand this paradox, keep in mind that under appropriate conditions, glacial ice, like the metamorphic rocks that we described in Chapter 7, can undergo **plastic deformation**. During this process, existing grains slowly change shape without breaking or new grains grow while old ones disappear.

FIGURE 18.3 Several types of glaciers form in mountainous areas.

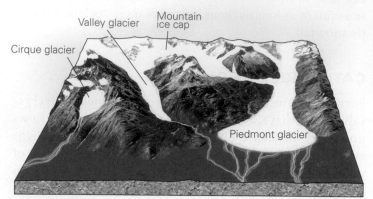

(a) Mountain glaciers are classified based on shape and position.

(b) Valley glaciers draining a mountain ice cap in Alaska.

(c) A valley glacier and cirque glaciers in Switzerland.

(d) A piedmont glacier near the coast of Alaska.

FIGURE 18.4 Antarctica is an ice-covered continent.

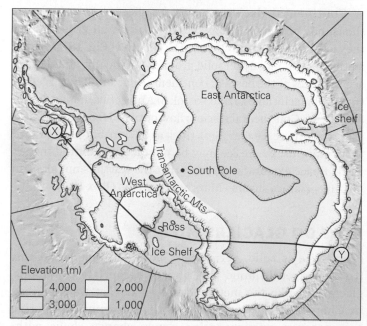

(a) A contour map of the Antarctic ice sheet. Valley glaciers carry ice from the ice sheet of East Antarctica down to the Ross Ice Shelf.

Simplistically, plastic deformation happens when, in response to stress, chemical bonds break and then immediately reattach in a somewhat different configuration. In silicate rocks, plastic deformation can occur only at depths in the crust where temperatures exceed a few hundred degrees and pressures become huge. In glacial ice, due to the weak bonds among water molecules in ice, plastic deformation can happen at depths of only about 60 m,

Did you ever wonder . . .
how a glacier moves?

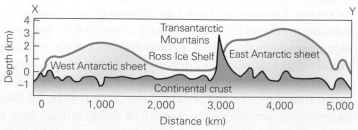

(b) A cross section X to Y of the Antarctic ice sheet. The Transantarctic Mountains separate East Antarctica from West Antarctica.

BOX 18.1 CONSIDER THIS...

Polar Ice Caps on Mars

Mars has white polar ice caps whose sizes change with the season, suggesting that they partially melt and then refreeze (**Fig. Bx18.1**). The question of what the ice caps consist of remained a puzzle until fairly recently. It now appears that the Martian ice caps consist mostly of H_2O ice mixed with a small amount of dust, and they attain a maximum thickness of 3 km. During the winter, atmospheric carbon dioxide freezes and covers the north polar cap with a 1-m-thick layer of dry ice. During the summer, this layer sublimates away. The south polar cap has an 8-m-thick blanket of dry ice, which doesn't disappear entirely in the summer. The difference between the north and south polar caps may reflect elevation—the south pole sits 6 km higher and therefore remains colder.

High-resolution photographs reveal that distinctive canyons, up to 10 km wide and 1 km deep, spiral outward from the center of the north polar ice cap. Why did this pattern form? Calculations suggest that if the ice sublimates on the sunny side of a crack and refreezes on the shady side, the crack will migrate sideways over time. If the cracks migrate more slowly closer to the pole, where it's colder, than they do farther away, they will naturally evolve into spirals.

FIGURE Bx18.1 The ice caps of Mars.

Ice caps during the winter

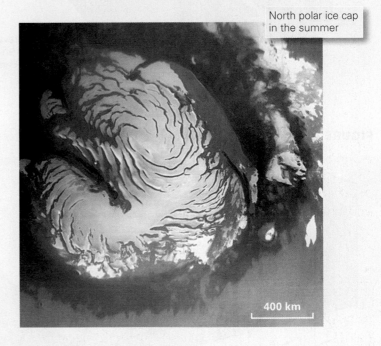

North polar ice cap in the summer

400 km

a depth known as the *brittle–plastic transition* for ice. Because glaciers can be hundreds of meters to several kilometers thick, most of the ice in a glacier can flow plastically. In *polar glaciers*, in which ice remains below its melting temperature throughout the year, plastic deformation takes place completely in the solid state. In *temperate glaciers*, where ice tends to be at or near its melting temperature during at least part of the year, plastic deformation also involves grains sliding past their neighbors on thin water films.

Above the brittle–plastic transition, glacial ice tends to be too brittle to flow easily. So, as the deeper parts of a glacier flow, the shallower parts may crack. A crack that opens into a downward-tapering gash is called a **crevasse** (**Fig. 18.5**). In large glaciers, crevasses can be hundreds of meters long, and

at the surface of the glacier, they may be up to 15 m across. People traversing glaciers must take safety precautions to avoid falling into crevasses, which may be hidden by snow drifts that could collapse under the weight of a snowmobile or even of a single person.

Why do glaciers move? Simply, the force of gravity can overcome the strength of ice (**Fig. 18.6**). A glacier flows in the direction in which its top surface slopes. Therefore, mountain glaciers flow down slopes or valleys, and continental ice sheets spread outward from their thickest point. To picture the movement of a continental ice sheet, imagine pouring honey on a tabletop. The honey spreads out until the puddle reaches an even thickness, and if you keep adding more honey to the center of the pile, it keeps flowing. In the case

FIGURE 18.5 Crevasses form in the upper layer of a glacier, in which the ice is brittle. Commonly, cracking takes place where the glacier bends while flowing over steps or ridges in its substrate.

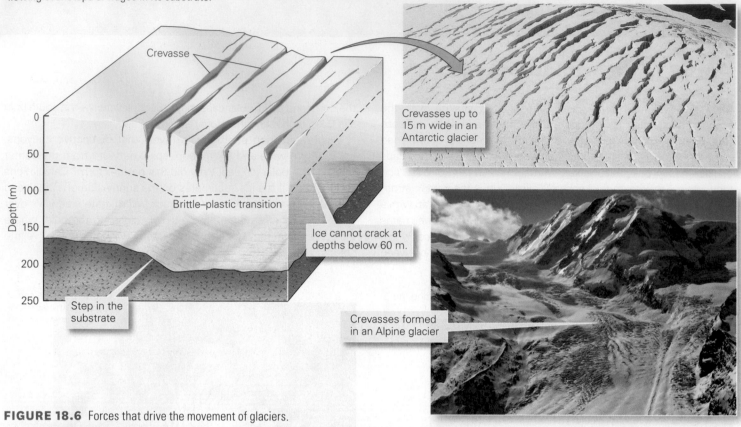

Crevasse

Brittle–plastic transition

Ice cannot crack at depths below 60 m.

Step in the substrate

Crevasses up to 15 m wide in an Antarctic glacier

Crevasses formed in an Alpine glacier

FIGURE 18.6 Forces that drive the movement of glaciers.

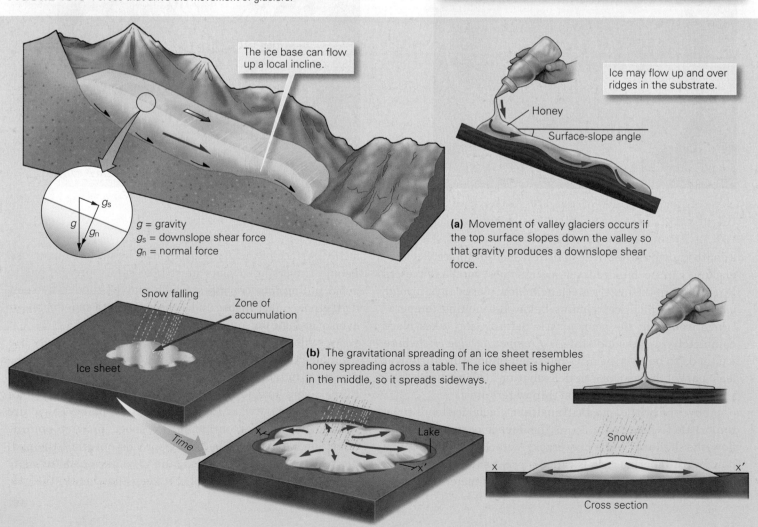

The ice base can flow up a local incline.

g = gravity
g_s = downslope shear force
g_n = normal force

Ice may flow up and over ridges in the substrate.

Honey

Surface-slope angle

(a) Movement of valley glaciers occurs if the top surface slopes down the valley so that gravity produces a downslope shear force.

Snow falling

Zone of accumulation

Ice sheet

Time

Lake

(b) The gravitational spreading of an ice sheet resembles honey spreading across a table. The ice sheet is higher in the middle, so it spreads sideways.

Snow

Cross section

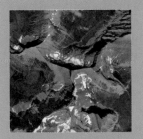
of a continental ice sheet, a thick pile of ice builds up, and gravity causes the top of the pile to push down on the ice at the base. Eventually, the basal ice can no longer support the weight of the overlying ice and begins to deform plastically. When this happens, the basal ice starts squeezing out to the side, carrying the overlying ice with it. The greater the volume of ice that builds up, the wider the ice sheet can become. Note that because the slope of a glacier's surface determines the direction that it flows, glaciers can move up and over ridges or hills on the land surface at their base.

Generally, glaciers flow at rates of 10 to 300 m per year. But not all parts of a glacier move at the same rate, for friction between rock and ice slows a glacier. Therefore, the interior of a valley glacier moves faster than its margins, and the top of a glacier moves faster than its base **(Fig. 18.7a)**. When glaciers flow into conditions where liquid water can accumulate and mix with sediment at the base of a glacier, a process called *basal sliding* takes place. During basal sliding, the glacier glides along on a wet sediment slurry without coming into frictional contact with bedrock and, as a consequence, may undergo a *surge*, meaning that it moves much faster than normal. During a surge, ice flow has been clocked at speeds of 10 to 110 m per day!

GLACIAL ADVANCE AND RETREAT

Glaciers resemble bank accounts: snowfall accumulates and adds to the account, while **ablation**—the removal of ice by melting (turning to liquid), sublimation (vaporizing directly), and *calving* (breaking off chunks)—subtracts from the account. Snowfall adds to the glacier in the *zone of accumulation*, whereas ablation subtracts in the *zone of ablation*. The boundary between these two zones is the **equilibrium line (Fig. 18.7b)**.

FIGURE 18.7 Flow velocities vary with location in a glacier. Overall, ice flows from the zone of accumulation to the toe.

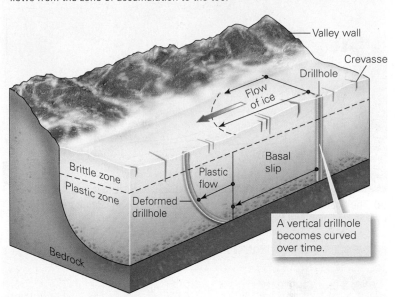

(a) Different parts of a glacier flow at different velocities due to friction with the substrate. The top and center regions flow fastest.

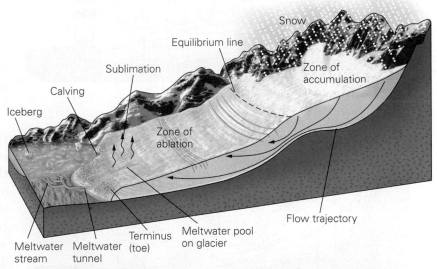

(b) The equilibrium line separates the zone of accumulation from the zone of ablation. Arrows indicate the flow path of ice within the glacier.

Whether a glacier gets larger or smaller over time depends on the balance between accumulation and ablation. If the rate at which ice builds up in the zone of accumulation exceeds the rate at which ablation occurs below the equilibrium line, then the glacier's *toe*, its leading edge, moves forward into previously unglaciated regions—such a change is called a **glacial advance (Fig. 18.8a, b)**. In mountain glaciers, the position of a toe moves downslope during an advance, and in continental glaciers, the toe moves outward, away from the glacier's origin. At the southern edge of a glacier, an advance causes the toe to move to lower latitudes. If the rate of ablation below

FIGURE 18.8 Glacial advance and retreat.

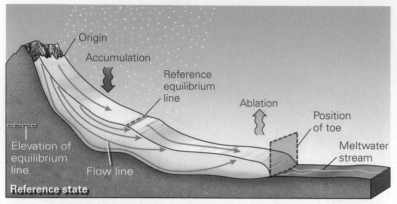

(a) The position of the toe represents a balance between addition of ice by accumulation and loss of ice by ablation.

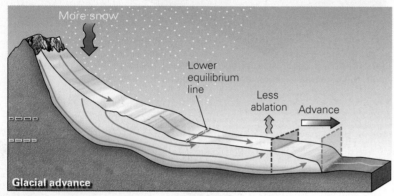

(b) If accumulation exceeds ablation, the glacier advances, the toe moves farther from the origin, and the ice thickens.

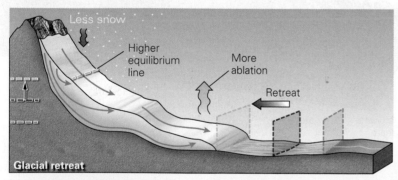

(c) If ablation exceeds accumulation, the glacier retreats and thins. The toe moves back, even though ice continues to flow toward the toe.

the equilibrium line equals the rate of accumulation, then the position of the toe remains fixed, even though ice continues to flow toward the toe. But if the rate of ablation exceeds the rate of accumulation, then the position of the toe moves back toward the origin of the glacier—such a change is called a **glacial retreat (Fig. 18.8c)**. During a mountain glacier's retreat, the position of the toe moves upslope, and during a continental glacier's retreat, the position of the toe moves back

toward the glacier's origin. This means that its southern margin moves to higher latitudes. It's important to realize that when a glacier retreats, only the position of the toe moves back toward the glacier's origin, for ice always flows toward the toe.

As glaciers flow, ice tends to follow curved trajectories, in cross section. This means that beneath the zone of accumulation, ice flows down toward the base of the glacier as new ice accumulates above it, whereas beneath the zone of ablation, ice gradually moves up toward the surface of the glacier, as overlying ice ablates. Because ice follows curving flow paths in the glacier, rocks picked up by ice at the base of the glacier may eventually reach the top of the glacier.

ICE IN THE SEA

On the moonless night of April 14, 1912, the ocean liner *Titanic* struck a large iceberg in the frigid North Atlantic. Its crew was convinced that they could see and avoid the biggest bergs and that smaller ones could not damage the steel hull of this brand new "unsinkable" vessel. Lookouts indeed detected the ghostly, floating mass of ice and alerted the ship's pilot, but the ship was unable to turn fast enough to avoid disaster. The force of the glancing blow split the steel hull, water gushed in, and less than 3 hours later, the ship disappeared beneath the surface and 1,500 people perished.

Where do icebergs, such as the one responsible for the *Titanic*'s demise, originate? At high latitudes, glaciers flow down to sea level. Glaciers whose toes lie in the water are called **tidewater glaciers**. Valley glaciers entering the sea become *ice tongues*, longer than they are wide, whereas continental glaciers entering the sea become broad, flat sheets known as *ice shelves*. In shallow water, glacial ice remains grounded **(Fig. 18.9a)**. But in deeper water, the ice floats such that about four-fifths of its mass lies below the water's surface. At the toe of a tidewater glacier, blocks of ice calve off and tumble into the water with an impressive splash. A free-floating chunk that rises 6 m above the water, and is at least 15 m long, can be called an *iceberg*. Since most of its mass lies below the surface of the sea, the base of a large iceberg may actually extend down for a few hundred meters below the water's surface **(Fig. 18.9b, c)**.

Not all ice floating in the sea originates in glaciers on land. In polar climates, the surface of the sea itself freezes, forming **sea ice (Fig. 18.9d)**. The Earth's north polar ice cap consists of sea ice, formed on the surface of the Arctic Ocean. Some sea ice floats freely, but some protrudes outward from the shore.

FIGURE 18.9 Ice shelves, tidewater glaciers, and sea ice.

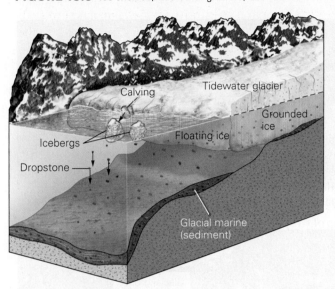

(a) Ice is grounded in shallow water but floats in deep water.

(b) This artist's rendition of an iceberg emphasizes that most of the ice is underwater.

(c) In summer, some of the sea ice of Antarctica breaks up to form tabular icebergs.

(d) In winter, sea ice covers most of the Arctic Ocean (left) and surrounds Antarctica (right).

TAKE-HOME MESSAGE

Glaciers form when buried snow lasts all year and turns to ice. Mountain glaciers form at high elevation and flow downslope. Ice sheets form in high latitudes and spread over continents. Flow of ice can be accomplished by plastic deformation or by basal gliding. The balance of accumulation and ablation determines whether a glacier advances or retreats. In polar regions, ice shelves, ice tongues, and sea ice cover large areas.

QUICK QUESTION Does ice actually flow uphill during a glacial retreat?

18.3 **Carving and Carrying by Ice**

THE PROCESS OF GLACIAL EROSION

During the last ice age, valley glaciers carved deep, steep-sided valleys into the Sierra Nevada of California. In the process, some granite domes were cut in half, leaving a rounded surface on one side and a steep cliff on the other. Half Dome, in Yosemite National Park, formed in this way **(Fig. 18.10a)**—its steep cliff has challenged many rock climbers. Such glacial erosion also produces the knife-edge ridges and pointed spires of high mountains **(Fig. 18.10b)** and broad expanses where rock outcrops have been stripped of overlying sediment and polished smooth **(Fig. 18.10c)**. In many places, the rock surface that is visible today was once in contact with ice, but

elsewhere, later rockfalls and river erosion have substantially modified the surface.

Glaciers pick up fragments of their substrate in several ways. During *glacial incorporation*, ice surrounds loose rock and carries it away, whereas during *glacial plucking*, a glacier breaks off fragments of bedrock. Plucking occurs when ice flows into joints that intersect the bedrock's surface. Freezing and thawing, along with the push of moving ice, causes joints to grow until a joint-bounded block of rock finally breaks free of its substrate and starts to move with the ice. As glaciers flow, sand and silt embedded in the ice act like the teeth of a giant rasp and grind away the substrate. This process, *glacial abrasion*, pulverizes rock into a fine powder known as *rock flour*, and yields shiny *glacially polished surfaces* on bedrock.

FIGURE 18.10 Products of glacial erosion. Ice is a very aggressive agent of erosion.

(a) Half Dome in Yosemite National Park, California.

(b) Rugged, glacially carved peaks in the Swiss Alps.

(c) Glacially polished outcrop in Central Park, New York City.

Glacial groove

(d) Glacial grooves in Victoria, British Columbia.

Clasts moving with the ice may also carve grooves or scratches, known as **glacial striations**, into bedrock **(Fig. 18.10d)**. Where carved by isolated sand grains into a polished surface, striations may look like scratches, with a width of less than a few millimeters and a length of tens of centimeters. In places where the base of the ice was carrying abundant gravel, the flow of ice may carve *glacial grooves* that can be up to a meter across, a meter deep, and tens of meters long. As you might expect, striations run parallel to the flow direction of the ice.

Let's now look more closely at erosional landscape features associated with a mountain glacier **(Fig. 18.11a)**. Freezing and thawing during the fall and spring help fracture the rock bordering the *head* of the glacier (the ice edge high in the mountains). This rock falls on the ice or gets picked up at the base of the ice, and moves downslope with the glacier. As a consequence, a bowl-shaped depression, or **cirque**, develops on the side of the mountain. If the ice later melts, a lake called a *tarn* may form at the base of the cirque. An **arête**, a residual knife-edge ridge of rock, separates two adjacent cirques, and a pointed mountain peak surrounded by at least three cirques is called a **horn**. The Matterhorn, a peak in Switzerland, is a particularly beautiful example of a horn; each of its faces began as a cirque **(Fig. 18.11b)**.

Glacial erosion drastically modifies the shape of a valley. To see how, compare a river-eroded valley with a glacially eroded valley. A topographic profile across a river in unglaciated mountains shows that a river typically flows down a **V-shaped valley**, with the river channel itself forming the point of the V. The V develops because river erosion occurs only in the channel, and mass wasting causes the valley slopes to approach the angle of repose. In contrast, a profile across a glacially eroded valley resembles a U, with steep walls and a curved floor. A **U-shaped valley** forms because the combined processes of glacial abrasion and plucking not only lower the floor of the valley but also erode its sides **(Fig. 18.11c)**.

Glacial erosion in mountains also modifies the intersections between tributaries and the trunk valley. In a river system, the trunk stream serves as the local base level for tributaries (see Chapter 14), so the mouths of the tributary valleys lie at the same elevation as the trunk valley. The ridges (spurs) between valleys taper to a point when they join the trunk valley floor. During glaciation, tributary glaciers flow down side valleys into a trunk glacier. But the trunk glacier cuts the floor of its valley down to a depth that far exceeds the depth cut by the tributary glaciers. Consequently, when the glaciers melt away, the mouths of the tributary valleys perch at a higher elevation than the floor of the trunk valley. Such side valleys are called **hanging valleys**. The water in post-glacial streams that flow down a hanging valley cascades over a spectacular waterfall to reach the post-glacial trunk stream **(Fig. 18.11d, e)**. As they

FIGURE 18.11 Features formed by the glacial erosion of a mountainous landscape.

Before glaciation, valleys are V-shaped, and tributary mouths emerge at the same elevation as the trunk stream.

Tributary valley

V-shaped valley

Trunk valley

Trunk valley

Tributary valley

During glaciation, the valleys fill with ice.

Time

After glaciation, the region contains U-shaped valleys, hanging valleys, truncated spurs, and horns.

Horn

Cirque   Arête

U-shaped valley

Hanging valley

Truncated spur

(a) Stages in the development of a glacially carved mountainous landscape.

(b) The Matterhorn, in the Swiss Alps, challenges climbers.

Profile

Profile

(c) A U-shaped valley in the Swiss Alps. The glacier that carved it has melted away. In contrast, a river carves a V-shaped valley, as shown in the inset.

(d) A waterfall spilling out of a U-shaped hanging valley in the Sierra Nevada.

(e) The dark, deep valley in the distance, near Zermatt, Switzerland, was carved by a Pleistocene valley glacier. Smaller glaciers still remain at higher elevations. When they melt, they will leave behind hanging valleys.

erode, trunk glaciers also remove the ends of ridges between valleys, producing *truncated spurs*.

Now let's look at the erosional features produced by continental ice sheets. To a large extent, these depend on the nature of the pre-glacial landscape. Where an ice sheet spreads over a region of low relief, such as the Canadian Shield, glacial erosion creates a vast region of polished, flat, striated surfaces. Where an ice sheet spreads completely over a hilly area, it smooths hills. Glacially eroded hills may become elongate in the direction of flow and may be asymmetric. Glacial rasping bevels the upstream part of the hill, to produce a gentle slope, whereas glacial plucking eats away at the downstream part, yielding a steep slope. Ultimately, the hill's profile resembles that of a sheep lying in a meadow—such a hill is called a **roche moutonnée**, from the French for sheep rock **(Fig. 18.12)**.

FJORDS: SUBMERGED GLACIAL VALLEYS

If the floor of a glacially carved valley lies below sea level along the coast or beneath the water table inland, the floor of the valley becomes submerged with water. Geologists refer to any glacially carved valley that has filled partially or entirely with water as a **fjord**. *Marine fjords,* or *coastal fjords,* occur along the coast and have filled with seawater, whereas *freshwater fjords* are inland lakes **(Fig. 18.13)**.

Spectacular examples of marine fjords can be found along the coasts of Norway, New Zealand, Chile, Alaska, and Greenland—in some cases, the walls of submerged U-shaped valleys rise straight from the sea as vertical cliffs up to 1,000 m high, and the water depth just offshore may exceed a few hundred meters. How do these particularly dramatic fjords develop? As noted earlier, where a valley glacier meets the sea, the glacier's base remains in contact with the ground until the water depth exceeds about four-fifths of the glacier's thickness. Further, during an ice age, water extracted from the sea becomes locked in the ice sheets on land, so sea level drops significantly. Therefore, the floors of valleys cut by coastal glaciers during the Pleistocene Ice Age could be cut

FIGURE 18.12 A roche moutonnée is an asymmetric bedrock hill shaped by the flow of glacial ice.

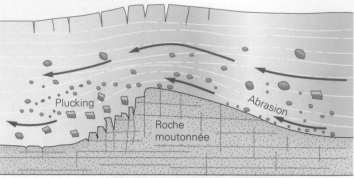

(a) Abrasion rasps the upstream side, and plucking carries away joint-bounded blocks on the downstream side.

(b) An example of a roche moutonnée. The glacier flowed from right to left.

FIGURE 18.13 One of the many spectacular fjords of Norway. The water is an arm of the sea that fills a glacially carved valley. Tourists are standing on Pulpit Rock (Preikestolen).

much deeper than present sea level. When sea level rose after ice-age glaciers melted, seawater submerged the valleys.

TAKE-HOME MESSAGE

A glacier scrapes up and plucks rock from its substrate and carries debris that falls on its surface. Glacial erosion polishes and scratches rock and carves distinctive landforms, such as U-shaped valleys, cirques, and horns. An elongate bay or lake formed when water partially fills a glacial valley is a marine fjord.

QUICK QUESTION Why do we find hanging valleys spilling waterfalls into trunk valleys in regions that have been eroded by mountain glaciers?

18.4 Deposition Associated with Glaciation

THE GLACIAL CONVEYOR

Glaciers, in general, do not consist of pure ice. All glaciers incorporate clasts at their base and sides, where ice comes in contact with its substrate. And in the case of valley glaciers, the ice also carries clasts that tumble down from bordering mountain slopes onto the surface of glaciers. All this debris, regardless of grain size, moves with the ice, so glaciers—like giant conveyor belts—transport large quantities of unsorted sediment in the direction of flow (**Fig. 18.14a, b**). Geologists

FIGURE 18.14 The glacial conveyor and the formation of moraines.

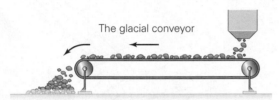

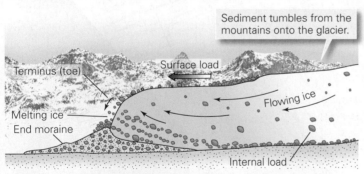

(a) Sediment falls on a glacier from bordering mountains and gets plucked up from below. Glaciers are like conveyor belts, moving sediment toward the toe of the glacier.

(b) This glacier in the French Alps carries lots of sediment.

(c) A medial moraine forms where lateral moraines of two valley glaciers merge.

Glaciers and Glacial Landforms

Continental ice sheet

Crevasses

Ice shelf

Higher sea level

Lower sea level

Dropstones

Iceberg

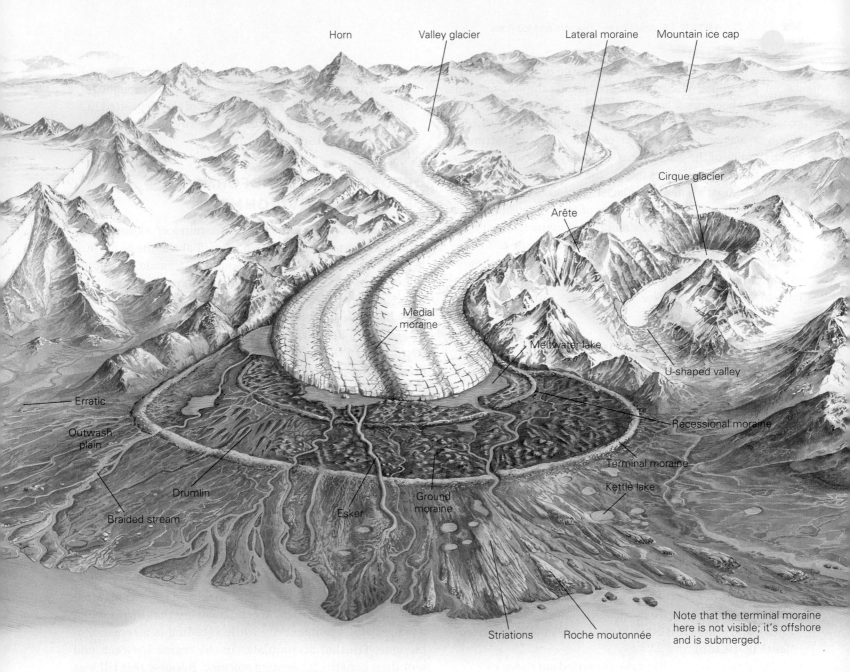

Horn Valley glacier Lateral moraine Mountain ice cap

Cirque glacier

Arête

Medial moraine

Meltwater lake

U-shaped valley

Erratic

Outwash plain

Recessional moraine

Drumlin

Terminal moraine

Esker Ground moraine

Kettle lake

Braided stream

Striations Roche moutonnée

Note that the terminal moraine here is not visible; it's offshore and is submerged.

Glaciers are rivers or sheets of ice that last all year and slowly flow. Continental ice sheets, vast glaciers up to a few kilometers thick, covered extensive areas of land during times when the Earth had a colder climate. At the peak of the last ice age, ice sheets covered almost all of Canada, much of the United States, northern Europe, and parts of Russia. Because ice sheets store so much of the Earth's water, sea level becomes lower during an ice age.

The upper part of a glacier is brittle and may crack to form crevasses. When an ice sheet reaches the sea, it becomes an ice shelf. Rocks that the glacier has plucked up along the way can be carried out to sea with the ice; when the ice melts, the rocks fall to the seafloor as dropstones. At the edge of the shelf, icebergs calve off and float away.

Glacial recession may happen when the climate warms, so ice melts away faster at the toe (terminus) of the glacier than

it can be added at the source. Consequences of glacial erosion and deposition remain when a glacier melts away. Erosion features include striations on bedrock and roches moutonnées. Deposition features include glacial moraines, glacial outwash, and esker deposits. Even when the toe remains fixed in position for a while, the ice continues to flow and molds underlying sediment into drumlins. Ice blocks buried in till melt to form kettle holes.

Mountain or alpine glaciers grow in mountainous areas because snow can last all year at high elevations. In the mountains, glaciers fill valleys or form ice caps. Sediment falling from the mountains creates lateral and medial moraines. Glaciers carve distinct landforms in the mountains, such as cirques, arêtes, horns, and U-shaped valleys.

refer to a pile of sediment either carried on or left behind by a glacier as a **moraine**. Sediment dropped on the side margins of a glacier becomes a *lateral moraine*, a stripe of debris adjacent to the valley wall on the top of the glacier (**Fig. 18.14c**). When a glacier melts, lateral moraines become stranded along the side of the glacially carved valley, like bathtub rings. Where two valley glaciers merge, the lateral moraines along the sides of the two merging glaciers join to become a *medial moraine*, a stripe down the interior of the composite glacier. Trunk glaciers formed by the merging of many tributary glaciers contain several medial moraines. Sediment left at the base of a glacier when the glacier melts away becomes *ground moraine*, and sediment transported to a glacier's toe accumulates to form an *end moraine*.

TYPES OF GLACIAL SEDIMENTARY DEPOSITS

Several different types of sediment can be deposited in glacial environments—all of these types together constitute **glacial drift**. The term dates from pre-Agassiz studies of glacial deposits, when geologists thought that the sediment had "drifted" into place during an immense flood. Specifically, glacial drift includes the following:

> *Glacial till* consists of sediment transported by ice and deposited beneath, at the side, or at the toe of a glacier in moraines. Till has not been sorted, because the solid ice of glaciers can carry clasts of all sizes (**Fig. 18.15a**).

> *Glacial erratics* are relatively large cobbles and boulders that have been dropped by a glacier. Some lie within till piles, and others rest directly on glacially polished surfaces (**Fig. 18.15b**).

> *Glacial marine* consists of clasts carried out to sea by icebergs. When the icebergs melt, the clasts sink to the seafloor and mix with marine sediment.

> *Glacial outwash* forms when meltwater streams carry and sort till. The clasts are deposited by a braided stream network to form a broad area of gravel and sandbars (**Fig. 18.15c**).

> *Loess* forms from clay and silt that were transported by strong *katabatic winds*. These develop when warmer air above ice-free land beyond the toe of a glacier rises and the cold, denser air from above the glacier rushes in to take its place. Deposits of loess tend to stick together, so erosion of the deposits yields steep escarpments (**Fig. 18.15d**).

> *Glacial lake-bed sediment* consists of very fine-grained sediment, including rock flour, that accumulates on the floor of a meltwater lake. Such sediment commonly contains varves. A **varve** is a pair of thin layers deposited during a single year—one layer consists of silt brought in during spring floods and the other of clay deposited in winter when the lake's surface freezes over and the water is still (**Fig. 18.15e**).

DEPOSITIONAL LANDFORMS OF GLACIAL ENVIRONMENTS

Picture a group of hunters wearing reindeer skins, gazing southward from the crest of an ice cliff at the toe of a continental glacier in what is now southern Canada. It's about 12,000 years ago, and the glacier has been receding for at least a millennium. The hunters would have been able to see a variety of landscape features, some formed by glacial erosion and some by deposition, due to moving ice and meltwater (**Geology at a Glance**, pp. 590–591). We've already described erosional features, so now let's focus on the depositional features of the landscape (**Fig. 18.16a, b**).

From their vantage point, the hunters would probably see a few curving ridges of sediment in the region between the glacier's toe and the horizon. Each of these ridges is an end moraine, formed from till deposited when the position of the glacier's toe remained in the same location for a while. As we've seen, ice keeps flowing to the toe, so the longer the toe stays in the same place, the bigger the end moraine becomes. Geologists refer to the end moraine at the farthest limit of glaciation as the **terminal moraine**. A ridge of sediment in the northeastern United States—a large terminal moraine built up during the most recent Ice Age—now underlies Long Island, New York, and Cape Cod, Massachusetts (**Fig. 18.16c**). An end moraine that forms when a glacier stalls temporarily while receding is a *recessional moraine*. Till that accumulates in the region between end moraines constitutes *ground moraine*. Because this till was deposited by moving ice, clasts within it may be aligned and scratched.

Locally, flowing glacial ice reshapes the underlying till into elongate hills known as **drumlins** (from the Gaelic word for small hill or ridge). Drumlins commonly occur in swarms and tend to be about 50 m high. Their long axis trends parallel to the flow direction of the glacier. Notably, drumlins taper in the direction of flow—a drumlin's upstream end is steeper than its downstream end (**Fig. 18.17a, b**).

Landscapes composed of glacial deposits tend to be hummocky (bumpy), consisting of small hills interspersed with depressions. In some places, the random ups and downs of a moraine surface are due to local variations in the amount of sediment supplied by the ice during deposition or removed subsequently by meltwater streams. But in other places, each depression on a moraine has a roughly circular

FIGURE 18.15 Sedimentation processes and deposits associated with glaciation. Glacial sediment is distinctive.

(a) This glacial till in Ireland is unsorted, because ice can carry sediment of all sizes.

(b) Glacial erratics resting on a glacially polished surface in Wyoming.

(c) Braided streams choked with glacial outwash in Alaska. The streams carry away finer sediment and leave the gravel behind.

(d) Thick loess deposits underlie parts of the prairie in Illinois.

(e) In the quiet water of an Alaskan glacial lake, fine-grained sediments accumulate. Varves in lake-bed sediment, now exposed in an outcrop near Puget Sound, Washington, reflect seasonal changes.

FIGURE 18.16 The formation of depositional landforms associated with continental glaciation.

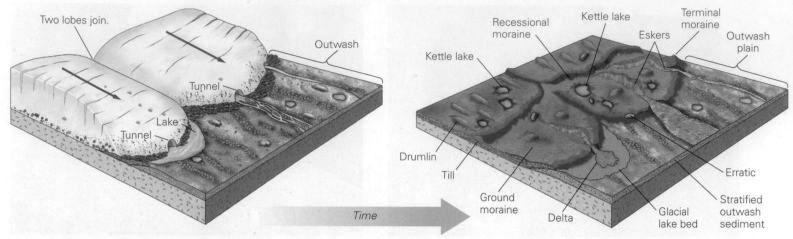

(a) The ice in continental glaciers flows toward the toe; sediment accumulates at the base and at the toe of the ice sheet.

(b) Several distinct depositional landforms form during glaciation; some developed under the ice and some at the toe.

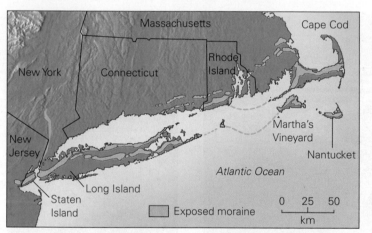

(c) Cape Cod, Long Island, and other landforms in the northeastern United States formed at the end of the continental ice sheet.

shape, as seen in map view. Such depressions, known as **kettle holes**, form when till-covered blocks of ice calve off the toe of a glacier; when the blocks melt, the depressions form **(Fig. 18.17c, d)**. Geologists refer to land surfaces spotted with many kettle holes separated by rounded hills or ridges of sediment as *knob-and-kettle topography* **(Fig. 18.17e)**. In places with a high water table, some kettles fill with water to become ponds.

As we've noted, ice does not directly deposit all of the sediment associated with glacial landscapes, for meltwater also carries and deposits sediment. Water-transported sediment, in contrast to till, tends to be sorted and stratified. Sediment deposited in meltwater tunnels beneath a glacier may remain as a sinuous ridge, known as an **esker**, when the glacier melts away **(Fig. 18.18)**. Braided meltwater streams that flow beyond the end of a glacier deposit layers of sand

and gravel that underlie *glacial outwash plains*. Meltwater may collect in a lake adjacent to the glacier's toe, to form an *ice-margin lake*, and additional meltwater lakes and swamps may form in low areas on the ground moraine. Sediments deposited in eskers and glacial outwash plains serve as important sources of sand and gravel for construction, and the fine sediment of former glacial lake beds may become fertile soil for agriculture.

TAKE-HOME MESSAGE

Glaciers, like giant conveyor belts, transport vast amounts of sediment. When the ice melts, the sediment accumulates as unsorted till. Meltwater streams and wind transport and sort the sediment to form outwash-plain gravels and loess deposits, respectively. Deposition by glaciers produces distinctive landforms, such as moraines, eskers, and kettle holes.

QUICK QUESTION How does knob-and-kettle topography develop?

18.5 Consequences of Continental Glaciation

ICE LOADING AND GLACIAL REBOUND

When a large ice sheet (more than 50 km in diameter) grows on a continent, its weight causes the surface of the lithosphere to sink. In other words, ice loading causes **glacial subsidence**.

FIGURE 18.17 Drumlins and knob-and-kettle topography.

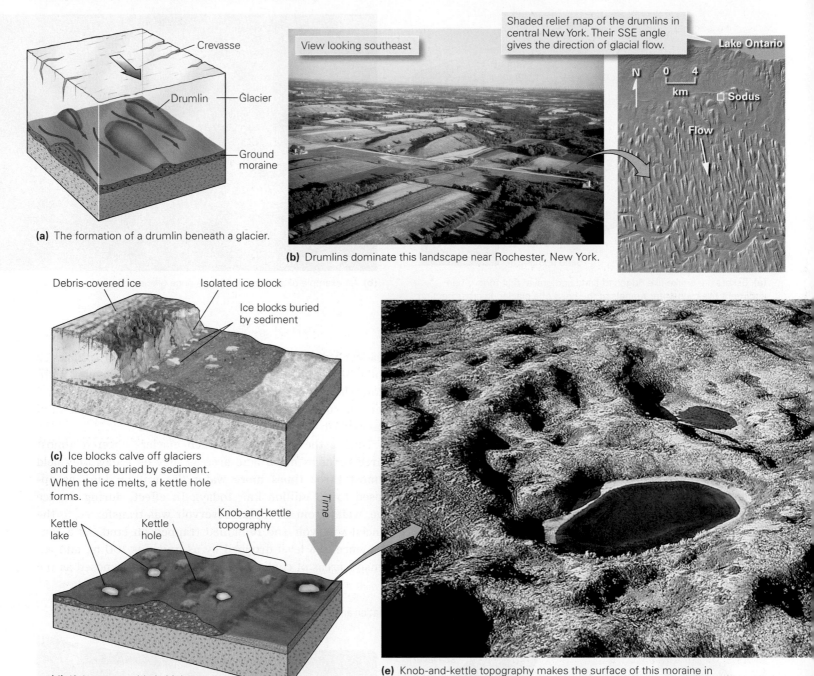

(a) The formation of a drumlin beneath a glacier.

View looking southeast

Shaded relief map of the drumlins in central New York. Their SSE angle gives the direction of glacial flow.

(b) Drumlins dominate this landscape near Rochester, New York.

(c) Ice blocks calve off glaciers and become buried by sediment. When the ice melts, a kettle hole forms.

(d) If the water table is high, kettle holes fill with water and turn into roughly circular lakes.

(e) Knob-and-kettle topography makes the surface of this moraine in Yellowstone Park, Wyoming, very hummocky.

Lithosphere, the relatively rigid outer shell of the Earth, can sink because the underlying asthenosphere is soft enough to flow slowly out of the way **(Fig. 18.19a)**. Because of ice loading, the rock surfaces underlying large areas of Antarctica's and Greenland's ice sheets now lie below sea level, so if the ice were instantly to melt away, these continents would be flooded by a shallow sea.

What happens when continental ice sheets melt? Gradually, the surface of the underlying continental lithosphere rises back up, by a process called **post-glacial rebound**. As this happens, the asthenosphere flows back underneath the lithosphere to fill the space. Because asthenosphere flows so slowly—at rates of a few millimeters per year—it may take thousands of years for ice-depressed continents

FIGURE 18.18 The formation of eskers.

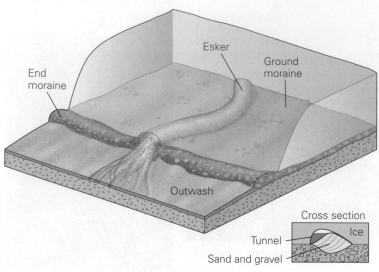

(a) Eskers are snake-like ridges of sand and gravel that form when sediment fills meltwater tunnels at the base of a glacier. In cross section (inset), wedges of sand accumulate in the tunnel.

(b) An example of an esker in an area once glaciated but now farmed.

to rebound completely. In fact, glacial rebound is still continuing in Canada and the northern United States (Fig. 18.19b).

SEA-LEVEL CHANGES: THE GLACIAL RESERVOIR

More of the Earth's surface and near-surface freshwater resides in glacial ice than in any other reservoir. In fact, glacial ice accounts for 2.15% of the Earth's total water supply (including oceans), while lakes, rivers, soil, and the atmosphere together contain only 0.03%. The melting of glacial ice transfers this water back into the ocean, causing

sea level to rise. In fact, if today's ice sheets in Antarctica and Greenland were to melt, large areas of the coastal plain along the east coast and Gulf Coast of North America would become submerged, as would much of the Ganges Delta of Bangladesh.

During the last ice age, when glaciers covered almost three times as much land area as they do today, they held almost three times more water (70 million km³, as opposed to 25 million km³ today). In effect, during the ice age, water from the ocean reservoir was transferred to the glacial reservoir and remained trapped on land. As a consequence, sea level dropped by as much as 100 m, and extensive areas of continental shelves became exposed as the

FIGURE 18.19 The concept of subsidence and rebound due to continental glaciation and deglaciation.

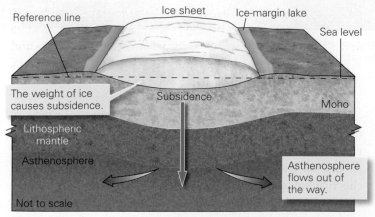

(a) The glacial load causes the lithosphere to sink.

(b) Rebound of the lithosphere produced uplifted beach terraces in Canada.

FIGURE 18.20 The link between sea level and global glaciation. Glaciers store water on land, so when glaciers grow, sea level falls, and when glaciers melt, sea level rises.

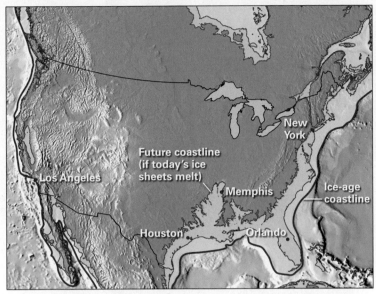

(a) The red line shows the coastline during the last ice age; much of the continental shelf was dry. If present-day ice sheets melt, coastal lands will flood.

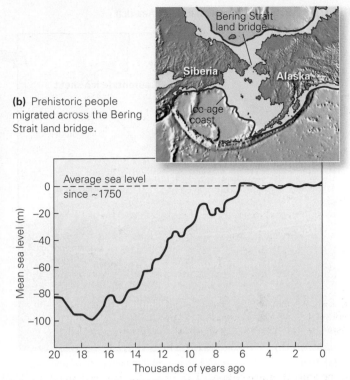

(b) Prehistoric people migrated across the Bering Strait land bridge.

(c) Sea-level rise between 17,000 and 7,000 B.C.E. was due to the melting of ice-age glaciers.

coastline migrated seaward, in places by more than 100 km **(Fig. 18.20)**. People and animals migrated into the newly exposed ice-age coastal plains. In fact, fishermen dragging their nets along the Atlantic Ocean floor off New England today occasionally recover human artifacts. The drop in sea level also produced dry-land access, known as land bridges, across the Bering Strait between North America and northeastern Asia and between Australia and Indonesia, providing migration routes for early people.

EFFECTS ON DRAINAGE AND LAKES

Continental glaciation can significantly modify the locations and character of rivers and streams draining the land. Locally, the growth of a glacier or the deposition of a moraine can block an individual stream. Diverted flow finds a new route and can carve new channels. By the time the glacier melts away, these new streams have become so well established that pre-glacial channels remain abandoned.

Erosion by glacial meltwater can carve valleys. Especially dramatic examples of this process have resulted when either ice dams or moraine dams holding back large lakes suddenly broke. In a matter of hours to days, the contents of the lakes drained, yielding an immense flood called an outburst flood or *glacial torrent*. Torrents can carve huge valleys and steep cliffs, strip the land of soil, and leave behind immense ripple

marks. For example, when the ice dam holding back Glacial Lake Missoula in Montana broke, it released an immense torrent—known as the *Great Missoula Flood*—that scoured eastern Washington, yielding a barren, soil-free landscape known as the *channeled scablands*. Recent evidence suggests that this process repeated several times. A torrent flowed down the channel of what is now the Illinois River when a moraine dam broke in northern Illinois, and carved a broad, steep-sided valley that is much too large to have been cut by the present-day Illinois River.

Large *ice-margin lakes* formed when the weight of continental ice sheets depressed the surface of the lithosphere sufficiently to generate depressions that filled with water. The largest known ice-margin lake covered portions of Manitoba and Ontario in south-central Canada and North Dakota and Minnesota in the United States **(Fig. 18.21a)**. This body of water, known as Glacial Lake Agassiz, existed between 11,700 and 9,000 years ago, a time during which the last ice age came to a close and the continental ice sheet retreated north. At its largest, the lake covered over 250,000 km², an area greater than that of all the present Great Lakes combined. Eventually, the ice sheet receded from the north shore of Glacial Lake Agassiz, so near the end of its life, the lake was surrounded by ice-free land. Field evidence suggests that the lake's demise came when it drained catastrophically, sending a torrent down what is now the St. Lawrence Seaway.

FIGURE 18.21 Ice-age lakes in North America.

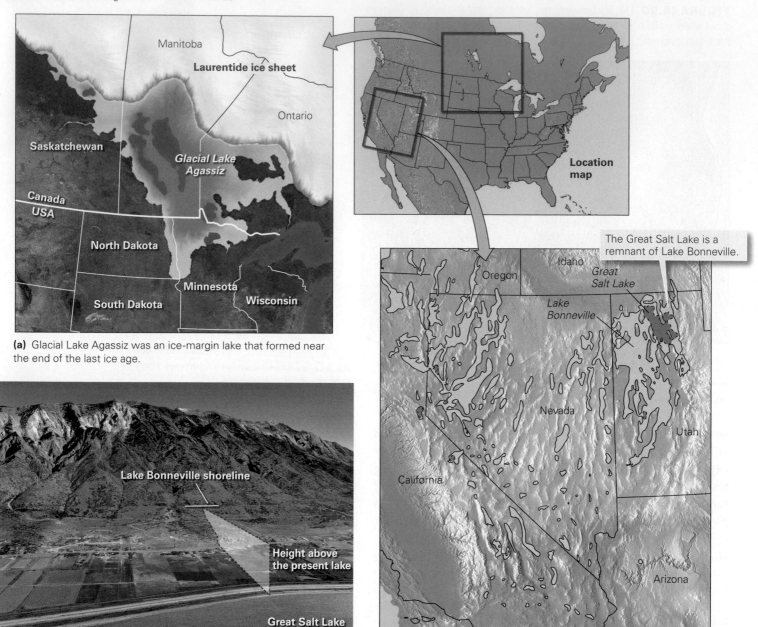

(a) Glacial Lake Agassiz was an ice-margin lake that formed near the end of the last ice age.

The Great Salt Lake is a remnant of Lake Bonneville.

(b) Pluvial lakes occurred throughout the Basin and Range Province during the last ice age due to the wetter climate. The largest of these was Lake Bonneville. Subtle horizontal terraces define the remnants of beaches, now over 100 m above the present level of the Great Salt Lake.

PLUVIAL FEATURES

During the Pleistocene Ice Age, regions to the south of continental ice sheets were wetter than they are today. Fed by enhanced rainfall, lakes accumulated in low-lying land at a great distance from the ice front. Many of these **pluvial lakes** (from the Latin word *pluvia*, meaning rain) flooded interior basins of the Basin and Range Province in Utah and Nevada, a region that is now a desert **(Fig. 18.21b)**. The largest such lake, Lake Bonneville, covered almost a

third of western Utah. When this lake suddenly drained after a natural dam broke, the former shoreline remained like a bathtub ring rimming the mountains near Salt Lake City. Today's Great Salt Lake is a small remnant of Lake Bonneville.

PERIGLACIAL ENVIRONMENTS

In regions adjacent to the fronts of continental ice sheets during the last ice age, the mean annual temperature stayed

FIGURE 18.22 Periglacial regions are not ice covered but do include substantial areas of permafrost.

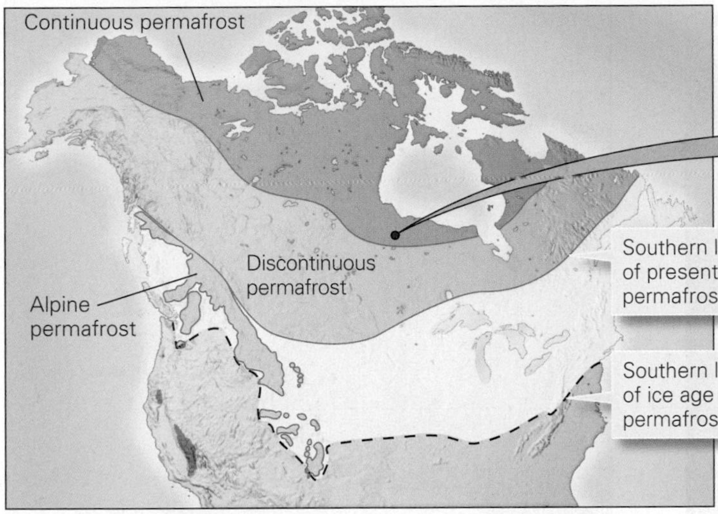

(a) The distribution of periglacial environments in North America.

(b) An example of patterned ground near a pond in Manitoba, Canada.

cold enough (below –5°C) that soil moisture and groundwater froze and the ground remained solid nearly all year. Regions underlain by such permanently frozen ground, or **permafrost**, are called *periglacial environments* (from the Greek word *peri*, meaning encircling), because periglacial environments develop around the edges of glacial environments **(Fig. 18.22a)**.

The upper few meters of permafrost may melt during the summer months, only to refreeze again when winter comes. As a consequence of the freeze-thaw process, the ground of some permafrost areas splits into pentagonal or hexagonal shapes, resulting in a landscape called **patterned ground (Fig. 18.22b)**. Permafrost presents a unique challenge to people who live or work in polar regions. For example, heat from a building may melt underlying permafrost, producing a mire into which the building settles. For this reason, buildings in permafrost regions must be placed on stilts, so that cold air can circulate beneath them to keep the ground underneath frozen.

TAKE-HOME MESSAGE

Growth of glaciers can affect many aspects of the Earth System. The weight of a growing continental ice sheet causes underlying lithosphere to subside. Later melting allows it to rebound. Continental ice sheets store so much water that their growth or melting affects global sea level. Sudden drainage of lakes related to glaciation leads to torrents that can scour the landscape.

QUICK QUESTION How did the landscape south of the Pleistocene ice sheet differ from the landscape of today?

18.6 The Pleistocene Ice Age

THE PLEISTOCENE GLACIERS

Today, most of the land surface in New York City lies hidden beneath concrete and steel, but in Central Park it's still possible to see land in a seminatural state. If you stroll through the park, you'll find that the top surfaces of many outcrops are smooth and polished, and in places they have been grooved and scratched (see Fig. 18.10c). Here and there, glacial erratics rest on the bedrock. You are seeing evidence that an ice sheet once scraped along this now-urban ground. Geologists estimate that the ice sheet that overrode the New York City area may have been 250 m thick, tall enough to overtop a 75-story building.

The fact that glaciated landscapes still decorate the surface of the Earth means that the last ice age occurred fairly recently during the Earth's history. Otherwise, these features would have eroded away. The ice age responsible for the glaciated landscapes of North America, Europe, and Asia happened during the Pleistocene Epoch, which began about 2.6 million years ago (Ma), so as we've noted earlier (see Chapter 11), this event is commonly known as the **Pleistocene Ice Age**.

Based on studying patterns of glacial striations and of the sources of erratics, geologists have developed an approximate idea of where the Pleistocene ice sheets originated and the directions in which the ice sheets flowed. In North America, major ice sheets began to grow in at least three locations **(Fig. 18.23)**—the *Labrador ice sheet* over northeastern Canada, the *Keewatin ice sheet* in northwestern Canada, and the *Baffin ice sheet* over Baffin Bay. These sheets, together with one or more smaller ones, merged to form the giant

FIGURE 18.23 Pleistocene ice sheets of the northern hemisphere.

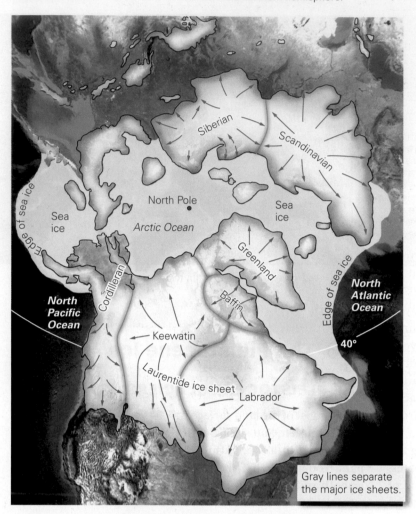

Gray lines separate the major ice sheets.

FIGURE 18.24 Climate belts during the Pleistocene.

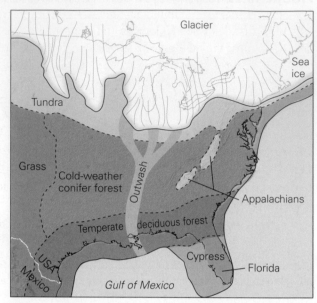

(a) Tundra covered parts of the United States, and southern states had forests like those of New England's today.

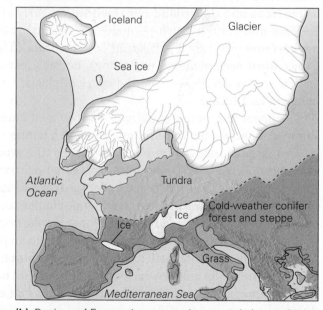

(b) Regions of Europe that support large populations today would have been barren tundra during the Pleistocene.

(c) Cold-adapted, now-extinct, large mammals roamed regions that are now temperate.

Laurentide ice sheet that covered all of Canada east of the Rocky Mountains and spread southward over the northern portion of the United States. The *Cordilleran ice sheet*, which originated in the mountains of western Canada, spread westward to the Pacific coast and eastward until it merged with the Laurentide ice sheet. Other Pleistocene ice sheets formed in Greenland, Scandinavia, Britain, central Asia, and Siberia.

In addition to continental ice sheets, sea ice in the northern hemisphere expanded during the Pleistocene to cover all of the Arctic Ocean and parts of the North Atlantic. In fact, sea ice surrounded Iceland and approached Scotland and also fringed most of western Canada and southeastern Alaska.

LIFE AND CLIMATE IN THE PLEISTOCENE WORLD

During the Pleistocene Ice Age, all climatic belts of the northern hemisphere shifted southward (**Fig. 18.24a, b**). Geologists can document this shift by examining fossil pollen, which can survive for thousands of years if preserved in the sediment of bogs. Presently, the southern boundary of

North America's *tundra*, a treeless region supporting only low shrubs, moss, and lichen capable of living on permafrost, lies at a latitude of 68° N. During the Pleistocene Ice Age, its southern limit moved down to 48° N. Much of the interior of the United States, which now has temperate, deciduous forest, harbored cold-weather spruce and pine forest. Ice-age climates also changed the distribution of rainfall on the planet. As we noted earlier, rainfall increased in North America, south of the glaciers, leading to the filling of pluvial lakes in Utah and Nevada. In contrast, rainfall decreased in equatorial regions, leading to shrinkage of the tropical rainforest. And because glaciers trapped so much water, as we have seen, sea level dropped.

Fossils also show that numerous species of now-extinct large mammals inhabited the Pleistocene world **(Fig. 18.24c)**. Giant mammoths and mastodons, relatives of the elephant, along with woolly rhinos, musk oxen, reindeer, giant ground sloths, bison, lions, saber-toothed cats, giant cave bears, and giant hyenas wandered forests and tundra in North America. Ancestral human species were already foraging in the woods by the beginning of the Pleistocene Epoch, and by the end, modern *Homo sapiens* lived on every continent except Antarctica and had discovered fire and invented tools.

TIMING OF THE PLEISTOCENE ICE AGE

Louis Agassiz assumed that only one ice age had affected the planet. But close examination of the stratigraphy of glacial deposits on land revealed that *paleosols* (ancient soil preserved in the stratigraphic record), as well as beds containing fossils of warmer-weather animals and plants, lay between distinct layers of glacial sediment. This observation indicated that between episodes of glacial deposition, continental ice sheets disappeared and temperate climates prevailed. In the second half of the 20th century, when modern methods for dating geological materials became available, the age differences among the different layers of glacial sediment were confirmed. Clearly, ice sheets advanced and then retreated more than once during the Pleistocene. Times during which the glaciers grew and covered substantial areas of the continents are called *glacial periods*, or *glaciations*, and times between glacial periods are called interglacial periods, or *interglacials*.

Using the continental sedimentary record, geologists recognized five Pleistocene glaciations in Europe and four in the midwestern United States (Wisconsinan, Illinoian, Kansan, and Nebraskan, named after the southernmost states in which their till was deposited; **Fig. 18.25)**. Since the 1980s, geologists no longer distinguish Nebraskan and Kansan, but instead group glacial sediments deposited before the Illinoian simply as "pre-Illinoian."

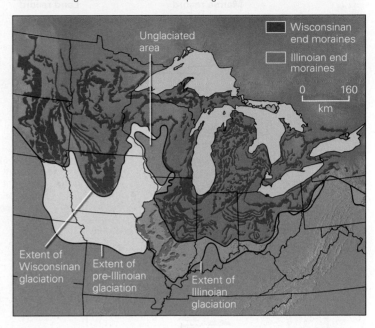

FIGURE 18.25 Pleistocene glacial deposits in the north-central United States. Curving moraines reflect the shape of glacial lobes.

During the 1960s, geologists realized that this traditional chronology of glaciations was an oversimplification. They discovered evidence of glaciations preserved in the stratigraphic record of Pleistocene marine sediment. By searching for layers containing glacially transported grains or for layers containing distinctive cold-water plankton, they determined that 20 to 30 glaciations took place during the Pleistocene Epoch. The stratigraphic record of the ice age that had been preserved on land turned out to be very incomplete.

Geologists refined their conclusions about the frequency of Pleistocene glaciations by examining the isotopic composition of fossil shells. Shells of many plankton species consist of calcite ($CaCO_3$). The oxygen in the shells includes two isotopes, a heavier one (^{18}O) and a lighter one (^{16}O). The ratio of these isotopes tells us about the water temperature in which the plankton grew, for as water gets colder, plankton incorporate a higher proportion of ^{18}O into their shells. The isotope record confirms that many distinct cold events occurred during the past 2.6 million years **(Fig. 18.26a)**.

OLDER ICE AGES DURING EARTH HISTORY

So far, we've focused on the Pleistocene Ice Age because of its importance in developing the Earth's present landscape. Was this the only ice age during Earth history, or did ice ages happen earlier in Earth history? To answer such questions, geologists study the stratigraphic record and search for ancient glacial deposits that have hardened into rock. These deposits, called **tillites**, consist of larger clasts distributed throughout

FIGURE 18.26 The timing of glaciations. Ice ages have occurred at several times in the geologic past.

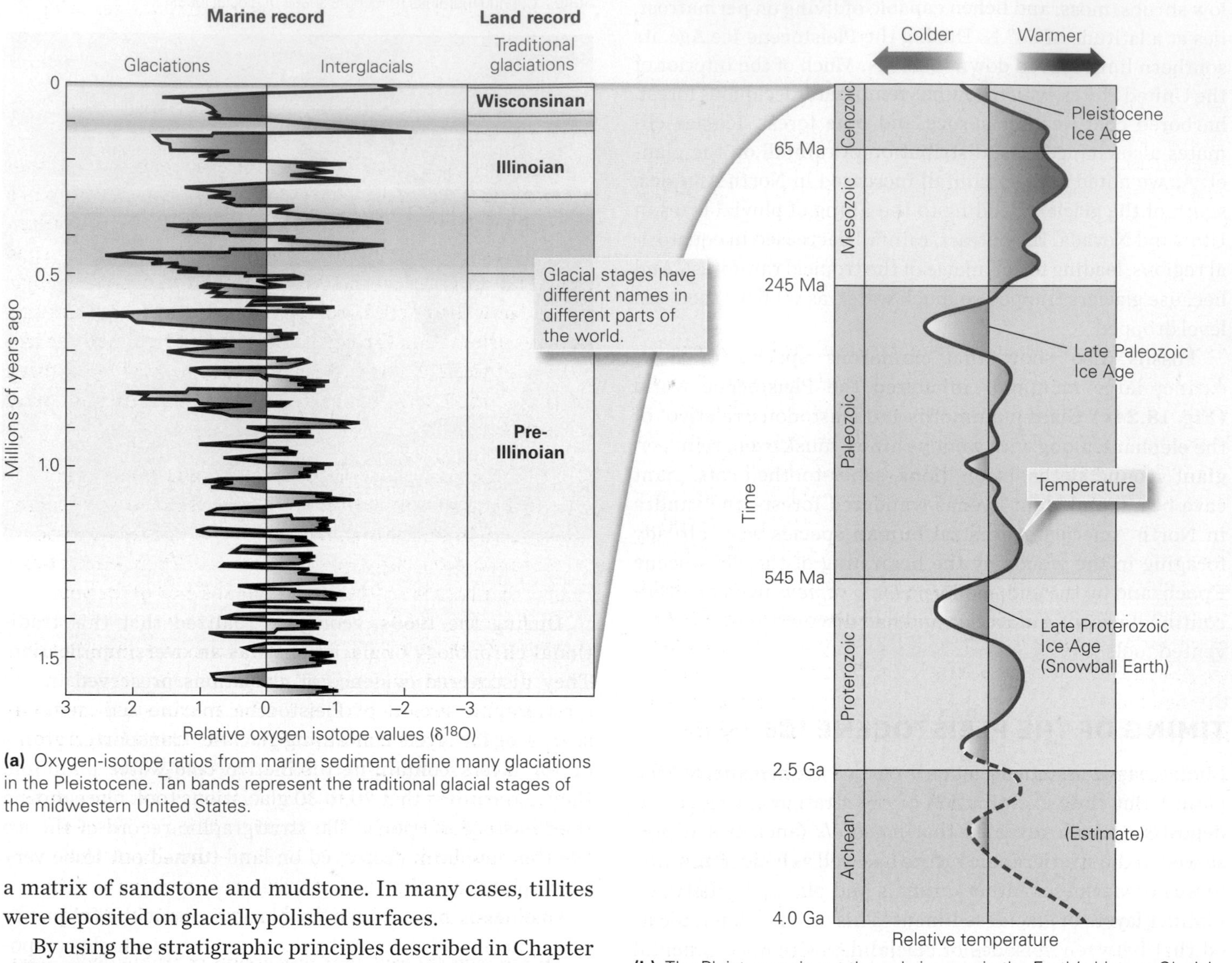

(a) Oxygen-isotope ratios from marine sediment define many glaciations in the Pleistocene. Tan bands represent the traditional glacial stages of the midwestern United States.

(b) The Pleistocene is not the only ice age in the Earth's history. Glacial advances happened in colder intervals of earlier eras, too.

a matrix of sandstone and mudstone. In many cases, tillites were deposited on glacially polished surfaces.

By using the stratigraphic principles described in Chapter 10, geologists have determined that tillites associated with continental glaciers were deposited during the late Paleozoic **(Fig. 18.26b)**. These are the deposits Alfred Wegener studied when he argued in favor of continental drift. Tillites have also yielded dates at the end of the Proterozoic Eon (about 600 to 700 Ma), near the beginning of the Proterozoic (about 2.2 Ga), and perhaps in the Archean Eon (about 2.7 Ga). Strata deposited at other times in Earth history do not contain widespread tillites. Thus, it appears that glacial advances and retreats have not occurred steadily throughout Earth history, but rather are restricted to four or five specific time intervals, or ice ages. Of particular note, some tillites of the late Proterozoic event were deposited at equatorial latitudes, suggesting that, for at least a short time, the

Did you ever wonder . . .
how many ice ages have happened during Earth history?

continents worldwide were largely glaciated, and the sea may have been covered worldwide by ice. Geologists refer to the ice-encrusted planet as **snowball Earth**.

TAKE-HOME MESSAGE

During the Pleistocene (2.6 Ma to 11 Ka), ice sheets advanced and retreated up to 30 times. The record on land is less complete than that in marine strata, for evidence of only four or five major glaciations is preserved on land. Ice ages also happened earlier in Earth history. During the late Proterozoic, ice may have covered all of snowball Earth.

QUICK QUESTION What is the evidence for multiple Pleistocene glaciations?

18.7 **The Causes of Ice Ages**

Ice ages occur only during restricted intervals of the Earth's history, hundreds of millions of years apart. During an ice age, glaciers advance and retreat periodically, with a frequency measured in tens of thousands to hundreds of thousands of years. Consequently, there must be both long-term and short-term controls on glaciation. What are they?

LONG-TERM CAUSES

As we noted earlier, the most recent ice age, the Pleistocene Ice Age, happened during the last 2.6 million years. The previous ice age took place near the end of the Paleozoic, with the most widespread glaciations occurring between about 305 and 325 Ma, the time when Pangaea existed. The ice age prior to that was between 630 and 780 Ma, before complex multicellular animals had appeared, and the ice age before that was 2100 to 2400 Ma, when the Earth was just beginning to have oxygen in its atmosphere. Clearly, ice ages happen only rarely, suggesting that their appearance may in some way link to major tectonic or evolutionary events during Earth history. Geologists have suggested many possible links, though all for now remain in the realm of speculation. Hypotheses under discussion suggest that ice ages may be triggered when the following occur:

> Plate movements place large areas of continents at high latitudes, so the land surface receives less solar energy and can become cold.

> High-latitude continents move over mantle upwelling zones, so land elevation overall rises to higher, cooler elevations.

> The global volume of mid-ocean ridges decreases so the capacity of ocean basins increases. When this happens, shallow seas, whose presence could moderate temperatures, disappear from the land.

> The growth of island arcs blocks ocean currents from carrying warm water to high-latitude regions, so the regions become cold enough to host ice-sheet formation.

> Organisms that extract CO_2 (a greenhouse gas) from the atmosphere flourish. The resulting lower CO_2 concentration would decrease the ability of the atmosphere to retain heat.

> The overall amount of chemical weathering on land increases, due to the uplift of mountain ranges. Weathering absorbs CO_2 and would decrease the ability of the atmosphere to retain heat.

> The amount of volcanic activity increases, for volcanic activity can put aerosols into the atmosphere that prevent solar energy from reaching the Earth.

> Uplift of mountain belts changes the pattern of atmospheric circulation, preventing warm air from reaching high latitudes.

SHORT-TERM CAUSES

Once an ice age has started, for whatever reason, continental glaciers advance and retreat many times, and there is a periodicity to these changes. Why? In 1920, Milutin Milanković, a Serbian astronomer and geophysicist, published work that provides the basis for an explanation. Milanković studied how the Earth's orbit changes shape and how its axis changes orientation through time, and he calculated the frequency of these changes. In particular, he evaluated three aspects of the Earth's movement around the Sun and found the following:

> The *eccentricity of the Earth's orbit* gradually changes from being more circular (low eccentricity) to being more elliptical in shape (high eccentricity) over a time period of around 100,000 years **(Fig. 18.27a)**. When the Earth's orbit has higher eccentricity, it spends part of the year farther from the Sun.

> The *tilt of the Earth's axis* varies between 22.5° and 24.5°, over a time period of 41,000 years **(Fig. 18.27b)**—we experience seasons because the Earth's axis is not perpendicular to the plane of its orbit, so when the tilt is greater, seasonal differences in high latitudes are greater.

> The *precession of the Earth's axis* changes over the course of about 23,000 years. The Earth wobbles, or *precesses*, like a spinning top **(Fig. 18.27c)**—this affects the relationship between the timing of the seasons and the position of the Earth along its orbit around the Sun.

Milanković showed that precession combines with variations in orbital eccentricity and tilt to affect both the total annual amount of *insolation* (exposure to the Sun's rays) and the seasonal distribution of insolation that the Earth receives at mid-latitudes to high latitudes by as much as 25%. According to Milanković, these changes are enough to cause periodic cooler summers on a time frame measured in tens of thousands of years **(Fig. 18.27d)**. When geologists began to study the climate record, they found climate cycles with the frequency predicted by Milanković. These climate cycles are now called **Milankovitch cycles**.

Milankovitch cycles, however, cannot be the whole story. Geologists suggest that several other factors may come into play in order to trigger or amplify a glacial advance, including these:

> *Changing albedo:* When snow remains on land throughout the year, or clouds form in the sky, the *albedo*

FIGURE 18.27 Milankovitch cycles influence the amount of insolation received at high latitudes.

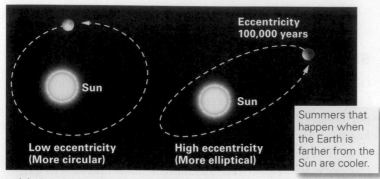

(a) Variations caused by changes in orbital shape.

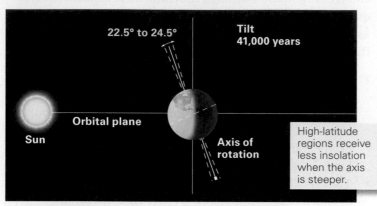

(b) Variations caused by changes in axial tilt.

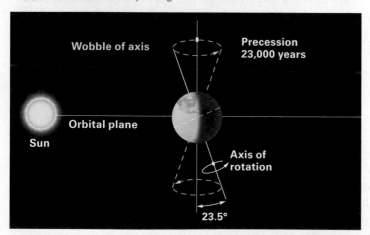

(c) Variations caused by the precession of the Earth's axis.

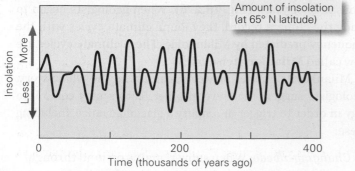

(d) Combining the effects of eccentricity, tilt, and precession produces distinct periods of more or less insolation.

(reflectivity) of the Earth increases, so the Earth's surface reflects incoming sunlight and thus becomes even cooler.

> *Interrupting the global heat conveyor:* As the climate cools, evaporation rates from the sea decrease, so seawater becomes less salty. Decreasing salinity might stop the system of thermohaline circulation that brings warm water to high latitudes (see Chapter 15). As a result, high-latitude regions may become even colder than otherwise.

> *Biological processes that change CO_2 concentration:* As we have noted, biological processes may amplify climate changes by altering the concentration of CO_2 in the atmosphere.

> *Solar variability:* Changes in the radiation output of the Sun could affect the amount of energy the Earth receives.

A MODEL FOR PLEISTOCENE ICE-AGE HISTORY

Long-Term Cooling in the Cenozoic Era Taking all of the above causes into account, geologists have proposed a scenario that may explain how the Pleistocene Ice Age began, and how a cycle of glacial advance and retreat takes place. The story begins in the Eocene Epoch, about 55 Ma **(Fig. 18.28)**. At that time, the Earth's climate was relatively balmy, even above the Arctic Circle. Near the end of the Middle Eocene (37 Ma), however, climate began to cool. This long-term climate change may have been caused, in part, by a change in the pattern of atmospheric circulation that happened when uplift of the Himalayas and Tibet diverted

FIGURE 18.28 Until recently, the Earth's atmosphere has been gradually cooling, overall, since the Cretaceous.

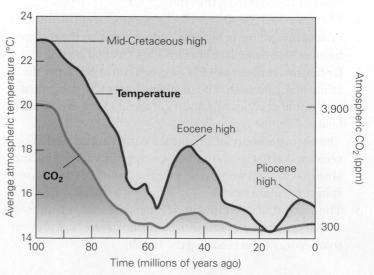

winds in a way that cooled the climate and exposed more rock to CO_2-absorbing chemical weathering reactions. In addition, plate motions opened the Southern Ocean and isolated Antarctica from warm currents, allowing the cold Antarctic Circumpolar Current to develop, which in turn may have cooled the global ocean.

So far, we've examined hypotheses that explain long-term cooling since about 40 Ma, but what caused the sudden birth of the Laurentide ice sheet growth at about 2.6 Ma? This event may coincide with other plate-tectonic events. For example, the gap between North and South America closed when the Isthmus of Panama grew and separated the waters of the Caribbean from those of the tropical Pacific for the first time. When this happened, warm currents that previously flowed out of the Caribbean into the Pacific were blocked and diverted northward to merge with the Gulf Stream. The Gulf Stream transfers warm water from the Caribbean, up the Atlantic Coast of North America, and ultimately to the British Isles. As the warm water moves up the Atlantic Coast, it generates warm, moisture-laden air that provides a source for the snow that falls over New England, eastern Canada, and Greenland. In other words, the Arctic has long been cold enough for ice caps, but until the Gulf Stream was diverted northward, by the growth of Panama, no source of moisture was available to make the abundant snow and ice needed for glacial growth.

Short-Term Advances and Retreats in the Pleistocene Once the Earth's climate had cooled overall, the Milankovitch cycles controlled periodic advances and retreats of the glaciers. To understand how, let's look at a possible case history of a single advance and retreat of the Laurentide ice sheet. (Note that such models remain the subject of vigorous debate.)

> *Stage 1:* When the Earth reaches a point in the Milankovitch cycle when the average mean temperature in temperate latitudes drops, not all winter snow melts away during the summer in northern Canada. The snow reflects sunlight, so the region grows still colder and even more snow can accumulate. When the snow gets deep enough, the base of the pile turns to ice, which begins to spread outward as a new continental glacier.

> *Stage 2:* The ice sheet continues to grow as more snow piles up in the zone of accumulation, and the atmosphere continues to cool because of the albedo effect. Note that increasing albedo is a *positive feedback*, meaning that its development enhances the effect that caused it. Eventually, however, the weight of the ice

loads the continent and makes it sink, so the elevation of the glacier's surface decreases, and it becomes warmer. Also, sea ice covers the ocean offshore, cutting off the source of moisture for snow, so the amount of snowfall diminishes. In effect, the glacial advance chokes on its own success—the rate of ablation begins to exceed the rate of accumulation, and the glacier begins to retreat.

> *Stage 3:* As the glacier retreats, albedo decreases, temperatures gradually increase, and the sea ice begins to melt. The supply of water to the atmosphere from evaporation increases once again, but with the warmer temperatures and lower elevations, this water precipitates as rain during the summer. The rain drastically accelerates the rate of ice melting, so the retreat progresses rapidly.

WILL THERE BE ANOTHER GLACIAL ADVANCE?

What does the future hold? Considering the periodicity of glacial advances and retreats during the Pleistocene Epoch, we may be living in an interglacial period. Pleistocene interglacials lasted about 10,000 years, and since the present interglacial began about 12,000 years ago, the time may be ripe for a new glaciation to begin. If a glacier on the scale of the Laurentide ice sheet were to develop, major cities and agricultural belts would be overrun by ice, and their inhabitants would have to migrate southward. Long before the ice front arrived, however, the climate would become so hostile that northern cities would already have been abandoned.

The Earth actually had a brush with ice-age conditions between the 1300s and the mid-1800s, when average annual temperatures in the northern hemisphere fell sufficiently for mountain glaciers to advance significantly. During this period, now known as the *Little Ice Age*, sea ice surrounded Iceland and canals froze in the Netherlands—it was during this time that ice skating became a popular pastime in the country **(Fig. 18.29a, b)**.

During the past 150 years, temperatures have warmed, and most mountain glaciers have retreated significantly **(Fig. 18.29c)**. Large slabs have been calving off Antarctic ice shelves. In fact, the Larsen Ice Shelf of Antarctica, an area larger than Rhode Island, disintegrated in 2002 over a period of only one month. Greenland's glaciers, in particular, are showing signs of accelerating retreat **(Fig. 18.30)**. Large meltwater ponds are forming on the surface of the ice sheet, and some of these drain abruptly through cracks to

FIGURE 18.29 The Little Ice Age and its demise. Glaciers that advanced between 1550 and 1850 have since retreated.

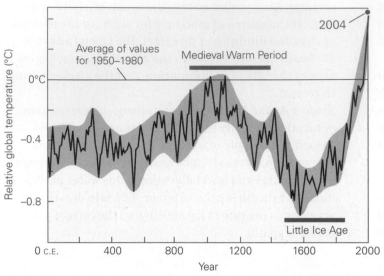

(a) A model of global temperature for the past 2,000 years. Overall trends display the Medieval Warm Period followed by the Little Ice Age. Since 1850, temperatures have warmed.

(b) Skaters (ca. 1600) on the frozen canals of the Netherlands during the Little Ice Age.

the base of the ice a kilometer below. The vast majority of researchers suggest that this global-warming trend is due to the increased CO_2 in the atmosphere from the burning of fossil fuels. Such global warming could conceivably cause a "super-interglacial" period, one that lasts much longer than the Milankovitch cycle predicts. The next chapter addresses the causes and consequences of such change.

TAKE-HOME MESSAGE

Ice ages may be triggered when the distribution of continents, the ocean currents, and the concentration of atmospheric CO_2 are appropriate. Advances and retreats during a given ice age are controlled by Milankovitch cycles, determined by periodic variations in Earth's orbital shape and rotation-axis orientation. We may be living in a super-interglacial period.

QUICK QUESTION How does positive feedback contribute to a glaciation during an ice age, and why does a glaciation eventually cease?

(c) During the Little Ice Age, a glacier filled this valley. In this 2003 photo, most of the glacier has vanished. Most of the retreat has happened in the last century.

FIGURE 18.30 Greenland's melting glaciers. Melting has accelerated in the last few decades.

(a) Lakes of meltwater accumulate on the surface of the ice sheet during the summer.

(c) Where glaciers meet the sea, huge masses calve off and crash into the water. This is happening so fast that the ice front is retreating.

(b) Lakes suddenly drain through cracks that carry the water to the base of the glacier, 1 km down. Addition of liquid water to the base allows the glacier to move faster, causing a "surge."

■ Melting ice

(d) The melting area has increased dramatically in recent years.

Chapter 18 Review

CHAPTER SUMMARY

> Glaciers are streams or sheets of recrystallized ice that survive for the entire year and flow in response to gravity. Mountain glaciers exist in high regions and fill cirques and valleys. Continental glaciers (ice sheets) spread over substantial areas of the continents.

> Glaciers form when snow accumulates over a long period of time. The snow turns first to firn and then to ice as it gets buried.

> Temperate glaciers melt during part of the year. Polar glaciers do not. Glaciers move by basal sliding or by plastic deformation of ice grains. In general, glaciers move tens of meters per year.

> Gravitational pull causes glaciers to flow in the direction of their overall surface slope.

> Whether the toe of a glacier stays fixed in position, advances, or retreats depends on the balance between the rate at which snow builds up in the zone of accumulation and the rate at which glaciers melt, calve, or sublimate in the zone of ablation.

> Icebergs break off glaciers that flow into the sea. Continental glaciers that flow into the sea along a coast make ice shelves. Sea ice forms where the ocean's surface freezes.

> As glacial ice flows over sediment, it incorporates clasts. The clasts embedded in glacial ice abrade the substrate and can polish bedrock and cut striations into it.

> Mountain glaciers carve numerous landforms, including cirques, arêtes, horns, U-shaped valleys, hanging valleys, and truncated spurs. Marine fjords are glacially carved valleys that filled with water when sea level rose after an ice age.

> Glaciers can transport sediment of all sizes. Till consists of unsorted sediment dropped by melting ice. Glacial marine collects underwater, by melting of icebergs. Streams sort till and deposit it as glacial outwash. Some of the finer sediment accumulates as lake-bed mud or as windblown loess.

> Glacial depositional landforms include moraines, knob-and-kettle topography, drumlins, eskers, meltwater lakes, and outwash plains. Lateral moraines accumulate along the sides of valley glaciers, medial moraines down the middle, and end moraines at a glacier's toe.

> Continental crust subsides as a result of ice loading. When the glacier melts away, the underlying lithosphere rebounds. When continental glaciers store substantial amounts of water, global sea level drops. When these glaciers melt, sea level rises.

> During past ice ages, the climate in regions south of the continental glaciers was wetter, and pluvial lakes formed. Permafrost exists in periglacial environments.

> During the Pleistocene Ice Age, large continental glaciers covered much of North America, Europe, and Asia.

> The stratigraphy of Pleistocene glacial deposits indicates that glaciers advanced and retreated many times during the ice age. The record of glaciations is more complete in oceanic sediment.

> Over the long term, plate tectonics and changes in the concentration of atmospheric CO_2 may set the stage for ice ages. Short-term advances and retreats may be caused by Milankovitch cycles.

GUIDE TERMS

ablation (p. 583)
arête (p. 586)
cirque (p. 586)
continental glacier (p. 579)
crevasse (p. 581)
drumlin (p. 592)
equilibrium line (p. 583)
erratic (p. 577)
esker (p. 594)
firn (p. 579)

fjord (p. 588)
glacial advance (p. 583)
glacial drift (p. 592)
glacial retreat (p. 584)
glacial striation (p. 586)
glacial subsidence (p. 594)
glacier (p. 579)
hanging valley (p. 586)
horn (p. 586)
kettle hole (p. 594)

Laurentide ice sheet (p. 600)
Milankovitch cycle (p. 603)
moraine (p. 592)
mountain glacier (p. 579)
patterned ground (p. 599)
permafrost (p. 599)
Pleistocene Ice Age (p. 599)
pluvial lake (p. 598)
post-glacial rebound (p. 595)

roche moutonnée (p. 588)
sea ice (p. 584)
snowball Earth (p. 602)
sublimate (p. 579)
terminal moraine (p. 592)
tidewater glacier (p. 584)
tillite (p. 601)
U-shaped valley (p. 586)
V-shaped valley (p. 586)
varve (p. 592)

GEOTOURS *THIS CHAPTER'S GEOTOURS WORKSHEET (R) FEATURES QUESTIONS AND GOOGLE EARTH SITES ON:*

....

> Continental glacier features in the northeastern United States
> Piedmont glacial features in Alaska

> Alpine/valley glacial features around the world

REVIEW QUESTIONS

The letters following each Review Question refer to the corresponding Learning Objective from the Chapter Opener.

1. What evidence did Louis Agassiz offer to support the idea that an ice age happened? **(D)**

2. How do mountain glaciers and continental glaciers differ? Describe the different types of mountain glaciers, and identify each in the image. **(A)**

3. Describe the transformation from snow to glacial ice. **(A)**

4. Describe the mechanisms that enable glaciers to move, and explain why glaciers move. Why do crevasses form? **(B)**

5. How fast do glaciers normally move? How fast can they move during a surge? **(B)**

6. Explain how the balance between ablation and accumulation controls advances and retreats. **(B)**

7. How can a glacier continue to flow toward its toe even though its toe is retreating? **(B)**

8. Explain how arêtes, cirques, and horns form and how they differ in shape. **(B)**

9. How does a glacier transform a V-shaped valley into a U-shaped valley? Discuss how hanging valleys form. **(B)**

10. Describe the various kinds of glacial deposits. Be sure to note the materials from which the deposits are made and the landforms that result from deposition. Which features are visible in this diagram? **(C)**

11. How does the lithosphere respond to the weight of glacial ice? How does sea level change during an ice age? **(B)**

12. How was the on-land chronology of glaciations developed? Why was it so incomplete? How was it modified with the study of marine sediment? **(D)**

13. Were there ice ages before the Pleistocene? If so, when? **(D)**

14. What are some of the long-term causes that lead to ice ages? What are the short-term causes that trigger glaciations and interglacials? **(E)**

ON FURTHER THOUGHT

15. If you fly over central Illinois during the early spring, you will see slight differences in soil color due to variations in moisture content—wetter soil is darker. These variations outline the shapes of polygons that are tens of meters across. What do these patterns represent? **(E)**

16. An unusual late Precambrian rock unit crops out in the Flinders Range, a small mountain belt in South Australia. This unit consists of clasts of granite and gneiss, in a wide range of sizes, suspended through a matrix of slate. The rock unit lies unconformably above a basement of granite and gneiss, and if you dig out the unconformity surface, you will find that it is polished and striated. What is the unusual rock? **(C)**

ONLINE RESOURCES

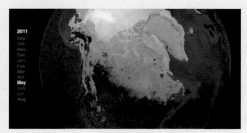

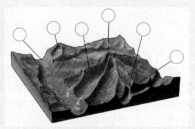

Animations This chapter features interactive animations that simulate glacial bodies under different conditions.

Videos This chapter features videos on rapid glacier change and studies of Greenland's retreating glaciers.

Smartwork5 This chapter features visual identification and labeling exercises on the effects of glacial systems on the landscape.

CHAPTER 19

GLOBAL CHANGE IN THE EARTH SYSTEM

By the end of this chapter, you should be able to . . .

A. explain why the Earth has been able to change in many ways over its history.

B. distinguish between unidirectional and cyclic changes and between gradual and catastrophic changes, and give examples of each.

C. describe how elements or compounds flow among various reservoirs during biogeochemical cycles.

D. characterize changes in the Earth's climate over geologic history, and discuss phenomena that play a role in regulating climate at different time scales.

E. discuss ways in which human society has significantly modified the Earth System, particularly since the industrial revolution.

F. evaluate evidence that climate has changed during the past few centuries, and describe the causes and consequences of this change, as well as possible methods of investigating it.

G. provide scenarios for changes in the Earth System in the near and very distant future.

19.1 Introduction

Did the Earth's surface look the same in the Jurassic Period as it does today? Definitely not! Two hundred million years ago, the North Atlantic Ocean was a narrow sea and the South Atlantic Ocean didn't exist at all, so most dry land lay within a single vast supercontinent **(Fig. 19.1)**. Today, both parts of the Atlantic are wide oceans, and the Earth has seven separate continents. Moreover, during the Jurassic, the call of the wild rumbled from the throats of dinosaurs, whereas today, mammals dominate the land. In essence, what we see of the Earth today is just a snapshot, an instant in the life story of a constantly changing planet. This idea arguably stands as geology's greatest philosophical contribution to humanity's understanding of our Universe.

Why has the Earth changed so much over geologic time, and why does it continue to change? Ultimately, change happens both because the Earth's internal heat keeps the asthenosphere weak enough to flow, and because the Sun's radiation can keep most of the Earth's surface at temperatures above the freezing point of water. Flow in the asthenosphere permits plate tectonics, which in turn leads to continental drift, volcanism, and mountain building. Solar heat keeps streams, glaciers, waves, and wind in motion, thereby causing erosion and deposition, and it also fuels photosynthesis. If the Earth did not have just the right mix of tectonic activity and solar heat, it would be a frozen dust bowl like Mars, a crater-pocked wasteland like the Moon, or a cloud-choked oven like Venus, and could not host life as we know it. Many of the changes that take place on the Earth reflect complex interactions among geologic and biological phenomena. For example, photosynthetic organisms changed the composition of the atmosphere by providing oxygen, and atmospheric composition, in turn, influences the nature of chemical weathering reactions that can take place in rocks.

We've referred to the global interconnecting web of physical and biological phenomena on the Earth as the *Earth System*. In this context, we can now define **global change**, in a general sense, as the modifications of physical and biological components in the Earth System over time.

Geologists distinguish among different types of global change on the basis of the rate or way in which change progresses. *Gradual change* takes place slowly, over long periods of geologic time (millions to billions of years); *catastrophic change* takes place relatively rapidly (seconds to millennia); *unidirectional change* involves transformations that never repeat; and *cyclic change* repeats the same steps over and over, though not necessarily with the same results or at the same rate.

In this chapter, we begin by reviewing examples of global change involving phenomena discussed earlier in this book. Then we introduce the concept of a *biogeochemical cycle*, the exchange of chemicals among various living and nonliving reservoirs. We'll see that some kinds of global change reflect modifications in the proportions of chemicals held in different reservoirs. Finally, we focus on *global climate change*, the transformations of the Earth's climate over time. We conclude this chapter, and this book, by considering hypotheses that describe the ultimate global change—the end of the Earth—in the very distant future.

FIGURE 19.1 The map of the Earth's surface changes over time because of plate motions.

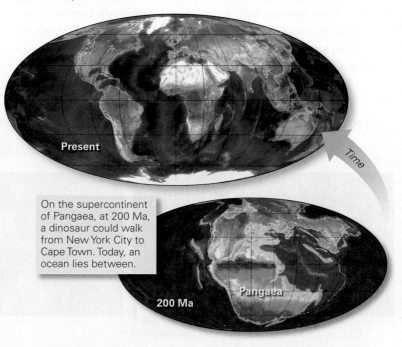

Present

Time

On the supercontinent of Pangaea, at 200 Ma, a dinosaur could walk from New York City to Cape Town. Today, an ocean lies between.

Pangaea

200 Ma

◀ (facing page) The buildings of New York City and adjacent New Jersey blanket what, only a few centuries ago, was a forest and wetlands bordering an estuary. About 20,000 years ago, the landscape we see here was covered by a vast ice sheet. Clearly, the Earth changes over time, and in recent centuries, humans have played a role in that change.

19.2 Unidirectional Changes

THE EVOLUTION OF THE SOLID EARTH

Recall from Chapter 1 that the Earth began as a fairly homogeneous mass, formed by the coalescence of planetesimals. The homogeneous proto-Earth did not last long, for within about 100 million years of its birth, when it had attained a diameter approaching 1,000 km, the planet began to melt, yielding liquid iron alloy that sank to the center to form the core. This process of **differentiation** represents a major unidirectional change in our planet—it produced a layered planet with an iron alloy core surrounded by a rocky mantle.

According to a widely held model, another protoplanet collided with the newborn Earth soon after its differentiation. This collision caused a catastrophic change, in that a significant portion of the Earth fragmented or vaporized, yielding an orbiting ring of debris that quickly coalesced to form the Moon **(Fig. 19.2)**. For a time after this collision, the Earth's mantle remained largely molten, and the planet's surface was a sea of magma. But cooling probably happened fairly quickly so that, according to recent research, the surface of the Earth had solidified and may even have hosted liquid water before 4.0 billion years ago (Ga).

The early surface of the Earth didn't last forever, though. Intense bombardment by asteroids and comets, between about 4.0 and 3.9 Ga, destroyed crust and sea that had existed prior to 3.9 Ga by pulverizing and melting it. Eventually, however, bombardment ceased, and our planet gradually cooled, permitting a new crust to stabilize at its surface, the oceans to form, and an early form of plate tectonics to begin operating. Over time, igneous activity produced lasting continental crust, blocks of which collided and sutured together, and by about 2.7 Ga, most of the continental crust that exists today had been distilled from the mantle. Differentiation, Moon formation, crust formation, and ocean formation all represent major unidirectional changes that took place early in the Earth's history.

THE EVOLUTION OF THE ATMOSPHERE AND OCEANS

Like its surface, the Earth's atmosphere has changed unidirectionally over time. Early on, it formed from gases released by volcanic activity. More gases arrived when comets collided with the new planet. Eventually, the Earth accumulated an early atmosphere composed dominantly of carbon dioxide (CO_2) and water (H_2O). Other gases, such as nitrogen (N_2), composed only a minor proportion of the early atmosphere. When the Earth's surface cooled, however, water condensed and fell as rain, collecting in low areas to form oceans. Permanent oceans have existed since about 3.8 Ga. Gradually, CO_2 dissolved in the oceans and also was absorbed by chemical-weathering reactions on land, so its concentration in the atmosphere decreased. Nitrogen, which doesn't react with other chemicals, was left behind. Consequently, the atmosphere's composition changed to become dominated by nitrogen. Photosynthetic organisms appeared early in the Archean, but it probably wasn't until between 2.5 and 2.0 Ga, the early Proterozoic, that oxygen became a significant proportion of the atmosphere. This unidirectional change is known as the *great oxygenation event*. Oxygen didn't reach breathable concentrations for another billion years.

> **Did you ever wonder...**
> if the Earth's atmosphere has always been breathable?

THE EVOLUTION OF LIFE

The fossil record indicates that life had appeared, in the form of single-celled archaea and bacteria, at least by 3.7 Ga, and

FIGURE 19.2 A popular model attributes the formation of the Moon to a collision with a planetesimal.

Time

A protoplanet collided with the Earth.

Debris sprayed into space.

The debris coalesced to form the Moon.

FIGURE 19.3 Because of evolution, organisms inhabiting the Earth today are not the same as those that inhabited the planet in the past.

The largest Mesozoic land animal was a dinosaur. The inset shows an elephant at the same scale.

The largest land animal today is the elephant.

that it has undergone unidirectional change (*evolution*) ever since (see Interlude E). Complex multicellular organisms appeared in the late Proterozoic. Shells appeared, and life diversified substantially at the beginning of the Cambrian, during an event called the *Cambrian explosion*.

Initially, multicellular life inhabited only the sea, and the land was completely barren. By the Middle Ordovician, plants had succeeded in invading the land, and animals followed in the Silurian. Early land animals included millipedes and scorpions. Reptiles appeared in the Carboniferous, and placental mammals in the Cretaceous. When dinosaurs became extinct, at the end of the Mesozoic, mammals became the largest land animals **(Fig. 19.3)**. While primates became a distinct order of mammals as early as 55 million years ago (Ma), our own species appeared only about 150,000 years ago. Once a species has gone extinct, it never reappears, and new species appear over time. Therefore, we can consider life evolution to be unidirectional change. As a result of evolution, life now inhabits regions from a few kilometers below the surface to a few kilometers above, yielding a diverse and complex biosphere.

CATASTROPHIC MASS-EXTINCTION EVENTS

Changes that happen on the Earth almost instantaneously are called *catastrophic changes*. For example, a volcanic explosion, an earthquake, a tsunami, or a landslide can change a local landscape in seconds or minutes. But such events affect only relatively small areas. Can such catastrophes happen on a global scale? In recent decades, geoscientists have come to the conclusion that the answer is yes. The stratigraphic record shows that the Earth's history so far includes several **mass-extinction**

events, during which large numbers of species abruptly vanished—these events define boundaries between geologic periods. A mass-extinction event decreases *biodiversity*—the number of different species that exist at a given time. It takes millions of years after a mass-extinction event for biodiversity to recover, and the new species that appear differ from those that vanished, for evolution is unidirectional.

Geologists speculate that some mass-extinction events reflect incredibly voluminous volcanic eruptions or the impact of a comet or an asteroid with the Earth. Either of these events could eject enough debris into the atmosphere to block sunlight. Without the warmth of the Sun, winter-like or night-like conditions would last for weeks to years, long enough to disrupt the food chain. In addition, either type of event could eject aerosols that would turn into global acid rain, scatter hot debris that would ignite forest fires, or give off chemicals that would make the ocean either toxic or overly nutritious and thus disrupt its ecosystems. At present, many geologists favor the hypothesis that the Permian-Triassic extinction was due to the eruption of 3 million km³ of basalt in Siberia, and that the Cretaceous-Paleogene (K-Pg) extinction was due to a giant meteorite impact.

> ### TAKE-HOME MESSAGE
>
> Since it first formed, our planet has undergone major changes that are unidirectional, in that they modified the Earth System forever and will never repeat. Examples include internal differentiation (core formation), Moon formation, atmospheric evolution, ocean formation, and life evolution. Catastrophic changes have taken place on occasion.
>
> ---
>
> **QUICK QUESTION** How are life evolution and atmospheric evolution linked?

19.3 Cyclic Changes

During *cyclic changes*, a sequence of stages may be repeated over time. Some cyclic changes are periodic in that the cycles happen with a definable frequency, but others are not. Below, we look at several examples of cyclic change—you'll see that some involve movements of physical components of the Earth, whereas others involve transfer of chemicals among both living and nonliving reservoirs.

THE SUPERCONTINENT CYCLE

Over the course of the Earth's history, the map of our planet's surface has changed constantly. During some intervals of geologic time, almost all continental crust lay within one or two supercontinents, and during other intervals, the planet's surface has hosted several smaller continents. The process of change during which a supercontinent forms by continental collision and later breaks apart by rifting is the *supercontinent cycle*. Geologists have found evidence that supercontinents existed at least three or four times during the past 3 billion years of Earth history—no two were alike. The most recent supercontinent, Pangaea, formed at 300 Ma, the end of the Paleozoic Era, and began to break apart by about 200 Ma, in the Mesozoic. The supercontinent cycle can take hundreds of millions of years to pass through a single complete cycle.

THE SEA-LEVEL CHANGE CYCLE

Global sea level rose and fell by as much as 300 m during the Phanerozoic, and likely did the same in the Precambrian.

When sea level rose, the shoreline migrated inland, resulting in a transgression during which low-lying plains became submerged. At times of particularly high sea level, shallow seas covered more than half of the Earth's continental area, and new layers of sediment buried broad portions of continents. When sea level fell, the continents became dry again. The lack of deposition or the occurrence of widespread erosion after a regression produced regional-scale unconformities. Geologists refer to the succession of strata deposited between two regional unconformities as a *sedimentary sequence*.

After studying sedimentary sequences around the world, geologists have pieced together a chart defining the history of global transgressions and regressions during the Phanerozoic Eon. The global sedimentary cycle chart may largely reflect the cycles of *eustatic* (worldwide) *sea-level change* (**Fig. 19.4**). Such sea-level changes may be due to a variety of factors, including advances and retreats of continental glaciers, changes in the volume of mid-ocean ridge systems, and changes in overall continental elevation and area. The chart of Figure 19.4 probably does not give us an exact image of sea-level change over time, however, because the sedimentary record reflects other factors as well, such as changes in sediment supply.

THE ROCK CYCLE

We learned early in this book that the crust of the Earth consists of three rock types: igneous, sedimentary, and metamorphic. Atoms making up the minerals of one rock type may later—after erosion and deposition, after metamorphism, or after melting—become part of another version of the same rock type or of a different rock type. In effect, rocks serve as reservoirs of atoms, and the atoms can move from reservoir to

FIGURE 19.4 This chart provides one interpretation of sea-level change during the past half billion years, based on the stratigraphic record.

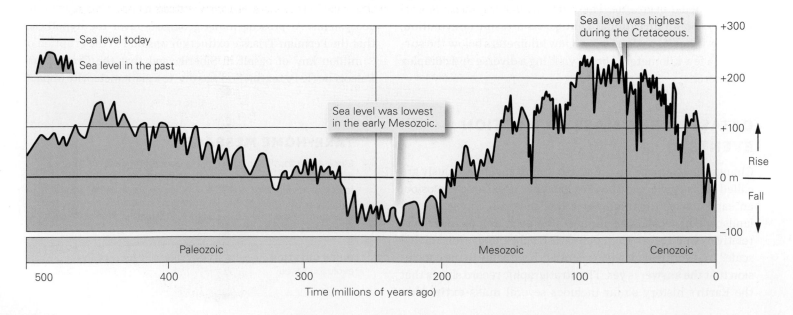

reservoir over time, a process called the *rock cycle*. Each stage in the rock cycle changes the Earth by redistributing and modifying material.

BIOGEOCHEMICAL CYCLES

A **biogeochemical cycle** involves the passage of chemicals among nonliving and living reservoirs in the Earth System, mostly on or near the planet's surface. Nonliving reservoirs include the atmosphere, rock, magma, soil, and the ocean, whereas living reservoirs include plants, animals, and microbes.

Some stages in a biogeochemical cycle may take only hours, some may take thousands of years, and others may take millions of years. The transfer of a chemical from reservoir to reservoir during these cycles doesn't really seem like a change in the Earth System in the way that the movement of continents or the metamorphism of rock seem like changes. In fact, for intervals of time, biogeochemical cycles attain a **steady-state condition**, meaning that the proportions of a chemical in different reservoirs remain fairly constant, even though there is a constant *flux* (flow) of the chemical among reservoirs. When we speak of global change in a biogeochemical cycle, we mean a change in the relative proportions of a chemical held in different reservoirs at a given time—in other words, a change in the steady-state condition. A great variety of chemicals (water, carbon, oxygen, sulfur, ammonia, phosphorus, and nitrogen) participate in biogeochemical cycles. Here, we look at two: water (H_2O) and carbon (C).

The Hydrologic Cycle As we learned in Interlude F, the hydrologic cycle involves the movement of water from reservoir to reservoir on or near the surface of the Earth. During the cycle, a chemical compound, H_2O, passes through both nonliving and living entities—the oceans, the atmosphere, surface water, groundwater, glaciers, soil, and living organisms. Global change in the hydrologic cycle occurs primarily when a change in global climate alters the ratio between the amount of water held in the ocean and the amount held in continental ice sheets. For example, during an ice age, water that had been stored in oceans moves into glacial reservoirs. When the continents become covered with ice, sea level drops. When the climate warms, water returns to the oceans, and sea level rises.

The Carbon Cycle Most carbon in the Earth's near-surface realm originally bubbled out of the mantle in the form of CO_2 gas released by volcanoes **(Fig. 19.5)**. Once it enters the atmosphere, carbon moves through various reservoirs of the Earth System through the **carbon cycle**. Some dissolves in seawater to form bicarbonate (HCO_3^-) ions, which may later eventually become incorporated in the shells of invertebrates or precipitate directly from water as $CaCO_3$. Atmospheric CO_2 can also be absorbed by photosynthetic organisms that convert it to sugar and other organic chemicals—this carbon enters the food chain and ultimately makes up the flesh, fat, and sinew of animals.

Of the carbon incorporated in organisms, some returns directly to the atmosphere (as CO_2) through the respiration of animals, the flatulence of animals (as methane, CH_4), or the decay of dead organisms. But some can be stored underground for long periods in fossil fuels (oil, gas, and coal), in organic shale, or in limestone ($CaCO_3$). This carbon can return to the atmosphere (as CO_2) through the burning of fossil fuels, production of cement, or the metamorphism of limestone. Or it can return to the sea after undergoing chemical weathering followed by dissolution in river water or groundwater.

FIGURE 19.5 In the carbon cycle, carbon moves among various reservoirs at or near the Earth's surface. Red arrows indicate release to the air, and green arrows indicate absorption from the air.

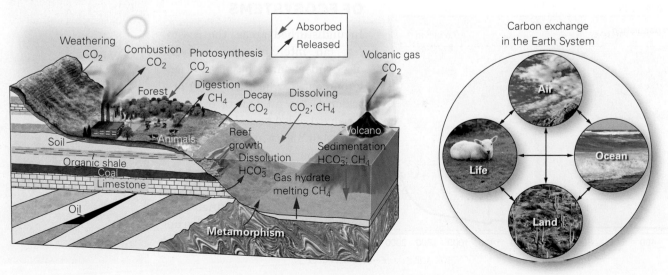

Substantial amounts of carbon may also be stored in methane hydrates (ice containing CH_4) buried in marine sediments. This carbon can be released as methane gas when methane hydrates melt.

TAKE-HOME MESSAGE

Some changes in the Earth System are cyclic, in that they have stages that may be repeated. Examples include the supercontinent cycle, the sea-level cycle, the rock cycle, the hydrologic cycle, and biogeochemical cycles. During the carbon cycle, carbon transfers among living and nonliving reservoirs.

QUICK QUESTION What evidence indicates that sea level rises and falls over time?

19.4 Human Impact on Land and Life

According to some researchers, perhaps only 2,000 to 20,000 people lived on the Earth after the catastrophic eruption of the Toba volcano about 70,000 years ago, and by the dawn of civilization, 4000 B.C.E., the population was still, at most, a few tens of millions. But by the beginning of the 19th century, revolutions in industrial methods, agriculture, medicine, and hygiene had substantially lowered death rates and raised living standards, so that the human population began to grow at accelerating rates, reaching 1 billion in 1850. It took only 80 years for the population to double after that, reaching 2 billion in 1930. Now the doubling time is only 44 years—the

FIGURE 19.6 The human population now doubles about every 44 years. The Black Death pandemic in the Middle Ages caused an abrupt drop that lasted for a few decades.

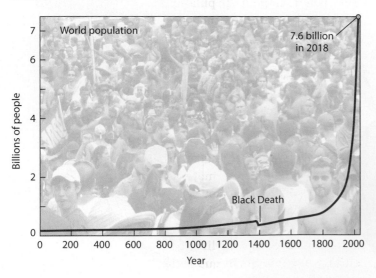

population passed the 6 billion mark just before the year 2000 and surpassed the 7.6 billion mark in 2018 **(Fig. 19.6)**.

As the population grows and the standard of living improves, per capita usage of resources increases. We use land for agriculture and grazing, forests for wood, rock and soil for construction, oil and coal for energy or plastics, and ores for metals (see Chapter 12). Without a doubt, our usage of resources has affected the Earth System profoundly, and thus humanity has become a major agent of global change. Here, we examine some of these *anthropogenic* (human-induced) *impacts*.

THE MODIFICATION OF LANDSCAPES

Every time we pick up a shovel and move a pile of rock or soil, we redistribute a portion of the Earth's crust. In the last century, the pace of Earth movement has accelerated, for now shovels in mines can move 300 m³ of rock in a single scoop, trucks can carry 200 tons of rock in a single load, and tankers can transport 500 million liters of oil during a single journey. In North America, human activity now moves more sediment each year than rivers do. Extracting rock during mining, building levees and dams along rivers or seawalls along the coast, and constructing highways and cities—all involve redistribution of Earth materials **(Fig. 19.7)**. In addition, people clear and plow fields, drain and fill wetlands, and pave over the land surface. All these activities change the landscape, the water table, and the supply of sediment.

THE MODIFICATION OF ECOSYSTEMS

In undisturbed areas, the *ecosystem* (an interconnected network of organisms and the physical environment in which they live) of a region is the product of evolution for an extended period of time. The ecosystem's flora and fauna include species that have adapted to living together in a particular climate and on the soil available. Human-caused deforestation, overgrazing, agriculture, and urbanization disrupt ecosystems and lead to decreased biodiversity.

SEE FOR YOURSELF...

Fields and Villages, China

LATITUDE
35°6'46.35" N

LONGITUDE
114°30'7.81" E

Look down from 4 km.

This landscape has been completely altered by human society. What was once forest or grassland has been divided into cultivated rectangles. In villages, houses and roads cover the surface. These changes affect both the carbon cycle and the hydrologic cycle.

FIGURE 19.7 Excavation, agriculture, and construction modify topography, drainage, infiltration, and ecology.

Humans move vast amounts of rock.

Agriculture eliminates diverse ecosystems.

Urbanization changes the water table.

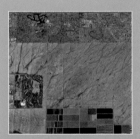
Archaeological studies have found that the earliest example of humans modifying an ecosystem occurred in the Stone Age, when hunters played a major role in causing mass extinctions of mammoths and other large mammals. Today, less than 5% of Europe and the United States retains its original ecosystems. Worldwide, tropical rainforests cover less than half the area that they covered before the dawn of civilization, and they are disappearing at a rate of about 1.8% per year **(Fig. 19.8a, b)**. Much of this loss comes from slash-and-burn agriculture, in which farmers and ranchers destroy forest to create open land for farming and grazing **(Fig. 19.8c)**.

Some changes humans make to the land have permanent consequences in a human time frame. The heavy rainfall of tropical regions removes nutrients from the soil of the fields produced by deforestation, making the soil useless in just a few years. Overgrazing by domesticated animals can remove vegetation so completely that some grasslands have undergone desertification. And urbanization replaces the natural land surface with concrete or asphalt, a process that completely destroys an ecosystem and radically changes the amount of rain that infiltrates the land surface to become groundwater.

POLLUTION

The environment has always contained contaminants such as soot, dust, chemical runoff, and waste produced by organisms. But when human populations grew, urbanization, industrial and agricultural activity, the production of electricity, and modern modes of transportation greatly increased both the quantity and diversity of contaminants that entered the air, surface water, and groundwater. These contaminants, or **pollution**, include both natural and synthetic materials in liquid, solid, and gaseous form. They have become a problem because the Earth System cannot naturally absorb or modify them quickly enough. Small quantities

FIGURE 19.8 The area of forests has been shrinking. For example, tropical rainforests are being logged or burned.

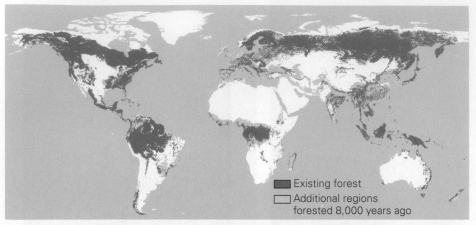

Existing forest

Additional regions forested 8,000 years ago

(a) Remaining forests today are much smaller than forests of 8,000 years ago.

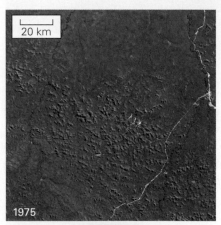

20 km

1975

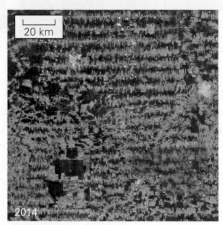

20 km

2014

(b) Comparison of satellite imagery highlights areas that have undergone deforestation, such as this one near Ariquemes, Brazil.

(c) Part of the Amazon rainforest's destruction is due to slash-and-burn agriculture.

of sewage can be absorbed by clay minerals in the soil or destroyed by bacterial metabolism, but large quantities overwhelm natural controls and can accumulate in destructive concentrations. Pollution of the Earth System can be considered a type of global change because it represents a redistribution and reformulation of materials. Some key problems associated with this change include the following:

> *Water contamination:* Humans dump a great variety of chemicals into surface water and groundwater. Examples include gasoline, other organic chemicals, radioactive waste, acids, fertilizers—the list could go on for pages. These chemicals not only can make water unsafe for human consumption, but also can affect biodiversity.

> *Destruction of habitat:* Agriculture, grazing, deforestation, urban sprawl, road construction, and other ways in which people change the surface of the land have resulted in the widespread destruction of habitats in which ecosystems thrived.

> *Smog:* The term was originally coined to refer to the dank, dark air that resulted when smoke from burning coal mixed with fog in London and other industrial cities. Another kind of smog, called *photochemical smog,* is the ozone-rich brown haze that blankets cities when exhaust from cars and trucks reacts with air in the presence of sunlight **(Fig. 19.9a)**.

> *Acid runoff and acid rain:* Dissolution of sulfide-containing minerals in ores or coal by groundwater or stream water makes the water into *acid runoff.* When rain passes through air that contains sulfur-containing aerosols (emitted from power plants or factories), the water dissolves the sulfur and yields **acid rain**. Wind can carry aerosols far from a power plant, so acid rain can damage a broad region.

> *Radioactive materials:* The by-products of nuclear weapons, nuclear energy, and some medical procedures transfer radioactive materials from rock to the Earth's surface environment.

> *Ozone depletion:* When emitted into the atmosphere, human-produced chemicals, most notably chlorofluorocarbons (CFCs), react with ozone in the stratosphere. This reaction, which happens most rapidly on the surfaces of tiny ice crystals in polar stratospheric clouds, destroys ozone molecules. It yields an *ozone hole* over high-latitude regions, particularly during the spring **(Fig. 19.9b)**. Note that the "hole" is not an area without any ozone, but is an area where atmospheric ozone has been decreased substantially. Ozone holes have dangerous consequences, for they affect the ability of the atmosphere to shield the Earth's surface from harmful ultraviolet radiation.

FIGURE 19.9 Consequences of adding pollutants to the atmosphere.

(a) Photochemical smog hazes the view in Beijing, on an otherwise sunny day.

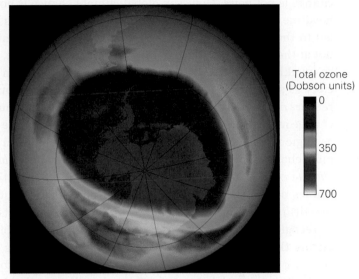

Total ozone
(Dobson units)

0

350

700

(b) The ozone hole over Antarctica in 2014. A Dobson unit is a measure of the amount of ozone in the atmosphere; red is more, purple is less.

TAKE-HOME MESSAGE

Human activities affect the landscape and ecosystems significantly by moving rock and soil, and by changing the character of land cover. Society also introduces contaminants into the environment at rates that cannot be accommodated by the Earth System, leading to pollution of land, air, surface water, and groundwater.

QUICK QUESTION What is the ozone hole, and why did it form?

19.5 Global Climate Change

WHAT IS CLIMATE CHANGE?

How often have you seen a news report proclaim "Record High Temperatures!" when thermometers register temperatures several degrees above "normal" for days on end. Do such headlines mean that the climate is changing? To start addressing this question, we first need to distinguish between two common terms: *weather* and *climate*.

Atmospheric conditions during a specific time interval at a given locality define the locality's **weather**. Weather may be "windy, rainy, and cool" in the morning, and "calm, sunny, and dry" in the afternoon. The term **climate**, in contrast, refers to an overall range of weather conditions, as well as the typical daily to seasonal variability of weather conditions,

as observed over a period of decades for a region. A region with a *tropical climate* tends to have hot, humid days all year, whereas a place with a *temperate climate* tends to have contrasting seasons and, therefore, a wide range of temperature and humidity over the course of a year. So a newspaper's headline about a hot spell or cold snap does not, in itself, mean that the climate is changing. But if a new set of conditions—say, an overall increase in average temperature, a rising snow line, a longer growing season, or a change in storm frequency or intensity—becomes the norm for a region, then climate change has taken place. And if such changes happen worldwide, then **global climate change** has occurred. An increase in global average atmospheric and sea-surface temperatures represents *global warming*, and a decrease represents *global cooling*.

METHODS OF STUDYING CLIMATE CHANGE

To gain insight into how global climate has already changed throughout Earth's long history, researchers study **paleoclimate** (past climate) by searching for climate indicators—such as depositional environments, fossil species, or geochemical signatures—that can be dated accurately. Paleoclimate studies give a sense of the magnitude of future change that might be possible.

Long-term global climate change, which happens in a time frame of tens to hundreds of millions of years, has been recognized since geologists first learned how to interpret the stratigraphic record. This record shows that depositional settings at a locality change significantly over time, so such change is not news. However, global climate change is now a headline, because research during the past few decades has led to the conclusion that this type of change is happening, not at the slow pace typical of most geologic phenomena, but so quickly that it will have significant effects on human society as soon as the next few decades. In other words, many readers of this book will likely see the effects of global climate change in their lifetimes. In this section, we first examine the geologic manifestations and possible causes of long-term climate change. Then, we turn our attention to *short-term global climate change*, meaning changes that take place over time frames of less than 1 million years. Of note, the resolution (detail) retained in the geologic record generally permits us to recognize only short-term changes that have taken place during the past few million years. This discussion sets the stage for the next section of the chapter, in which we focus on *contemporary global warming*, meaning changes that have taken place during the past few centuries and may affect lifestyles during this century.

LONG-TERM CLIMATE CHANGE

The stratigraphic record shows that the Earth's surface temperature has stayed between the freezing point of water and the boiling point of water for almost the entire time since the beginning of the Archean, for strata of all ages contain water-deposited sediment. Characteristics of strata reveal that at some times in the past, the Earth's atmosphere has been significantly warmer than it is today, whereas at other times, it has been significantly cooler. The warmer periods have come to be known as **greenhouse periods** (or *hothouse periods*), and the colder ones are called **icehouse periods (Fig. 19.10)**. (The more familiar term, *ice age*, refers to the specific times during an icehouse period when the Earth was cold enough for ice sheets to advance and cover substantial areas of the continents.) During greenhouse periods, even lands at polar latitudes were largely free of ice. During icehouse periods, glaciers covered land at mid-latitudes. Of note, during the late Proterozoic icehouse period, land at equatorial latitudes became ice covered, and this planet may have frozen over entirely to become "snowball Earth."

What causes long-term global climate change? The answer may lie in the complex relationships among the various components of the Earth System, as described earlier. For example, the positions of continents, due to plate movements, influence the climate by controlling the pattern of oceanic currents, which redistribute heat around the planet's surface. Continental movement also determines whether the land lies at high or low latitudes (and thus how much solar radiation strikes it), and whether or not large continental interior

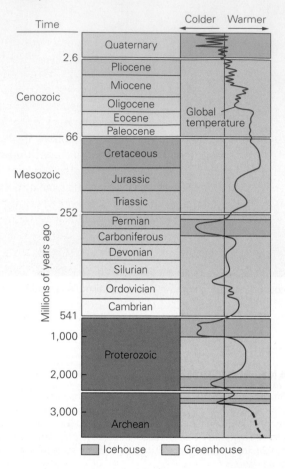

FIGURE 19.10 The Earth has experienced icehouse and greenhouse periods at various times during Earth history.

regions exist where extremely cold winter temperatures can develop. Changes in the concentration of *greenhouse gases* also play an important role. Greenhouse gases, such as CO_2 and CH_4, constitute only a small proportion of the atmosphere, but they play a major role in regulating the temperature of the atmosphere **(Box 19.1)**. Over geologic time, the concentration of these gases can change as a consequence of many factors, including: life evolution (such as the appearance of photosynthetic organisms, which remove CO_2); the degree to which carbon-containing materials (coal, oil, and limestone) form and become buried and, therefore, able to trap the carbon in underground reservoirs; and whether plate interactions have caused the uplift of mountain ranges, for uplift exposes rock to weathering reactions that involve greenhouse gases and can affect atmospheric circulation. Changes in solar radiation may also change, as the Sun itself evolves.

SHORT-TERM CLIMATE CHANGE

The detail of paleoclimate information retained in the stratigraphic record decreases as rocks get older, partly because

BOX 19.1 CONSIDER THIS...

The Role of Greenhouse Gases and Feedback Mechanisms

The Sun constantly bathes the Earth in visible and ultraviolet light. Some of this energy reflects off the atmosphere or off the Earth's surface, so that our planet shines when viewed from space, and some gets absorbed by gases in the atmosphere. The remaining visible and ultraviolet light reaches the surface and gets absorbed by rocks, sediments, soils, or water. These materials heat up and then release the energy back into the atmosphere in the form of infrared radiation (thermal energy) that heads back upward. If the Earth had no atmosphere, all of this infrared radiation would escape back into space. But our planet does have an atmosphere, and certain gases in the air absorb infrared radiation and re-radiate it. Some of the re-radiated energy heads up into space, but some heads downward and warms the lower atmosphere **(Fig. Bx19.1)**. In effect, these special gases trap thermal energy and keep the lower atmosphere warm, somewhat as glass traps heat in a greenhouse. The overall trapping process is called the **greenhouse effect**, and gases (such as H_2O, CO_2, CH_4, and N_2O) that cause it are called **greenhouse gases**. An increase in the concentration of greenhouse gases in the atmosphere will cause the atmosphere to become warmer, whereas a decrease in the concentration of these gases will allow the atmosphere to become cooler.

Atmospheric warming or cooling due to changes in the concentration of greenhouse gases can be amplified by positive or negative **feedback mechanisms**. *Negative feedback* slows a process down or even reverses it, whereas *positive feedback* enhances a process and amplifies its consequences. Let's consider an example of a feedback mechanism involving the greenhouse effect. Imagine that global average atmospheric temperature increases due to an increase in CO_2. This increase will, in turn, cause the oceans to warm and evaporate more, so more water transfers into the atmospheric reservoir globally, causing still more warming—this is a positive feedback. Warming may also cause permafrost in arctic regions and methane hydrates on the seafloor to melt. This would release CH_4 into the atmosphere, which would cause even more warming, another example of positive feedback.

FIGURE Bx19.1 The greenhouse effect traps thermal energy in the atmosphere.

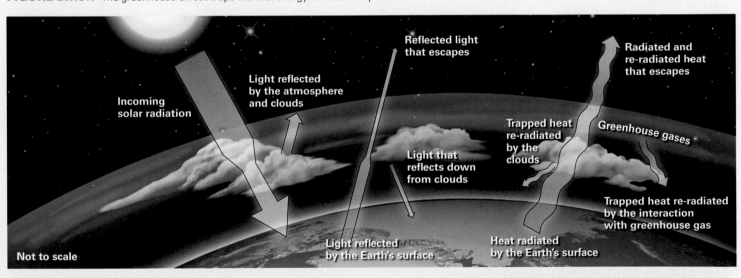

Reflected light that escapes

Radiated and re-radiated heat that escapes

Light reflected by the atmosphere and clouds

Incoming solar radiation

Trapped heat re-radiated by the clouds

Greenhouse gases

Light that reflects down from clouds

Trapped heat re-radiated by the interaction with greenhouse gas

Light reflected by the Earth's surface

Heat radiated by the Earth's surface

Not to scale

methods to provide age constraints have a range of uncertainty, and partly because geologic phenomena such as compaction, diagenesis, and metamorphism can destroy the record. Thus, geologists can obtain a detailed record of paleoclimate changes for only the past few thousand to few million years. The following data sources provide this record:

> *Microfossils:* In Pleistocene and Holocene deposits, a record of climate change can come from the study of fossil

plankton and pollen. Studies of plankton shells preserved in marine deposits at a given location, for example, can define variations in seawater temperature over time at the location, for plankton shells have distinctive shapes that allow species to be identified accurately, and the species that live at a locality reflect the water temperature. Similarly, studies of delicate pollen grains preserved in lake deposits have allowed researchers to determine how the type of forest at a given latitude changed in association

FIGURE 19.11 Change in the assemblage of pollen in sediment indicates a shift in climate.

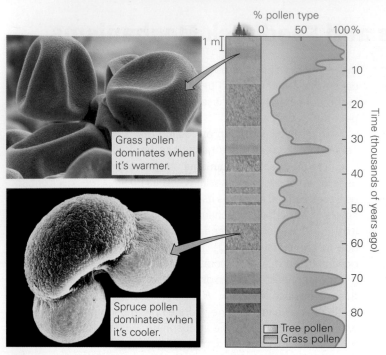

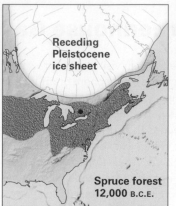
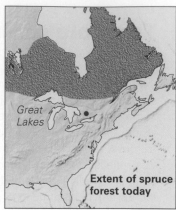

(b) Based on studying the pollen record, geologists discovered that spruce forests (green) grew farther south 12,000 years ago than they do today.

(a) The proportion of tree pollen relative to grass pollen can change in a sedimentary sequence over time. Researchers plot changes in the proportion of pollen types over time by examining samples from a column of sediment.

with the advance and retreat of continental glaciers **(Fig. 19.11)**.

> *Oxygen-isotope ratios:* Researchers have found that the ratio of ^{18}O to ^{16}O—two isotopes of oxygen that are preserved in the calcite ($CaCO_3$) or silica (SiO_2) molecules of shells or in the water (H_2O) molecules of glacial ice—depends on global climate at the time the shells or ice formed. Therefore, by measuring the $^{18}O/^{16}O$ ratio in fossil shells in a succession of layers in marine sediment or in a succession of layers in glacial ice, researchers can quantify how global climate has changed over time. For example, during times of cooler climates, when continental glaciers grow, the $^{18}O/^{16}O$ ratio in shells tends to be higher, and the $^{18}O/^{16}O$ ratio in glacial ice (formed at high latitudes) tends to be lower.

Researchers have now obtained ice cores down to a depth of almost 3 km in Antarctica and Greenland; these cores provide a record of global climate change for the past 800,000 years **(Fig. 19.12a, b)**. Drill cores of marine sediments extend the paleoclimate record back over millions of years **(Fig. 19.12c)**.

> *Growth rings:* If you've ever looked at a tree stump, you will have noticed the concentric rings visible in the wood. In general, each ring represents one year of growth, and the thickness of the ring indicates the rate of growth in a given year. Trees grow faster during warmer, wetter years

and more slowly during cold, dry years. Thus, the succession of ring widths provides a record of climate during the lifetime of the tree. Growth rings in corals and shells may provide similar information.

As we discussed in Chapter 18, paleoclimate data led to the realization that continental glaciers advanced and retreated up to 30 times in the northern hemisphere during the Pleistocene Ice Age. Each advance represents an interval of global cooling, and each retreat represents an interval of global warming. If we focus on the last 15,000 years, we see trends of cooling or warming that last thousands of years, within which some cooling or warming events may last as long as a few centuries **(Fig. 19.13a)**. Specifically, the time between about 15,000 and 10,500 B.C.E. was a warming period during which the last ice-age glaciers retreated. This warming trend was followed by the *Younger Dryas*, an interval of cooler temperatures named for an Arctic flower that became widespread during the interval. Then temperatures increased again, reaching a peak at 5,000 to 6,000 years ago, a period called the *Holocene maximum*, when average temperatures peaked at about 2°C above temperatures of today. This warming led to increased evaporation and therefore to increased precipitation, making the Middle East unusually wet and fertile—conditions that may have contributed to the rise of civilization in Mesopotamia and Egypt. The temperature dipped to a low about 3,000 years ago, before returning to a high during the Middle Ages, a time called the *Medieval Warm Period*. During this time, Vikings established self-supporting agricultural settlements along the coast of Greenland **(Fig. 19.13b)**. The temperature dropped again from 1500 C.E. to about 1800 C.E., a period known as the *Little Ice Age*, when Alpine glaciers advanced and the canals of the Netherlands froze over in winter **(Fig. 19.13c;** see Fig. 18.29). Overall, the climate has warmed

FIGURE 19.12 The proportion of oxygen isotopes transferred between reservoirs during evaporation or precipitation depends on temperature. The $^{18}O/^{16}O$ ratio can be studied in glacial ice (H_2O) and fossil shells ($CaCO_3$).

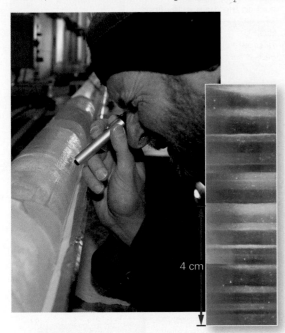

(a) A researcher examines an ice core in the field. Lab photos reveal annual layers.

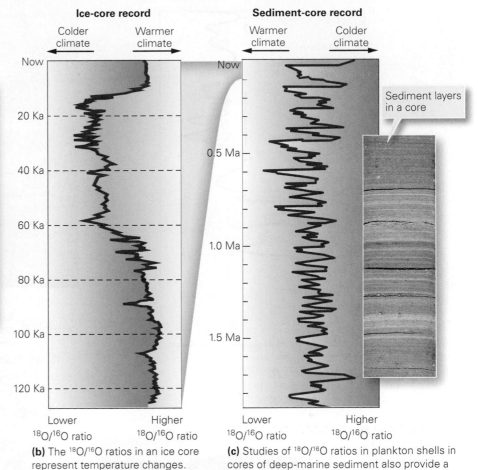

(b) The $^{18}O/^{16}O$ ratios in an ice core represent temperature changes.

(c) Studies of $^{18}O/^{16}O$ ratios in plankton shells in cores of deep-marine sediment also provide a climate record.

since the end of the Little Ice Age, and today it is as warm or warmer than it was during the Medieval Warm Period.

Geologists propose that several factors contributed to short-term warming and cooling events apparent in the near-term paleoclimate record:

> *Changes in the Earth's orbit and tilt:* As Milanković first recognized in 1920, the change in tilt of the Earth's axis, the Earth's precession cycle, and changes in the eccentricity of its orbit together cause variation in the amount of summer insolation at high latitudes (see Chapter 18). These changes correlate with observed ups and downs in atmospheric and oceanic temperature.

> *Changes in ocean currents:* Recent studies suggest that the configuration of currents can change quite quickly, and that this configuration affects the climate.

> *Large eruptions of volcanic aerosols:* Not all of the sunlight that reaches the Earth penetrates its atmosphere and warms the ground. Some is reflected by the atmosphere. The degree of reflectivity, or **albedo**, of the atmosphere increases if the concentration of volcanic aerosols in the atmosphere increases.

> *Changes in surface albedo:* Regional-scale changes in the nature of continental vegetation cover, or the proportion of snow and ice on the Earth's surface, or the sudden deposition of reflective volcanic ash affect our planet's

albedo. Increasing surface albedo causes cooling, whereas decreasing albedo causes warming.

> *Fluctuations in solar radiation:* The amount of energy produced by the Sun varies with the *sunspot cycle*. This cycle involves the appearance of large numbers of sunspots (black spots thought to be magnetic storms on the Sun's surface) about every 9 to 11.5 years. Whether longer-term cycles exist, capable of contributing to short-term climate change, has not yet been determined.

> *Fluctuation in cosmic rays:* Some researchers suggest that changes in the rate of influx of cosmic rays affect climate, perhaps by generating clouds. According to this concept, cosmic rays striking the atmosphere produce clusters of ions that become condensation nuclei around which water molecules congregate, leading to the formation of droplets making up clouds. This concept remains controversial.

> *Abrupt changes in concentrations of greenhouse gases:* A sudden change in greenhouse gas concentration in the atmosphere could affect climate. One such change might

FIGURE 19.13 Climate during the Holocene. Measurements suggest that temperature has varied significantly.

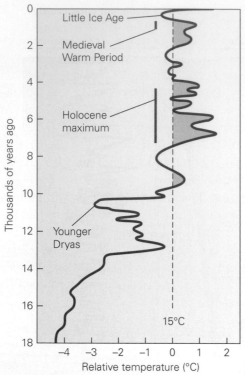

(a) There have been several temperature highs and lows during the Holocene.

At the end of the Little Ice Age, this tributary glacier in France reached the main valley floor.

Today, glaciers extend only partway down the side valleys.

(c) Glaciers advanced in Europe during the Little Ice Age.

(b) Vikings settled in Greenland during the Medieval Warm Period, when the climate was warm enough for agriculture.

happen if sea temperature warmed. This could cause sudden melting of a large amount of the methane hydrate that crystallized in sediment on the seafloor, releasing CH_4 to the atmosphere to cause warming. Sudden melting of permafrost could have a similar effect, for the resulting rot of organic material preserved in the permafrost produces CH_4. Algal blooms and reforestation absorb CO_2 and, therefore, could cause cooling.

TAKE-HOME MESSAGE

Geologic study shows that, over the long term, climate on the Earth alternates between greenhouse and icehouse conditions. Factors including life evolution, plate movements, and plate interactions likely cause this long-term change. Study of the near-term record shows that short-term changes in climate can also take place. These changes may relate to the Milankovitch cycle, to sudden release of methane, or to volcanic activity.

QUICK QUESTION How does continental drift contribute to climate change?

19.6 Contemporary Global Warming

OBSERVED CHANGES IN CO₂ AND CH₄

We've seen that greenhouse gases, most notably CO_2 and CH_4, play a major role in the regulation of the Earth's surface temperatures. Without these gases, the Earth could not be a home for life. Both gases cycle through various biogeochemical reservoirs of the Earth System, and the fluxes among reservoirs influence the amount in any given reservoir at any given time. For most of geologic time, the concentration of these gases in the atmosphere was governed by natural processes—volcanic eruptions, life-evolution events, forest fires, sea-level rise or fall, orogenic uplift and related weathering, warming and cooling due to the Milankovitch cycle, changes in solar activity or cosmic-ray flux, and even meteor impact. But beginning around 8,000 years ago, human activities began to modify the environment significantly, first with the invention of agriculture and animal husbandry, and then, during the past two centuries, with the spread of industrialization and new construction techniques.

How does agriculture affect the carbon cycle? In many regions, farming replaces high-biomass forests with low-biomass crops; in other words, deforestation returns carbon to the air. Farming also tends to extract carbon that had been incorporated in soil and puts it back into the air. Planting rice in soggy paddies produces methane, as vegetation in the bogs decays. Husbandry replaces forest with low-biomass grazing land, and animals produce methane as they digest their food.

Burning fossil fuels oxidizes vast quantities of carbon that took millions of years to accumulate and had been stored underground for millions of years more in oil, gas, or coal to produce CO_2 that mixes into the atmosphere. Heating calcite ($CaCO_3$) to produce the lime (CaO) of concrete takes carbon that had been locked in limestone and produces CO_2 that also mixes into the air.

FIGURE 19.14 Changes in carbon dioxide (CO_2) and methane (CH_4) concentrations in the atmosphere over time.

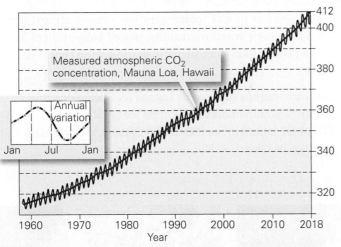

(a) Monthly measurements begin in 1960. There is an annual cycle, related to seasons, but the overall increase is clear.

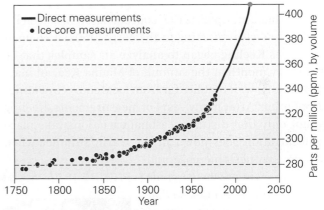

(b) Since the industrial revolution, CO_2 concentrations have steadily increased.

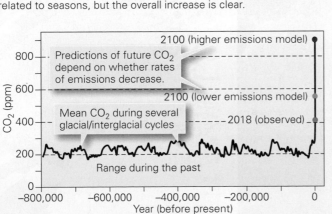

(c) In studies of ice cores from glaciers, researchers find that CO_2 concentrations varied between 180 and 300 ppm throughout glacial advances and retreats. The current value is far above this range.

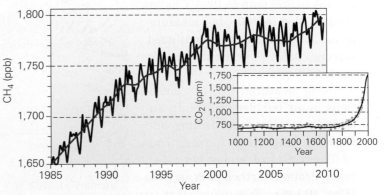

(d) Concentrations of methane, another greenhouse gas, have been increasing, too.

It may seem strange that human society has such a significant effect on the Earth's greenhouse gases. But comparing the quantity of greenhouse gases produced by humans to those produced by volcanoes demonstrates the significance of these gases. Specifically, researchers estimate that all volcanic eruptions together in a given year—including submarine and subaerial eruptions—emit about 0.15 to 0.26 billion tons of CO_2. By comparison, human activities emit about 35 billion tons of CO_2 every year, about 135 times as much. This means that people now produce more CO_2 in three days than do all of the volcanoes on the Earth in a typical year. Studies of carbon isotopes also indicate that the carbon comes from organic, not volcanic sources. We are transferring huge amounts of CO_2 from the surface and subsurface reservoirs of the Earth System, and are sending it into the atmosphere.

Of the anthropogenic CO_2 that humans send into the atmosphere, 85% comes from burning fossil fuels and producing cement, while the remaining 15% is a consequence of deforestation. About 40% to 50% of this gas dissolves in the ocean, reacts with rocks during chemical weathering, or gets incorporated into plants during photosynthesis. The remainder stays in the atmosphere and mixes thoroughly with other gases.

Evidence that atmospheric CO_2 concentration has been increasing became clear starting in the early 1960s. A chemist named Charles Keeling began to analyze air samples that he collected every month at the summit of Mauna Kea, an inactive volcano in Hawaii that lies far from local sources of atmospheric pollution. After many years of measurements, Keeling showed not only that distinct seasonal variations take place (CO_2 concentration goes down in the warm summer when rates of photosynthesis increase, and it goes up in the winter when organic matter dies and decays), but also that the average annual concentration of CO_2 was rising steadily (Fig. 19.14a)! Specifically, the average yearly CO_2 concentration went from 320 ppm in 1965 to 360 ppm in 1995. Charles Keeling died in 2005, but measurements at Mauna Kea have continued. During 2018, CO_2 concentration exceeded 412 ppm.

Using measured CO_2 concentrations in air bubbles trapped in ancient ice obtained by drilling into glaciers, researchers have extended the record of atmospheric CO_2 concentration further back in time (Fig. 19.14b). They found that in 1750, CO_2 concentration was only 280 ppm, and thus, atmospheric CO_2 concentration has increased by about 40% since the beginning of the industrial revolution. Studies of gas bubbles trapped in the glaciers of Antarctica allow researchers to extend the record of atmospheric CO_2 concentrations back almost 800,000 years into the past. This record demonstrates that during the alternating glacial and interglacial periods of the Late Pleistocene, CO_2 concentrations varied between about 180 and 300 ppm (Fig. 19.14c). Thus, the increases since the beginning of the industrial revolution are beyond the range of natural fluctuations that occurred during the last 800,000 years. Methane (CH_4) concentrations have also risen measurably over the past two centuries (Fig. 19.14d).

Have CO_2 concentrations ever been higher than they are today? Yes. For example, the CO_2 concentration during the Eocene climatic optimum (49 to 56 Ma) was over 1,000 ppm, about 2.5 times what it is today. But whether rates of change have ever been comparable to what researchers have measured in the past 200 years remains uncertain.

OBSERVATIONS OF CLIMATE CHANGE

The fundamental principle of the greenhouse effect requires that an increase in CO_2 concentration causes atmospheric warming. Is such warming taking place? Researchers have published thousands of observations indicating that it is. For example:

> Large ice shelves, such as the Larsen Ice Shelf along the Antarctic Peninsula, and the Ayles Ice Shelf along Ellesmere Island in northernmost Canada, are breaking up (Fig. 19.15a).

FIGURE 19.15 Visible examples of changes in Earth's ice cover.

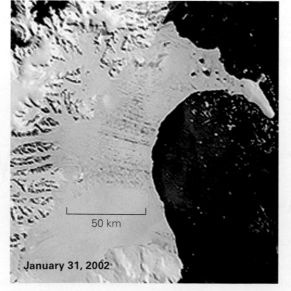
January 31, 2002

March 7, 2002

(a) In 2002, a large portion of the Larsen Ice Shelf of Antarctica disintegrated over the course of a month.

50 km

> The area covered by sea ice in the Arctic Ocean has decreased substantially. Some estimates place the rate of ice-cover loss at about 3% per decade. This observation leads to the prediction that it may be possible to sail across the Arctic Ocean within decades **(Fig.19.15b)**.

The average age of sea ice has also decreased significantly.

> The Greenland ice sheet has been melting at an accelerating pace **(Fig. 19.15c)**. Studies suggest that the rate of ice loss has increased from 90 to 220 km³ per year in the

FIGURE 19.15 (*continued*)

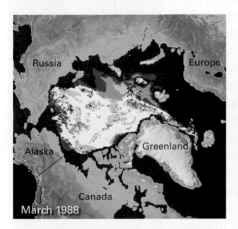

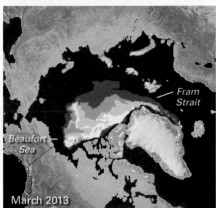

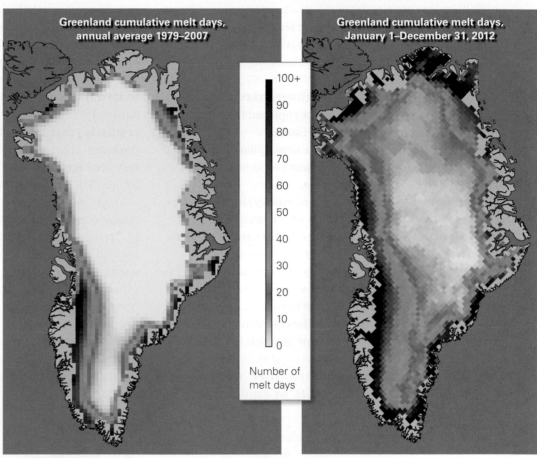

(b) The coverage and age of sea ice in the Arctic Ocean has decreased substantially during the past 25 years.

(c) The number of days in a year during which melting of the Greenland ice sheet takes place has increased.

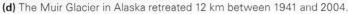

(d) The Muir Glacier in Alaska retreated 12 km between 1941 and 2004.

last 10 years and that, in places, the sheet is thinning by about 1 m per year.

> Valley glaciers worldwide have been retreating rapidly, so that areas that were once ice covered are now bare. The change is truly dramatic in many locations **(Fig. 19.15d)**. Of the 150 glaciers that were in Glacier National Park, Montana, when the park was founded in 1910, fewer than 25 remain, and according to models, the last glaciers there will disappear within 30 years.

> Including both the melting of polar ice sheets and the melting of mountain glaciers, geologists estimate that, worldwide, glacial volumes are diminishing at a rate of about 400 km³ every year.

> The area of permafrost in high latitudes has decreased substantially, and melt ponds have replaced frozen land across Siberia, Alaska, and arctic Canada.

> Average annual water vapor in the atmosphere has been increasing due to evaporation of warmer seas, possibly leading to more severe hurricanes.

> Biological phenomena that are sensitive to climate are being disrupted. For example, the time when sap in the maple trees of the northeastern United States starts to flow has changed; the mosquito line (the elevation at which mosquitoes can survive) has risen substantially; plant hardiness zones in North America have migrated northward **(Fig. 19.16)**; and the average weight of polar bears has been decreasing, because the bears can no longer walk over pack ice to reach their hunting grounds in the sea.

Directly measuring temperatures over time can be challenging, because weather varies so significantly with location, even on a daily basis. To determine average temperatures, researchers must use statistical methods to analyze temperature measurements obtained at recording stations around the world. Results of such analyses indicate that warming is happening. They show that global mean atmospheric temperature has risen by almost 1°C during the last century and is higher now than it has been at any time during the past 2,000 years **(Fig. 19.17)**. Although this change may seem relatively small, the total temperature change between the last ice age and now was only 3° to 5°C. A small change in average temperature may have major consequences. Significantly, temperature change varies around the world—some regions appear to be warming more than others, and some areas have been cooling. Because of the increase in atmospheric temperature, average measured values of near-surface ocean-water temperatures have also been rising **(Fig. 19.18)**.

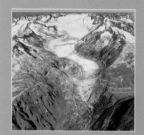
FIGURE 19.16 Plant hardiness zones have been migrating northward.

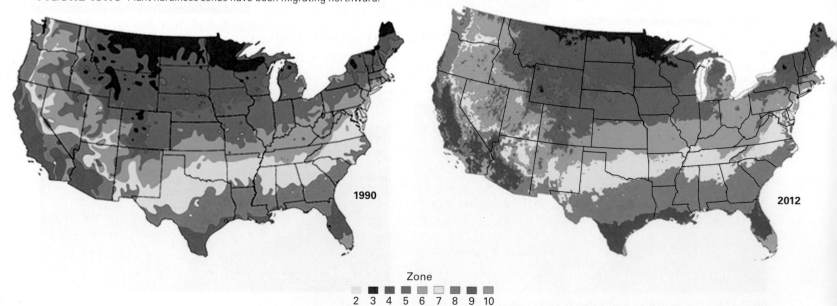

1990

2012

Zone

2 3 4 5 6 7 8 9 10

FIGURE 19.17 Measurements of global warming and global temperature anomalies.

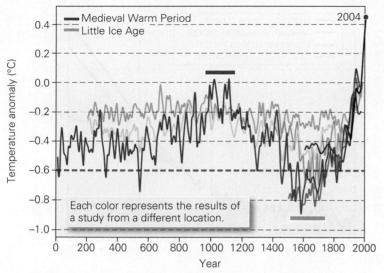

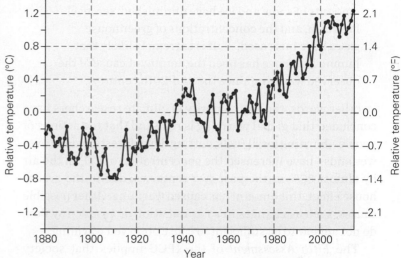

(a) Reconstructions of temperature during the past 2,000 years. Each color represents a published estimate from a research team. The black dashed line is the average of direct measurements.

(b) Graph showing the change in annual global average temperature (combined land and ocean) since 1880 relative to the average for the 20th century.

INTERPRETATIONS AND POTENTIAL CONSEQUENCES OF CLIMATE CHANGE

Climate data are not easy to obtain or interpret. In an attempt to make sense of overwhelming volumes of sometimes contradictory data and interpretations relevant to climate change, in 1988 a group of leading scientists founded the *Intergovernmental Panel on Climate Change* (IPCC), whose purpose is to evaluate published climate studies from a broad perspective and provide assessments of conclusions from the studies. These assessments have societal significance because climate change can have major

implications for society and for global policy decisions. The IPCC, sponsored by the World Meteorological Organization and the United Nations, summarizes its conclusions in a report that is revised and published every 5 years. The language describing the likelihood that global warming is happening, and that humans have contributed significantly to causing it, has become progressively less tentative in successive editions of the report. The *Fifth Assessment Report*, published in 2014, states,

> Warming of the climate system is unequivocal, and since the 1950s, many of the observed changes

FIGURE 19.18 The global annual average temperature of shallow ocean water relative to the average temperature between 1961 and 2006.

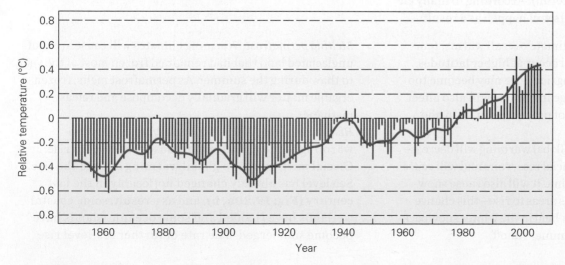

are unprecedented over decades to millennia. The atmosphere and ocean have warmed, the amounts of snow and ice have diminished, sea level has risen, and the concentrations of greenhouse gases have increased.... It is extremely likely that human influence has been the dominant cause of the observed warming since the mid-20th century.

In other words, the vast majority of climate researchers have concluded that global warming is real, and that the actions of people—burning fossil fuels, cutting down forests, paving over wetlands—have increased the concentration of CO_2 in the air by almost 50% in the past two centuries; due to the greenhouse effect, this amount has caused warming. Other possible causes, such as changes in solar radiation or cosmic-ray flux, do not appear to be sufficient to cause observed warming.

The *Fifth Assessment* of the IPCC implies that society faces the challenge of either slowing climate change or dealing with its consequences. The effects of global warming over the coming decades to centuries remain the subject of intense debate because predictions depend on computer models, and not all researchers agree on how to construct or interpret these models. The role of clouds in climate change, for example, remains poorly understood and inadequately addressed. But newer analyses, running on faster computers, provide increasingly reliable constraints on future trends.

In a worst-case scenario, computer models predict that global warming will continue into the future at the present rate, so that by 2050—within the lifetime of many readers of this book—the average annual temperature will have increased in some parts of the world by 1.5°C to 2°C. At these rates, by the end of the century, temperatures could be almost 4°C warmer **(Fig. 19.19a)**, and by 2150, global temperatures may be 5°C to 11°C warmer than at present—the warmest since the Eocene, 40 million years ago. Models predict that warming will not be the same everywhere—the greatest impact will be in the Arctic **(Fig. 19.19b)**. According to many climate models, the following events will happen as a result:

> *Shift in climate belts:* As the climate overall warms, temperate and desert regions will occur at higher latitudes. Thus, regions that now host agriculture may become too dry to be farmable. The change in climate will also affect the amount and distribution of precipitation (rain and snow) that falls.

> *Ice retreat and snow-line rise:* Global warming will decrease the volume of water held in glacial reservoirs and will cause areas covered by sea ice to shrink. It will also cause snow-line elevations in mountainous areas to rise—this change will affect the amount of water held in the winter snowpack and therefore will decrease summer runoff.

FIGURE 19.19 Model predictions of future global warming.

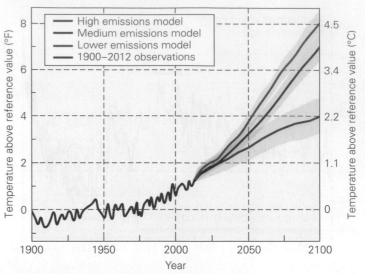

(a) Model calculations of global warming. Different models assume different rates of CO_2 emissions. All suggest significant global temperature increase by 2100.

2070–2100 prediction vs. 1960–1990 average

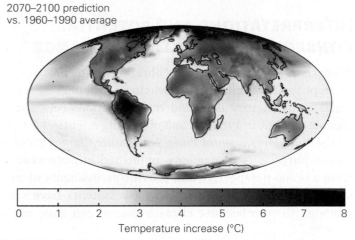

(b) Global warming does not mean that all locations warm by the same amount. This map shows a model prediction of how temperature increases may vary with location for the time period 2070–2100.

> *Melting permafrost:* Warming climates will cause areas of unglaciated land that had remained frozen most of the year to thaw during the summer. As permafrost melts, frozen organic matter will gradually decompose and release methane.

> *Rise in sea level:* Water expands when heated, so warming of the global ocean causes sea level to rise. Glacial melting due to global warming adds to the rise. Sea level has already changed noticeably in the last century **(Fig. 19.20a, b)**, and as a result, some coastal wetlands have flooded and some oceanic islands have become submerged. The rate of further sea-level rise

FIGURE 19.20 Sea-level rise of the recent past and possible sea-level rise of the future.

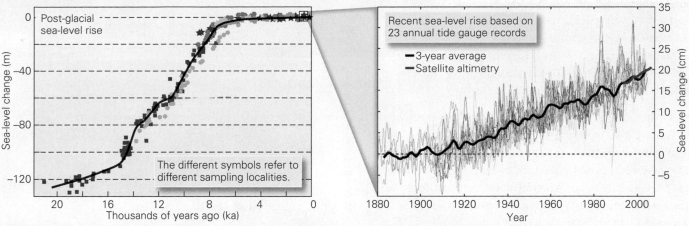

(a) Sea level rose rapidly after the last ice age due to melting of continental glaciers.

(b) Tide gauges document sea-level rise over the past 130 years.

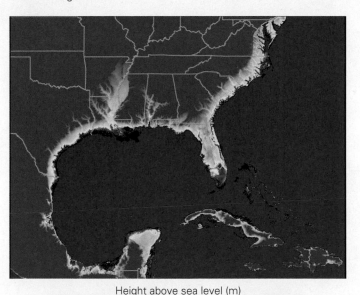

Height above sea level (m)

0 1 2 3 5 8 12 20 35 60 80

(c) The red and black areas of this map are so low that they could be flooded if sea level rises a few meters.

depends on the rate of global warming. Models suggest that by 2100, it may rise by an additional 20 to 60 cm **(Fig. 19.20c)**. If seas rise a meter or two, the result would be to inundate regions of the world where 20% of the human population currently lives (**Geology at a Glance**, pp. 632–633).

> *Stronger storms:* An increase in average ocean and atmospheric temperatures may lead to increased evaporation from the oceans. The additional moisture might nourish stronger hurricanes. Some climate models predict that global warming could change global weather patterns or cause more intense flooding or drought.

> *Increases in wildfires:* Warmer temperatures may lead to an increase in the frequency of wildfires because the moisture content of plants will be lower.

> *Interruption of thermohaline circulation:* Ocean currents play a major role in transferring heat across latitudes. According to some models, if global warming melts enough polar ice, the resulting freshwater will dilute surface ocean water at high latitudes. This water would not sink, so thermohaline circulation would be shut off (see Chapter 15), preventing the oceans from conveying heat.

The potential changes described above, along with the results of other studies that estimate the large economic cost of global warming, imply that the issue needs to be addressed seriously and soon. But what can be done? In 1997, at a summit meeting in Japan, 160 nations signed the *Kyoto Accord*, which declared that society must slow the input of greenhouse gases into the atmosphere. The goals of the Kyoto Accord were amplified, in modified form, after the 2015 U.N. Climate Change Conference, when 195 countries (all but three that attended the conference) signed the *Paris Climate Accord* in 2016. The United States withdrew from the accord in 2017. The accord encourages countries to reduce greenhouse-gas emissions and to begin to develop ways to accommodate the climate change now under way.

Society can decrease greenhouse-gas emissions by switching to power-generation systems that do not burn fossil fuels, and by increasing energy conservation. Emissions to the atmosphere might also be decreased by *carbon capture and sequestration* (*CCS*). This process forces the CO_2 produced by power plants down injection wells into a permeable rock layer underground. Needless to say, climate-change mitigation efforts involve political, economic, and lifestyle decisions, so they remain controversial in some countries.

Consequences of Sea-Level Change

Antarctica

If Antarctica suddenly became ice free, most of West Antarctica would be underwater.

A large volume of water resides in the glaciers of Antarctica and Greenland. If all this ice were to melt suddenly, these two continents would become mostly dry land. (Parts would remain submerged until glacial rebound takes place, because subsidence due to the weight of ice has pushed the crust's surface below sea level.) Transfer of water from the glacial reservoir back to the oceanic reservoir would cause a sea-level rise of about 65–70 m (215–230 feet). Even if only 10% of this rise were to happen, coastal cities would flood.

San Francisco (after 7-m rise)

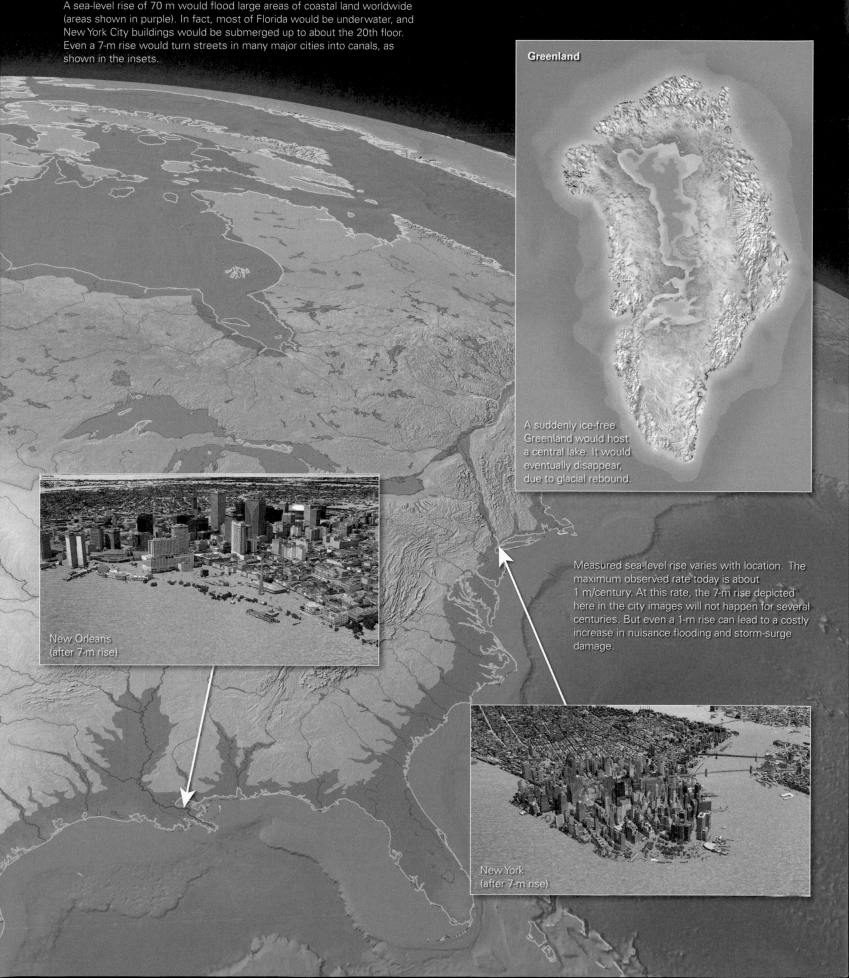

A sea-level rise of 70 m would flood large areas of coastal land worldwide (areas shown in purple). In fact, most of Florida would be underwater, and New York City buildings would be submerged up to about the 20th floor. Even a 7-m rise would turn streets in many major cities into canals, as shown in the insets.

Greenland

A suddenly ice-free Greenland would host a central lake. It would eventually disappear, due to glacial rebound.

New Orleans (after 7-m rise)

Measured sea-level rise varies with location. The maximum observed rate today is about 1 m/century. At this rate, the 7-m rise depicted here in the city images will not happen for several centuries. But even a 1-m rise can lead to a costly increase in nuisance flooding and storm-surge damage.

New York (after 7-m rise)

TAKE-HOME MESSAGE

Growing evidence indicates that greenhouse gases (CO_2 and CH_4) being introduced into the atmosphere by human activities are causing global warming of both the air and sea that might not otherwise have happened. This change is associated with glacial melting, shifting climate belts, sea-level rise, and other phenomena.

QUICK QUESTION How do researchers develop predictions of future climate change?

19.7 The Future of the Earth

In the geologic near term, the future of the Earth probably depends largely on human activities. Whether society achieves **sustainable growth** (an ability to prosper within the constraints of the Earth System) will depend on our own foresight and ingenuity. Projecting thousands of years into the future, we might well wonder if the Earth will return to ice-age conditions, with glaciers growing over major cities and the continental shelf becoming dry land, or if the ice age is over for good because of anthropogenic global warming. No one really knows for sure.

If we project millions of years into the future, it is clear that the map of the planet will change significantly because of the continuing activity of plate tectonics. For example, during the next 50 million years or so, the Atlantic Ocean will probably become bigger, the Pacific Ocean will shrink, and the western part of California will migrate northward. Eventually, Australia may crush against the southern margin of Asia, and the islands of Indonesia will be flattened in between.

Predicting the map of the Earth beyond that point is hard because we can't know where new subduction zones will develop. Perhaps subduction of the Pacific Plate will lead to the collision of the Americas with Asia to produce a new supercontinent ("Amasia"). A subduction zone will eventually form on one side (or both sides) of the Atlantic Ocean, and the ocean will be consumed. As a consequence, the eastern margin of the Americas will collide with the western margin of Europe or Africa. The sites of major cities—New York, Miami, Rio de Janeiro, Buenos Aires, and London—will be incorporated in a collisional mountain belt and probably will be subjected to metamorphism and igneous intrusion before being uplifted and eroded. Shallow seas may once again cover the interiors of continents and then later retreat, and glaciers may once again cover the continents—it happened in the past, so it could happen again!

And if the past is the key to the future, we *Homo sapiens* might not be around to watch our cities enter the rock cycle. Biological evolution may have introduced new species to the biosphere. Perhaps, 100 million years from now, the stratigraphic record of our time will be several centimeters of strata containing anomalous isotopes and unusual chemicals, trace fossils of concrete structures, and a record of widespread extinctions. In fact, some researchers suggest that we are now facing the *sixth extinction*, a mass extinction due to the current, ongoing, rapid destruction of global ecosystems. Because anthropogenic global change appears to be progressing so rapidly now, some researchers use the term *Anthropocene* in reference to the geologic time interval during which humans have had a profound impact on the Earth System.

And what of the end of the Earth? Geologic catastrophes resulting from asteroid and comet collisions will undoubtedly occur in the future as they have in the past. We can't predict when the next strike will come, but unless the object can be diverted, the Earth will experience another radical readjustment of surface conditions. It's not likely, however, that such collisions will destroy our planet. Rather, astronomers predict that the end of the Earth will occur some 5 billion years from now, when the Sun begins to run out of nuclear fuel. When this happens, outward-directed thermal pressure caused by fusion reactions will no longer be able to prevent the Sun from collapsing inward because of the immense gravitational pull of its mass. Were the Sun a few times larger than it is, the collapse would trigger a supernova explosion that would blast matter out into space to form a new nebula surrounding a black hole. But the Sun is not that large, so the thermal energy generated when its interior collapses inward will heat the gases of its outer layers sufficiently to cause them to expand. As a result, the Sun will become a *red giant*, a huge star whose radius will grow beyond the orbit of the Earth **(Fig. 19.21)**. Our planet will then vaporize, and its atoms will join an expanding ring of gas—the ultimate global change. When that happens, the atoms that once formed the Earth and all its inhabitants throughout geologic time might eventually be incorporated in a future solar system, where the cycle of planetary formation and evolution will begin anew.

TAKE-HOME MESSAGE

In the near term, the Earth's surface will be affected by decisions of human society. Over longer time scales, the map of the Earth will change due to plate interactions and sea-level change.

QUICK QUESTION What might happen to the atoms that make up the Earth long after the red-giant stage of Solar System evolution?

FIGURE 19.21 In about 5 billion years, the Sun will become a red giant.

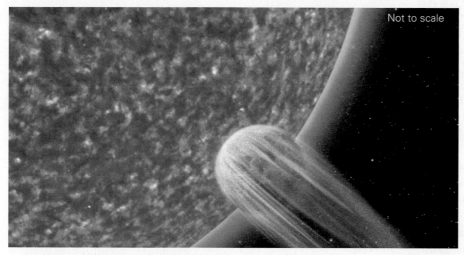

Not to scale

(a) As the surface of the red giant approaches the Earth, the planet will evaporate like a giant comet.

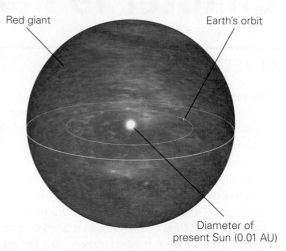

Red giant   Earth's orbit

Diameter of present Sun (0.01 AU)

(b) At full size, the red giant formed from the Sun may have a radius greater than that of the Earth's orbit, so the Earth's atoms will become part of the star.

ANOTHER VIEW The delta plain of the Yangtze River, near Shanghai, was once a landscape of marshes, shrubs, and meandering distributaries. Now it hosts rice paddies, housing, canals, and tree plantations, a manifestation of society's impact on the landscape.

Chapter 19 Review

CHAPTER SUMMARY

> We refer to the global interconnecting web of physical and biological phenomena on our planet as the Earth System. Global change involves the transformations or modifications of physical and biological components of the Earth System through time.

> Unidirectional change results in transformations that never repeat, whereas cyclic change involves repetition of the same steps over and over, although the end results may not always be the same.

> Examples of unidirectional change include the gradual evolution of the solid Earth from a homogeneous collection of planetesimals to a layered planet, the formation of the oceans, the gradual change in the composition of the atmosphere, and the evolution of life.

> Mass extinction, a catastrophic change in biodiversity, may be caused by the impact of a comet or asteroid or by intense volcanic activity.

> Examples of physical cycles that take place on the Earth include the supercontinent cycle, the sea-level cycle, and the rock cycle.

> A biogeochemical cycle involves the passage of a chemical among nonliving and living reservoirs. Examples include the hydrologic cycle and the carbon cycle. Global change occurs when factors change the relative proportions of the chemical in different reservoirs.

> During the last two centuries, humans have changed landscapes; modified ecosystems; and added pollutants to the land, air, and water at rates faster than the Earth System can process.

> Studies of long-term climate change show that, at times in the past, the Earth experienced greenhouse (warmer) periods; at other times, there were icehouse (cooler) periods. Factors leading to long-term climate change include the positions of continents, volcanic activity, the uplift of land, and the addition or removal of CO_2, an important greenhouse gas.

> Short-term climate change can be seen in the near-term record of the last million years and can be studied through micropaleontology, oxygen-isotope ratios, bubbles in ice, and growth rings in trees.

> During only the past 15,000 years, we see that the climate has warmed and cooled a few times. Causes of short-term climate change include the Milankovitch cycle, fluctuations in solar radiation and cosmic rays, changes in albedo, and changes in ocean currents.

> The addition of CO_2 and CH_4 to the atmosphere causes global warming, which could shift climate belts and lead to a rise in sea level. Sources for the increase include fossil fuel burning, cement production, and deforestation.

> In the future, in addition to climate change, the Earth will witness a continued rearrangement of continents resulting from plate tectonics and will likely suffer the impact of asteroids and comets. The end of the Earth may come in about 5 billion years, when the Sun runs out of fuel and becomes a red giant.

GUIDE TERMS

acid rain (p. 618)
albedo (p. 623)
biogeochemical cycle (p. 615)
carbon cycle (p. 615)
climate (p. 619)

differentiation (p. 612)
feedback mechanism (p. 621)
global change (p. 611)
global climate change (p. 619)
greenhouse effect (p. 621)

greenhouse gas (p. 621)
greenhouse (hothouse) period (p. 620)
icehouse period (p. 620)
mass-extinction event (p. 613)

paleoclimate (p. 619)
pollution (p. 617)
steady-state condition (p. 615)
sustainable growth (p. 634)
weather (p. 619)

 GEOTOURS *THIS CHAPTER'S GEOTOURS WORKSHEET (S) FEATURES QUESTIONS AND GOOGLE EARTH SITES ON:*

> Effects of global warming   > Effects of deforestation   > Water use in arid regions   > Preservation of natural habitats   > Sea-level change

REVIEW QUESTIONS

The letters following each Review Question refer to the corresponding Learning Objective from the Chapter Opener.

1. What does the term *Earth System* refer to? **(C)**

2. How have the Earth's interior, crust, and atmosphere changed since the planet first formed? **(A)**

3. Does the map of the Earth stay the same over time? Why? **(A)**

4. What processes control the rise and fall of sea level on the Earth? **(B)**

5. What events can cause catastrophic change in the Earth System? **(B)**

6. What are pollutants, and why do they pose a problem? What is the ozone hole, and why did it form? **(E)**

7. Describe the reservoirs that play a role in the carbon cycle and how carbon transfers among reservoirs. Identify the reservoirs on the diagram. **(C)**

8. How have humans changed the solid Earth? **(E)**

9. What are the possible causes of long-term climate change? What factors explain short-term climate change? **(D)**

10. Explain the role of greenhouse gases in regulating climate. **(D)**

11. Contrast icehouse and greenhouse conditions. Did icehouse conditions happen prior to the most recent ice age? **(D)**

12. How do paleoclimatologists study ancient climate change? **(D)**

13. What does this graph tell us about the change in CO_2 content since 1960? **(E)**

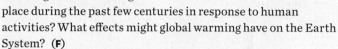

14. What evidence do researchers use to argue that global warming has been taking place during the past few centuries in response to human activities? What effects might global warming have on the Earth System? **(F)**

15. What areas of the Earth are expected to increase most in temperature? What are possible consequences of this temperature rise? **(F)**

16. What are some likely scenarios for the long-term future of the Earth? **(G)**

ON FURTHER THOUGHT

17. If global warming continues, how will the distribution of grain crops change? Might this change affect national economies, and if so, how? **(E)**

18. Currently, tropical rainforests are being cut down at a rate of 1.8% per year. At this rate, how many more years will the forests survive? **(E)**

ONLINE RESOURCES

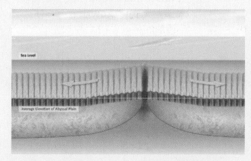

Animations This chapter features animations on global sea-level change, including glacial ice volume, thermal expansion, seafloor-spreading rate, and oceanic volcanic plateaus.

Videos This chapter features real-world videos on rising CO_2 levels, deforestation, ice sheet stratigraphy, and more.

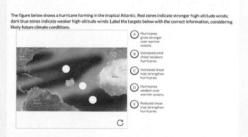

Smartwork5 This chapter features questions on the causes and impacts of long- and short-term climate change and the factors that control and impact regional climate conditions.

ADDITIONAL CHARTS

METRIC CONVERSION CHART

Length

1 kilometer (km) = 0.6214 mile (mi)
1 meter (m) = 1.094 yards = 3.281 feet
1 centimeter (cm) = 0.3937 inch
1 millimeter (mm) = 0.0394 inch
1 mile (mi) = 1.609 kilometers (km)
1 yard = 0.9144 meter (m)
1 foot = 0.3048 meter (m)
1 inch = 2.54 centimeters (cm)

Area

1 square kilometer (km^2) = 0.386 square mile (mi^2)
1 square meter (m^2) = 1.196 square yards (yd^2)
= 10.764 square feet (ft^2)
1 square centimeter (cm^2) = 0.155 square inch (in^2)
1 square mile (mi^2) = 2.59 square kilometers (km^2)
1 square yard (yd^2) = 0.836 square meter (m^2)
1 square foot (ft^2) = 0.0929 square meter (m^2)
1 square inch (in^2) = 6.4516 square centimeters (cm^2)

Volume

1 cubic kilometer (km^3) = 0.24 cubic mile (mi^3)
1 cubic meter (m^3) = 264.2 gallons
= 35.314 cubic feet (ft^3)
1 liter (l) = 1.057 quarts
= 33.815 fluid ounces
1 cubic centimeter (cm^3) = 0.0610 cubic inch (in^3)
1 cubic mile (mi^3) = 4.168 cubic kilometers (km^3)
1 cubic yard (yd^3) = 0.7646 cubic meter (m^3)
1 cubic foot (ft^3) = 0.0283 cubic meter (m^3)
1 cubic inch (in^3) = 16.39 cubic centimeters (cm^3)

Mass

1 metric ton = 2,205 pounds
1 kilogram (kg) = 2.205 pounds
1 gram (g) = 0.03527 ounce
1 pound (lb) = 0.4536 kilogram (kg)
1 ounce (oz) = 28.35 grams (g)

Pressure

1 kilogram per
square centimeter (kg/cm^2)* = 0.96784 atmosphere (atm)
= 0.98066 bar
= 9.8067 × 10^4 pascals (Pa)
1 bar = 0.1 megapascals (Mpa)
= 1.0 × 10^5 pascals (Pa)
= 29.53 inches of mercury (in a barometer)
= 0.98692 atmosphere (atm)
= 1.02 kilograms per square centimeter (kg/cm^2)
1 pascal (Pa) = 1 kg/m/s^2
1 pound per square inch = 0.06895 bars
= 6.895 × 10^3 pascals (Pa)
= 0.0703 kilogram per square
centimeter

Temperature

To change from Fahrenheit (F) to Celsius (C):

$$°C = \frac{(°F - 32°)}{1.8}$$

To change from Celsius (C) to Fahrenheit (F):

$$°F = (°C \times 1.8) + 32°$$

To change from Celsius (C) to Kelvin (K):

$$K = °C + 273.15$$

To change from Fahrenheit (F) to Kelvin (K):

$$K = \frac{(°F - 32°)}{1.8} + 273.15$$

*Note: Because kilograms are a measure of mass whereas pounds are a unit of weight, pressure units incorporating kilograms assume a given gravitational constant (g) for Earth. In reality, the gravitational for Earth varies slightly with location.

Alkali metals

Transition elements (metals)

Nonmetals

Inert gases

Legend:
Symbol — He
Atomic number — 2
Name — Helium
Atomic weight — 4.002

1	2	3	4	5	6	7	8	9	10	11	12	13	14	15	16	17	18
H 1 Hydrogen 1.007																	He 2 Helium 4.002
Li 3 Lithium 6.941	Be 4 Beryllium 9.0121											B 5 Boron 10.811	C 6 Carbon 12.011	N 7 Nitrogen 14.006	O 8 Oxygen 15.999	F 9 Fluorine 18.998	Ne 10 Neon 20.179
Na 11 Sodium 22.989	Mg 12 Magnesium 24.305											Al 13 Aluminum 26.981	Si 14 Silicon 28.085	P 15 Phosphorus 30.973	S 16 Sulfur 32.066	Cl 17 Chlorine 35.452	Ar 18 Argon 39.948
K 19 Potassium 39.098	Ca 20 Calcium 40.078	Sc 21 Scandium 44.955	Ti 22 Titanium 47.88	V 23 Vanadium 50.941	Cr 24 Chromium 51.996	Mn 25 Manganese 54.938	Fe 26 Iron 55.847	Co 27 Cobalt 58.933	Ni 28 Nickel 58.693	Cu 29 Copper 63.546	Zn 30 Zinc 65.39	Ga 31 Gallium 69.723	Ge 32 Germanium 72.61	As 33 Arsenic 74.921	Se 34 Selenium 78.96	Br 35 Bromine 79.904	Kr 36 Krypton 83.80
Rb 37 Rubidium 85.467	Sr 38 Strontium 87.62	Y 39 Yttrium 88.905	Zr 40 Zirconium 91.224	Nb 41 Niobium 92.906	Mo 42 Molybdenum 95.94	Tc 43 Technetium 98.907	Ru 44 Ruthenium 101.07	Rh 45 Rhodium 102.905	Pd 46 Palladium 106.42	Ag 47 Silver 107.868	Cd 48 Cadmium 112.411	In 49 Indium 114.82	Sn 50 Tin 118.710	Sb 51 Antimony 121.757	Te 52 Tellurium 127.60	I 53 Iodine 126.904	Xe 54 Xenon 131.29
Cs 55 Cesium 132.905	Ba 56 Barium 137.327	La 57 Lanthanum 138.905	Hf 72 Hafnium 178.49	Ta 73 Tantalum 180.947	W 74 Tungsten 183.85	Re 75 Rhenium 186.207	Os 76 Osmium 190.2	Ir 77 Iridium 192.22	Pt 78 Platinum 195.08	Au 79 Gold 196.966	Hg 80 Mercury 200.59	Tl 81 Thallium 204.383	Pb 82 Lead 207.2	Bi 83 Bismuth 208.980	Po 84 Polonium 208.982	At 85 Astatine 209.987	Rn 86 Radon 222.017
Fr 87 Francium 223.019	Ra 88 Radium 226.025	Ac 89 Actinium 227.027															

Lanthanides / Actinides:

Ce 58 Cerium 140.115	Pr 59 Praseodymium 140.907	Nd 60 Neodymium 144.24	Pm 61 Promethium 144.912	Sm 62 Samarium 150.36	Eu 63 Europium 151.965	Gd 64 Gadolinium 15725	Tb 65 Terbium 158.925	Dy 66 Dysprosium 162.50	Ho 67 Holmium 164.930	Er 68 Erbium 167.26	Tm 69 Thulium 168.934	Yb 70 Ytterbium 173.04	Lu 71 Lutetium 174.967
Th 90 Thorium 232.038	Pa 91 Protactinium 231.035	U 92 Uranium 238.028	Np 93 Neptunium 237.048	Pu 94 Plutonium 244.064	Am 95 Americium 243.061	Cm 96 Curium 247.070	Bk 97 Berkelium 247.070	Cf 98 Californium 251.079	Es 99 Einsteinium 252.083	Fm 100 Fermium 257.095	Md 101 Mendelevium 258.10	No 102 Nobelium 259.100	Lr 103 Lawrencium 262.11

The modern periodic table of the elements. Each column groups elements with related properties. For example, inert gases are listed in the column on the right. Metals are found in the central and left parts of the chart.

GLOSSARY

a'a   A lava flow with a rubbly surface.

ablation   The removal of ice at the toe of a glacier by melting, sublimation (the evaporation of ice into water vapor), and/or calving.

absolute plate velocity   The movement of a plate relative to a fixed point in the mantle.

abyssal plain   A broad, relatively flat region of the ocean that lies at least 4.5 km below sea level.

accretionary prism   A wedge-shaped mass of sediment and rock scraped off the top of a downgoing plate and accreted onto the overriding plate at a convergent plate margin.

accretion disk   A pancake-shaped accumulation of matter that forms as a nebula and then transforms into a star and its associated planets; the bulbous central zone of the accretion disk becomes the star.

acid rain   Precipitation in which air pollutants react with water to make a weak acid that then falls from the sky.

active continental margin   A continental margin that coincides with a plate boundary.

active margin   A continental margin that is also a plate boundary.

active volcano   A volcano that has erupted within the past few centuries and will likely erupt again.

aerosols   Tiny solid particles or liquid droplets that remain suspended in the atmosphere for a long time.

aftershocks   The series of smaller earthquakes that follow a major earthquake.

agent of erosion   Natural entities that remove material from the Earth's surface, and transport it elsewhere; examples include rivers, glaciers, and the wind.

albedo   The reflectivity of a surface.

alluvial fan   A gently sloping apron of sediment dropped by an ephemeral stream at the base of a mountain in arid or semiarid regions.

alluvium   Sorted sediment deposited by a stream.

Alpine-Himalayan chain   The largest orogenic belt on the Earth today, formed by collisions of the former Gondwana continents with the southern margins of Europe and Asia.

angle of repose   The angle of the steepest slope that a pile of uncemented material can attain without collapsing from the pull of gravity.

angular unconformity   An unconformity in which the strata below were tilted or folded before the unconformity developed; strata below the unconformity therefore have a different tilt than strata above.

annual probability   The likelihood that a flood of a given size or larger will happen at a specified locality during any given year.

Anthropocene   A term used informally in reference to the past few thousand years, to emphasize that during this time, human society has modified the Earth System significantly.

anticline   A fold with an arch-like shape in which the limbs dip away from the hinge.

apparent polar-wander path   A path on the globe along which a magnetic pole appears to have wandered over time; in fact, the continents drift, while the magnetic pole stays fairly-fixed.

aquifer   Sediment or rock that transmits water easily.

aquitard   Sediment or rock that does not transmit water easily and therefore retards the motion of the water.

Archean Eon   The middle Precambrian eon (4.0–2.5 Ga).

Archimedes' principle   The mass of the water displaced by a block of material equals the mass of the whole block of material.

arête   A residual knife-edge ridge of rock that separates two adjacent cirques.

arkose   A clastic sedimentary rock containing both quartz and feldspar grains.

artesian spring   A location where the ground surface intersects a natural fracture (joint) that taps a confined aquifer in which the pressure can drive the water to the surface.

artesian well   A well in which water rises on its own.

artificial levee   A man-made retaining wall to hold back a river from flooding.

ash   *See* Volcanic ash.

assimilation   The process of magma contamination in which blocks of wall rock fall into a magma chamber and dissolve.

asteroid   One of the fragments of solid material, left over from planet formation or produced by collision of planetesimals, that resides between the orbits of Mars and Jupiter.

asthenosphere   The layer of the mantle that lies between 100–150 km and 350 km deep; the asthenosphere is relatively soft and can flow when acted on by force.

atom   The smallest piece of an element that has the properties of the element; it consists of a nucleus surrounded by an electron cloud.

atomic mass   The amount of matter in an atom; roughly, it is the sum of the number of protons plus the number of neutrons in the nucleus.

atomic number   The number of protons in the nucleus of a given element.

axial (plane) surface   The imaginary surface plane that encompasses the hinges of successive layers of a fold.

backwash   The gravity-driven flow of water back down the slope of a beach.

banded-iron formation (BIF) Iron-rich sedimentary layers consisting of alternating gray beds of iron oxide and red beds of iron-rich chert.

bar (1) A sheet or elongate lens or mound of alluvium; (2) a unit of air-pressure measurement approximately equal to 1 atm.

base level The lowest elevation a stream channel's floor can reach at a given locality.

basin A fold or depression shaped like a right-side-up bowl.

Basin and Range Province A broad, Cenozoic continental rift that has affected a portion of the western United States in Nevada, Utah, and Arizona; in this province, tilted fault blocks form ranges, and alluvium-filled valleys are basins.

batholith A vast composite, intrusive, igneous rock body up to several hundred km long and 100 km wide, formed by the intrusion of numerous plutons in the same region.

bathymetry Variation in depth.

beach A gently sloping fringe of sediment along the shore.

beach erosion The removal of beach sand caused by wave action and longshore currents.

beach profile A cross section illustrating the shape of a beach's surface.

bearing The compass heading of a line.

bed An individual layer of sediment or sedimentary rock in a deposit.

bedding Layering or stratification in sedimentary rocks.

bed load Large particles, such as sand, pebbles, or cobbles, that bounce or roll along a streambed.

bedrock Rock still attached to the Earth's crust.

Big Bang nucleosynthesis The formation of new atomic nuclei (mostly hydrogen and helium) during the Big Bang.

Big Bang theory A theory suggested by scientists in which a cataclysmic explosion represents the formation of the Universe; before this event, all matter and all energy were packed into one volume-less point.

biochemical sedimentary rock Sedimentary rock formed from material (such as shells) produced by living organisms.

biodiversity The number of different species that exist at a given time.

biofuel Gas or liquid fuel made from plant material (biomass). Examples of biofuel include alcohol (from fermented sugar), biodiesel from vegetable oil, and wood.

biogenic material Substances that originated within living organisms.

biogeochemical cycle The exchange of chemicals between living and nonliving reservoirs in the Earth System.

biomarker A molecule or set of molecules that could only have been produced by living organisms; their presence can be used for correlation and for determining the time at which life on the Earth began.

biomass The amount of organic material in a specified volume.

bioremediation The injection of oxygen and nutrients into a contaminated aquifer to foster the growth of bacteria that will ingest or break down contaminants.

biosphere The region of the Earth and atmosphere inhabited by life; this region stretches from a few km below the Earth's surface to a few km above.

black smoker The cloud of suspended minerals formed where hot water spews out of a vent along a mid-ocean ridge; the dissolved sulfide components of the hot water instantly precipitate when the water mixes with seawater and cools.

block Large, angular pyroclastic fragments consisting of volcanic rock, broken up during the eruption.

blowout A deep, bowl-like depression scoured out of desert terrain by a turbulent vortex of wind.

body waves Seismic waves that pass through the interior of the Earth.

bomb A stream-lined block of rock ejected by a volcano while still hot; it gets shaped as it flies through the air.

Bowen's reaction series The sequence in which different silicate minerals crystallize during the progressive cooling of a melt.

braided stream A sediment-choked stream consisting of entwined subchannels.

breccia Coarse sedimentary rock consisting of angular fragments; or rock broken into angular fragments by faulting.

brittle deformation The cracking and fracturing of a material subjected to stress.

burial metamorphism Metamorphism due only to the consequences of very deep burial.

butte A medium-sized, flat-topped hill in an arid region.

caldera A large circular depression with steep walls and a fairly-flat floor, formed after an eruption as the center of the volcano collapses into the drained magma chamber below.

Cambrian explosion The remarkable diversification of life, indicated by the fossil record, that occurred at the beginning of the Cambrian Period.

canyon A trough or valley with steeply sloping walls, cut into the land by a stream.

capacity (of a stream) The total quantity of sediment a stream can carry.

capillary fringe The thin subsurface layer in which water molecules seep up from the water table by capillary action to fill pores.

carbonate rocks Rocks containing calcite and/or dolomite.

carbon cycle the progressive transfer of carbon, from reservoir to reservoir in the Earth System.

cast Sediment that preserves the shape of a shell it once filled before the shell dissolved or mechanically weathered away.

cave An opening in the side of a cliff, or completely underground, which remains dark all day long; larger underground caves are also known as caverns.

cement Mineral material that precipitates from water and fills the spaces between grains, holding the grains together.

cementation The phase of lithification in which cement, consisting of minerals that precipitate from groundwater, partially or completely fills the spaces between clasts and attaches each grain to its neighbor.

chain reaction A self-perpetuating process in a nuclear reaction, whereby neutrons released during the fission trigger more fission.

channel A trough dug into the ground surface by flowing water.

chemical sedimentary rocks Sedimentary rocks made up of minerals that precipitate directly from water solution.

chemical weathering The process in which chemical reactions alter or destroy minerals when rock comes in contact with water solutions and/or air.

chimney (1) A conduit in a magma chamber in the shape of a long vertical pipe through which magma rises and erupts at the surface; (2) an isolated column of rock in an arid region.

chron The time interval between successive magnetic reversals.

cinder cone A subaerial volcano consisting of a cone-shaped pile of tephra whose slope approaches the angle of repose for tephra.

cirque A bowl-shaped depression carved by a glacier on the side of a mountain.

clast A fragment or grain produced by the physical or chemical weathering of a pre-existing rock.

clastic (detrital) sedimentary rock Sedimentary rock consisting of cemented-together detritus derived from the weathering of preexisting rock.

cliff (scarp) retreat The change in the position of a cliff face caused by erosion.

climate The average weather conditions, along with the range of conditions, of a region over a year.

closure temperature The temperature at which ions can no longer move in or out of a mineral quickly; when a rock drops below the closure temperature for a mineral, the mineral can be used for isotopic dating.

coal A black, organic rock consisting of greater than 50% carbon; it forms from the buried and altered remains of plant material.

coal gasification The process of producing relatively clean-burning gases from solid coal.

coal rank A measurement of the carbon content of coal; higher-rank coal forms at higher temperatures.

coal reserve The quantities of discovered, but not yet mined, coal in sedimentary rock of the continents.

coastal fjord A glacially carved valley that became submerged when sea level rose, so it is now an elongate bay of the sea.

coastal plain Low-relief regions of land adjacent to the coast.

coastal wetland A flat-lying coastal area that floods during high tide and drains during low tide, and hosts salt-resistant plants.

color The characteristic of a material due to the spectrum of light emitted or reflected by the material, as perceived by eyes or instruments.

columnar jointing A type of fracturing that yields roughly hexagonal columns of basalt; columnar joints form when a dike, sill, or lava flow cools.

comet A ball of ice and dust, probably remaining from the formation of the Solar System, that orbits the Sun.

compaction The phase of lithification in which the pressure of the overburden on the buried rock squeezes out water and air that was trapped between clasts, and the clasts press tightly together.

competence (of a stream) The maximum particle size a stream can carry.

compression A push or squeezing felt by a body.

compressional waves Waves in which particles of material move back and forth parallel to the direction in which the wave itself moves.

cone of depression The downward-pointing, cone-shaped surface of the water table in a location where the water table is experiencing drawdown because of pumping at a well.

conglomerate Very coarse-grained sedimentary rock consisting of rounded clasts.

contact metamorphism *See* Thermal metamorphism.

contaminant plume A cloud of contaminated groundwater that moves away from the source of the contamination.

continental crust The crust beneath the continents.

continental drift The hypothesis that continents have moved and are still moving slowly across the Earth's surface.

continental glacier A vast sheet of ice that spreads over thousands of square km of continental crust.

continental rift A linear belt along which continental lithosphere stretches and pulls apart.

continental rifting The process by which a continent stretches and splits along a belt; if it is successful, rifting separates a larger continent into two smaller continents separated by a divergent boundary.

continental shelf A broad, shallowly submerged fringe of a continent; ocean-water depth over the continental shelf is generally less than 200 meters; the widest continental shelves occur over passive margins.

conventional reserve A volume of oil or gas in a reservoir rock within a trap; it can be pumped relatively easily from the reservoir rock.

convergent margin *See* Convergent plate boundary.

convergent (consuming) boundary A boundary at which two plates move toward each other so that one plate sinks (subducts) beneath the other; only oceanic lithosphere can subduct.

coral reef A mound of coral and coral debris forming a region of shallow water.

core The dense, iron-rich center of the Earth.

core-mantle boundary An interface 2,900 km below the Earth's surface separating the mantle and core.

Coriolis effect The deflection of objects, winds, and currents on the surface of the Earth owing to the planet's rotation.

correlation The process of defining the age relations between the strata at one locality and the strata at another.

cosmology The study of the overall structure of the Universe.

crater (1) A circular depression at the top of a volcanic mound; (2) a depression formed by the impact of a meteorite.

craton A long-lived block of durable continental crust commonly found in the stable interior of a continent.

cratonic platform A province in the interior of a continent in which Phanerozoic strata bury most of the underlying Precambrian rock.

creep The gradual downslope movement of regolith.

crevasse A large crack that develops by brittle deformation in the top 60 m of a glacier.

cross bed Internal laminations in a bed, inclined at an angle to the main bedding; cross beds are a relict of the slip face of dunes or ripples.

crust   The rock that makes up the outermost layer of the Earth.

crustal root   Low-density crustal rock that protrudes downward beneath a mountain range.

cryosphere   The realm of the Earth System that consists of frozen water (ice); it includes glaciers, sea ice, and permafrost.

crystal   A single, continuous piece of a mineral bounded by flat surfaces that formed naturally as the mineral grew.

crystal face   The flat surfaces of a crystal, formed during the crystal's growth.

crystal habit   The general shape of a crystal or cluster of crystals that grew unimpeded.

crystal structure   The arrangement of atoms in a crystal.

crystalline igneous rock   An igneous rock that consists of minerals that grew when a melt solidified, and eventually interlock like pieces of a jigsaw puzzle.

crystalline material   A substance in which atoms are arranged in a crystalline lattice.

current   (1) A well-defined stream of ocean water; (2) the moving flow of water in a stream.

cycle   A series of interrelated events or steps that occur in succession and can be repeated, perhaps indefinitely.

Darcy's (disappearing stream) law   A mathematical equation stating that a volume of water, passing through a specified area of material at a given time, depends on the material's permeability and hydraulic gradient.

daughter atom (isotope)   The decay product of radioactive decay.

debris fall   A mass movement event in which fragments of various sizes, including large chunks and fine sediment, free fall down a slope.

debris flow (debris slide)   A downslope movement of mud mixed with larger rock fragments.

debris slide   A sudden downslope movement of material consisting only of regolith.

deflation   The process of lowering the land surface by wind abrasion.

delta   A wedge of sediment formed at a river mouth when the running water of the stream enters standing water, the current slows, the stream loses competence, and sediment settles out.

deposition   The process by which sediment settles out of a transporting medium.

depositional environment   A setting in which sediments accumulate; its character (fluvial, deltaic, reef, glacial, etc.) reflects local conditions.

depositional landform   A landform resulting from the deposition of sediment where the medium carrying the sediment evaporates, slows down, or melts.

desert   A region so arid that it contains no permanent streams, except for those that bring water in from elsewhere, and has very sparse vegetation cover.

desertification   The process of transforming nondesert areas into desert.

desert pavement   A mosaic-like stone surface forming the ground in a desert.

desert varnish   A dark, rusty-brown coating of iron oxide and magnesium oxide that accumulates on the surface of the rock.

diagenesis   All of the physical, chemical, and biological processes that transform sediment into sedimentary rock and that alter the rock after the rock has formed.

differential stress   A condition causing a material to experience a push or pull in one direction of a greater magnitude than the push or pull in another direction; in some cases, differential stress can result in shearing.

differentiation (of a planet)   A process early in a planet's history during which dense iron alloy melted and sank downward to form the core, leaving less-dense mantle behind.

dike   A tabular (wall-shaped) intrusion of rock that cuts across the layering of country rock.

dimension stone   An intact block of granite or marble to be used for architectural purposes.

dip   The angle of a plane's slope as measured in a vertical plane perpendicular to the strike.

dipole   A magnetic field with a north and south pole, like that of a bar magnet.

directional drilling   The process of controlling the trajectory of a drill bit to make sure that the drill hole goes exactly where desired.

discharge   The volume of water in a conduit or channel passing a point in 1 second.

discharge area   A location where groundwater flows back up to the surface and may emerge at springs.

disconformity   An unconformity parallel to the two sedimentary sequences it separates.

displacement (offset)   The amount of movement or slip across a fault plane.

dissolved load   Ions dissolved in a stream's water.

distillation column   A vertical pipe in which crude oil is separated into several components.

distributaries   The fan of small streams formed where a river spreads out over its delta.

divergent boundary   A boundary at which two lithosphere plates move apart from each other; they are marked by mid-ocean ridges.

domain   In the context of discussing the classification of organisms, it is the highest rank in a taxonomic hierarchy; there are three domains (Bacteria; Archaea; Eukarya).

dome   Folded or arched layers with the shape of an overturned bowl.

dormant volcano   A volcano that has not erupted for hundreds to thousands of years but does have the potential to erupt again in the future.

downslope movement   The tumbling or sliding of rock and sediment from higher elevations to lower ones.

drainage divide   A highland or ridge that separates one watershed from another.

drainage network (basin)   An array of interconnecting streams that together drain an area.

drainage reversal   When the overall direction of flow in a drainage network becomes the opposite of what it once had been.

drilling mud   A slurry of water mixed with clay that oil drillers use to cool a drill bit and flush rock cuttings up and out of the hole.

drumlin A streamlined, elongate hill formed when a glacier overrides glacial till.

dune A pile of sand generally formed by deposition from the wind.

dynamic metamorphism Metamorphism that occurs as a consequence of shearing alone, with no change in temperature or pressure.

dynamo A power plant generator in which water or wind power spins an electrical conductor around a permanent magnet.

dynamothermal metamorphism Metamorphism that involves heat, pressure, and shearing.

Earth materials A general term for the great variety of substances that make up this planet.

earthquake A vibration caused by the sudden breaking or frictional sliding of rock in the Earth.

earthquake early warning system A communications network that provides an alert within microseconds after the first earthquake waves arrive at a seismograph near the epicenter, but before damaging vibrations reach population centers.

earthquake intensity A representation of the strength of an earthquake, based on the amount of damage due to the event and on people's perception of ground shaking during the event; intensity is measured with the Mercalli scale; intensity decreases with increasing distance from the epicenter.

earthquake magnitude A representation of the energy released by an earthquake, as indicated by the amplitude of specific seismic waves as they would be recorded by a seismometer at a set distance from the epicenter.

EarthScope A study, funded by the US National Science Foundation and conducted between about 2004 and 2019, that involved the installation of large arrays of seismometers and other instruments to study the internal structure of the Earth and, in places, deformation of the land surface of the United States.

Earth System The global interconnecting web of physical and biological phenomena involving the solid Earth, the hydrosphere, and the atmosphere.

ecliptic The plane defined by a planet's orbit.

effusive eruption An eruption that yields mostly lava, not ash.

elastic behavior A response of a material to stress, during which the material changes shape. The amount of change depends on the magnitude of stress, and the change disappears when stress is removed. The behavior occurs when chemical bonds bend or stretch, but do not break.

elastic-rebound theory The concept that earthquakes happen because stress builds up, causing rock adjacent to a fault to bend elastically until breaking and slip on a fault occurs; the slip relaxes the elastic bending and decreases stress.

element A material consisting entirely of one kind of atom; elements cannot be subdivided or changed by chemical reactions.

energy The capacity to do work.

energy density The amount of energy contained by a unit volume of material.

eon The largest subdivision of geologic time.

epeirogeny An event of epeirogenic movement; the term is usually used in reference to the formation of broad mid-continent domes and basins.

ephemeral (intermittent) stream A stream whose bed lies above the water table, so that the stream flows only when the rate at which water enters the stream from rainfall or meltwater exceeds the rate at which water infiltrates the ground below.

epicenter The point on the surface of the Earth directly above the focus of an earthquake.

epoch An interval of geologic time representing the largest subdivision of a period.

equant A term for a grain that has the same dimensions in all directions.

equilibrium line (of a glacier) The boundary between the zone of accumulation and the zone of ablation.

equipotential surface In the context of discussing groundwater, it is an imaginary surface that groundwater will rise to in a stand pipe.

era An interval of geologic time representing the largest subdivision of the Phanerozoic Eon.

erosion The grinding away and removal of the Earth's surface materials by moving water, air, or ice.

erosional landform A landform that results from the breakdown and removal of rock or sediment.

erratic A boulder or cobble that was picked up by a glacier and deposited hundreds of kilometers away from the outcrop from which it detached.

eruption The release of lava and/or pyroclastic debris from a volcanic vent.

esker A ridge of sorted sand and gravel that snakes across a ground moraine; the sediment of an esker was deposited in subglacial meltwater tunnels.

estuary An inlet in which seawater and river water mix; created when a coastal valley is flooded because of either rising sea level or land subsidence.

evaporate To change state from liquid to vapor.

evaporite Thick salt deposits that form as a consequence of precipitation from saline water.

evapotranspiration The sum of evaporation from bodies of water and the ground surface and transpiration from plants and animals.

evolution *See* Theory of Evolution by Natural Selection.

exhumation The process (involving uplift and erosion) that returns deeply buried rocks to the surface.

expanding Universe theory The theory that the whole Universe must be expanding because galaxies in every direction seem to be moving away from us.

explosive eruptions Violent volcanic eruptions that produce clouds and avalanches of pyroclastic debris.

external energy In the context of the Earth System, this is the energy that comes to the Earth from the Sun.

external process A geomorphologic process—such as downslope movement, erosion, or deposition—that is the consequence of gravity or of the interaction between the solid Earth and its fluid envelope (air and water). Energy for these processes comes from gravity and sunlight.

extinction The death of the last members of a species so that there are no parents to pass on their genetic traits to offspring.

extinct volcano   A volcano that was active in the past but has now shut off entirely and will not erupt in the future.

extraordinary fossil   A rare fossilized relict, or trace, of the soft part of an organism.

extrusive igneous rock   Rock that forms by the freezing of lava above ground, after it flows or explodes out (extrudes) onto the surface and comes into contact with the atmosphere or ocean.

facet (of a gem)   The ground and polished surface of a gem, produced by a gem cutter using a grinding lap.

failure surface   A weak surface that forms the base of a landslide.

fault   A fracture on which one body of rock slides past another.

fault scarp   A small step on the ground surface where one side of a fault has moved vertically with respect to the other.

feedback mechanism   A condition that arises when the consequence of a phenomenon influences the phenomenon itself.

firn   Compacted granular ice (derived from snow) that forms where snow is deeply buried; if buried more deeply, firn turns into glacial ice.

fissure   A conduit in a magma chamber in the shape of a long crack through which magma rises and erupts at the surface.

fjord   A deep, glacially carved, U-shaped valley flooded by rising sea level.

flash flood   A flood that occurs during unusually intense rainfall or as the result of a dam collapse, during which the floodwaters rise very fast.

flood   An event during which the volume of water in a stream becomes so great that it covers areas outside the stream's normal channel.

flood basalt   Vast sheets of basalt that spread from a volcanic vent over an extensive surface of land; they may form where a rift develops above a continental hot spot, and where lava is particularly hot and has low viscosity.

flood-hazard map   A representation of a portion of the Earth's surface that is designed to show how the danger of flooding varies with location.

floodplain   The flat land on either side of a stream that becomes covered with water during a flood.

focus   The location where a fault slips during an earthquake (hypocenter).

fold   A bend or wrinkle of rock layers or foliation; folds form as a consequence of ductile deformation.

foliation   Layering formed as a consequence of the alignment of mineral grains, or of compositional banding in a metamorphic rock.

foreshocks   The series of smaller earthquakes that precede a major earthquake.

fossil   The remnant, or trace, of an ancient living organism that has been preserved in rock or sediment.

fossil assemblage   A group of fossil species found in a specific sequence of sedimentary rock.

fossilization   The process of forming a fossil.

fractional crystallization   The process by which a magma becomes progressively more silicic as it cools, because early-formed crystals settle out.

fracture zone   A narrow band of vertical fractures in the ocean floor; fracture zones lie roughly at right angles to a mid-ocean ridge, and the actively slipping part of a fracture zone is a transform fault.

fragmental igneous rock   A rock consisting of igneous chunks and/or shards that are packed together, welded together, or cemented together after having solidified.

friction   Resistance to sliding on a surface.

fuel   A substance that can be used to produce energy.

Ga   Billions of years ago (abbreviation).

galaxy   An immense system of hundreds of billions of stars.

gas   Matter that consists of atoms or molecules that are not attached to each other; a gas fills a container that contains it; air is an example of a gas.

gem   A finished (cut and polished) gemstone ready to be set in jewelry.

gemstone   A mineral that has special value because it is rare and people consider it beautiful.

geocentric model   An ancient Greek idea suggesting that the Earth sat motionless in the center of the Universe while stars and other planets and the Sun orbited around it.

geochronology   The science of dating geologic events in years.

geoid   A reference surface representing the elevations, worldwide, at which gravitational potential energy is the same.

geologic column   A composite stratigraphic chart that represents the entirety of the Earth's history.

geologic contact   The surface between two distinct geologic units.

geologic cross section   A depiction of contacts in the subsurface as represented by their traces on an imaginary vertical slice into the Earth.

geologic map   A map showing the distribution of rock units and structures across a region.

geologic time   The span of time since the formation of the Earth.

geologic time scale   A scale that describes the intervals of geologic time.

geologist   A scientist who specializes in studying the Earth.

geology (geoscience)   The study of the Earth, including our planet's composition, behavior, and history.

geophysics   The subdiscipline of geology focused on the quantitative analysis and modeling of physical characteristics of the Earth; it includes the study of earthquakes, gravity, and magnetism.

geosphere   In the context of the Earth System, it is the solid part of the Earth from the surface to the center.

geothermal gradient   The rate of change in temperature with depth.

geothermal region   A region of current or recent volcanism in which magma or very hot rock heats up groundwater, which may discharge at the surface in the form of hot springs and/or geysers.

geyser   A fountain of steam and hot water that erupts periodically from a vent in the ground in a geothermal region.

giant planets   The four outer, or Jovian, planets of our Solar System, which are significantly larger than the rest of the planets and consist largely of gas and/or ice.

glacial advance The forward movement of a glacier's toe when the supply of snow exceeds the rate of ablation.

glacial drift Sediment deposited in glacial environments.

glacial retreat The movement of a glacier's toe back toward the glacier's origin; glacial retreat occurs if the rate of ablation exceeds the rate of supply.

glacial striation Grooves or scratches cut into bedrock when clasts embedded in the moving glacier act like the teeth of a giant rasp.

glacial subsidence The sinking of the surface of a continent caused by the weight of an overlying glacial ice sheet.

glacier A river or sheet of ice that slowly flows across the land surface and lasts all year long.

glass A solid in which atoms are not arranged in an orderly pattern.

glassy igneous rock Igneous rock consisting entirely of glass, or of tiny crystals surrounded by a glass matrix.

global change The transformations or modifications of both physical and biological components of the Earth System through time.

global circulation The movement of volumes of air in paths that ultimately take it around the planet.

global climate change Transformations or modifications in the Earth's climate over time.

global positioning system (GPS) A satellite system people can use to measure rates of movement of the Earth's crust relative to one another, or simply to locate their position on the Earth's surface.

gneiss A compositionally banded metamorphic rock typically composed of alternating dark- and light-colored layers.

Gondwana A supercontinent that consisted of today's South America, Africa, Antarctica, India, and Australia. (Also called Gondwanaland.)

grade (of an ore) The concentration of a useful metal in an ore—the higher the concentration, the higher the grade.

grain A fragment of a mineral crystal or of a rock.

granite A coarse-grained, intrusive, silicic igneous rock.

gravity The attractive force that one mass exerts on another; the magnitude depends on the size of the objects and the distance between them.

gravity anomaly A value of gravitational pull that is greater than or lesser than the pull predicted by the geoid.

great oxygenation event (GOE) The time in Earth's history, about 2.4 Ga, when the concentration of oxygen in the atmosphere increased dramatically.

greenhouse (hothouse) period Relatively warm global climate leading to the rising of sea level for an interval of geologic time.

greenhouse effect The trapping of heat in the Earth's atmosphere by carbon dioxide and other greenhouse gases, which absorb infrared radiation; somewhat analogous to the effect of glass in a greenhouse.

greenhouse gases Atmospheric gases, such as carbon dioxide and methane, that regulate the Earth's atmospheric temperature by absorbing infrared radiation.

groundwater Water that resides under the surface of the Earth, mostly in pores or cracks of rock or sediment.

groundwater contamination Addition of chemicals or microbes (e.g., from agricultural and industrial activities, and landfills or septic tanks) to the groundwater supply.

guyot A seamount that had a coral reef growing on top of it, so that it is now flat-crested.

Hadean Eon The oldest of the Precambrian eons; the time between the Earth's origin and the formation of the first rocks that have been preserved.

half-life The time it takes for half of a group of a radioactive element's isotopes to decay.

hand specimen A piece of rock, about the size of a fist, that can be collected for study.

hanging valley A glacially carved tributary valley whose floor lies at a higher elevation than the floor of the trunk valley.

hardness (of a mineral) A measure of the relative ability of a mineral to resist scratching; it represents the resistance of bonds in the crystal structure from being broken.

headward erosion The process by which a stream channel lengthens up its slope as the flow of water increases.

headwaters The beginning point of a stream.

heat Thermal energy resulting from the movement of molecules.

heliocentric model An idea proposed by Greek philosophers around 250 B.C.E. suggesting that all heavenly objects including the Earth orbited the Sun.

hinge The portion of a fold where curvature is greatest.

Holocene Epoch The period of geologic time since the last glaciation.

Holocene climatic maximum The period from 5,000 to 6,000 years ago, when Holocene temperatures reached a peak.

horn A pointed mountain peak surrounded by at least three cirques.

hornfels Rock that undergoes metamorphism simply because of a change in temperature, without being subjected to differential stress.

hot spot A location at the base of the lithosphere, at the top of a mantle plume, where temperatures can cause melting.

hot spring A spring that emits water ranging in temperature from about 30°C to 104°C.

Hubbert's Peak The high point on a graph of production vs. time; the concept that we can define Hubbert's Peak for a resource emphasizes that supplies of resources are limited.

hydraulic head The potential energy available to drive the flow of a given volume of groundwater at a location; it can be measured as an elevation above a reference.

hydrocarbon A chain-like or ring-like molecule made of hydrogen and carbon atoms; petroleum and natural gas are hydrocarbons.

hydrocarbon generation A process in which oil shale warms to temperatures of greater than about 90°C so kerogen molecules transform into oil and natural gas molecules.

hydrocarbon reserve A known supply of oil and gas held underground.

hydrofracturing (fracking) A process by which drillers generate new fractures or open preexisting ones underground, by pumping a high-pressure fluid into a portion of the drill hole, in order to increase the permeability of surrounding hydrocarbon-bearing rocks.

hydrologic cycle The continual passage of water from reservoir to reservoir in the Earth System.

hydrosphere The Earth's water, including surface water (lakes, rivers, and oceans), groundwater, and liquid water in the atmosphere.

hydrothermal metamorphism When very hot water passes through the crust and causes metamorphism of rock.

hypothesis An idea that has the potential to explain a phenomenon; a hypothesis must be rigorously tested if it is to eventually become a theory.

icehouse period A period of time when the Earth's temperature was cooler than it is today and ice ages could occur.

igneous rock Rock that forms when hot molten rock (magma or lava) cools and freezes solid.

index fossil A fossil of an organism that lived during a relatively short period of time over a relatively large area of the Earth, and can be used for stratigraphic correlation.

inequant A general adjective indicating that the dimensions of an object are not the same in all directions; for example, an object that is longer than it is wide is inequant.

internal energy Energy in the Earth System that comes from within the planet, due to residual heat from the Earth's formation, or radioactive decay.

internal process A process in the Earth System, such as plate motion, mountain building, or volcanism, ultimately caused by the Earth's internal heat.

intraplate earthquakes Earthquakes that occur away from plate boundaries.

intrusive igneous rock Rock formed by the freezing of magma underground.

isostasy (isostatic equilibrium) The condition that exists when the buoyancy force pushing lithosphere up equals the gravitational force pulling lithosphere down.

isotopes Different versions of a given element that have the same atomic number but different atomic weights.

isotopic dating Another term for radiometric dating, meaning the determination of the numerical age of rocks and minerals.

joints Naturally formed cracks in rocks.

karst landscape A region underlain by caves in limestone bedrock; the collapse of the caves creates a landscape of sinkholes separated by higher topography, or of limestone spires separated by low areas.

kerogen The waxy molecules into which the organic material in shale transforms on reaching about 100°C. At higher temperatures, kerogen transforms into oil.

kettle hole A circular depression in the ground made when a block of ice calves off the toe of a glacier, becomes buried by till, and later melts.

kingdom In the context of taxonomy, it is second highest rank of life (beneath domain); examples of kingdoms include Animalia, Plantae, and Fungi.

laccolith A blister-shaped igneous intrusion that forms when magma injects between layers underground in a manner that pushes overlying layers upward to form a dome.

lag deposit The coarse sediment left behind in a desert after wind erosion removes the finer sediment.

lahar A thick slurry formed when volcanic ash and debris mix with water, either in rivers or from rain or melting snow and ice on the flank of a volcano.

landform A particular land-surface shape at a location.

landscape The overall shape and character of the land surface in a region.

landslide A sudden movement of rock and debris down a non-vertical slope.

lapilli Any pyroclastic particle that is 2 to 64 mm in diameter (i.e., marble-sized); the particles can consist of frozen lava clots, pumice fragments, or ash clumps.

large igneous province (LIP) A region in which huge volumes of lava and/or ash erupted over a relatively short interval of geologic time.

Laurentia A continent in the early Paleozoic Era composed of today's North America and Greenland.

Laurentide ice sheet An ice sheet that spread over northeastern Canada during the Pleistocene ice age(s).

lava Molten rock that has flowed out onto the Earth's surface.

lava flows Sheets or mounds of lava that flow onto the ground surface or sea floor in molten form and then solidify.

lava fountain An eruption of lava at a volcano, during which lava spurts into the air while still molten.

lava tube The empty space left when a lava tunnel drains; this happens when the surface of a lava flow solidifies while the inner part of the flow continues to stream downslope.

limb (of fold) The side of a fold, showing less curvature than at the hinge.

liquid Matter that can flow to conform to the shape of the container that holds it.

lithification The transformation of loose sediment into solid rock through compaction and cementation.

lithosphere The relatively rigid, nonflowable, outer 100- to 150-km-thick layer of the Earth, constituting the crust and the top part of the mantle.

longitudinal profile A cross-sectional image showing the variation in elevation along the length of a river.

longshore current The flow of water parallel to the shore just off a coast, because of the diagonal movement of waves toward the shore.

longshore drift The movement of sediment laterally along a beach; it occurs when waves wash up a beach diagonally.

lower mantle The deepest section of the mantle, stretching from 670 km down to the core-mantle boundary.

low-velocity zone The asthenosphere underlying oceanic lithosphere in which seismic waves travel more slowly, probably because rock has partially melted.

luster The way a mineral surface scatters light.

Ma Millions of years ago (abbreviation).

macrofossil A fossil large enough to be seen with the naked eye.

magma Molten rock beneath the Earth's surface.

magma chamber A space below ground filled with magma.

magnetic anomaly The difference between the expected strength of the Earth's magnetic field at a certain location and the actual measured strength of the field at that location.

magnetic declination The angle between the direction a compass needle points at a given location and the direction of true north.

magnetic field The region affected by the force emanating from a magnet.

magnetic inclination The angle between a magnetic needle free to pivot on a horizontal axis and a horizontal plane parallel to the Earth's surface.

magnetic-reversal chronology The history of magnetic reversals through geologic time.

magnetosphere The region protected from the electrically charged particles of the solar winds by the Earth's magnetic field.

mainshock The largest earthquake during a succession of related earthquakes; foreshocks, which precede the mainshock, and aftershocks that come after the mainshock, all release much less energy than the mainshock.

manganese nodules Lumpy accumulations of manganese-oxide minerals precipitated onto the sea floor.

mantle The thick layer of rock below the Earth's crust and above the core.

marble A metamorphic rock composed of calcite and transformed from a protolith of limestone.

mare The broad, darker areas on the Moon's surface; they consist of flood basalts that erupted over 3 billion years ago and spread out across the Moon's lowlands.

mass-extinction event A time when vast numbers of species abruptly vanish.

mass movement (mass wasting) The gravitationally caused downslope transport of rock, regolith, snow, or ice.

mass-transfer cycle The progressive movement of material from one reservoir in the Earth System to another.

meander A snake-like curve along a stream's course.

meandering stream A reach of stream containing many meanders (snake-like curves).

melt Molten (liquid) rock.

meltdown The melting of the fuel rods in a nuclear reactor that occurs if the rate of fission becomes too fast and the fuel rods become too hot.

metaconglomerate A metamorphic rock produced by metamorphism of a conglomerate; typically, it contains flattened pebbles and cobbles.

metal A solid composed almost entirely of atoms of metallic elements; it is generally opaque, shiny, smooth, malleable, and can conduct electricity.

metamorphic aureole The region around a pluton, stretching tens to hundreds of meters out, in which heat transferred into the country rock and metamorphosed the country rock.

metamorphic facies A set of metamorphic mineral assemblages indicative of metamorphism under a specific range of pressures and temperatures.

metamorphic foliation A fabric defined by parallel surfaces or layers that develop in a rock as a result of metamorphism; schistocity and gneissic layering are examples.

metamorphic grade A representation of the intensity of metamorphism, meaning the amount or degree of metamorphic change.

metamorphic mineral New minerals that grow in place within a solid rock under metamorphic temperatures and pressures.

metamorphic rock Rock that forms when preexisting rock changes into new rock as a result of an increase in pressure and temperature and/or shearing under elevated temperatures; metamorphism occurs without the rock first becoming a melt or a sediment.

metamorphic texture A distinctive arrangement of mineral grains produced by metamorphism.

metamorphic zone The region between two metamorphic isograds, typically named after an index mineral found within the region.

metamorphism The process by which one kind of rock transforms into a different kind of rock.

metasomatism The process by which a rock's overall chemical composition changes during metamorphism because of reactions with hot water that bring in or remove elements.

meteor A streak of bright, glowing gas created as a meteoroid vaporizes in the atmosphere due to friction.

meteorite A piece of rock or metal alloy that fell from space and landed on the Earth.

microfossil A fossil that can be seen only with a microscope or an electron microscope.

mid-ocean ridge A 2-km-high submarine mountain belt that forms along a divergent oceanic plate boundary.

migmatite A rock formed when gneiss is heated high enough so that it begins to partially melt, creating layers, or lenses, of new igneous rock that mix with layers of the relict gneiss.

Milankovitch cycles Climate cycles that occur over tens to hundreds of thousands of years because of changes in the Earth's orbit and tilt.

mineral classes Groups of minerals distinguished from each other on the basis of chemical composition.

mineral resource An accumulation of a useful ore, in which valuable elements are sufficiently concentrated to be worth mining.

mineralogy The study of minerals and their characteristics.

mixture A material consisting of two or more substances that can be separated mechanically (i.e., without chemical reactions).

Modified Mercalli Intensity (MMI) Scale An earthquake characterization scale based on the amount of damage that the earthquake causes.

Moho The seismic-velocity discontinuity that defines the boundary between the Earth's crust and mantle. Named for Andrija Mohorovičić.

Mohs hardness scale A list of ten minerals in a sequence of relative hardness, with which other minerals can be compared.

mold A cavity in sedimentary rock left behind when a shell that once filled the space weathers out.

molecule The smallest piece of a compound that has the properties of the compound; it consists of two or more atoms attached by chemical bonds.

moment magnitude scale   A modern scale for measuring the relative size of earthquakes that involves studying the amplitude of waves on a seismograph, along with other parameters.

monocline   A fold in the land surface whose shape resembles that of a carpet draped over a stair step.

moon   A sizable solid body locked in orbit around a planet.

moraine   A sediment pile composed of till deposited by a glacier.

morphology   The form or shape of an object (or the study of form or shape); for example, fossil shells can be classified based on the morphology of their shells.

mountain belt   An elongate band of mountains, formed as the result of an orogeny.

mountain building   The process of generating a mountain range.

mountain (alpine) glacier   A glacier that exists in or adjacent to a mountainous region.

mouth   The outlet of a stream where it discharges into another stream, a lake, or a sea.

mudstone   Very fine-grained sedimentary rock that will not easily split into sheets.

natural hazard   A natural feature of the environment that can cause injury to living organisms and/or damage to buildings and the landscape.

natural levees   A pair of low ridges that appear on either side of a stream and develop as a result of the accumulation of sediment deposited naturally during flooding.

natural selection   The process by which the fittest organisms survive to pass on their characteristics to the next generation.

nebula   A cloud of gas or dust in space.

nebular theory of planet formation   The concept that planets grow out of rings of gas, dust, and ice surrounding a newborn star.

nonconformity   A type of unconformity at which sedimentary rocks overlie basement (older intrusive igneous rocks and/or metamorphic rocks).

normal fault   A fault in which the hanging-wall block moves down the slope of the fault.

normal polarity   Polarity in which the paleomagnetic dipole has the same orientation as it does today.

nuclear fusion   The process by which the nuclei of atoms fuse together, thereby creating new, larger atoms.

nuclear reactor   The part of a nuclear power plant where the fission reactions occur.

nuclear waste   The radioactive material produced as a by-product in a nuclear plant that must be disposed of carefully due to its dangerous radioactivity.

numerical age   (in older literature, "absolute age") The age of a geologic feature given in years.

oasis   A verdant region surrounded by desert, occurring at a place where natural springs provide water at the surface.

oblique-slip fault   A fault in which sliding occurs diagonally along the fault plane.

Oil Age   The period of human history, including our own, so named because the economy depends on oil.

oil seep   A location where oil bubbles out of the ground on its own, without pumping.

oil shale   An organic shale containing abundant kerogen.

oil window   The narrow range of temperatures under which oil can form in a source rock.

olistotrome   A large, submarine slump block, buried and preserved.

Oort Cloud   A cloud of icy objects, left over from Solar System formation, that orbit the Sun in a region outside of the heliosphere.

ordinary well   A well whose base penetrates below the water table and can thus provide water.

ore   Rock containing native metals or a concentrated accumulation of ore minerals.

ore deposit   An economically significant accumulation of ore.

ore minerals   Minerals that have metal in high concentrations and in a form that can be easily extracted.

organic chemical   A carbon-containing compound that occurs in living organisms, or that resembles such compounds; it consists of carbon atoms bonded to hydrogen atoms along with varying amounts of oxygen, nitrogen, and other chemicals.

organic coast   A coast along which living organisms control landforms along the shore.

organic sedimentary rock   Sedimentary rock (such as coal) formed from carbon-rich relicts of organisms.

orogen (orogenic belt)   A linear range of mountains.

orogenic collapse   The process in which mountains begin to collapse under their own weight and spread out laterally.

orogeny   A mountain-building event.

outcrop   An exposure of bedrock.

oxbow lake   A meander that has been cut off yet remains filled with water.

pahoehoe   A lava flow with a surface texture of smooth, glassy, ropelike ridges.

paleomagnetism   The record of ancient magnetism preserved in rock.

paleontologist   A scientist who specializes in studying and interpreting fossils.

paleontology   The study of ancient life and its evolution as recorded by fossils.

paleopole   The supposed position of the Earth's magnetic pole in the past, with respect to a particular continent.

Pangaea   A supercontinent that assembled at the end of the Paleozoic Era.

parent atom (isotope)   A radioactive isotope that undergoes decay.

partial melt   The magma formed when the lower-melting-temperature component of a rock has melted.

partial melting   The melting in a rock of the minerals with the lowest melting temperatures, while other minerals remain solid.

passive continental margin   A continental margin that does not coincide with a plate boundary, and therefore does not display seismicity.

passive margin   (see passive continental margin)

patterned ground   A polar landscape in which the ground splits into pentagonal or hexagonal shapes.

peat   Compacted and partially decayed vegetation accumulating beneath a swamp.

perched water table A quantity of groundwater that lies above the regional water table because an underlying lens of impermeable rock or sediment prevents the water from sinking down to the regional water table.

period An interval of geologic time representing a subdivision of a geologic era.

permafrost Permanently frozen ground.

permanent stream A stream that flows year-round because its bed lies below the water table, or because more water is supplied from upstream than can infiltrate the ground.

permeability The degree to which a material allows fluids to pass through it via an interconnected network of pores and cracks.

permineralization The fossilization process in which plant material becomes transformed into rock by the precipitation of silica from groundwater.

petrified wood Wood that has undergone permineralization and has turned into agate; growth rings and cell walls may remain visible in samples.

petroglyph Drawings formed by chipping into the desert varnish of rocks to reveal the lighter rock beneath.

Phanerozoic Eon The most recent eon, an interval of time from 542 Ma to the present.

photomicrograph A photograph of a thin section taken through a microscope.

photovoltaic cell A devise capable of transforming solar energy directly into electricity.

phyllite A fine-grained metamorphic rock with a foliation caused by the preferred orientation of very fine-grained mica.

phylogenetic tree A chart representing the ideas of paleontologists showing which groups of organisms radiated from which ancestors.

phylogeny The study of how a species of organism evolves.

physical weathering The process in which intact rock breaks into smaller grains or chunks.

pillow lava Mafic lava that extruded underwater to form blob-like shapes, or pillows, which typically have a glassy rind.

planet An object that orbits a star, is roughly spherical, and has cleared its neighborhood of other objects.

planetesimal Tiny, solid pieces of rock and metal that collect in a planetary nebula and eventually accumulate to form a planet.

plastic deformation The deformational process in which mineral grains behave like plastic and, when compressed or sheared, become flattened or elongate without cracking or breaking.

plate One of about 20 distinct pieces of the relatively rigid lithosphere.

plate boundary The border between two adjacent lithosphere plates.

plate tectonics *See* Theory of plate tectonics.

playa The flat, typically salty lake bed that remains when all the water evaporates in drier times; forms in desert regions.

Pleistocene Ice Age The ice age that began about 2.6 Ma, and involves many advances and retreats of continental glaciers.

plunge In the context of defining the orientation of a line, plunge refers to the angle between the line and horizontal, as measured in a vertical plane.

pluton An irregular or blob-shaped intrusion; can range in size from tens of m across to tens of km across.

pluvial lake A lake formed to the south of a continental glacier as a result of enhanced rainfall during an ice age.

point bar A wedge-shaped deposit of sediment on the inside bank of a meander.

polarity subchron The time interval between magnetic reversals if the interval is of short duration (less than 200,000 years long).

pollution Natural and synthetic contaminant materials introduced to the Earth's environment by the activities of humans.

polymorphs Two minerals that have the same chemical composition but a different crystal lattice structure.

pore A small, open space within sediment or rock.

porosity The total volume of empty space (pore space) in a material, usually expressed as a percentage.

Portland cement Cement made by mechanically mixing limestone, sandstone, and shale in just the right proportions, before heating in a kiln, to provide the correct chemical make-up of cement.

post-glacial rebound The rise of the surface of a continent after an ice sheet has melted away, so that isostasy is reestablished.

pothole A bowl-shaped depression carved into the floor of a stream by a long-lived whirlpool carrying sand or gravel.

Precambrian The interval of geologic time between the Earth's formation about 4.57 Ga and the beginning of the Phanerozoic Eon 542 Ma.

preferred orientation The parallelism of inequant grains in a metamorphic rock.

preservation potential The likelihood that an organism will be preserved as a fossil.

pressure Force per unit area, or the "push" acting on a material in cases where the push (compressional stretch) is the same in all directions.

Proterozoic Eon The most recent of the Precambrian eons (2,500–541 Ma).

protolith The original rock from which a metamorphic rock formed.

protoplanet A body that grows by the accumulation of planetesimals but has not yet become big enough to be called a planet.

protoplanetary disk The plate-shaped region of gas and dust, surrounding the newborn Sun, from which the planets formed.

protostar A dense body of gas that is collapsing inward because of gravitational forces and that may eventually become a star.

P-waves Compressional seismic waves that move through the body of the Earth.

P-wave shadow zone A band between 103° and 143° from an earthquake epicenter, as measured along the circumference of the Earth, inside which P-waves do not arrive at seismograph stations.

pyroclastic debris Fragmented material that sprayed out of a volcano and landed on the ground or sea floor in solid form.

pyroclastic flow A fast-moving avalanche that occurs when hot volcanic ash and debris mix with air and flow down the side of a volcano.

pyroclastic rock Rock made from fragments that were blown out of a volcano during an explosion and were then packed or welded together.

quartzite A metamorphic rock composed of quartz and transformed from a protolith of quartz sandstone.

radioactive decay The process by which a radioactive atom undergoes fission or releases particles, thereby being transformed into a new element.

radiometric dating (also called numerical dating) The science of determining the age of materials in years by measuring the ratio of parent radioactive atoms to daughter product atoms in the material.

rapids A reach of a stream in which water becomes particularly turbulent; as a consequence, waves develop on the surface of the stream.

recharge area A location where water enters the ground and infiltrates down to the water table.

recurrence interval The average time between events of a given size or magnitude; the term is commonly used to give a sense of the frequency of earthquakes or of flooding.

reflection What happens when energy bounces off a boundary; when this happens, the incoming angle and outgoing angle are equal.

refraction The bending of a ray as it passes through a boundary between two different materials.

refractory materials Substances that have a relatively high melting point and tend to exist in solid form.

regional metamorphism *See also* Dynamothermal metamorphism; metamorphism of a broad region, usually the result of deep burial during an orogeny.

regolith Any kind of unconsolidated debris that covers bedrock.

regression The seaward migration of a shoreline caused by a lowering of sea level.

relative age The age of one geologic feature with respect to another.

relative plate velocity The movement of one lithosphere plate with respect to another.

relief The difference in elevation between adjacent high and low regions on the land surface.

reserve A known occurrence of a resource in sufficient quantities and concentration to make it potentially worth extracting.

reservoir A region that contains a volume of material; for example, in the context of the Earth System, the ocean is the largest reservoir of liquid water on the Earth, and in the context of discussing hydrocarbons, a porous sandstone may be a good reservoir for oil or gas.

reservoir rock Rock with high porosity and permeability, so it can contain an abundant amount of easily accessible oil.

residence time The average length of time that a substance stays in a particular reservoir.

reversed polarity Polarity in which the paleomagnetic dipole points north.

reverse fault A steeply dipping fault on which the hanging-wall block slides up.

Richter scale A scale that defines earthquakes on the basis of the amplitude of the largest ground motion recorded on a seismogram.

ridge-push force A process in which gravity causes the elevated lithosphere at a mid-ocean ridge axis to push on the lithosphere that lies farther from the axis, making it move away.

rifting The process by which continental lithosphere stretches horizontally and thins vertically.

ripple mark Relatively small elongated ridges that form on a sedimentary bed surface at right angles to the direction of current flow.

riprap Loose boulders or concrete piled together along a beach to absorb wave energy before it strikes a cliff face.

roche moutonnée A glacially eroded hill that becomes elongate in the direction of flow and asymmetric; glacial rasping smooths the upstream part of the hill into a gentle slope, while glacial plucking erodes the downstream edge into a steep slope.

rock A coherent, naturally occurring solid, consisting of an aggregate of minerals or a mass of glass.

rock composition The chemical makeup of a rock, as represented by the proportions of different minerals that it contains.

rock cycle The succession of events that results in the transformation of Earth materials from one rock type to another, then another, and so on.

rockfall A mass of rock that separates from a cliff, typically along a joint, and then free-falls downslope.

rockslide A sudden downslope movement of rock.

rogue wave Waves that are two to five times the size of most of the large waves passing a locality in a given time interval.

Runoff The water the flows on the surface of the Earth to drain the land; it includes streamflow and sheetflow.

R-waves (Rayleigh waves) Surface seismic waves that cause the ground to ripple up and down, like water waves in a pond.

salinity The degree of concentration of salt in water.

saltation The movement of a sediment in which grains bounce along their substrate, knocking other grains into the water column (or air) in the process.

salt dome A rising bulbous dome of salt that bends up the adjacent layers of sedimentary rock.

sand spit An area where the beach stretches out into open water across the mouth of a bay or estuary.

sandstone Coarse-grained sedimentary rock consisting almost entirely of quartz.

schist A medium-to-coarse-grained metamorphic rock that possesses schistosity.

science The systematic study of natural phenomena via observation, computation, experiment, and modeling.

scientific law A concise statement that completely describes a natural relationship or phenomenon; it does not, however, explain the phenomenon.

scientific method A sequence of steps for systematically analyzing scientific problems in a way that leads to verifiable results.

scouring A process by which running water removes loose fragments of sediment from a streambed.

seafloor spreading The gradual widening of an ocean basin as new oceanic crust forms at a mid-ocean ridge axis and then moves away from the axis.

sea ice Ice formed by the freezing of the surface of the sea.

seal rock A relatively impermeable rock, such as shale, salt, or unfractured limestone, that lies above a reservoir rock and stops the oil from rising further.

seamount An isolated submarine mountain.

seasonal floods Floods that appear almost every year during seasons when rainfall is heavy or when winter snows start to melt.

sediment An accumulation of loose mineral grains, such as boulders, pebbles, sand, silt, or mud, that are not cemented together.

sediment liquefaction When pressure in the water in the pores push sediment grains apart so that they become surrounded by water and no longer rest against each other, and the sediment becomes able to flow like a liquid.

sedimentary basin A depression, created as a consequence of subsidence, that fills with sediment.

sedimentary rock Rock that forms either by the cementing together of fragments broken off preexisting rock or by the precipitation of mineral crystals out of water solutions at or near the Earth's surface.

sedimentary structure A geometry or arrangement of material in sediment or sedimentary rock that formed during or shortly after deposition, not in response to later tectonic stress; examples include cross beds and mudcracks.

seismic belts (seismic zones) The relatively narrow strips of crust on the Earth under which most earthquakes occur.

seismicity Earthquake activity.

seismic ray The changing position of an imaginary point on a wave front as the front moves through rock.

seismic-reflection profile A cross-sectional view of the crust made by measuring the reflection of artificial seismic waves off boundaries between different layers of rock in the crust.

seismic tomography Analysis by sophisticated computers of global seismic data in order to create a three-dimensional image of variations in seismic-wave velocities within the Earth.

seismic velocity The speed at which seismic waves travel.

seismic-velocity discontinuity A boundary in the Earth at which seismic velocity changes abruptly.

seismic (earthquake) waves Waves of energy emitted at the focus of an earthquake.

seismic zone A region in which earthquakes happen fairly frequently; a seismic belt is an elongate seismic zone.

seismogram The record of an earthquake produced by a seismometer.

seismologist A scientist who specializes in the study of earthquakes, or in the study of how seismic waves characterize the interior of the Earth.

seismometer (seismograph) An instrument that can record the ground motion from an earthquake.

shale Very fine-grained sedimentary rock that breaks into thin sheets.

shale gas Gas extracted directly from a source rock (organic shale).

shale oil Oil extracted directly from a source rock.

shatter cones Small, cone-shaped fractures formed by the shock of a meteorite impact.

shear waves Seismic waves in which particles of material move back and forth perpendicular to the direction in which the wave itself moves.

sheetwash A film of water less than a few mm thick that covers the ground surface during heavy rains.

shield An older, interior region of a continent.

shield volcano A subaerial volcano with a broad, gentle dome, formed either from low-viscosity basaltic lava or from large pyroclastic sheets.

shock metamorphism The changes that can occur in a rock due to the passage of a shock wave, generally resulting from a meteorite impact.

shore The region of land adjacent to a body of water.

shoreline The boundary between the water and land.

silicate rock Rock composed of silicate minerals.

silicates (silicate minerals) Minerals built from silicon-oxygen tetrahedra arranged in chains, sheets, or 3-D networks; they make up most of the Earth's crust and mantle.

siliceous sedimentary rock Sedimentary rock that contains abundant quartz.

silicon-oxygen tetrahedron The SiO_4^{4-} anionic group, in which four oxygen atoms surround a single silicon atom, thereby defining the corners of a tetrahedron.

sill A nearly horizontal tabletop-shaped tabular intrusion that occurs between the layers of country rock.

siltstone Fine-grained sedimentary rock generally composed of very small quartz grains.

sinkhole A circular depression in the land that forms when an underground cavern collapses.

slab-pull force The force that downgoing plates (or slabs) apply to oceanic lithosphere at a convergent margin.

slate Fine-grained, low-grade metamorphic rock, formed by the metamorphism of shale.

slickensides The polished surface of a fault caused by slip on the fault; lineated slickensides also have grooves that indicate the direction of fault movement.

slope failure The downslope movement of material on an unstable slope.

slow-onset flood A flood that takes days to weeks to develop; these include seasonal floods that cover flood plains and delta plains.

slump A semi-coherent volume of regolith that slipped down a slope above a spoon-shaped failure surface at slow to moderate speed.

slumping   Downslope movement in which a mass of regolith detaches from its substrate along a spoon-shaped, sliding surface and slips downward semi-coherently.

snow avalanche   Rapid downslope movement of a mass of snow; typically, the movement transforms the snow into a turbulent cloud.

snowball Earth   A model proposing that, at times during Earth history, glaciers covered all land, and the entire ocean surface froze.

soil   Sediment that has undergone changes at the surface of the Earth, including reaction with rainwater and the addition of organic material.

soil erosion   The removal of soil by wind and runoff.

soil horizon   Distinct zones within a soil, distinguished from each other by factors such as chemical composition and organic content.

soil moisture   Underground water that wets the surface of the mineral grains and organic material making up soil, but lies above the water table.

soil order   A given type of soil in a common soil classification scheme; for example, an aridisol is a soil order formed in very dry climates.

soil profile   A vertical sequence of distinct zones of soil.

Solar System   Our Sun and all the materials that orbit it (including planets, moons, asteroids, Kuiper Belt objects, and Oort Cloud objects).

solid   A material that can maintain its shape indefinitely.

solifluction   The type of creep characteristic of tundra regions; during the summer, the uppermost layer of permafrost melts, and the soggy, weak layer of ground then flows slowly downslope in overlapping sheets.

sorting   (1) The range of clast sizes in a collection of sediment; (2) the degree to which sediment has been separated by flowing currents into different-sized fractions.

source rock   A rock (organic-rich shale) containing the raw materials from which hydrocarbons eventually form.

specific gravity   A number representing the density of a mineral, as specified by the ratio between the weight of a volume of the mineral and the weight of an equal volume of water.

speleothem   A formation that grows in a limestone cave by the accumulation of travertine precipitated from water solutions dripping in a cave or flowing down the wall of a cave.

spring   A natural outlet from which groundwater flows up onto the ground surface.

stalactite   An icicle-like cone that grows from the ceiling of a cave as dripping water precipitates limestone.

stalagmite   An upward-pointing cone of limestone that grows when drips of water hit the floor of a cave.

star   An object in the Universe in which fusion reactions occur pervasively, producing vast amounts of energy; our Sun is a star.

steady-state condition   The condition when proportions of a chemical in different reservoirs remain fairly constant even though there is a constant flux (flow) of the chemical among the reservoirs.

stellar nucleosynthesis   The production of new, larger atoms by fusion reactions in stars; the process generates more massive elements that were not produced by the Big Bang.

stellar wind   The stream of atoms emitted from a star into space.

stick-slip behavior   Stop-start movement along a fault plane caused by friction, which prevents movement until stress builds up sufficiently.

stoping   A process by which magma intrudes; blocks of wall rock break off and then sink into the magma.

storm surge   Excess seawater driven landward by wind during a storm; the low atmospheric pressure beneath the storm allows sea level to rise locally, increasing the surge.

strain   The change in shape of an object in response to deformation (i.e., as a result of the application of a stress).

strata   A succession of several layers or beds together.

strategic mineral   A mineral containing elements, typically metals, of strategic importance to technology.

stratigraphic column   A cross-section diagram of a sequence of strata summarizing information about the sequence.

stratigraphic formation   A recognizable layer of a specific sedimentary rock type or set of rock types, deposited during a certain time interval, that can be traced over a broad region.

stratigraphic group (group)   Several adjacent stratigraphic formations in a succession.

stratovolcano   A large, cone-shaped subaerial volcano consisting of alternating layers of lava and tephra.

streak   The color of the powder produced by pulverizing a mineral on an unglazed ceramic plate.

stream   A ribbon of water that flows in a channel.

stream gradient   The slope of a stream's channel in the downstream direction.

stream piracy   A process that happens when headward erosion by one stream causes the stream to intersect the course of another stream and capture its flow.

stream rejuvenation   The renewed downcutting of a stream into a floodplain or peneplain, caused by a relative drop of the base level.

stream terrace   When a stream downcuts through the alluvium of a floodplain so that a new, lower floodplain develops and the original floodplain becomes a step-like platform.

stress   The push, pull, or shear that a material feels when subjected to a force; formally, the force applied per unit area over which the force acts.

strike   The compass orientation of a horizontal line on a plane.

strike-slip fault   A fault in which one block slides horizontally past another (and therefore parallel to the strike line), so there is no relative vertical motion.

stromatolite   Layered mounds of sediment formed by cyanobacteria; cyanobacteria secrete a mucous-like substance to which sediment sticks, and as each layer of cyanobacteria gets buried by sediment, it colonizes the surface of the new sediment, building a mound upward.

subchron   *See* **polarity subchron.**

subduction   The process by which one oceanic plate bends and sinks down into the asthenosphere beneath another plate.

sublimation The evaporation of ice directly into vapor without first forming a liquid.

submarine canyon A narrow, steep canyon that dissects a continental shelf and slope.

subsidence The vertical sinking of the Earth's surface in a region, relative to a reference plane.

subsoil The B-horizon, or zone of accumulation, in a soil; it underlies the topsoil.

supernova A short-lived, very bright object in space that results from the cataclysmic explosion marking the death of a very large star; the explosion ejects large quantities of matter into space to form new nebulae.

superplume A huge mantle plume.

surface load (bed load) Sediment that rolls and bounce along the ground (under the air) or along a stream bed (under water).

surface waves Seismic waves that travel along the Earth's surface.

suspended load Tiny solid grains carried along by a stream without settling to the floor of the channel.

sustainable growth The ability of society to prosper without depleting the supply of natural resources, and without destroying the environment.

suture The contact defining the boundary of what were two separate crustal blocks, prior to collision.

swamp A wetland dominated by trees.

swash The upward surge of water that flows up a beach slope when breakers crash onto the shore.

S-waves Seismic shear waves that pass through the body of the Earth.

S-wave shadow zone A band between 103° and 180° from the epicenter of an earthquake inside of which S-waves do not arrive at seismograph stations.

syncline A trough-shaped fold whose limbs dip toward the hinge.

talus A sloping apron of fallen rock along the base of a cliff.

tar sand Sandstone reservoir rock in which less viscous oil and gas molecules have either escaped or been eaten by microbes, so that only tar remains.

taxonomy The study and classification of the relationships among different forms of life.

temperature A measure of the hotness or coldness of a material.

tephra Unconsolidated accumulations of pyroclastic grains.

terminal moraine The end moraine at the farthest limit of glaciation.

terrestrial planets Planets that are of comparable size and character to the Earth and consist of a metallic core surrounded by a rock mantle.

theory A scientific idea supported by an abundance of evidence that has passed many tests and failed none.

theory of evolution by natural selection The idea that species change over time, new species appear, and old species disappear, due to the survival of the fittest.

theory of plate tectonics The theory that the outer layer of the Earth (the lithosphere) consists of separate plates that move with respect to one another.

thermal energy The total kinetic energy in a material due to the vibration and movement of atoms in the material.

thermal metamorphism Metamorphism caused by heat conducted into country rock from an igneous intrusion.

thermohaline circulation The rising and sinking of water driven by contrasts in water density, which is due in turn to differences in temperature and salinity; this circulation involves both surface and deep-water currents in the ocean.

thin section A 3/100-mm-thick slice of rock that can be examined with a petrographic microscope.

thrust fault A gently dipping reverse fault; the hanging-wall block moves up the slope of the fault.

tidal range The difference in sea level between high tide and low tide at a given point.

tide The daily rising or falling of sea level at a given point on the Earth.

tidewater glacier A glacier that has entered the sea along a coast.

till A mixture of unsorted mud, sand, pebbles, and larger rocks deposited by glaciers.

tillite A rock formed from hardened ancient glacial deposits and consisting of larger clasts distributed through a matrix of sandstone and mudstone.

topographic profile A line representing the intersection of the land surface with an imaginary vertical plane at a locality.

topography Variations in elevation.

topsoil The top soil horizons, which are typically dark and nutrient-rich.

transform fault A fault marking a transform plate boundary; along mid-ocean ridges, transform faults are the actively slipping segment of a fracture zone between two ridge segments.

transform boundary A boundary at which one lithosphere plate slips laterally past another.

transgression The inland migration of shoreline resulting from a rise in sea level.

transition zone The middle portion of the mantle, from 400 to 670 km deep, in which there are several jumps in seismic velocity.

travel time The time that it takes for a seismic wave to travel from the focus of an earthquake to a seismometer along a given ray path.

travel-time curve A graph that plots the time since an earthquake began on the vertical axis and the distance to the epicenter on the horizontal axis.

travertine A rock composed of crystalline calcium carbonate ($CaCO_3$) formed by chemical precipitation from groundwater that has seeped out at the ground surface.

trench A deep, elongate trough bordering a volcanic arc; a trench defines the trace of a convergent plate boundary.

tributary A smaller stream that flows into a larger stream.

triple junction A point where three lithosphere plate boundaries intersect.

tropical cyclone A large spiral-shaped rotating storm that forms over the ocean in tropical latitudes; the categories includes hurricanes, typhoons, and cyclones.

tsunami A large wave along the sea surface triggered by an earthquake or large submarine slump.

tuff A pyroclastic igneous rock composed of volcanic ash and fragmented pumice, formed when accumulations of the debris cement together.

turbidite A graded bed of sediment built up at the base of a submarine slope and deposited by turbidity currents.

turbidity current A submarine avalanche of sediment and water that speeds down a submarine slope.

unconformity A boundary between two different rock sequences representing an interval of time during which new strata were not deposited and/or were eroded.

unconventional reserve A supply of oil or gas that cannot be easily pumped; it includes forms of hydrocarbons that are too viscous to pump, or occur in impermeable rock; examples include tar sand, oil shale, shale oil, and shale gas.

uniformitarianism (the principle of) (see principle of uniformitarianism)

Universe All of space and all the matter and energy within it.

uplift (n. geology) The upward vertical movement of the Earth's surface.

upper mantle The uppermost section of the mantle, reaching down to a depth of 400 km.

U-shaped valley A steep-walled valley shaped by glacial erosion into the form of a U.

vacuum Space that contains very little matter in a given volume (e.g., a region in which air has been removed).

valley A trough with sloping walls, cut into the land by a stream.

varve A pair of thin layers of glacial lake-bed sediment, one consisting of silt brought in during the spring floods and the other of clay deposited during the winter when the lake's surface freezes over and the water is still.

vein A seam of minerals that forms when dissolved ions carried by water solutions precipitate in cracks.

ventifact (faceted rock) A desert rock whose surface has been faceted by the wind.

viscosity The resistance of material to flow.

volatiles (volatile materials) Elements or compounds such as H_2O and CO_2 that evaporate at relatively low temperatures and can exist in gaseous forms at the Earth's surface.

volcanic arc A curving chain of active volcanoes formed adjacent to a convergent plate boundary.

volcanic ash Tiny glass shards formed when a fine spray of exploded lava freezes instantly upon contact with the atmosphere.

volcano (1) A vent from which melt from inside the Earth spews out onto the planet's surface; (2) a mountain formed by the accumulation of extrusive volcanic rock.

V-shaped valley A valley whose cross-sectional shape resembles the shape of a V; the valley probably has a river running down the point of the V.

Wadati-Benioff zone A sloping band of seismicity defined by intermediate- and deep-focus earthquakes that occur in the downgoing slab of a convergent plate boundary.

waterfall A place where water drops over an escarpment.

watershed The region that collects water that feeds into a given drainage network.

water table The boundary, approximately parallel to the Earth's surface, that separates substrate in which groundwater fills the pores from substrate in which air fills the pores.

wave A disturbance that transmits energy from one point to another in the form of periodic motions.

wave base The depth, approximately equal in distance to half a wavelength in a body of water, beneath which there is no wave movement.

wave refraction (ocean) The bending of waves as they approach a shore so that their crests make no more than a 5° angle with the shoreline.

weather Local-scale conditions as defined by temperature, air pressure, relative humidity, and wind speed.

weathering The processes that break up and corrode solid rock, eventually transforming it into sediment.

well A hole in the ground dug or drilled in order to obtain water.

xenolith A relict of wall rock surrounded by intrusive rock when the intrusive rock freezes.

yardang A mushroom-like column with a resistant rock perched on an eroding column of softer rock; created by wind abrasion in deserts where a resistant rock overlies softer layers of rock.

zone of accumulation (1) The layer of regolith in which new minerals precipitate out of water passing through, thus leaving behind a load of fine clay; (2) the area of a glacier in which snowfall adds to the glacier.

zone of leaching The layer of regolith in which water dissolves ions and picks up very fine clay; these materials are then carried downward by infiltrating water.

PHOTO CREDITS

P. Jacobs/JLM Visuals; **p. 95 (top center left):** Stephen Marshak; **p. 95 (top center right):** Richard P. Jacobs/JLM Visuals96 (bottom); **p. 95 (bottom center left and right):** Stephen Marshak; **p. 95 (bottom left):** Richard P. Jacobs/JLM Visuals; **p. 95 (bottom right):** Scientifica/Visuals Unlimited; **p. 97 (top center):** 1992 Jeff Scovil; **p. 97 (top right):** Marli Miller/Visuals Unlimited, Inc.; **p. 97 (top left):** Richard P. Jacobs/JLM Visuals; **p. 97 (bottom left):** Richard P. Jacobs/JLM Visuals; **p. 97 (bottom center):** Richard P. Jacobs/JLM Visuals; **p. 97 (bottom right):** Thomas Hunn/Visuals Unlimited, Inc.; **p. 98 (top left):** Ann Bryant/Geology.com; **p. 98 (top right):** Arco Images GmbH/Alamy; **p. 98 (center):** On display at the Harvard Museum of Natural History, courtesy Mineralogical and Geological Museum, Photo by Stephen Marshak © President & Fellows, Harvard College; **p. 98 (bottom):** Stephen Marshak; **p. 99 (right):** Construction Photography/Alamy; **p. 99 (left):** Farbled/Dreamstime.com; **p. 99 (inset):** Dennis Kunkel/Science Source; **p. 101 (bottom left):** Jason Pineau/Getty Images; **p. 101 (top left):** Ken Lucas/Visuals Unlimited; **p. 101 (right):** © Petra Diamonds Limited; **p. 102 (bottom left):** Images provided by Google Earth mapping services/NASA, © DigitalGlobe, © Terra Metrics, © GeoEye, © Europa Technologies, Copyright 2014; **p. 102 (bottom right):** Images provided by Google Earth mapping services/NASA, © DigitalGlobe, © Terra Metrics, © GeoEye, © Europa Technologies, Copyright 2014; **p. 102 (top):** The Hope Diamond/Smithsonian Institution, Washington DC, US/Bridgeman Images; **p. 103 (top):** Albert Copley/Visuals Unlimited; **p. 103 (inset):** Albert Copley/Visuals Unlimited; **p. 105:** © 1996 Jeff Scovil.

INTERLUDE A

Page 106: Photos 12/Alamy; **p. 108 (top center):** Courtesy David W. Houseknecht, USGS; **p. 108 (top left):** Stephen Marshak; **p. 108 (bottom left):** sciencephotos/Alamy; **p. 108 (bottom center):** Courtesy of Kent Ratajeski, Dept. of Geology and Geophysics, University of Wisconsin, Madison; **p. 109 (all):** Stephen Marshak; **p. 110 (all):** Stephen Marshak; **p. 112 (all except bottom right):** Stephen Marshak; **p. 112 (bottom right):** Scienics & Science/Alamy; **p. 113:** Stephen Marshak.

CHAPTER 4

Page 114: Stephen Marshak; **p. 116 (top left):** Google Earth; **p. 116 (top right):** J. D. Griggs/U. S. Geological Survey; **p. 116 (bottom right):** Stephen Marshak; **p. 116 (bottom left):** Stephen Marshak; **p. 117 (top):** Stephen Marshak; **p. 117 (bottom):** Stephen Marshak; **p. 122 (top right):** Stephen Marshak; **p. 122 (bottom left):** Stephen Marshak; **p. 122 (bottom right):** Stephen Marshak; **p. 122 (top left):** USGS; **p. 123 (left):** Stephen Marshak; **p. 123 (right):** Stephen Marshak; **p. 125 (top):** Stephen Marshak; **p. 125 (bottom right):** Stephen Marshak; **p. 126:** Images provided by Google Earth mapping services/NASA, © DigitalGlobe, © Terra Metrics, © GeoEye, © Europa Technologies, Copyright 2014; **p. 127**

(inset): Stephen Marshak; **p. 127 (bottom left):** Stephen Marshak; **p. 128 (right):** Doug Sokell/Visuals Unlimited; **p. 128 (left inset):** Dr. Kent Ratajeski; **p. 128 (right inset):** Dr. Matthew Genge; **p. 128 (center):** Mark A. Schneider/Science Source; **p. 128 (center inset):** Omphacite. 2006. Wikimedia: http://en.wikipedia.org/wiki/Public_domain; **p. 128 (left):** Stephen Marshak; **p. 131 (top left):** geoz/Alamy; **p. 131 (bottom left):** Joyce Photographics/Science Source; **p. 131 (center left):** Mark A. Schneider/Science Source; **p. 131 (center right):** Siim Sepp/Alamy; **p. 131 (top right):** Stephen Marshak; **p. 131 (bottom left):** Stephen Marshak; **p. 131 (bottom right):** Stephen Marshak; **p. 131 (bottom right):** Wally Eberhart/Getty Images; **p. 134:** imageBROKER/Alamy Stock Photo; **p. 135:** Stephen Marshak; **p. 137 (top):** Stephen Marshak; **p. 137 (center):** Stephen Marshak.

CHAPTER 5

Page 138: Westend61 GmbH /Alamy Stock Photo; **p. 140 (top right):** Jack Repcheck; **p. 140 (bottom left):** Stephen Marshak; **p. 140 (bottom right):** Stephen Marshak; **p. 140 (top left):** Yale Center for British Art, Paul Mellon Collection/Bridgeman Images; **p. 141 (top left):** Robert Francis/Agefotostock; **p. 141 (top right):** Stephen Marshak; **p. 141 (center right):** Stephen Marshak; **p. 141 (bottom left):** Stephen Marshak; **p. 141 (bottom right):** Stephen Marshak; **p. 141 (center left):** USGS; **p. 142 (bottom):** Images provided by Google Earth mapping services/NASA, © DigitalGlobe, © Terra Metrics, © GeoEye, © 2017; **p. 142 (left):** Stephen Marshak; **p. 142 (right):** Stephen Marshak; **p. 143 (top):** Marli Miller/Visuals Unlimited; **p. 143 (top):** Stephen Marshak; **p. 143 (bottom right):** Stephen Marshak; **p. 144 (left):** AP Photo; **p. 144 (inset):** Stephen Marshak; **p. 144 (right):** Stephen Marshak; **p. 145 (top left):** AF/Getty Images; **p. 145 (top right):** AFP/Getty Images; **p. 145 (bottom right):** Anthony Phelps/Reuters/Newscom; **p. 145 (center left):** Photo by Suzanne MacLachlan, British Ocean Sediment Core Research Facility, National Oceanography Centre, Southampton; **p. 145 (bottom left):** Stephen Marshak; **p. 145 (bottom center):** Stephen Marshak; **p. 145 (center right):** USGS; **p. 147 (top):** Sunshine Pics/Alamy; **p. 147 (bottom):** USGS; **p. 148:** Marli Miller/Visuals Unlimited; **p. 149 (top right):** Google EarthImage Landsat/Copernicus/Data LDEO-Columbia, NSF, NOAA/Data SIO, NOAA, U.S. Navey, NGA, GEBCO; **p. 149 (top left):** Marli Miller/Visuals Unlimited; **p. 149 (inset):** Robert Harding World Imagery/Alamy; **p. 150 (top right):** AFP/Getty Images; **p. 150 (top left):** USGS; **p. 150 (bottom):** USGS; **p. 151 (top right):** © Tom Pfeiffer/www.volcanodiscovery.com; **p. 151 (bottom right):** Corbis Historical/Getty Images; **p. 151 (bottom left):** USGS; **p. 153 (top);** Lyn Topinka/USGS; **p. 153 (bottom):** Stephen Marshak; **p. 157 (top right):** Image courtesy of Submarine Ring of Fire 2002 Exploration, NOAA–OE; **p. 157 (bottom left and right):** Images provided by Google Earth mapping services/NASA, © DigitalGlobe, © Terra Metrics, © GeoEye, © Europa Technologies, Copyright 2014; **p. 159:**

Stephen Marshak; **p. 160 (bottom):** Bob Rauber; **p. 160 (top):** Images provided by Google Earth mapping services/NASA, © DigitalGlobe, © Terra Metrics, © GeoEye, © 2017; **p. 161 (top left):** USGS; **p. 161 (top center left):** Vittoriano Rastelli/Getty Images; **p. 161(bottom center left):** AP Photo; **p. 161 (bottom left):** Stephen Marshak; **p. 161 (top right):** Stocktrek Images, Inc./Alamy Stock Photo; **p. 161 (top center right):** Reuters/Cristobal Saavedra/Newscom; **p. 161 (bottom center right):** Magnus T. Gudmundson, University of Iceland; **p. 161 (bottom right):** USGS; **p. 163:** Peter Turnley/Getty Images; **p. 164:** Stephen Marshak; **p. 165 (bottom):** Planetary Visions/NERC-COMET/JAXA/ESA; **p. 167 (right):** Sigurgeir Jonasson//Getty Images; **p. 167 (left):** Vittoriano Rastelli/Getty Images; **p. 168:** Gail Mooney/ Corbis/VCG/Getty Images; **p. 169 (bottom):** Carlos Gutierrez/Reuters; **p. 169 (top left):** NASA; **p. 169 (top right):** NASA; **p. 169 (center left):** NASA; **p. 169 (top center):** NASA/JPL; **p. 169 (center):** NASA/JPL.

INTERLUDE B

Page 172: Emma Marshak; **p. 174 (all):** Stephen Marshak; **p. 175 (all):** Stephen Marshak; **p. 176 (all):** Stephen Marshak; **p. 177:** Stephen Marshak; **p. 179 (all):** Stephen Marshak; **p. 180 (all):** Stephen Marshak; **p. 181 (all):** Stephen Marshak; **p. 182:** Stephen Marshak; **p. 184:** U.S. Department of Agriculture; **p. 185 (both);** Stephen Marshak; **p. 186 (both):** Stephen Marshak.

CHAPTER 6

Page 188: Stephen Marshak; **p. 189;** Stephen Marshak; **p. 190:** Stephen Marshak; **p. 192 (all):** Stephen Marshak; **p. 193 (all):** Stephen Marshak; **p. 194 (all):** Stephen Marshak; **p. 195 (both):** Stephen Marshak; **p. 196 (inset):** Stephen Marshak; **p. 196:** Visuals Unlimited; **p. 197 (all):** Stephen Marshak; **p. 198 (all except bottom right):** Stephen Marshak; **p. 198 (bottom right):** Images provided by Google Earth mapping services/NASA, © DigitalGlobe, © Terra Metrics, © GeoEye, © 2017; **p. 199:** Stephen Marshak; **p. 200 (top):** Canada Photos/Alamy; **p. 200 (bottom):** 1980 Grand Canyon Natural History Association; **p. 200 (inset):** Stephen Marshak; **p. 201 (both):** Stephen Marshak; **p. 202 (top left):** IMAGINA Photography/Alamy; **p. 202 (bottom left):** Stephen Marshak; **p. 202 (bottom right):** Stephen Marshak; **p. 203 (inset):** Marli Miller/Visuals Unlimited; **p. 203 (top right):** Stephen Marshak; **p. 203 (bottom left):** Stephen Marshak; **p. 203 (bottom right):** Stephen Marshak; **p. 204:** Images provided by Google earth mapping services/NASA, © DigitalGlobe, © Terra Metrics, © GeoEye, © Europa Technologies, Copyright 2015; **p. 205 (top left):** Emma Marshak; **p. 205 (top right):** Marli Miller/Visuals Unlimited; **p. 205 (bottom right):** Polar-TREC/Arctic Research Consortium of the United States (ARCUS); **p. 205 (top center):** Stephen Marshak; **p. 205 (center right):** Stephen Marshak; **p. 205 (bottom left):** Stephen Marshak; **p. 206:** Stephen Marshak; **p. 208:** Images provided by Google earth mapping services/NASA, DigitalGlobe, © Terra Metrics, © GeoEye, © Europa Technologies, Copyright 2015; **p. 209 (right):** Images provided by Google Earth mapping services/NASA, © DigitalGlobe, © Terra Metrics, © GeoEye, © Europa Technologies, Copyright 2017; **p. 209 (left):** Stephen Marshak; **p. 210 (top right):** David Wall/Alamy Stock Photo; **p. 210 (bottom right):** G. R. "Dick" Roberts ©/Natural Sciences Image Library; **p. 210 (bottom left):** The Natural History Museum/Alamy Stock Photo; **p. 213:** Stephen Marshak.

CHAPTER 7

Page 216: Stephen Marshak; **p. 218 (top left):** Stephen Marshak; **p. 218 (top right):** Stephen Marshak; **p. 218 (inset):** Stephen Marshak; **p. 218 (bottom left):** Visuals Unlimited; **p. 218 (bottom right):** Kurt Freihauf; **p. 222 (top right):** Emma Marshak; **p. 222 (top left):** Stefano Clemente/Alamy Stock Photo; **p. 222 (bottom):** Stephen Marshak; **p. 223 (all):** Stephen Marshak; **p. 224 (both):** Stephen Marshak; **p. 225 (all):** Stephen Marshak; **p. 230 (all):** Stephen Marshak; **p. 231 (left):** Images provided by Google earth mapping services/NASA, © DigitalGlobe, © Terra Metrics, © GeoEye, © Europa Technologies, Copyright 2015; **p. 231 (right):** Images provided by Google earth mapping services/NASA, © DigitalGlobe, © Terra Metrics, © GeoEye, © Europa Technologies, Copyright 2017; **p. 234:** Images provided by Google earth mapping services/NASA, © DigitalGlobe, © Terra Metrics, © GeoEye, © Europa Technologies, Copyright 2017; **p. 235 (both):** Stephen Marshak.

INTERLUDE C

Page 238 (all): Stephen Marshak.

CHAPTER 8

Page 246: Kyodo News via Getty Images; **p. 248 (top):** AFP/Getty Images; **p. 248 (right):** AP Photo/Kyodo News; **p. 248 (center):** JIJI Press/AFP/Getty Images; **p. 250 (both):** Photo Courtesy of Paul "Kip" Otis-Diehl, USMC, 29 Palms, CA; **p. 253:** Peltzer et al. (1999), Evidence of nonlinear elasticity of the Crust. Science, v. 286. Copyright © 1999, AAAS; **p. 254:** Images provided by Google Earth mapping services/NASA, ©DIgitalGlobe, ©TerraMetrics, ©GeoEye, ©EuropaTechnologies. Copyright 2017; **p. 265 (bottom):** AP Photo/Paul Sakuma; **p. 265 (top):** Library of Congress/Getty Images; **p. 267 (left):** Anna Kompanek/CIPE; **p. 267 (right):** Omar Havana/Getty Images; **p. 268:** New Madrid earthquake woodcut from Deven's Our First Century (1877)/Wikimedia Commons; **p. 270 (top):** AP Photo; **p. 270 (bottom center):** M. Celebi, U.S. Geographical Survey; **p. 270 (top center):** Pacific Press Service/Alamy; **p. 270 (bottom):** Reuters/Newscom; **p. 271 (bottom left):** AP Photo/New Zealand Herald, Geoff Sloan; **p. 271 (top left):** Barry Lewis/ Alamy; **p. 271 (top right):** Bettmann/Getty Images; **p. 271 (center left):** NOAA/National Geophysical Data Center (NGDC); **p. 272 (top left):** Karl V.

Steinbrugge Collection, University of California, Berkeley; **p. 272 (center right):** National Geophysical Data Center (NGDC); **p. 273:** Images provided by Google Earth mapping/ services/ NASA, © DigitalGlobe, © Terra Metrics, © GeoEye, © Europa Technologies, Copyright 2017; **p. 276 (top left):** AFP/Getty Images; **p. 276 (bottom right, both):** Ikonos images Copyright Centre for Remote Images, Sensing and Processing, National University of Singapore and Space Imaging; **p. 276 (top right):** Photo by David Rydevik. 2004. Wikimedia; **p. 277 (inset):** Air Photo Service/Reuters/ Landov; **p. 277 (top):** AP Photo/Kyodo News; **p. 277 (bottom):** EPA/The Tokyo Electric Power Company/Landov; **p. 278 (top):** Reuters/Eduardo Munoz; **p. 278 (bottom left):** Reuters/Eduardo Munoz; **p. 278 (bottom right):** USGS; **p. 282 (center right):** NOAA/NOA Center for Tsunami Research; **p. 282 (bottom right):** Stephen Marshak; **p. 283:** Aurora Photos/Alamy Stock Photo.

INTERLUDE D

Page 286: USGS; **p. 295 (bottom left):** Figures provided courtesy F. Lemoine & J. Frawley, NASA Goddard Space Flight Center; **p. 295 (bottom right):** Figures provided courtesy F. Lemoine & J. Frawley, NASA Goddard Space Flight Center; **p. 295 (top right):** Courtesy of Greg Moore, University of Hawaii and Nathan Bangs, University of Texas; **p. 295 (top left):** Courtesy of Sercel and CGG Veritas/Shell Oil Company; **p. 296:** USGS; **p. 298:** USGS.

CHAPTER 9

Page 300: Stephen Marshak; **pp. 302-303:** NOAA/ETO-PO1382; **p. 304 (all):** Stephen Marshak; **p. 305 (both):** Stephen Marshak; **p. 307:** Images provided by Google Earth mapping services/NASA, ©DigitalGlobe, © Terra Metrics, © GeoEye, © Europa Technologies. Copyright 2017; **p. 308 (left):** Russ Bishop/Alamy Stock Photo; **p. 308 (center):** Stephen Marshak; **p. 308 (right):** Stephen Marshak; **p. 309:** Stephen Marshak; **p. 311 (bottom):** Lloyd Cluff/Getty Images; **p. 311 (top):** Stephen Marshak; **p. 311 (center):** USGS; **p. 312 (bottom):** Images provided by Google Earth mapping services/NASA, ©DigitalGlobe, © Terra Metrics, © GeoEye, © Europa Technologies. Copyright 2017; **p. 312 (top left):** Stephen Marshak; **p. 312 (top center):** Stephen Marshak; **p. 312 (top right):** Stephen Marshak; **p. 314 (bottom):** Landsat/USGS; **p. 314 (top left):** Stephen Marshak; **p. 314 (top center):** Stephen Marshak; **p. 314 (top right):** Stephen Marshak; **p. 314 (bottom center):** Stephen Marshak; **p. 315 (top):** Stephen Marshak; **p. 315 (bottom):** Stephen Marshak; **p. 316:** Stephen Marshak; **p. 317:** Images provided by Google Earth mapping services/NASA, ©DigitalGlobe, © Terra Metrics, © GeoEye, © Europa Technologies. Copyright 2017; **p. 318 (top left):** Images provided by Google Earth mapping services/NASA, © DigitalGlobe, © Terra Metrics, © GeoEye, © Europa Technologies, Copyright 2014; **p. 318 (bottom):** Images provided by Google Earth mapping services/ NASA, ©DigitalGlobe, © Terra Metrics, © GeoEye, © Europa Technologies. Copyright 2017; **p. 318 (top right):** Stephen Marshak; **p. 319:** Image from Google Earth; data map from: http://all-geo.org/highlyallochthonous/2013/02/friday-focal-mechanism-the-himalayas-long-tectonic-shadow/, http://onlinelibrary.wiley.com/doi/10.1029/2005JB004120/abstract;jsessionid=DCF2B17B34D889A35B10ADCD8C64A520.f03t01; **p. 322:** Stephen Marshak; **p. 324:** Stephen Marshak; **p. 325 (both):** Stephen Marshak.

INTERLUDE E

Page 330: Stephen Marshak; **p. 331 (all):** Stephen Marshak; **p. 334 (top center):** Dirk Wiersma/Science Source; **p. 334 (bottom center):** Kevin Schafer/Getty Images; **p. 334 (top left):** Sovfoto/UIG via Getty Images; **p. 334 (top right):** Stephen Marshak; **p. 334 (bottom left):** Stephen Marshak; **p. 334 (bottom right):** Stephen Marshak; **p. 335 (left):** Stephen Marshak; **p. 335 (right):** UCL Micropaleontology Collections, UCL Museums & Collections; **p. 336 (left):** Courtesy of Senckenberg, Messel Research Department; **p. 336 (right):** Humboldt-Universitat zu Berlin Museum fur Naturkunde. Photo by W. Harre; **p. 337:** Illustration by Karen Carr and Karen Carr Studio, Inc. © Smithsonian Institution.

CHAPTER 10

Page 342: Stephen Marshak; **p. 344 (both):** Stephen Marshak; **p. 345 (both):** Stephen Marshak; **p. 348:** Stephen Marshak; **p. 349:** Stephen Marshak; **p. 351:** Stephen Marshak; **p. 352:** Images provided by Google Earth mapping services/ NASA, © DigitalGlobe, © Terra Metrics, © GeoEye, © Europa Technologies. Copyright 2017; **p. 353 (left):** USGS; **p. 353 (right):** Paul Karabinos, Williams College and USGS; **p. 355:** Images provided by Google Earth mapping services/NASA, © DigitalGlobe, © Terra Metrics, © GeoEye, © Europa Technologies. Copyright 2017; **p. 357 (all):** Stephen Marshak; **p. 361:** Stephen Marshak; **p. 362 (right):** Damon Runberg, USGS; **p. 362 (left):** Stephen Marshak; **p. 365:** USGS National Center of EROS and NASA, Landsat Project Science Office; **p. 367:** Stephen Marshak.

CHAPTER 11

Page 368: Stephen Marshak; **p. 370 (left):** artwork Copyright Don Dixon/ www.cosmographica.com; **p. 370 (right):** artwork Copyright Don Dixon/ www.cosmographica.com; **p. 372 (left):** Courtesy of Dr. J. William Schopf/UCLA; **p. 372 (right):** Stephen Marshak; **p. 373:** USGS; **p. 376 (bottom left):** Courtesy of Dr. Paul Hoffman, Harvard University; **p. 376 (top left):** De Agostini Picture Library Universal Images Group/ Newscom; **p. 376 (top right):** Stephen Marshak; **p. 377 (both):** Ronald C. Blakey; Colorado Plateau Geosystems, Inc.; **p. 378:** Tom McHugh/Science Source; **p. 379 (bottom):** Edward B. Daeschler; **p. 379 (top):** Ronald C. Blakey; Colorado Plateau Geosystems, Inc.; **p. 380:** Ronald C. Blakey; Colorado Plateau

Geosystems, Inc.; **p. 381 (right):** Science Photo Library/Alamy Stock Photo; **p. 382 (right):** Images provided by Google Earth mapping services/NASA, © DigitalGlobe, © Terra Metrics, © GeoEye, © Europa Technologies, Copyright 2017; **p. 382 (bottom left):** Richard Bizley; **p. 382 (top inset):** Ronald C. Blakey; Colorado Plateau Geosystems, Inc.; **p. 383:** Ronald C. Blakey; Colorado Plateau Geosystems, Inc.; **p. 384 (left):** Ronald C. Blakey; Colorado Plateau Geosystems, Inc.; **p. 385 (center):** Images provided by Google Earth mapping services/NASA/ © 2015; **p. 385 (top):** NASA, JPL; **p. 385 (inset):** Ronald C. Blakey; Colorado Plateau Geosystems, Inc.; **p. 385 (bottom):** Virgil L. Shaprton, Lunar and Planetary Institute; **p. 386:** Images provided by Google Earth mapping services/NASA, © DigitalGlobe, © Terra Metrics, © GeoEye, © Europa Technologies, Copyright 2017; **p. 388 (all):** Ronald C. Blakey; Colorado Plateau Geosystems, Inc.; **p. 390:** Stephen Marshak; **p. 391:** Jacques Descloitres, MODIS Team, NASA Visible Earth.

CHAPTER 12

Page 394: Stephen Marshak; **p. 401:** Data courtesy of Fugro. Credit: Virtual Seismic Atlas; http://www.seismicatlas.org; **p. 403 (top and center):** Stephen Marshak; **p. 403:** Calvin Larsen/Science Source; **p. 404:** Images provided by Google Earth mapping services/NASA, © DigitalGlobe, © Terra Metrics, © GeoEye, © Europa Technologies, Copyright 2017; **p. 407:** Andrew Harrer/Bloomberg via Getty Images; **p. 409 (bottom):** Department of Natural Resources, Alaska; **p. 409 (top):** Field Museum Library/Getty Images; **p. 411 (bottom right):** Cultura Creative (RF)/Alamy; **p. 411 (top left):** Stephen Marshak; **p. 411 (bottom left):** Stephen Marshak; **p. 412:** Ron Chapple/Dreamstime; **p. 414:** Images provided by Google Earth mapping services/NASA, © DigitalGlobe, © Terra Metrics, © GeoEye, © Europa Technologies, Copyright 2017; **p. 415:** Education Images/UIG via Getty Images; **p. 416 (all):** Stephen Marshak; **p. 419 (left):** NASA; **p. 419 (right):** NASA; **p. 420 (right):** AFP/Getty Images; **p. 420 (left):** AP Photo/Stapleton; **p. 421 (inset):** Layne Kennedy/Getty Images; **p. 421:** Stephen Marshak; **p. 422 (left):** Richard P. Jacobs/JLM Visuals; **p. 422 (right):** Stephen Marshak; **p. 424:** Stephen Marshak; **p. 425 (right):** Images provided by Google Earth mapping services/NASA, © DigitalGlobe, © Terra Metrics, © GeoEye, © Europa Technologies, Copyright 2017; **p. 425 (top left):** Stephen Marshak; **p. 425 (bottom left):** Stephen Marshak; **p. 429 (bottom right):** Stephen Marshak; **p. 429 (top left):** Richard P. Jacobs/JLM Visuals; **p. 429 (top right):** Stephen Marshak; **p. 429 (bottom left):** Stephen Marshak; **p. 431 (right):** Bill Brooks/Alamy Stock Photo; **p. 431 (left):** Doug Sokell/Visuals Unlimited.

INTERLUDE F

Page 434: Stephen Marshak; **p. 436 (top left):** Stephen Marshak; **p. 436 (top right):** Stephen Marshak; **p. 436 (center left):** Stephen Marshak; **p. 436 (center right):** Stephen Marshak; **p. 436 (bottom left):** G. R. 'Dick' Roberts © Natural Sciences Image Library; **p. 436 (bottom right):** Julie Dermansky/Corbis via Getty Images; **p. 439 (background):** Stephen Marshak; **p. 439 (left):** Stephen Marshak; **p. 439 (right):** Stephen Marshak; **p. 443 (top left):** JSC/NASA; **p. 443 (top right):** Dr. David Smith, NASA Goddard Space Flight Center/ MOLA Science Team; **p. 443 (bottom left):** NASA; **p. 443 (bottom right):** NASA; **p. 444 (left):** ESA/DLR/FU Berlin; **p. 444 (right):** NASA.

CHAPTER 13

Page 446: Markus Hell/Tareom Aerials; **p. 447 (top):** Lloyd Cluff/Getty Images; **p. 447 (bottom):** Lloyd Cluff/Getty Images; **p. 448 (left):** Stephen Marshak; **p. 448 (right):** Marli Miller; **p. 449 (top left):** Stephen Marshak; **p. 449 (bottom left):** Stephen Marshak; **p. 449 (bottom right):** Stephen Marshak; **p. 450 (top left):** Shana Reis/EPA; **p. 450 (top right):** Cascades Volcano Observatory/USGS; **p. 450 (bottom left):** Stephen Marshak; **p. 450 (bottom right):** Stephen Marshak; **p. 451 (top left):** Images provided by Google Earth mapping services/NASA, © DigitalGlobe, © Terra Metrics, © GeoEye, © Europa Technologies, Copyright 2015; **p. 451 (bottom right):** Images provided by Google Earth mapping services/ NASA, © DigitalGlobe, © Terra Metrics, © GeoEye, © Europa Technologies, Copyright 2015; **p. 452 (top left):** Bob Schuster, USGS; **p. 452 (right):** National Geographic Image Collection/ Alamy; **p. 452 (bottom left):** Ron Varela/Ventura County Star; **p. 453:** Stephen Marshak; **p. 454 (top left):** mediacolors/ Alamy Stock Photo; **p. 454 (top right):** Alaska Stock/Alamy; **p. 454 (bottom left):** Stephen Marshak; **p. 454 (bottom right):** Stephen Marshak; **p. 455 (top):** Jerome Neufeld and Stephen Morris, nonlinear Physica University of Toronto; **p. 455 (bottom left):** USGS/Barry W. Eakins; **p. 455 (bottom right):** USGS, Geologic Investigations Series I-2809 by Barry W. Eakins, Joel E. Robinson, Toshiya Kanamatsu, Jiro Naka, John R. Smith, Eiichi Takahashi, and David A. Clague; **p. 456:** Stephen Marshak; **p. 461:** Breck P. Kent/JLM Visuals; **p. 465:** Stephen Marshak.

CHAPTER 14

Page 468: Stephen Marshak; **p. 469 (top):** Helen H. Richardson/The Denver Post/ Getty Images; **p. 469 (center):** John Gibson/Getty Images; **p. 469 (bottom):** Andy Clark/Reuters/ Newscom; **p. 471:** Stephen Marshak; **p. 472:** Images provided by Google Earth mapping services/NASA, © DigitalGlobe, © Terra Metrics, © GeoEye, © Europa Technologies, Copyright 2015; **p. 473 (left):** Stephen Marshak; **p. 473 (right):** Stephen Marshak; **p. 475 (left):** Stephen Marshak; **p. 475 (right):** © Ron Niebrugge; **p. 476 (left):** Stephen Marshak; **p. 476 (right):** Stephen Marshak; **p. 478 (left):** Stephen Marshak; **p. 478 (right):** Stephen Marshak; **p. 479 (left):** Images provided by Google Earth mapping services/NASA, © DigitalGlobe,

© Terra Metrics, © GeoEye, © Europa Technologies, Copyright 2017; **p. 479 (top right):** Stephen Marshak; **p. 479 (bottom right):** Stephen Marshak; **p. 480 (top right):** Stephen Marshak; **p. 480 (bottom right):** Stephen Marshak; **p. 481 (bottom left):** Marli Miller; **p. 481 (top):** Stephen Marshak; **p. 481 (bottom):** Stephen Marshak; **p. 482 (bottom right):** NASA/GSFC/meti/ersdac/jaros, and US/Japan ASTER Science Team; **p. 484:** Google Earth; **p. 487 (left, both):** NASA images created by Jesse Allen, Earth Observatory, using data provided courtesy of the Landsat Project Science Office- Copyright 2008; **p. 487 (bottom right):** United Nations, Evan Schneider; **p. 487 (top right):** Missouri Department of Transportation; **p. 488 (left):** Reuters/Yoray Cohen/Eilat Rescue Unit; **p. 488 (right):** USGS; **p. 489:** Courtesy of Johnstown Area Heritage Association; **p. 490 (top):** Stephen Marshak; **p. 490 (bottom left):** Stephen Marshak; **p. 490 (bottom center):** Stephen Marshak; **p. 490 (bottom right):** Stephen Marshak; **p. 491 (center):** Stephen Marshak; **p. 491 (top right):** Stephen Marshak; **p. 493 (top left):** Stephen Marshak; **p. 493 (top right):** Stephen Marshak; **p. 493 (bottom right):** AP Photo/The News-Star, Margaret Croft; **p. 495 (top):** Photo courtesy of the Bureau of Reclamation; **p. 495 (bottom):** Stephen Marshak.

CHAPTER 15

Page 498: Stephen Marshak; **p. 503 (left):** Stephen Marshak; **p. 503 (right):** http://en.wikipedia.org/wiki/File:MtSt-Michel_avion.jpg, pd; **p. 504 (bottom right):** Stephen Marshak; **p. 505:** Stephen Marshak; **p. 506:** NASA; **p. 508:** Los Alamos National Laboratory; **p. 510 (left):** Stephen Marshak; **p. 510 (right):** Stephen Marshak; **p. 511 (left):** NASA; **p. 511 (right):** Stephen Marshak; **p. 512 (top right):** G. R. 'Dick' Roberts © Natural Sciences Image Library; **p. 512 (left center):** Stephen Marshak; **p. 512 (bottom left):** Stephen Marshak; **p. 512 (bottom right):** Emma Marshak; **p. 513 (top right):** Images provided by Google Earth mapping services/NASA, © DigitalGlobe, © Terra Metrics, © GeoEye, © Europa Technologies, Copyright 2015; **p. 513 (bottom right):** Cody Duncan/Alamy; **p. 514 (bottom right):** Images provided by Google Earth mapping services/NASA, © DigitalGlobe, © Terra Metrics, © GeoEye, © Europa Technologies, Copyright 2015; **p. 514 (top left):** Stephen Marshak; **p. 514 (top right):** Stephen Marshak; **p. 514 (bottom right):** Images provided by Google Earth mapping services/NASA, © DigitalGlobe, © Terra Metrics, © GeoEye, © Europa Technologies, Copyright 2015; **p. 515 (right):** Stephen Marshak; **p. 515 (left):** Stephen Marshak; **p. 515 insert:** Steve Bloom Images/Alamy; **p. 520:** Images provided by Google Earth mapping services/NASA, © DigitalGlobe, © Terra Metrics, © GeoEye, © Europa Technologies, Copyright 2017; **p. 521 (left, both):** USGS; **p. 521 (right):** Stephen Marshak; **p. 522:** Stephen Marshak; **p. 523:** AP Photo; **p. 524:** NASA; **p. 525 (top left):** NOAA; **p. 525 (top right):** AP Photo/Susan Walsh; **p. 525 (center):** U.S. Geological Survey (2007); **p. 525 (bottom left):** POOL/AFP/Getty Images; **p. 525 (bottom right):** Stephen Marshak; **p. 526:** AP Photo/David J. Phillip; **p. 527:** Stephen Marshak.

CHAPTER 16

Page 530: Stephen Marshak; **p. 531 (top left):** USGS; **p. 531 (right):** Images provided by Google Earth mapping Services/NASA, © DigitalGlobe, © Terra Metrics, © GeoEye, © Europa Technologies, Copyright 2017; **p. 532:** Images provided by Google Earth mapping services/NASA, © DigitalGlobe, © Terra Metrics, © GeoEye, © Europa Technologies, Copyright 2015; **p. 533 (center left):** Photo courtesy of Eric Prokacki and Jim Best, University of Illinois; **p. 533 (bottom left):** Stephen Marshak; **p. 539:** Stephen Marshak; **p. 541:** Stephen Marshak; **p. 542 (top right):** Allan Tuchman; **p. 542 (center left):** Stephen Marshak; **p. 542 (center right):** Emma Marshak; **p. 542 (bottom left):** Thom Foley; **p. 542 (bottom right):** Stephen Marshak; **p. 548:** Stephen Marshak; **p. 549 (top right):** Stephen Marshak; **p. 549 (bottom right):** Stephen Marshak; **p. 549 (left):** Image provided by Google Earth mapping services/NASA, © Digital Globe, © Terra Metrics, © GeoEye, © Europa Technologies. Copyright 2015; **p. 550 (top right):** Lois Kent; **p. 550 (bottom left):** Paul F. Hudson, University of Texas; **p. 550 (bottom right):** ML Sinibaldi/Media Bakery; **p. 551 (top left):** Lois Kent; **p. 551 (top right):** Ashley Cooper/Getty Images; **p. 551 (bottom right):** Stephen Marshak; **p. 552 (left):** Stephen Marshak; **p. 552 (right):** River in the Mist and Evening Light, 1987 (hanging scroll, ink & colour on paper), Keran, Li (1907–89)/Private collection/Photo © Christie's Images/Bridgeman Images; **p. 553:** Stephen Marshak.

CHAPTER 17

Page 556: Stephen Marshak; **p. 557:** Stephen Marshak; **p. 558:** Stephen Marshak; **p. 559:** Professor Andrew Danderfer; **p. 560 (bottom left):** Stephen Marshak; **p. 560 (top right):** Stephen Marshak; **p. 560 (center right):** Stephen Marshak; **p. 561 (left):** Stephen Marshak; **p. 561 (right):** Stephen Marshak; **p. 562 (top right):** Daniel J Bryant/Getty Images; **p. 562 (bottom left):** Stephen Marshak; **p. 563 (left):** Stephen Marshak; **p. 563 (right):** Images provided by Google Earth mapping services/NASA, © DigitalGlobe, © Terra Metrics, © GeoEye, © Europa Technologies, Copyright 2015; **p. 564 (top left):** Stephen Marshak; **p. 564 (top right):** Stephen Marshak; **p. 564 (bottom):** Stephen Marshak; **p. 565 (top left):** Stephen Marshak; **p. 565 (top right):** Stephen Marshak; **p. 565 (inset):** Stephen Marshak; **p. 565 (bottom right):** Images provided by Google Earth mapping services/NASA, © DigitalGlobe, © Terra Metrics, © GeoEye, © Europa Technologies, Copyright 2015; **p. 568 (right):** Stephen Marshak; **p. 569 (bottom left):** Stephen Marshak; **p. 569 (inset):** Images provided by Google Earth mapping services/NASA, © DigitalGlobe, © Terra Metrics, © GeoEye, © Europa Technologies, Copyright 2015; **p. 570 (top):** Stephen Marshak;

p. 570 (top inset): Stephen Marshak; **p. 570 (bottom):** Stephen Marshak; **p. 570 (bottom inset):** Stephen Marshak; **p. 571 (bottom right):** Stephen Marshak; **p. 571 (top left):** Images provided by Google Earth mapping services/NASA, © DigitalGlobe, © Terra Metrics, © GeoEye, © Europa Technologies, Copyright 2015; **p. 572:** JB Russell/Panos Pictures; **p. 573 (top):** Image courtesy Jacques Descloitres MODIS Rapid Response Team; **p. 573 (bottom):** United States Department of Agriculture.

CHAPTER 18

Page 576: Stephen Marshak; **p. 577:** Stephen Marshak; **p. 578 (bottom left):** Ted Spiegel/National Geographic Creative; **p. 578 (top left):** Shutterstock; **p. 578 (top center):** Stephen Marshak; **p. 578 (top right):** Stephen Marshak; **p. 578 (center left):** Emma Marshak; **p. 580 (left):** Stephen Marshak; **p. 580 (right):** Stephen Marshak; **p. 581 (left):** NASA-JPL; **p. 581 (right):** NASA; **p. 582 (bottom right):** National Geophysical Data Center/NOAA; **p. 582 (top inset):** Stephen Marshak; **p. 583:** Images provided by Google Earth mapping services/ NASA, © DigitalGlobe, © Terra Metrics, © GeoEye, © Europa Technologies, Copyright 2015; **p. 585 (top center):** Newspix/ Getty Images; **p. 585 (top right):** Stephen Marshak; **p. 585 (bottom, both):** ESA; **p. 586 (top left):** Shutterstock; **p. 586 (top right):** Stephen Marshak; **p. 586 (bottom left):** Stephen Marshak; **p. 586 (bottom right):** Stephen Marshak; **p. 587 (top left):** Stephen Marshak; **p. 587 (top inset):** Stephen Marshak; **p. 587 (top right):** Stephen Marshak; **p. 587 (bottom left):** 1986 Keith S. Walklet/Quietworks; **p. 587 (bottom right):** Stephen Marshak; **p. 588 (center right):** Marli Miler; **p. 588 (bottom right):** Wolfgang Meier/zefa/Getty Images; **p. 588 (left):** Images provided by Google Earth mapping services/NASA, © DigitalGlobe, © Terra Metrics, © GeoEye, © Europa Technologies, Copyright 2015; **p. 589 (both):** Stephen Marshak; **p. 593 (top left):** Stephen Marshak; **p. 593 (top right):** Stephen Marshak; **p. 593 (center inset):** Stephen Marshak; **p. 593 (center left):** Stephen Marshak; **p. 593 (center right):** Stephen Marshak; **p. 593 (bottom right):** Kevin Schafer/Alamy; **p. 593 (bottom left):** Stephen Marshak; **p. 595 (top):** Stephen Marshak; **p. 595 (top inset):** Image provided by Google Earth mapping services/NASA, © Digital-Globe, © Terra Metrics, © GeoEye, © EuropaTechnologies, Copyright 2017; **p. 595 (bottom right):** Glenn Oliver/Visuals Unlimited; **p. 596 (top):** Grambo Photography/Getty Images; **p. 596 (bottom):** Mike Beauregard/Creative Commons Attribution 2.0 Generic; **p. 598:** Image provided by Google Earth mapping services/NASA, © DigitalGlobe, © Terra Metrics,

© GeoEye, © EuropaTechnologies, Copyright 2017; **p. 599:** Lynda Dredge/Geological Survey of Canada; **p. 600:** Detail of mural by Charles R. Knight, American Museum of Natural History, #4950 (5), Photo by Denis Finnin; **p. 606 (top):** Prisma Archivo/Alamy Stock PhotoDR64G8; **p. 606 (bottom):** Stephen Marshak; **p. 607 (top left):** NaturePL/Superstock; **p. 607 (top right):** Roger J. Braithwaite; **p. 607 (center left):** Michael Melford/Getty Images; **p. 607 (bottom, both):** NASA.

CHAPTER 19

Page 610: Stephen Marshak; **p. 611 (both):** Ronald C. Blakey; Colorado Plateau Geosystems, Inc.; **p. 613:** Photo by Felix Andrews, 2005, https://creativecommons.org/licenses/by-sa/ 3.0/deed.en; **p. 615 (right, all):** Stephen Marshak; **p. 616 (left):** Stephen Marshak; **p. 616 (right):** Images provided by Google Earth mapping services/NASA, © DigitalGlobe, © Terra Metrics, © GeoEye, © Europa Technologies, Copyright 2015; **p. 617 (top left):** Courtesy P&H Mining Equipment; **p. 617 (top center):** Richard Hamilton Smith/Getty Images; **p. 617 (top right):** Stephen Marshak; **p. 617 (bottom left):** Images provided by Google Earth mapping services/NASA, © DigitalGlobe, © Terra Metrics, © GeoEye, © Europa Technologies, Copyright 2015; **p. 617 (bottom right):** Images provided by Google Earth mapping services/NASA, © DigitalGlobe, © Terra Metrics, © GeoEye, © Europa Technologies, Copyright 2015; **p. 618 (center, both):** Images provided by Google Earth mapping services/NASA, © DigitalGlobe, © Terra Metrics, © GeoEye, © Europoa Technologies, Copyright 2014; **p. 618 (bottom):** Nigel Dickinson/Alamy; **p. 619 (left):** Stephen Marshak; **p. 619 (right):** NASA; **p. 622 (top left):** Bob Sacha/Getty Images; **p. 622 (bottom left):** ISM/ medicalimages.com; **p. 623 (left):** Karim Agabi/Science Source; **p. 623 (right inset):** NOAA; **p. 623 (left inset):** Lamont-Doherty Earth Observatory; **p. 624 (top right):** akg–images; **p. 624 (bottom left):** Wikimedia Commons: https://en.wikipedia.org/wiki/File:Hvalsey_Church.jpg; **p. 624 (bottom right):** Stephen Marshak; **p. 626 (both):** NASA/Goddard Space Flight Center Scientific Visualization Studio; **p. 627 (top right, both):** National Snow and Ice Data Center; **p. 627 (bottom left):** USGS; **p. 627 (bottom right):** USGS photograph by Bruce Molnia; **p. 628:** Images provided by Google Earth mapping services/NASA, © DigitalGlobe, © Terra Metrics, © GeoEye, © Europa Technologies, Copyright 2015; **pp. 632-633 (all):** Images provided by Google Earth mapping services/NASA, © DigitalGlobe, © Terra Metrics, © GeoEye, © Europa Technologies, Copyright 2017; **p. 635:** Stephen Marshak.

INDEX

Note: Page numbers in *italics* refer to illustrations, tables, and figures. Page numbers in **boldface** refer to guide terms.

100-year flood, 492, *494*

A

A-horizon, 182
a'a' lava flow, *141,* **142**
abandoned meander, 482, *490*
ablation, **583,** *583, 584*
abrasion, erosion by, 474
absolute age, 344
absolute plate velocity, **80,** *80*
abyssal plains, 33, *33,* 50, 64, 66, 500, *501*
Acadian orogeny, 378, *379*
acceleration, 17
accreted terranes, 317, 318–19, *319*
accretion, **317**
accretion disk, **21**
accretionary coasts, 517
accretionary disk, *25*
accretionary lapilli, *145,* 146
accretionary prisms, *68,* **69,** *264, 500, 519*
    blueschist formation in, 226, *233*
acid mine runoff, 418, 431, *431,* 618
acid rain, **418,** 550, **618**
acid runoff, 618
active continental margins, *500,* 501, *501*
active faults, 250, 309
active margins, **63,** *500,* 501, *501*
active sand layer, 511
active volcanoes, **164**
adit, 428
advection, 22, *22*
aerosols, **143**
Africa, *375, 376,* 377, *377,* 379, 381, 383, 386, 485
African Rift, 266
aftershocks, earthquake, **251,** 266, 269
Agassiz, Louis, 577, *577,* 601
agate, 197, *198*
"Age of Dinosaurs," 356, 365
"Age of Mammals," 356, 365, 390

agents of erosion, **438**
agents of metamorphism, 217
aggregate, 429
agriculture
    flooding and, 495
    greenhouse gases and, 625
    irrigation *see* irrigation
    landscape modification by, 438, 617, *617*
    overuse of water and, 495
    pollution of coastal waters, 521–22
    slash-and-burn, 617, *618*
air-fall tuff, 147
air pollution, 418, 431
air pressure, 31
Alaska, *4, 143, 151, 157, 454, 576, 593*
    fjords of, 588
    glaciers in, 579, *580*
Alaska, 1964 Good Friday earthquake in, 254, 265, 270, *271*
    Turnagain Heights liquefaction disaster in, 270–71, *272*
albedo
    glacial advance and changing, 603, 605
    global climate change and, **623**
Alberta 2013 flood, 469, *469*
Aleutian volcanic arc, *157*
alfisol soils, 184, *185*
Alleghanian orogeny, 379, *380, 381*
alloy, 34, *34*
alluvial aprons, *567*
alluvial-fan environments, 204–5, *205*
alluvial fans, **205,** *205, 240,* **480,** *480, 563, 564, 566, 567*
alluvium, **475,** 563
alluvium-filled valleys, 478, *479*
Alpine Fault, 70, 264
alpine glaciers
    see mountain glaciers
Alpine-Himalayan Chain, **386,** *386*
Alps, 302, 386, *436,* 577, *578,* 579, *586, 589*
    avalanche in Austrian Alps, 451–52, *454*
aluminum oxide, 424, 429
aluminum smelting, 421
Alvarez, Luis, 385

Alvarez, Walter, 385
Amalfi Coast, Italy, 509
Amasia, 634
Amazon rain forest, *618*
Amazon River, 485
    discharge of, 472, 474
amber, 102
    fossils preserved in, 333, *334,* 335
ambient geothermal energy, 414
American Falls of Niagara Falls, *480*
amethyst, *84,* 85
ammonites, 337, *337*
amphibians, 365, 380, *389*
amphibole, *97,* 99, 124
amplitude, 257, 261
Amsden Shale, *461*
Amundsen, Roald, 189
Anak Krakatau, *152*
Anatolian faults, 265
Ancestral Rockies, 379, *380*
Andes, *68,* 317, *318,* 387, *468,* 516, 579
Andes Mountains, *68,* 157, 317, *318, 478,* 485
    lahars in, 451
andesite, 129, *131*
andesitic lava flows, 139, *140,* 142–43, 144, 146
Andrew, Hurricane, 526
angiosperms, 383, 387
angle of repose, 146, *149,* **457,** *457,* 572
angular unconformities, **348,** *349*
angularity, 191
anhedral grains, 91, *94*
animal attack, 177
animal life
    evolution of *see* evolution
    extinction of *see* extinction
    fossils *see* fossils
    physical weathering of rock by, 177
Animalia, 336, *337*
anions, 86, 98
annual probability (of earthquakes), 279
annual probability (of flooding), **492**
anoxic environment, 332, 335
Antarctic Bottom Water, *506*
Antarctic ice sheet, 626
Antarctic Peninsula, 626

I-1